DELMAR'S
Standard
Textbook of
Electricity

FIFTH EDITION

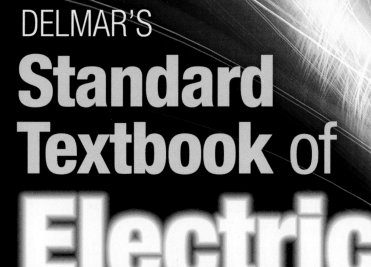

Stephen L. Herman

DELMAR
CENGAGE Learning™

Australia • Brazil • Japan • Korea • Mexico • Singapore • Spain • United Kingdom • United States

DELMAR
CENGAGE Learning™

Delmar's Standard Textbook of Electricity, 5th Edition
Stephen L. Herman

Vice President, Career and Professional Editorial: Dave Garza

Director of Learning Solutions: Sandy Clark

Acquisitions Editor: Stacy Masucci

Managing Editor: Larry Main

Senior Product Manager: John Fisher

Senior Editorial Assistant: Andrea Timpano

Vice President, Career and Professional Marketing: Jennifer Baker

Marketing Director: Deborah Yarnell

Marketing Manager: Kathryn Hall

Associate Marketing Manager: Mark Pierro

Production Director: Wendy Troeger

Production Manager: Mark Bernard

Content Project Manager: Barbara LeFleur

Senior Art Director: David Arsenault

Technology Project Manager: Christopher Catalina

Cover images:
Energy-efficient light bulb:
© iStockphoto/Eltoddo

Bright wave on black:
Buslik, 2011. Used under license from Shutterstock.com

For product information and technology assistance, contact us at
Professional Group Cengage Learning Customer & Sales Support,
1-800-354-9706
For permission to use material from this text or product,
submit all requests online at **cengage.com/permissions.**
Further permissions questions can be e-mailed to
permissionrequest@cengage.com.

Library of Congress Control Number: 2010932363

ISBN-13: 978-1-1115-3915-3

ISBN-10: 1-1115-3915-4

Delmar
5 Maxwell Drive
Clifton Park, NY 12065-2919
USA

Cengage Learning is a leading provider of customized learning solutions with office locations around the globe, including Singapore, the United Kingdom, Australia, Mexico, Brazil and Japan. Locate your local office at: **international.cengage.com/region**

Cengage Learning products are represented in Canada by Nelson Education, Ltd.

For your lifelong learning solutions, visit **delmar.cengage.com**

Visit our corporate website at **cengage.com.**

Printed in the U.S.A.
1 2 3 4 5 X X 12 11 10

Contents

SECTION IV
Small Sources of Electricity   356

SECTION X
Transformers   744

UNIT 28
Single-Phase Transformers   745

UNIT 29
Three-Phase Transformers   802

SECTION XI
DC Machines   842

UNIT 30
DC Generators   843

UNIT 31
DC Motors   883

APPENDIX A

APPENDIX B

APPENDIX C

APPENDIX D

APPENDIX E

APPENDIX F

To my wife, Debbie, God's greatest gift to me.

Preface

Intended Use

Delmar's Standard Textbook of Electricity, 5th edition, is intended for students in electrical trade programs at high schools and community colleges, as well as those in industry training. It assumes that the reader has had no prior knowledge of electricity but also provides enough comprehensive coverage to be used as a reference tool for experienced electricians.

Subject & Approach

The content itself is presented as a blend of the practical and theoretical. It not only explains the different concepts relating to electrical theory but also provides many practical examples of how to do many of the common tasks the industrial electrician must perform. An extensive art program containing full color photographs and line drawings, as well as the inclusion of practical exercises for the student, also serve to further clarify theoretical concepts.

Design of Text

The subject matter has been divided into 34 separate units—each designed to "stand alone." The "stand alone" concept permits the information to be presented in almost any sequence the instructor desires, as teaching techniques vary from one instructor to another. The information is also presented in this manner to allow students and instructors quick reference on a particular subject.

Math Level

The math level has been kept to basic algebra and trigonometry, and Appendix B contains a section of electrical formulas—all divided into groups that are related to a particular application. Unit 15 of the text provides an introduction to basic trigonometry and vectors for those students weak in the subject.

A Note about Calculations

Delmar's Standard Textbook of Electricity, 5th edition, like all other scientific texts, contains numerous mathematical equations and calculations. Students often become concerned if their answers to problems are not exactly the same as the solutions given in the text. The primary reason for a discrepancy is the rounding off of values. Different scientific calculators carry out numbers to different places, depending on the manufacturer and model. Some calculators carry numbers to 8 places, some to 10 places, and some to 12 places. There may also be times when numbers that are reentered into the calculator are carried to only 2 or 3 decimal places of accuracy. For example, the numbers shown below will be multiplied with a calculator that carries numbers out to 8 places of accuracy:

$$3.21 \times 34.6 \times 4.32 \times 0.021 \times 3.098 \times 0.467$$

The answer is 14.577480.

The same problem will again be multiplied, but this time each answer will be reentered before it is multiplied by the next number. Each time the answer is reentered, it will be rounded off to 3 places after the decimal. If the fourth number after the decimal is 5 or greater, the third decimal place will be rounded up. If the fourth number is less than 5, it will be rounded down. The answer is 14.577405.

The same set of numbers will again be multiplied, but this time each answer will be reentered after rounding off the number to one place after the decimal. The answer is 14.617100.

Notice that all three answers are different, but all three are essentially correct. The most accurate answer is 14.577480, and the least accurate answer is 14.617100. Although these answers may look substantially different, they are within approximately 1% of each other.

Another consideration is problems that contain multiple steps. The more steps it takes to solve a problem, the more chance there is for inaccuracy. In most instances in this text, the answers were left in the display of the calculator, which permits the greatest degree of accuracy. When numbers had to be reentered, they were taken to 3 places of accuracy. When you work a problem in this text and your answer is different, consider the degree of difference before concluding that your answer is incorrect.

New to this Edition

The fifth edition of *Delmar's Standard Textbook of Electricity* continues to remain true to the comprehensive nature and visually appealing style that are its trademark features but will now offer more emphasis on the practical approach to electrical theory. New to this edition:

- *Explanation of the American Wire Gauge measurement used throughout industry*

- *Extended coverage of the effects of temperature on conductor resistance*

- *Coverage of fuel cells*

- *The addition of constant-current transformers*

- *Coverage of parallel transformer connections*

- *Energy saving "Green Tips" where applicable*

- *New Introduction*
  "Electrical Occupations" contains information about electrical personnel, building codes, and solar and wind energy.

Features of The Text

- *"Safety Overview"*
  At the beginning of Section I, Safety Overview provides information on general safety rules, personal protective equipment, potential job hazards, lock-out/tag-out procedures, GFCI, Grounding—and more! Students are acquainted with the all important safety concerns applicable to working in a lab and on the work site.

- *"Cautions"*
  Author highlights text where students should be aware of potential risks in working with various types of electrical equipment.

Caution: The **ammeter**, unlike the voltmeter, is a very low-impedance device. The ammeter is used to measure current and must be connected in series with the load to permit the load to limit the current flow *(Figure 10–13)*. ■

- *Math Presentation*
  Section on vectors in Unit 17 is presented earlier in the text in Unit 15, *Basic Trigonometry,* providing a foundation for students as they work through math equations.

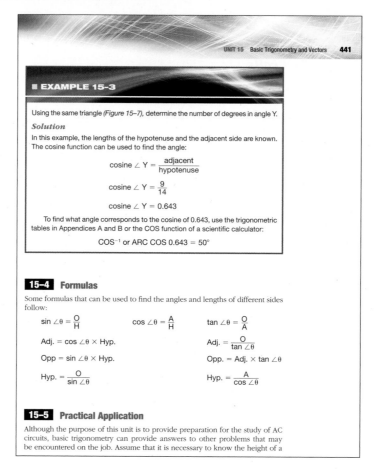

■ EXAMPLE 15–3

Using the same triangle *(Figure 15–7)*, determine the number of degrees in angle Y.

Solution

In this example, the lengths of the hypotenuse and the adjacent side are known. The cosine function can be used to find the angle:

$$\text{cosine} \angle Y = \frac{\text{adjacent}}{\text{hypotenuse}}$$

$$\text{cosine} \angle Y = \frac{9}{14}$$

$$\text{cosine} \angle Y = 0.643$$

To find what angle corresponds to the cosine of 0.643, use the trigonometric tables in Appendices A and B or the COS function of a scientific calculator:

$$\text{COS}^{-1} \text{ or ARC COS } 0.643 = 50°$$

15–4   Formulas

Some formulas that can be used to find the angles and lengths of different sides follow:

$$\sin \angle \theta = \frac{O}{H} \qquad \cos \angle \theta = \frac{A}{H} \qquad \tan \angle \theta = \frac{O}{A}$$

$$\text{Adj.} = \cos \angle \theta \times \text{Hyp.} \qquad \text{Adj.} = \frac{O}{\tan \angle \theta}$$

$$\text{Opp} = \sin \angle \theta \times \text{Hyp.} \qquad \text{Opp.} = \text{Adj.} \times \tan \angle \theta$$

$$\text{Hyp.} = \frac{O}{\sin \angle \theta} \qquad \text{Hyp.} = \frac{A}{\cos \angle \theta}$$

15–5   Practical Application

Although the purpose of this unit is to provide preparation for the study of AC circuits, basic trigonometry can provide answers to other problems that may be encountered on the job. Assume that it is necessary to know the height of a

- *"Why You Need to Know"*

    Boxed articles at the beginning of each unit explain to students the importance of learning the material presented in each unit, and how it may apply to actual job situations.

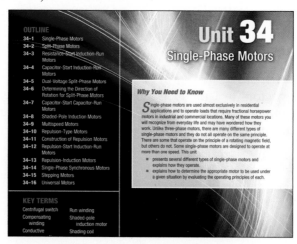

- *"Practical Applications"*
  Word problems step the students through potential situations on the job and encourage them to develop critical thinking skills.

> **Practical Applications**
>
> An office building uses a bank of 63 lead-acid cells connected in series with a capacity of 80 amp-hours each to provide battery backup for their computers. The lead-acid cells are to be replaced with nickel-metal hydride cells with a capacity of 40 amp-hours each. How many nickel-metal hydride cells will be required to replace the lead-acid cells and how should they be connected? ∎

- *DVD Correlation*
  Units are highlighted where material can be viewed on the accompanying DVD series, providing another source of learning for the student: **DC Electrical Theory, AC Electrical Theory, Single-PhaseTransformers & Electrical Machines, Three-Phase Circuits & Electrical Machines**

- *Text Design*
  A fresh design creates a text that makes it even easier to navigate through content, serving to facilitate learning for students.

- *New, Up-to-Date Art*
  Approximately 32 new four-color photos and line illustrations combined bring text up to date, keeping students aware of the latest technology in the industry.

- *Dedication to Technical Accuracy and Consistency*
  Text was thoroughly reviewed for technical accuracy and consistency, ensuring existing errors were corrected, enabling students to readily grasp more difficult concepts.

Supplement Package

- *Lab-Volt Manual* provides experiments for students to test and troubleshoot key concepts presented in the text, using Lab-volt equipment. (Order #: 1-1115-3916-2).
  Also available: The Complete Laboratory Manual for Electricity, by Steve Herman. This manual is designed to be conducted with common lab equipment. (Order #: 1-4283-2430-5).

- ***Instructor Resource (CD-ROM for Instructors)***
  (Order #: 1-1115-3916-2).
  Instructor Guide contains answers to all review questions and practical applications contained within the text, as well as practice exams.
  - *PowerPoint* presentations provide a thorough review of all major concepts presented in each unit, featuring four-color photos and line illustrations from the text. The fifth edition contains numerous Power-Point presentations not available before.
  - *Computerized Testbank* offered in *ExamView 4.0* contains approximately 700 questions for instructors to test student knowledge as they progress through the text. Allows instructors to edit the exams and add their own questions.
  - *Image Library* consists of all the images from the text in electronic format, allowing instructors to create their own classroom presentations.
  - *Video Clips* drawn from each video provide key lessons from the series.
  - *Instructors Guide & Solutions to Lab-Volt Manual* is in Word format.

  To access additional course materials including CourseMate, please visit www.cengagebrain.com. At the CengageBrain.com home page, search for the ISBN of your title (from the back cover of your book) using the search box at the top of the page. This will take you to the product page where these resources can be found.

- ***A DVD Set*** brings important concepts to life through easy-to-understand explanations and examples, professional graphics and animations, and a necessary emphasis on safety. Videos run approximately 20 minutes. The DVDs are interactive and provide test questions and remediation.

  DC Electrical Theory DVD (4 videos) includes Basic Electricity, Series & Parallel Circuits, Combination Circuits, and Small Sources of Electricity.

  AC Electrical Theory DVD (5 videos) includes Alternating Current, Inductance, Capacitors, Capacitors in AC Circuits, and Series Circuits.

  Single-Phase Transformers & Electrical Machines DVD (4 videos) includes Single-Phase Transformers; DC Machines; Single-Phase Motors, Part I; Single-Phase Motors, Part II.

  Three-Phase Circuits & Electrical Machines DVD (4 videos) includes Three-Phase Circuits; Three-Phase Transformers; Three-Phase Motors, Part I; Three-Phase Motors, Part II.

- ***Blackboard supplement*** features include chapter objectives, practice tests, glossary, and links to relevant websites. (Order #: 1-1115-3918-9).

Electrical Course Notes: This is a 6 panel brochure outlining the most common key concepts and formulas used when studying electrical theory. Order #: 1-1115-3923-5

CourseMate for Delmar's Standard Textbook of Electricity, 5E: This interactive and assignable web based solutions includes a CLeBook, Unit slides in PowerPoint, quizzes, animations, glossary, and engagement tracker.

A Note about the Lab Manuals

The two laboratory manuals, entitled *Experiments in Electricity for Use with Lab-Volt EMS Equipment* and *The Complete Laboratory Manual for Electricity, 3E,* provide extensive opportunities for students to apply what they have learned. Both manuals contain multiple hands-on experiments for each unit of the textbook and have been extensively field-tested to ensure that all the experiments will work as planned. The engineers at Lab-Volt conducted each of the experiments in *Experiments in Electricity for Use with Lab-Volt EMS Equipment,* and, following their testing, Lab-Volt has endorsed this manual. It is the manual they recommend to their customers. *The Complete Laboratory Manual for Electricity,* was field-tested at the Shreveport-Bossier Regional Technical School under the direction of Richard Cameron.

About the Author

Stephen L. Herman has been both a teacher of industrial electricity and an industrial electrician for many years. His formal training was obtained at Catawba Valley Technical College in Hickory, North Carolina. Mr. Herman has worked as a maintenance electrician for Superior Cable Corp. and as a class "A" electrician for National Liberty Pipe and Tube Co. During those years of experience, Mr. Herman learned to combine his theoretical knowledge of electricity with practical application. The books he has authored reflect his strong belief that a working electrician must have a practical knowledge of both theory and experience to be successful.

Mr. Herman was the Electrical Installation and Maintenance instructor at Randolph Technical College in Asheboro, North Carolina, for 9 years. After a return to industry, he became the lead instructor of the Electrical Technology Curriculum at Lee College in Baytown, Texas. He retired from Lee College after 20 years of service and, at present, resides in Pittsburg, Texas, with his wife. He continues to stay active in the industry, write, and update his books.

Acknowledgments

The author and publisher would like to express thanks to those reviewers who provided insightful feedback throughout the development of the fifth edition of this text:

James Blackett, Thomas Nelson Community College, Hampton, VA

James Cipollone, Antelope Valley Community College, Lancaster, CA

Eduardo Del Toro, MacArthur High School/Independent Electrical Contractors, San Antonio, TX

Randy Ludington, Guilford Community College, Greensboro, NC

J.C. Morrow, Hopkinsville Community College, Hopkinsville, KY

Robert B. Meyers, Jr., Harrisburg Area Community College, Harrisburg, PA

Larry Pogoler, LA Trade Tech College, Los Angeles, CA

Elmer Tepper, Gateway Community College, Phoenix, AZ

Dean Senter, Pratt Community College, Pratt, KS

Justin Shores, Antelope Valley Community College, Lancaster, CA

Raul Vasquez, Independent Electrical Contractors, San Antonio, TX

Introduction

Electrical Occupations

Organization of the Industry

The electrical industry is one of the largest in the United States and Canada. In 2008, electricians held about 692,000 jobs. Electrical contracting firms employed about 65% of the wage and salaried workers. The remainder worked as electricians in other related industries. About 9% of the electricians were self-employed. The opportunity for employment and advancement as an electrician is one of the highest of any industry. Basically, the entire country runs on electricity. Industry, commercial locations, and homes all employ electricity as the main source of power. It has been estimated that between 2008 and 2018 the need for qualified electricians will increase at a rate of about 12%. That represents an annual increase of over 8000 electricians over the next 10 years. The lay-off rate of electricians is one of the lowest of any occupation. If industry operates, it will require electricians to keep it running.

Electrical Personnel

Electricians can generally be divided into several categories, depending on their specific area of employment. Each of these categories may require special skills.

- **Construction**
  Electricians working in the construction industry generally require a basic knowledge of electrical theory and an extensive knowledge of *National Electrical Code* requirements and wiring practices. Electricians in the construction area can generally be divided into helpers, journeymen, and masters. Many states require tests for journeymen and master levels.

- **Industrial Electricians**
  Industrial electricians are generally concerned with maintaining equipment that has already been installed. Electricians in an industrial environment require an extensive knowledge of electrical theory and *National Electrical Code* requirements for installation of motors, capacitor banks, and transformers. Industrial electricians should also possess a basic knowledge of electronics and

electronic devices. Modern industry employs many electronic devices, such as variable frequency drives, solid state controls for direct current motors, and programmable logic controllers. Another area of concern for most industrial electricians is motor controls. Motor control systems are generally either relay logic or electronic in the form of programmable logic controllers or distributive control systems.

- ### *Instrumentation Technicians*
  Instrumentation technicians calibrate and maintain devices that sense such quantities as temperature, pressure, liquid level, flow rate, and others. These people should have an extensive knowledge of electrical theory, especially as it pertains to low-voltage and closed-loop systems.

- ### *Related Industries*
  The fields related to the electrical industry are too numerous to mention but include air conditioning and refrigeration, aircraft electronics, automotive, cable TV, broadcast media, energy and utilities, and home appliance and repair, as well as many, many others. The opportunity for employment in the electrical field is almost unlimited.

Union and Nonunion Employees

The largest percentage of electricians are nonunion employees. Many construction electricians receive training at various trade and technical schools. Some employers also sponsor apprenticeship programs. Apprenticeship-type programs generally require the electrician to work on the job as well as attend classes. The advantage to apprenticeship training is that it permits a person to earn money while he or she attends class. The disadvantage is that it can create an extremely busy schedule. Most industrial electricians, and those in related fields, require special training at a trade or technical school.

The largest electrician's union is the International Brotherhood of Electrical Workers (IBEW). The construction electricians who belong to the IBEW generally receive apprenticeship-type training for an organization called the National Joint Apprenticeship Training Committee (NJATC). Union electricians who work in related fields generally belong to unions organized for their particular industry, such as United Auto Workers or United Steel Workers.

Apprentices, whether union or nonunion, attend classes several hours a week and work on the job under the supervision of a journeyman. Most journeymen have completed their apprenticeship training and a set number of hours of practical work, and are required to pass an examination to become a journeyman. Journeymen work under the supervision of a master electrician. The master is ultimately responsible for the work performed and is answerable to the architect or owner. Most states require not only that a master pass a very rigorous examination but also be bonded for a particular sum of money, depending on the size of the job he or she bids on.

Ethics

Probably the greatest document concerning ethical behavior was given to a man named Moses on top of a mountain several thousand years ago and is called the Ten Commandments. Ethics are the principles by which behavior is judged to be right or wrong. There is an old saying stating that the best advertisement is word of mouth. This type of advertisement, however, can be a two-edged sword. People who do poor work, charge for work that was not done, make promises that are never kept, and cheat people at every opportunity gain a reputation that eventually catches up with them.

People who do an honest day's work for an honest wage, keep promises, and deal fairly with other people gain a reputation that will lead to success. Many years ago I worked for a man who had a business of rebuilding engines. He charged about twice the going rate of any other person in town and had more business than he could handle. I once asked him how he could charge more than anyone else and still have more business than anyone else. His answer was simple. He said, "There are two ways by which a business can be known. One is as the cheapest in town and the other is as the best in town. I'm the best in town." Most people are willing to pay more for a person that has a reputation for doing quality work and dealing honestly with customers.

Appearance

Appearance plays a major role in how a person is perceived. The old saying that first impressions are the most important is true. This doesn't mean that formal office attire is required to make a good impression on a prospective customer, but a professional person is expected to look professional. A person who wears clean work clothes and drives a relatively clean vehicle makes a much better impression than someone who shows up in filthy clothes with shirttail hanging out and pants sagging almost to the knees.

Communication

Communication skills are extremely important on any job. These skills can be divided into several areas such as speaking, listening, and writing.

Speaking: Speaking well is probably one of the most important skills for obtaining a successful career in any field. Generally, one of the first impressions you make concerns your ability to speak properly. Even though slang is widely used among friends, family, and the media, a person who uses proper English gives the impression of being educated, informed, and professional.

The ability to speak also involves communicating with people on the job, whether that person is a journeyman or an employer. The ability to explain clearly how a job was done or why it was done a certain way is also important, as it is often necessary to communicate with people who have no knowledge of the electrical field. The ability to explain to a homeowner

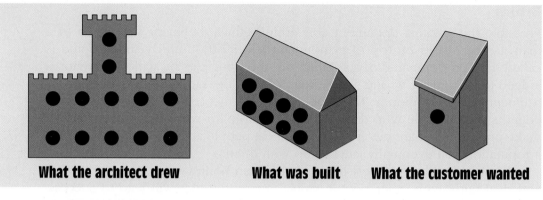

What the architect drew   What was built   What the customer wanted

Delmar/Cengage Learning

FIGURE OCCUPATIONS 1   Listening to the customer can save money and time.

why a receptacle or switch should or should not be placed in a particular location is important.

Listening: Listening is probably the most understated skill concerning communication. You should not only listen to what a person wants but also make sure you *understand* what he or she is saying. Not understanding what a person wants can lead to extremely costly mistakes. The most costly work is that which has to be redone because of a misunderstanding. An example of how misunderstandings can lead to costly mistakes is shown in *Figure Occupations 1.*

Writing: Many jobs require the electrician to fill out work reports that can include a description of the job, the materials used, and the time required to complete the job. This is especially true of a person in charge of other workers, such as a journeyman.

Maintenance electricians in an industrial environment generally submit a report on the maintenance performed on a particular machine. The report commonly includes the particular machine, the problem encountered, the materials necessary for repair, and the time spent in troubleshooting and repair.

Working on a Team

Teamwork is essential on most construction jobs. The typical construction job may include people that pour the concrete foundation; carpenters; brick masons; stone masons; plumbers; landscapers; people that install flooring and carpet; air-conditioning and refrigeration contractors; and, of course, electricians. One of the key elements to a successful team effort is communication. If conduit is to be run under the slab, it is better to communicate with the people doing the foundation and inform them that conduit needs to be run before the slab is poured.

Be respectful of other trades. If an electrical outlet box is in the way of a sewer line, the plumber may ask that it be moved. It is much easier to move an outlet box than it is to reroute a sewer line. If electrical boxes are to be placed in an outside brick wall, ask the brick mason how he would like the box to be placed. A little respect for other trades plus communication can solve many problems before they happen.

If possible, help other people. If you are already in an attic and the air-conditioning contractor asks whether you would be willing to do a small job that would save him time and effort, it is good working relations to do so. Grudges and hard feelings do not happen in a work setting where kindness is practiced.

Building Codes

Many cities, counties, and states have their own building codes that supersede the *National Electrical Code*. The *National Electrical Code* is law only if the local authority has adopted it as law. Always check local codes before beginning a construction project. Local codes often specify the manner in which wiring is to be installed and the size or type of wire that must be used for a particular application.

Green Building

"Green building" basically means making buildings more energy efficient. This can encompass many areas of the construction such as using "low E" energy-efficient windows, adding extra insulation, adding solar collectors to assist the water heater, and installing solar panels and/or wind generators to assist the electrical service. For the electrician, it may be installing larger wire than necessary to help overcome voltage drop, or installing energy-efficient appliances such as heat pump–type water heaters. These water heaters use about half the amount of power of a standard electric water heater. Energy-efficient appliances are generally identified by an Energy Star label. Energy Star is a government-backed symbol awarded to products that are considered energy efficient. Energy Star was established to reduce greenhouse gas emissions and other pollutants caused by inefficient use of energy, and to aid consumers in identifying and purchasing energy-efficient products that will save money without sacrificing performance, features, or comfort.

Before a product can receive an Energy Star label, it must meet certain requirements set forth in Energy Star product specifications:

- Product categories must produce significant energy savings nationwide.

- Qualified products must deliver the features and performance demanded by customers as well as increase energy efficiency.

- If the qualified product cost more than a conventional, less-efficient counterpart, purchasers must be able to recover their investment in increased energy efficiency through utility bill savings, within a reasonable period of time.

- Energy efficiency must be achievable through broadly available, nonproprietary technologies offered by more than one manufacturer.

- Product energy consumption and performance must be measurable and verified with testing.

- Labeling should effectively differentiate products and be clearly visible to purchasers.

Solar Energy

One of the primary sources of green energy is solar power. Solar energy is the primary source of heating water in many countries and can be as simple as a dark colored container mounted on the roof of a structure, *Figure Occupations 2.* Other types of solar water heaters involve a solar collector, a special tank that contains a heat exchanger, and related equipment, *Figure Occupations 3.* Most of these types of water heaters contain backup electric heating elements for cloudy weather when the solar collector cannot supply enough energy to heat the water.

Delmar/Cengage Learning

FIGURE OCCUPATIONS 2 Solar water heaters mounted on a roof.

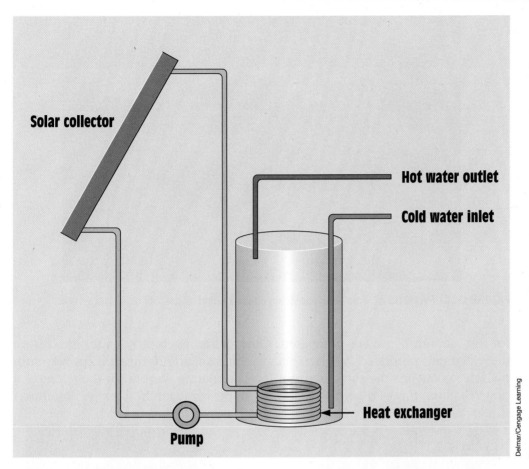

FIGURE OCCUPATIONS 3 Some solar water heaters use a solar panel and special tank with a heat exchanger.

Some solar systems generate electricity and are generally called PV (photo-voltaic) systems. In these types of systems solar panels are mounted on the roof of a dwelling or in an open area on the ground, *Figure Occupations 4*. Pho-tovoltaic cells generate direct current, which must be changed into alternating current by an inverter, *Figure Occupations 5*. The home remains connected to the utility company at all times. The solar panels augment the incoming power to help reduce the energy supplied by the utility company. There are various methods of supplying power to the utility company, depending on the require-ments of the utility company and state laws. Some systems cause the electric meter to run backward during times that the solar panels are producing more energy than is being supplied by the utility company. Other systems require the use of two separate meters, *Figure Occupations 6*. One records the amount of power supplied by the utility company and the other records the amount of power supplied by the solar cells. The utility company then purchases the power from the homeowner or in some cases gives the homeowner credit

FIGURE OCCUPATIONS 4   Four solar panels are often mounted on the roof or in an open area.

for the amount of power generated. Other systems employ batteries to store the electricity produced by the solar panels. An uninterruptable power supply (UPS) converts the direct current into alternating current. In the event of a power failure, the UPS continues to supply power from the storage batteries.

FIGURE OCCUPATIONS 5
Inverter changes the direct current produced by the solar cells into alternating current.

FIGURE OCCUPATIONS 6   One meters records the power supplied by the utility company, and another records the amount of power supplied by the solar panels.

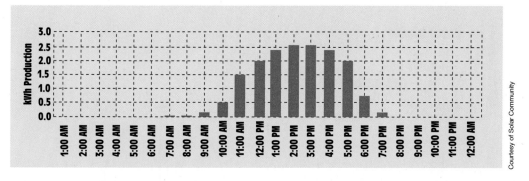

Courtesy of Solar Community

FIGURE OCCUPATIONS 7   Generation of electric power over a 24-hour period.

The amount of electricity produced by the solar panels is directly proportional to the intensity of sunlight striking the panels. The graph shown in *Figure Occupations* 7 illustrates the power output over a 24-hour period. The information was gathered during the month of March. Solar cells have a very long life span, generally considered to be 50 years or more. Most manufacturers of solar panels cover the cells with a material that is designed to remain clear in direct sunlight and is strong enough to withstand the average hail storm. Solar panels connect cells in series and parallel to obtain the desired voltage and current capacity.

Regardless of the type of system, there are generally specific procedures that must be followed during the installation of solar systems. Special circuit breakers designed for direct current and high amperage interrupt capability are often required. Manufacturers' recommendations as well as national and local electrical codes should be followed.

Wind Power

Another widely used form of "Green" energy is wind. Wind is actually a product of solar energy. The Sun heats different areas of the Earth's surface at different rates. Hot air rises at a faster rate than cool air. As the hot air rises, cool air rushes in to replace the void left by the rising hot air, and wind is created. Air has mass, and moving air can contain a lot of energy. Wind generators convert the kinetic energy of moving air into electricity. Wind energy increases by the cube of the speed, which means that each time the wind speed doubles, the amount of energy increases eight times. This is the reason that the shape of many automobiles is designed to move through the air with less friction. The wind resistance of an automobile traveling at 60 miles per hour will be eight times greater than when traveling at 30 miles per hour.

Wind generators are often referred to as wind turbines. There are two basic designs of wind turbines, the horizontal axis and the vertical axis, *Figure*

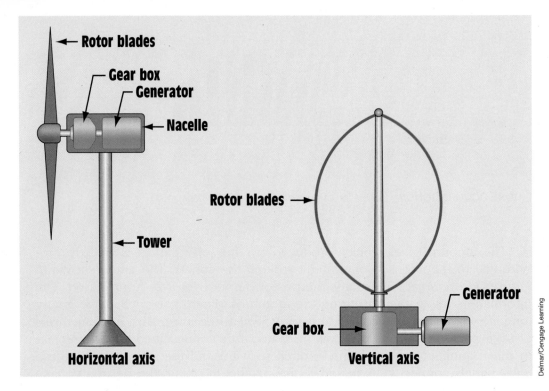

Delmar/Cengage Learning

FIGURE OCCUPATIONS 8   Wind turbine types.

Occupations 8. Vertical axis turbines are often called "egg beaters." The main advantage of vertical axis turbines is that they are omnidirectional, meaning that they will operate regardless of wind direction. Although horizontal axis turbines must turn to face the wind, they are mostly used for producing electricity. The size of wind turbines can vary greatly depending on the amount of electricity they are intended to produce. Utility scale turbines used in land-based wind farms, *Figure Occupations 9,* can have rotor diameters that range

Delmar/Cengage Learning

FIGURE OCCUPATIONS 9   Wind farm.

from 50 meters (164 feet) to 90 meters (295 feet). The tower height is generally the same as the rotor diameter. Utility wind turbines generally feed the electricity they produce directly into the power grid to aid other electricity-generating plants that use fossil fuels.

Wind turbines intended for residential or small commercial use are much smaller. Most have rotor diameters of 8 meters (26 feet) or smaller and are mounted on towers of 40 meters (131 feet) or less. As with solar installations, wind-powered systems can be installed to connect directly to the power grid or to charge a bank of batteries. An inverter is used to convert the direct current of the batteries into alternating current to supply the home. Inverters used to couple the wind turbine to the power line must be able to maintain a steady power flow with varying wind speeds and varying voltages. They must also be able to shape the waveform to that of a sine wave, *Figure Occupations 10*. Similar to solar installations, some wind-powered systems cause the electric

Courtesy of Diversified Technology, Inc.

FIGURE OCCUPATIONS 10   12kW Wind Power Inverter

meter to run backward when it is producing more power than is required by the home. Some utility companies will give credit for the amount of power generated, and some will purchase the power from the customer. Other utilities require the use of two separate meters to determine the amount of wind power produced.

As with solar systems, when installing a wind-powered system, manufacturers' instructions and utility requirements should be followed. Before installing a wind-powered system, check to make certain that the area has a high enough average wind speed to justify the cost of the system.

Lighting

Electric lighting began in 1879 when Thomas Edison invented the first incandescent lamp. He employed the use of a carbon filament that was heated to a temperature that produced a dim light by today's standards. In 1906, the incandescent lamp was improved by replacing the carbon filament with one made of tungsten. Tungsten could be heated to a much higher temperature and therefore could produce a much brighter light. Incandescent lamps today still use tungsten filaments. Incandescent lamps have the advantage of being inexpensive to purchase, but they also have a disadvantage in that they are very energy inefficient. These lamps are basically room heaters that produce light as a byproduct. At best, incandescent lamps are about 5% efficient, which means that a 100-watt lamp actually produces about 95 watts of heat and about 5 watts of light. They consume about 400% more energy to produce the same amount light as a standard fluorescent lamp.

Light is measured in *lumens*. The lumens, a metric measure of light intensity as perceived by the human eye, is based on the English measurement of a candela. Basically, a light source that uniformly radiates 1 candela in all directions is equal to 4π lumens. Lighting efficiency is measured by the lumens produced by 1 watt of electricity (lumens per watt). The chart in *Figure Occupations 11* lists the average lumens per watt for different types of lighting. The actual light output per watt can vary greatly for each type of lamp, depending on many conditions such as temperature, age, wattage, and so on. The range is listed for each type.

The chart indicates that some types are much more energy efficient than others, but all are not suited for use inside buildings. High-pressure sodium is the most efficient, but it has a very orange color, making it unsuitable for many applications. These lamps are generally used in outdoor applications such as parking lots and street lamps. Metal halide is also very efficient and has a near white color. These lamps are often used in large buildings like factories, warehouses, and commercial locations such as building supply stores. Florescent lighting is probably the type most used for homes, office buildings, and retail stores. Compact fluorescent lamps are replacing incandescent lamps

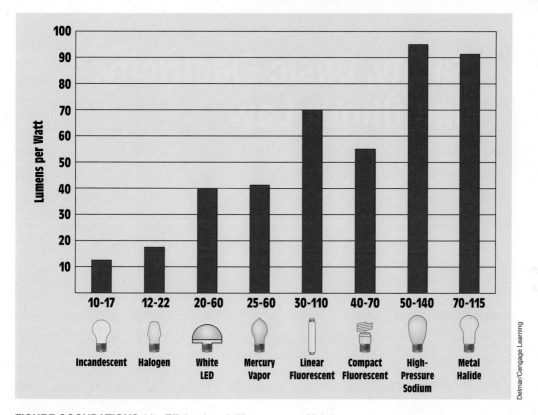

Delmar/Cengage Learning

FIGURE OCCUPATIONS 11   Efficiencies of different types of lighting.

in many homes. Although compact fluorescent lamps have an initial cost that is greater than incandescent lamps, they use about one-fourth the energy to produce a similar amount of light, and their average life expectancy is about 10 times longer. Because compact fluorescent lamps are more energy efficient, they produce less heat for the same amount of light, reducing the load on air-conditioning systems. Over the life expectancy of the lamp, the average cost of the compact fluorescent lamp will be less than a similar incandescent lamp.

Summary

The electrical field offers many avenues that can lead to success. Most electricians work in the construction industry, but many are employed as maintenance technicians in industry and other related fields. The demand for qualified electricians is expected to increase at a rate of over 8000 new jobs per year over the next 10 years. The lay-off rate for electricians is one of the lowest in the country. Electricity is the power that operates homes, businesses, and industry. If industry runs, it will require electricians to keep it running.

Safety, Basic Electricity, and Ohm's Law

Safety Overview

KEY TERMS

Artificial respiration

Cardiopulmonary resuscitation (CPR)

Confined spaces

De-energized circuit

Disconnect

Energized circuit

Fibrillation

Fire-retardant clothing

Horseplay

Idiot proofing

Lockout and tagout

Material safety data sheets (MSDS)

Meter

Milliamperes (mA)

Occupational Safety and Health Administration (OSHA)

Scaffolds

Why You Need to Know

Safety is the job of each individual. You should be concerned not only with your own safety but also with the safety of others around you. This is especially true for persons employed in the electrical field. Some general rules should be followed when working with electric equipment or circuits.

Delmar/Cengage Learning

Objectives

After studying this unit, you should be able to

- state basic safety rules.

- describe the effects of electric current on the body.

- discuss the origin and responsibilities of OSHA.

- discuss material safety data sheets.

- discuss lockout and tagout procedures.

- discuss types of protective clothing.

- explain how to properly place a straight ladder against a structure.

- discuss different types of scaffolds.

- discuss classes of fires.

- discuss ground-fault circuit interrupters.

- discuss the importance of grounding.

S-1   General Safety Rules

Never Work on an Energized Circuit If the Power Can Be Disconnected

When possible, use the following three-step check to make certain that power is turned off:

1. Test the **meter** on a known live circuit to make sure the meter is operating.

2. Test the circuit that is to become the **de-energized circuit** with the meter.

3. Test the meter on the known live circuit again to make certain the meter is still operating.

Install a warning tag at the point of disconnection so people will not restore power to the circuit. If possible, use a lock to prevent anyone from turning the power back on.

Think

Of all the rules concerning safety, this one is probably the most important. No amount of safeguarding or **idiot proofing** a piece of equipment can protect a person as well as taking time to think before acting. Many technicians have been killed by supposedly "dead" circuits. Do not depend on circuit breakers, fuses, or someone else to open a circuit. Test it yourself before you touch it. If you are working on high-voltage equipment, use insulated gloves and meter probes to measure the voltage being tested. *Think* before you touch something that could cost you your life.

Avoid Horseplay

Jokes and **horseplay** have a time and place, but not when someone is working on an electric circuit or a piece of moving machinery. Do not be the cause of someone's being injured or killed, and do not let someone else be the cause of your being injured or killed.

Do Not Work Alone

This is especially true when working in a hazardous location or on a live circuit. Have someone with you who can turn off the power or give **artificial respiration** and/or **cardiopulmonary resuscitation (CPR).** Several electric shocks can cause breathing difficulties and can cause the heart to go into fibrillation.

Work with One Hand When Possible

The worst kind of electric shock occurs when the current path is from one hand to the other, which permits the current to pass directly through the heart. A person can survive a severe shock between the hand and foot that would cause death if the current path were from one hand to the other.

Learn First Aid

Anyone working on electric equipment, especially those working with voltages greater than 50 volts, should make an effort to learn first aid. A knowledge of first aid, especially CPR, may save your own or someone else's life.

Avoid Alcohol and Drugs

The use of alcohol and drugs has no place on a work site. Alcohol and drugs are not only dangerous to users and those who work around them; they also cost industry millions of dollars a year. Alcohol and drug abusers kill thousands of people on the highways each year and are just as dangerous on a work site as they are behind the wheel of a vehicle. Many industries have instituted testing policies to screen for alcohol and drugs. A person who tests positive generally receives a warning the first time and is fired the second time.

S–2 Effects of Electric Current on the Body

Most people have heard that it is not the voltage that kills but the current. This is true, but do not be misled into thinking that voltage cannot harm you. Voltage is the force that pushes the current though the circuit. It can be compared to the pressure that pushes water through a pipe. The more pressure available, the greater the volume of water flowing through the pipe. Students often ask how much current will flow through the body at a particular voltage. There is no easy answer to this question. The amount of current that can flow at a particular voltage is determined by the resistance of the current path. Different people have different resistances. A body has less resistance on a hot day when sweating, because salt water is a very good conductor. What one eats and drinks for lunch can have an effect on the body's resistance, as can the length of the current path. Is the current path between two hands or from one hand to one foot? All these factors affect body resistance.

Figure S–1 illustrates the effects of different amounts of current on the body. This chart is general—some people may have less tolerance to electricity and others may have a greater tolerance.

A current of 2 to 3 **milliamperes (mA)** (0.002 to 0.003 amperes) usually causes a slight tingling sensation, which increases as current increases and becomes very noticeable at about 10 milliamperes (0.010 amperes). The tingling sensation is very painful at about 20 milliamperes. Currents between 20 and 30 milliamperes cause a person to seize the line and be unable to let go of the circuit. Currents between 30 and 40 milliamperes cause muscular paralysis, and those between 40 and 60 milliamperes cause breathing difficulty. When the current increases to about 100 milliamperes, breathing is extremely difficult. Currents from 100 to 200 milliamperes generally cause death because the heart usually goes into **fibrillation,** a condition in which the heart begins to "quiver" and the pumping action stops. Currents above 200 milliamperes cause the heart to squeeze shut. When the current is removed, the heart usually returns to a normal pumping action. This is the operating principle of a defibrillator. The voltage considered to be the most dangerous to work with is 120 volts, because that generally causes a current flow of between 100 and 200 milliamperes through most people's bodies. Large amounts of current can cause severe electric burns that are often very serious because they occur on the inside of the body. The exterior of the body may not look seriously burned, but the inside may be severely burned.

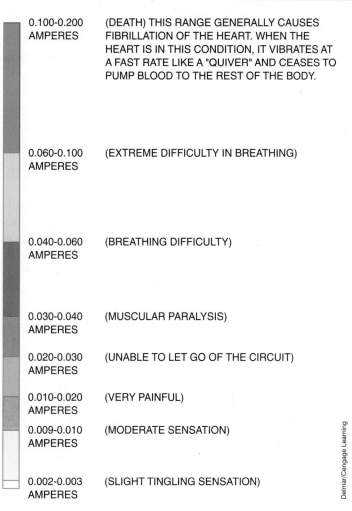

	0.100-0.200 AMPERES	(DEATH) THIS RANGE GENERALLY CAUSES FIBRILLATION OF THE HEART. WHEN THE HEART IS IN THIS CONDITION, IT VIBRATES AT A FAST RATE LIKE A "QUIVER" AND CEASES TO PUMP BLOOD TO THE REST OF THE BODY.
	0.060-0.100 AMPERES	(EXTREME DIFFICULTY IN BREATHING)
	0.040-0.060 AMPERES	(BREATHING DIFFICULTY)
	0.030-0.040 AMPERES	(MUSCULAR PARALYSIS)
	0.020-0.030 AMPERES	(UNABLE TO LET GO OF THE CIRCUIT)
	0.010-0.020 AMPERES	(VERY PAINFUL)
	0.009-0.010 AMPERES	(MODERATE SENSATION)
	0.002-0.003 AMPERES	(SLIGHT TINGLING SENSATION)

Delmar/Cengage Learning

FIGURE S–1   The effects of electric current on the body.

S-3   On the Job

OSHA

OSHA is an acronym for **Occupational Safety and Health Administration,** U.S. Department of Labor. Created by congress in 1971, its mission is to ensure safe and healthful workplaces in the United States. Since its creation, workplace fatalities have been cut in half, and occupational injury and illness rates have declined by 40%. Enforcement of OSHA regulations is the responsibility of the Secretary of Labor.

OSHA standards cover many areas, such as the handling of hazardous materials, fall protection, protective clothing, and hearing and eye protection. Part 1910, Subpart S, deals mainly with the regulations concerning electrical safety. These regulations are available in books and can be accessed at the OSHA website on the Internet at *http://www.osha.org.*

Hazardous Materials

It may become necessary to deal with some type of hazardous material. A hazardous material or substance is any substance to which exposure may result in adverse effects on the health or safety of employees. Hazardous materials may be chemical, biological, or nuclear. OSHA sets standards for dealing with many types of hazardous materials. The required response is determined by the type of hazard associated with the material. Hazardous materials are required to be listed as such. Much information concerning hazardous materials is generally found on **material safety data sheets (MSDS).** (A sample MSDS is included at the end of the unit.) If you are working in an area that contains hazardous substances, always read any information concerning the handling of the material and any safety precautions that should be observed. After a problem exists is not the time to start looking for information on what to do.

Some hazardous materials require a hazardous materials (HAZMAT) response team to handle any problems. A HAZMAT team is any group of employees designated by the employer who are expected to handle and control an actual or potential leak or spill of a hazardous material. They are expected to work in close proximity to the material. A HAZMAT team is not always a fire brigade, and a fire brigade may not necessarily have a HAZMAT team. On the other hand, a HAZMAT team may be part of a fire brigade or fire department.

Employer Responsibilities

Section 5(a)1 of the Occupational Safety and Health Act basically states that employers must furnish each of their employees a place of employment that is free of recognized hazards that are likely to cause death or serious injury. This places the responsibility for compliance on employers. Employers must identify hazards or potential hazards within the work site and eliminate them, control them, or provide employees with suitable protection from them. It is the employee's responsibility to follow the safety procedures set up by the employer.

To help facilitate these safety standards and procedures, OSHA requires that an employer have a competent person oversee implementation and enforcement of these standards and procedures. This person must be able to recognize unsafe or dangerous conditions and have the authority to correct or eliminate them. This person also has the authority to stop work or shut down a work site until safety regulations are met.

MSDS

MSDS stands for material safety data sheets, which are provided with many products. They generally warn users of any hazards associated with the product. They outline the physical and chemical properties of the product; list precautions that should be taken when using the product; and list any potential health hazards, storage consideration, flammability, reactivity, and, in some instances, radioactivity. They sometimes list the name, address, and telephone number of the manufacturer; the MSDS date and emergency telephone numbers; and, usually, information on first aid procedures to use if the product is swallowed or comes in contact with the skin. Safety data sheets can be found on many home products such as cleaning products, insecticides, and flammable liquids.

Trenches

It is often necessary to dig trenches to bury conduit. Under some conditions, these trenches can be deep enough to bury a person if a cave-in should occur. Safety regulations for the shoring of trenches is found in OSHA Standard 1926, Subpart P, App C, titled "Timber Shoring for Trenches." These procedures and regulations are federally mandated and must be followed. Some general safety rules also should be followed:

1. Do not walk close to trenches unless it is necessary. This can cause the dirt to loosen and increase the possibility of a cave-in.

2. Do not jump over trenches if it is possible to walk around them.

3. Place barricades around trenches *(Figure S–2)*.

4. Use ladders to enter and exit trenches.

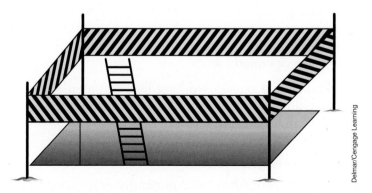

Delmar/Cengage Learning

FIGURE S–2 Place a barricade around a trench and use a ladder to enter and exit the trench.

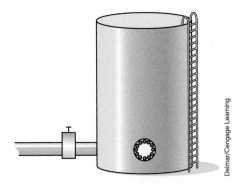

FIGURE S–3 A confined space is any space having a limited means of entrance or exit.

Confined Spaces

Confined spaces have a limited means of entrance or exit *(Figure S–3)*. They can be very hazardous workplaces, often containing atmospheres that are extremely harmful or deadly. Confined spaces are very difficult to ventilate because of their limited openings. It is often necessary for a worker to wear special clothing and use a separate air supply to work there. OSHA Section 12, "Confined Space Hazards," lists rules and regulations for working in a confined space. In addition, many industries have written procedures that must be followed when working in confined spaces. Some general rules include the following:

1. Have a person stationed outside the confined space to watch the person or persons working inside. The outside person should stay in voice or visual contact with the inside workers at all times. He or she should check air sample readings and monitor oxygen and explosive gas levels.

2. The outside person should never enter the space, even in an emergency, but should contact the proper emergency personnel. If he or she should enter the space and become incapacitated, there would be no one available to call for help.

3. Use only electric equipment and tools that are approved for the atmosphere found inside the confined area. It may be necessary to obtain a burning permit to operate tools that have open brushes and that spark when they are operated.

4. As a general rule, a person working in a confined space should wear a harness with a lanyard that extends to the outside person, so the outside person could pull him or her to safety if necessary.

Lockout and Tagout Procedures

Lockout and tagout procedures are generally employed to prevent someone from energizing a piece of equipment by mistake. This could apply to switches, circuit breakers, or valves. Most industries have their own internal

Delmar/Cengage Learning

FIGURE S–4   Safety tag used to tagout equipment.

Delmar/Cengage Learning

FIGURE S–5   The equipment can be locked out by several different people.

policies and procedures. Some require that a tag similar to the one shown in *Figure S–4* be placed on the piece of equipment being serviced; some also require that the equipment be locked out with a padlock. The person performing the work places the lock on the equipment and keeps the key in his or her possession. A device that permits the use of multiple padlocks and a safety tag is shown in *Figure S–5*. This is used when more than one person is working on the same piece of equipment. Violating lockout and tagout procedures is considered an extremely serious offense in most industries and often results in immediate termination of employment. As a general rule, there are no first-time warnings.

After locking out and tagging a piece of equipment, it should be tested to make certain that it is truly de-energized before working on it. A simple three-step procedure is generally recommended for making certain that a piece of electric equipment is de-energized. A voltage tester or voltmeter that has a high enough range to safely test the voltage is employed. The procedure is as follows:

1. Test the voltage tester or voltmeter on a known energized circuit to make certain the tester is working properly.

2. Test the circuit you intend to work on with the voltage tester or voltmeter to make sure that it is truly de-energized.

3. Test the voltage tester or voltmeter on a known energized circuit to make sure that the tester is still working properly.

This simple procedure helps to eliminate the possibility of a faulty piece of equipment indicating that a circuit is de-energized when it is not.

S–4   Protective Clothing

Maintenance and construction workers alike are usually required to wear certain articles of protective clothing, dictated by the environment of the work area and the job being performed.

Head Protection

Some type of head protection is required on almost any work site. A typical electrician's hard hat, made of nonconductive plastic, is shown in *Figure S–6*. It has a pair of safety goggles attached that can be used when desired or necessary.

Delmar/Cengage Learning

FIGURE S–6   Typical electrician's hard hat with attached safety goggles.

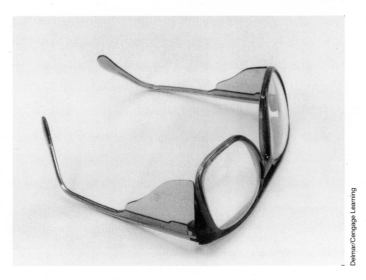

Delmar/Cengage Learning

FIGURE S–7   Safety glasses provide side protection.

Eye Protection

Eye protection is another piece of safety gear required on almost all work sites. Eye protection can come in different forms, ranging from the goggles shown in *Figure S–6* to the safety glasses with side shields shown in *Figure S–7*. Common safety glasses may or may not be prescription glasses, but almost all provide side protection *(Figure S–7)*. Sometimes a full face shield may be required.

Hearing Protection

Section III, Chapter 5, of the OSHA Technical Manual includes requirements concerning hearing protection. The need for hearing protection is based on the ambient sound level of the work site or the industrial location. Workers are usually required to wear some type of hearing protection when working in certain areas, usually in the form of earplugs or earmuffs.

Fire-Retardant Clothing

Special clothing made of fire-retardant material is required in some areas, generally certain industries as opposed to all work sites. **Fire-retardant clothing** is often required for maintenance personnel who work with high-power sources such as transformer installations and motor-control centers. An arc flash in a motor-control center can easily catch a person's clothes on fire. The typical motor-control center can produce enough energy during an arc flash to kill a person 30 feet away.

FIGURE S–8 Leather gloves with rubber inserts.

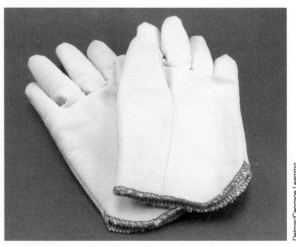

FIGURE S–9 Kevlar gloves protect against cuts.

Gloves

Another common article of safety clothing is gloves. Electricians often wear leather gloves with rubber inserts when it is necessary to work on energized circuits *(Figure S–8)*. These gloves are usually rated for a certain amount of voltage. They should be inspected for holes or tears before they are used. Kevlar gloves *(Figure S–9)* help protect against cuts when stripping cable with a sharp blade.

Safety Harness

Safety harnesses provide protection from falling. They buckle around the upper body with leg, shoulder, and chest straps; and the back has a heavy metal D-ring *(Figure S–10)*. A section of rope approximately 6 feet in length, called

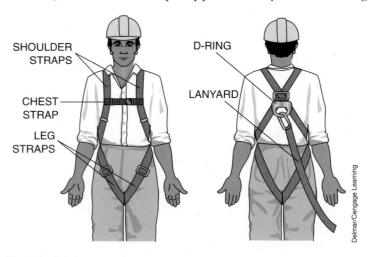

SHOULDER STRAPS

CHEST STRAP

LEG STRAPS

D-RING

LANYARD

FIGURE S–10 Typical safety harness.

a lanyard, is attached to the D-ring and secured to a stable structure above the worker. If the worker falls, the lanyard limits the distance he or she can drop. A safety harness should be worn:

1. When working more than 6 feet above the ground or floor

2. When working near a hole or drop-off

3. When working on high scaffolding

A safety harness is shown in *Figure S–11.*

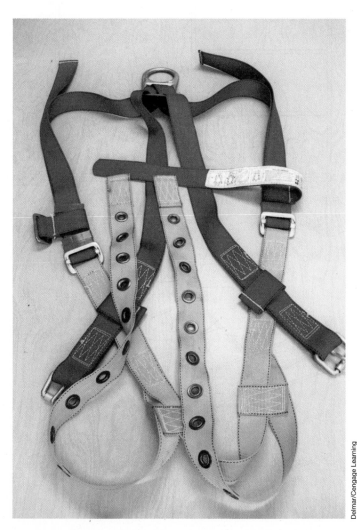

Delmar/Cengage Learning

FIGURE S–11   Safety harness.

S–5   Ladders and Scaffolds

It is often necessary to work in an elevated location. When this is the case, ladders or scaffolds are employed. **Scaffolds** generally provide the safest elevated working platforms. They are commonly assembled on the work site from standard sections *(Figure S–12)*. The bottom sections usually contain adjustable feet that can be used to level the sections. Two end sections are connected by X braces that form a rigid work platform *(Figure S–13)*. Sections of scaffolding are stacked on top of each other to reach the desired height.

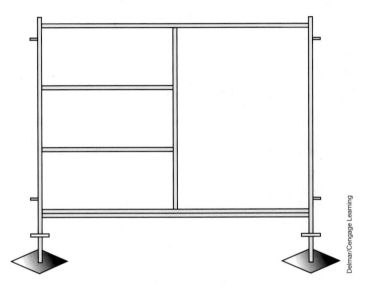

Delmar/Cengage Learning

FIGURE S–12   Typical section of scaffolding.

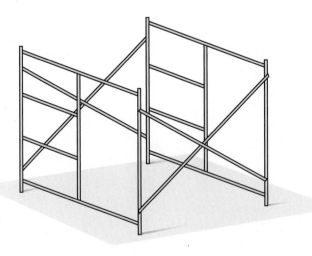

Delmar/Cengage Learning

FIGURE S–13   X braces connect scaffolding sections together.

Rolling Scaffolds

Rolling scaffolds are used in areas that contain level floors, such as inside a building. The major difference between a rolling scaffold and those discussed previously is that it is equipped with wheels on the bottom section that permit it to be moved from one position to another. The wheels usually contain a mechanism that permits them to be locked after the scaffold is rolled to the desired location.

Hanging or Suspended Scaffolds

Hanging or suspended scaffolds are suspended by cables from a support structure. They are generally used on the sides of buildings to raise and lower workers by using hand cranks or electric motors.

Straight Ladders

Ladders can be divided into two main types, straight and step. Straight ladders are constructed by placing rungs between two parallel rails *(Figure S–14)*. They generally contain safety feet on one end that help prevent the ladder from slipping. Ladders used for electrical work are usually wood or fiberglass; aluminum ladders are avoided because they conduct electricity. Regardless of the type of ladder used, you should check its load capacity before using it. This information is found on the side of the ladder. Load capacities of 200 pounds, 250 pounds, and 300 pounds are common. Do not use a ladder that does not have enough load capacity to support your weight plus the weight of your tools and the weight of any object you are taking up the ladder with you.

Delmar/Cengage Learning

FIGURE S–14 Straight ladder.

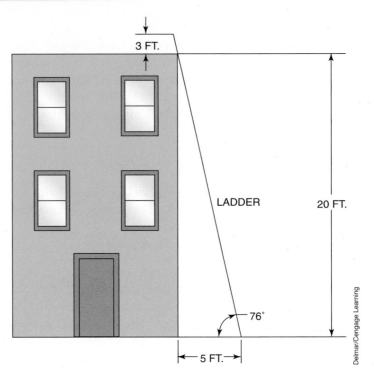

FIGURE S–15   A ladder should be placed at an angle of approximately 76°.

Straight ladders should be placed against the side of a building or other structure at an angle of approximately 76° *(Figure S–15)*. This can be accomplished by moving the base of the ladder away from the structure a distance equal to one-fourth the height of the ladder. If the ladder is 20 feet high, it should be placed 5 feet from the base of the structure. If the ladder is to provide access to the top of the structure, it should extend 3 feet above the structure.

Step Ladders

Step ladders are self-supporting, constructed of two sections hinged at the top *(Figure S–16)*. The front section has two rails and steps, the rear portion two rails and braces. Like straight ladders, step ladders are designed to withstand a certain load capacity. Always check the load capacity before using a ladder. As a general rule, ladder manufacturers recommend that the top step not be used because of the danger of becoming unbalanced and falling. Many people mistakenly think the top step is the top of the ladder, but it is actually the last step before the ladder top.

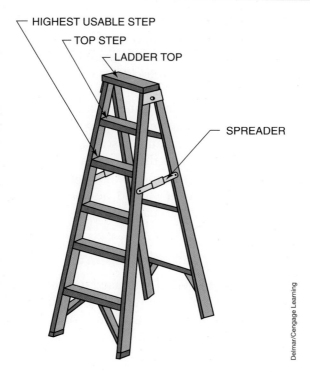

HIGHEST USABLE STEP
TOP STEP
LADDER TOP
SPREADER

Delmar/Cengage Learning

FIGURE S–16   Typical step ladder.

S–6   Fires

For a fire to burn, it must have three things: fuel, heat, and oxygen. Fuel is anything that can burn, including materials such as wood, paper, cloth, combustible dusts, and even some metals. Different materials require different amounts of heat for combustion to take place. If the temperature of any material is below its combustion temperature, it will not burn. Oxygen must be present for combustion to take place. If a fire is denied oxygen, it will extinguish.

Fires are divided into four classes: A, B, C, and D. Class A fires involve common combustible materials such as wood or paper. They are often extinguished by lowering the temperature of the fuel below the combustion temperature. Class A fire extinguishers often use water to extinguish a fire. A fire extinguisher listed as Class A only should never be used on an electrical fire.

Class B fires involve fuels such as grease, combustible liquids, or gases. A Class B fire extinguisher generally employs carbon dioxide (CO_2), which greatly lowers the temperature of the fuel and deprives the fire of oxygen. Carbon dioxide extinguishers are often used on electrical fires, because they do not destroy surrounding equipment by coating it with a dry powder.

Class C fires involve energized electric equipment. A Class C fire extinguisher usually uses a dry powder to smother the fire. Many fire extinguishers can be used on multiple types of fires; for example, an extinguisher labeled ABC could be used on any of the three classes of fire. The important thing to remember is never to use an extinguisher on a fire for which it is not rated. Using a Class A extinguisher filled with water on an electrical fire could be fatal.

Class D fires consist of burning metal. Spraying water on some burning metals can actually cause the fire to increase. Class D extinguishers place a powder on top of the burning metal that forms a crust to cut off the oxygen supply to the metal. Some metals cannot be extinguished by placing powder on them, in which case the powder should be used to help prevent the fire from spreading to other combustible materials.

S–7   Ground-Fault Circuit Interrupters

Ground-fault circuit interrupters (GFCI) are used to prevent people from being electrocuted. They work by sensing the amount of current flow on both the ungrounded (hot) and grounded (neutral) conductors supplying power to a device. In theory, the amount of current in both conductors should be equal but opposite in polarity *(Figure S–17)*. In this example, a current of 10 amperes flows in both the hot and neutral conductors.

A ground fault occurs when a path to ground other than the intended path is established *(Figure S–18)*. Assume that a person comes in contact with a defective electric appliance. If the person is grounded, a current path can be

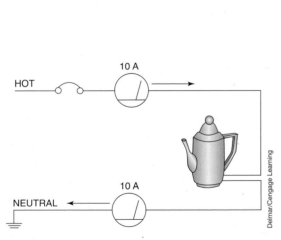

FIGURE S–17   The current in both the "hot" and neutral conductors should be the same, but flowing in opposite directions.

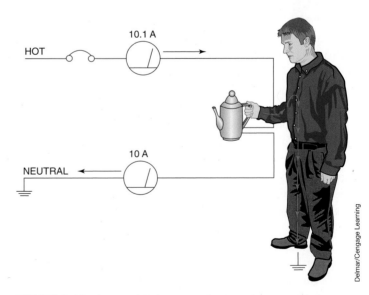

FIGURE S–18   A ground fault occurs when a path to ground other than the intended path is established.

established through the person's body. In the example shown in *Figure S–18*, it is assumed that a current of 0.1 ampere is flowing through the person. This means that the hot conductor now has a current of 10.1 amperes, but the neutral conductor has a current of only 10 amperes. The GFCI is designed to detect this current difference to protect personnel by opening the circuit when it detects a current difference of approximately 5 milliamperes (0.005 ampere). The *National Electrical Code® (NEC®) 210.8* lists places where ground-fault protection is required in dwellings. The *National Electrical Code* and *NEC* are registered trademarks of the National Fire Protection Association, Quincy, MA.

GFCI Devices

Several devices can be used to provide ground-fault protection, including the ground-fault circuit breaker *(Figure S–19)*. The circuit breaker provides ground-fault protection for an entire circuit, so any device connected to the circuit is ground-fault protected. A second method of protection, ground-fault receptacles *(Figure S–20)*, provide protection at the point of attachment. They have some

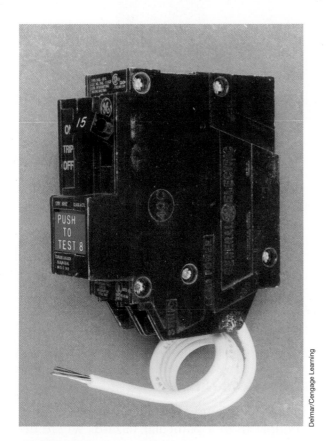

FIGURE S–19   Ground-fault circuit breaker.

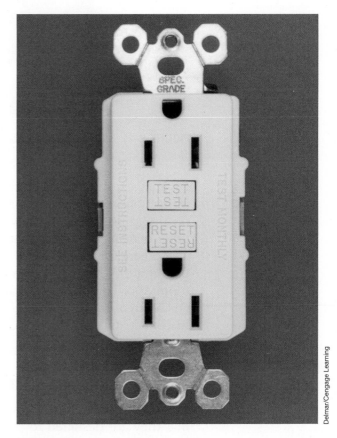

FIGURE S–20   Ground-fault receptacle.

Delmar/Cengage Learning

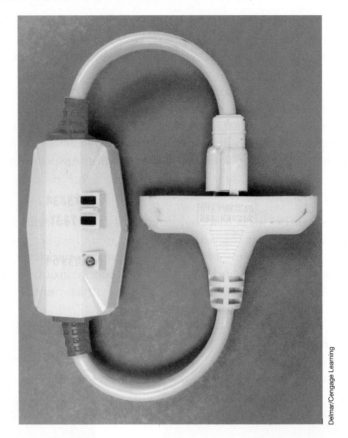

Delmar/Cengage Learning

FIGURE S–21 Ground-fault extension.

advantages over the GFCI circuit breaker. They can be connected so that they protect only the devices connected to them and do not protect any other outlets on the same circuit, or they can be connected so they provide protection to other outlets. Another advantage is that, because they are located at the point of attachment for the device, there is no stray capacitance loss between the panel box and the equipment being protected. Long wire runs often cause nuisance tripping of GFCI circuit breakers. A third ground-fault protective device is the GFCI extension cord *(Figure S–21)*. It can be connected into any standard electric outlet, and any devices connected to it are then ground-fault protected.

S–8 Arc-Fault Circuit Interrupters (AFCIs)

Arc-fault circuit interrupters are similar to ground fault circuit interrupters in that they are designed to protect people from a particular hazard. Where the ground fault interrupter is designed to protect against electrocution, the arc-fault interrupter is intended to protect against fire. Studies have shown that one-third of electrical related fires are caused by an arc-fault condition. At present, the *National Electrical Code* requires that arc-fault circuit interrupters be used on all 120-volt, single-phase, 15- and 20-ampere circuits installed in dwelling units supplying power to family rooms, dining rooms, living rooms, parlors, libraries, dens, bedrooms, sunrooms, recreation rooms, closets, hallways, or similar rooms or areas.

An arc-fault is a plasma flame that can develop temperatures in excess of 6000°C (10,832°F). Arc faults occur when an intermittent gap between two conductors or a conductor and ground permits current to "jump" between the two conductive surfaces. There are two basic types of arc faults, the parallel and the series.

Parallel Arc Faults

Parallel arc faults are caused by two conductors becoming shorted together *(Figure S–22)*. A prime example of this is when the insulation of a lamp cord or extension cord has become damaged and permits the two conductors to short together. The current in this type of fault is limited by the resistance of the conductors in the circuit. The current in this type of fault is generally much higher than the rated current of a typical thermomagnetic circuit breaker. A continuous short will usually cause the circuit breaker to trip almost immediately because it will activate the magnetic part of the circuit breaker, but an intermittent short may take some time to heat the thermal part of the circuit

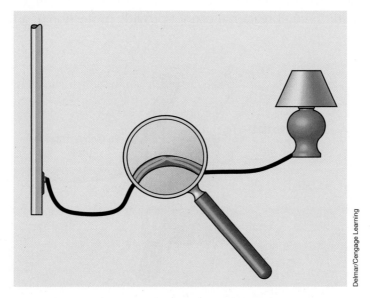

Delmar/Cengage Learning

FIGURE S–22   Parallel arc faults are caused by two conductors touching.

breaker enough to cause it to trip open. Thermal/magnetic type circuit breakers are generally effective in protecting against this type of arc fault, but cords with small-size conductors, such as lamp and small extension cords, can add enough resistance to the circuit to permit the condition to exist long enough to produce sufficient heat to start a fire.

Parallel arc faults can be more hazardous than series arc faults because they generate a greater amount of heat. Arc faults of this type often cause hot metal to be ejected into combustible material. Parallel arc faults, however, generally produce peak currents that are well above the normal current rating of a circuit breaker. This permits the electronic circuits in the arc-fault circuit interrupter to detect them very quickly and trip the breaker in a fraction of a second.

Series Arc Faults

Series arc faults are generally caused by loose connections. A loose screw on an outlet terminal, or an improperly made wire nut connection, is a prime example of this type of problem. They are called series arc faults because the circuit contains some type of current-limiting resistance connected in series with the arc *(Figure S–23)*. Although the amount of electrical energy converted into heat is less than that of a parallel arc fault, series arc faults can be more dangerous. The fact that the current is limited by some type of load keeps the current below the thermal and magnetic trip rating of a common thermo/magnetic circuit breaker. Because the peak arc current is never greater than the normal steady current flow, series arcing is more difficult to detect than parallel arcing.

When the current of an arc remains below the normal range of a common thermomagnetic circuit breaker, it cannot provide protection. If a hair dryer, for

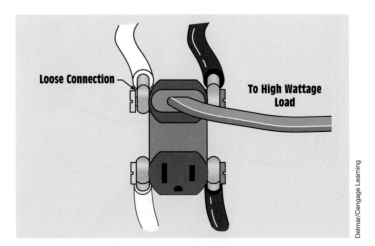

FIGURE S–23   Series arc-faults are generally caused by bad connections.

example, normally has a current draw of 12 amperes, but the wall outlet has a loose screw at one terminal so that the circuit makes connection only half of the time, the average circuit current is 6 amperes. This is well below the trip rating of a common circuit breaker. A 6-ampere arc, however, can produce a tremendous amount of heat in a small area.

Arc-Fault Detection

There are conditions where arcing in an electric circuit is normal, such as these:

- Turning a light switch on or off
- Switching on or off of a motor relay
- Plugging in an appliance that is already turned on
- Changing a light bulb with the power turned on
- Arcing caused by motors that contain a commutator and brushes

The arc-fault circuit interrupter is designed to be able to distinguish between normally occurring arcs and an arc fault. An arc caused by a toggle switch being used to turn a light on or off will produce a current spike of short duration, as shown in *Figure S–24*. An arc fault, however, is an intermittent

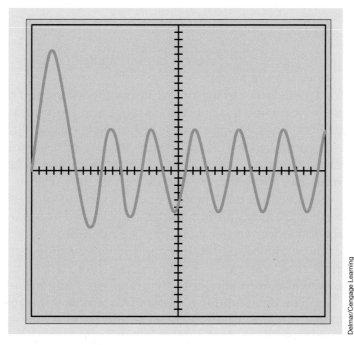

FIGURE S–24 Current spike produced by turning a light on or off.

Delmar/Cengage Learning

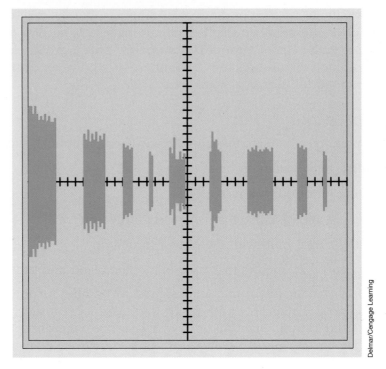

FIGURE S–25 Waveform produced by typical arc fault.

connection and will generally produce current spikes of various magnitudes and lengths of time *(Figure S–25)*.

   In order for an arc-fault circuit interrupter to determine the difference between a normally occurring arc and an arc fault, a microprocessor and other related electronic components are employed to detect these differences. The AFCI contains current and temperature sensors as well as a microprocessor and nonvolatile (retains its information when power is switched off) memory. The current and temperature sensors permit the AFCI to operate as a normal circuit breaker in the event of a circuit overload or short circuit. The microprocessor continuously monitors the current and compares the waveform to information stored in the memory. The microprocessor is monitoring the current for the magnitude, duration, and length of time between pulses, not for a particular waveform. For this reason, there are some appliances that can produce waveforms similar to that of an arc fault and may cause the AFCI to trip. Appliances containing motors that employ the use of brushes and a commutator, such as vacuum cleaners and hand drills, will produce a similar waveform.

FIGURE S–26   Arc-fault circuit breaker.

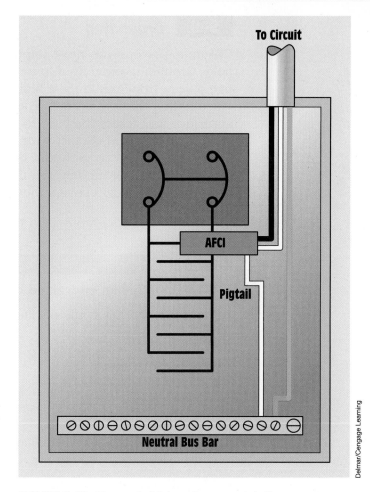

FIGURE S–27   The arc-fault interrupter connects in the same manner as a ground-fault interrupter.

Connecting an Arc-Fault Circuit Interrupter

The AFCI is connected in the same manner as a ground fault circuit breaker. The AFCI contains a white pigtail *(Figure S–26)* that is connected to the neutral bus bar in the panel box. Both the neutral and hot or ungrounded conductors of the branch circuit are connected to the arc-fault circuit breaker. The circuit breaker contains a silver-colored and a brass-colored screw. The neutral or white wire of the branch circuit is inserted under the silver screw, and the black wire is inserted under the brass screw *(Figure S–27)*. A rocker switch located on the front of the AFCI permits the breaker to be tested for both short and arc condition. In addition to the manual test switch, the microprocessor performs a self-test about once every 10 minutes.

S–9 Grounding

Grounding is one the most important safety considerations in the electrical field. Grounding provides a low resistance path to ground to prevent conductive objects from existing at a high potential. Many electric appliances are provided with a three-wire cord. The third prong is connected to the case of the appliance and forces the case to exist at ground potential. If an ungrounded conductor comes in contact with the case, the grounding conductors conduct the current directly to ground. The third prong on a plug should never be cut off or defeated. Grounding requirements are far too numerous to list in this chapter, but *NEC 250* covers the requirements for the grounding of electrical systems.

Summary

- Never work on an **energized circuit** if the power can be **disconnected**.
- The most important rule of safety is to think.
- Avoid horseplay.
- Do not work alone.
- Work with one hand when possible.
- Learn first aid and CPR.
- A current of 100 to 200 milliamperes passing through the heart generally causes death.
- The mission of OSHA is to ensure safe and healthy workplaces in the United States.
- Avoid using alcohol and drugs in the workplace.
- Do not walk close to trenches unless it is necessary.
- Do not jump over trenches if it is possible to walk around them.
- Place barricades around trenches.
- Use ladders to enter and exit trenches.
- When working in confined spaces, an outside person should keep in constant contact with people inside the space.
- Lockout and tagout procedures are used to prevent someone from energizing a circuit by mistake.
- Scaffolds generally provide the safest elevated working platforms.

- The bottom of a straight ladder should be placed at a distance from the wall that is equal to one fourth the height of the ladder.

- Fires can be divided into four classes: Class A is common items such as wood and paper; Class B is grease, liquids, and gases; Class C is energized electric equipment; and Class D is metals.

- Ground-fault circuit interrupters are used to protect people from electric shock.

- GFCI protectors open the circuit when approximately 5 milliamperes of ground-fault current are sensed.

- Arc-fault interrupters protect against electrical fires by sensing an arc-fault condition. Arc-fault circuit interrupters employ a microprocessor to sense an arc-fault condition.

- *NEC 250* lists requirements for grounding electrical systems.

Review Questions

1. What is the most important rule of electrical safety?

2. Why should a person work with only one hand when possible?

3. What range of electric current generally causes death?

4. What is fibrillation of the heart?

5. What is the operating principle of a defibrillator?

6. Who is responsible for enforcing OSHA regulations?

7. What is the mission of OSHA?

8. What is an MSDS?

9. A padlock is used to lock out a piece of equipment. Who should have the key?

10. A ladder is used to reach the top of a building 16 feet tall. What distance should the bottom of the ladder be placed from the side of the building?

11. What is a ground fault?

12. What is the approximate current at which a ground-fault detector will open the circuit?

13. Name three devices used to provide ground-fault protection.

14. What type of fire is Class B?

15. What section of the *NEC* covers grounding?

Section 1	Identity of Material
Trade Name	OATEY HEAVY DUTY CLEAR LO-VOC PVC CEMENT
Product Numbers	31850, 31851, 31853, 31854
Formula	PVC Resin in Solvent Solution
Synonyms	PVC Plastic Pipe Cement
Firm Name & Mailing Address	OATEY CO., 4700 West 160th Street, P.O. Box 35906 Cleveland, Ohio 44135, U.S.A. http://www.oatey.com
Oatey Phone Number	1-216-267-7100
Emergency Phone Numbers	For Emergency First Aid call 1-303-623-5716 COLLECT. For chemical transportation emergencies ONLY, call Chemtrec at 1-800-424-9300
Prepared By	Charles N. Bush, Ph.D.

Section 2	Hazardous Ingredients		
Ingredients	%	*Cas Number*	*Sec 313*
Acetone	0–5%	67-64-1	No
Amorphous Fumed Silica (Nonhazardous)	1–3%	112945-52-5	No
Proprietary (Nonhazardous)	5–15%	N/A	No
PVC Resin (Nonhazardous)	10–16%	9002-86-2	No
Cyclohexanone	5–15%	108-94-1	No
Tetrahydrofuran (See SECTION 11)	30–50%	109-99-9	No
Methyl Ethyl Ketone	20–35%	78-93-3	Yes

Section 3	Known Hazards Under U.S. 29 CFR 1910.1200				
Hazards	*Yes*	*No*	*Hazards*	*Yes*	*No*
Combustible Liquid		x	Skin Hazard	x	
Flammable Liquid	x		Eye Hazard	x	
Pyrophoric Material		x	Toxic Agent	x	
Explosive Material		x	Highly Toxic Agent		x
Unstable Material		x	Sensitizer		x

TABLE S–1 Heavy Duty Clear LO-VOC PVC Cement

Hazards	Yes	No	Hazards	Yes	No
Water Reactive Material		x	Kidney Toxin	x	
Oxidizer		x	Reproductive Toxin	x	
Organic Peroxide		x	Blood Toxin		x
Corrosive Material		x	Nervous System Toxin	x	
Compressed Gas		x	Lung Toxin	x	
Irritant	x		Liver Toxin	x	
Carcinogen NTP/IARC/ OSHA (see SECTION 11)		x			

Section 4	Emergency and First Aid Procedures—Call 1-303-623-5716 Collect
Skin	If irritation arises, wash thoroughly with soap and water. Seek medical attention if irritation persists. Remove dried cement with Oatey Plumber's Hand Cleaner or baby oil.
Eyes	If material gets into eyes or if fumes cause irritation, immediately flush eyes with water for 15 minutes. If irritation persists, seek medical attention.
Inhalation	Move to fresh air. If breathing is difficult, give oxygen. If not breathing, give artificial respiration. Keep victim quiet and warm. Call a poison control center or physician immediately. If respiratory irritation occurs and does not go away, seek medical attention.
Ingestion	**DO NOT INDUCE VOMITING.** This product may be aspirated into the lungs and cause chemical pneumonitis, a potentially fatal condition. Drink water and call a poison control center or physician immediately. Avoid alcoholic beverages. Never give anything by mouth to an unconscious person.

Section 5	Fire Fighting Measures
Precautions	Do not use or store near heat, sparks, or flames. Do not smoke when using. Vapors may accumulate in low places and may cause flash fires.
Special Fire Fighting Procedures	**FOR SMALL FIRES:** Use dry chemical, CO_2, water or foam extinguisher. **FOR LARGE FIRES:** Evacuate area and call Fire Department immediately.

TABLE S–1 Continued

Section 6	Accidental Release Measures
Spill or Leak Procedures	Remove all sources of ignition and ventilate area. Stop leak if it can be done without risk. Personnel cleaning up the spill should wear appropriate personal protective equipment, including respirators if vapor concentrations are high. Soak up spill with absorbent material such as sand, earth or other noncombusting material. Put absorbent material in covered, labeled metal containers. Contaminated absorbent material may pose the same hazards as the spilled product. See Section 13 for disposal information.
Section 7	Handling and Storage
Precautions	**HANDLING & STORAGE:** Keep away from heat, sparks and flames; store in cool, dry place. **OTHER:** Containers, even empties, will retain residue and flammable vapors.
Section 8	Exposure Controls/Personal Protection
Protective Equipment Types	**EYES:** Safety glasses with side shields. **RESPIRATORY:** NIOSH-approved canister respirator in absence of adequate ventilation. **GLOVES:** Rubber gloves are suitable for normal use of the product. For long exposures to pure solvents, chemical-resistant gloves may be required. **OTHER:** Eye wash and safety shower should be available.
Ventilation	**LOCAL EXHAUST:** Open doors & windows. Exhaust ventilation capable of maintaining emissions at the point of use below PEL. If used in enclosed area, use exhaust fans. Exhaust fans should be explosion-proof or set up in a way that flammable concentrations of solvent vapors are not exposed to electrical fixtures or hot surfaces.

Section 9	Physical and Chemical Properties			
NFPA Hazard Signal	Health 2	Stability 1	Flammability 3	Special None
HMIS Hazard Signal	Health 3	Stability 1	Flammability 4	Special None
Boiling Point	151 °F/66 °C			
Melting Point	N/A			
Vapor Pressure	145 mmHg @ 20 °C			

TABLE S–1 Continued

Vapor Density (Air = 1)	2.5
Volatile Components	70–80%
Solubility In Water	Negligible
PH	N/A
Specific Gravity	0.95 +/–0.015
Evaporation Rate	(BUAC = 1) = 5.5 – 8.0
Appearance	Clear Liquid
Odor	Ether-Like
Will Dissolve In	Tetrahydrofuran
Material Is	Liquid

Unit 1
Atomic Structure

Why You Need to Know

Atoms are the building blocks of the universe, and all matter is composed of them. One component of an atom is the electron, and all electrical quantities, such as voltage, current, and watts, are based on other electrical units that are derived from the measurement of electrons. A basic understanding of electron flow will remove the "mystery" of electricity and will start you on a path to a further understanding of electrical theory. This unit explains

- how electricity is produced and how those sources are divided in alternating current (AC) and direct current (DC) for utilization.
- why some materials are conductors and others are insulators.
- why a conductor becomes warm as a current flows through it.

KEY TERMS

Alternating current (AC)
Atom
Atomic number
Attraction
Bidirectional
Conductor
Direct current (DC)
Electron
Electron orbit
Element
Insulators
Matter
Molecules
Negative
Neutron
Nucleus
Positive
Proton
Repulsion
Semiconductors
Unidirectional
Valence electrons

Objectives

After studying this unit, you should be able to

■ list the three principal parts of an atom.

■ state the law of charges.

■ discuss centripetal force.

■ discuss the differences between conductors and insulators.

Preview

Electricity is the driving force that provides most of the power for the industrialized world. It is used to light homes, cook meals, heat and cool buildings, drive motors, and supply the ignition for most automobiles. The technician who understands electricity can seek employment in almost any part of the world.

Electric sources are divided into two basic types, **direct current (DC)** and **alternating current (AC).** Direct current is **unidirectional,** which means that it flows in only one direction. The first part of this text is mainly devoted to the study of direct current. Alternating current is **bidirectional,** which means that it reverses its direction of flow at regular intervals. The latter part of this text is devoted mainly to the study of alternating current. ■

1–1   Early History of Electricity

Although the practical use of electricity has become common only within the last hundred years, it has been known as a force for much longer. The Greeks were the first to discover electricity about 2500 years ago. They noticed that when amber was rubbed with other materials, it became charged with an unknown force that had the power to attract objects such as dried leaves, feathers, bits of cloth, or other lightweight materials. The Greeks called amber *elektron*. The word *electric* was derived from it and meant "to be like amber," or to have the ability to attract other objects.

This mysterious force remained little more than a curious phenomenon until about 2000 years later, when other people began to conduct experiments. In the early 1600s, William Gilbert discovered that amber was not the only material that could be charged to attract other objects. He called materials that could be charged *electriks* and materials that could not be charged *nonelektriks*.

About 300 years ago, a few men began to study the behavior of various charged objects. In 1733, a Frenchman named Charles DuFay found that a piece of charged glass would repel some charged objects and attract others. These men soon learned that the force of **repulsion** was just as important as the force

LIST A	LIST B
Glass (rubbed on silk)	Hard rubber (rubbed on wool)
Glass (rubbed on wool or cotton)	Block of sulfur (rubbed on wool or fur)
Mica (rubbed on cloth)	Most kinds of rubber (rubbed on cloth)
Asbestos (rubbed on cloth or paper)	Sealing wax (rubbed on silk, wool, or fur)
Stick of sealing wax (rubbed on wool)	Mica (rubbed on dry wool)
	Amber (rubbed on cloth)

Delmar/Cengage Learning

FIGURE 1–1 List of charged materials.

of **attraction.** From these experiments, two lists were developed *(Figure 1–1)*. It was determined that any material in list A would attract any material in list B, that all materials in list A would repel each other, and that all materials in list B would repel each other *(Figure 1–2)*. Various names were suggested for the materials in lists A and B. Any opposite-sounding names could have been chosen, such as east and west, north and south, male and female. Benjamin

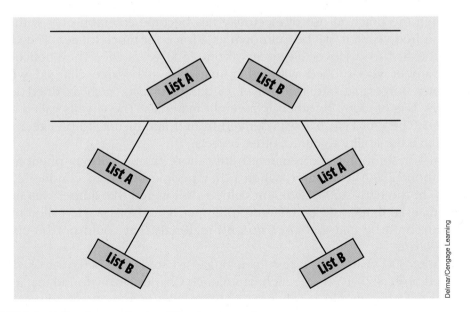

Delmar/Cengage Learning

FIGURE 1–2 Unlike charges attract and like charges repel.

Franklin named the materials in list A **positive** and the materials in list B **negative.** These names are still used today. The first item in each list was used as a standard for determining whether a charged object was positive or negative. Any object repelled by a piece of glass rubbed on silk would have a positive charge, and any item repelled by a hard rubber rod rubbed on wool would have a negative charge.

1-2   Atoms

Understanding electricity necessitates starting with the study of atoms. The **atom** is the basic building block of the universe. All **matter** is made from a combination of atoms. Matter is any substance that has mass and occupies space. Matter can exist in any of three states: solid, liquid, or gas. Water, for example, can exist as a solid in the form of ice, as a liquid, or as a gas in the form of steam *(Figure 1–3)*. An **element** is a substance that cannot be chemically divided into two or more simpler substances. A table listing both natural and artificial elements is shown in *Figure 1–4*. An atom is the smallest part of an element. The three principal parts of an atom are the **electron,** the **neutron,**

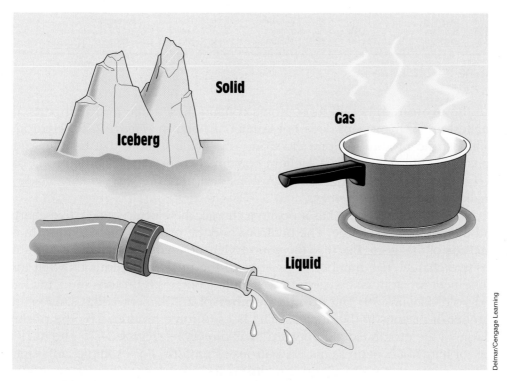

Delmar/Cengage Learning

FIGURE 1–3   Water can exist in three states, depending on temperature and pressure.

ATOMIC NUMBER	NAME	VALENCE ELECTRONS	SYMBOL	ATOMIC NUMBER	NAME	VALENCE ELECTRONS	SYMBOL	ATOMIC NUMBER	NAME	VALENCE ELECTRONS	SYMBOL
1	Hydrogen	1	H	37	Rubidium	1	Rb	73	Tantalum	2	Ta
2	Helium	2	He	38	Strontium	2	Sr	74	Tungsten	2	W
3	Lithium	1	Li	39	Yttrium	2	Y	75	Rhenium	2	Re
4	Beryllum	2	Be	40	Zirconium	2	Zr	76	Osmium	2	Os
5	Boron	3	B	41	Niobium	1	Nb	77	Iridium	2	Ir
6	Carbon	4	C	42	Molybdenum	1	Mo	78	Platinum	1	Pt
7	Nitrogen	5	N	43	Technetium	2	Tc	79	Gold	1	Au
8	Oxygen	6	O	44	Ruthenium	1	Ru	80	Mercury	2	Hg
9	Fluorine	7	F	45	Rhodium	1	Rh	81	Thallium	3	Tl
10	Neon	8	Ne	46	Palladium	–	Pd	82	Lead	4	Pb
11	Sodium	1	Na	47	Silver	1	Ag	83	Bismuth	5	Bi
12	Magnesium	2	Ma	48	Cadmium	2	Cd	84	Polonium	6	Po
13	Aluminum	3	Al	49	Indium	3	In	85	Astatine	7	At
14	Silicon	4	Si	50	Tin	4	Sn	86	Radon	8	Rd
15	Phosphorus	5	P	51	Antimony	5	Sb	87	Francium	1	Fr
16	Sulfur	6	S	52	Tellurium	6	Te	88	Radium	2	Ra
17	Chlorine	7	Cl	53	Iodine	7	I	89	Actinium	2	Ac
18	Argon	8	A	54	Xenon	8	Xe	90	Thorium	2	Th
19	Potassium	1	K	55	Cesium	1	Cs	91	Protactinium	2	Pa
20	Calcium	2	Ca	56	Barium	2	Ba	92	Uranium	2	U
21	Scandium	2	Sc	57	Lanthanum	2	La				
22	Titanium	2	Ti	58	Cerium	2	Ce		Artificial Elements		
23	Vanadium	2	V	59	Praseodymium	2	Pr				
24	Chromium	1	Cr	60	Neodymium	2	Nd	93	Neptunium	2	Np
25	Manganese	2	Mn	61	Promethium	2	Pm	94	Plutonium	2	Pu
26	Iron	2	Fe	62	Samarium	2	Sm	95	Americium	2	Am
27	Cobalt	2	Co	63	Europium	2	Eu	96	Curium	2	Cm
28	Nickel	2	Ni	64	Gadolinium	2	Gd	97	Berkelium	2	Bk
29	Copper	1	Cu	65	Terbium	2	Tb	98	Californium	2	Cf
30	Zinc	2	Zn	66	Dysprosium	2	Dy	99	Einsteinium	2	E
31	Gallium	3	Ga	67	Holmium	2	Ho	100	Fermium	2	Fm
32	Germanium	4	Ge	68	Erbium	2	Er	101	Mendelevium	2	Mv
33	Arsenic	5	As	69	Thulium	2	Tm	102	Nobelium	2	No
34	Selenium	6	Se	70	Ytterbium	2	Yb	103	Lawrencium	2	Lw
35	Bromine	7	Br	71	Lutetium	2	Lu				
36	Krypton	8	Kr	72	Hafnium	2	Hf				

Delmar/Cengage Learning

FIGURE 1–4 Table of elements.

and the **proton.** Although most atoms contain these three principal parts, the smallest atom, hydrogen, does not contain a neutron *(Figure 1–5)*. Hydrogen contains one proton and one electron. The smallest atom that contains neutrons is helium *(Figure 1–6)*. Helium contains two protons, two neutrons, and two electrons. It is theorized that protons and neutrons are actually made of smaller particles called *quarks*.

Notice that the proton has a positive charge, the electron a negative charge, and the neutron no charge. The neutrons and the protons combine to form the **nucleus** of the atom. Because the neutron has no charge, the nucleus has a net positive charge. The number of protons in the nucleus determines what kind of element an atom is. Oxygen, for example, contains 8 protons in its nucleus, and gold contains 79. The **atomic number** of an element is the same as the number of protons in the nucleus. The lines of force produced by the positive charge of the proton extend outward in all directions *(Figure 1–7)*. The nucleus may or may not contain as many neutrons as protons. For example, an atom of helium contains 2 protons and 2 neutrons in its nucleus, whereas an atom of copper contains 29 protons and 35 neutrons *(Figure 1–8)*.

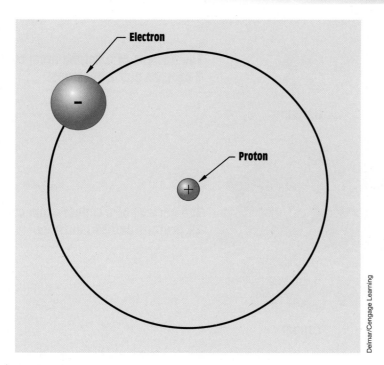

FIGURE 1–5   Hydrogen contains one proton and one electron.

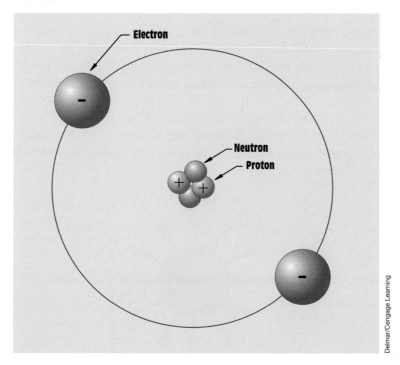

FIGURE 1–6   Helium contains two protons, two neutrons, and two electrons.

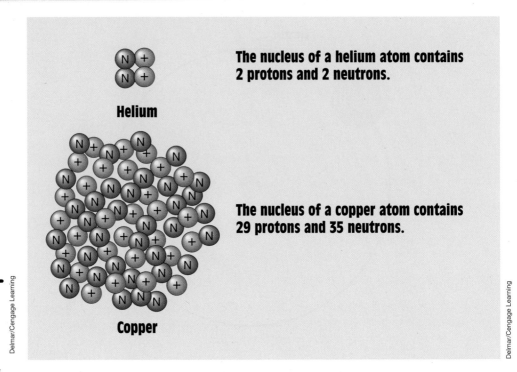

The nucleus of a helium atom contains 2 protons and 2 neutrons.

Helium

The nucleus of a copper atom contains 29 protons and 35 neutrons.

Copper

Proton

FIGURE 1–7 The lines of force extend outward.

FIGURE 1–8 The nucleus may or may not contain the same number of protons and neutrons.

The electron orbits the outside of the nucleus. Notice in *Figure 1–5* that the electron is shown to be larger than the proton. Actually, an electron is about three times as large as a proton. The estimated size of a proton is 0.07 trillionth of an inch in diameter, and the estimated size of an electron is 0.22 trillionth of an inch in diameter. Although the electron is larger in size, the proton weighs about 1840 times more. Imagine comparing a soap bubble with a piece of buckshot. Compared with the electron, the proton is a very massive particle. Because the electron exhibits a negative charge, the lines of force come in from all directions *(Figure 1–9)*.

Electron

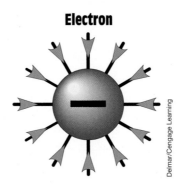

FIGURE 1–9 The lines of force come inward.

1-3   The Law of Charges

Understanding atoms necessitates first understanding a basic law of physics that states that ***opposite charges attract and like charges repel.*** In *Figure 1–10,* which illustrates this principle, charged balls are suspended from strings. Notice that the two balls that contain opposite charges are attracted to each other. The two positively charged balls and the two negatively charged balls repel each other. The reason for this is that lines of force can never cross each other. The outward-going lines of force of a positively charged object combine with the inward-going lines of force of a negatively charged object *(Figure 1–11).* This combining produces an attraction between the two objects. If two objects with like charges come close to

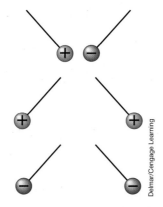

FIGURE 1–10   Unlike charges attract and like charges repel.

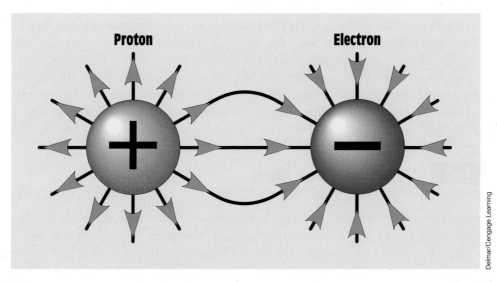

FIGURE 1–11   Unlike charges attract each other.

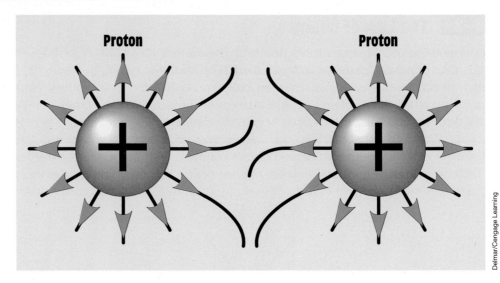

FIGURE 1–12 Like charges repel each other.

each other, the lines of force repel *(Figure 1–12)*. Because the nucleus has a net positive charge and the electron has a negative charge, the electron is attracted to the nucleus.

Because the nucleus of an atom is formed from the combination of protons and neutrons, one may ask why the protons of the nucleus do not repel each other because they all have the same charge. Two theories attempt to explain this. The first theory asserted that the force of gravity held the nucleus together. Neutrons, like protons, are extremely massive particles. It was first theorized that the gravitational attraction caused by their mass overcame the repelling force of the positive charges. By the mid-1930s, however, it was known that the force of gravity could not hold the nucleus together. According to Coulomb's law, the electromagnetic force in helium is about 1.1×10^{36} times greater than the gravitational force as determined by Newton's law. In 1947, the Japanese physicist Hideki Yukawa identified a subatomic particle that acts as a mediator to hold the nucleus together. The particle is a quark known as a *gluon*. The force of the gluon is about 10^2 times stronger than the electromagnetic force.

1–4   Structure of the Atom

In 1808, a scientist named John Dalton proposed that all matter was composed of atoms. Although the assumptions that Dalton used to prove his theory were later found to be factually incorrect, the idea that all matter is composed of atoms was adopted by most of the scientific world. Then in 1897, J.J. Thomson discovered the electron. Thomson determined that electrons have a negative charge and that they have very little mass compared to the atom. He

proposed that atoms have a large positively charged massive body with negatively charged electrons scattered throughout it. Thomson also proposed that
the negative charge of the electrons exactly balanced the positive charge of the
large mass, causing the atom to have a net charge of zero. Thomson's model of
the atom proposed that electrons existed in a random manner within the atom,
much like firing BBs from a BB gun into a slab of cheese. This was referred to
as the plum pudding model of the atom.

In 1913, a Danish scientist named Neils Bohr presented the most accepted
theory concerning the structure of an atom. In the Bohr model, electrons exist
in specific or "allowed" orbits around the nucleus in much the same way that
planets orbit the Sun *(Figure 1–13)*. The orbit in which the electron exists is

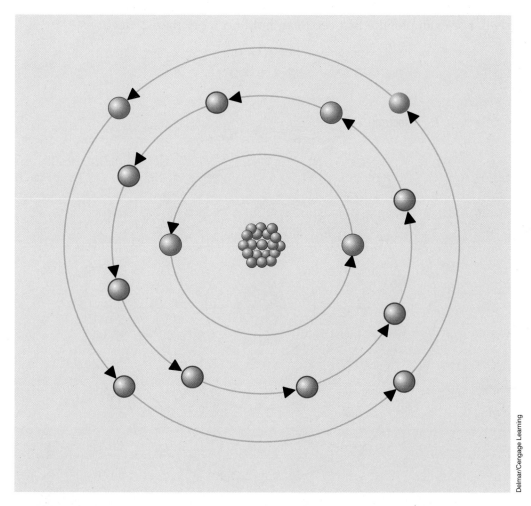

Delmar/Cengage Learning

FIGURE 1–13 Bohr's model of the atom proposed that electrons orbit the nucleus in much the same
way that planets orbit the Sun.

determined by the electron's mass times its speed times the radius of the orbit. These factors must equal the positive force of the nucleus. In theory there can be an infinite number of allowed orbits.

When an electron receives enough energy from some other source, it "quantum jumps" into a higher allowed orbit. Electrons, however, tend to return to a lower allowed orbit. When this occurs, the electron emits the excess energy as a single photon of electromagnetic energy.

1–5 Electron Orbits

Each **electron orbit** of an atom contains a set number of electrons *(Figure 1–14)*. The number of electrons that can be contained in any one orbit, or shell, is found by the formula ($2N^2$). The letter N represents the number of the orbit, or shell. For example, the first orbit can hold no more than 2 electrons:

$$2 \times (1)^2 \text{ or}$$
$$2 \times 1 = 2$$

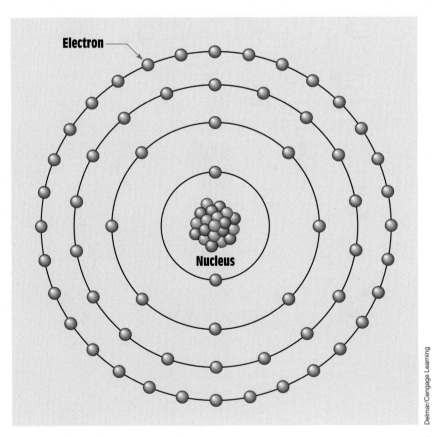

FIGURE 1–14 Electron orbits.

Delmar/Cengage Learning

The second orbit can hold no more than 8 electrons:

$$2 \times (2)^2 \text{ or}$$
$$2 \times 4 = 8$$

The third orbit can contain no more than 18 electrons:

$$2 \times (3)^2 \text{ or}$$
$$2 \times 9 = 18$$

The fourth and fifth orbits cannot hold more than 32 electrons. Thirty-two is the maximum number of electrons that can be contained in any orbit:

$$2 \times (4)^2 \text{ or}$$
$$2 \times 16 = 32$$

Although atoms are often drawn flat, as illustrated in *Figure 1–14*, electrons orbit the nucleus in a spherical fashion, as shown in *Figure 1–15*. Electrons travel at such a high rate of speed that they form a shell around the nucleus. For this reason, electron orbits are often referred to as *shells*.

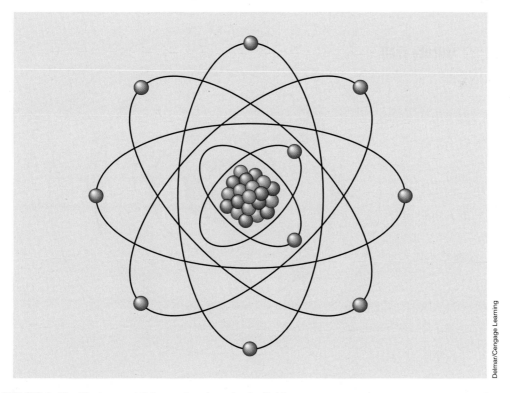

Delmar/Cengage Learning

FIGURE 1–15   Electrons orbit the nucleus in a circular fashion.

1-6   Valence Electrons

The outer shell of an atom is known as the *valence shell*. Any electrons located in the outer shell of an atom are known as **valence electrons** *(Figure 1–16)*. The valence shell of an atom cannot hold more than eight electrons. The valence electrons are of primary concern in the study of electricity because these electrons explain much of electrical theory. A **conductor,** for instance, is made from a material that contains between one and three valence electrons. Atoms with one, two, or three valence electrons are unstable and can be made to give up these electrons with little effort. Conductors are materials that permit electrons to flow through them easily. When an atom has only one or two valence electrons, these electrons are loosely held by the atom and are easily given up for current flow. Silver, copper, and gold all contain one valence electron and are excellent conductors of electricity. Silver is the best natural conductor of electricity, followed by copper, gold, and

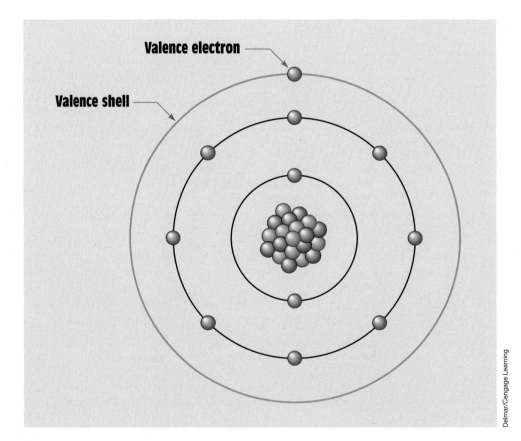

FIGURE 1–16   The electrons located in the outer orbit of an atom are valence electrons.

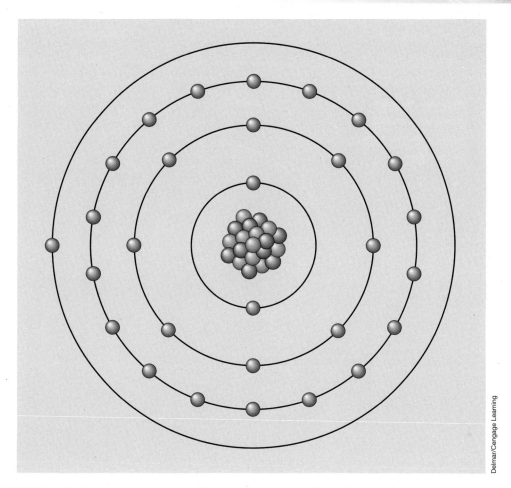

FIGURE 1–17 A copper atom contains 29 electrons and has 1 valence electron.

Delmar/Cengage Learning

aluminum. An atom of copper is shown in *Figure 1–17*. Although it is known that atoms containing few valence electrons are the best conductors, it is not known why some of these materials are better conductors than others. Copper, gold, platinum, and silver all contain only one valence electron. Silver, however, conducts electricity more readily than any of the others. Aluminum, which contains three valence electrons, is a better conductor than platinum, which contains only one valence electron.

1–7 Electron Flow

Electrical current is the flow of electrons. There are several theories concerning how electrons are made to flow through a conductor. One theory is generally referred to as the *bump theory*. It states that current flow is produced when an

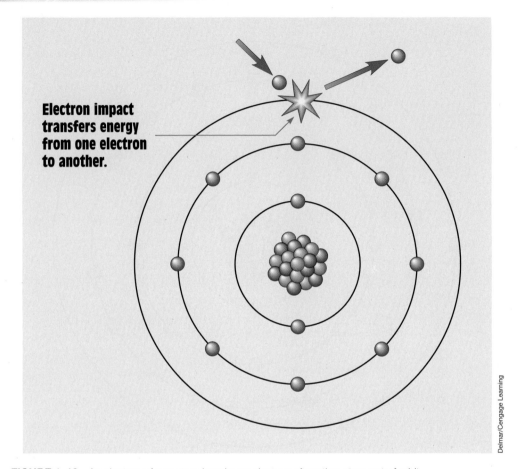

Electron impact transfers energy from one electron to another.

Delmar/Cengage Learning

FIGURE 1–18   An electron of one atom knocks an electron of another atom out of orbit.

electron from one atom knocks electrons of another atom out of orbit. *Figure 1–18* illustrates this action. When an atom contains only one valence electron, that electron is easily given up when struck by another electron. The striking electron gives its energy to the electron being struck. The striking electron may settle into orbit around the atom, and the electron that was struck moves off to strike another electron. This same effect can be seen in the game of pool. If the moving cue ball strikes a stationary ball exactly right, the energy of the cue ball is given to the stationary ball. The stationary ball then moves off with the cue ball's energy, and the cue ball stops moving *(Figure 1–19)*. The additional energy causes the electron to move out of orbit and become a free electron. After traveling a short distance, the electron enters the valence orbit of a different atom. When it returns to orbit, some or all of the gained energy is released in the form of heat, which is why conductors become warm when

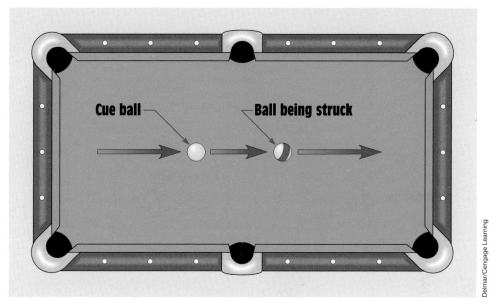

Cue ball

Ball being struck

Delmar/Cengage Learning

FIGURE 1–19 The energy of the cue ball is given to the ball being struck.

current flows through them. If too much current flows in a conductor, it may become hot enough to cause a fire.

If an atom containing two valence electrons is struck by a moving electron, the energy of the striking electron is divided between the two valence electrons *(Figure 1–20)*. If the valence electrons are knocked out of orbit, they contain only half the energy of the striking electron. This effect can also be seen in the game of pool *(Figure 1–21)*. If a moving cue ball strikes two stationary balls at the same time, the energy of the cue ball is divided between the two stationary balls. Both stationary balls will move but with only half the energy of the cue ball.

Other theories deal with the fact that all electric power sources produce a positive terminal and a negative terminal. The negative terminal is created by causing an excess of electrons to form at that terminal, and the positive terminal is created by removing a large number of electrons from that terminal *(Figure 1–22)*. Different methods can be employed to produce the excess of electrons at one terminal and deficiency of electrons at the other, but when a circuit is completed between the two terminals, negative electrons are repelled away from the negative terminal and attracted to the positive *(Figure 1–23)*. The greater the difference in the number of electrons between the negative and positive terminals, the greater the force of repulsion and attraction.

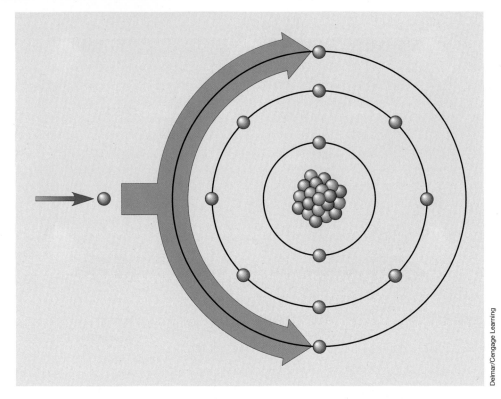

FIGURE 1–20 The energy of the striking electron is divided.

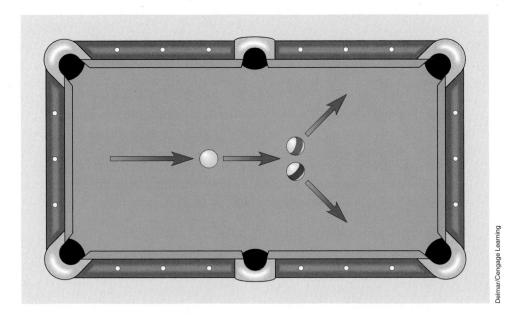

FIGURE 1–21 The energy of the cue ball is divided between the two other balls.

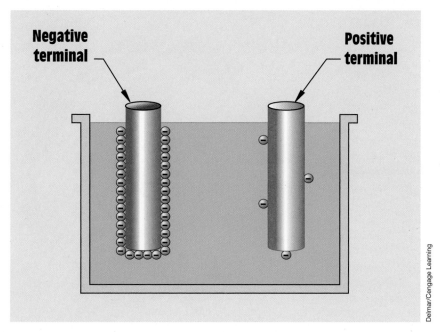

FIGURE 1–22   All electrical power sources produce a positive and a negative terminal.

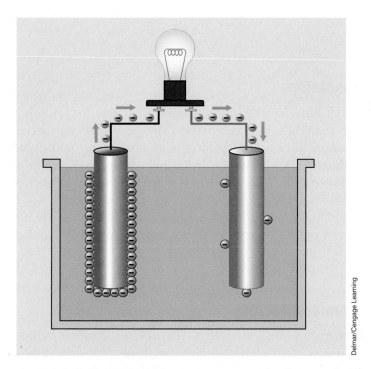

FIGURE 1–23   Completing a circuit between the positive and negative terminals causes electrons to be repelled from the negative terminal and attracted to the positive terminal.

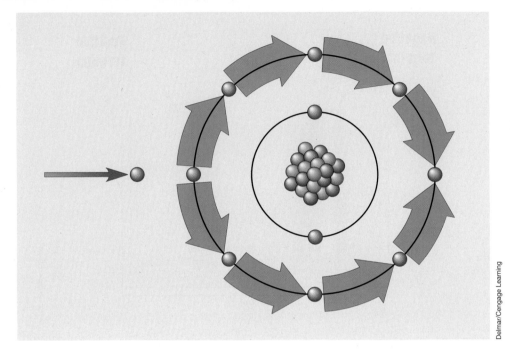

FIGURE 1–24 The energy of the striking electron is divided among the eight electrons.

1–8   Insulators

Materials containing seven or eight valence electrons are known as **insulators.** Insulators are materials that resist the flow of electricity. When the valence shell of an atom is full or almost full, the electrons are held tightly and are not given up easily. Some good examples of insulator materials are rubber, plastic, glass, and wood. *Figure 1–24* illustrates what happens when a moving electron strikes an atom containing eight valence electrons. The energy of the moving electron is divided so many times that it has little effect on the atom. Any atom that has seven or eight valence electrons is extremely stable and does not easily give up an electron.

1–9   Semiconductors

Semiconductors are materials that are neither good conductors nor good insulators. They contain four valence electrons *(Figure 1–25)* and are characterized by the fact that as they are heated, their resistance decreases. Heat has the opposite effect on conductors, whose resistance *increases* with an

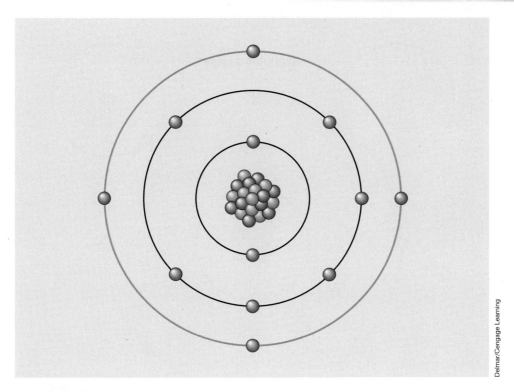

Delmar/Cengage Learning

FIGURE 1–25 Semiconductors contain four valence electrons.

increase of temperature. Semiconductors have become extremely important in the electrical industry since the invention of the transistor in 1947. All solid-state devices such as diodes, transistors, and integrated circuits are made from combinations of semiconductor materials. The two most common materials used in the production of electronic components are silicon and germanium. Of the two, silicon is used more often because of its ability to withstand heat. Before any pure semiconductor can be used to construct an electronic device, it must be mixed or "doped" with an impurity.

1–10 Molecules

Although all matter is made from atoms, atoms should not be confused with **molecules,** which are the smallest part of a compound. Water, for example, is a compound, not an element. The smallest particle of water is a molecule made of two atoms of hydrogen and one atom of oxygen, H_2O *(Figure 1–26).* If the molecule of water is broken apart, it becomes two hydrogen atoms and one oxygen atom and is no longer water.

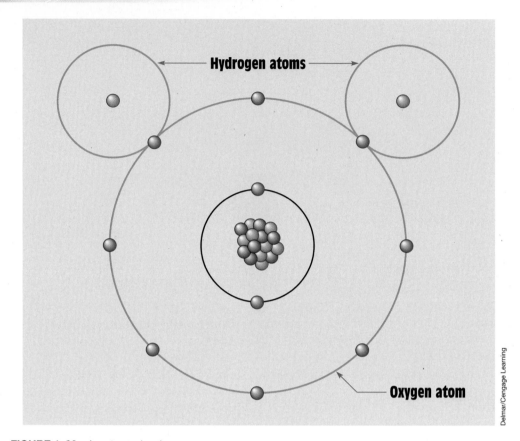

FIGURE 1–26   A water molecule.

Delmar/Cengage Learning

1–11   Methods of Producing Electricity

So far in this unit, it has been discussed that electricity is a flow of electrons. There are six basic methods for producing electricity:

1. Magnetism
2. Chemical action
3. Pressure
4. Heat
5. Friction
6. Light

Of the six methods listed, magnetism is the most common method used to produce electricity. Electromagnetic induction is the operating principle of all generators and alternators. These principles are covered fully later in this text.

The second most common method of producing electricity is chemical action. The chemical production of electricity involves the movement of entire ions instead of just electrons. The principles of conduction in liquids are discussed in Unit 12 and Unit 13.

The production of electricity by pressure involves the striking, bending, or twisting of certain crystals. This effect is referred to as the *piezo* electric effect. The word *piezo* is derived from a Greek word meaning "pressure."

Producing electricity with heat is referred to as the *Seebeck* effect. The Seebeck effect is the operating principle of thermocouples. Thermocouples are discussed in Unit 13.

Static charges are probably the best example of producing electricity by friction. A static charge occurs when certain materials are rubbed together and electrons are transferred from one object to the other. Static electricity is discussed in Unit 3.

Producing electricity from light involves the use of particles called *photons*. In theory, photons are massless particles of pure energy. Photons can be produced when electrons are forced to change to a lower energy level. This is the operating principle of gas-filled lights such as sodium vapor, mercury vapor, and so on. Electricity can be produced by photons when they strike a semiconductor material. The energy of the photon is given to an electron, forcing it to move out of orbit. This is the operating principle of *photovoltaic* devices called *solar cells*. Solar cells are discussed in Unit 13. Other photo-operated devices are *photoemissive* and *photoconductive*. Photoemissive devices include photodiodes, phototransistors, photoSCRs, and so on. These devices are generally used to sense light when the speed of operation is imperative. Photoconductive devices change resistance with a change of light. The most common photoconductive device is the cad cell *(Figure 1–27)*. Cad cells exhibit a resistance

Delmar/Cengage Learning

FIGURE 1–27 Cad cell.

of about 50 ohms (Ω) in direct sunlight and several hundred thousand ohms in darkness.

1–12 Electrical Effects

With the exception of friction, electricity can be used to cause the same effects that produce it:

1. Magnetism

2. Chemical reactions

3. Pressure

4. Heat

5. Light

Anytime an electric current flows through a conductor, a magnetic field is created around the conductor. This principle is discussed in Unit 4.

Electricity can be used to produce certain chemical reactions, such as electroplating. Electroplating is accomplished by placing a base metal and a pure metal in a chemical solution. An object can be copper plated, for example, by placing a base metal object and a piece of pure copper in a solution of cuprous cyanide. The object to be plated is connected to the negative electrode, and the pure copper is connected to the positive electrode. Atoms of copper are transferred through the solution and deposited on the object to be plated.

Another example of electricity producing chemical reactions can be seen in the process of *electrolysis*. Electrolysis is the process of separating elements electrically. These principles are discussed in Unit 12.

Just as the twisting or bending of certain crystals can produce electricity, electricity can cause certain crystals to bend or twist. When electricity is applied to a certain size and shape of quartz crystal, the crystal vibrates at a certain rate. This principle has been used in crystal radios for many years. If an electric current is applied to a piece of Rochelle salt crystal, the crystal vibrates. This is the operating principle of a crystal earphone *(Figure 1–28)*.

As discussed previously in this unit, when electrons enter a valence orbit, heat is often produced. This is the reason that conductors become warm as current flows through them. This is also the operating principle of many heat-producing devices such as electric ranges, electric irons, electric heaters, and so on.

Light is produced when electrons move to a lower orbit and produce a photon. When electric current is applied to certain conductors, they not only become hot, but they also emit photons of light. Incandescent lamps use this principle of operation. When electric current is applied to the filament of an incandescent lamp, most of the electrical energy is converted into heat, but part of it produces photons of light. Incandescent lamps, however, are very

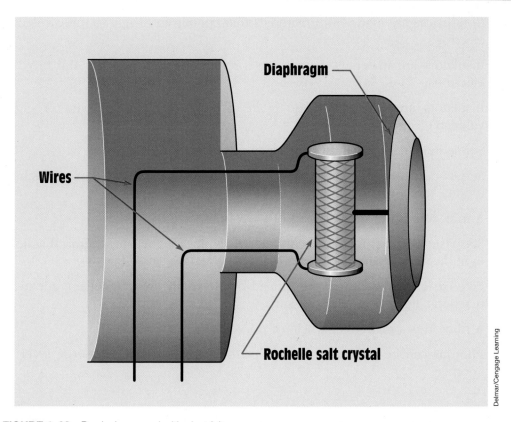

Wires

Diaphragm

Rochelle salt crystal

Delmar/Cengage Learning

FIGURE 1–28 Producing sound with electricity.

inefficient. The typical 100-watt lamp produces about 95 watts of heat and 5 watts of light. Other lighting sources such as sodium vapor, mercury vapor, and fluorescent are much more efficient. Some semiconductor devices can be used to produce light without heat. Light-emitting diodes are a good example of these devices.

Summary

- The atom is the smallest part of an element.
- The three principal parts of an atom are the proton, the electron, and the neutron.
- Protons have a positive charge, electrons a negative charge, and neutrons no charge.
- Valence electrons are located in the outer orbit of an atom.
- Conductors are materials that provide an easy path for electron flow.

- Conductors are made from materials that contain from one to three valence electrons.

- Insulators are materials that do not provide an easy path for the flow of electrons.

- Insulators are generally made from materials containing seven or eight valence electrons.

- Semiconductors contain four valence electrons.

- Semiconductors are used in the construction of all solid-state devices such as diodes, transistors, and integrated circuits.

- A molecule is the smallest part of a compound.

- Six basic methods for producing electricity are magnetism, chemical action, light, heat, pressure, and friction.

- Five basic effects that can be caused by electricity are magnetism, chemical reactions, light, heat, and pressure.

- A photon is a massless particle of pure energy.

- Photons can be produced when electrons move from one energy level to another.

Review Questions

1. What are the three principal parts of an atom, and what charge does each carry?

2. How many times larger is an electron than a proton?

3. How many times more does a proton weigh than an electron?

4. State the law of charges.

5. What force keeps an electron in orbit around the nucleus of an atom?

6. How many valence electrons are generally contained in materials used for conductors?

7. How many valence electrons are generally contained in materials used for insulators?

8. What is electricity?

9. What is a gluon?

10. It is theorized that protons and neutrons are actually formed from a combination of smaller particles. What are these particles called?

Unit 2

Electrical Quantities and Ohm's Law

KEY TERMS

Ampere (A)
British thermal unit (Btu)
Complete path
Conventional current flow theory
Coulomb (C)
Electromotive force (EMF)
Electron flow theory
Grounding conductor
Horsepower (hp)
Impedance
Joule
Neutral conductor
Ohm (Ω)
Ohm's law
Potential difference
Power
Resistance
Volt (V)
Watt (W)

Why You Need to Know

*A*nyone working in the electrical field must know and understand the basic units used to measure electric power. To accomplish this, it is important to understand how electricity works and is evaluated. In order to work with electric components and control devices in the field, you must know the electrical measuring terms and how to apply these terms. This unit presents

- the difference between voltage and current. You will discover, for example, that voltage is actually the force that pushes the electrons through a conductor. Voltage cannot flow, but it can cause current to flow.
- current. This is a quantity of electrons moving through a conductor within a certain length of time. Current, or amperes, is the actual amount of electricity that flows through the circuit. If it were possible to cut a wire and catch electricity in a container, you would have a container full of electrons.
- watts. This is a measure of power. It is basically the rate at which electrical energy is being converted into some other form. It is also the measurement used by the power company to charge its customers for the amount of energy consumed.
- Ohm's law. This is the basis for all electrical calculations. Ohm's law is a method of mathematically determining electrical quantities when other quantities are known. It is possible, for example, to determine the amount of current that will flow in a circuit if the resistance in the circuit and the voltage applied to it are known.
- a discussion of the similarity of electricity and water systems.

Delmar/Cengage Learning

Objectives

After studying this unit, you should be able to

- define a coulomb.
- define an ampere.
- define a volt.
- define an ohm.
- define a watt.
- calculate different electrical values using Ohm's law.
- discuss different types of electric circuits.
- select the proper Ohm's law formula from a chart.

Preview

Electricity has a standard set of values. Before one can work with electricity, one must know these values and how to use them. Because the values of electrical measurement have been standardized, they are understood by everyone who uses them. For instance, carpenters use a standard system for measuring length, such as the inch, foot, meter, or centimeter. Imagine what a house would look like that was constructed by two carpenters who used different lengths of measure for an inch or foot. The same holds true for people who work with electricity. The standards of measurement must be the same for everyone. Meters should be calibrated to indicate the same quantity of current flow or voltage or resistance. A volt, an ampere, or an ohm is the same everywhere in the world. ■

2–1 The Coulomb

A **coulomb** is a quantity measurement for electrons. One coulomb contains 6.25×10^{18}, or 6,250,000,000,000,000,000 electrons. To better understand the number of electrons contained in a coulomb, think of comparing one second to 200 billion years. Because the coulomb is a quantity measurement, it is similar to a quart, a gallon, or a liter. It takes a certain amount of liquid to equal a liter, just as it takes a certain amount of electrons to equal a coulomb.

The coulomb is named for a French scientist who lived in the 1700s named Charles Augustin de Coulomb. Coulomb experimented with electrostatic charges and developed a law dealing with the attraction and repulsion of these forces. The law, known as ***Coulomb's law of electrostatic charges, states that the force of electrostatic attraction or repulsion is directly proportional to the product of the two charges and inversely proportional to the square of the distance between them.*** The number of electrons contained in the coulomb was determined by the average charge of an electron. The symbol for coulomb is the letter *C*. It is the System Internationale (SI) unit of electric charge. A coulomb is defined as the charge transferred by a current of 1 ampere in one second.

2-2   The Ampere

The **ampere** is named for André Ampère, a scientist who lived from the late 1700s to the early 1800s. Ampère is most famous for his work dealing with electromagnetism, which is discussed in a later unit. The ampere (A) is equal to 1 coulomb per second. Notice that the definition of an ampere involves a quantity measurement, the coulomb, and a time measurement, the second. One ampere of current flows through a wire when 1 coulomb flows past a point in one second *(Figure 2–1)*. The ampere is a measurement of the amount of electricity that is flowing through a circuit. In a water system, it would be comparable to gallons per minute or gallons per second *(Figure 2–2)*. The letter *I*, which stands for intensity of current, and the letter *A,* which stands for ampere, are both used to represent current flow in algebraic formulas. This text uses the letter *I* in formulas to represent current.

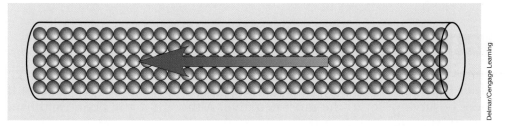

Delmar/Cengage Learning

FIGURE 2–1   One ampere equals one coulomb per second.

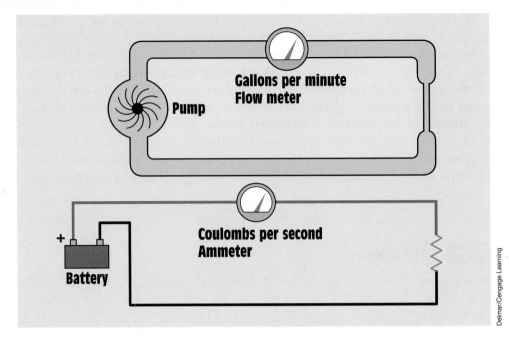

FIGURE 2–2   Current in an electric circuit can be compared to flow rate in a water system.

2–3   The Electron Flow Theory

There are actually two theories concerning current flow. One theory is known as the **electron flow theory** and states that because electrons are negative particles, current flows from the most negative point in the circuit to the most positive. The electron flow theory is the more widely accepted as being correct and is used throughout this text.

2–4   The Conventional Current Flow Theory

The second theory, known as the **conventional current flow theory,** is older than the electron flow theory and states that current flows from the most positive point to the most negative. Although it has been established almost to a certainty that the electron flow theory is correct, the conventional current flow theory is still widely used for several reasons. Most electronic circuits use the negative terminal as ground or common. When the negative terminal is used as ground, the positive terminal is considered to be above ground, or hot. It is easier for most people to think of something flowing down rather than up, or from a point above ground to

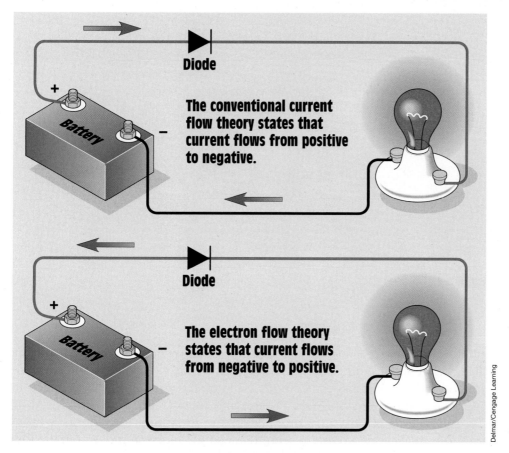

Diode

The conventional current flow theory states that current flows from positive to negative.

Battery

Diode

The electron flow theory states that current flows from negative to positive.

Battery

Delmar/Cengage Learning

FIGURE 2–3   Conventional current flow theory and electron flow theory.

ground. An automobile electric system is a good example of this type of circuit. Most people consider the positive battery terminal to be the hot terminal.

Many people who work in the electronics field prefer the conventional current flow theory because all the arrows on the semiconductor symbols point in the direction of conventional current flow. If the electron flow theory is used, it must be assumed that current flows against the arrow *(Figure 2–3).* Another reason that many people prefer using the conventional current flow theory is that most electronic schematics are drawn in a manner that assumes that current flows from the more positive to the more negative source. In *Figure 2–4,* the positive voltage point is shown at the top of the schematic and the negative (ground) is shown at the bottom. When tracing the flow of current through a circuit, most people find it easier to go from top to bottom than from bottom to top.

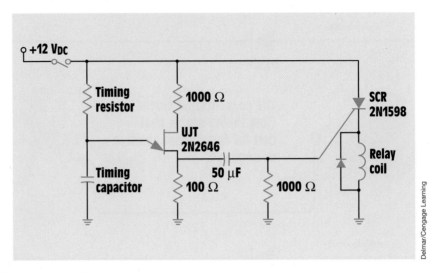

FIGURE 2–4 On-delay timer.

2–5 Speed of Current

To determine the speed of current flow through a wire, one must first establish exactly what is being measured. As stated previously, current is a flow of electrons through a conductive substance. Assume for a moment that it is possible to remove a single electron from a wire and identify it by painting it red. If it were possible to observe the progress of the identified electron as it moved from atom to atom, it would be seen that a single electron moves rather slowly *(Figure 2–5)*. It is estimated that a single electron moves at a rate of about 3 inches per hour at 1 ampere of current flow.

Another factor that must be considered is whether the circuit is DC, AC, or radio waves. Radio waves move at approximately the speed of light, which is

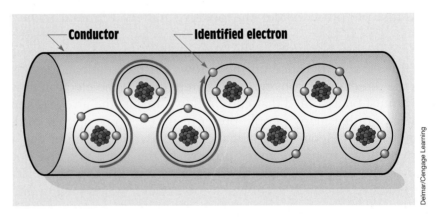

FIGURE 2–5 Electrons moving from atom to atom.

186,000 miles per second or 300,000,000 meters per second. The velocity of AC through a conductor is less than the speed of light because magnetic fields travel more slowly in material dielectrics than they do through free air. The formula shown can be used to calculate the wavelength of a signal traveling through a conductor. Wavelength is the distance that current travels during one AC cycle. Wavelength is discussed more fully in Unit 15.

$$L = \frac{984 \, V}{f}$$

where

$$L = \text{length in feet}$$
$$V = \text{velocity factor}$$
$$f = \text{frequency in megahertz (MHz)}$$

The velocity factor is determined by the type of conductor. *Table 2–1* gives the velocity factor for several different types of coaxial cables and parallel conductors.

How many feet would a 5-MHz signal travel through a conductor with a velocity factor of 0.66 during one AC cycle?

$$L = \frac{984 \times 0.66}{5}$$

$$L = 129.888 \text{ feet}$$

Description or Type Number	Velocity Factor	Characteristic Impedance	Capacitance per Foot
Coaxial Cable			
RG–8A/U	0.66	53	29.5 pF
RG–58A/U	0.66	53	28.5 pF
RG–17A/U	0.66	50	30 pF
RG–11A/U	0.66	75	20.5 pF
RG59A/U	0.66	73	21 pF
Parallel Conductors			
Air Insulated	0.975	200–600	—
214–023	0.71	75	20 pF
214–056	0.82	300	5.8 pF
214–076	0.84	300	3.9 pF
214–022	0.85	300	3 pF

TABLE 2–1   Data for Different Types of Transmission Lines

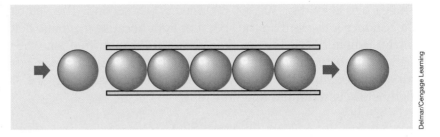

Delmar/Cengage Learning

FIGURE 2–6 When a ball is pushed into one end, another ball is forced out the other end. This basic principle causes the instantaneous effect of electric impulses.

In a DC circuit, the impulse of electricity can appear to be faster than the speed of light. Assume for a moment that a pipe has been filled with table-tennis balls *(Figure 2–6)*. If a ball is forced into the end of the pipe, the ball at the other end will be forced out. Each time a ball enters one end of the pipe, another ball is forced out the other end. This principle is also true for electrons in a wire. There are billions of electrons in a wire. If an electron enters one end of a wire, another electron is forced out the other end. Assume that a wire is long enough to be wound around the earth 10 times. If a power source and a switch were connected at one end of the wire and a light at the other end *(Figure 2–7)*, the light would turn on the moment the switch was closed. It would take light approximately 1.3 seconds to travel around the earth 10 times.

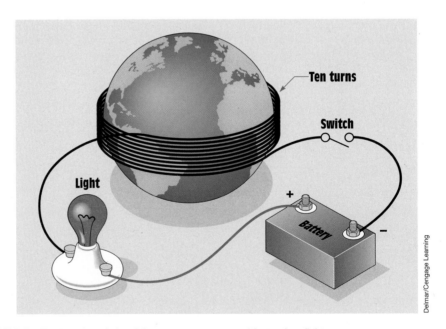

Delmar/Cengage Learning

FIGURE 2–7 The impulse of electricity can appear to travel faster than light.

2-6   Basic Electric Circuits

A **complete path** must exist before current can flow through a circuit *(Figure 2–8)*. A complete circuit is often referred to as a *closed circuit*, because the power source, conductors, and load form a closed loop. In *Figure 2–8*, a lamp is used as the load. The load offers resistance to the circuit and limits the amount of current that can flow. If the switch is opened, there is no longer a closed loop and no current can flow. This is often referred to as an incomplete, or open, circuit.

Another type of circuit is the short circuit, which has very little or no resistance. It generally occurs when the conductors leading from and back to the power source become connected *(Figure 2–9)*. In this example, a separate current path has been established that bypasses the load. Because the load is the device that limits the flow of current, when it is bypassed, an excessive amount of current can flow. Short circuits generally cause a fuse to blow or a circuit breaker to open. If the circuit has not been protected by a fuse or circuit breaker, a short circuit can damage equipment, melt wires, and start fires.

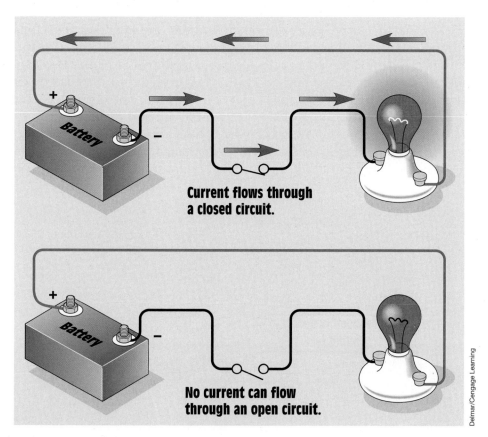

Current flows through a closed circuit.

No current can flow through an open circuit.

Delmar/Cengage Learning

FIGURE 2–8   Current flows only through a closed circuit.

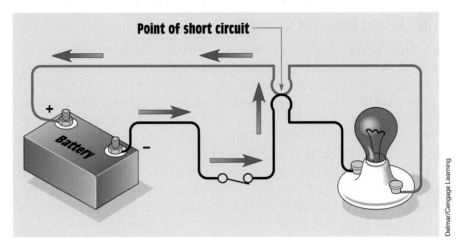

FIGURE 2–9   A short circuit bypasses the load and permits too much current to flow.

Another type of circuit, one that is often confused with a short circuit, is a grounded circuit. Grounded circuits can also cause an excessive amount of current flow. They occur when a path other than the one intended is established to ground. Many circuits contain an extra conductor called the **grounding conductor.** A typical 120-volt appliance circuit is shown in *Figure 2–10*. In this circuit, the ungrounded, or hot, conductor is connected to the fuse or circuit breaker. The hot conductor supplies power to the load. The grounded conductor, or **neutral conductor,** provides the return path and completes the circuit back to the power source. The grounding conductor is generally connected to the case of the appliance to provide a low-resistance path to ground. Although both the neutral and grounding conductors are grounded at the power source,

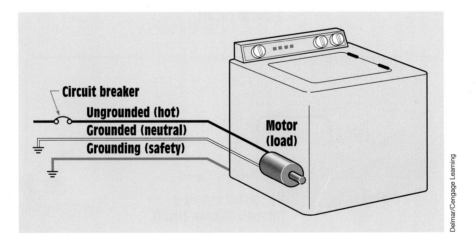

FIGURE 2–10   120-V appliance circuit.

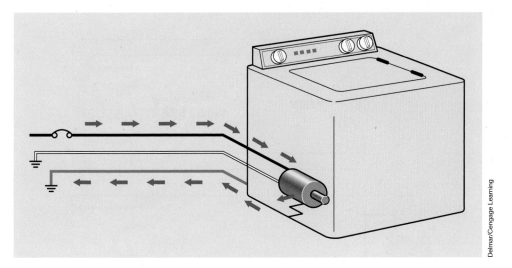

Delmar/Cengage Learning

FIGURE 2–11   The grounding conductor provides a low-resistance path to ground.

the grounding conductor is not considered to be a circuit conductor, because current will flow through the grounding conductor only when a circuit fault develops. In normal operation, current flows through the hot and neutral conductors only.

The grounding conductor is used to help prevent a shock hazard in the event that the ungrounded, or hot, conductor comes in contact with the case or frame of the appliance *(Figure 2–11)*. This condition can occur in several ways. In this example, assume that the motor winding becomes damaged and makes connection to the frame of the motor. Because the frame of the motor is connected to the frame of the appliance, the grounding conductor provides a circuit path to ground. If enough current flows, the circuit breaker will open. Without a grounding conductor connected to the frame of the appliance, the frame would become hot (in the electrical sense) and anyone touching the case and a grounded point, such as a water line, would complete the circuit to ground. The resulting shock could be fatal. For this reason, the grounding prong of a plug should never be cut off or bypassed.

2–7   The Volt

Voltage is defined as the potential difference between two points of a conducting wire carrying a constant current of 1 ampere when the power dissipated between these points is 1 watt. Voltage is also referred to as **potential difference** or **electromotive force (EMF).** It is the force that pushes the electrons through a wire and is often referred to as electrical pressure. A **volt** is the amount of potential necessary to cause 1 coulomb to produce 1 joule of work. One thing to remember is that voltage cannot flow. Voltage in an electrical

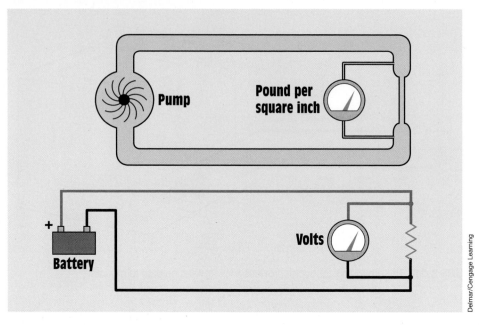

Delmar/Cengage Learning

FIGURE 2–12 Voltage in an electric circuit can be compared to pressure in a water system.

circuit is like pressure in a water system *(Figure 2–12)*. To say that voltage flows through a circuit is like saying that pressure flows through a pipe. Pressure can push water through a pipe, and it is correct to say that water flows through a pipe, but it is not correct to say that pressure flows through a pipe. The same is true for voltage. Voltage pushes current through a wire, but voltage cannot flow through a wire.

Voltage is often thought of as the potential to do something. For this reason it is frequently referred to as potential, especially in older publications and service manuals. Voltage must be present before current can flow, just as pressure must be present before water can flow. A voltage, or potential, of 120 volts is present at a common wall outlet, but there is no flow until some device is connected and a complete circuit exists. The same is true in a water system. Pressure is present, but water cannot flow until the valve is opened and a path is provided to a region of lower pressure. The letter E, which stands for EMF, or the letter V, which stands for volt, can be used to represent voltage in an algebraic formula. This text uses the letter E to represent voltage in an algebraic formula.

2–8 The Ohm

An **ohm** is the unit of **resistance** to current flow. It was named after the German scientist Georg S. Ohm. The symbol used to represent an ohm, or resistance, is the Greek letter omega (Ω). The letter R, which stands for resistance, is used to

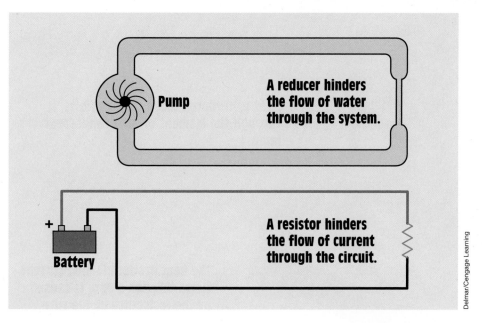

Pump

A reducer hinders
the flow of water
through the system.

+

Battery

A resistor hinders
the flow of current
through the circuit.

Delmar/Cengage Learning

FIGURE 2–13   A resistor in an electric circuit can be compared to a reducer in a water system.

represent ohms in an algebraic formula. An ohm is the amount of resistance that allows 1 ampere of current to flow when the applied voltage is 1 volt. Without resistance, every electric circuit would be a short circuit. All electric loads, such as heating elements, lamps, motors, transformers, and so on, are measured in ohms. In a water system, a reducer can be used to control the flow of water; in an electric circuit, a resistor can be used to control the flow of electrons *(Figure 2–13)*.

To understand the effect of resistance on an electric circuit, imagine a person running along a beach. As long as the runner stays on the hard, compact sand, he or she can run easily along the beach. Likewise, current can flow easily through a good conductive material, such as a copper wire. Now imagine that the runner wades out into the water until it is knee deep. He or she will no longer be able to run along the beach as easily because of the resistance of the water. Now imagine that the runner wades out into the water until it is waist deep. His or her ability to run along the beach will be hindered to a greater extent because of the increased resistance of the water against his or her body. The same is true for resistance in an electric circuit. The higher the resistance, the greater the hindrance to current flow.

Another fact an electrician should be aware of is that any time current flows through a resistance, heat is produced *(Figure 2–14)*. That is why a wire becomes warm when current flows through it. The elements of an electric range become hot, and the filament of an incandescent lamp becomes extremely hot because of resistance.

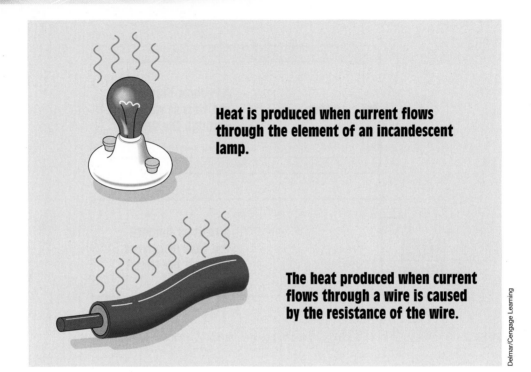

Heat is produced when current flows through the element of an incandescent lamp.

The heat produced when current flows through a wire is caused by the resistance of the wire.

Delmar/Cengage Learning

FIGURE 2–14 Heat is produced when current flows through resistance.

Another term similar in meaning to resistance is **impedance.** Impedance is most often used in calculations of AC rather than DC. Impedance is discussed to a greater extent later in this text.

GREEN TIP: Larger wire size will result in less resistance. Less resistance reduces the voltage drop and consequently the amount of power loss due to heating the conductor. ■

2–9 The Watt

Wattage is a measure of the amount of power that is being used in a circuit. The **watt** was named in honor of the English scientist James Watt. In an algebraic formula, wattage is generally represented either by the letter P, for power, or W, for watts. It is proportional to the amount of voltage and the amount of current flow. To understand watts, return to the example of the water system. Assume that a water pump has a pressure of 120 pounds per square inch (PSI) and causes a flow rate of 1 gallon per second. Now assume that this water is

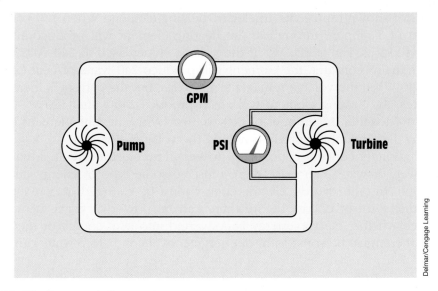

FIGURE 2–15 Force equals flow rate times pressure.

used to drive a turbine, as shown in *Figure 2–15*. The turning force or torque developed by the turbine is proportional to the amount of water flow and the pressure forcing it against the turbine blades. If the pressure is increased and the flow rate remains constant, the water will strike the turbine blades with greater force and the torque will increase. If the pressure remains constant and a greater volume of water is permitted to flow, the turbine blades will be struck by more pounds of water in the same amount of time and torque will again increase. As you can see, the torque developed by the turbine is proportional to both the pressure and the flow rate of the water.

The power of an electric circuit is very similar. *Figure 2–16* shows a resistor connected to a circuit with a voltage of 120 volts and a current flow of 1 ampere.

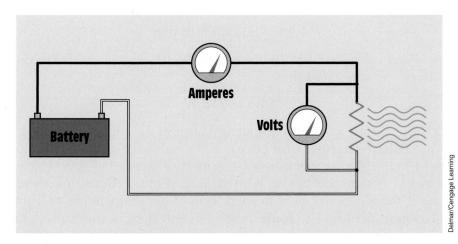

FIGURE 2–16 Amperes times volts equals watts.

The resistor shown represents an electric heating element. When 120 volts force a current of 1 ampere through it, the heating element will produce 120 watts of heat (120 V × 1 A = 120 W). If the voltage is increased to 240 volts, but the current remains constant, the element will produce 240 watts of heat (240 V × 1 A = 240 W). If the voltage remains at 120 volts, but the current is increased to 2 amperes, the heating element will again produce 240 watts (120 V × 2 A = 240 W). Notice that the amount of power used by the heating element is determined by the amount of current flow and the voltage driving it.

An important concept concerning **power** in an electric circuit is that before true power, or watts, can exist, there must be some type of energy change or conversion. In other words, electric energy must be changed or converted into some other form of energy before there can be power or watts. It makes no difference whether electric energy is converted into heat energy or mechanical energy; there must be some form of energy conversion before watts can exist.

2–10 Other Measures of Power

The watt is not the only unit of power measure. Many years ago, James Watt decided that in order to sell his steam engines, he would have to rate their power in terms that the average person could understand. He decided to compare his steam engines to the horses he hoped his engines would replace. After experimenting, Watt found that the average horse working at a steady rate could do 550 foot-pounds of work per second. A foot-pound (ft-lb) is the amount of force required to raise a 1 pound weight 1 foot. This rate of doing work is the definition of a **horsepower (hp):**

$$1 \text{ hp} = 550 \text{ ft-lb/s}$$

Horsepower can also be expressed as 33,000 foot-pounds per minute (550 ft-lb + 60 s = 33,000):

$$1 \text{ hp} = 33,000 \text{ ft-lb/min}$$

It was later calculated that the amount of electric energy needed to produce one horsepower was 746 watts:

$$1 \text{ hp} = 746 \text{ W}$$

Another measure of energy frequently used in the English system of measure is the **British thermal unit (Btu).** A Btu is defined as the amount of heat required to raise the temperature of 1 pound of water 1 degree Fahrenheit. In the metric system, the calorie is used instead of the Btu to measure heat. A calorie is the amount of heat needed to raise the temperature of 1 gram of water 1 degree Celsius. The **joule** is the SI equivalent of the watt. A joule is defined as 1 newton-meter. A newton is a force of 100,000 dynes, or about

1 Horsepower =	746 watts
1 Horsepower =	550 ft-lb/s
1 Watt =	0.00134 horsepower
1 Watt =	3.412 Btu/hr
1 Wattsecond =	1 joule
1 Btu-hr =	0.293 watts
1 Cal/s =	4.19 watts
1 Ft-lb/s =	1.36 watts
1 Btu =	1050 joules
1 Joule =	0.2388 cal
1 Cal =	4.187 joules

Delmar/Cengage Learning

FIGURE 2–17 Common power units.

3-½ ounces, and a meter is about 39 inches. The joule can also be expressed as the amount of work done by 1 coulomb flowing through a potential of 1 volt, or as the amount of work done by 1 watt for 1 second:

$$1 \text{ joule} = 1 \text{ wattsecond}$$

The chart in *Figure 2–17* gives some common conversions for different quantities of energy. These quantities can be used to calculate different values.

■ EXAMPLE 2-1

An elevator must lift a load of 4000 lb to a height of 50 ft in 20 s. How much horsepower is required to operate the elevator?

Solution

Find the amount of work that must be performed, and then convert that to horsepower:

$$4,000 \text{ lb} \times 50 \text{ ft} = 200,000 \text{ ft-lb}$$

$$\frac{200,000 \text{ ft-lb}}{20 \text{ s}} = 10,000 \text{ ft-lb/s}$$

$$\frac{10,000 \text{ ft-lb/s}}{550 \text{ ft-lb/s}} = 18.18 \text{ hp}$$

■ EXAMPLE 2-2

A water heater contains 40 gallons of water. Water weighs 8.34 lb per gallon. The present temperature of the water is 68°F. The water must be raised to a temperature of 160°F in one hour. How much power will be required to raise the water to the desired temperature?

Solution

First determine the weight of the water in the tank, because a Btu is the amount of heat required to raise the temperature of 1 pound of water 1 degree Fahrenheit:

$$40 \text{ gal} \times 8.34 \text{ lb per gal} = 333.6 \text{ lb}$$

The second step is to determine how many degrees of temperature the water must be raised. This amount will be the difference between the present temperature and the desired temperature:

$$160°F - 68°F = 92°F$$

The amount of heat required in Btu will be the product of the pounds of water and the desired increase in temperature:

$$333.6 \text{ lb} \times 92°F = 30{,}691.2 \text{ lb-degrees or Btu}$$

$$1 \text{ W} = 3.412 \text{ Btu/hr}$$

Therefore,

$$\frac{30.691 \text{ Btu}}{3.412 \text{ Btu/hr/w}} = 8995.1 \text{w}$$

2-11   Ohm's Law

In its simplest form, **Ohm's law** states that *it takes 1 volt to push 1 ampere through 1 ohm.* Ohm discovered that all electric quantities are proportional to each other and can therefore be expressed as mathematical formulas. He found that if the resistance of a circuit remained constant and the voltage increased, there was a corresponding proportional increase of current. Similarly, if the resistance remained constant and the voltage decreased, there would be a proportional decrease of current. He also found that if the voltage remained constant and the resistance increased, there would be a decrease of current; and if the voltage remained constant and the resistance decreased, there would be an increase of current. This finding led Ohm to the conclusion that *in a DC* *circuit, the current is directly proportional to the voltage and inversely proportional to the resistance.*

Because Ohm's law is a statement of proportion, it can be expressed as an algebraic formula when standard values such as the volt, the ampere, and the ohm are used. The three basic Ohm's law formulas are

$$E = I \times R$$

$$I = \frac{E}{R}$$

$$R = \frac{E}{I}$$

where

$$E = \text{EMF, or voltage}$$

$$I = \text{intensity of current, or amperage}$$

$$R = \text{resistance}$$

The first formula states that the voltage can be found if the current and resistance are known. Voltage is equal to amperes multiplied by ohms. For example, assume that a circuit has a resistance of 50 ohms and a current flow through it of 2 amperes. The voltage connected to this circuit is 100 volts.

$$E = I \times R$$

$$E = 2\ \text{A} \times 50\ \Omega$$

$$E = 100\ \text{V}$$

The second formula states that the current can be found if the voltage and resistance are known. In the example shown, 120 volts are connected to a resistance of 30 ohms. The amount of current flow will be 4 amperes.

$$I = \frac{E}{R}$$

$$I = \frac{120\ \text{V}}{30\ \Omega}$$

$$I = 4\ \text{A}$$

The third formula states that if the voltage and current are known, the resistance can be found. Assume that a circuit has a voltage of 240 volts and a current flow of 10 amperes. The resistance in the circuit is 24 ohms.

$$R = \frac{E}{I}$$

$$R = \frac{240\ \text{V}}{10\ \text{A}}$$

$$R = 24\ \Omega$$

FIGURE 2–18 Chart for finding values of voltage, current, and resistance.

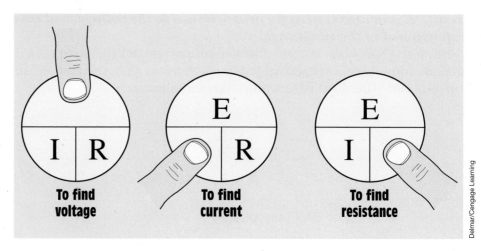

FIGURE 2–19 Using the Ohm's law chart.

Figure 2–18 shows a simple chart that can be a great help when trying to remember an Ohm's law formula. To use the chart, cover the quantity that is to be found. For example, if the voltage, E, is to be found, cover the E on the chart. The chart now shows the remaining letters IR *(Figure 2–19)*; thus, E = I × R. The same method reveals the formulas for current (I) and resistance (R).

A larger chart, which shows the formulas needed to find watts as well as voltage, amperage, and resistance, is shown in *Figure 2–20*. The letter

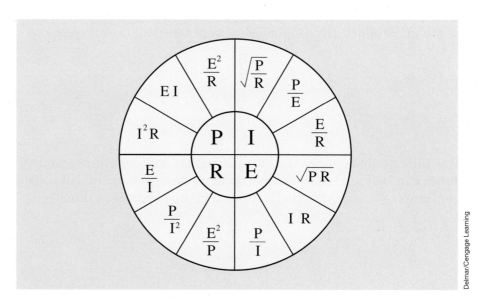

FIGURE 2–20 Formula chart for finding values of voltage, current, resistance, and power.

P (power) is used to represent the value of watts. Notice that this chart is divided into four sections and that each section contains three different formulas. To use this chart, select the section containing the quantity to be found and then choose the proper formula from the given quantities.

■ EXAMPLE 2-3

An electric iron is connected to 120 V and has a current draw of 8 A. How much power is used by the iron?

Solution

The quantity to be found is watts, or power. The known quantities are voltage and amperage. The proper formula to use is shown in *Figure 2–21*.

$$P = EI$$
$$P = 120\text{ V} \times 8\text{ A}$$
$$P = 960\text{ W}$$

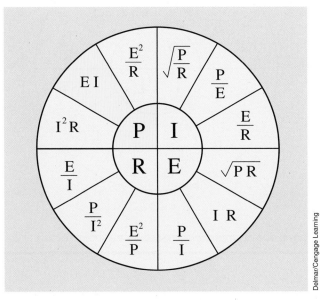

FIGURE 2–21   Finding power when voltage and current are known.

Delmar/Cengage Learning

■ EXAMPLE 2-4

An electric hair dryer has a power rating of 1000 W. How much current will it draw when connected to 120 V?

Solution

The quantity to be found is amperage, or current. The known quantities are power and voltage. To solve this problem, choose the formula shown in *Figure 2–22*.

$$I = \frac{P}{E}$$

$$I = \frac{1000 \text{ W}}{120 \text{ V}}$$

$$I = 8.333\text{A}$$

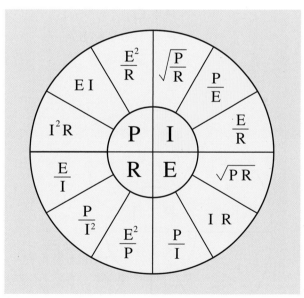

FIGURE 2–22 Finding current when power and voltage are known.

■ EXAMPLE 2-5

An electric hotplate has a power rating of 1440 W and a current draw of 12 A. What is the resistance of the hotplate?

Solution

The quantity to be found is resistance, and the known quantities are power and current. Use the formula shown in *Figure 2–23*.

$$R = \frac{P}{I^2}$$

$$R = \frac{1440 \text{ W}}{12 \text{ A} \times 12 \text{ A}}$$

$$R = \frac{1440 \text{ W}}{144 \text{ A}}$$

$$R = 10 \text{ } \Omega$$

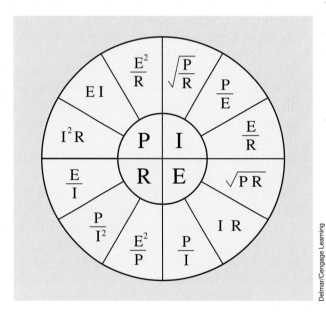

FIGURE 2–23   Finding resistance when power and current are known.

Kilo	1000
Hecto	100
Deka	10
Base unit	1
Deci	$\frac{1}{10}$ or 0.1
Centi	$\frac{1}{100}$ or 0.01
Milli	$\frac{1}{1000}$ or 0.001

Delmar/Cengage Learning

FIGURE 2–24   Standard metric prefixes.

2–12   Metric Prefixes

Metric prefixes are used in the electrical field just as they are in most other scientific fields. A special type of notation, known as *engineering notation*, is used in electrical measurements. Engineering notation is similar to scientific notation except that engineering notation is in steps of 1000 instead of 10. The chart in *Figure 2–24* shows standard metric prefixes. The first step above the base unit is deka, which means 10. The second prefix is hecto, which means 100, and the third prefix is kilo, which means 1000. The first prefix below the base unit is deci, which means $\frac{1}{10}$; the second prefix is centi, which means $\frac{1}{100}$; and the third is milli, which means $\frac{1}{1000}$.

The chart in *Figure 2–25* shows prefixes used in engineering notation. The first prefix above the base unit is kilo, or 1000; the second prefix is mega, or 1,000,000; and the third prefix is giga, or 1,000,000,000. Notice that each prefix is 1000 times greater than the previous prefix. The chart also shows that the first prefix below the base unit is milli, or $\frac{1}{1000}$; the second is micro, represented by the Greek mu (μ), or $\frac{1}{1,000,000}$; and the third is nano, or $\frac{1}{1,000,000,000}$.

Metric prefixes are used in almost all scientific measurements for ease of notation. It is much simpler to write a value such as 10 MΩ than it is to write 10,000,000 ohms, or to write 0.5 ns than to write 0.000,000,000, 5 second. Once the metric prefixes have been learned, measurements such as 47 kilohms (kΩ) or 50 milliamperes (mA) become commonplace to the technician.

The SI System

Note that the term *metric* is commonly used to indicate a system that employs measurements that increase or decrease in steps of 10. The prefixes just discussed are commonly referred to as *metric* units of measure. These prefixes are actually part of the SI (System Internationale) system that was adopted for

ENGINEERING UNIT	SYMBOL	MULTIPLY BY	
Tera	T	1,000,000,000,000	$\times 10^{12}$
Giga	G	1,000,000,000	$\times 10^{9}$
Mega	M	1,000,000	$\times 10^{6}$
Kilo	k	1,000	$\times 10^{3}$
Base unit		1	
Milli	m	0.001	$\times 10^{-3}$
Micro	μ	0.000,001	$\times 10^{-6}$
Nano	n	0.000,000,001	$\times 10^{-9}$
Pico	p	0.000,000,000,001	$\times 10^{-12}$

FIGURE 2–25   Standard prefixes of engineering notation.

use in the United States in 1960 by the 11th General Conference on Weights and Measures (abbreviated CGPM from the official French name *Conference Generale des Poids et Mesures*). The intention of the SI system is to provide a worldwide standard of weights and measures. The SI system uses seven base and two supplementary units that are regarded as dimensionally independent *(Figure 2–26)*. From these base and supplementary units, other units have been derived. Some of these units commonly used in the electrical field are shown in *Figure 2–27*.

Quantity	Unit	Symbol
Length	Meter	m
Mass	Kilogram	kg
Time	Second	s
Electric current	Ampere	A
Thermodynamic temperature	Kelvin	K
Amount of substance	Mole	mol
Luminous intensity	Candela	cd
Phase angle	Radian	rad
Solid angle	Sterdian	sr

FIGURE 2–26   SI base and supplementary units.

Quantity	Unit	Symbol	Formula
Frequency	Hertz	Hz	$1/s$
Force	Newton	N	$(kg \cdot m)/s^2$
Pressure, stress	Pascal	Pa	N/m^2
Energy, work, quantity of heat	Joule	J	$N \cdot m$
Power, radiant flux	Watt	W	J/s
Quantity of electricity, charge	Coulomb	C	$A \cdot s$
Electric potential, electromotive force	Volt	V	W/A
Capacitance	Farad	F	C/V
Electric resistance	Ohm	Ω	V/A
Conductance	Siemens	S	A/V
Magnetic flux	Weber	Wb	$V \cdot s$
Magnetic flux density	Tesla	T	Wb/m^2
Inductance	Henry	H	Wb/A
Luminous flux	Lumen	lm	$cd \cdot sr$
Illuminance	Lux	lx	lm/m^2

FIGURE 2–27   Derived SI units.

Summary

- A coulomb is a quantity measurement of electrons.
- An ampere (A) is 1 coulomb per second.
- The letter *I*, which stands for intensity of current flow, is normally used in Ohm's law formulas.
- Voltage is referred to as electric pressure, potential difference, or electromotive force. An E or a V can be used to represent voltage in Ohm's law formulas.
- An ohm (Ω) is a measurement of resistance (R) in an electric circuit.
- The watt (W) is a measurement of power in an electric circuit. It is represented by either a W or a P (power) in Ohm's law formulas.
- Electric measurements are generally expressed in engineering notation.
- Engineering notation differs from scientific notation in that it uses steps of 1000 instead of steps of 10.
- Before current can flow, there must be a complete circuit.
- A short circuit has little or no resistance.

Review Questions

1. What is a coulomb?

2. What is an ampere?

3. Define voltage.

4. Define ohm.

5. Define watt.

6. An electric heating element has a resistance of 16 Ω and is connected to a voltage of 120 V. How much current will flow in this circuit?

7. How many watts of heat are being produced by the heating element in Question 6?

8. A 240-V circuit has a current flow of 20 A. How much resistance is connected in the circuit?

9. An electric motor has an apparent resistance of 15 Ω. If 8 A of current are flowing through the motor, what is the connected voltage?

10. A 240-V air-conditioning compressor has an apparent resistance of 8 Ω. How much current will flow in the circuit?

11. How much power is being used by the motor in Question 10?

12. A 5-kW electric heating unit is connected to a 240-V line. What is the current flow in the circuit?

13. If the voltage in Question 12 is reduced to 120 V, how much current would be needed to produce the same amount of power?

14. Is it less expensive to operate the electric heating unit in Question 12 on 240 V or 120 V?

Practical Applications

You are an electrician on the job. The electrical blueprint shows that eight 500-W lamps are to be installed on the same circuit. The circuit voltage is 277 V and is protected by a 20-A circuit breaker. A continuous-use circuit can be loaded to only 80% of its rating. Is a 20-A circuit large enough to carry this load? ■

Practical Applications

You have been sent to a new home. The homeowner reports that sometimes the electric furnace trips the 240-V, 60-A circuit breaker connected to it. Upon examination, you find that the furnace contains three 5000-W heating elements designed to turn on in stages. For example, when the thermostat calls for heat, the first 5000-W unit turns on. After some period of time, the second unit will turn on, and then, after another time delay, the third unit will turn on. What do you think the problem is, and what would be your recommendation for correcting it? Explain your answer. ■

Practical Applications

You are an electrician installing the wiring in a new home. The homeowner desires that a ceiling fan with light kits be installed in five different rooms. Each fan contains a light kit that can accommodate four 60-watt lamps. Each fan motor draws a current of 1.8 amperes when operated on high speed. It is assumed that each fan can operate more than three hours at a time and therefore must be considered a continuous-duty device. The fans are to be connected to a 15-ampere circuit. Because the devices are continuous duty, the circuit current must be limited to 80% of the continuous connected load. How many fans can be connected to a single 15-ampere circuit? How many circuits will be required to supply power to all five fans? ■

Practical Applications

A homeowner is installing a swimming pool. You have been asked to install a circuit to operate a 600-watt underwater light and a circulating pump. The motor nameplate reveals that the pump has a current draw of 8.5 amperes. The devices are considered continuous duty. Can the power to operate both of these devices be supplied by a single 20-ampere circuit? ■

Practice Problems

Ohm's Law

Fill in the missing values.

Volts (E)	Amperes (I)	Ohms (R)	Watts (P)
153 V	0.056 A		
	0.65 A	470 Ω	
24 V			124 W
	0.00975 A		0.035 W
		6.8 kΩ	0.86 W
460 V		72 Ω	
48 V	1.2 A		
	154 A	0.8 Ω	
277 V			760 W
	0.0043 A		0.0625 W
		130 kΩ	0.0225 W
96 V		2.2 kΩ	

Unit 3
Static Electricity

Why You Need to Know

Many processes and devices use static electricity in a productive way. Copy machines, for example, could not operate except for the principles governing static electricity. Other devices such as electronic air cleaners and paint spray operations employ static electricity. The concepts of static electricity and how it can be used or prevented are important to anyone in the electrical field. To fully understand static electricity, this unit presents

- a demonstration of electron flow and how, by adding electrons, an object is negatively charged and how, by removing electrons, an object becomes positively charged.
- an explanation of electron flow in thunderclouds and lightning. This natural element is probably the greatest example of static electricity, and in this unit you can see how it applies to the theory of positive and negative charges in nature.

KEY TERMS

Electroscope
Electrostatic charges
Lightning
Lightning arrestor
Lightning bolts
Lightning rods
Nuisance static charges
Precipitators
Selenium
Static
Thundercloud
Useful static charges

Objectives

After studying this unit, you should be able to

■ discuss the nature of static electricity.

■ use an electroscope to determine unknown charges.

■ discuss lightning protection.

■ list nuisance charges of static electricity.

■ list useful charges of static electricity.

Preview

S tatic electric charges occur often in everyday life. Almost everyone has received a shock after walking across a carpet and then touching a metal object or after sliding across a car seat and touching the door handle. Almost everyone has combed their hair with a hard rubber or plastic comb and then used the comb to attract small pieces of paper or other lightweight objects. Static electric charges cause clothes to stick together when they are taken out of a clothes dryer. Lightning is without doubt the greatest display of a static electric discharge. ■

3–1   Static Electricity

Although static charges can be a nuisance *(Figure 3–1)*, or even dangerous, they can also be beneficial. Copy machines, for example, operate on the principle of static electricity. The manufacture of sandpaper also relies on the application of static electricity. Grains of sand receive a static charge to

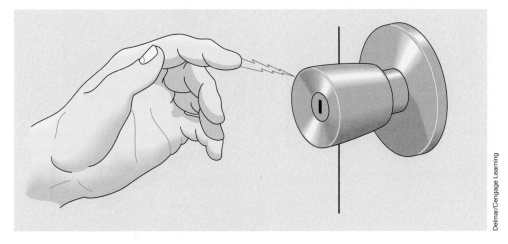

FIGURE 3–1   Static electric charges can cause a painful shock.

make them stand apart and expose a sharper edge *(Figure 3–2)*. Electronic air filters—**precipitators**—use static charges to attract small particles of smoke, dust, and pollen *(Figure 3–3)*. The precipitator uses a high-voltage DC power supply to provide a set of wires with a positive charge and a set of plates with a negative charge. As a blower circulates air through the unit, small particles

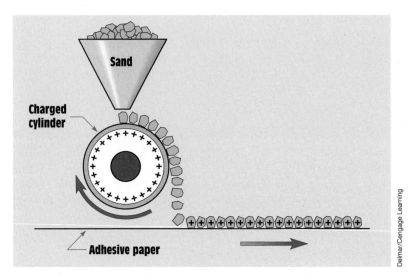

FIGURE 3–2   Grains of sand receive a charge to help them stand apart.

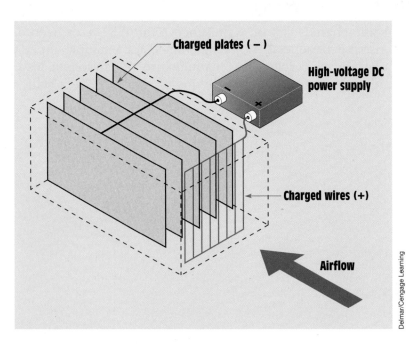

FIGURE 3–3   Electronic air cleaner.

receive a positive charge as they move across the charged wires. The charged particles are then attracted to the negative plates. The negative plates hold the particles until the unit is turned off and the plates are cleaned.

The word **static** means not moving or sitting still. Static electricity refers to electrons that are sitting still and not moving. Static electricity is therefore a charge and not a current. **Electrostatic charges** are built up on insulator materials because insulators are the only materials that can hold the electrons stationary and keep them from flowing to a different location. A static charge can be built up on a conductor only if the conductor is electrically insulated from surrounding objects. A static charge can be either positive or negative. If an object has a lack of electrons, it has a positive charge; and if it has an excess of electrons, it has a negative charge.

3–2   Charging an Object

The charge that accumulates on an object is determined by the materials used to produce the charge. If a hard rubber rod is rubbed on a piece of wool, the wool deposits excess electrons on the rod and gives it a negative charge. If a glass rod is rubbed on a piece of wool, electrons are removed from the rod, thus producing a positive charge on the rod *(Figure 3–4)*.

3–3   The Electroscope

An early electric instrument that can be used to determine the polarity of the electrostatic charge of an object is the **electroscope** *(Figure 3–5)*. An electroscope is a metal ball attached to the end of a metal rod. The other end of the rod is attached to two thin metal leaves. The metal leaves are inside a transparent container that permits the action of the leaves to be seen. The metal rod is insulated from the box. The metal leaves are placed inside a container so that air currents cannot affect their movement.

Before the electroscope can be used, it must first be charged. This is done by touching the ball with an object that has a known charge. For this example, assume that a hard rubber rod has been rubbed on a piece of wool to give it a negative charge. When the rubber rod is wiped against the metal ball, excess electrons are deposited on the metal surface of the electroscope. Because both of the metal leaves now have an excess of electrons, they repel each other, as shown in *Figure 3–6*.

Testing an Object

A charged object can now be tested to determine whether it has a positive or negative polarity. Assume that a ballpoint pen is charged by rubbing the plastic body through a person's hair. Now bring the pen close to but not touching

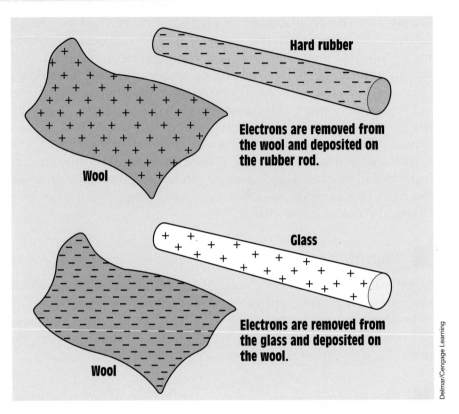

Hard rubber

Electrons are removed from the wool and deposited on the rubber rod.

Wool

Glass

Electrons are removed from the glass and deposited on the wool.

Wool

FIGURE 3–4 Producing a static charge.

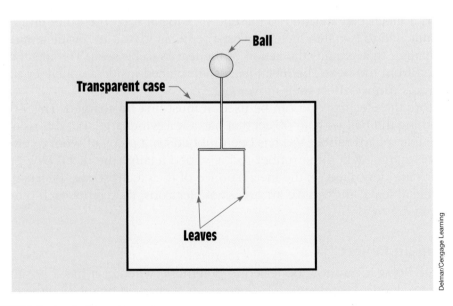

Ball

Transparent case

Leaves

FIGURE 3–5 An electroscope.

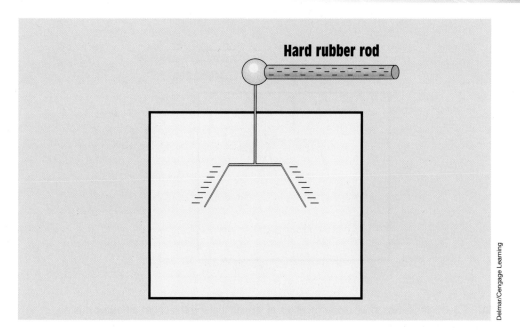

FIGURE 3–6 The electroscope is charged with a known static charge.

the ball and observe the action of the leaves. If the pen has taken on a negative charge, the leaves will move farther apart, as shown in *Figure 3–7*. The field caused by the negative electrons on the pen repels electrons from the ball. These electrons move down the rod to the leaves, causing the leaves to

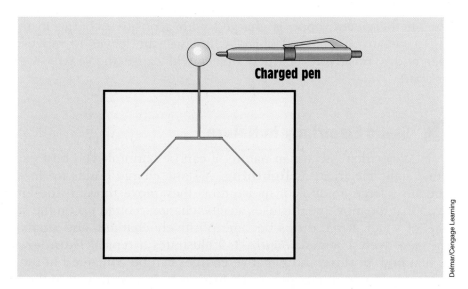

FIGURE 3–7 The leaves are deflected farther apart, indicating that the object has a negative charge.

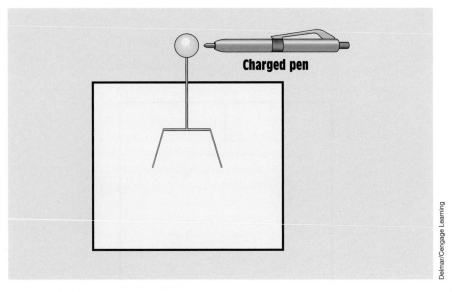

Charged pen

Delmar/Cengage Learning

FIGURE 3–8   The leaves move closer together, indicating that the object has a positive charge.

become more negative and to repel each other more, forcing the leaves to move farther apart.

If the pen has a positive charge, the leaves will move closer together when the pen is moved near the ball *(Figure 3–8)*. This action is caused by the positive field of the pen attracting electrons. When electrons are attracted away from the leaves, they become less negative and move closer together. If the electroscope is charged with a positive charge in the beginning, a negatively charged object will cause the leaves to move closer together and a positively charged object will cause the leaves to move farther apart.

3–4   Static Electricity in Nature

When static electricity occurs in nature, it can be harmful. The best example of natural static electricity is **lightning.** A static charge builds up in clouds that contain a large amount of moisture as they move through the air. It is theorized that the movement causes a static charge to build up on the surface of drops of water. Large drops become positively charged, and small drops become negatively charged. *Figure 3–9* illustrates a typical **thundercloud.** Notice that both positive and negative charges can be contained in the same cloud. Most lightning discharges, or **lightning bolts,** occur within the cloud. Lightning discharges can also take place between different clouds, between a

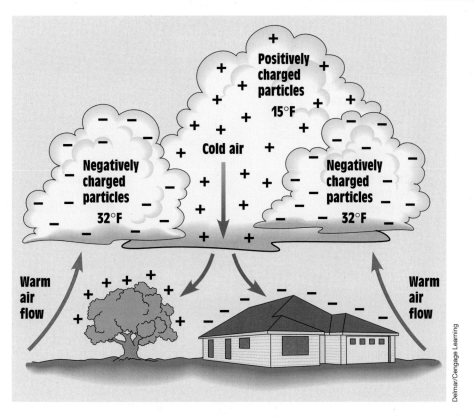

FIGURE 3–9   The typical thundercloud contains both negatively and positively charged particles.

cloud and the ground, and between the ground and the cloud *(Figure 3–10)*. Whether a lightning bolt travels from the cloud to the ground or from the ground to the cloud is determined by which contains the negative and which the positive charge. Current always flows from negative to positive. If a cloud is negative and an object on the ground is positive, the lightning discharge travels from the cloud to the ground. If the cloud has a positive charge and the object on the ground has a negative charge, the discharge travels from the ground to the cloud. A lightning bolt has an average voltage of about 15,000,000 volts.

Lightning Protection

Lightning rods are sometimes used to help protect objects from lightning. Lightning rods work by providing an easy path to ground for current flow. If the protected object is struck by a lightning bolt, the lightning rod bleeds the lightning discharge to ground before the protected object can be harmed *(Figure 3–11)*. Lightning rods were invented by Benjamin Franklin.

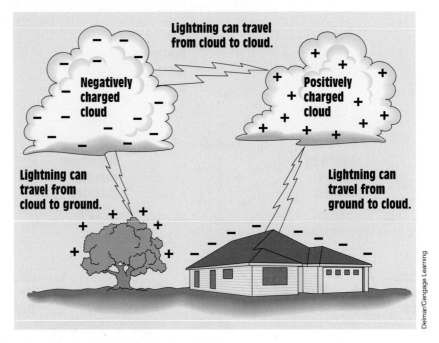

FIGURE 3–10 Lightning travels from negative to positive.

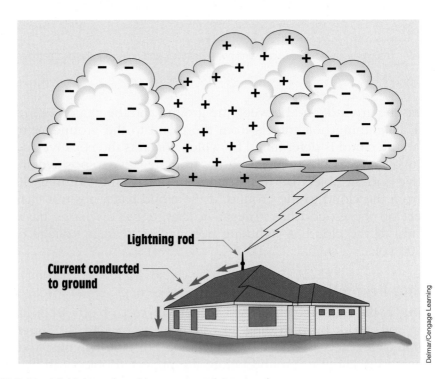

FIGURE 3–11 A lightning rod provides an easy path to ground.

Another device used for lightning protection is the **lightning arrestor.** The lightning arrestor works in a manner very similar to the lightning rod except that it is not designed to be struck by lightning itself and it does not provide a direct path to ground. The lightning arrestor is grounded at one end, and the other end is brought close to but not touching the object to be protected. If the protected object is struck, the high voltage of the lightning arcs across to the lightning arrestor and bleeds to ground.

Power lines are often protected by lightning arrestors that exhibit a very high resistance at the normal voltage of the line. If the power line is struck by lightning, the increase of voltage causes the resistance of the arrestor to decrease and conduct the lightning discharge to ground.

3–5   Nuisance Static Charges

Static charges are sometimes a nuisance. Some examples of **nuisance static charges** are listed here:

1. *The static charge that accumulates on automobiles as they move through dry air.* These static charges can cause dangerous conditions under certain circumstances. For that reason, trucks carrying flammable materials such as gasoline or propane use a drag chain. One end of the drag chain is attached to the frame of the vehicle, and the other end drags the ground. The chain is used to provide a path to ground while the vehicle is moving and to prevent a static charge from accumulating on the body of the vehicle.

2. *The static charge that accumulates on a person's body as he or she walks across a carpet.* This charge can cause a painful shock when a metal object is touched and it discharges in the form of an electric spark. Most carpets are made from man-made materials that are excellent insulators such as nylon. In the winter, the heating systems of most dwellings remove moisture from the air and cause the air to have a low humidity. The dry air combined with an insulating material provides an excellent setting for the accumulation of a static charge. This condition can generally be eliminated by the installation of a humidifier. A simple way to prevent the painful shock of a static discharge is to hold a metal object, such as a key or coin, in one hand. Touch the metal object to a grounded surface, and the static charge will arc from the metal object to ground instead of from your finger to ground.

3. *The static charge that accumulates on clothes in a dryer.* This static charge is caused by the clothes moving through the dry air. The greatest static charges generally are built up on man-made fabrics because they are the best insulators and retain electrons more readily than natural fabrics such as cotton or wool.

3–6   Useful Static Charges

Not all static charges are a nuisance. Some examples of **useful static charges** follow:

1. Static electricity is often used in spray painting. A high-voltage grid is placed in front of the spray gun. This grid has a positive charge. The object to be painted has a negative charge *(Figure 3–12)*. As the droplets of paint pass through the grid, the positive charge causes electrons to be removed from the paint droplets. The positively charged droplets are attracted to the negatively charged object. This static charge helps to prevent waste of the paint and at the same time produces a uniform finish.

2. Another device that depends on static electricity is the dry copy machine. The copy machine uses an aluminum drum coated with **selenium** *(Figure 3–13)*. Selenium is a semiconductor material that changes its conductivity with a change of light intensity. When selenium is in the presence of light, it has a very high conductivity. When it is in darkness, it has a very low conductivity.

A high-voltage wire located near the drum causes the selenium to have a positive charge as it rotates *(Figure 3–14)*. The drum is in darkness when it is charged. An image of the material to be copied is reflected on the drum by

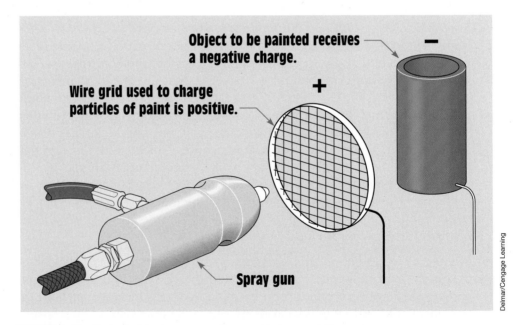

FIGURE 3–12   Static electric charges are often used in spray painting.

FIGURE 3–13   The drum of a copy machine is coated with selenium.

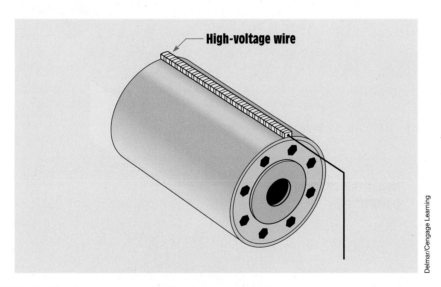

FIGURE 3–14   The drum receives a positive charge.

a system of lenses and mirrors *(Figure 3–15)*. The light portions of the paper reflect more light than the dark portions. When the reflected light strikes the drum, the conductivity of the selenium increases greatly, and negative electrons from the aluminum drum neutralize the selenium charge at that point. The dark area of the paper causes the drum to retain a positive charge.

A dark powder that has a negative charge is applied to the drum *(Figure 3–16)*. The powder is attracted to the positively charged areas on the drum. The powder on the neutral areas of the drum falls away.

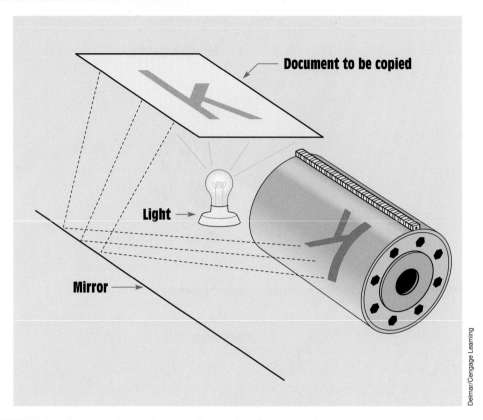

FIGURE 3–15   The image is transferred to the selenium drum.

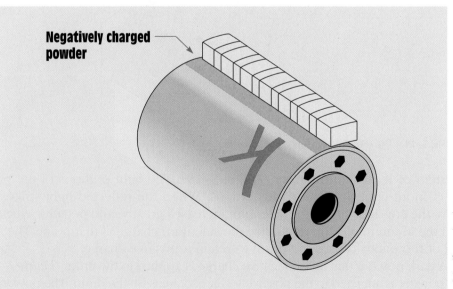

FIGURE 3–16   Negatively charged powder is applied to the positively charged drum.

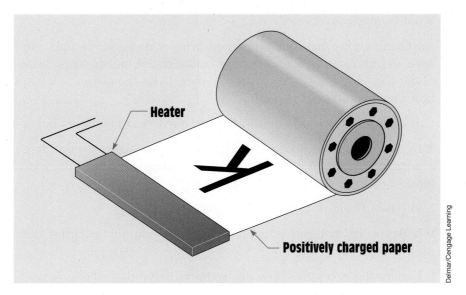

FIGURE 3-17 The negatively charged powder is attracted to the positively charged paper.

A piece of positively charged paper passes under the drum *(Figure 3–17)* and attracts the powder from the drum. The paper then passes under a heating element, which melts the powder into the paper and causes the paper to become a permanent copy of the original.

Summary

- The word *static* means not moving.
- An object can be positively charged by removing electrons from it.
- An object can be negatively charged by adding electrons to it.
- An electroscope is a device used to determine the polarity of an object.
- Static charges accumulate on insulator materials.
- Lightning is an example of a natural static charge.

Review Questions

1. Why is static electricity considered to be a charge and not a current?
2. If electrons are removed from an object, is the object positively or negatively charged?
3. Why do static charges accumulate on insulator materials only?

4. What is an electroscope?

5. An electroscope has been charged with a negative charge. An object with an unknown charge is brought close to the electroscope. The leaves of the electroscope come closer together. Does the object have a positive or a negative charge?

6. Can one thundercloud contain both positive and negative charges?

7. A thundercloud has a negative charge, and an object on the ground has a positive charge. Will the lightning discharge be from the cloud to the ground or from the ground to the cloud?

8. Name two devices used for lightning protection.

9. What type of material is used to coat the aluminum drum of a copy machine?

10. What special property does this material have that makes it useful in a copy machine?

Unit 4
Magnetism

KEY TERMS

Ampere-turns
Demagnetized
Electromagnets
Electron spin
    patterns
Flux
Flux density
Left-hand rule
Lines of flux
Lodestones

Magnetic domains
Magnetic molecules
Magnetomotive
    force (mmf)
Permanent magnets
Permeability
Reluctance
Residual magnetism
Saturation

Why You Need to Know

Magnetism and electricity are inseparable. If you have one, the other can be produced. A list of devices that operate on magnetism is almost endless. Everything from a simple compass to the largest electric motor in industry operates on magnetism. This unit presents those basic principles and discusses the most common terms used to measure magnetism:

- flux density.
- reluctance.
- ampere-turns.
- the left-hand rule regarding electromagnets.

These concepts appear many times in the study of electricity.

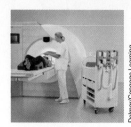

Delmar/Cengage Learning

Objectives

After studying this unit, you should be able to

- discuss the properties of permanent magnets.

- discuss the difference between the axis poles of the earth and the magnetic poles of the earth.

- discuss the operation of electromagnets.

- determine the polarity of an electromagnet when the direction of the current is known.

- discuss the different systems used to measure magnetism.

- define terms used to describe magnetism and magnetic quantities.

Preview

Magnetism is one of the most important phenomena in the study of electricity. It is the force used to produce most of the electrical power in the world. The force of magnetism has been known for over 2000 years. It was first discovered by the Greeks when they noticed that a certain type of stone was attracted to iron. This stone was first found in Magnesia in Asia Minor and was named magnetite. In the Dark Ages, the strange powers of the magnet were believed to be caused by evil spirits or the devil. ■

4–1   The Earth Is a Magnet

The first compass was invented when it was noticed that a piece of magnetite, a type of stone that is attracted to iron, placed on a piece of wood floating in water always aligned itself north and south *(Figure 4–1)*. Because they are always able to align themselves north and south, natural magnets became known as "leading stones" or **lodestones.** The reason that the lodestone aligned itself north and south is because the earth itself contains magnetic poles. *Figure 4–2* illustrates the positions of the true North and South Poles, or the axis, of the earth and the positions of the magnetic poles. Notice that *magnetic* north is not located at the true North Pole of the earth. This is the reason that navigators must distinguish between true north and magnetic north. The angular difference between the two is known as the *angle of declination*. Although the illustration shows the magnetic lines of force to be only on each side of the earth, the lines actually surround the entire earth like a magnetic shell.

Also notice that the magnetic north pole is located near the southern polar axis and the magnetic south pole is located near the northern polar

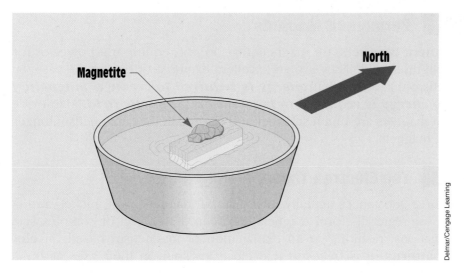

FIGURE 4–1   The first compass.

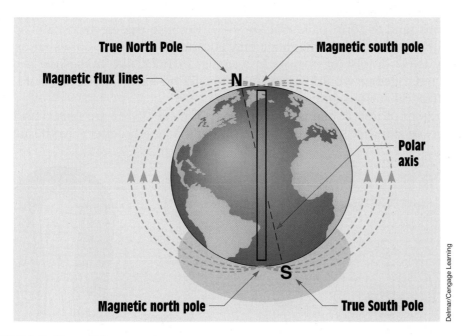

FIGURE 4–2   The earth is a magnet.

axis. The reason that the *geographic* poles (axes) are called north and south is because the north pole of a compass needle points in the direction of the north geographic pole. Because unlike magnetic poles attract, the north magnetic pole of the compass needle is attracted to the south magnetic pole of the earth.

4–2   Permanent Magnets

Permanent magnets are magnets that do not require any power or force to maintain their field. They are an excellent example of one of the basic laws of magnetism that states that *energy is required to create a magnetic field, but no energy is required to maintain a magnetic field.* Man-made permanent magnets are much stronger and can retain their magnetism longer than natural magnets.

4–3   The Electron Theory of Magnetism

Only three substances actually form natural magnets: iron, nickel, and cobalt. Why these materials form magnets has been the subject of complex scientific investigations, resulting in an explanation of magnetism based on **electron spin patterns.** It is believed that electrons spin on their axes as they orbit around the nucleus of the atom. This spinning motion causes each electron to become a tiny permanent magnet. Although all electrons spin, they do not all spin in the same direction. In most atoms, electrons that spin in opposite directions tend to form pairs *(Figure 4–3)*. Because the electron pairs spin in opposite directions, their magnetic effects cancel each other out as far as having any effect on distant objects. In a similar manner, two horseshoe magnets connected together would be strongly attracted to each other but would have little effect on surrounding objects *(Figure 4–4)*.

An atom of iron contains 26 electrons. Of these 26, 22 are paired and spin in opposite directions, canceling each other's magnetic effect. In the next to

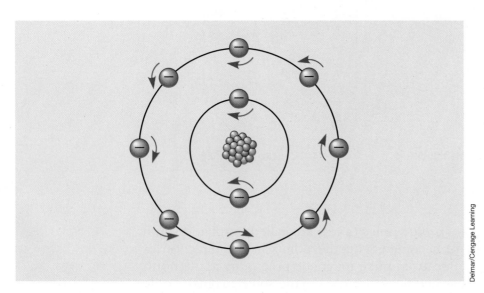

FIGURE 4–3   Electron pairs generally spin in opposite directions.

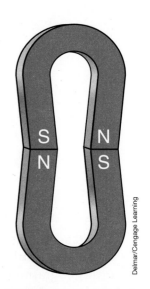

FIGURE 4–4   Two horseshoe magnets attract each other.

the outermost shell, however, 4 electrons are not paired and spin in the same direction. These 4 electrons account for the magnetic properties of iron. At a temperature of 1420°F, or 771.1°C, the electron spin patterns rearrange themselves and iron loses its magnetic properties.

When the atoms of most materials combine to form molecules, they arrange themselves in a manner that produces a total of eight valence electrons. The electrons form a spin pattern that cancels the magnetic field of the material. When the atoms of iron, nickel, and cobalt combine, however, the magnetic field is not canceled. Their electrons combine so that they share valence electrons in such a way that their spin patterns are in the same direction, causing their magnetic fields to add instead of cancel. The additive effect forms regions in the molecular structure of the metal called **magnetic domains** or **magnetic molecules.** These magnetic domains act like small permanent magnets.

A piece of nonmagnetized metal has its molecules in a state of disarray *(Figure 4–5)*. When the metal is magnetized, its molecules align themselves in an orderly pattern *(Figure 4–6)*. In theory, each molecule of a magnetic

Delmar/Cengage Learning

FIGURE 4–5   The atoms are disarrayed in a piece of nonmagnetized metal.

Delmar/Cengage Learning

FIGURE 4–6   The atoms are aligned in an orderly fashion in a piece of magnetized metal.

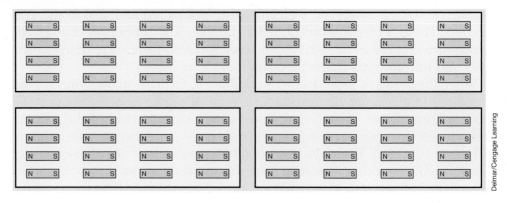

FIGURE 4–7   When a magnet is cut apart, each piece becomes a separate magnet.

material is itself a small magnet. If a permanent magnet is cut into pieces, each piece is a separate magnet *(Figure 4–7)*.

4–4   Magnetic Materials

Magnetic materials can be divided into three basic classifications:

- *Ferromagnetic materials* are metals that are easily magnetized. Examples of these materials are iron, nickel, cobalt, and manganese.

- *Paramagnetic materials* are metals that can be magnetized, but not as easily as ferromagnetic materials. Some examples of paramagnetic materials are platinum, titanium, and chromium.

- *Diamagnetic materials* are either metal or nonmetal materials that cannot be magnetized. The magnetic lines of force tend to go around them instead of through them. Some examples of these materials are copper, brass, and antimony.

Some of the best materials for the production of permanent magnets are alloys. One of the best permanent magnet materials is Alnico 5, which is made from a combination of aluminum, nickel, cobalt, copper, and iron. Another type of permanent magnet material is made from a combination of barium ferrite and strontium ferrite. Ferrites can have an advantage in some situations because they are insulators and not conductors. They have a resistivity of approximately 1,000,000 ohm-centimeters. Barium ferrite and strontium ferrite can be powdered. The powder is heated to the melting point and then rolled and heat treated. This treatment changes the grain structure and magnetic properties of the material. The new type of material has a property more like stone than metal and is known as a *ceramic magnet.* Ceramic magnets can be powdered and mixed with rubber, plastic, or liquids. Ceramic magnetic materials mixed with liquids can be used to make magnetic ink, which is used on checks. Another frequently used magnetic material is iron oxide, which is used to make magnetic recording tape and computer disks.

4–5 Magnetic Lines of Force

Magnetic lines of force are called **flux.** The symbol used to represent flux is the Greek letter phi (Φ). Flux lines can be seen by placing a piece of cardboard on a magnet and sprinkling iron filings on the cardboard. The filings will align themselves in a pattern similar to the one shown in *Figure 4–8*. The pattern produced by the iron filings forms a two-dimensional figure, but the flux lines actually surround the entire magnet *(Figure 4–9)*. Magnetic **lines of flux** repel each other and never cross. Although magnetic lines of flux do not flow, it is assumed they are in north to south direction.

A basic law of magnetism states that ***unlike poles attract and like poles repel.*** *Figure 4–10* illustrates what happens when a piece of cardboard is placed over two magnets with their north and south poles facing each other and iron filings are sprinkled on the cardboard. The filings form a pattern showing that the magnetic lines of flux are attracted to each other. *Figure 4–11* illustrates the pattern formed by the iron filings when the cardboard is placed over two magnets with like poles facing each other. The filings show that the magnetic lines of flux repel each other.

If the opposite poles of two magnets are brought close to each other, they are attracted to each other *(Figure 4–12)*. If like poles of the two magnets are brought together, they repel each other.

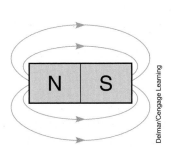

FIGURE 4–8 Magnetic lines of force are called lines of flux.

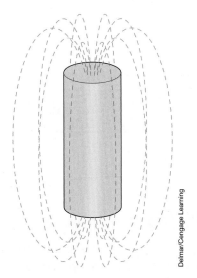

FIGURE 4–9 Magnetic lines of flux surround the entire magnet.

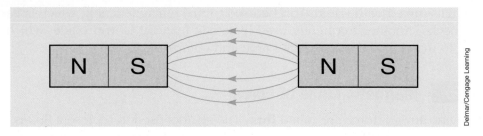

FIGURE 4–10   Opposite magnetic poles attract each other.

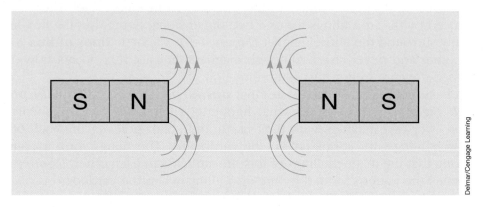

FIGURE 4–11   Like magnetic poles repel each other.

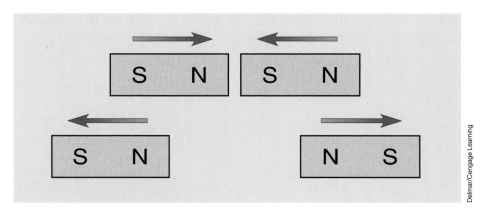

FIGURE 4–12   Opposite poles of a magnet attract, and like poles repel.

4–6   Electromagnetics

A basic law of physics states that ***whenever an electric current flows through a conductor, a magnetic field is formed around the conductor.*** **Electromagnets** depend on electric current flow to produce a magnetic field. They are generally designed to produce a magnetic field only as long as the current is flowing; they do not retain their magnetism when current flow stops. Electromagnets operate on the principle that current flowing through a conductor produces a magnetic field around the conductor *(Figure 4–13)*. If the conductor is wound into a coil *(Figure 4–14)*, the magnetic lines of flux add to produce a stronger magnetic field. A coil with 10 turns of wire produces a magnetic field that is 10 times as strong as the magnetic field around a single conductor.

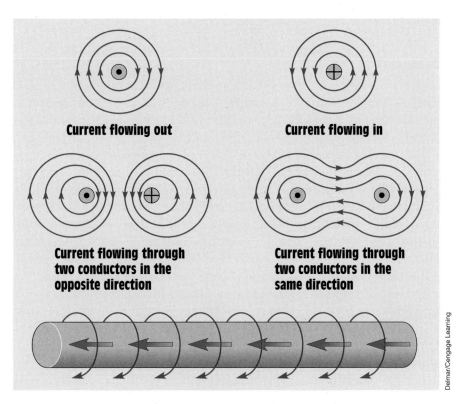

FIGURE 4–13   Current flowing through a conductor produces a magnetic field around the conductor.

FIGURE 4–14   Winding the wire into a coil increases the strength of the magnetic field.

Another factor that affects the strength of an electromagnetic field is the amount of current flowing through the wire. An increase in current flow causes an increase in magnetic field strength. The two factors that determine the number of flux lines produced by an electromagnet are the number of turns of wire and the amount of current flow through the wire. The strength of an electromagnet is proportional to its **ampere-turns.** Ampere-turns are determined by multiplying the number of turns of wire by the current flow.

Core Material

Coils can be wound around any type of material to form an electromagnet. The base material is called the *core material.* When a coil is wound around a nonmagnetic material such as wood or plastic, it is known as an *air-core magnet.* When a coil is wound around a magnetic material such as iron or soft steel, it is known as an *iron-core magnet.* The addition of magnetic material to the center of the coil can greatly increase the strength of the magnet. If the core material causes the magnetic field to become 10 times stronger, the core material has a **permeability** of 10 *(Figure 4–15).* Permeability is a measure of a material's ability to become magnetized. The number of flux lines produced is proportional to the ampere-turns. The magnetic core material provides an easy path for the flow of magnetic lines in much the same way a conductor provides an easy path for the flow of electrons. This increased permeability permits the flux lines to be concentrated in a smaller area, which increases the number of flux lines per square inch or per square centimeter. In a similar manner, a person using a garden hose with an adjustable nozzle attached can adjust the nozzle to spray the water in a fine mist that covers a large area or in a concentrated stream that covers a small area.

Another common magnetic measurement is **reluctance.** Reluctance is resistance to magnetism. A material such as soft iron or steel has a high permeability and low reluctance because it is easily magnetized. A material such as copper has a low permeability and high reluctance.

If the current flow in an electromagnet is continually increased, the magnet eventually reaches a point where its strength increases only slightly with an increase in current. When this condition occurs, the magnetic material is at a point of **saturation.** Saturation occurs when all the molecules of the magnetic material are lined up. Saturation is similar to pouring 5 gallons of water into a 5-gallon bucket. Once the bucket is full, it simply cannot hold any more water. If it became necessary to construct a stronger magnet, a larger piece of core material would be required.

When the current flow through the coil of a magnet is stopped, some magnetism may be left in the core material. The amount of magnetism left in a material after the magnetizing force has stopped is called **residual magnetism.** If the residual magnetism of a piece of core material is hard to remove, the

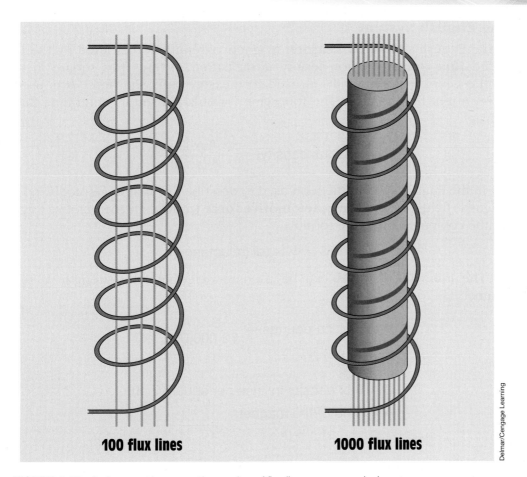

100 flux lines   **1000 flux lines**

Delmar/Cengage Learning

FIGURE 4–15   An iron core increases the number of flux lines per square inch.

material has a high coercive force. *Coercive force* is a measure of a material's ability to retain magnetism. A high coercive force is desirable in materials that are intended to be used as permanent magnets. A low coercive force is generally desirable for materials intended to be used as electromagnets. Coercive force is measured by determining the amount of current flow through the coil in the direction opposite to that required to remove the residual magnetism. Another term that is used to describe a material's ability to retain magnetism is *retentivity*.

4–7   Magnetic Measurement

The terms used to measure the strength of a magnetic field are determined by the system that is being used. Three different systems are used to measure magnetism: the English system, the CGS system, and the MKS or SI system.

The English System

In the English system of measure, magnetic strength is measured in a term called **flux density.** Flux density is measured in lines per square inch. As the Greek letter phi (Φ) is used to measure flux, the letter B is used to represent flux density. The following formula is used to determine flux density:

$$B \text{ (flux density)} = \frac{\Phi \text{ (flux lines)}}{A \text{ (area)}}$$

In the English system, the term used to describe the total force producing a magnetic field, or flux, is **magnetomotive force (mmf).** Magnetomotive force can be computed using the formula

$$mmf = \Phi \times rel \text{ (reluctance)}$$

The following formula can be used to determine the strength of the magnet:

$$\text{Pull (in pounds)} = \frac{B \times A}{72,000,000}$$

where

$$B = \text{flux density in lines per square inch}$$
$$A = \text{area of the magnet.}$$

The CGS System

In the CGS (centimeter-gram-second) system of measurement, one magnetic line of force is known as a *maxwell.* A gauss represents a magnetic force of 1 maxwell per square centimeter. In the English or SI system, magnetomotive force is measured in ampere-turns. In the CGS system, gilberts are used to represent the same measurement. Because the main difference between these two systems of measurement is that they use different units of measure, a conversion factor can be employed to help convert one set of units to another:

$$1 \text{ gilbert} = 1.256 \text{ ampere-turns}$$

In the CGS system, a standard called the *unit magnetic pole* is used. In *Figure 4–16,* two magnets are separated by a distance of 1 centimeter. These magnets repel each other with a force of 1 dyne. The dyne is a very weak unit of force. One dyne is equal to $\frac{1}{27,800}$ of an ounce, or it requires

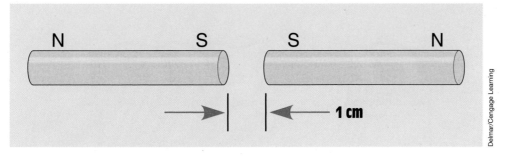

FIGURE 4–16   A unit magnetic pole produces a force of one dyne.

27,800 dynes to equal a force of 1 ounce. When the two magnets separated by a distance of 1 centimeter exert a force of 1 dyne, they are considered to be a unit magnetic pole. Magnetic force can then be determined using the formula

$$\text{Force (in dynes)} = \frac{M_1 \times M_2}{D^2}$$

where

M_1 = strength of first magnet in unit magnetic poles

M_2 = strength of second magnet in unit magnetic poles

D = distance between the poles in centimeters

The MKS or SI System

The MKS or SI system employs the units of measure of the MKS (meter-kilogram-second) system. In the SI system, the unit of force is the newton. One newton is equal to 0.2248 pounds, or it requires 4.448 newtons to equal a force of 1 pound. The weber is used to measure magnetic flux. One weber equals 100,000,000 lines of flux or 10^8 maxwells.

4–8   Magnetic Polarity

The polarity of an electromagnet can be determined using the **left-hand rule.** When the fingers of the left hand are placed around the windings in the direction of electron current flow, the thumb points to the north magnetic pole *(Figure 4–17).* If the direction of current flow is reversed, the polarity of the magnetic field also reverses.

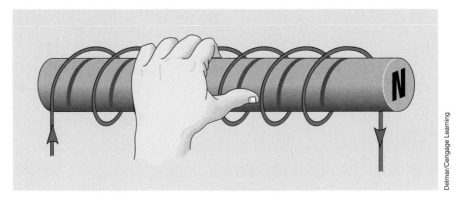

Delmar/Cengage Learning

FIGURE 4–17   The left-hand rule can be used to determine the polarity of an electromagnet.

4–9   Demagnetizing

When an object is to be **demagnetized,** its molecules must be disarranged as they are in a nonmagnetized material. This can be done by placing the object in the field of a strong electromagnet connected to an AC line. Because the magnet is connected to AC, the polarity of the magnetic field reverses each time the current changes direction. The molecules of the object to be demagnetized are therefore aligned first in one direction and then in the other. If the object is pulled away from the AC magnetic field, the effect of the field becomes weaker as the object is moved farther away *(Figure 4–18)*. The weakening of the magnetic field causes the molecules of the object to

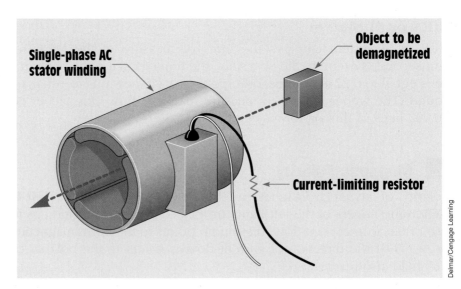

Delmar/Cengage Learning

FIGURE 4–18   Demagnetizing an object.

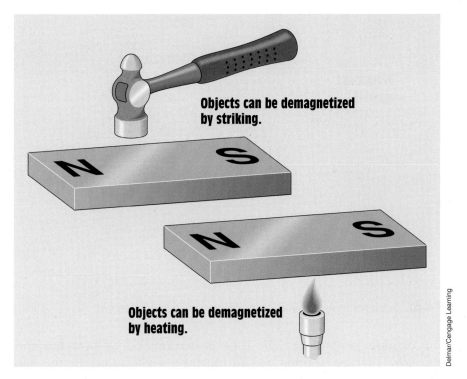

Objects can be demagnetized by striking.

Objects can be demagnetized by heating.

Delmar/Cengage Learning

FIGURE 4–19 Other methods for demagnetizing objects.

be left in a state of disarray. The ease or difficulty with which an object can be demagnetized depends on the strength of the AC magnetic field and the coercive force of the object.

An object can be demagnetized in two other ways *(Figure 4–19)*. If a magnetized object is struck, the vibration often causes the molecules to rearrange themselves in a disordered fashion. It may be necessary to strike the object several times. Heating also demagnetizes an object. When the temperature becomes high enough, the molecules rearrange themselves in a disordered fashion.

4–10 Magnetic Devices

A list of devices that operate on magnetism would be very long indeed. Some of the more common devices are electromagnets, measuring instruments, inductors, transformers, and motors.

The Speaker

The speaker is a common device that operates on the principle of magnetism *(Figure 4–20)*. The speaker produces sound by moving a cone; the movement

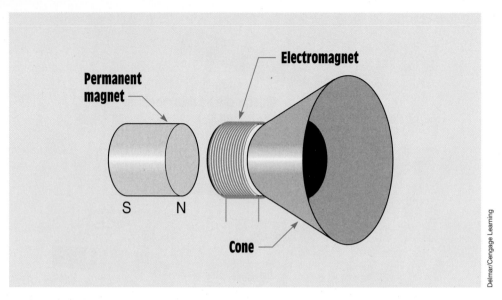

FIGURE 4–20 A speaker uses both an electromagnet and a permanent magnet.

causes a displacement of air. The tone is determined by how fast the cone vibrates. Low or bass sounds are produced by vibrations in the range of 20 cycles per second. High sounds are produced when the speaker vibrates in the range of 20,000 cycles per second.

The speaker uses two separate magnets. One is a permanent magnet, and the other is an electromagnet. The permanent magnet is held stationary, and the electromagnet is attached to the speaker cone. When current flows through the coil of the electromagnet, a magnetic field is produced. The polarity of the field is determined by the direction of current flow. When the electromagnet has a north polarity, it is repelled away from the permanent magnet, causing the speaker cone to move outward and displace air. When the current flow reverses through the coil, the electromagnet has a south polarity and is attracted to the permanent magnet. The speaker cone then moves inward and again displaces air. The number of times per second that the current through the coil reverses determines the tone of the speaker.

Summary

- Early natural magnets were known as lodestones.
- The earth has a north and a south magnetic pole.
- The magnetic poles of the earth and the axes poles are not the same.
- Like poles of a magnet repel each other, and unlike poles attract each other.

- Some materials have the ability to become better magnets than others.

- Three basic types of magnetic material are

  a. Ferromagnetic

  b. Paramagnetic

  c. Diamagnetic

- When current flows through a wire, a magnetic field is created around the wire.

- The direction of current flow through the wire determines the polarity of the magnetic field.

- The strength of an electromagnet is determined by the ampere-turns.

- The type of core material used in an electromagnet can increase its strength.

- Three different systems are used to measure magnetic values:

  a. The English system

  b. The CGS system

  c. The SI system

- An object can be demagnetized by placing it in an AC magnetic field and pulling it away, by striking, and by heating.

Review Questions

1. Is the north magnetic pole of the earth a north polarity or a south polarity?

2. What were early natural magnets known as?

3. The south pole of one magnet is brought close to the south pole of another magnet. Will the magnets repel or attract each other?

4. How can the polarity of an electromagnet be determined if the direction of current flow is known?

5. Define the following terms:

   Flux density
   Permeability
   Reluctance
   Saturation
   Coercive force
   Residual magnetism

6. A force of 1 ounce is equal to how many dynes?

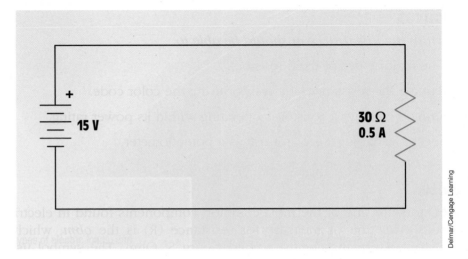

FIGURE 5–1 Resistor used to limit the flow of current.

The second principal function of resistors is to produce a **voltage divider.** The three resistors shown in *Figure 5–2* are connected in series with a 17.5-volt battery. If the leads of a voltmeter were connected between different points in the circuit, it would indicate the following voltages:

A to B, 1.5 V

A to C, 7.5 V

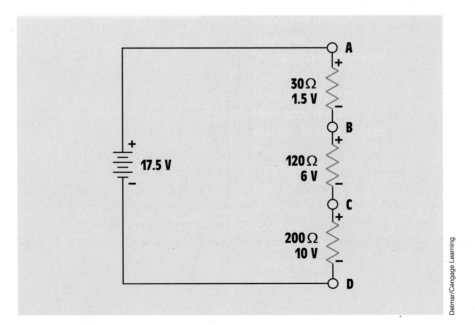

FIGURE 5–2 Resistors used as a voltage divider.

A to D, 17.5 V

B to C, 6 V

B to D, 16 V

C to D, 10 V

By connecting resistors of the proper value, almost any voltage desired can be obtained. Voltage dividers were used to a large extent in vacuum-tube circuits many years ago. Voltage divider circuits are still used today in applications involving field-effect transistors (FETs) and in multirange voltmeter circuits.

5–2   Fixed Resistors

Fixed resistors have only one ohmic value, which cannot be changed or adjusted. There are several different types of fixed resistors. One of the most common types of fixed resistors is the **composition carbon resistor.** Carbon resistors are made from a compound of carbon graphite and a resin bonding material. The proportions of carbon and resin material determine the value of resistance. This compound is enclosed in a case of nonconductive material with connecting leads *(Figure 5–3)*.

Carbon resistors are very popular for most applications because they are inexpensive and readily available. They are made in standard values that range from about 1 ohm to about 22 megohms (MΩ), and they can be obtained in power ratings of $\frac{1}{8}$, $\frac{1}{4}$, $\frac{1}{2}$, 1, and 2 watts. The power rating of the resistor is indicated by its size. A $\frac{1}{2}$-watt resistor is approximately $\frac{3}{8}$ inch in length and $\frac{1}{8}$ inch in diameter. A 2-watt resistor has a length of approximately $\frac{11}{16}$ inch

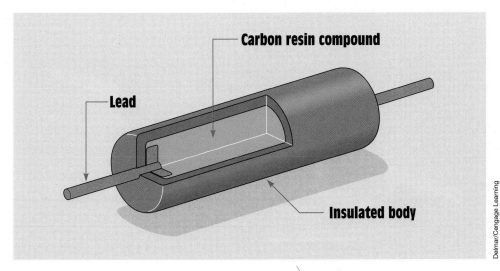

Delmar/Cengage Learning

FIGURE 5–3   Composition carbon resistor.

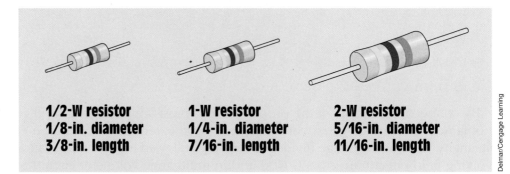

1/2-W resistor
1/8-in. diameter
3/8-in. length

1-W resistor
1/4-in. diameter
7/16-in. length

2-W resistor
5/16-in. diameter
11/16-in. length

FIGURE 5–4 Power rating is indicated by size.

and a diameter of approximately $\frac{5}{16}$ inch *(Figure 5–4)*. The 2-watt resistor is larger than the $\frac{1}{2}$-watt or 1-watt because it must have a larger surface area to be able to dissipate more heat. Although carbon resistors have a lot of desirable characteristics, they have one characteristic that is not desirable. Carbon resistors will change their value with age or if they are overheated. Carbon resistors generally increase instead of decrease in value.

Metal Film Resistors

Another type of fixed resistor is the **metal film resistor.** Metal film resistors are constructed by applying a film of metal to a ceramic rod in a vacuum *(Figure 5–5)*. The resistance is determined by the type of metal used to form the film and the thickness of the film. Typical thicknesses for the film are

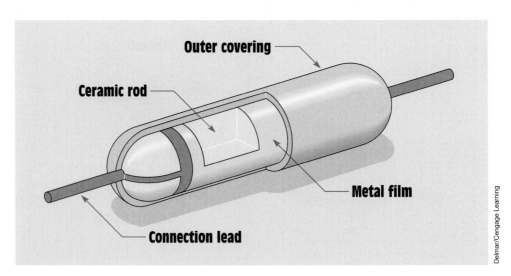

Outer covering

Ceramic rod

Metal film

Connection lead

FIGURE 5–5 Metal film resistor.

from 0.00001 to 0.00000001 inch. Leads are then attached to the film coating, and the entire assembly is covered with a coating. These resistors are superior to carbon resistors in several respects. Metal film resistors do not change their value with age, and their **tolerance** is generally better than carbon resistors. Tolerance indicates the plus and minus limits of a resistor's ohmic value. Carbon resistors commonly have a tolerance range of 20%, 10%, or 5%. Metal film resistors generally range in tolerance from 2% to 0.1%. The disadvantage of the metal film resistor is that it costs more.

Carbon Film Resistors

Another type of fixed resistor that is constructed in a similar manner is the **carbon film resistor.** This resistor is made by coating a ceramic rod with a film of carbon instead of metal. Carbon film resistors are less expensive to manufacture than metal film resistors and can have a higher tolerance rating than composition carbon resistors.

Metal Glaze Resistors

The **metal glaze resistor** is also a fixed resistor, similar to the metal film resistor. This resistor is made by combining metal with glass. The compound is then applied to a ceramic base as a thick film. The resistance is determined by the amount of metal used in the compound. Tolerance ratings of 2% and 1% are common.

Wire-Wound Resistors

Wire-wound resistors are fixed resistors that are made by winding a piece of resistive wire around a ceramic core *(Figure 5–6)*. The resistance of a wire-wound resistor is determined by three factors:

1. the type of material used to make the resistive wire

2. the diameter of the wire

3. the length of the wire

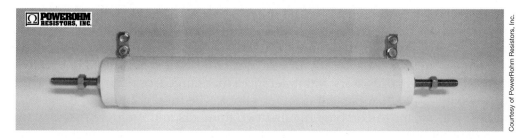

FIGURE 5–6   Wire-wound resistor.

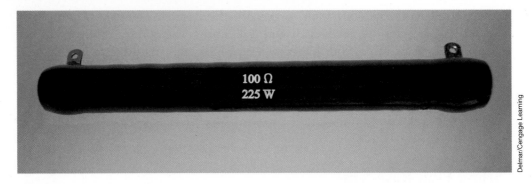

FIGURE 5-7   Wire-wound resistor with hollow core.

Wire-wound resistors can be found in various case styles and sizes. These resistors are generally used when a high power rating is needed. Wire-wound resistors can operate at higher temperatures than any other type of resistor. A wire-wound resistor that has a hollow center is shown in *Figure 5–7*. This type of resistor should be mounted vertically and not horizontally. The center of the resistor is hollow for a very good reason. When the resistor is mounted vertically, the heat from the resistor produces a chimney effect and causes air to circulate through the center *(Figure 5–8)*. This increase of airflow dissipates heat at a faster rate to help keep the resistor from overheating. The disadvantage of wire-wound resistors is that they are expensive and generally require a large amount of space for mounting. They can also exhibit an amount of inductance in circuits that operate at high frequencies. This added inductance can cause problems to the rest of the circuit. Inductance is covered in later units.

5-3   Color Code

The value of a resistor can often be determined by the **color code.** Many resistors have bands of color that are used to determine the resistance value, tolerance, and in some cases reliability. The color bands represent numbers. Each color represents a different numerical value *(Table 5–1)*. The chart shown in *Figure 5–9* lists the color and the number value assigned to each color. The resistor shown beside the color chart illustrates how to determine the value of a resistor. Resistors can have from three to five bands of color. Resistors that have a tolerance of ±20% have only three color bands. Most resistors contain four bands of color. For resistors with tolerances that range from ±10% to ±2%, the first two color bands represent number values. The third color band is the multiplier. This means to combine the first two numbers and multiply the resulting two-digit number by the power of 10 indicated by the value of the third band. The fourth band indicates the tolerance. For example, assume a resistor has

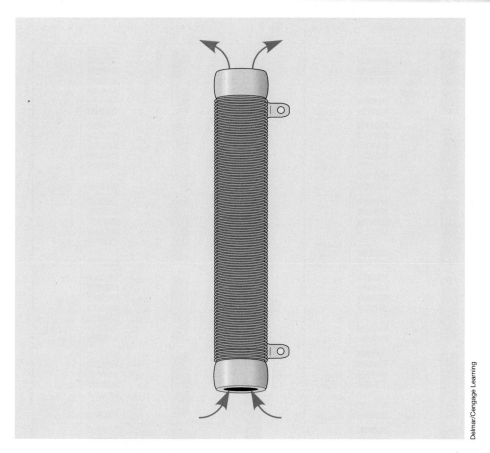

Delmar/Cengage Learning

FIGURE 5–8   Airflow helps cool the resistor.

Color	Value	Tolerance	
Black	0	No color	± 20%
Brown	1	Silver	± 10%
Red	2	Gold	± 5%
Orange	3	Red	± 2%
Yellow	4	Brown	± 1%
Green	5		
Blue	6		
Violet	7		
Gray	8		
White	9		

TABLE 5–1   Colors and Numeric Values.

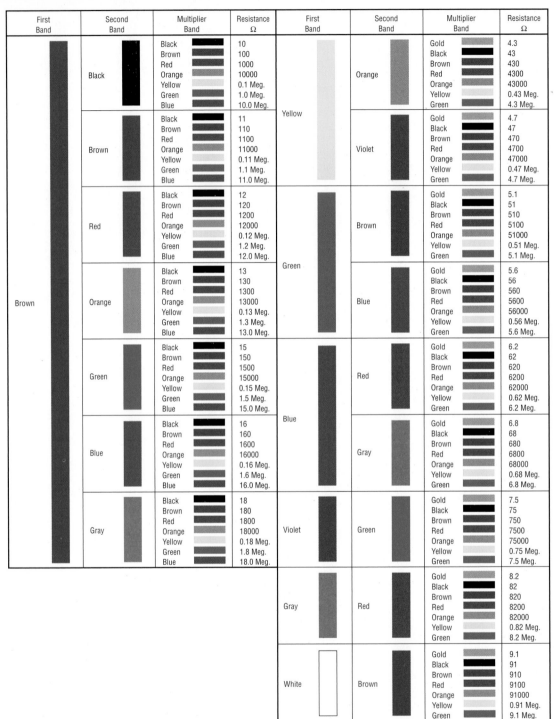

FIGURE 5–9 Resistor color code chart.

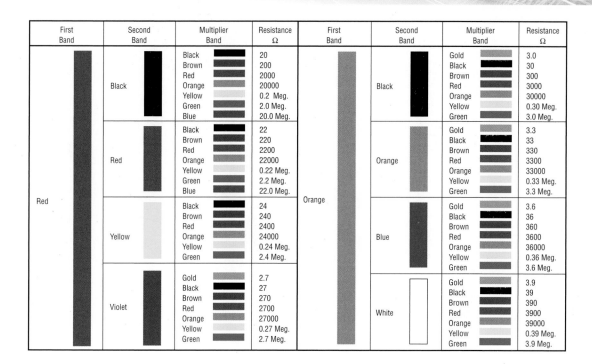

First Band	Second Band	Multiplier Band	Resistance Ω
Red	Black	Black	20
		Brown	200
		Red	2000
		Orange	20000
		Yellow	0.2 Meg.
		Green	2.0 Meg.
		Blue	20.0 Meg.
	Red	Black	22
		Brown	220
		Red	2200
		Orange	22000
		Yellow	0.22 Meg.
		Green	2.2 Meg.
		Blue	22.0 Meg.
	Yellow	Black	24
		Brown	240
		Red	2400
		Orange	24000
		Yellow	0.24 Meg.
		Green	2.4 Meg.
	Violet	Gold	2.7
		Black	27
		Brown	270
		Red	2700
		Orange	27000
		Yellow	0.27 Meg.
		Green	2.7 Meg.

First Band	Second Band	Multiplier Band	Resistance Ω
Orange	Black	Gold	3.0
		Black	30
		Brown	300
		Red	3000
		Orange	30000
		Yellow	0.30 Meg.
		Green	3.0 Meg.
	Orange	Gold	3.3
		Black	33
		Brown	330
		Red	3300
		Orange	33000
		Yellow	0.33 Meg.
		Green	3.3 Meg.
	Blue	Gold	3.6
		Black	36
		Brown	360
		Red	3600
		Orange	36000
		Yellow	0.36 Meg.
		Green	3.6 Meg.
	White	Gold	3.9
		Black	39
		Brown	390
		Red	3900
		Orange	39000
		Yellow	0.39 Meg.
		Green	3.9 Meg.

Standard Color Code

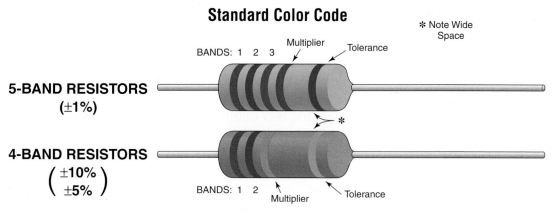

✻ Note Wide Space

BANDS: 1 2 3   Multiplier   Tolerance

5-BAND RESISTORS
(±1%)

✻

4-BAND RESISTORS
$\left(\begin{array}{c}\pm 10\% \\ \pm 5\%\end{array}\right)$

BANDS: 1 2   Multiplier   Tolerance

Band 1 1st Digit		Band 2 2nd Digit		Band 3 (if used) 3rd Digit		Multiplier		Resistance Tolerance	
Color	Digit	Color	Digit	Color	Digit	Color	Multiplier	Color	Tolerance
Black	0	Black	0	Black	0	Black	1	Silver	±10%
Brown	1	Brown	1	Brown	1	Brown	10	Gold	± 5%
Red	2	Red	2	Red	2	Red	100	Brown	± 1%
Orange	3	Orange	3	Orange	3	Orange	1,000		
Yellow	4	Yellow	4	Yellow	4	Yellow	10,000		
Green	5	Green	5	Green	5	Green	100,000		
Blue	6	Blue	6	Blue	6	Blue	1,000,000		
Violet	7	Violet	7	Violet	7	Silver	0.01		
Gray	8	Gray	8	Gray	8	Gold	0.1		
White	9	White	9	White	9				

FIGURE 5–9 Continued.

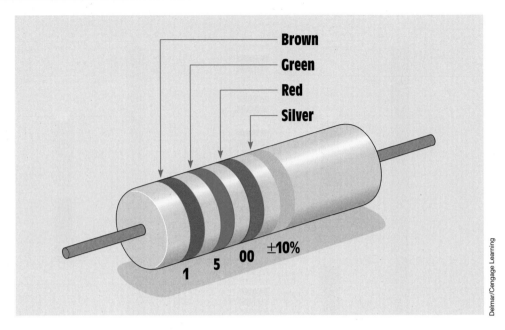

FIGURE 5–10 Determining resistor values using the color code.

color bands of brown, green, red, and silver *(Figure 5–10)*. The first two bands represent the numbers 1 and 5 (brown is 1 and green is 5). The third band is red, which has a number value of 2. The number 15 should be multiplied by 10^2, or 100. The value of the resistor is 1500 ohm. Another method, which is simpler to understand, is to add the number of zeros indicated by the multiplier band to the combined first two numbers. The multiplier band in this example is red, which has a numeric value of 2. Add two zeros to the first two numbers. The number 15 becomes 1500.

The fourth band is the tolerance band. The tolerance band in this example is silver, which means ±10%. This resistor should be 1500 ohm plus or minus 10%. To determine the value limits of this resistor, find 10% of 1500 ohm:

$$1500 \ \Omega \times 0.10 = 150 \ \Omega$$

The value can range from 1500 Ω + 10%, or 1500 Ω + 150 Ω = 1650 Ω, to 1500 Ω − 10% or 1500 Ω − 150 Ω = 1350 Ω.

Resistors that have a tolerance of ±1%, as well as some military resistors, contain five bands of color.

Gold and Silver as Multipliers

The colors gold and silver are generally found in the fourth band of a resistor, but they can be used in the multiplier band also. When the color gold is used

■ **EXAMPLE 5-1**

The resistor shown in *Figure 5–11* contains the following bands of color:

First band = brown

Second band = black

Third band = black

Fourth band = brown

Fifth band = brown

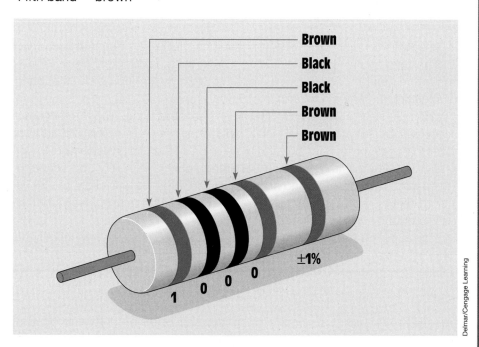

Brown
Black
Black
Brown
Brown

±1%

1   0   0   0

Delmar/Cengage Learning

FIGURE 5–11   Determining the value of a ±1% resistor.

Solution

The brown fifth band indicates that this resistor has a tolerance of ±1%. To determine the value of a 1% resistor, the first three bands are numbers, and the fourth band is the multiplier. In this example, the first band is brown, which has a number value of 1. The next two bands are black, which represents a number value of 0. The fourth band is brown, which means add one 0 to the first three numbers. The value of this resistor is 1000 Ω ±1%.

■ EXAMPLE 5-2

A five-band resistor has the following color bands:

First band = red

Second band = orange

Third band = violet

Fourth band = red

Fifth band = brown

Solution

The first three bands represent number values. Red is 2, orange is 3, and violet is 7. The fourth band is the multiplier; in this case, red represents 2. Add two zeros to the number 237. The value of the resistor is 23,700 Ω. The fifth band is brown, which indicates a tolerance of ±1%.

Military resistors often have five bands of color also. These resistors are read in the same manner as a resistor with four bands of color. The fifth band can represent different things. A fifth band of orange or yellow is used to indicate reliability. Resistors with a fifth band of orange have a reliability good enough to be used in missile systems, and a resistor with a fifth band of yellow can be used in space-flight equipment. A military resistor with a fifth band of white indicates the resistor has solderable leads.

Resistors with tolerance ratings ranging from 0.5% to 0.1% generally have their values printed directly on the resistor.

as the multiplier band, it means to divide the combined first two numbers by 10. If silver is used as the multiplier band, it means to divide the combined first two numbers by 100. For example, assume a resistor has color bands of orange, white, gold, and gold. The value of this resistor is 3.9 ohm with a tolerance of ±5% (orange = 3; white = 9; gold means to divide 39 by 10 = 3.9; and gold in the fourth band means ±5% tolerance).

5-4　Standard Resistance Values of Fixed Resistors

Fixed resistors are generally produced in standard values. The higher the tolerance value, the fewer resistance values available. Standard resistor values are listed in the chart shown in *Figure 5–12*. In the column under 10%, only

STANDARD RESISTANCE VALUES (Ω)									
0.1% 0.25%		0.1% 0.25%		0.1% 0.25%		0.1% 0.25%		0.1% 0.25%	
0.5%	1%	0.5%	1%	0.5%	1%	0.5%	1%	0.5%	1%
10.0	10.0	17.2	–	29.4	29.4	50.5	–	86.6	86.6
10.1	–	17.4	17.4	29.8	–	51.1	51.1	87.6	–
10.2	10.2	17.6	–	30.1	30.1	51.7	–	88.7	88.7
10.4	–	17.8	17.8	30.5	–	52.3	52.3	89.8	–
10.5	10.5	18.0	–	30.9	30.9	53.0	–	90.9	90.9
10.6	–	18.2	18.2	31.2	–	53.6	53.6	92.0	–
10.7	10.7	18.4	–	31.6	31.6	54.2	–	93.1	93.1
10.9	–	18.7	18.7	32.0	–	54.9	54.9	94.2	–
11.0	11.0	18.9	–	32.4	32.4	55.6	–	95.3	95.3
11.1	–	19.1	19.1	32.8	–	56.2	56.2	96.5	–
11.3	11.3	19.3	–	33.2	33.2	56.9	–	97.6	97.6
11.4	–	19.6	19.6	33.6	–	57.6	57.6	98.8	–
11.5	11.5	19.8	–	34.0	34.0	58.3	–		
11.7	–	20.0	20.0	34.4	–	59.0	59.0		
11.8	11.8	20.3	–	34.8	34.8	59.7	–		
12.0	–	20.5	20.5	35.2	–	60.4	60.4		
12.1	12.1	20.8	–	35.7	35.7	61.2	–		
12.3	–	21.0	21.0	36.1	–	61.9	61.9		
12.4	12.4	21.3	–	36.5	36.5	62.6	–		
12.6	–	21.5	21.5	37.0	–	63.4	63.4		
12.7	12.7	21.8	–	37.4	37.4	64.2	–	2%,5%	10%
12.9	–	22.1	22.1	37.9	–	64.9	64.9	10	10
13.0	13.0	22.3	–	38.3	38.3	65.7	–	11	–
13.2	–	22.6	22.6	38.8	–	66.5	66.5	12	12
13.3	13.3	22.9	–	39.2	39.2	67.3	–	13	–
13.5	–	23.2	23.2	39.7	–	68.1	68.1	15	15
13.7	13.7	23.4	–	40.2	40.2	69.0	–	16	–
13.8	–	23.7	23.7	40.7	–	69.8	69.8	18	18
14.0	14.0	24.0	–	41.2	41.2	70.6	–	20	–
14.2	–	24.3	24.3	41.7	–	71.5	71.5	22	22
14.3	14.3	24.6	–	42.2	42.2	72.3	–	24	–
14.5	–	24.9	24.9	42.7	–	73.2	73.2	27	27
14.7	14.7	25.2	–	43.2	43.2	74.1	–	30	–
14.9	–	25.5	25.5	43.7	–	75.0	75.0	33	33
15.0	15.0	25.8	–	44.2	44.2	75.9	–	36	–
15.2	–	26.1	26.1	44.8	–	76.8	76.8	39	39
15.4	15.4	26.4	–	45.3	45.3	77.7	–	43	–
15.6	–	26.7	26.7	45.9	–	78.7	78.7	47	47
15.8	15.8	27.1	–	46.4	46.4	79.6	–	51	–
16.0	–	27.4	27.4	47.0	–	80.6	80.6	56	56
16.2	16.2	27.7	–	47.5	47.5	81.6	–	62	–
16.4	–	28.0	28.0	48.1	–	82.5	82.5	68	68
16.5	16.5	28.4	–	48.7	48.7	83.5	–	75	–
16.7	–	28.7	28.7	49.3	–	84.5	84.5	82	82
16.9	16.9	29.1	–	49.9	49.9	85.6	–	91	–

Delmar/Cengage Learning

FIGURE 5–12   Standard resistance values.

12 values of resistors are listed. These standard values, however, can be multiplied by factors of 10. Notice that one of the standard values listed is 33 ohm. There are also standard values in 10% resistors of 0.33, 3.3, 330, 3300, 33,000, 330,000, and 3,300,000 ohm. The 2% and 5% column shows 24 resistor values, and the 1% column list 96 values. All the values listed in the chart can be multiplied by factors of 10 to obtain other resistance values.

5–5   Power Ratings

Resistors also have a power rating in watts that should not be exceeded or the resistor will be damaged. The amount of heat that must be dissipated by (given off to the surrounding air) the resistor can be determined by the use of one of the following formulas:

$$P = I^2R$$

$$P = \frac{E^2}{R}$$

$$P = EI$$

■ EXAMPLE 5-3

The resistor shown in *Figure 5–13* has a value of 100 Ω and a power rating of ½ W. If the resistor is connected to a 10-V power supply, will it be damaged?

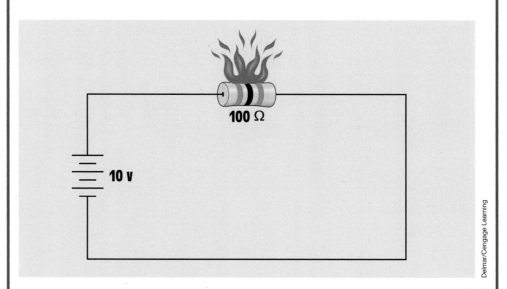

100 Ω

10 V

Delmar/Cengage Learning

FIGURE 5–13   Exceeding the power rating causes damage to the resistor.

Solution

Using the formula $P = \frac{E^2}{R}$, determine the amount of heat that will be dissipated by the resistor.

$$P = \frac{E^2}{R}$$

$$P = \frac{10\text{ V} \times 10\text{ V}}{100\ \Omega}$$

$$P = \frac{100\text{ V}}{100\ \Omega}$$

$$P = 1\text{ W}$$

Because the resistor has a power rating of ½ W, and the amount of heat that will be dissipated is 1 W, the resistor will be damaged.

5-6 Variable Resistors

A **variable resistor** is a resistor whose values can be changed or varied over a range. Variable resistors can be obtained in different case styles and power ratings. *Figure 5–14* illustrates how a variable resistor is constructed. In this example, a resistive wire is wound in a circular pattern, and a sliding tap makes contact with the wire. The value of resistance can be adjusted between one end of the resistive wire and the sliding tap. If the resistive wire has a total value of 100 ohms, the resistor can be set between the values of 0 and 100 ohms.

A variable resistor with three terminals is shown in *Figure 5–15*. This type of resistor has a wiper arm inside the case that makes contact with the resistive element. The full resistance value is between the two outside terminals, and the wiper arm is connected to the center terminal. The resistance between the center terminal and either of the two outside terminals can be adjusted by turning the shaft and changing the position of the wiper arm. Wire-wound variable resistors of this type can be obtained also *(Figure 5–16)*. The advantage of the wire-wound type is a higher power rating.

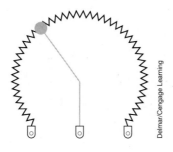

FIGURE 5–14 Variable resistor.

Delmar/Cengage Learning

Courtesy of Rockwell Automation

FIGURE 5–15   Variable resistors with three terminals.

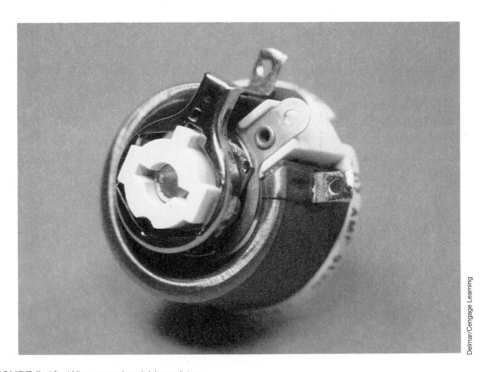

Delmar/Cengage Learning

FIGURE 5–16   Wire-wound variable resistor.

Delmar/Cengage Learning

FIGURE 5–17 Multiturn variable resistor.

The resistor shown in *Figure 5–15* can be adjusted from its minimum to maximum value by turning the control approximately three-quarters of a turn. In some types of electric equipment, this range of adjustment may be too coarse to allow for sensitive adjustments. When this becomes a problem, a multiturn resistor *(Figure 5–17)* can be used. **Multiturn variable resistors** operate by moving the wiper arm with a screw of some number of turns. They generally range from 3 turns to 10 turns. If a 10-turn variable resistor is used, it will require 10 turns of the control knob to move the wiper from one end of the resistor to the other end instead of three-quarters of a turn.

Variable Resistor Terminology

Variable resistors are known by several common names. The most popular name is **pot,** which is shortened from the word **potentiometer.** Another common name is **rheostat.** A rheostat is actually a variable resistor that has two terminals. They are used to adjust the current in a circuit to a certain value. A potentiometer is a variable resistor that has three terminals. Potentiometers can be used as rheostats by only using two of their three terminals. A potentiometer describes how a variable resistor is used rather than some specific type of resistor. The word *potentiometer* comes from the word *potential,* or voltage. A potentiometer is a variable resistor used to provide a variable voltage, as shown in *Figure 5–18.* In this example, one end of a variable resistor is connected to +12 volts and the other end is connected to ground. The middle terminal, or

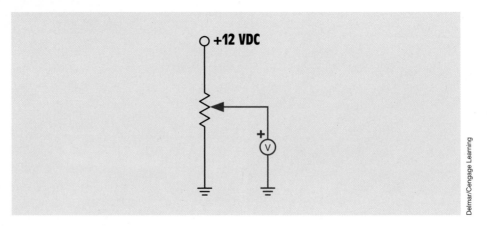

FIGURE 5–18   Variable resistor used as a potentiometer.

wiper, is connected to the positive terminal of a voltmeter and the negative lead is connected to ground. If the wiper is moved to the upper end of the resistor, the voltmeter will indicate a potential of 12 volts. If the wiper is moved to the bottom, the voltmeter will indicate a value of 0 volts. The wiper can be adjusted to provide any value of voltage between 12 and 0 volts.

5–7   Schematic Symbols

Electrical schematics use symbols to represent the use of a resistor. Unfortunately, the symbol used to represent a resistor is not standard. *Figure 5–19* illustrates several schematic symbols used to represent both fixed and variable resistors.

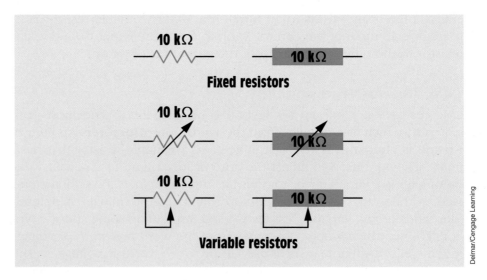

FIGURE 5–19   Schematic symbols used to represent resistors.

Summary

- Resistors are used in two main applications: as voltage dividers and to limit the flow of current in a circuit.

- The value of fixed resistors cannot be changed.

- There are several types of fixed resistors, such as composition carbon, metal film, and wire-wound.

- Carbon resistors change their resistance with age or if overheated.

- Metal film resistors never change their value but are more expensive than carbon resistors.

- The advantage of wire-wound resistors is their high power ratings.

- Resistors often have bands of color to indicate their resistance value and tolerance.

- Resistors are produced in standard values. The number of values between 0 and 100 ohms is determined by the tolerance.

- Variable resistors can change their value within the limit of their full value.

- A potentiometer is a variable resistor used as a voltage divider.

Review Questions

1. Name three types of fixed resistors.

2. What is the advantage of a metal film resistor over a carbon resistor?

3. What is the advantage of a wire-wound resistor?

4. How should tubular wire-wound resistors be mounted and why?

5. A 0.5-W, 2000-Ω resistor has a current flow of 0.01 A through it. Is this resistor operating within its power rating?

6. A 1-W, 350-Ω resistor is connected to 24 V. Is this resistor operating within its power rating?

7. A resistor has color bands of orange, blue, yellow, and gold. What are the resistance and tolerance of this resistor?

8. A 10,000-Ω resistor has a tolerance of 5%. What are the minimum and maximum values of this resistor?

9. Is 51,000 Ω a standard value for a 5% resistor?

10. What is a potentiometer?

Practical Applications

*Y*ou are an electrician on the job. It is decided that the speed of a large DC motor is to be reduced by connecting a resistor in series with its armature. The DC voltage applied to the motor is 250 V, and the motor has a full-load armature current of 50 A. Your job is to reduce the armature current to 40 A at full load by connecting the resistor in series with the armature. What value of resistance should be used, and what is the power rating of the resistor? ■

Practical Applications

*Y*ou are working on an electronic circuit. The circuit current is 5 mA. A resistor is marked with the following bands: brown, black, red, gold. A voltmeter measures a voltage drop of 6.5 V across the resistor. Is this resistor within its tolerance rating? ■

Practical Applications

A homeowner uses a 100-watt incandescent lamp as a heater in an outside well pump house to protect the pump from freezing in cold weather. Unfortunately, however, the lamp can burn out and leave the pump unprotected. You have been asked to install a heater that will not burn out and leave the pump unprotected. You have available a 100-watt, 150-ohm wire-wound resistor. Can this resistor be connected to the 120-volt source without damage to the resistor? If so, what would be the power output of the resistor? ■

Practical Applications

Y ou have determined that a 4700-ohm, ½-watt resistor on an electronic circuit board is defective. Assuming room permits, can the resistor be replaced with a 4700-ohm, 1-watt resistor without damage to the rest of the board, or will the higher wattage resistor generate excessive heat that could damage other components? ■

Practice Problems

Resistors

Fill in the missing values.

1st Band	2nd Band	3rd Band	4th Band	Value	%Tolerance
Red	Yellow	Brown	Silver		
				6800 Ω	5
Orange	Orange	Orange	Gold		
				12 Ω	2
Brown	Green	Silver	Silver		
				1.8 MΩ	10
Brown	Black	Yellow	None		
				10 kΩ	5
Violet	Green	Black	Red		
				4.7 kΩ	20
Gray	Red	Green	Red		
				5.6 Ω	2

Basic Electric Circuits

Unit 6
Series Circuits

KEY TERMS

Chassis ground
Circuit breakers
Earth ground
Fuses
Ground point
Series circuit
Voltage drop
Voltage polarity

Why You Need to Know

*I*t is important to learn the rules governing the values of resistance, voltage, current, and power in a series circuit. Knowledge of series circuits is essential in understanding more complex circuits that are encountered throughout an electrician's career. There are some series circuits in everyday use, such as many street lighting systems, but knowledge of series circuits is most useful in understanding how components in combination circuits relate to each other. Combination circuits contain components or branches that are connected in both series and parallel. To apply the rules of a series circuit, this unit

- provides a basic understanding of how voltage drop impacts devices that are connected in series.
- explains why each series-connected device increases the total opposition to current flow.
- provides an understanding of how current flows through a series circuit.

Delmar/Cengage Learning

Objectives

After studying this unit, you should be able to

- discuss the properties of series circuits.
- list three rules for solving electrical values of series circuits.
- calculate values of voltage, current, resistance, and power for series circuits.
- calculate the values of voltage drop in a series circuit using the voltage divider formula.

Preview

Electric circuits can be divided into three major types: series, parallel, and combination. Combination circuits are circuits that contain both series and parallel paths. The first type discussed is the series circuit. ∎

6–1   Series Circuits

A **series circuit** is a circuit that has only one path for current flow *(Figure 6–1)*. Because there is only one path for current flow, the current is the same at any point in the circuit. Imagine that an electron leaves the negative terminal of the battery. This electron must flow through each resistor before it can complete the circuit to the positive battery terminal.

One of the best examples of a series-connected device is a fuse or circuit breaker *(Figure 6–2)*. Because **fuses** and **circuit breakers** are connected in series with the rest of the circuit, all the circuit current must flow through them. If the current becomes excessive, the fuse or circuit breaker will open and disconnect the rest of the circuit from the power source.

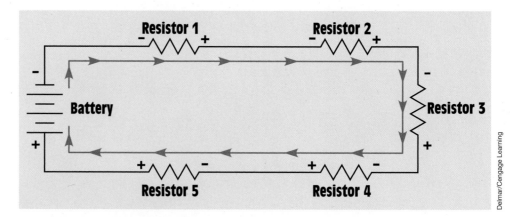

FIGURE 6–1   A series circuit has only one path for current flow.

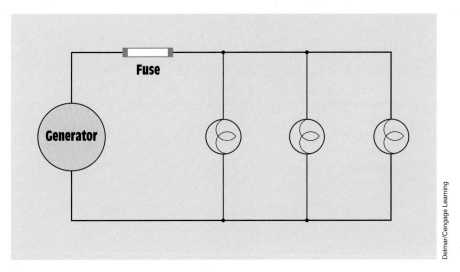

FIGURE 6–2   All the current must flow through the fuse.

6–2   Voltage Drops in a Series Circuit

Voltage is the force that pushes the electrons through a resistance. The amount of voltage required is determined by the amount of current flow and resistance. If a voltmeter is connected across a resistor *(Figure 6–3)*, the amount of voltage necessary to push the current through that resistor is indicated by the meter. This amount is known as **voltage drop.** It is similar to pressure drop in a water system. ***In a series circuit, the sum of all the voltage drops across all the resistors***

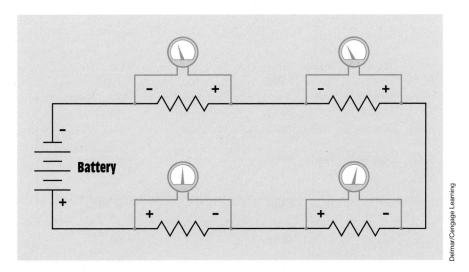

FIGURE 6–3   The voltage drops in a series circuit must equal the applied voltage.

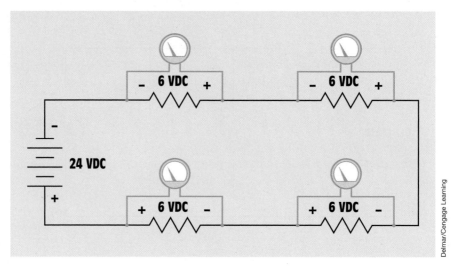

FIGURE 6–4 The voltage drop across each resistor is proportional to its resistance.

must equal the voltage applied to the circuit. The amount of voltage drop across each resistor is proportional to its resistance and the circuit current.

In the circuit shown in *Figure 6–4*, four resistors are connected in series. It is assumed that all four resistors have the same value. The circuit is connected to a 24-volt battery. Because all the resistors have the same value, the voltage drop across each will be 6 volts (24 V/4 resistors = 6 V). Note that all four resistors will have the same voltage drop only if they all have the same value. The circuit shown in *Figure 6–5* illustrates a series circuit comprising resistors

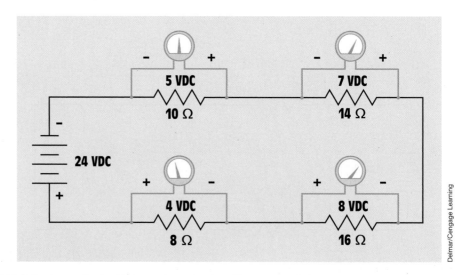

FIGURE 6–5 Series circuit with four resistors having different voltage drops.

of different values. Notice that the voltage drop across each resistor is proportional to its resistance. Also notice that the sum of the voltage drops is equal to the applied voltage of 24 volts.

6–3   Resistance in a Series Circuit

Because only one path exists for the current to flow through a series circuit, it must flow through each resistor in the circuit *(Figure 6–1)*. Each resistor limits or impedes the flow of current in the circuit. Therefore, the total amount of resistance to current flow in a series circuit is equal to the sum of the resistances in that circuit.

6–4   Calculating Series Circuit Values

Three rules can be used with Ohm's law for finding values of voltage, current, resistance, and power in any series circuit:

1. The current is the same at any point in the circuit.

2. The total resistance is the sum of the individual resistors.

3. The applied voltage is equal to the sum of the voltage drops across all the resistors.

The circuit shown in *Figure 6–6* shows the values of current flow, voltage drop, and resistance for each of the resistors. The total resistance (R_T) of

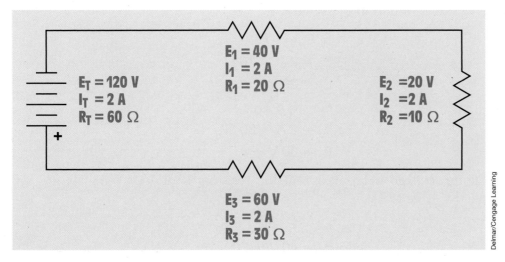

$E_T = 120$ V
$I_T = 2$ A
$R_T = 60$ Ω

$E_1 = 40$ V
$I_1 = 2$ A
$R_1 = 20$ Ω

$E_2 = 20$ V
$I_2 = 2$ A
$R_2 = 10$ Ω

$E_3 = 60$ V
$I_3 = 2$ A
$R_3 = 30$ Ω

Delmar/Cengage Learning

FIGURE 6–6   Series circuit values.

the circuit can be found by adding the values of the three resistors—resistance adds:

$$R_T = R_1 + R_2 + R_3$$
$$R_T = 20\ \Omega + 10\ \Omega + 30\ \Omega$$
$$R_T = 60\ \Omega$$

The amount of current flow in the circuit can be found by using Ohm's law:

$$I = \frac{E}{R}$$
$$I = \frac{120\ V}{60\ \Omega}$$
$$I = 2\ A$$

A current of 2 amperes flows through each resistor in the circuit:

$$I_T = I_1 = I_2 = I_3$$

Because the amount of current flowing through resistor R_1 is known, the voltage drop across the resistor can be found using Ohm's law:

$$E_1 = I_1 \times R_1$$
$$E_1 = 2\ A \times 20\ \Omega$$
$$E_1 = 40\ V$$

In other words, it takes 40 volts to push 2 amperes of current through 20 ohms of resistance. If a voltmeter were connected across resistor R_1, it would indicate a value of 40 volts *(Figure 6–7)*. The voltage drop across resistors R_2 and R_3 can be found in the same way:

$$E_2 = I_2 \times R_2$$
$$E_2 = 2\ A \times 10\ \Omega$$
$$E_2 = 20\ V$$
$$E_3 = I_3 \times R_3$$
$$E_3 = 2\ A \times 30\ \Omega$$
$$E_3 = 60\ V$$

If the voltage drop across all the resistors is added, it equals the total applied voltage (E_T):

$$E_T = E_1 + E_2 + E_3$$
$$E_T = 40\ V + 20\ V + 60\ V$$
$$E_T = 120\ V$$

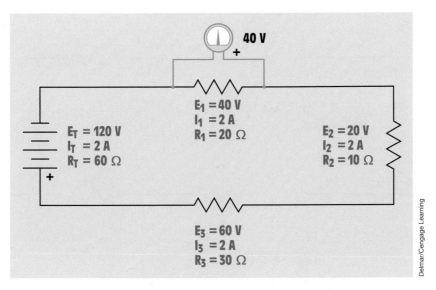

Delmar/Cengage Learning

FIGURE 6–7 The voltmeter indicates a voltage drop of 40 volts.

6–5 Solving Circuits

In the following problems, circuits that have missing values are shown. The missing values can be found by using the rules for series circuits and Ohm's law.

■ EXAMPLE 6-1

The first step in finding the missing values in the circuit shown in *Figure 6–8* is to find the total resistance (R_T). This can be done using the second rule of series circuits, which states that resistances add to equal the total resistance of the circuit:

$$R_T = R_1 + R_2 + R_3 + R_4$$
$$R_T = 100 \ \Omega + 250 \ \Omega + 150 \ \Omega + 300 \ \Omega$$
$$R_T = 800 \ \Omega$$

Now that the total voltage and the total resistance are known, the current flow through the circuit can be found using Ohm's law:

$$I = \frac{E}{R}$$
$$I = \frac{40 \ V}{800 \ \Omega}$$
$$I = 0.050 \ A$$

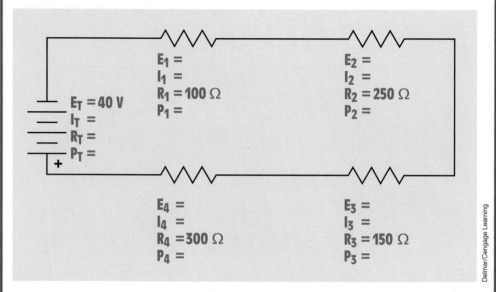

FIGURE 6–8 Series circuit, Example 1.

The first rule of series circuits states that current remains the same at any point in the circuit. Therefore, 0.050 A flows through each resistor in the circuit *(Figure 6–9)*. The voltage drop across each resistor can now be found using Ohm's law *(Figure 6–10)*:

$$E_1 = I_1 \times R_1$$
$$E_1 = 0.050 \text{ A} \times 100 \text{ }\Omega$$
$$E_1 = 5 \text{ V}$$
$$E_2 = I_2 \times R_2$$
$$E_2 = 0.050 \text{ A} \times 250 \text{ }\Omega$$
$$E_2 = 12.5 \text{ V}$$
$$E_3 = I_3 \times R_3$$
$$E_3 = 0.050 \text{ A} \times 150 \text{ }\Omega$$
$$E_3 = 7.5 \text{ V}$$
$$E_4 = I_4 \times R_4$$
$$E_4 = 0.050 \text{ A} \times 300 \text{ }\Omega$$
$$E_4 = 15 \text{ V}$$

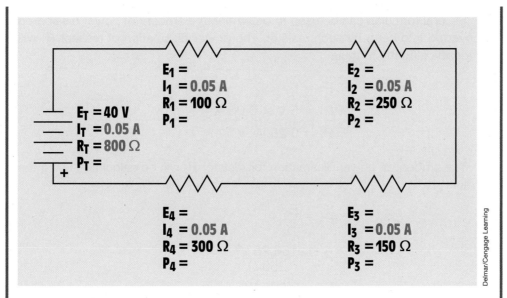

FIGURE 6–9 The current is the same at any point in a series circuit.

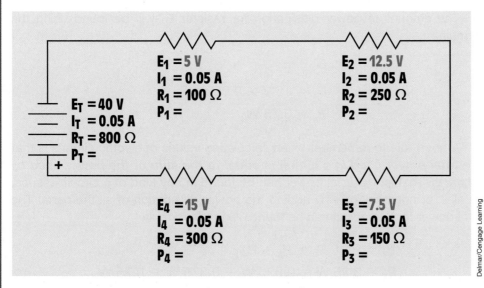

FIGURE 6–10 The voltage drop across each resistor can be found using Ohm's law.

Several formulas can be used to determine the amount of power dissipated (converted into heat) by each resistor. The power dissipation of resistor R_1 will be found using the formula

$$P_1 = E_1 \times I_1$$
$$P_1 = 5\text{ V} \times 0.05\text{ A}$$
$$P_1 = 0.25\text{ W}$$

The amount of power dissipation for resistor R_2 will be calculated using the formula

$$P_2 = \frac{E_2{}^2}{R_2}$$
$$P_2 = \frac{156.25\text{ V}^2}{250\ \Omega}$$
$$P_2 = 0.625\text{ W}$$

The amount of power dissipation for resistor R_3 will be calculated using the formula

$$P_3 = I_3{}^2 \times R_3$$
$$P_3 = 0.0025\text{ A}^2 \times 150\ \Omega$$
$$P_3 = 0.375\text{ W}$$

The amount of power dissipation for resistor R_4 will be found using the formula

$$P_4 = E_4 \times I_4$$
$$P_4 = 15\text{ V} \times 0.05\text{ A}$$
$$P_4 = 0.75\text{ W}$$

A good rule to remember when calculating values of electric circuits is that **the total power used in a circuit is equal to the sum of the power used by all parts.** That is, the total power can be found in any kind of a circuit—series, parallel, or combination—by adding the power dissipation of all the parts. The total power for this circuit can be found using the formula

$$P_T = P_1 + P_2 + P_3 + P_4$$
$$P_T = 0.25\text{ W} + 0.625\text{ W} + 0.375\text{ W} + 0.75\text{ W}$$
$$P_T = 2\text{ W}$$

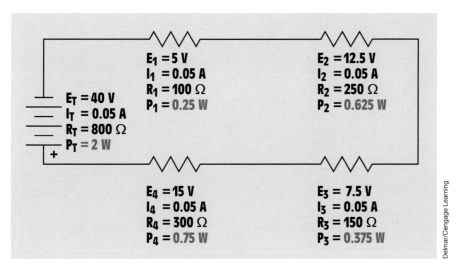

FIGURE 6–11   The final values for the circuit in Example 1.

Now that all the missing values have been found *(Figure 6–11),* the circuit can be checked by using the third rule of series circuits, which states that voltage drops add to equal the applied voltage:

$$E_T = E_1 + E_2 + E_3 + E_4$$
$$E_T = 5\ V + 12.5\ V + 7.5\ V + 15\ V$$
$$E_T = 40\ V$$

■ EXAMPLE 6-2

The second circuit to be solved is shown in *Figure 6–12.* In this circuit, the total resistance is known, but the value of resistor R_2 is not. The second rule of series circuits states that resistances add to equal the total resistance of the circuit. Because the total resistance is known, the missing resistance of R_2 can be found by adding the values of the other resistors and subtracting their sum from the total resistance of the circuit *(Figure 6–13):*

$$R_2 = R_T - (R_1 + R_3 + R_4)$$
$$R_2 = 6000\ \Omega - (1000\ \Omega + 2000\ \Omega + 1200\ \Omega)$$
$$R_2 = 6000\ \Omega - 4200\ \Omega$$
$$R_2 = 1800\ \Omega$$

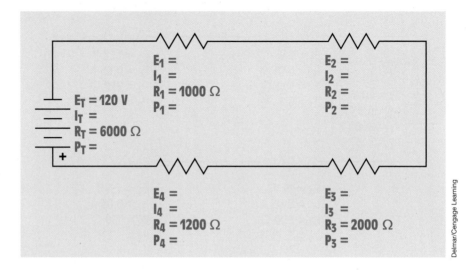

FIGURE 6–12 Series circuit, Example 2.

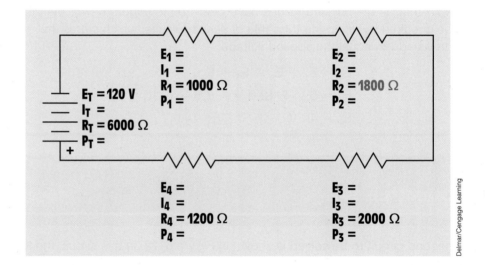

FIGURE 6–13 The missing resistor value.

The amount of current flow in the circuit can be found using Ohm's law:

$$I = \frac{E}{R}$$

$$I = \frac{120\ V}{6000\ \Omega}$$

$$I = 0.020\ A$$

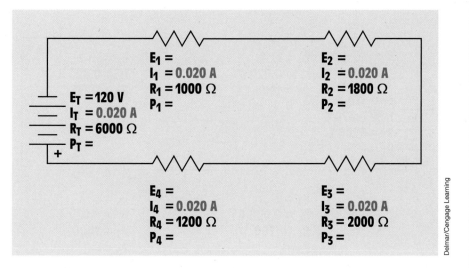

FIGURE 6–14 The current is the same through each circuit element.

Because the amount of current flow is the same through all elements of a series circuit *(Figure 6–14),* the voltage drop across each resistor can be found using Ohm's law *(Figure 6–15):*

$$E_1 = I_1 \times R_1$$
$$E_1 = 0.020 \, A \times 1000 \, \Omega$$
$$E_1 = 20 \, V$$
$$E_2 = I_2 \times R_2$$
$$E_2 = 0.020 \, A \times 1800 \, \Omega$$
$$E_2 = 36 \, V$$
$$E_3 = I_3 \times R_3$$
$$E_3 = 0.020 \, A \times 2000 \, \Omega$$
$$E_3 = 40 \, V$$
$$E_4 = I_4 \times R_4$$
$$E_4 = 0.020 \, A \times 1200 \, \Omega$$
$$E_4 = 24 \, V$$

The third rule of series circuits can be used to check the answers:

$$E_T = E_1 + E_2 + E_3 + E_4$$
$$E_T = 20 \, V + 36 \, V + 40 \, V + 24 \, V$$
$$E_T = 120 \, V$$

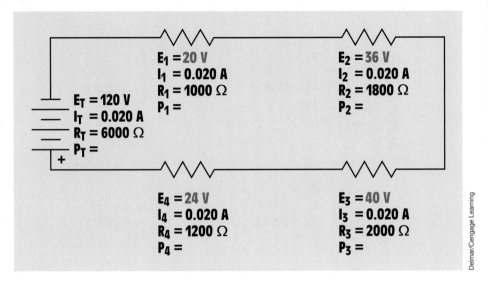

FIGURE 6–15 Voltage drops across each resistor.

The amount of power dissipation for each resistor in the circuit can be calculated using the same method used to solve the circuit in Example 1. The power dissipated by resistor R_1 is calculated using the formula

$$P_1 = E_1 \times I_1$$
$$P_1 = 20 \text{ V} \times 0.02 \text{ A}$$
$$P_1 = 0.4 \text{ W}$$

The amount of power dissipation for resistor R_2 is found by using the formula

$$P_2 = \frac{E_2{}^2}{R_2}$$
$$P_2 = \frac{1296 \text{ V}^2}{1800 \text{ }\Omega}$$
$$P_2 = 0.72 \text{ W}$$

The power dissipation of resistor R_3 is found using the formula

$$P_3 = I_3{}^2 \times R_3$$
$$P_3 = 0.0004 \text{ A}^2 \times 2000 \text{ }\Omega$$
$$P_3 = 0.8 \text{ W}$$

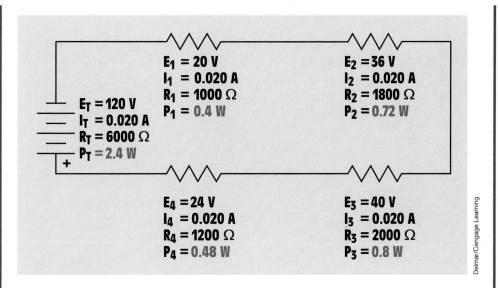

$E_1 = 20$ V
$I_1 = 0.020$ A
$R_1 = 1000$ Ω
$P_1 = 0.4$ W

$E_2 = 36$ V
$I_2 = 0.020$ A
$R_2 = 1800$ Ω
$P_2 = 0.72$ W

$E_T = 120$ V
$I_T = 0.020$ A
$R_T = 6000$ Ω
$P_T = 2.4$ W

$E_4 = 24$ V
$I_4 = 0.020$ A
$R_4 = 1200$ Ω
$P_4 = 0.48$ W

$E_3 = 40$ V
$I_3 = 0.020$ A
$R_3 = 2000$ Ω
$P_3 = 0.8$ W

Delmar/Cengage Learning

FIGURE 6–16   The remaining unknown values for the circuit in Example 2.

The power dissipation of resistor R_4 is calculated using the formula

$$P_4 = E_4 \times I_4$$
$$P_4 = 24 \text{ V} \times 0.02 \text{ A}$$
$$P_4 = 0.48 \text{ W}$$

The total power is calculated using the formula

$$P_T = E_T \times I_T$$
$$P_T = 120 \text{ V} \times 0.02 \text{ A}$$
$$P_T = 2.4 \text{ W}$$

The circuit with all calculated values is shown in *Figure 6–16.*

■ EXAMPLE 6-3

In the circuit shown in *Figure 6–17,* resistor R_1 has a voltage drop of 6.4 V, resistor R_2 has a power dissipation of 0.102 W, resistor R_3 has a power dissipation of 0.154 W, resistor R_4 has a power dissipation of 0.307 W, and the total power consumed by the circuit is 0.768 W.

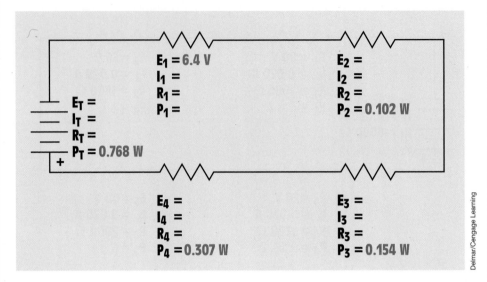

Delmar/Cengage Learning

FIGURE 6–17   Series circuit, Example 3.

The only value that can be found with the given quantities is the amount of power dissipated by resistor R_1. Because the total power is known and the power dissipated by the three other resistors is known, the power dissipated by resistor R_1 can be found by subtracting the power dissipated by resistors R_2, R_3, and R_4 from the total power used in the circuit:

$$P_1 = P_T - (P_2 + P_3 + P_4)$$

or

$$P_1 = P_T - P_2 - P_3 - P_4$$
$$P_1 = 0.768 \text{ W} - 0.102 \text{ W} - 0.154 \text{ W} - 0.307 \text{ W}$$
$$P_1 = 0.205 \text{ W}$$

Now that the amount of power dissipated by resistor R_1 and the voltage drop across R_1 are known, the current flow through resistor R_1 can be found using the formula

$$I = \frac{P}{E}$$
$$I = \frac{0.205 \text{ W}}{6.4 \text{ V}}$$
$$I = 0.032 \text{ A}$$

Because the current in a series circuit must be the same at any point in the circuit, it must be the same through all circuit components *(Figure 6–18).*

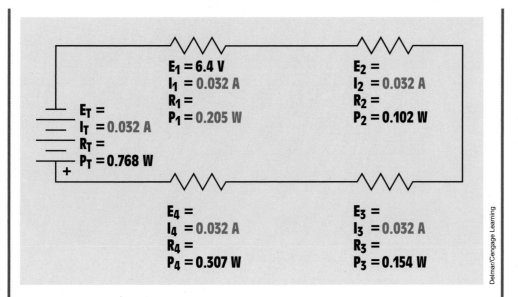

FIGURE 6–18   The current flow in the circuit in Example 3.

Now that the power dissipation of each resistor and the amount of current flowing through each resistor are known, the voltage drop of each resistor can be calculated *(Figure 6–19)*:

$$E_2 = \frac{P_2}{I_2}$$

$$E_2 = \frac{0.102 \text{ W}}{0.032 \text{ A}}$$

$$E_2 = 3.188 \text{ V}$$

$$E_3 = \frac{P_3}{I_3}$$

$$E_3 = \frac{0.154 \text{ W}}{0.032 \text{ A}}$$

$$E_3 = 4.813 \text{ V}$$

$$E_4 = \frac{P_4}{I_4}$$

$$E_4 = \frac{0.307 \text{ W}}{0.032 \text{ A}}$$

$$E_4 = 9.594 \text{ V}$$

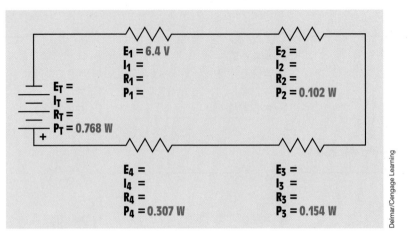

Delmar/Cengage Learning

FIGURE 6–19 Voltage drops across each resistor.

Ohm's law can now be used to find the ohmic value of each resistor in the circuit *(Figure 6–20):*

$$R_1 = \frac{E_1}{I_1}$$

$$R_1 = \frac{6.4 \text{ V}}{0.032 \text{ A}}$$

$$R_1 = 200 \text{ } \Omega$$

$$R_2 = \frac{E_2}{I_2}$$

$$R_2 = \frac{3.188 \text{ V}}{0.032 \text{ A}}$$

$$R_2 = 99.625 \text{ } \Omega$$

$$R_3 = \frac{E_3}{I_3}$$

$$R_3 = \frac{4.813 \text{ V}}{0.032 \text{ A}}$$

$$R_3 = 150.406 \text{ } \Omega$$

$$R_4 = \frac{E_4}{I_4}$$

$$R_4 = \frac{9.594 \text{ V}}{0.032 \text{ A}}$$

$$R_4 = 299.813 \text{ } \Omega$$

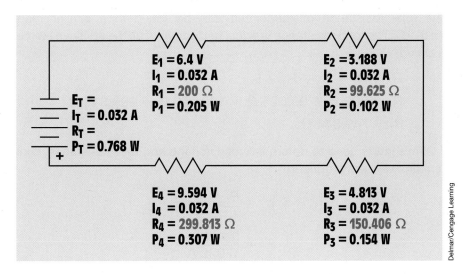

FIGURE 6–20   The ohmic value of each resistor.

The voltage applied to the circuit can be found by adding the voltage drops across the resistor *(Figure 6–21):*

$$E_T = E_1 + E_2 + E_3 + E_4$$
$$E_T = 6.4 \text{ V} + 3.188 \text{ V} + 4.813 \text{ V} + 9.594 \text{ V}$$
$$E_T = 23.995 \text{ V}$$

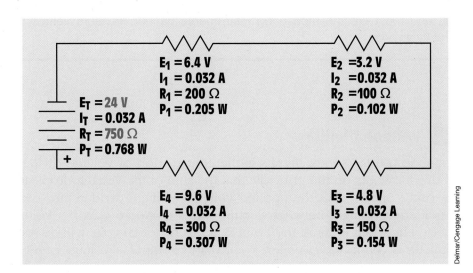

FIGURE 6–21   The applied voltage and the total resistance.

The total resistance of the circuit can be found in a similar manner *(Figure 6–21)*. The total resistance is equal to the sum of all the resistive elements in the circuit:

$$R_T = R_1 + R_2 + R_3 + R_4$$
$$R_T = 200 \ \Omega + 99.625 \ \Omega + 150.406 \ \Omega + 299.813 \ \Omega$$
$$R_T = 749.844 \ \Omega$$

If Ohm's law is used to determine total voltage and total resistance, slightly different answers are produced:

$$E_T = \frac{P_T}{I_T}$$

$$E_T = \frac{0.768 \ W}{0.032 \ A}$$

$$E_T = 24 \ V$$

$$R_T = \frac{E_T}{I_T}$$

$$R_T = \frac{24 \ V}{0.032 \ A}$$

$$R_T = 750 \ \Omega$$

The slight difference in answers is caused by the rounding off of values. Although there is a small difference between the answers, they are within 1% of each other. This small difference has very little effect on the operation of the circuit, and most electric measuring instruments cannot measure values this accurately anyway.

6–6   Voltage Dividers

One common use for series circuits is the construction of voltage dividers. A voltage divider works on the principle that the sum of the voltage drops across a series circuit must equal the applied voltage. Voltage dividers are used to provide different voltages between certain points *(Figure 6–22)*. If a voltmeter is connected between Point A and Point B, a voltage of 20 volts will be seen. If the voltmeter is connected between Point B and Point D, a voltage of 80 volts will be seen.

Voltage dividers can be constructed to produce any voltage desired. For example, assume that a voltage divider is connected to a source of 120 volts

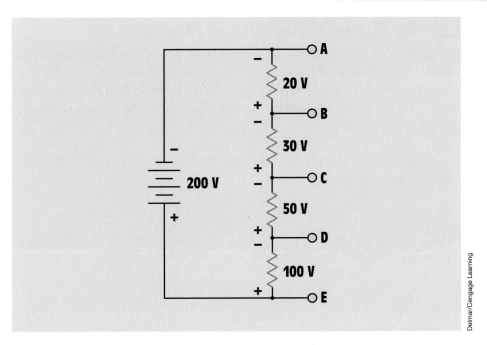

FIGURE 6–22 Series circuit used as a voltage divider.

and is to provide voltage drops of 36 volts, 18 volts, and 66 volts. Notice that the sum of the voltage drops equals the applied voltage. The next step is to decide how much current is to flow through the circuit. Because there is only one path for current flow, the current will be the same through all the resistors. In this circuit, a current flow of 15 milliamperes (0.015 A) will be used. The resistance value of each resistor can now be determined:

$$R = \frac{E}{I}$$

$$R_1 = \frac{36 \text{ V}}{0.015 \text{ A}}$$

$$R_1 = 2.4 \text{ k}\Omega \ (2400 \ \Omega)$$

$$R_2 = \frac{18 \text{ V}}{0.015 \text{ A}}$$

$$R_2 = 1.2 \text{ k}\Omega \ (1200 \ \Omega)$$

$$R_3 = \frac{66 \text{ V}}{0.015 \text{ A}}$$

$$R_3 = 4.4 \text{ k}\Omega \ (4400 \ \Omega)$$

6-7 The General Voltage Divider Formula

Another method of determining the voltage drop across series elements is to use the general voltage divider formula. Because the current flow through a series circuit is the same at all points in the circuit, the voltage drop across any particular resistance is equal to the total circuit current times the value of that resistor:

$$E_X = I_T \times R_X$$

The total circuit current is proportional to the source voltage (E_T) and the total resistance of the circuit:

$$I_T = \frac{E_T}{R_T}$$

If the value of I_T is substituted for E_T/R_T in the previous formula, the expression now becomes:

$$E_X = \left(\frac{E_T}{R_T}\right)R_X$$

If the formula is rearranged, it becomes what is known as the general voltage divider formula:

$$E_X = \left(\frac{R_X}{R_T}\right)E_T$$

The voltage drop across any series component (E_X) can be calculated by substituting the value of R_X for the resistance value of that component when the source voltage and total resistance are known.

■ EXAMPLE 6-4

Three resistors are connected in series to a 24-V source. Resistor R_1 has a resistance of 200 Ω, resistor R_2 has a value of 300 Ω, and resistor R_3 has a value of 160 Ω. What is the voltage drop across each resistor?

Solution

Find the total resistance of the circuit:

$$R_T = R_1 + R_2 + R_3$$
$$R_T = 200\ \Omega + 300\ \Omega + 160\ \Omega$$
$$R_T = 660\ \Omega$$

Now use the voltage divider formula to calculate the voltage drop across each resistor:

$$E_1 = \left(\frac{R_1}{R_T}\right) E_T$$

$$E_1 = \left(\frac{200\ \Omega}{660\ \Omega}\right) 24\ V$$

$$E_1 = 7.273\ V$$

$$E_2 = \left(\frac{R_2}{R_T}\right) E_T$$

$$E_2 = \left(\frac{300\ \Omega}{660\ \Omega}\right) 24\ V$$

$$E_2 = 10.909\ V$$

$$E_3 = \left(\frac{R_3}{R_T}\right) E_T$$

$$E_3 = \left(\frac{160\ \Omega}{660\ \Omega}\right) 24\ V$$

$$E_3 = 5.818\ V$$

6-8 Voltage Polarity

It is often necessary to know the polarity of the voltage dropped across a resistor. **Voltage polarity** can be determined by observing the direction of current flow through the circuit. In the circuit shown in *Figure 6–22*, it will be assumed that the current flows from the negative terminal of the battery to the positive terminal. Point A is connected to the negative battery terminal, and Point E is connected to the positive terminal. If a voltmeter is connected across Terminals A and B, Terminal B will be positive with respect to A. If a voltmeter is connected across Terminals B and C, however, Terminal B will be negative with respect to Terminal C. Notice that Terminal B is closer to the negative terminal of the battery than Terminal C is. Consequently, electrons flow through the resistor in a direction that makes Terminal B more negative than C. Terminal C would be negative with respect to Terminal D for the same reason.

6–9 Using Ground as a Reference

Two symbols are used to represent ground *(Figure 6–23)*. The symbol shown in *Figure 6–23(A)* is an **earth ground** symbol. It symbolizes a **ground point** that is made by physically driving an object such as a rod or a pipe into the ground. The symbol shown in *Figure 6–23(B)* symbolizes a **chassis ground.** This is a point that is used as a common connection for other parts of a circuit, but it is not actually driven into the ground. Although the symbol shown in *Figure 6–23(B)* is the accepted symbol for a chassis ground, the symbol shown in *Figure 6–23(A)* is often used to represent a chassis ground also.

An excellent example of using a chassis ground as a common connection can be found in the electric system of an automobile. The negative terminal of the battery is grounded to the frame or chassis of the vehicle. The frame of the automobile is not connected directly to earth ground; it is insulated from the ground by rubber tires. In the case of an automobile electric system, the chassis of the vehicle is the negative side of the circuit. An electric circuit using ground as a common connection point is shown in *Figure 6–24*. This circuit is an electronic burglar alarm. Notice the numerous ground points in the schematic. In practice, when the circuit is connected, all the ground points will be connected together.

In voltage divider circuits, ground is often used to provide a common reference point to produce voltages that are above and below ground *(Figure 6–25)*. An above-ground voltage is a voltage that is positive with respect to ground. A below-ground voltage is negative with respect to ground. In *Figure 6–25*, one terminal of a zero-center voltmeter is connected to ground. If the probe is connected to Point A, the pointer of the voltmeter gives a negative indication for voltage. If the probe is connected to Point B, the pointer indicates a positive voltage.

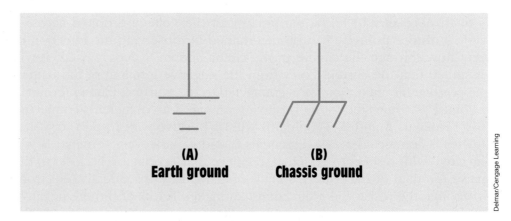

(A)
Earth ground

(B)
Chassis ground

Delmar/Cengage Learning

FIGURE 6–23 Ground symbols.

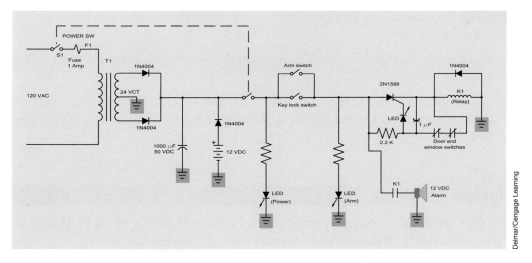

FIGURE 6–24   Burglar alarm with battery backup.

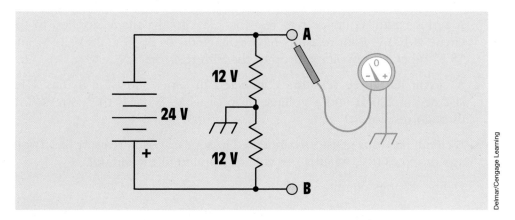

FIGURE 6–25   A common ground used to produce above- and below-ground voltage.

Summary

- Series circuits have only one path for current flow.

- The individual voltage drops in a series circuit can be added to equal the applied voltage.

- The current is the same at any point in a series circuit.

- The individual resistors can be added to equal the total resistance of the circuit.

- Fuses and circuit breakers are connected in series with the devices they are intended to protect.

- The total power in any circuit is equal to the sum of the power dissipated by all parts of the circuit.

- When the source voltage and the total resistance are known, the voltage drop across each element can be calculated using the general voltage divider formula.

Review Questions

1. A series circuit has individual resistor values of 200 Ω, 86 Ω, 91 Ω, 180 Ω, and 150 Ω. What is the total resistance of the circuit?

2. A series circuit contains four resistors. The total resistance of the circuit is 360 Ω. Three of the resistors have values of 56 Ω, 110 Ω, and 75 Ω. What is the value of the fourth resistor?

3. A series circuit contains five resistors. The total voltage applied to the circuit is 120 V. Four resistors have voltage drops of 35 V, 28 V, 22 V, and 15 V. What is the voltage drop of the fifth resistor?

4. A circuit has three resistors connected in series. Resistor R_2 has a resistance of 220 Ω and a voltage drop of 44 V. What is the current flow through resistor R_3?

5. A circuit has four resistors connected in series. If each resistor has a voltage drop of 60 V, what is the voltage applied to the circuit?

6. Define a series circuit.

7. State the three rules for series circuits.

8. A series circuit has resistance values of 160 Ω, 100 Ω, 82 Ω, and 120 Ω. What is the total resistance of this circuit?

9. If a voltage of 24 V is applied to the circuit in Question 8, what will be the total amount of current flow in the circuit?

10. Referring to the circuit described in Questions 8 and 9, determine the voltage drop across each of the resistors.

   160 Ω, _____ V

   100 Ω, _____ V

   82 Ω, _____ V

   120 Ω, _____ V

11. A series circuit contains the following values of resistors:

   $R_1 = 510\ \Omega \quad R_2 = 680\ \Omega \quad R_3 = 390\ \Omega \quad R_4 = 750\ \Omega$

   Assume a source voltage of 48 V. Use the general voltage divider formula to calculate the voltage drop across each of the resistors.

   $E_1 = $ _____ V $E_2 = $ _____ V $E_3 = $ _____ V $E_4 = $ _____ V

Practical Applications

A 12-V DC automobile head lamp is to be used on a fishing boat with a 24-V power system. The head lamp is rated at 50 W. A resistor is to be connected in series with the lamp to permit it to operate on 24 V. What should be the resistance and power rating of the resistor? ■

Practical Applications

Three wire-wound resistors have the following values: 30 Ω, 80 Ω, and 100 Ω. Each resistor has a voltage rating of 100 V. If these three resistors are connected in series, can they be connected to a 240-V circuit without damage to the resistors? Explain your answer. ■

Practical Applications

You are an electrician working in an industrial plant. A circuit contains eight incandescent lamps connected in series across 480 volts. One lamp has burned out, and you must determine which one is defective. You have available a voltmeter, ammeter, and ohmmeter. Which meter would you use to determine which lamp is defective in the shortest possible time? Explain how you would use this meter and why. ■

Practical Applications

direct current motor is connected to a 250-volt DC supply. The armature has a current draw of 165 amperes when operating at full load. You have been assigned the task of connecting two resistors in the armature circuit to provide speed control for the motor. When both resistors are connected in the circuit, the armature current is to be limited to 50% of the full-load current draw. When only one resistor is connected in the circuit, the armature current is to be limited to 85% of full-load current. Determine the ohmic value and minimum power rating of each resistor. Refer to *Figure 6–26*. When both switches S_1 and S_2 are open (off), both resistors are connected in the armature circuit, limiting current to 50% of its normal value. When switch S_1 is closed, it causes the current to bypass resistor 1. Resistor 2 now limits the current to 85% of the full-load current. When both switches S_1 and S_2 are closed, all resistance is bypassed, and the armature is connected to full power. ■

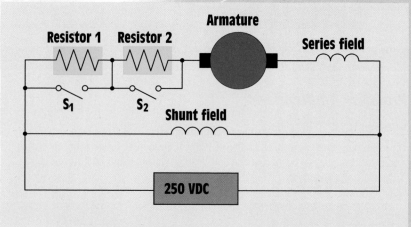

Delmar/Cengage Learning

FIGURE 6–26   Determine the resistance and power rating of the two series resistors.

Practice Problems

Series Circuits

1. Using the three rules for series circuits and Ohm's law, solve for the missing values.

E_T 120 V	E_1	E_2	E_3	E_4	E_5
I_T	I_1	I_2	I_3	I_4	I_5
R_T	R_1 430 Ω	R_2 360 Ω	R_3 750 Ω	R_4 1000 Ω	R_5 620 Ω
P_T	P_1	P_2	P_3	P_4	P_5

E_T	E_1	E_2	E_3 11 V	E_4	E_5
I_T	I_1	I_2	I_3	I_4	I_5
R_T	R_1	R_2	R_3	R_4	R_5
P_T 0.25 W	P_1 0.03 W	P_2 0.0825 W	P_3	P_4 0.045 W	P_5 0.0375 W

E_T 340 V	E_1 44 V	E_2 94 V	E_3 60 V	E_4 40 V	E_5
I_T	I_1	I_2	I_3	I_4	I_5
R_T	R_1	R_2	R_3	R_4	R_5
P_T	P_1	P_2	P_3	P_4	P_5 0.204 W

2. Use the general voltage divider formula to calculate the values of voltage drop for the following series-connected resistors. Assume a source voltage of 120 V.

   $R^1 = 1K$ Ω $R_2 = 2.2K$ Ω $R_3 = 1.8K$ Ω $R_4 = 1.5K$ Ω

   $E_1 =$ _____ V $E_2 =$ _____ V $E_3 =$ _____ V $E_4 =$ _____ V

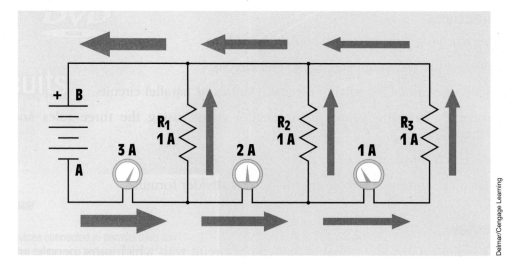

FIGURE 7–1 Parallel circuits provide more than one path for current flow.

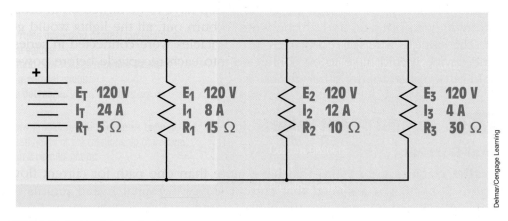

FIGURE 7–2 Parallel circuit values.

seen that each resistor is connected directly to the power source. A second rule for parallel circuits states that ***the voltage drop across any branch of a parallel circuit is the same as the applied voltage.*** For this reason, most electric circuits in homes are connected in parallel. Each lamp and receptacle is supplied with 120 volts *(Figure 7–3)*.

Total Resistance

In the circuit shown in *Figure 7–4,* three separate resistors have values of 15 ohms, 10 ohms, and 30 ohms. The total resistance of the circuit, however, is 5 ohms. ***The total resistance of a parallel circuit is always less than the resistance of the lowest value resistor, or branch, in the circuit.*** Each

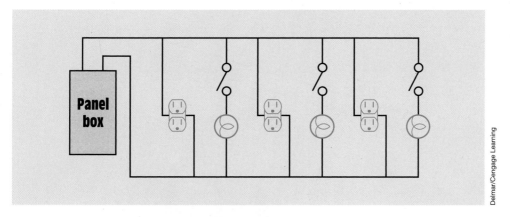

FIGURE 7–3 Lights and receptacles are connected in parallel.

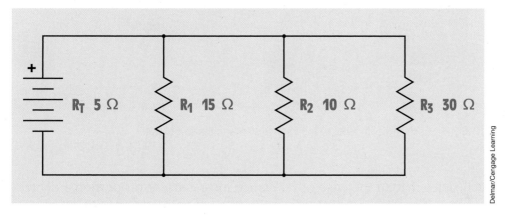

FIGURE 7–4 Total resistance is always less than the resistance of any single branch.

time another element is connected in parallel, there is less opposition to the flow of current through the entire circuit. Imagine a water system consisting of a holding tank, a pump, and return lines to the tank *(Figure 7–5)*. Although large return pipes have less resistance to the flow of water than small pipes, the small pipes do provide a return path to the holding tank. Each time another return path is added, regardless of size, there is less overall resistance to flow and the rate of flow increases.

That concept often causes confusion concerning the definition of **load** among students of electricity. Students often think that an increase of resistance constitutes an increase of load. An increase of current, not resistance, results in an increase of load. In laboratory exercises, students often see the circuit current increase each time a resistive element is connected to the circuit, and they conclude that an increase of resistance must therefore cause an increase of current. That conclusion is, of course, completely contrary to Ohm's law,

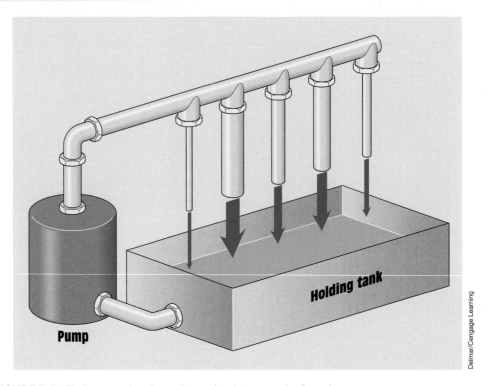

Delmar/Cengage Learning

FIGURE 7–5   Each new path reduces the total resistance to the flow of water.

which states that an increase of resistance must cause a proportional decrease of current. The false concept that an increase of resistance causes an increase of current can be overcome once the student understands that if the resistive elements are being connected in parallel, the circuit resistance is actually being decreased and not increased.

7–2   Parallel Resistance Formulas

Resistors of Equal Value

Three formulas can be used to determine the total resistance of a parallel circuit. The first formula can be used only when all the resistors in the circuit are of equal value. This formula states that ***when all resistors are of equal value, the total resistance is equal to the value of one individual resistor, or branch, divided by the number (N) of resistors or branches.***

$$R_T = \frac{R}{N}$$

For example, assume that three resistors, each having a value of 24 ohms, are connected in parallel *(Figure 7–6)*. The total resistance of this circuit can

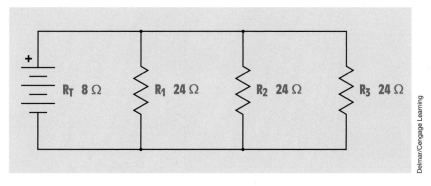

Delmar/Cengage Learning

FIGURE 7–6   Finding the total resistance when all resistors have the same value.

be found by dividing the resistance of one single resistor by the total number of resistors:

$$R_T = \frac{R}{N}$$

$$R_T = \frac{24\ \Omega}{3}$$

$$R_T = 8\ \Omega$$

Product over Sum

The second formula used to determine the total resistance in a parallel circuit divides the product of pairs of resistors by their sum sequentially until only one pair is left. This is commonly referred to as the product-over-sum method for finding total resistance.

$$R_T = \frac{R_1 \times R_2}{R_1 + R_2}$$

In the circuit shown in *Figure 7–7,* three branches having resistors with values of 20 ohms, 30 ohms, and 60 ohms are connected in parallel. To find the total resistance of the circuit using the product-over-sum method, find the total resistance of any two branches in the circuit *(Figure 7–8):*

$$R_T = \frac{R_2 \times R_3}{R_2 + R_3}$$

$$R_T = \frac{30\ \Omega \times 60\ \Omega}{30\ \Omega + 60\ \Omega}$$

$$R_T = \frac{1800\ \Omega}{90\ \Omega}$$

$$R_T = 20\ \Omega$$

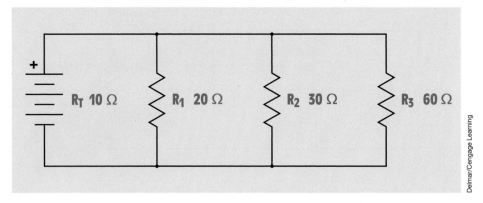

FIGURE 7–7 Finding the total resistance of a parallel circuit by dividing the product of two resistors by their sum.

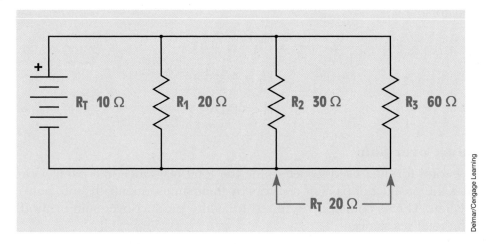

FIGURE 7–8 The total resistance of the last two branches.

The total resistance of the last two resistors in the circuit is 20 ohms. This 20 ohms, however, is connected in parallel with a 20-ohms resistor. The total resistance of the last two resistors is now substituted for the value of R_1 in the formula, and the value of the first resistor is substituted for the value of R_2 *(Figure 7–9)*:

$$R_T = \frac{R_1 \times R_2}{R_1 + R_2}$$

$$R_T = \frac{20\ \Omega \times 20\ \Omega}{20\ \Omega + 20\ \Omega}$$

$$R_T = \frac{400\ \Omega}{40\ \Omega}$$

$$R_T = 10\ \Omega$$

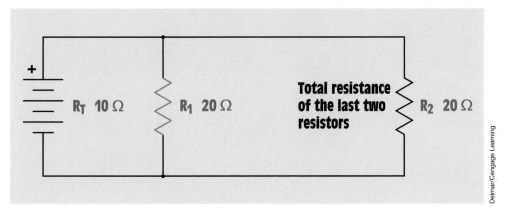

FIGURE 7–9 The total value of the first two resistors is used as Resistor 2.

Reciprocal Formula

The third formula used to find the total resistance of a parallel circuit is

$$\frac{1}{R_T} = \frac{1}{R_1} + \frac{1}{R_2} + \frac{1}{R_3} + \frac{1}{R_N}$$

Notice that this formula actually finds the reciprocal of the total resistance, instead of the total resistance. To make the formula equal to the total resistance, it can be rewritten as follows:

$$R_T = \frac{1}{\dfrac{1}{R_1} + \dfrac{1}{R_2} + \dfrac{1}{R_3} + \dfrac{1}{R_N}}$$

The value R_N stands for the number of resistors in the circuit. If the circuit has 25 resistors connected in parallel, for example, the last resistor in the formula would be R_{25}.

This formula is known as the reciprocal formula. The reciprocal of any number is that number divided into 1. The reciprocal of 4, for example, is 0.25 because $\frac{1}{4} = 0.25$. Another rule of parallel circuits is that ***the total resistance of a parallel circuit is the reciprocal of the sum of the reciprocals of the individual branches.*** A modified version of this formula is used in several different applications to find values other than resistance. Some of those other formulas are covered later.

Before the invention of handheld calculators, the slide rule was often employed to help with the mathematical calculations in electrical work. At that time, the product-over-sum method of finding total resistance was the most popular. Since the invention of calculators, however, the reciprocal formula has become the most popular because scientific calculators have a reciprocal key (1/X), which makes calculating total resistance using the reciprocal method very easy.

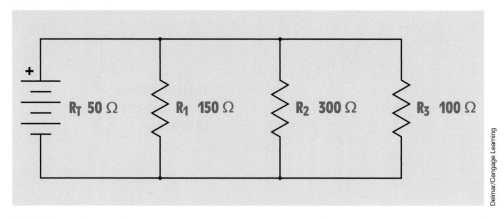

FIGURE 7–10 Finding the total resistance using the reciprocal method.

In *Figure 7–10*, three resistors having values of 150 ohms, 300 ohms, and 100 ohms are connected in parallel. The total resistance can be found using the reciprocal formula:

$$R_T = \frac{1}{\dfrac{1}{R_1} + \dfrac{1}{R_2} + \dfrac{1}{R_3}}$$

$$R_T = \frac{1}{\dfrac{1}{150\ \Omega} + \dfrac{1}{300\ \Omega} + \dfrac{1}{100\ \Omega}}$$

$$R_T = \frac{1}{(0.006667 + 0.003333 + 0.01)\dfrac{1}{\Omega}}$$

$$R_T = \frac{1}{(0.02)\dfrac{1}{\Omega}}$$

$$R_T = 50\ \Omega$$

To find the total resistance of the previous example using a scientific calculator, press the following keys. Note that the calculator automatically carries each answer to the maximum number of decimal places. This increases the accuracy of the answer.

$$\boxed{1}\ \boxed{5}\ \boxed{0}\ \boxed{1/x}\ \boxed{+}\ \boxed{3}\ \boxed{0}\ \boxed{0}\ \boxed{1/x}\ \boxed{+}\ \boxed{1}\ \boxed{0}\ \boxed{0}\ \boxed{1/x}\ \boxed{=}\ \boxed{1/x}$$

Note that this is intended to illustrate how total parallel resistance can be determined using many scientific calculators. Some calculators may require a different key entry or pressing the equal key at the end.

■ EXAMPLE 7–1

In the circuit shown in *Figure 7–11,* three resistors having values of 300 Ω, 200 Ω, and 600 Ω are connected in parallel. The total current flow through the circuit is 0.6 A. Find all the missing values in the circuit.

Solution

The first step is to find the total resistance of the circuit. The reciprocal formula is used:

$$R_T = \frac{1}{\dfrac{1}{R_1} + \dfrac{1}{R_2} + \dfrac{1}{R_3}}$$

$$R_T = \frac{1}{\dfrac{1}{300\ \Omega} + \dfrac{1}{200\ \Omega} + \dfrac{1}{600\ \Omega}}$$

$$R_T = \frac{1}{(0.00333 + 0.0050 + 0.00167)\dfrac{1}{\Omega}}$$

$$R_T = \frac{1}{(0.01)\dfrac{1}{\Omega}}$$

$$R_T = 100\ \Omega$$

Now that the total resistance of the circuit is known, the voltage applied to the circuit can be found by using the total current value and Ohm's law:

$$E_T = I_T \times R_T$$
$$E_T = 0.6\ A \times 100\ \Omega$$
$$E_T = 60\ V$$

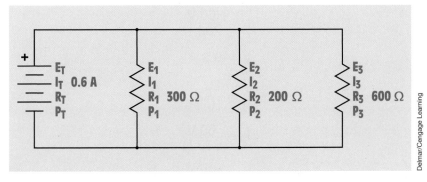

FIGURE 7–11   Parallel circuit, Example 1.

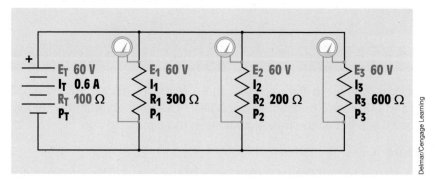

Delmar/Cengage Learning

FIGURE 7–12 The voltage is the same across all branches of a parallel circuit.

One of the rules for parallel circuits states that the voltage drops across all the parts of a parallel circuit are the same as the total voltage. Therefore, the voltage drop across each resistor is 60 V *(Figure 7–12):*

$$E_T = E_1 = E_2 = E_3$$

Because the voltage drop and the resistance of each resistor are known, Ohm's law can be used to determine the amount of current flow through each resistor *(Figure 7–13):*

$$I_1 = \frac{E_1}{R_1}$$

$$I_1 = \frac{60 \text{ V}}{300 \text{ } \Omega}$$

$$I_1 = 0.2 \text{ A}$$

$$I_2 = \frac{E_2}{R_2}$$

$$I_2 = \frac{60 \text{ V}}{200 \text{ } \Omega}$$

$$I_2 = 0.3 \text{ A}$$

$$I_3 = \frac{E_3}{R_3}$$

$$I_3 = \frac{60 \text{ V}}{600 \text{ } \Omega}$$

$$I_3 = 0.1 \text{ A}$$

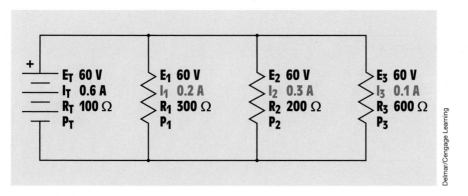

Delmar/Cengage Learning

FIGURE 7–13   Ohm's law is used to calculate the amount of current through each branch.

The amount of power (W) used by each resistor can be found by using Ohm's law. A different formula is used to find the amount of electrical energy converted into heat by each of the resistors:

$$P_1 = \frac{E_1^{\,2}}{R_1}$$

$$P_1 = \frac{60 \text{ V} \times 60 \text{ V}}{300 \text{ }\Omega}$$

$$P_1 = \frac{3600 \text{ V}^2}{300 \text{ }\Omega}$$

$$P_1 = 12 \text{ W}$$

$$P_2 = I_2^{\,2} \times R_2$$

$$P_2 = 0.3 \text{ A} \times 0.3 \text{ A} \times 200 \text{ }\Omega$$

$$P_2 = 0.09 \text{ A}^2 \times 200 \text{ }\Omega$$

$$P_2 = 18 \text{ W}$$

$$P_3 = E_3 \times I_3$$

$$P_3 = 60 \text{ V} \times 0.1 \text{ A}$$

$$P_3 = 6 \text{ W}$$

In Unit 6, it was stated that the total amount of power in a circuit is equal to the sum of the power used by all the parts. This is true for any type of circuit. Therefore, the total amount of power used by this circuit can be found by taking the sum of the power used by all the resistors *(Figure 7–14)*:

$$P_T = P_1 + P_2 + P_3$$

$$P_T = 12 \text{ W} + 18 \text{ W} + 6 \text{ W}$$

$$P_T = 36 \text{ W}$$

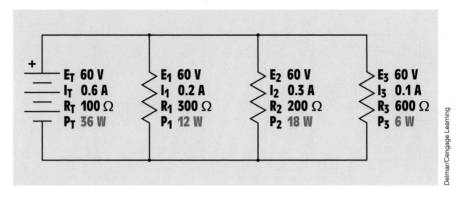

FIGURE 7–14   The amount of power used by the circuit.

■ EXAMPLE 7-2

In the circuit shown in *Figure 7–15,* three resistors are connected in parallel. Two of the resistors have a value of 900 Ω and 1800 Ω. The value of Resistor R_2 is unknown. The total resistance of the circuit is 300 Ω. Resistor R_2 has a current flow of 0.2 A. Find the missing circuit values.

Solution

The first step in solving this problem is to find the missing resistor value. This can be done by changing the reciprocal formula as shown:

$$\frac{1}{R_2} = \frac{1}{R_T} - \frac{1}{R_1} - \frac{1}{R_3}$$

or

$$R_2 = \frac{1}{\dfrac{1}{R_T} - \dfrac{1}{R_1} - \dfrac{1}{R_3}}$$

One of the rules for parallel circuits states that the total resistance is equal to the reciprocal of the sum of the reciprocals of the individual resistors. Therefore, the reciprocal of any individual resistor is equal to the reciprocal of the difference

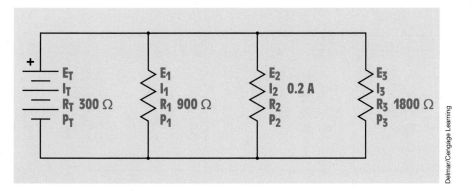

FIGURE 7–15   Parallel circuit, Example 2.

between the reciprocal of the total resistance and the sum of the reciprocals of the other resistors in the circuit:

$$R_2 = \cfrac{1}{\cfrac{1}{R_T} - \cfrac{1}{R_1} - \cfrac{1}{R_3}}$$

$$R_2 = \cfrac{1}{\cfrac{1}{300\ \Omega} - \cfrac{1}{900\ \Omega} - \cfrac{1}{1800\ \Omega}}$$

$$R_2 = \cfrac{1}{(0.003333 - 0.001111 - 0.0005556)\,\dfrac{1}{\Omega}}$$

$$R_2 = \cfrac{1}{(0.001666)\,\dfrac{1}{\Omega}}$$

$$R_2 = 600\ \Omega$$

Now that the resistance of Resistor R_2 has been found, the voltage drop across Resistor R_2 can be determined using the current flow through the resistor and Ohm's law *(Figure 7–16)*:

$$E_2 = I_2 \times R_2$$

$$E_2 = 0.2\ A \times 600\ \Omega$$

$$E_2 = 120\ V$$

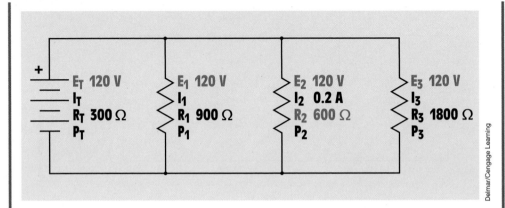

FIGURE 7–16 The missing resistor and voltage values.

If 120 V is dropped across Resistor R_2, the same voltage is dropped across each component of the circuit:

$$E_2 = E_T = E_1 = E_3$$

Now that the voltage drop across each part of the circuit is known and the resistance is known, the current flow through each branch can be determined using Ohm's law *(Figure 7–17)*:

$$I_T = \frac{E_T}{R_T}$$

$$I_T = \frac{120 \text{ V}}{300 \text{ }\Omega}$$

$$I_T = 0.4 \text{ A}$$

$$I_1 = \frac{E_1}{R_1}$$

$$I_1 = \frac{120 \text{ V}}{1800 \text{ }\Omega}$$

$$I_1 = 0.1333 \text{ A}$$

$$I_3 = \frac{E_3}{R_3}$$

$$I_3 = \frac{120 \text{ V}}{1800 \text{ }\Omega}$$

$$I_3 = 0.0667 \text{ A}$$

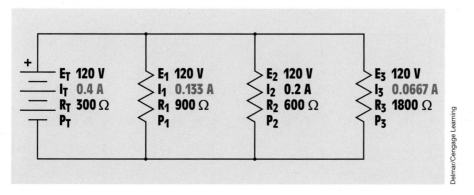

Delmar/Cengage Learning

FIGURE 7–17   Determining the current using Ohm's law.

The amount of power used by each resistor can be found using Ohm's law *(Figure 7–18)*:

$$P_1 = \frac{E_1^{\,2}}{R_1}$$

$$P_1 = \frac{120\ V \times 120\ V}{900\ \Omega}$$

$$P_1 = \frac{14,400\ V^2}{900\ \Omega}$$

$$P_1 = 16\ W$$

$$P_2 = I_2^{\,2} \times R_2$$

$$P_2 = 0.2\ A \times 0.2\ A \times 600\ \Omega$$

$$P_2 = 0.04\ A^2 \times 600\ \Omega$$

$$P_2 = 24\ W$$

$$P_3 = E_3 \times I_3$$

$$P_3 = 120\ V \times 0.0667\ A$$

$$P_3 = 8.004\ W$$

$$P_T = E_T \times I_T$$

$$P_T = 120\ V \times 0.4\ A$$

$$P_T = 48\ W$$

If the wattage values of the three resistors are added to calculate total power for the circuit, it will be seen that their total is 48.004 W instead of the calculated 48 W. The small difference in answers is caused by the rounding off of other values. In this instance, the current of Resistor R_3 was rounded from 0.066666666 to 0.0667.

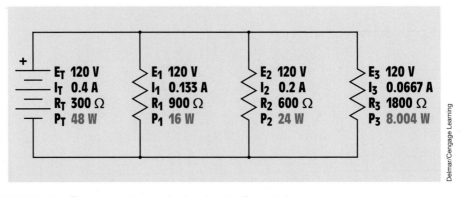

FIGURE 7–18 The values of power for the circuit in Example 2.

■ EXAMPLE 7-3

In the circuit shown in *Figure 7–19,* three resistors are connected in parallel. Resistor R_1 is producing 0.075 W of heat, R_2 is producing 0.45 W of heat, and R_3 is producing 0.225 W of heat. The circuit has a total current of 0.05 A.

Solution

Because the amount of power dissipated by each resistor is known, the total power for the circuit can be found by finding the sum of the power used by the components:

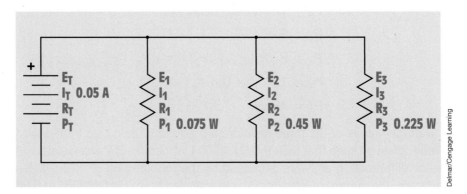

FIGURE 7–19 Parallel circuit, Example 3.

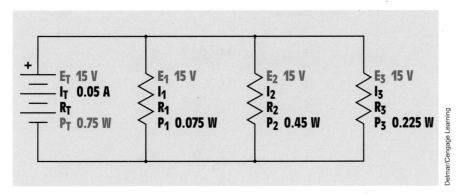

FIGURE 7–20   The applied voltage for the circuit.

$$P_T = P_1 + P_2 + P_3$$
$$P_T = 0.075 \text{ W} + 0.45 \text{ W} + 0.225 \text{ W}$$
$$P_T = 0.75 \text{ W}$$

Now that the amount of total current and total power for the circuit are known, the applied voltage can be found using Ohm's law *(Figure 7–20):*

$$E_T = \frac{P_T}{I_T}$$

$$E_T = \frac{0.75 \text{ W}}{0.05 \text{ A}}$$

$$E_T = 15 \text{ V}$$

The amount of current flow through each resistor can now be found using Ohm's law *(Figure 7–21):*

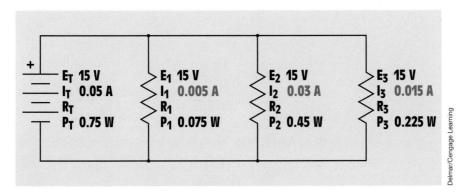

FIGURE 7–21   The current through each branch.

$$I_1 = \frac{P_1}{E_1}$$

$$I_1 = \frac{0.075 \text{ W}}{15 \text{ V}}$$

$$I_1 = 0.005 \text{ A}$$

$$I_2 = \frac{P_2}{E_2}$$

$$I_2 = \frac{0.45 \text{ W}}{15 \text{ V}}$$

$$I_2 = 0.03 \text{ A}$$

$$I_3 = \frac{P_3}{E_3}$$

$$I_3 = \frac{0.225 \text{ W}}{15 \text{ V}}$$

$$I_3 = 0.015 \text{ A}$$

All resistance values for the circuit can now be found using Ohm's law *(Figure 7–22):*

$$R_1 = \frac{E_1}{I_1}$$

$$R_1 = \frac{15 \text{ V}}{0.005 \text{ A}}$$

$$R_1 = 3000 \ \Omega$$

$$R_2 = \frac{E_2}{I_2}$$

$$R_2 = \frac{15 \text{ V}}{0.03 \text{ A}}$$

$$R_2 = 500 \ \Omega$$

$$R_3 = \frac{E_3}{I_3}$$

$$R_3 = \frac{15 \text{ V}}{0.015 \text{ A}}$$

$$R_3 = 1000 \ \Omega$$

$$R_T = \frac{E_T}{I_T}$$

$$R_T = \frac{15 \text{ V}}{0.05 \text{ A}}$$

$$R_T = 300 \ \Omega$$

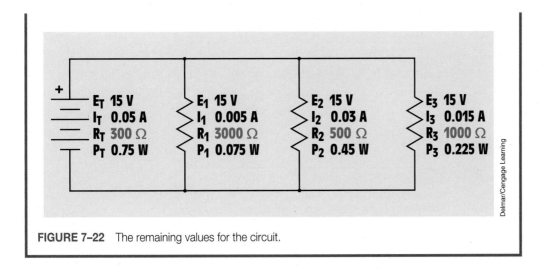

FIGURE 7–22   The remaining values for the circuit.

Current Dividers

All parallel circuits are **current dividers** *(Figure 7–23)*. As previously discussed in this unit, the sum of the currents in a parallel circuit must equal the total current. Assume that a current of 1 ampere enters the circuit at Point A. This 1 ampere of current will divide between Resistors R_1 and R_2, and then recombine at Point B. The amount of current that flows through each resistor is inversely proportional to the resistance value. A greater amount of current will

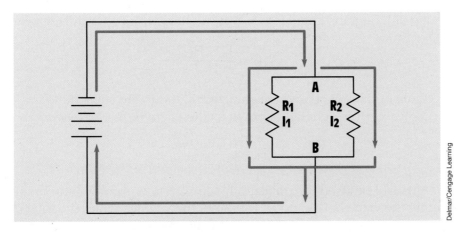

FIGURE 7–23   Parallel circuits are current dividers.

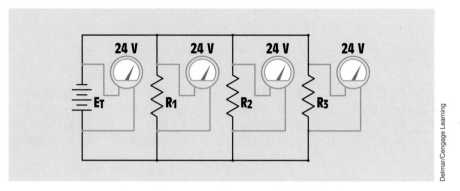

FIGURE 7–24 The voltage is the same across all branches of a parallel circuit.

flow through a low-value resistor, and less current will flow through a high-value resistor. In other words, the amount of current flowing through each resistor is inversely proportional to its resistance.

In a parallel circuit, the voltage across each branch must be equal *(Figure 7–24)*. Therefore, the current flow through any branch can be calculated by dividing the source voltage (E_T) by the resistance of that branch. The current flow through Branch 1 can be calculated using the formula

$$I_1 = \frac{E_T}{R_1}$$

It is also true that the total circuit voltage is equal to the product of the total circuit current and the total circuit resistance:

$$E_T = I_T \times R_T$$

If the value of ET is substituted for ($I_T \times R_T$) in the previous formula, it becomes

$$I_1 = \frac{I_T \times R_T}{R_1}$$

If the formula is rearranged and the values of I_1 and R_1 are substituted for I_X and R_X, it becomes what is generally known as the *current divider formula:*

$$I_X = \left(\frac{R_T}{R_X}\right) I_T$$

This formula can be used to calculate the current flow through any branch by substituting the values of I_X and R_X for the branch values when the total circuit current and the resistance are known. In the circuit shown in *Figure 7–25,*

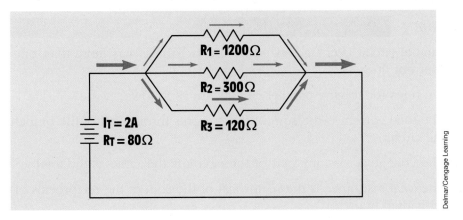

Delmar/Cengage Learning

FIGURE 7–25   The current divides through each branch of a parallel circuit.

Resistor R_1 has a value of 1200 ohms, Resistor R_2 has a value of 300 ohms, and Resistor R_3 has a value of 120 ohms, producing a total resistance of 80 ohms for the circuit. It is assumed that a total current of 2 amperes flows in the circuit. The amount of current flow through Resistor R_1 can be found using the formula

$$I_1 = \left(\frac{R_T}{R_1}\right) I_T$$

$$I_1 = \left(\frac{80\ \Omega}{1200\ \Omega}\right) 2\ A$$

$$I_1 = 0.133\ A$$

The current flow through each of the other resistors can be found by substituting in the same formula:

$$I_2 = \left(\frac{R_T}{R_2}\right) I_T$$

$$I_2 = \left(\frac{80\ \Omega}{300\ \Omega}\right) 2\ A$$

$$I_2 = 0.533\ A$$

$$I_3 = \left(\frac{R_T}{R_3}\right) I_T$$

$$I_3 = \left(\frac{80\ \Omega}{120\ \Omega}\right) 2\ A$$

$$I_3 = 1.333\ A$$

Summary

- A Parallel circuit is characterized by the fact that it has more than one path for current flow.

- Three rules for solving parallel circuits are as follows:

  a. The total current is the sum of the currents through all of the branches of the circuit.

  b. The voltage across any part of the circuit is the same as the total voltage.

  c. The total resistance is the reciprocal of the sum of the reciprocals of each individual branch.

- Circuits in homes are connected in parallel.

- The total power in a parallel circuit is equal to the sum of the power dissipation of all the components.

- Parallel circuits are current dividers.

- The current flowing through each branch of a parallel circuit can be calculated when the total resistance and the total current are known.

- The amount of current flow through each branch of a parallel circuit is inversely proportional to its resistance.

Review Questions

1. What characterizes a parallel circuit?

2. Why are circuits in homes connected in parallel?

3. State three rules concerning parallel circuits.

4. A parallel circuit contains four branches. One branch has a current flow of 0.8 A, another has a current flow of 1.2 A, the third has a current flow of 0.25 A, and the fourth has a current flow of 1.5 A. What is the total current flow in the circuit?

5. Four resistors having a value of 100 Ω each are connected in parallel. What is the total resistance of the circuit?

6. A parallel circuit has three branches. An ammeter is connected in series with the output of the power supply and indicates a total current flow of 2.8 A. If Branch 1 has a current flow of 0.9 A and Branch 2 has a current flow of 1.05 A, what is the current flow through Branch 3?

7. Four resistors having values of 270 Ω, 330 Ω, 510 Ω, and 430 Ω are connected in parallel. What is the total resistance in the circuit?

8. A parallel circuit contains four resistors. The total resistance of the circuit is 120 Ω. Three of the resistors have values of 820 Ω, 750 Ω, and 470 Ω. What is the value of the fourth resistor?

9. A circuit contains a 1200-Ω, a 2200-Ω, and a 3300-Ω resistor connected in parallel. The circuit has a total current flow of 0.25 A. How much current flows through each of the resistors?

Practical Applications

You have been hired by a homeowner to install a ceiling fan and light kit in a living room. The living room luminaire (light fixture) being used at the present time contains two 60-W lamps. After locating the circuit in the panel box, you find that the circuit is protected by a 15-A circuit breaker and is run with 14 AWG copper wire. After turning on all lights connected to this circuit, you have a current draw of 8.6 A and a voltage of 120 V. The ceiling fan light kit contains four 60-W lamps, and the fan has a maximum current draw of 1.6 A. Can this fan be connected to the existing living room circuit? Recall that a continuous-use circuit should not be loaded to more than 80% of its rated capacity. ■

Practical Applications

You are employed in a large industrial plant. A 480-V, 5000-W heater is used to melt lead in a large tank. It has been decided that the heater is not sufficient to raise the temperature of the lead to the desired level. A second 5000-W heater is to be installed on the same circuit. What will be the circuit current after installation of the second heater, and what is the minimum size circuit breaker that can be used if this is a continuous-duty circuit? ■

Practical Applications

You are an electrician. You have been asked by a homeowner to install a lighted mirror in a bathroom. The mirror contains eight 40-watt lamps. Upon checking the service panel you discover that the bathroom circuit is connected to a single 120-volt, 20-ampere circuit breaker. At the present time, the circuit supplies power to an electric wall heater rated at 1000 watts, a ceiling fan with a light kit, and a light fixture over the mirror. The fan motor has a full-load current draw of 3.2 amperes and the light kit contains three 60-watt lamps. The light fixture presently installed over the mirror contains four 60-watt lamps. The homeowner asked whether the present light fixture over the mirror can be replaced by the lighted mirror. Assuming all loads are continuous, can the present circuit supply the power needed to operate all the loads without overloading the circuit? ■

Practical Applications

A car lot uses incandescent lamps to supply outside lighting during the night. There are three strings of lamps connected to a single 20-ampere circuit. Each string contains eight lamps. What is the largest standard lamp that can be used without overloading the circuit? Standard size lamps are 25 watt, 40 watt, 60 watt, 75 watt, and 100 watt. ■

Practice Problems

Parallel Circuits

Using the rules for parallel circuits and Ohm's law, solve for the missing values.

1.

E_T	E_1	E_2	E_3	E_4
I_T 0.942 A	I_1	I_2	I_3	I_4
R_T	R_1 680 Ω	R_2 820 Ω	R_3 470 Ω	R_4 330 Ω
P_T	P_1	P_2	P_3	P_4

2.

E_T	E_1	E_2	E_3	E_4
I_T 0.00639 A	I_1	I_2 0.00139 A	I_3 0.00154 A	I_4 0.00115 A
R_T	R_1	R_2	R_3	R_4
P_T	P_1 0.640 W	P_2	P_3	P_4

3.

E_T	E_1	E_2	E_3	E_4
I_T	I_1	I_2	I_3 3.2 A	I_4
R_T 3.582 Ω	R_1 16 Ω	R_2 10 Ω	R_3	R_4 20 Ω
P_T	P_1	P_2	P_3	P_4

4.

E_T	E_1	E_2	E_3	E_4
I_T	I_1	I_2	I_3	I_4
R_T	R_1 82 kΩ	R_2 75 kΩ	R_3 56 kΩ	R_4 62 kΩ
P_T 3.436 W	P_1	P_2	P_3	P_4

5. A parallel circuit contains the following resistor values:

   $R_1 = 360\ \Omega$ $R_2 = 470\ \Omega$ $R_3 = 300\ \Omega$

   $R_4 = 270\ \Omega$ $I_T = 0.05$ A

   Find the following missing values:

   $R_T =$ _____ Ω $I_1 =$ _____ A $I_2 =$ _____ A

   $I_3 =$ _____ A $I_4 =$ _____ A

6. A parallel circuit contains the following resistor values:

   $R_1 = 270$K Ω $R_2 = 360$K Ω $R_3 = 430$K Ω

   $R_4 = 100$K Ω $I_T = 0.006$ A

   Find the following missing values:

   $R_T =$ _____ Ω $I_1 =$ _____ A $I_2 =$ _____ A

   $I_3 =$ _____ A $I_4 =$ _____ A

Unit 8 DVD VIDEO

Combination Circuits

Why You Need to Know

*U*nderstanding how values of voltage, current, and resistance relate to each other in combination circuits is essential to understanding how all electric circuits work. Although it is not a common practice for electricians on the job to use a calculator to calculate values of voltage and current, they must have an understanding of how these electrical values relate to each other in different kinds of circuits. An understanding of Ohm's law and basic circuits is also essential in being able to troubleshoot problems that occur in an electric circuit. In this unit

- the rules for series circuits are employed for those components connected in series, and parallel rules are used for components connected in parallel.
- you are shown how to reduce a complex combination circuit to a simple series or parallel circuit and then apply circuit rules and Ohm's law to determine the values of voltage, current, and power for the different components.

OUTLINE

KEY TERMS

Combination circuit
Node
Parallel block
Redraw
Reduce
Simple parallel circuit
Trace the current path

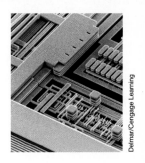

Objectives

After studying this unit, you should be able to

■ define a combination circuit.

■ list the rules for parallel circuits.

■ list the rules for series circuits.

■ solve combination circuits using the rules for parallel circuits, the rules for series circuits, and Ohm's law.

Preview

Combination circuits contain a combination of both series and parallel elements. To determine which components are in parallel and which are in series, trace the flow of current through the circuit. Remember that a series circuit has only one path for current flow and a parallel circuit has more than one path for current flow. ■

8–1   Combination Circuits

A simple combination circuit is shown in *Figure 8–1*. It is assumed that the current in *Figure 8–1* flows from Point A to Point B. To identify the series and parallel elements, **trace the current path.** All the current in the circuit must flow through the first resistor, R_1. R_1 is therefore in series with the rest of the circuit. When the current reaches the junction point of R_2 and R_3, however, it splits. A junction point such as this is often referred to as a **node.** Part of the current flows through R_2, and part flows through R_3. These two resistors are in parallel. Because this circuit contains both series and parallel elements, it is a **combination circuit.**

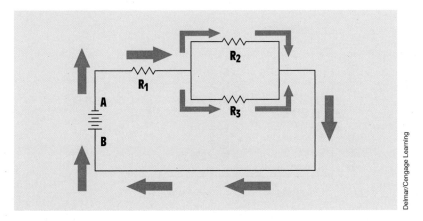

FIGURE 8–1   A simple combination circuit.

8-2 Solving Combination Circuits

The circuit shown in *Figure 8–2* contains four resistors with values of 325 ohms, 275 ohms, 150 ohms, and 250 ohms. The circuit has a total current flow of 1 ampere. In order to determine which resistors are in series and which are in parallel, trace the path for current flow through the circuit. When the path of current flow is traced, it can be seen that current can flow by two separate paths from the negative terminal to the positive terminal. One path is through R_1 and R_2, and the other path is through R_3 and R_4. These two paths are therefore in parallel. However, the same current must flow through R_1 and R_2. So these two resistors are in series. The same is true for R_3 and R_4.

To solve the unknown values in a combination circuit, use series circuit rules for those sections of the circuit that are connected in series and parallel circuit rules for those sections connected in parallel. The circuit rules are as follows:

Series Circuits

1. The current is the same at any point in the circuit.

2. The total resistance is the sum of the individual resistances.

3. The applied voltage is equal to the sum of the voltage drops across all the resistors.

Parallel Circuits

1. The voltage drop across any branch of a parallel circuit is the same as the applied voltage.

2. The total current flow is equal to the sum of the currents through all of the circuit branches.

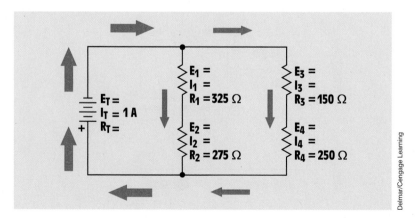

FIGURE 8–2 Tracing the current paths through a combination circuit.

3. The total resistance is equal to the reciprocal of the sum of the reciprocals of the branch resistances.

8-3   Simplifying the Circuit

The circuit shown in *Figure 8–2* can be reduced or simplified to a **simple parallel circuit** *(Figure 8–3)*. Because R_1 and R_2 are connected in series, their values can be added to form one equivalent resistor, $R_{c(1\&2)}$, which stands for a combination of Resistors 1 and 2. The same is true for R_3 and R_4. Their values are added to form $R_{c(3\&4)}$. Now that the circuit has been reduced to a simple parallel circuit, the total resistance can be found:

$$R_T = \cfrac{1}{\cfrac{1}{R_{c(1\&2)}} + \cfrac{1}{R_{c(3\&4)}}}$$

$$R_T = \cfrac{1}{\cfrac{1}{600 \; \Omega} + \cfrac{1}{400 \; \Omega}}$$

$$R_T = \cfrac{1}{(0.0016667 + 0.0025)\dfrac{1}{\Omega}}$$

$$R_T = \cfrac{1}{(0.0041667)\dfrac{1}{\Omega}}$$

$$R_T = 240 \; \Omega$$

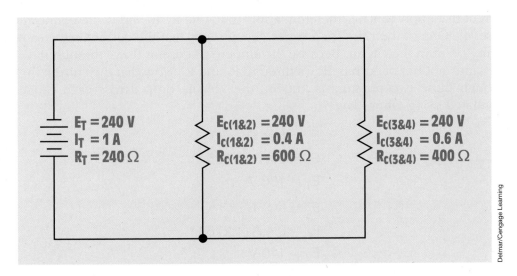

$E_T = 240 \text{ V}$
$I_T = 1 \text{ A}$
$R_T = 240 \; \Omega$

$E_{C(1\&2)} = 240 \text{ V}$
$I_{C(1\&2)} = 0.4 \text{ A}$
$R_{C(1\&2)} = 600 \; \Omega$

$E_{C(3\&4)} = 240 \text{ V}$
$I_{C(3\&4)} = 0.6 \text{ A}$
$R_{C(3\&4)} = 400 \; \Omega$

FIGURE 8–3   Simplifying the combination circuit.

Now that the total resistance has been found, the other circuit values can be calculated. The applied voltage can be found using Ohm's law:

$$E_T = I_T \times R_T$$
$$E_T = 1\,A \times 240\,\Omega$$
$$E_T = 240\,V$$

One of the rules for parallel circuits states that the voltage is the same across each branch of the circuit. For this reason, the voltage drops across $R_{c(1\&2)}$ and $R_{c(3\&4)}$ are the same. Because the voltage drop and the resistance are known, Ohm's law can be used to find the current flow through each branch:

$$I_{c(1\&2)} = \frac{E_{c(1\&2)}}{R_{c(1\&2)}}$$

$$I_{c(1\&2)} = \frac{240\,V}{600\,\Omega}$$

$$I_{c(1\&2)} = 0.4\,A$$

$$I_{c(3\&4)} = \frac{E_{c(3\&4)}}{R_{c(3\&4)}}$$

$$I_{c(3\&4)} = \frac{240\,V}{400\,\Omega}$$

$$I_{c(3\&4)} = 0.6\,A$$

These values can now be used to solve the missing values in the original circuit. $R_{c(1\&2)}$ is actually a combination of R_1 and R_2. The values of voltage and current that apply to $R_{c(1\&2)}$ therefore apply to R_1 and R_2. R_1 and R_2 are connected in series. One of the rules for a series circuit states that the current is the same at any point in the circuit. Because 0.4 ampere of current flows through $R_{c(1\&2)}$, the same amount of current flows through R_1 and R_2. Now that the current flow through these two resistors is known, the voltage drop across each can be calculated using Ohm's law:

$$E_1 = I_1 \times R_1$$
$$E_1 = 0.4\,A \times 325\,\Omega$$
$$E_1 = 130\,V$$

$$E_2 = I_2 \times R_2$$
$$E_2 = 0.4\,A \times 275\,\Omega$$
$$E_2 = 110\,V$$

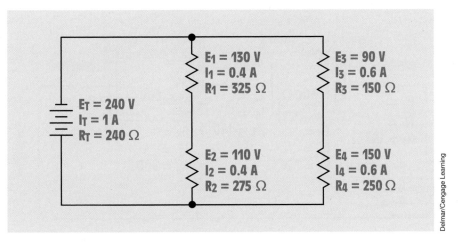

FIGURE 8–4 All the missing values for the combination circuit.

These values of voltage and current can now be added to the circuit in *Figure 8–2* to produce the circuit shown in *Figure 8–4*.

The values of voltage and current for $R_{c(3\&4)}$ apply to R_3 and R_4. The same amount of current that flows through $R_{c(3\&4)}$ flows through R_3 and R_4. The voltage across these two resistors can now be calculated using Ohm's law:

$$E_3 = I_3 \times R_3$$
$$E_3 = 0.6 \text{ A} \times 150 \text{ }\Omega$$
$$E_3 = 90 \text{ V}$$

$$E_4 = I_4 \times R_4$$
$$E_4 = 0.6 \text{ A} \times 250 \text{ }\Omega$$
$$E_4 = 150 \text{ V}$$

■ EXAMPLE 8–1

Solve the combination circuit shown in *Figure 8–5*.

Solution

The first step in finding the missing values is to trace the current path through the circuit to determine which resistors are in series and which are in parallel. All the current must flow through R_1. R_1 is therefore in series with the rest of the

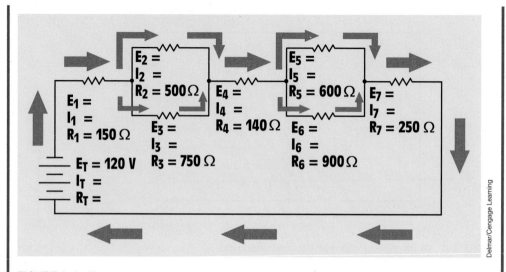

FIGURE 8–5   Tracing the flow of current through the combination circuit.

circuit. When the current reaches the junction of R_2 and R_3, it divides, and part flows through each resistor. R_2 and R_3 are in parallel. All the current must then flow through R_4, which is connected in series, to the junction of R_5 and R_6. The current path is divided between these two resistors. R_5 and R_6 are connected in parallel. All the circuit current must then flow through R_7.

The next step in solving this circuit is to **reduce** it to a simpler circuit. If the total resistance of the first **parallel block** formed by R_2 and R_3 is found, this block can be replaced by a single resistor:

$$R_{(2,\,3)} = \frac{1}{\dfrac{1}{R_2} + \dfrac{1}{R_3}}$$

$$R_{(2,\,3)} = \frac{1}{\dfrac{1}{500\ \Omega} + \dfrac{1}{750\ \Omega}}$$

$$R_{(2,\,3)} = \frac{1}{(0.002 + 0.0013333)\dfrac{1}{\Omega}}$$

$$R_{(2,\,3)} = \frac{1}{(0.0033333)\dfrac{1}{\Omega}}$$

$$R_{(2,\,3)} = 300\ \Omega$$

The equivalent resistance of the second parallel block can be calculated in the same way:

$$R_{(5, 6)} = \frac{1}{\dfrac{1}{R_5} + \dfrac{1}{R_6}}$$

$$R_{(5, 6)} = \frac{1}{\dfrac{1}{600 \ \Omega} + \dfrac{1}{900 \ \Omega}}$$

$$R_{(5, 6)} = \frac{1}{(0.0016667 + 0.0011111) \dfrac{1}{\Omega}}$$

$$R_{(5, 6)} = \frac{1}{(0.00277778) \dfrac{1}{\Omega}}$$

$$R_{(5, 6)} = 360 \ \Omega$$

Now that the total resistance of the second parallel block is known, you can **redraw** the circuit as a simple series circuit as shown in *Figure 8–6*. The first parallel block has been replaced with a single resistor of 300 Ω labeled $R_{c(2\&3)}$, and the second parallel block has been replaced with a single 360-Ω resistor labeled $R_{c(5\&6)}$. Ohm's law can be used to find the missing values in this series circuit.

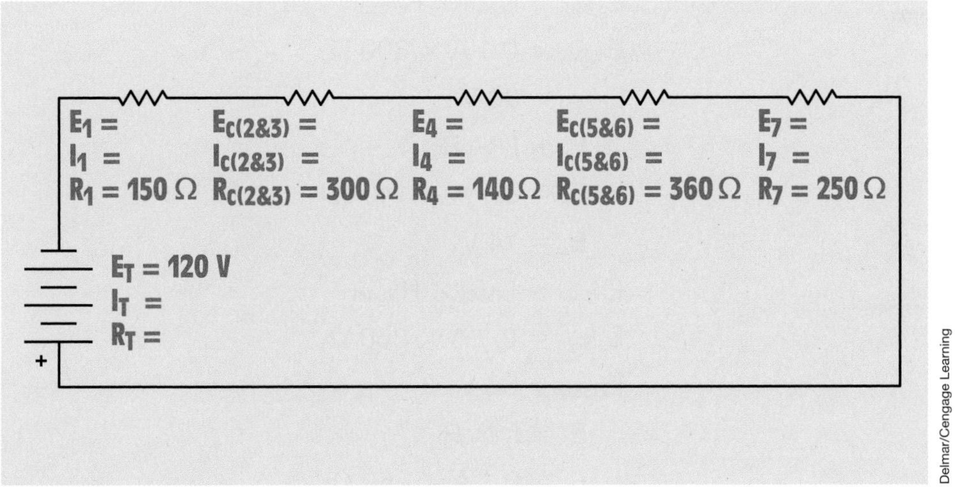

$E_1 =$	$E_{c(2\&3)} =$	$E_4 =$	$E_{c(5\&6)} =$	$E_7 =$
$I_1 =$	$I_{c(2\&3)} =$	$I_4 =$	$I_{c(5\&6)} =$	$I_7 =$
$R_1 = 150 \ \Omega$	$R_{c(2\&3)} = 300 \ \Omega$	$R_4 = 140 \ \Omega$	$R_{c(5\&6)} = 360 \ \Omega$	$R_7 = 250 \ \Omega$

$E_T = 120$ V
$I_T =$
$R_T =$

Delmar/Cengage Learning

FIGURE 8–6 Simplifying the combination circuit.

One of the rules for series circuits states that the total resistance of a series circuit is equal to the sum of the individual resistances. R_T can be calculated by adding the resistances of all the resistors:

$$R_T = R_1 + R_{c(2\&3)} + R_4 + R_{c(5\&6)} + R_7$$
$$R_T = 150 \ \Omega + 300 \ \Omega + 140 \ \Omega + 360 \ \Omega + 250 \ \Omega$$
$$R_T = 1200 \ \Omega$$

Because the total voltage and total resistance are known, the total current flow through the circuit can be calculated:

$$I_T = \frac{E_T}{R_T}$$
$$I_T = \frac{120 \ V}{1200 \ \Omega}$$
$$I_T = 0.1 \ A$$

The first rule of series circuits states that the current is the same at any point in the circuit. The current flow through each resistor is therefore 0.1 A. The voltage drop across each resistor can now be calculated using Ohm's law:

$$E_1 = I_1 \times R_1$$
$$E_1 = 0.1 \ A \times 150 \ \Omega$$
$$E_1 = 15 \ \Omega$$
$$E_{c(2\&3)} = I_{c(2\&3)} \times R_{c(2\&3)}$$
$$E_{c(2\&3)} = 0.1 \ A \times 300 \ \Omega$$
$$E_{c(2\&3)} = 30 \ V$$
$$E_4 = I_4 \times R_4$$
$$E_4 = 0.1 \ A \times 140 \ \Omega$$
$$E_4 = 14 \ V$$
$$E_{c(5\&6)} = I_{c(5\&6)} \times R_{c(5\&6)}$$
$$E_{c(5\&6)} = 0.1 \ A \times 360 \ \Omega$$
$$E_{c(5\&6)} = 36 \ V$$
$$E_7 = I_7 \times R_7$$
$$E_7 = 0.1 \ A \times 250 \ \Omega$$
$$E_7 = 25 \ V$$

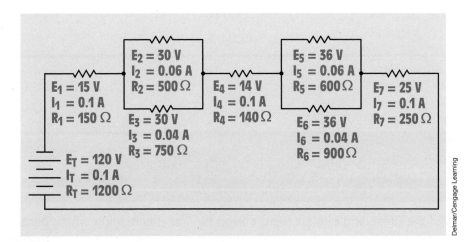

$E_1 = 15$ V $E_{c(2\&3)} = 30$ V $E_4 = 14$ V $E_{c(5\&6)} = 36$ V $E_7 = 25$ V
$I_1 = 0.1$ A $I_{c(2\&3)} = 0.1$ A $I_4 = 0.1$ A $I_{c(5\&6)} = 0.1$ A $I_7 = 0.1$ A
$R_1 = 150\ \Omega$ $R_{c(2\&3)} = 300\ \Omega$ $R_4 = 140\ \Omega$ $R_{c(5\&6)} = 360\ \Omega$ $R_7 = 250\ \Omega$

$E_T = 120$ V
$I_T = 0.1$ A
$R_T = 1200\ \Omega$

FIGURE 8–7 The simplified circuit with all values solved.

The series circuit with all solved values is shown in *Figure 8–7*. These values can now be used to solve missing parts in the original circuit.

$R_{c(2\&3)}$ is actually the parallel block containing R_2 and R_3. The values for $R_{c(2\&3)}$ therefore apply to this parallel block. One of the rules for a parallel circuit states that the voltage drop of a parallel circuit is the same at any point in the circuit. Because 30 V are dropped across $R_{c(2\&3)}$, the same 30 V are dropped across R_2 and R_3 *(Figure 8–8)*. The current flow through these resistors can now be

$E_2 = 30$ V
$I_2 = 0.06$ A
$R_2 = 500\ \Omega$

$E_5 = 36$ V
$I_5 = 0.06$ A
$R_5 = 600\ \Omega$

$E_1 = 15$ V
$I_1 = 0.1$ A
$R_1 = 150\ \Omega$

$E_4 = 14$ V
$I_4 = 0.1$ A
$R_4 = 140\ \Omega$

$E_7 = 25$ V
$I_7 = 0.1$ A
$R_7 = 250\ \Omega$

$E_3 = 30$ V
$I_3 = 0.04$ A
$R_3 = 750\ \Omega$

$E_6 = 36$ V
$I_6 = 0.04$ A
$R_6 = 900\ \Omega$

$E_T = 120$ V
$I_T = 0.1$ A
$R_T = 1200\ \Omega$

FIGURE 8–8 All values solved for the combination circuit.

calculated using Ohm's law:

$$I_2 = \frac{E_2}{R_2}$$

$$I_2 = \frac{30 \text{ V}}{500 \text{ } \Omega}$$

$$I_2 = 0.06 \text{ A}$$

$$I_3 = \frac{E_3}{R_3}$$

$$I_3 = \frac{30 \text{ V}}{750 \text{ } \Omega}$$

$$I_3 = 0.04 \text{ A}$$

The values of $R_{c(5\&6)}$ can be applied to the parallel block composed of R_5 and R_6. $E_{c(5\&6)}$ is 36 volts. This is the voltage drop across R_5 and R_6. The current flow through these two resistors can be calculated using Ohm's law:

$$I_5 = \frac{E_5}{R_5}$$

$$I_5 = \frac{36 \text{ V}}{600 \text{ } \Omega}$$

$$I_5 = 0.06 \text{ A}$$

$$I_6 = \frac{E_6}{R_6}$$

$$I_6 = \frac{36 \text{ V}}{900 \text{ } \Omega}$$

$$I_6 = 0.04 \text{ A}$$

■ EXAMPLE 8-2

Both of the preceding circuits were solved by first determining which parts of the circuit were in series and which were in parallel. The circuits were then re-duced to a simple series or parallel circuit. This same procedure can be used for

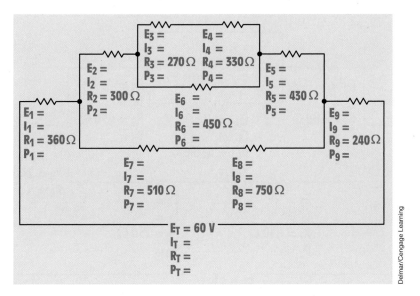

FIGURE 8–9 A complex combination circuit.

any combination circuit. The circuit shown in *Figure 8–9* is reduced to a simpler circuit first. Once the values of the simple circuit are found, they can be placed back in the original circuit to find other values.

Solution

The first step is to reduce the top part of the circuit to a single resistor. This part consists of R_3 and R_4. Because these two resistors are connected in series with each other, their values can be added to form one single resistor. This combination will form R_{c1} *(Figure 8–10):*

$$R_{c1} = R_3 + R_4$$
$$R_{c1} = 270\ \Omega + 330\ \Omega$$
$$R_{c1} = 600\ \Omega$$

The top part of the circuit is now formed by R_{c1} and R_6. These two resistors are in parallel with each other. If their total resistance is calculated, they can be changed into one single resistor with a value of 257.143 Ω. This combination becomes R_{c2} *(Figure 8–11):*

$$R_T = \cfrac{1}{\cfrac{1}{600\ \Omega} + \cfrac{1}{450\ \Omega}}$$
$$R_T = 257.143\ \Omega$$

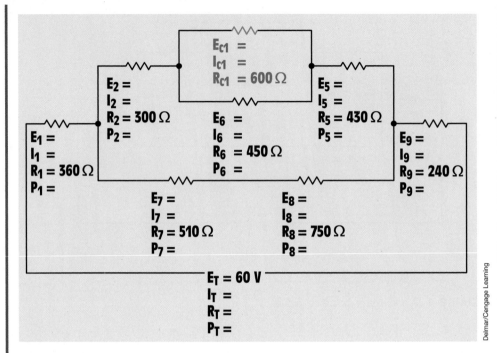

FIGURE 8–10 R_1 and R_2 are combined to form R_{c1}.

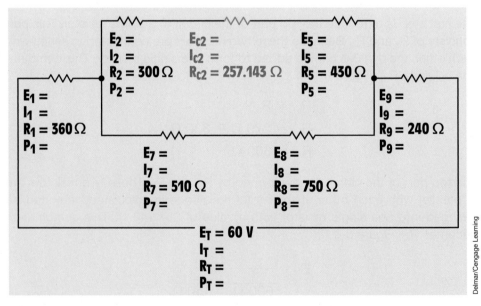

FIGURE 8–11 R_{c1} and R_6 are combined to form R_{c2}.

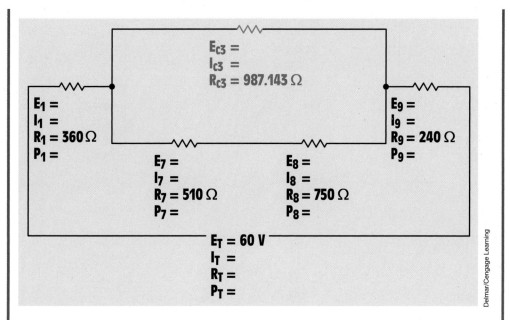

FIGURE 8–12 R_2, R_{c2}, and R_5 are combined to form R_{c3}.

The top of the circuit now consists of R_2, R_{c2}, and R_5. These three resistors are connected in series with each other. They can be combined to form R_{c3} by adding their resistances together *(Figure 8–12)*:

$$R_{c3} = R_2 + R_{c2} + R_5$$
$$R_{c3} = 300\ \Omega + 257.143\ \Omega + 430\ \Omega$$
$$R_{c3} = 987.143\ \Omega$$

R_7 and R_8 are connected in series with each other also. These two resistors are added to form R_{c4} *(Figure 8–13)*:

$$R_{c4} = R_7 + R_8$$
$$R_{c4} = 510\ \Omega + 750\ \Omega$$
$$R_{c4} = 1260\ \Omega$$

R_{c3} and R_{c4} are connected in parallel with each other. Their total resistance can be calculated to form R_{c5} *(Figure 8–14)*:

$$R_{c5} = \cfrac{1}{\cfrac{1}{987.143\ \Omega} + \cfrac{1}{1260\ \Omega}}$$
$$R_{c5} = 553.503\ \Omega$$

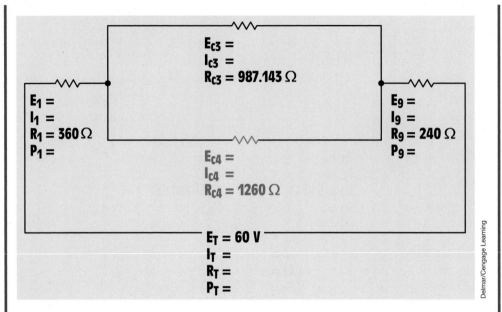

FIGURE 8–13 R_7 and R_8 are combined to form R_{c4}.

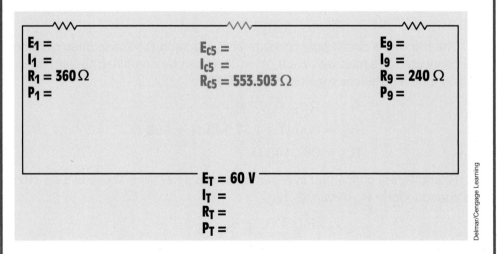

FIGURE 8–14 R_{c3} and R_{c4} are combined to form R_{c5}.

The circuit has now been reduced to a simple series circuit containing three resistors. The total resistance of the circuit can be calculated by adding R_1, R_{c5}, and R_9:

$$R_T = R_1 + R_{c5} + R_9$$
$$R_T = 360 \ \Omega + 553.503 \ \Omega + 240 \ \Omega$$
$$R_T = 1153.503 \ \Omega$$

Now that the total resistance and total voltage are known, the total circuit current and total circuit power can be calculated using Ohm's law:

$$I_T = \frac{E_T}{R_T}$$

$$I_T = \frac{60 \text{ V}}{1153.503 \ \Omega}$$

$$I_T = 0.052 \text{ A}$$

$$P_T = E_T \times I_T$$

$$P_T = 60 \text{ V} \times 0.052 \text{ A}$$

$$P_T = 3.12 \text{ W}$$

Ohm's law can now be used to find the missing values for R_1, R_{c5}, and R_9 (Figure 8–15):

$$E_1 = I_1 \times R_1$$

$$E_1 = 0.052 \text{ A} \times 360 \ \Omega$$

$$E_1 = 18.72 \text{ V}$$

$$P_1 = E_1 \times I_1$$

$$P_1 = 18.72 \text{ V} \times 0.052 \text{ A}$$

$$P_1 = 0.973 \text{ W}$$

$$E_{c5} = I_{c5} \times R_{c5}$$

$$E_{c5} = 0.052 \text{ A} \times 553.503 \ \Omega$$

$$E_{c5} = 28.782 \text{ V}$$

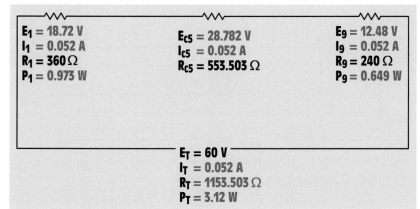

$E_1 = 18.72$ V	$E_{c5} = 28.782$ V	$E_9 = 12.48$ V
$I_1 = 0.052$ A	$I_{c5} = 0.052$ A	$I_9 = 0.052$ A
$R_1 = 360 \ \Omega$	$R_{c5} = 553.503 \ \Omega$	$R_9 = 240 \ \Omega$
$P_1 = 0.973$ W		$P_9 = 0.649$ W

$E_T = 60$ V
$I_T = 0.052$ A
$R_T = 1153.503 \ \Omega$
$P_T = 3.12$ W

FIGURE 8–15   Missing values are found for the first part of the circuit.

$$E_9 = I_9 \times R_9$$

$$E_9 = 0.052 \text{ A} \times 240 \text{ } \Omega$$

$$E_9 = 12.48 \text{ V}$$

$$P_9 = E_9 \times I_9$$

$$P_9 = 12.48 \text{ V} \times 0.052 \text{ A}$$

$$P_9 = 0.649 \text{ W}$$

R_{c5} is actually the combination of R_{c3} and R_{c4}. The values of R_{c5} therefore apply to R_{c3} and R_{c4}. Since these two resistors are connected in parallel with each other, the voltage drop across them will be the same. Each will have the same voltage drop as R_{c5} *(Figure 8–16)*. Ohm's law can now be used to find the remaining values of R_{c3} and R_{c4}:

$$I_{c4} = \frac{E_{c4}}{R_{c4}}$$

$$I_{c4} = \frac{28.782 \text{ V}}{1260 \text{ } \Omega}$$

$$I_{c4} = 0.0228 \text{ A}$$

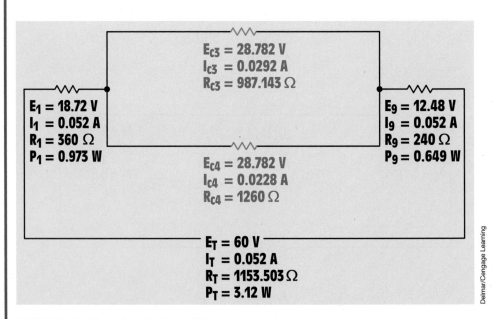

FIGURE 8–16   The values for R_{c3} and R_{c4}.

Delmar/Cengage Learning

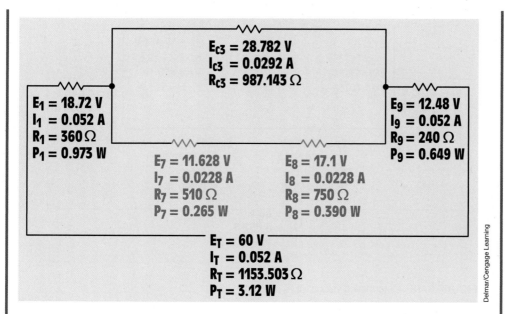

$E_{C3} = 28.782$ V
$I_{C3} = 0.0292$ A
$R_{C3} = 987.143$ Ω

$E_1 = 18.72$ V
$I_1 = 0.052$ A
$R_1 = 360$ Ω
$P_1 = 0.973$ W

$E_9 = 12.48$ V
$I_9 = 0.052$ A
$R_9 = 240$ Ω
$P_9 = 0.649$ W

$E_7 = 11.628$ V
$I_7 = 0.0228$ A
$R_7 = 510$ Ω
$P_7 = 0.265$ W

$E_8 = 17.1$ V
$I_8 = 0.0228$ A
$R_8 = 750$ Ω
$P_8 = 0.390$ W

$E_T = 60$ V
$I_T = 0.052$ A
$R_T = 1153.503$ Ω
$P_T = 3.12$ W

FIGURE 8–17   The values for R_7 and R_8.

Delmar/Cengage Learning

$$I_{c3} = \frac{E_{c3}}{R_{c3}}$$

$$I_{c3} = \frac{28.782 \text{ V}}{987.143 \text{ Ω}}$$

$$I_{c3} = 0.0292 \text{ A}$$

R_{c4} is the combination of R_7 and R_8. The values of R_{c4} apply to R_7 and R_8. Because R_7 and R_8 are connected in series with each other, the current flow will be the same through both *(Figure 8–17)*. Ohm's law can now be used to calculate the remaining values for these two resistors:

$$E_7 = I_7 \times R_7$$
$$E_7 = 0.0228 \text{ A} \times 510 \text{ Ω}$$
$$E_7 = 11.628 \text{ V}$$

$$P_7 = E_7 \times I_7$$
$$P_7 = 11.628 \text{ V} \times 0.0228 \text{ A}$$
$$P_7 = 0.265 \text{ W}$$

$$E_8 = I_8 \times R_8$$
$$E_8 = 0.0228 \text{ A} \times 750 \text{ Ω}$$
$$E_8 = 17.1 \text{ V}$$

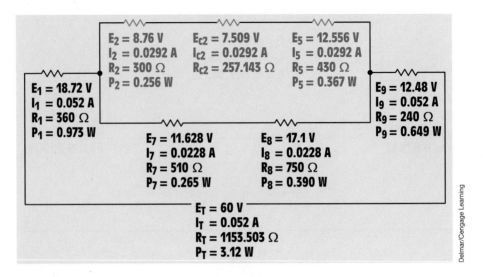

FIGURE 8–18   Determining values for R_1, R_{c2}, and R_5.

$$P_8 = E_8 \times I_8$$
$$P_8 = 17.1 \text{ V} \times 0.0228 \text{ A}$$
$$P_8 = 0.390 \text{ W}$$

R_{c3} is the combination of R_2, R_{c2}, and R_5. Because these resistors are connected in series with each other, the current flow through each will be the same as the current flow through R_{c3}. The remaining values can now be calculated using Ohm's law *(Figure 8–18)*:

$$E_{c2} = I_{c2} \times R_{c2}$$
$$E_{c2} = 0.0292 \text{ A} \times 257.143 \text{ }\Omega$$
$$E_{c2} = 7.509 \text{ V}$$

$$E_2 = I_2 \times R_2$$
$$E_2 = 0.0292 \text{ A} \times 300 \text{ }\Omega$$
$$E_2 = 8.76 \text{ V}$$

$$P_2 = E_2 \times I_2$$
$$P_2 = 8.76 \text{ A} \times 0.292 \text{ A}$$
$$P_2 = 0.256 \text{ W}$$

$$E_5 = I_5 \times R_5$$
$$E_5 = 0.0292 \text{ A} \times 430 \text{ }\Omega$$
$$E_5 = 12.556 \text{ V}$$

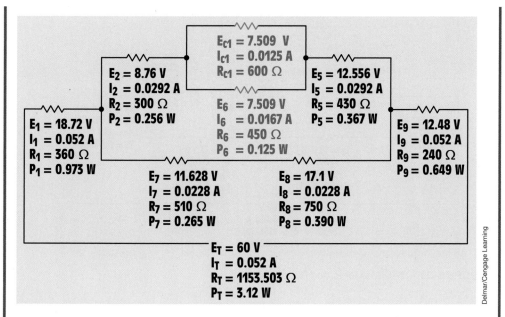

FIGURE 8–19   The values of R_{c1} and R_6.

$$P_5 = E_5 \times I_5$$
$$P_5 = 12.556 \text{ V} \times 0.0292 \text{ A}$$
$$P_5 = 0.367 \text{ W}$$

R_{c2} is the combination of R_{c1} and R_6. R_{c1} and R_6 are connected in parallel and will therefore have the same voltage drop as R_{c2}. Ohm's law can be used to calculate the remaining values for R_{c2} and R_6 *(Figure 8–19)*:

$$I_{c1} = \frac{E_{c1}}{R_{c1}}$$

$$I_{c1} = \frac{7.509 \text{ V}}{600 \text{ } \Omega}$$

$$I_{c1} = 0.0125 \text{ A}$$

$$I_6 = \frac{E_6}{R_6}$$

$$I_6 = \frac{7.509 \text{ V}}{450 \text{ } \Omega}$$

$$I_6 = 0.0167 \text{ A}$$

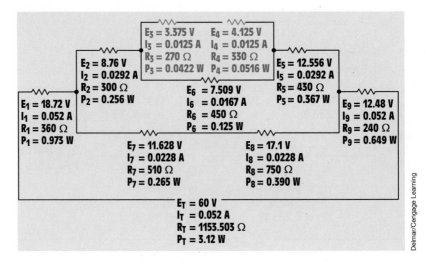

FIGURE 8–20 The values for R_3 and R_4.

$$P_6 = E_6 \times I_6$$
$$P_6 = 7.509 \text{ V} \times 0.0167 \text{ A}$$
$$P_6 = 0.125 \text{ W}$$

R_{c1} is the combination of R_3 and R_4. Because these two resistors are connected in series, the amount of current flow through R_{c1} will be the same as the flow through R_3 and R_4. The remaining values of the circuit can now be found using Ohm's law *(Figure 8–20):*

$$E_3 = I_3 \times R_3$$
$$E_3 = 0.0125 \text{ A} \times 270 \text{ } \Omega$$
$$E_3 = 3.375 \text{ V}$$

$$P_3 = E_3 \times I_3$$
$$P_3 = 3.375 \text{ V} \times 0.0125 \text{ A}$$
$$P_3 = 0.0422 \text{ W}$$

$$E_4 = I_4 \times R_4$$
$$E_4 = 0.0125 \text{ A} \times 330 \text{ } \Omega$$
$$E_4 = 4.125 \text{ V}$$

$$P_4 = E_4 \times I_4$$
$$P_4 = 4.125 \text{ V} \times 0.0125 \text{ A}$$
$$P_4 = 0.0516 \text{ W}$$

Summary

- Combination circuits are circuits that contain both series and parallel branches.

- The three rules for series circuits are as follows:

  a. The current is the same at any point in the circuit.

  b. The total resistance is the sum of the individual resistances.

  c. The applied voltage is equal to the sum of the voltage drops across all the resistors.

- The three rules for parallel circuits are as follows:

  a. The voltage drop across any branch of a parallel circuit is the same as the applied voltage.

  b. The total current is the sum of the individual currents through each path in the circuit (current adds).

  c. The total resistance is the reciprocal of the sum of the reciprocals of the branch resistances.

- When solving combination circuits, it is generally easier if the circuit is reduced to simpler circuits.

Review Questions

1. Refer to *Figure 8–2*. Replace the values shown with the following. Solve for all the unknown values.

   I_T = 0.6 A
   R_1 = 470 Ω
   R_2 = 360 Ω
   R_3 = 510 Ω
   R_4 = 430 Ω

2. Refer to *Figure 8–5*. Replace the values shown with the following. Solve for all the unknown values.

   E_T = 63 V
   R_1 = 1000 Ω
   R_2 = 2200 Ω
   R_3 = 1800 Ω
   R_4 = 910 Ω
   R_5 = 3300 Ω
   R_6 = 4300 Ω
   R_7 = 860 Ω

3. Refer to the circuit shown in *Figure 8–2*. Redraw the circuit and use the following values:

$E_T = 12$ V
$R_1 = 270$ Ω
$R_2 = 510$ Ω
$R_3 = 470$ Ω
$R_4 = 330$ Ω

Assume that an ammeter indicates a total circuit current of 15 mA. A voltmeter indicates the following voltage drops across each resistor:

$E_1 = 12$ V
$E_2 = 0$ V
$E_3 = 7$ V
$E_4 = 5$ V

What is the most likely problem with this circuit?

4. Refer to the circuit shown in *Figure 8–22*. The circuit has an applied voltage of 24 V and the resistors have values as follows:

$R_1 = 1$ kΩ
$R_2 = 300$ Ω
$R_3 = 750$ Ω
$R_4 = 1$ kΩ

An ammeter and a voltmeter indicate the following values:

$I_T = 42.5$ mA
$I_1 = 24$ mA
$E_1 = 24$ V
$I_2 = 18.5$ mA
$E_2 = 5.5$ V
$I_3 = 0$ A
$E_3 = 18.5$ V
$I_4 = 18.5$ mA
$E_4 = 18.5$ V

What is the most likely problem with this circuit?

5. Refer to *Figure 8–21*. Assume that the resistors have the following values:

$R_1 = 150$ Ω
$R_2 = 120$ Ω
$R_3 = 47$ Ω
$R_4 = 220$ Ω

Assume that an ohmmeter connected across the entire circuit indicates a value of 245 Ω. Does this reading indicate that there is a problem with the circuit and, if so, what is the most likely problem?

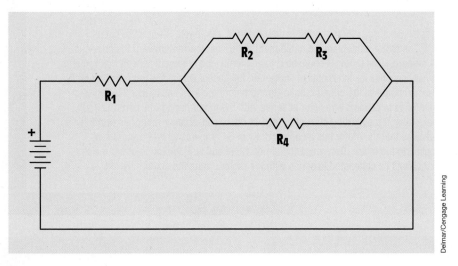

FIGURE 8–21 Series-parallel circuit.

Practical Applications

A circuit contains a 1000-Ω and a 300-Ω resistor connected in series with each other. The total circuit resistance must be reduced to a value of 1150 Ω. Using standard resistance values for 5% resistors discussed in Unit 5, explain how to change the present resistance value to the value desired. ∎

Practical Applications

Two resistors are connected in series. One resistor has a resistance of 1000 Ω, and the other resistor value is not known. The circuit is connected to a 50-V power supply, and there is a current of 25 mA flowing in the circuit. If a 1500-Ω resistor is connected in parallel with the unknown resistor, how much current will flow in this circuit? ∎

Practical Applications

A single-phase electric motor is connected to a 240-volt, 30-ampere circuit. The motor nameplate indicates that the full-load current draw of the motor is 21 amperes. The motor is connected to an inertia load (a large flywheel) that requires several seconds for the motor to accelerate full speed. Due to the long starting time, the motor causes the circuit breaker to trip before the load has reached full speed. Your job is to connect a resistor in series with the motor during the starting period that will limit the starting current to 90% of the circuit-breaker rating. An ammeter indicates that the motor has a current draw of 64 amperes when power is first applied to the motor. Determine the ohmic value and wattage rating of the resistor that should be connected in series with the motor during the starting period. ■

Practical Applications

The hot water for the heating system for a small factory is supplied by a single boiler. The boiler is located in a small building separate from the factory. The power supplied to the boiler is 240-volt single-phase without a neutral conductor. At the present time, the small building housing the boiler has no inside lighting. The business owner desires that four 100-watt lamps be installed inside the building. All lamps are to be connected to a single switch. The lamps are to be connected to the present 240-volt service inside the building. How would you accomplish this task? Explain your answer. ■

Practice Problems

Series-Parallel Circuits

Refer to the circuit shown in *Figure 8–21* to solve the following problems.

1. Find the unknown values in the circuit if the applied voltage is 75 V and the resistors have the following values:

$R_1 = 1.5 \ k\Omega$ $R_2 = 910 \ \Omega$ $R_3 = 2 \ k\Omega$ $R_4 = 3.6 \ k\Omega$

I_T _____ E_1 _____ E_2 _____ E_3 _____ E_4 _____

R_T _____ I_1 _____ I_2 _____ I_3 _____ I_4 _____

2. Find the unknown values in the circuit if the applied voltage is 350 V and the resistors have the following values:

$R_1 = 22\ k\Omega$ $R_2 = 18\ k\Omega$ $R_3 = 12\ k\Omega$ $R_4 = 30\ k\Omega$

I_T _____ E_1 _____ E_2 _____ E_3 _____ E_4 _____

R_T _____ I_1 _____ I_2 _____ I_3 _____ I_4 _____

3. Find the unknown values in the circuit if the applied voltage is 18 V and the resistors have the following values:

$R_1 = 82\ \Omega$ $R_2 = 160\ \Omega$ $R_3 = 220\ \Omega$ $R_4 = 470\ \Omega$

I_T _____ E_1 _____ E_2 _____ E_3 _____ E_4 _____

R_T _____ I_1 _____ I_2 _____ I_3 _____ I_4 _____

Parallel-Series Circuits

Refer to the circuit shown in *Figure 8–22* to solve the following problems.

4. Find the unknown values in the circuit if the total current is 0.8 A and the resistors have the following values:

$R_1 = 1.5\ k\Omega$ $R_2 = 910\ \Omega$ $R_3 = 2\ k\Omega$ $R_4 = 3.6\ k\Omega$

E_T _____ E_1 _____ E_2 _____ E_3 _____ E_4 _____

R_T _____ I_1 _____ I_2 _____ I_3 _____ I_4 _____

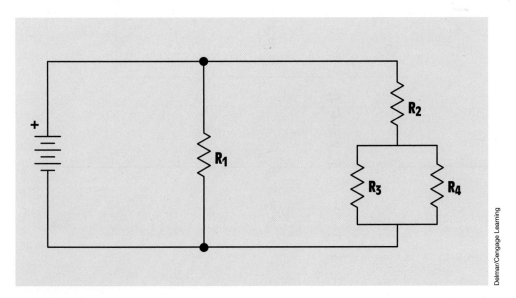

FIGURE 8–22 A parallel-series circuit.

5. Find the unknown values in the circuit if the total current is 0.65 A and the resistors have the following values:

$R_1 = 22 \text{ k}\Omega$ $R_2 = 15 \text{ k}\Omega$ $R_3 = 22 \text{ k}\Omega$ $R_4 = 33 \text{ k}\Omega$

E_T _____ E_1 _____ E_2 _____ E_3 _____ E_4 _____

R_T _____ I_1 _____ I_2 _____ I_3 _____ I_4 _____

6. Find the unknown values in the circuit if the total current is 1.2 A and the resistors have the following values:

$R_1 = 75 \ \Omega$ $R_2 = 47 \ \Omega$ $R_3 = 220 \ \Omega$ $R_4 = 160 \ \Omega$

E_T _____ E_1 _____ E_2 _____ E_3 _____ E_4 _____

R_T _____ I_1 _____ I_2 _____ I_3 _____ I_4 _____

Combination Circuits

Refer to the circuit shown in *Figure 8–23* to solve the following problems.

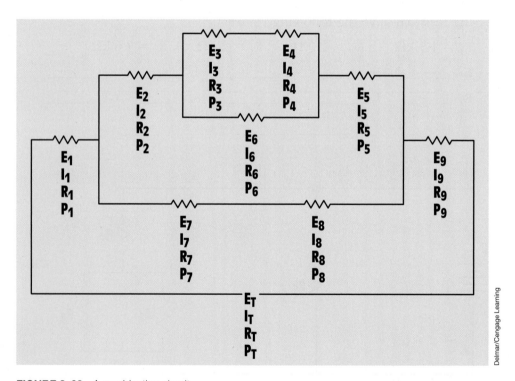

FIGURE 8–23   A combination circuit.

7.

E_T 250 V	E_1	E_2	E_3	E_4
I_T	I_1	I_2	I_3	I_4
R_T	R_1 220 Ω	R_2 500 Ω	R_3 470 Ω	R_4 280 Ω
P_T	P_1	P_2	P_3	P_4

E_5	E_6	E_7	E_8	E_9
I_5	I_6	I_7	I_8	I_9
R_5 400 Ω	R_6 500 Ω	R_7 350 Ω	R_8 450 Ω	R_9 300 Ω
P_5	P_6	P_7	P_8	P_9

8.

E_T	E_1	E_2	E_3	E_4 1.248 V
I_T	I_1	I_2	I_3	I_4
R_T	R_1	R_2	R_3	R_4
P_T 0.576 W	P_1 0.0806 W	P_2 0.0461 W	P_3 0.00184 W	P_4

E_5	E_6	E_7	E_8	E_9
I_5	I_6	I_7	I_8	I_9
R_5	R_6	R_7	R_8	R_9
P_5 0.0203 W	P_6 0.00995 W	P_7 0.0518 W	P_8 0.0726 W	P_9 0.288 W

Unit 9

Kirchhoff's Laws, Thevenin's, Norton's, and Superposition Theorems

Why You Need to Know

*T*here are some circuits that are difficult to solve using Ohm's Law. Circuits that contain more than one power source, for example, can be solved using Kirchhoff's voltage and current laws. Thevenin's and Norton's theorems can be employed to simplify the calculations needed to solve electrical values in certain types of circuits. The super-position theorem combines the principles of Thevenin's and Norton's theorems to simplify circuit calculations. In this unit

- Kirchhoff's voltage and current laws are presented.
- you are shown how to reduce a circuit to a simple Thevenin equivalent circuit that contains a single voltage source and resistor.
- you are shown how to reduce a circuit to a simple Norton equivalent circuit that contains a single current source and resistor.
- you will be shown how to solve a complex circuit using the superposition theorem.

OUTLINE

KEY TERMS

Kirchhoff's current law
Kirchhoff's voltage law
Norton's theorem
Superposition theorem
Thevenin's theorem

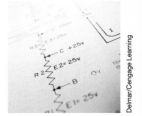

Objectives

After studying this unit, you should be able to

- state Kirchhoff's voltage and current laws.

- solve problems using Kirchhoff's laws.

- discuss Thevenin's theorem.

- find the Thevenin equivalent voltage and resistance values for a circuit network.

- discuss Norton's theorem.

- find the Norton equivalent current and resistance values for a circuit network.

- solve circuits using the superposition theorem.

Preview

Thus far in this text, electrical values for series, parallel, and combination circuits have been calculated using Ohm's law. There are some circuits, however, where Ohm's law either cannot be used or it would be very difficult to use to find unknown values. Some circuits do not have clearly defined series or parallel connections. When this is the case, Kirchhoff's laws are often employed because they can be used to solve any type of circuit. This is especially true for circuits that contain more than one source of power.

Also discussed in this unit are Thevenin's and Norton's theorems. These theorems are used to simplify circuit networks, making it easier to find electrical quantities for different values of load resistances. ■

9–1 Kirchhoff's Laws

Kirchhoff's laws were developed by a German physicist named Gustav R. Kirchhoff. In 1847, Kirchhoff stated two laws for dealing with voltage and current relationships in an electric circuit:

1. ***The algebraic sum of the voltage sources and voltage drops in a closed circuit must equal zero.***

2. ***The algebraic sum of the currents entering and leaving a point must equal zero.***

These two laws are actually two of the rules used for series and parallel circuits discussed earlier in this text. The first law is actually the series circuit rule that states the sum of the voltage drops in a series circuit must equal the applied voltage. The second law is the parallel circuit rule that states that the total current will be the sum of the currents through all the circuit branches.

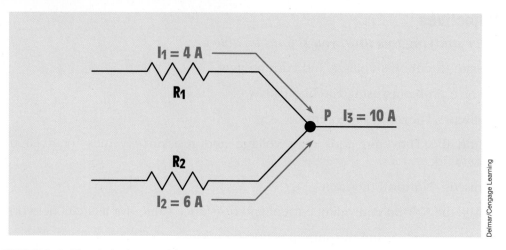

FIGURE 9–1   The algebraic sum of the currents entering and leaving a point must equal zero.

Kirchhoff's Current Law

Kirchhoff's current law states that the algebraic sum of the currents entering and leaving any particular point must equal zero. The proof of this law lies in the fact that if more current entered a particular point than left, some type of charge would have to develop at that point. Consider the circuit shown in *Figure 9–1*. Four amperes of current flow through R_1 to point P, and 6 amperes of current flow through R_2 to point P. The current leaving point P is the sum of the two currents or 10 amperes. Kirchhoff's current law, however, states that the algebraic sum of the currents must equal zero. When using Kirchhoff's law, current entering a point is considered to be positive and current leaving a point is considered to be negative.

$$4\,A + 6\,A - 10\,A = 0\,A$$

Currents I_1 and I_2 are considered to be positive because they enter point P. Current I_3 is negative because it leaves point P.

A second circuit that illustrates Kirchhoff's current law can be seen in *Figure 9–2*. Consider what happens to the current at point B. Two amperes of current flow into point B from R_1. The current splits at point B, part flowing to R_2 and part to R_4 through R_6. The current entering point B is positive, and the two currents leaving point B are negative.

$$I_1 - I_2 - I_4 = 0$$
$$2\,A - 0.8\,A - 1.2\,A = 0\,A$$

Now consider the currents at point E. There is 0.8 ampere of current entering point E from R_2, and 1.2 amperes of current enter point E from R_4 through R_6. Two amperes of current leave point E and flow through R_3.

$$0.8\,A + 1.2\,A - 2\,A = 0\,A$$

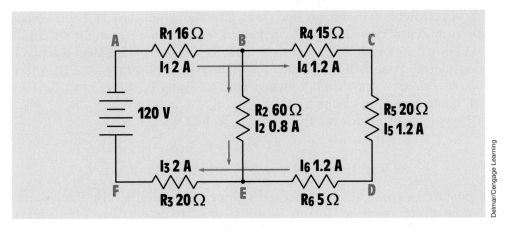

FIGURE 9-2 The current splits to separate branches.

Kirchhoff's Voltage Law

Kirchhoff's voltage law is very similar to the current law in that the algebraic sum of the voltages around any closed loop must equal zero. Before determining the algebraic sum of the voltages, it is necessary to first determine which end of the resistive element is positive and which is negative. To make this determination, assume a direction of current flow and mark the end of the resistive element where current enters and where current leaves. It will be assumed that current flows from negative to positive. Therefore, the point at which current enters a resistor will be marked negative, and the point where current leaves the resistor will be marked positive. The voltage drops and the polarity markings have been added to each of the resistors in the example circuit shown in *Figure 9-2*. The amended circuit is shown in *Figure 9-3*.

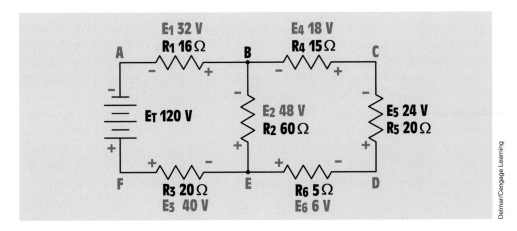

FIGURE 9-3 Marking resistor elements.

To use Kirchhoff's voltage law, start at some point and add the voltage drops around any closed loop. Be certain to return to the starting point. In the circuit shown in *Figure 9–3*, there are actually three separate closed loops to be considered. Loop ACDF contains the voltage drops E_1, E_4, E_5, E_6, E_3, and E_T (120-volt source). Loop ABEF contains voltage drops E_1, E_2, E_3, and E_T. Loop BCDE contains voltage drops E_4, E_5, E_6, and E_2.

The voltage drops for the first loop are as follows:

$$-E_1 - E_4 - E_5 - E_6 - E_3 + E_T = 0$$
$$-32\ V - 18\ V - 24\ V - 6\ V - 40\ V + 120\ V = 0\ V$$

The positive or negative sign for each number is determined by the assumed direction of current flow. In this example, it is assumed that current leaves point A and returns to point A. Current leaving point A enters R_1 at the negative end. Therefore, the voltage is considered to be negative ($-32\ V$). The same is true for R_4, R_5, R_6, and R_3. The current enters the voltage source at the positive end, however. Therefore, E_T is assumed to be positive.

For the second loop, it is assumed that current will leave point A and return to point A through R_1, R_2, R_3, and the voltage source. The voltage drops are as follows:

$$-E_1 - E_2 - E_3 + E_T = 0$$
$$-32\ V - 48\ V - 40\ V + 120\ V = 0\ V$$

The current path for the third loop assumes that current leaves point B and returns to point B. The current will flow through R_4, R_5, R_6, and R_2.

$$-E_4 - E_5 - E_6 + E_2 = 0$$
$$-18\ V - 24\ V - 6\ V + 48\ V = 0\ V$$

Solving Problems with Kirchhoff's Laws

Up to this point, the values of voltage and current have been given to illustrate how Kirchhoff's laws are applied to a circuit. If the values of voltage and current had to be known before Kirchhoff's laws could be used, however, there would be little need for them. In the circuit shown in *Figure 9–4*, the voltage drops and currents are not shown. This circuit contains three resistors and two voltage sources. Although there are three separate loops in this circuit, only two are needed to find the missing values. The two loops used are ABEF and CBED as shown in *Figure 9–4*. The resistors have been marked positive and negative to correspond with the assumed direction of current flow. For the first loop, it is assumed that current leaves point A and returns to point A. The equation for this loop is

$$-E_1 - E_3 + E_{S1} = 0$$
$$-E_1 - E_3 + 60\ V = 0\ V$$

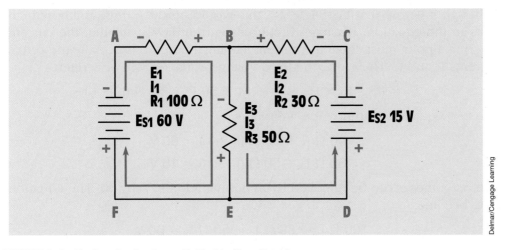

FIGURE 9–4 Finding circuit values with Kirchhoff's voltage law.

For the second loop, it is assumed that current leaves point C and returns to point C. The equation for the second loop is

$$-E_2 - E_3 + E_{S2} = 0$$
$$-E_2 - E_3 + 15\,V = 0\,V$$

To simplify these two equations, the whole numbers are moved to the other side of the equal sign. This is done in the first equation by subtracting 60 volts from both sides. The equation now becomes

$$-E_1 - E_3 = -60\,V$$

In the second equation, 15 volts is subtracted from both sides. The equation becomes

$$-E_2 - E_3 = -15\,V$$

The equations can be further simplified by removing the negative signs. To do this, both equations are multiplied by negative one (-1). The two equations now become

$$E_1 + E_3 = 60\,V$$
$$E_2 + E_3 = 15\,V$$

According to Ohm's law, the voltage drop across any resistive element is equal to the amount of current flowing through the element times its resistance ($E = I \times R$). In order to solve the equations presented, it is necessary to change the values of E_1, E_2, and E_3 to their Ohm's law equivalents:

$$E_1 = I_1 \times R_1 = I_1 \times 100\,\Omega = 100\,\Omega\,I_1$$
$$E_2 = I_2 \times R_2 = I_2 \times 30\,\Omega = 30\,\Omega\,I_2$$

Although it is true that $E_3 = I_3 \times R_3$, this would produce three unknown currents in the equation. Because Kirchhoff's current law states that the currents entering a point must equal the current leaving a point, I_3 is actually the sum of currents I_1 and I_2. Therefore, the third voltage equation will be written

$$E_3 = (I_1 + I_2) \times R_3 = (I_1 + I_2) \times 50 \ \Omega = 50 \ \Omega \ (I_1 + I_2)$$

The two equations can now be written as

$$100 \ \Omega \ I_1 + 50 \ \Omega \ (I_1 + I_2) = 60 \text{ V}$$
$$30 \ \Omega \ I_2 + 50 \ \Omega \ (I_1 + I_2) = 15 \text{ V}$$

The parentheses can be removed by multiplying I_1 and I_2 by 50. The equations now become

$$100 \ \Omega \ I_1 + 50 \ \Omega \ I_1 + 50 \ \Omega \ I_2 = 60 \text{ V}$$
$$30 \ \Omega \ I_2 + 50 \ \Omega \ I_1 + 50 \ \Omega \ I_2 = 15 \text{ V}$$

After gathering terms, the equations become

$$150 \ \Omega \ I_1 + 50 \ \Omega \ I_2 = 60 \text{ V}$$
$$50 \ \Omega \ I_1 + 80 \ \Omega \ I_2 = 15 \text{ V}$$

In order to solve these equations, it is necessary to solve them as simultaneous equations. To solve simultaneous equations, it is necessary to eliminate unknowns until there is only one unknown left. An equation cannot be solved if there is more than one unknown. Multiplying the bottom equation by negative 3 (-3) will result in 50 Ω I_1 becoming -150 Ω I_1. The positive 150 Ω I_1 in the top equation and the negative 150 Ω I_1 now cancel each other, eliminating one unknown.

$$-3(50 \ \Omega \ I_1 + 80 \ \Omega \ I_2 = 15 \text{ V})$$
$$-150 \ \Omega \ I_1 - 240 \ \Omega \ I_2 = -45 \text{ V}$$

The two equations can now be added:

$$150 \ \Omega \ I_1 + 50 \ \Omega \ I_2 = 60 \text{ V}$$
$$-150 \ \Omega \ I_1 - 240 \ \Omega \ I_2 = -45 \text{ V}$$

The positive 150 ohms I_1 and the negative 150 ohms I_1 cancel each other, leaving -190 ohms I_2.

$$-190 \ \Omega \ I_2 = 15 \text{ V}$$

Dividing both sides of the equation by -190 produces the answer for I_2:

$$I_2 = -0.0789 \text{ A}$$

The negative answer for I_2 indicates that the assumed direction of current flow was incorrect. Current actually flows through the circuit as shown in *Figure 9–5*.

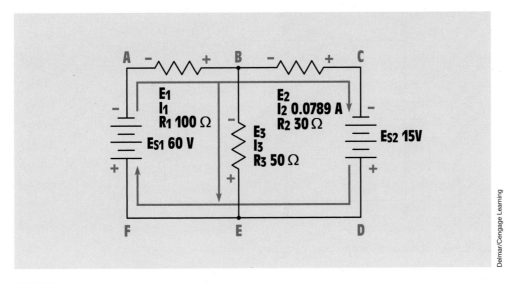

FIGURE 9–5   Actual direction of current flow.

Now that the value of I_2 is known, that answer can be substituted in either of the equations to find I_1.

$$150\ \Omega\ I_1 + 50\ \Omega\ (-0.0789\ A) = 60\ V$$
$$150\ \Omega\ I_1 - 3.945\ A = 60\ V$$

Now add +3.945 to both sides of the equation:

$$150\ \Omega\ I_1 = 63.945\ V$$
$$I_1 = 0.426\ A$$

There is 0.426 ampere leaving point A, flowing through R_1, and entering point B. At point B, 0.0789 ampere branches to point C through R_2 and the remainder of the current branches to point E through R_3. The value for I_3, therefore, can be found by subtracting 0.0789 ampere from 0.426 ampere:

$$I_3 = I_1 - I_2$$
$$I_3 = 0.426\ A - 0.0789\ A$$
$$I_3 = 0.347\ A$$

The voltage drops across each resistor can now be determined using Ohm's law:

$$E_1 = 0.426\ A \times 100\ \Omega$$
$$E_1 = 42.6\ V$$

$$E_2 = 0.0789\ A \times 30\ \Omega$$
$$E_2 = 2.367\ V$$

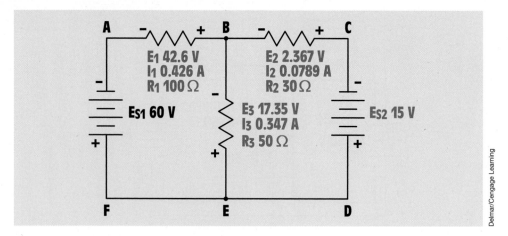

FIGURE 9–6   All circuit values have been calculated.

$$E_3 = 0.347 \text{ A} \times 50 \text{ }\Omega$$
$$E_3 = 17.35 \text{ V}$$

The values for the entire circuit are shown in *Figure 9–6*. To check the answers, add the voltages around the loops. The answers should total zero. Loop BCDE is added first.

$$-E_2 - E_{S2} + E_3 = 0$$
$$-2.367 \text{ V} - 15 \text{ V} + 17.35 \text{ V} = -0.017 \text{ V}$$

(The slight negative voltage in the answer is caused by rounding off values.) The second loop checked is ABEF:

$$-E_1 - E_3 + E_{S1} = 0$$
$$-42.6 \text{ V} - 17.35 \text{ V} + 60 \text{ V} = 0.05 \text{ V}$$

The third loop checked is ACDF:

$$-E_1 - E_2 - E_{S2} + E_{S1} = 0$$
$$-42.6 \text{ V} - 2.367 \text{ V} - 15 \text{ V} + 60 = 0.033 \text{ V}$$

9–2   Thevenin's Theorem

Thevenin's theorem was developed by a French engineer named M. L. Thevenin. It is *used to simplify a circuit network into an equivalent circuit, which contains a single voltage source and series resistor* (*Figure 9–7*). Imagine a black box that contains an unknown circuit and two

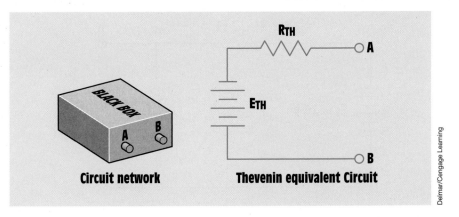

FIGURE 9–7 Thevenin's theorem reduces a circuit network to a single power source and a single series resistor.

output terminals labeled A and B. The output terminals exhibit some amount of voltage and some amount of internal impedance.

Thevenin's theorem reduces the circuit inside the black box to a single source of power and a series resistor equivalent to the internal impedance. The equivalent Thevenin circuit assumes the output voltage to be the open circuit voltage with no load connected. The equivalent Thevenin resistance is the open circuit resistance with no power source connected. Imagine the Thevenin circuit shown in *Figure 9–7* with the power source removed and a single conductor between Terminal B and the equivalent resistor *(Figure 9–8)*. If an ohmmeter were to be connected across Terminals A and B, the equivalent Thevenin resistance would be measured.

Calculating the Thevenin Values

The circuit shown in *Figure 9–9* has a single power source of 24 volts and two resistors connected in series. R_1 has a value of 2 ohms, and R_2 has

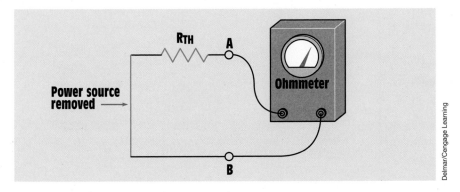

FIGURE 9–8 Equivalent Thevenin resistance.

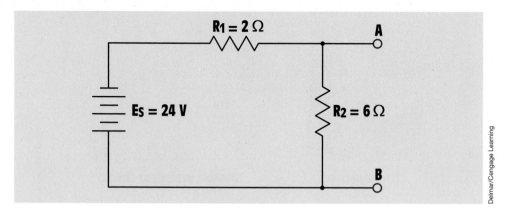

FIGURE 9–9 Determining the Thevenin equivalent circuit.

a value of 6 ohms. The Thevenin equivalent circuit is calculated across Terminals A and B. To do this, determine the voltage drop across R_2 because it is connected directly across Terminals A and B. The voltage drop across R_2 is the open circuit voltage of the equivalent Thevenin circuit when no load is connected across Terminals A and B. Because R_1 and R_2 form a series circuit, a total of 8 ohms are connected to the 24-volt power source. This produces a current flow of 3 amperes through R_1 and R_2 (24 V/8 Ω = 3 A). Because 3 amperes of current flow through R_2, a voltage drop of 18 volts will appear across it (3 A $\times$ 6 Ω = 18 V). The equivalent Thevenin voltage for this circuit is 18 volts.

To determine the equivalent Thevenin resistance, disconnect the power source and replace it with a conductor *(Figure 9–10)*. In this circuit, R_1 and

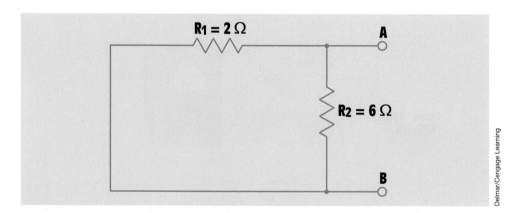

FIGURE 9–10 Determining the Thevenin equivalent resistance.

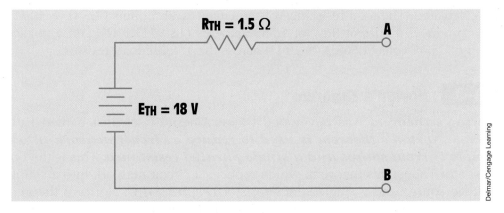

FIGURE 9–11 The Thevenin equivalent circuit.

R_2 are connected in parallel with each other. The total resistance can now be determined using one of the formulas for finding parallel resistance:

$$R_T = \frac{R_1 \times R_2}{R_1 + R_2}$$

$$R_T = \frac{2\ \Omega \times 6\ \Omega}{2\ \Omega + 6\ \Omega}$$

$$R_T = 1.5\ \Omega$$

The Thevenin equivalent circuit is shown in *Figure 9–11*.

Now that the Thevenin equivalent of the circuit is known, the voltage and current values for different load resistances can be quickly calculated. Assume, for example, that a load resistance of 10 ohms is connected across Terminals A and B *(Figure 9–12)*. The voltage and current values for the circuit can now easily

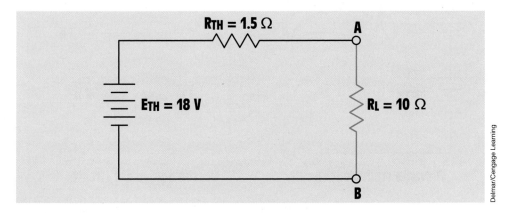

FIGURE 9–12 A 10-ohm load resistor is connected across Terminals A and B.

be calculated. The total resistance of the circuit is 11.5 ohms (1.5 Ω + 10 Ω). This produces a current flow of 1.565 amperes (18 V/11.5 Ω), and a voltage drop of 15.65 volts (1.565 A $\times$ 10 Ω) across the 10-ohm load resistor.

9–3   Norton's Theorem

Norton's theorem was developed by an American scientist named E. L. Norton. ***Norton's theorem is used to reduce a circuit network into a simple current source and a single parallel resistance.*** This is the opposite of Thevenin's theorem, which reduces a circuit network into a simple voltage source and a single series resistor *(Figure 9–13)*. Norton's theorem assumes a source of current that is divided among parallel branches. A source of current is often easier to work with, especially when calculating values for parallel circuits, than a voltage source, which drops voltages across series elements.

Current Sources

Power sources can be represented in one of two ways, as a voltage source or as a current source. Voltage sources are generally shown as a battery with a resistance connected in series with the circuit to represent the internal resistance of the source. This is the case when using Thevenin's theorem. Voltage sources are rated with some amount of voltage such as 12 volts, 24 volts, and so on.

Power sources can also be represented by a current source connected to a parallel resistance that delivers a certain amount of current such as 1 ampere, 2 amperes, 3 amperes, and so on. Assume that a current source is rated at

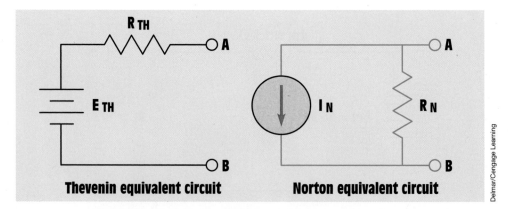

Thevenin equivalent circuit   Norton equivalent circuit

FIGURE 9–13   The Thevenin equivalent circuit contains a voltage source and series resistance. The Norton equivalent circuit contains a current source and parallel resistance.

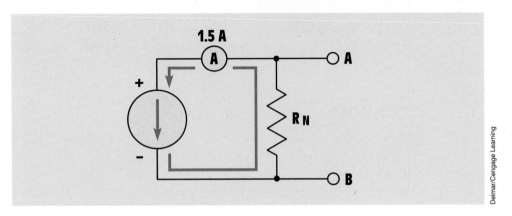

FIGURE 9–14   The current source supplies a continuous 1.5 amperes.

1.5 amperes *(Figure 9–14)*. This means that 1.5 amperes will flow from the power source regardless of the circuit connected. In the circuit shown in *Figure 9–14*, 1.5 amperes flow through R_N.

Determining the Norton Equivalent Circuit

The same circuit used previously to illustrate Thevenin's theorem is used to illustrate Norton's theorem. Refer to the circuit shown in *Figure 9–15*. In this basic circuit, a 2-ohm and a 6-ohm resistor are connected in series with a 24-volt power source. To determine the Norton equivalent of this circuit, imagine a short circuit placed across Terminals A and B *(Figure 9–16)*. Because this places the short circuit directly across R_2, that resistance is eliminated from the circuit and a resistance of 2 ohms is left connected in series with the voltage

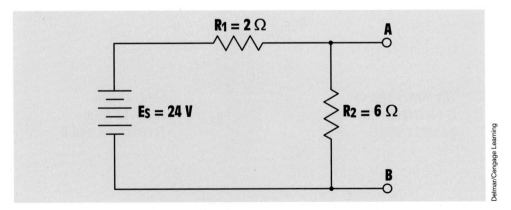

FIGURE 9–15   Determining the Norton equivalent circuit.

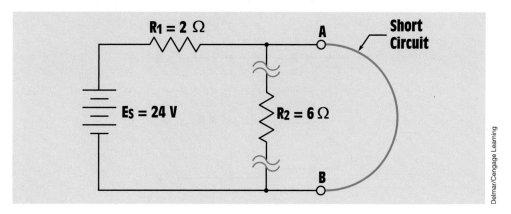

FIGURE 9–16 Shorting Terminals A and B eliminates the 6-ohm resistor.

source. The next step is to determine the amount of current that can flow through this circuit. This current value is known as I_N:

$$I_N = \frac{E_s}{R_1}$$

$$I_N = \frac{24\ V}{2\ \Omega}$$

$$I_N = 12\ A$$

I_N, or 12 amperes, is the amount of current available in the Norton equivalent circuit.

The next step is to find the equivalent parallel resistance, R_N, connected across the current source. To do this, remove the short circuit across Terminals A and B. Now replace the power source with a short circuit just as was done in determining the Thevenin equivalent circuit *(Figure 9–17)*. The circuit

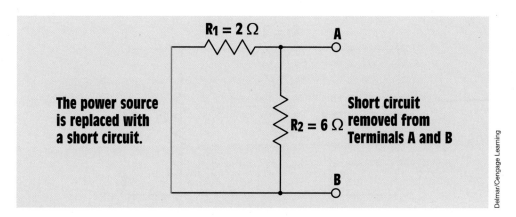

FIGURE 9–17 Determining the Norton equivalent resistance.

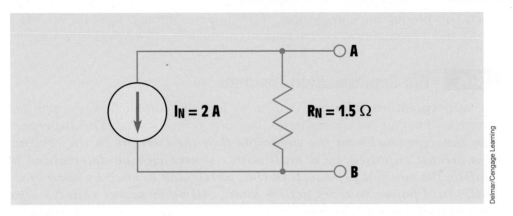

FIGURE 9–18   Equivalent Norton circuit.

now has a 2-ohm and a 6-ohm resistor connected in parallel. This produces an equivalent resistance of 1.5 ohms. The Norton equivalent circuit, shown in *Figure 9–18,* is a 1.5-ohm resistor connected in parallel with a 12-ampere current source.

Now that the Norton equivalent for the circuit has been calculated, any value of resistance can be connected across Terminals A and B and the electrical values calculated quickly. Assume that a 6-ohm load resistor, R_L, is connected across Terminals A and B *(Figure 9–19)*. The 6-ohm load resistor is connected in parallel with the Norton equivalent resistance of 1.5 ohms. This produces a total resistance of 1.2 ohms for the circuit. In a Norton equivalent circuit, it is assumed that the Norton equivalent current, I_N, flows at all times. In this circuit, the Norton equivalent current is 12 amperes. Therefore, a current of 12 amperes flows through the 1.2-ohm resistance. This produces a voltage drop of 14.4 volts across the resistance ($E = 12\text{ A} \times 1.2\ \Omega$). Because the resistors shown in *Figure 9–19* are connected in parallel, 14.4 volts are dropped across

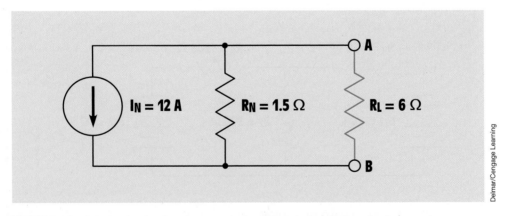

FIGURE 9–19   A 6-ohm load resistor is connected to the equivalent Norton circuit.

each. This produces a current flow of 9.6 amperes through R_N (14.4 V/1.5 Ω = 9.6 A) and a current flow of 2.4 amperes through R_L (14.4 V/6 Ω = 2.4 A).

9-4 The Superposition Theorem

The **superposition theorem** is used to find the current flow through any branch of a circuit containing more than one power source. *The superposition theorem works on the principle that the current in any branch of a circuit supplied by a multipower source can be determined by finding the current produced in that particular branch by each of the individual power sources acting alone. All other power sources must be replaced by a resistance equivalent to their internal resistances. The total current flow through the branch is the algebraic sum of the individual currents produced by each of the power sources.*

■ EXAMPLE 9-1

An example circuit is shown in *Figure 9–20*. In this example, the circuit contains two voltage sources. The amount of current flowing through R_2 is determined using the superposition theorem. The circuit can be solved by following a procedure through several distinct steps.

Solution

■ STEP 1

Reduce all but one of the voltage sources to zero by replacing it with a short circuit, leaving any internal series resistance. Reduce the current source to zero

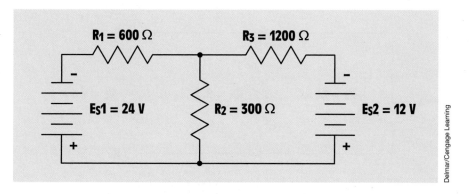

FIGURE 9–20   A circuit with two power sources.

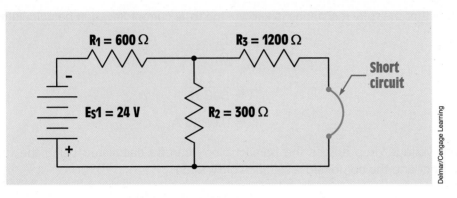

FIGURE 9–21   Voltage source E_S2 is replaced with a short circuit.

by replacing it with an open circuit, leaving any internal parallel resistance. Voltage source E_S2 will be shorted *(Figure 9–21)*. The circuit now exists as a simple combination circuit with R_1 connected in series with R_2 and R_3, which are in parallel with each other *(Figure 9–22)*. The total resistance of this circuit can now be found by finding the total resistance of the two resistors connected in parallel, R_2 and R_3, and adding them to R_1:

$$R_T = R_1 + \left(\cfrac{1}{\cfrac{1}{R_2} + \cfrac{1}{R_3}} \right)$$

$$R_T = 600 \ \Omega + \left(\cfrac{1}{\cfrac{1}{300 \ \Omega} + \cfrac{1}{1200 \ \Omega}} \right)$$

$$R_T = 840 \ \Omega$$

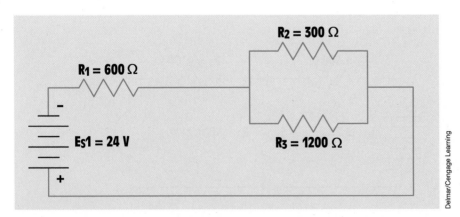

FIGURE 9–22   The circuit is reduced to a simple combination circuit.

Now that the total resistance is known, the total current flow can be calculated:

$$I_T = \frac{E_T}{R_T}$$

$$I_T = \frac{24\ V}{840\ \Omega}$$

$$I_T = 0.0286\ A$$

The voltage drop across the parallel block can be calculated using the total current and the combined resistance of R_2 and R_3:

$$E_{COMBINATION} = 0.0286\ A \times 240\ \Omega$$

$$E_{COMBINATION} = 6.864\ V$$

The current flowing through R_2 can now be calculated:

$$I_2 = \frac{6.864\ V}{300\ \Omega}$$

$$I_2 = 0.0229\ A$$

Notice that the current is flowing through R_2 in the direction of the arrow in *Figure 9–23*.

■ STEP 2

Find the current flow through R_2 by shorting voltage source E_S1. Power is now supplied by voltage source E_S2 *(Figure 9–24)*. In this circuit, R_1 and R_2 are in parallel with each other. R_3 is in series with R_1 and R_2.

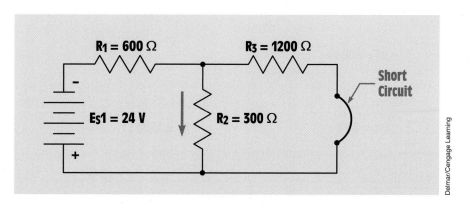

FIGURE 9–23   Current flows through the resistor in the direction shown.

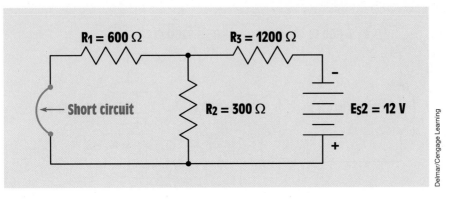

FIGURE 9–24 Voltage source E_S2 is shorted.

$$R_T = R_3 + \left(\cfrac{1}{\cfrac{1}{R_1} + \cfrac{1}{R_2}} \right)$$

$$R_T = 1200\ \Omega + 200\ \Omega$$

$$R_T = 1400\ \Omega$$

The total current flow in the circuit can now be determined:

$$I_T = \frac{E_T}{R_T}$$

$$I_T = \frac{12\ V}{1400\ \Omega}$$

$$I_T = 0.00857\ A$$

The amount of voltage drop across the parallel combination can now be calculated:

$$E_C = 0.00857\ A \times 200\ \Omega$$

$$E_C = 1.714\ V$$

The amount of current flow through R_2 can now be calculated using Ohm's law:

$$I_2 = \frac{1.714\ V}{300\ \Omega}$$

$$I_2 = 0.00571\ A$$

Notice that the current flowing through R_2 is in the same direction as in the previous circuit (Figure 9–25).

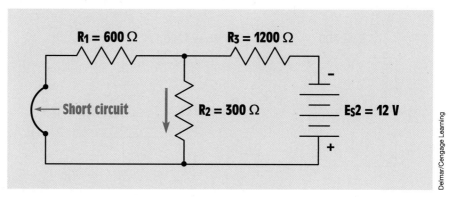

FIGURE 9–25 Current flows in the same direction.

■ STEP 3

The next step is to find the algebraic sum of the two currents. Because both currents flow through R_2 in the same direction, the two currents are added:

$$I_{2(TOTAL)} = 0.0229 \text{ A} + 0.00571 \text{ A}$$
$$I_{2(TOTAL)} = 0.0286 \text{ A}$$

■ EXAMPLE 9-2

Example circuit 2 is shown in *Figure 9–26*. This circuit contains a voltage source of 30 V and a current source of 0.2 A. The amount of current flowing through R_2 will be calculated.

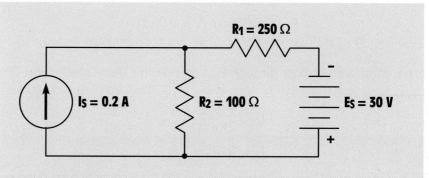

FIGURE 9–26 Example circuit 2 contains a current source and a voltage source.

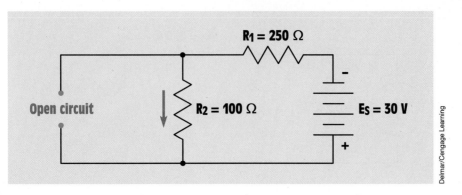

FIGURE 9–27   The current source is replaced with an open circuit.

Solution

▪ STEP 1

The first step is to find the current flow through R_2 using the voltage source only. This is done by replacing the current source with an open circuit *(Figure 9–27)*. Notice the direction of current flow through R_2.

When the current source is replaced with an open circuit, R_1 and R_2 become connected in series with each other. The total resistance is the sum of the two resistances:

$$R_T = R_1 + R_2$$
$$R_T = 250\ \Omega + 100\ \Omega$$
$$R_T = 350\ \Omega$$

Now that the total resistance is known, the total current flow in the circuit can be found using Ohm's law:

$$I_T = \frac{30\ V}{350\ \Omega}$$
$$I_T = 0.0857\ A$$

Because the current flow must be the same in all points of a series circuit, the same amount of current flows through resistor R_2:

$$I_T = I_2$$
$$I_2 = 0.0857\ A$$

▪ STEP 2

The next step is to find the amount of current flow through R_2 that would be supplied by the current source only. This can be done by replacing the voltage

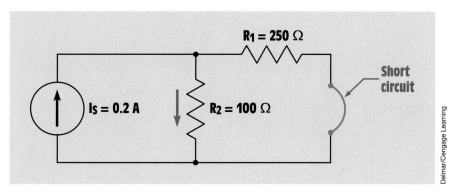

FIGURE 9–28   The voltage source is replaced with a short circuit.

source with a short circuit *(Figure 9–28)*. Notice the direction of current flow through R_2.

When the voltage source is removed and replaced with a short circuit, R_1 and R_2 become connected in parallel with each other. The current flow through R_2 will be calculated using the current divider formula:

$$I_2 = \left(\frac{R_1}{R_1 + R_2}\right) \times I_S$$

$$I_2 = \frac{250\ \Omega}{350\ \Omega} \times 0.2\ A$$

$$I_2 = 0.143\ A$$

■ STEP 3

The total amount of current flow through R_2 can now be determined by finding the algebraic sum of both currents:

$$I_{2(TOTAL)} = 0.0857\ A + 0.143\ A$$

$$I_{2(TOTAL)} = 0.229\ A$$

■ EXAMPLE 9-3

In the third example, a circuit contains two resistors and two current sources *(Figure 9–29)*. The amount of current flowing through R_2 will be determined.

■ STEP 1

Remove one of the current sources from the circuit and replace it with an open circuit. In this example, current source $I_S 2$ is removed first *(Figure 9–30)*. R_1 is

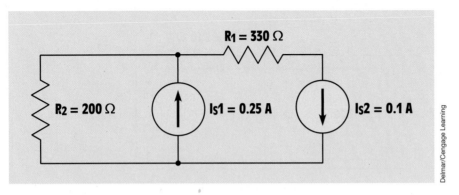

FIGURE 9–29 The circuit contains two current sources.

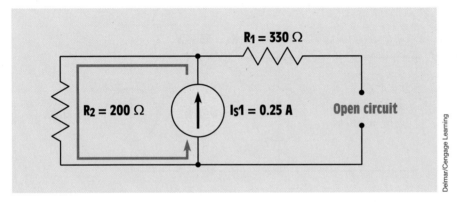

FIGURE 9–30 Current source I_s2 is replaced with an open circuit.

now removed from the circuit and the entire 0.25 A of current flows through R_2. Notice the direction of current flow through the resistor.

■ STEP 2
The next step is to replace current source I_s1 with an open circuit and determine the amount and direction of current flow through resistor R_2 produced by current source I_s2 *(Figure 9–31)*.

When I_s1 is replaced with an open circuit, R_1 and R_2 become connected in series with each other. Because the current is the same in a series circuit, both resistors have a current flow of 0.1 A through them.

■ STEP 3
Now that the amount and the direction of current flow through R_2 for both current sources are known, the total current flow can be determined by adding the two currents together. Because the currents flow in opposite directions, the

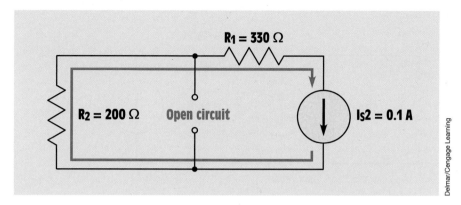

FIGURE 9–31 Current source I_S1 is replaced with an open circuit.

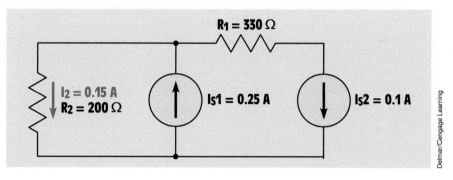

FIGURE 9–32 The amount and the direction of current flow through R_2 have been determined.

algebraic sum will be the difference of the two currents and the direction will be determined by the greater *(Figure 9–32)*.

$$I_{2(TOTAL)} = 0.25\ A - 0.1\ A$$
$$I_{2(TOTAL)} = 0.15\ A$$

Summary

- Kirchhoff's laws can be used to solve any type of circuit.
- Kirchhoff's voltage law states that the algebraic sum of the voltage drops and voltage sources around any closed path must equal zero.
- Kirchhoff's current law states that the algebraic sum of the currents entering and leaving any point must equal zero.

- Kirchhoff's laws can be used to solve unknown values for circuits that contain more than one power source.

- When using Kirchhoff's laws it is generally necessary to solve simultaneous equations.

- Thevenin's theorem involves reducing a circuit network to a simple voltage source and series resistance.

- The Thevenin equivalent voltage is the open circuit voltage across two points.

- To determine the Thevenin equivalent resistance, replace the voltage source with a short circuit.

- Norton's theorem involves reducing a circuit network to a current source and parallel resistance.

- The Norton equivalent current is determined by shorting the output terminals.

- The Norton equivalent resistance is determined by replacing the current source with a short circuit.

- The superposition theorem is used to find the current flow through any branch of a circuit containing more than one power source.

Review Questions

1. State Kirchhoff's voltage law.

2. State Kirchhoff's current law.

3. What is the purpose of Thevenin's and Norton's theorems?

4. When using Kirchhoff's current law, do the currents entering a point carry a positive or negative sign?

5. When using Kirchhoff's current law, if a negative answer is found for current flow, what does this indicate?

To answer Questions 6 through 10, refer to the circuit shown in *Figure 9–33*.

6. How much current flows through R_1?

$I_1 = $ _____

7. How much current flows through R_2?

$I_2 = $ _____

8. How much voltage is dropped across R_1?

$E_1 = $ _____

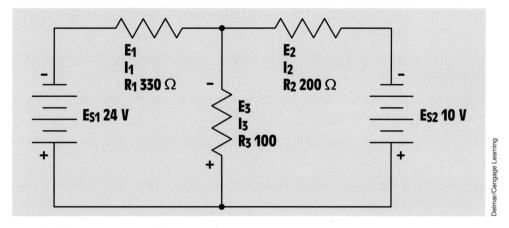

FIGURE 9–33 Review Questions 6 through 10.

9. How much voltage is dropped across R_2?

$E_2 = $ _____

10. How much voltage is dropped across R_3?

$E_3 = $ _____

To answer Questions 11 through 15, refer to the circuit shown in *Figure 9–34*.

11. In the circuit shown in *Figure 9–34*, assume that R_1 has a resistance of 4 Ω and R_2 has a resistance of 20 Ω. Battery E_S has a voltage of 48 V. What is the Thevenin equivalent voltage for this circuit across Terminals A and B?

$E_{THEV} = $ _____

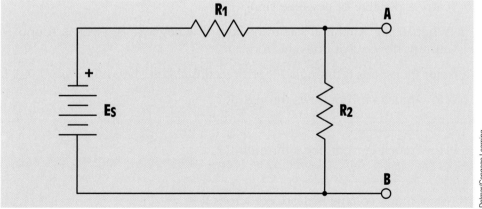

FIGURE 9–34 Review Questions 11 through 15.

12. What is the equivalent Thevenin resistance for the circuit described in Question 11?

R_{THEV} _____ Ω

13. In the circuit shown in *Figure 9–34*, assume that R_1 has a resistance of 2.5 Ω and R_2 has a resistance of 16 Ω. Power source E_S has a voltage of 20 V. As measured across Terminals A and B, what would be the equivalent Norton current for this circuit?

$I_{NORTON} =$ _____

14. What is the equivalent Norton resistance for the circuit described in Question 13?

$R_{NORTON} =$ _____ Ω

15. Assume that an 8-Ω load resistance is connected across Terminals A and B for the circuit described in Question 13. How much current will flow through the load resistance?

$I_{LOAD} =$ _____

Practice Problems

To solve the following Kirchhoff's laws problems, refer to the circuit shown in *Figure 9–35*.

1. $E_{S1} = 12$ V $E_1 =$ _____ $E_2 =$ _____ $E_3 =$ _____

   $E_{S2} = 32$ V $I_1 =$ _____ $I_2 =$ _____ $I_3 =$ _____

   $R_1 = 680$ Ω $R_2 = 1000$ Ω $R_3 = 500$ Ω

2. $E_{S1} = 3$ V $E_1 =$ _____ $E_2 =$ _____ $E_3 =$ _____

   $E_{S2} = 1.5$ V $I_1 =$ _____ $I_2 =$ _____ $I_3 =$ _____

   $R_1 = 200$ Ω $R_2 = 120$ Ω $R_3 = 100$ Ω

3. $E_{S1} = 6$ V $E_1 =$ _____ $E_2 =$ _____ $E_3 =$ _____

   $E_{S2} = 60$ V $I_1 =$ _____ $I_2 =$ _____ $I_3 =$ _____

   $R_1 = 1.6$ kΩ $R_2 = 1.2$ kΩ $R_3 = 2.4$ kΩ

To answer the following questions, refer to the circuit shown in *Figure 9–34*.

4. Find the Thevenin equivalent voltage and resistance across Terminals A and B.

$E_S = 32$ V $R_1 = 4$ Ω $R_2 = 6$ Ω $E_{TH} =$ _____ $R_{TH} =$ _____

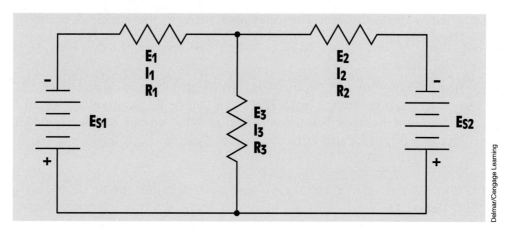

FIGURE 9–35 Kirchhoff's law practice problem circuit.

5. $E_S = 18$ V $R_1 = 2.5\ \Omega$ $R_2 = 12\ \Omega$ $E_{TH} =$ _____ $R_{TH} =$ _____

6. Find the Norton equivalent current and resistance across Terminals A and B.

   $E_S = 10$ V $R_1 = 3\ \Omega$ $R_2 = 7\ \Omega$ $I_N =$ _____ $R_N =$ _____

7. $E_S = 48$ V $R_1 = 12\ \Omega$ $R_2 = 64\ \Omega$ $I_N =$ _____ $R_N =$ _____

Meters and Wire Sizes

Unit 10
Measuring Instruments

KEY TERMS

Ammeter

Ammeter shunt

Analog meters

Ayrton shunt

Bridge circuit

Clamp-on ammeter

Current transformer

D'Arsonval
   movement

Galvanometers

Moving-coil meter

Multirange
   ammeters

Multirange
   voltmeters

Ohmmeter

Oscilloscope

Voltmeter

Wattmeter

Wheatstone bridge

Why You Need to Know

*M*easuring instruments are the eyes of the electrician. An understanding of how measuring instruments operate is very important to anyone working in the electrical field. They provide the electrician with the ability to evaluate problems on the job through the use of technical tools. They also enable an electrician to correctly determine electrical values of voltage, current, resistance, power, and many others. In this unit you will learn

- how different types of meters are constructed. This knowledge will aid you in knowing how the meters are to be connected in a circuit and how to interpret their readings.
- the differences between analog and digital meters, as well as how to interpret the waveforms shown on an oscilloscope. Oscilloscopes must be used to observe specific voltage patterns that would render a conventional voltmeter useless.

Objectives

After studying this unit, you should be able to

- discuss the operation of a d'Arsonval meter movement.
- discuss the operation of a moving-iron type of movement.
- connect a voltmeter to a circuit.
- connect and read an analog multimeter.
- connect an ammeter to a circuit.
- measure resistance using an ohmmeter.
- interpret waveforms shown on the display of an oscilloscope.
- connect a wattmeter into a circuit.

Preview

Anyone desiring to work in the electrical and electronics field must become proficient with the common instruments used to measure electrical quantities. These instruments are the voltmeter, the ammeter, and the ohmmeter. Without meters, it would be impossible to make meaningful interpretations of what is happening in a circuit. Meters can be divided into two general types: analog and digital. ■

10–1   Analog Meters

Analog meters are characterized by the fact that they use a pointer and scale to indicate their value *(Figure 10–1)*. There are different types of analog meter movements. One of the most common is the **d'Arsonval movement** shown in *Figure 10–2*. This type of movement is often referred to as a **moving-coil meter.** A coil of wire is suspended between the poles of a permanent magnet. The coil is suspended either by jeweled movements similar to those used in watches or by taut bands. The taut-band type offers less turning friction than the jeweled movement. These meters can be made to operate on very small amounts of current and often are referred to as **galvanometers.**

Principle of Operation

Analog meters operate on the principle that like magnetic poles repel each other. As current passes through the coil, a magnetic field is created around the coil. The direction of current flow through the meter is such that the same polarity of magnetic pole is created around the coil as that of the permanent magnet. This like polarity causes the coil to be deflected away from the pole

Courtesy of Simpson Electric

FIGURE 10–1 An analog meter.

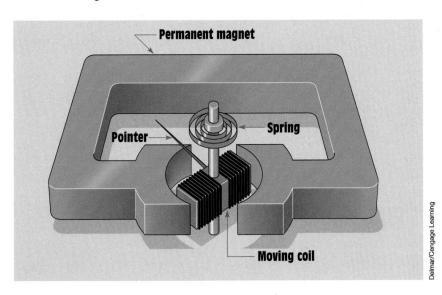

Delmar/Cengage Learning

FIGURE 10–2 Basic d'Arsonval meter movement.

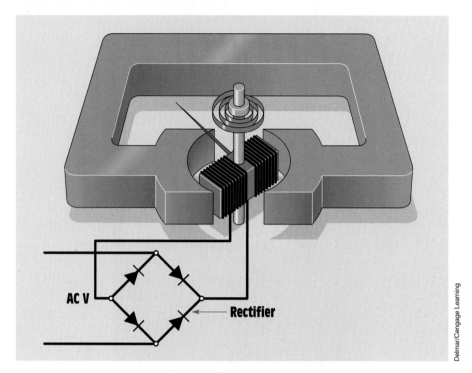

FIGURE 10–3   Rectifier changes AC voltage into DC voltage.

of the magnet. A spring is used to retard the turning of the coil. The distance the coil turns against the spring is proportional to the strength of the magnetic field developed in the coil. If a pointer is added to the coil and a scale is placed behind the pointer, a meter movement is created.

Because the turning force of this meter depends on the repulsion of magnetic fields, it will operate on DC only. If AC is connected to the moving coil, the magnetic polarity will change 60 times per second and the net turning force will be zero. For this reason, a DC voltmeter will indicate zero if connected to an AC line. When this type of movement is to be used to measure AC values, the current must be rectified, or changed into DC, before it is applied to the meter *(Figure 10–3)*.

10–2   The Voltmeter

The **voltmeter** is designed to be connected directly across the source of power. *Figure 10–4* shows a voltmeter being used to test the voltage of a battery. Notice that the leads of the meter are connected directly across the source of voltage. A voltmeter can be connected directly across the power

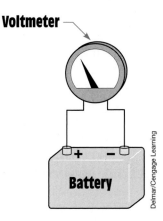

Voltmeter

Battery

Delmar/Cengage Learning

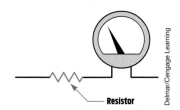

Resistor

Delmar/Cengage Learning

FIGURE 10–4   A voltmeter connects directly across the power source.

FIGURE 10–5   A resistor connects in series with the meter.

source because it has a very high resistance connected in series with the meter movement *(Figure 10–5)*. The industrial standard for a voltmeter is 20,000 ohms per volt for DC and 5000 ohms per volt for AC. Assume the voltmeter shown in *Figure 10–5* is an AC meter and has a full-scale range of 300 volts. The meter circuit (meter plus resistor) would therefore have a resistance of 1,500,000 ohms (300 V × 5000 Ω per volt = 1,500,000 Ω).

Calculating the Resistor Value

Before the resistor value can be calculated, the operating characteristics of the meter must be known. It will be assumed that the meter requires a current of 50 microamperes and a voltage of 1 volt to deflect the pointer full scale. These are known as the *full-scale values* of the meter.

When the meter and resistor are connected to a source of voltage, their combined voltage drop must be 300 volts. Because the meter has a voltage drop of 1 volt, the resistor must have a drop of 299 volts. The resistor and meter are connected in series with each other. In a series circuit, the current flow must be the same in all parts of the circuit. If 50 microamperes of current flow are required to deflect the meter full scale, then the resistor must have a current of 50 microamperes flowing through it when it has a voltage drop of 299 volts. The value of resistance can now be calculated using Ohm's law:

$$R = \frac{E}{I}$$

$$R = \frac{299 \text{ V}}{0.000050 \text{ A}}$$

$$R = 5.98 \text{ M}\Omega \ (5,980,000 \ \Omega)$$

10–3   Multirange Voltmeters

Most voltmeters are **multirange voltmeters,** which means that they are designed to use one meter movement to measure several ranges of voltage. For example, one meter may have a selector switch that permits full-scale ranges to be selected. These ranges may be 3 volts full scale, 12 volts full scale, 30 volts full scale, 60 volts full scale, 120 volts full scale, 300 volts full scale, and 600 volts full scale. Meters are made with that many scales so that they will be as versatile as possible. If it is necessary to check for a voltage of 480 volts, the meter can be set on the 600-volt range. It would be very difficult, however, to check a 24-volt system on the 600-volt range. If the meter is set on the 30-volt range, it is simple to test for a voltage of 24 volts. The meter shown in *Figure 10–6* has multirange selection for voltage.

Courtesy of Triplett Corp.

FIGURE 10–6   Volt-ohm-milliampere meter with multirange selection.

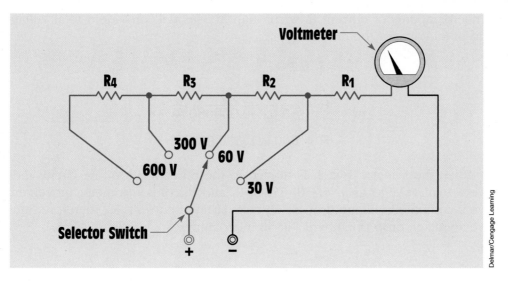

FIGURE 10–7   A rotary selector switch is used to change the full-range setting.

When the selector switch of this meter is turned, steps of resistance are inserted in the circuit to increase the range or are removed from the circuit to decrease the range. The meter shown in *Figure 10–7* has four range settings for full-scale voltage: 30 volts, 60 volts, 300 volts, and 600 volts. Notice that when the higher voltage settings are selected, more resistance is inserted in the circuit.

Calculating the Resistor Values

The values of the four resistors shown in *Figure 10–7* can be determined using Ohm's law. Assume that the full-scale values of the meter are 50 microamperes and 1 volt. The first step is to determine the value for R_1, which is used to provide a full-scale value of 30 volts. R_1, therefore, must have a voltage drop of 29 volts when a current of 50 microamperes is flowing through it.

$$R = \frac{E}{I}$$

$$R = \frac{29 \text{ V}}{0.000050 \text{ A}}$$

$$R = 5.80 \text{ k}\Omega \ (580{,}000 \ \Omega)$$

When the selector switch is moved to the second position, the meter circuit should have a total voltage drop of 60 volts. The meter movement and R_1 have a total voltage drop of 30 volts, so R_2 must have a voltage drop of 30 volts when

50 microamperes of current flow through it. This will provide a total voltage drop of 60 volts for the entire circuit:

$$R = \frac{E}{I}$$

$$R = \frac{30\ V}{50\ \mu A\ (0.000050\ A)}$$

$$R = 600\ k\Omega\ (600,000\ \Omega)$$

When the selector switch is moved to the third position, the circuit must have a total voltage drop of 300 volts. R_1 and R_2, plus the meter movement, have a combined voltage drop of 60 volts at rated current. R_3, therefore, must have a voltage drop of 240 volts at 50 microamperes.

$$R = \frac{E}{I}$$

$$R = \frac{240\ V}{50\ \mu A}$$

$$R = 4.8\ M\Omega\ (4,800,000\ \Omega)$$

When the selector switch is moved to the fourth position, the circuit must have a total voltage drop of 600 volts at rated current. Because R_1, R_2, and R_3, plus the meter movement produce a voltage drop of 300 volts at rated current, R_4 must have a voltage drop of 300 volts when 50 microamperes of current flow through it.

$$R = \frac{E}{I}$$

$$R = \frac{300\ V}{50\ \mu A}$$

$$R = 6\ M\Omega\ (6,000,000\ \Omega)$$

10–4   Reading a Meter

Learning to read the scale of a multimeter takes time and practice. Most people use meters every day without thinking about it. A common type of meter used daily by most people is shown in *Figure 10–8*. The meter illustrated is a speed-ometer similar to those seen in automobiles. This meter is designed to measure speed. It is calibrated in miles per hour (mph). The speedometer shown has a full-scale value of 80 miles per hour. If the pointer is positioned as shown in *Figure 10–8*, most people would know instantly that the speed of the automobile is 55 miles per hour.

FIGURE 10–8 A speedometer.

FIGURE 10–9 A fuel gauge.

Figure 10–9 illustrates another common meter used by most people. This meter is used to measure the amount of fuel in the tank of the automobile. Most people can glance at the pointer of the meter and know that the meter is indicating that there is one quarter of a tank of fuel remaining. Now assume that the tank has a capacity of 20 gallons. The meter is indicating that 5 gallons of fuel remain in the tank.

Learning to read the scale of a multimeter is similar to learning to read a speedometer or fuel gauge. The meter scale shown in *Figure 10–10* has several scales used to measure different quantities and values. The top of the scale is used to measure resistance, or ohms. Notice that the scale begins on the left at

FIGURE 10–10 Typical multimeter scale.

infinity and ends at zero on the right. Ohmmeters are covered later in this unit. The second scale is labeled AC–DC and is used to measure voltage. Notice that this scale has three different full-scale values. The top scale is 0–300, the second scale is 0–60, and the third scale is 0–12. The scale used is determined by the setting of the range control switch. The third set of scales is labeled "AC amperes." This scale is used with a clamp-on ammeter attachment that can be used with some meters. The last scale is labeled "dBm," which is used to measure decibels.

Reading a Voltmeter

Notice that the three voltmeter scales use the primary numbers 3, 6, and 12 and are in multiples of 10 of these numbers. Because the numbers are in multiples of 10, it is easy to multiply or divide the readings in your head by moving a decimal point. Remember that any number can be multiplied by 10 by moving the decimal point one place to the right, and any number can be divided by 10 by moving the decimal point one place to the left. For example, if the selector switch were set to permit the meter to indicate a voltage of 3 volts full scale, the 300-volt scale would be used, and the reading would be divided by 100. The reading can be divided by 100 by moving the decimal point two places to the left. In *Figure 10–11,* the pointer is indicating a value of 250. If the selector switch is set for 3 volts full scale, moving the decimal point two places to the left will give a reading of 2.5 volts. If the selector switch were set for a full-scale

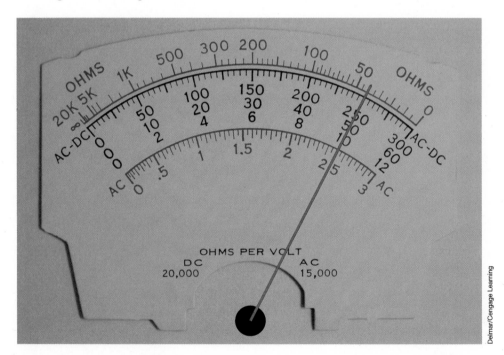

FIGURE 10–11   The meter indicates a value of 250.

value of 30 volts, the meter shown in *Figure 10–11* would be indicating a value of 25 volts. That reading is obtained by dividing the scale by 10 and moving the decimal point one place to the left.

Now assume that the meter has been set to have a full-scale value of 600 volts. The pointer in *Figure 10–12* is indicating a value of 44. Because the full-scale value of the meter is set for 600 volts, use the 60-volt range and multiply the reading on the meter by 10 by moving the decimal point one place to the right. The correct reading becomes 440 volts.

Three distinct steps should be followed when reading a meter. These steps are especially helpful for someone who has not had a great deal of experience reading a multimeter. The steps are as follows:

1. *Determine what the meter indicates.* Is the meter set to read a value of DC voltage, DC current, AC voltage, AC current, or ohms? It is impossible to read a meter if you do not know what the meter is used to measure.

2. *Determine the full-scale value of the meter.* The advantage of a multimeter is that it can measure a wide range of values and quantities. After it has been determined what quantity the meter is set to measure, it must then be determined what the range of the meter is.

There is a great deal of difference in reading when the meter is set to indicate a value of 600 volts full scale and when it is set for 30 volts full scale.

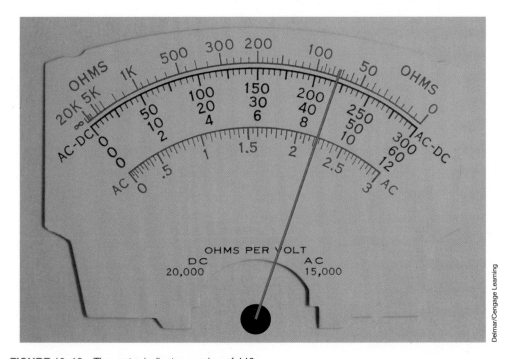

FIGURE 10–12 The meter indicates a value of 440.

3. *Read the meter.* The last step is to determine what the meter is indicating. It may be necessary to determine the value of the hash marks on the meter face for the range for which the selector switch is set. If the meter in *Figure 10–10* is set for 300 volts full scale, each hash mark has a value of 5 volts. If the full-scale value of the meter is 60 volts, however, each hash mark has a value of 1 volt.

10–5   The Ammeter

 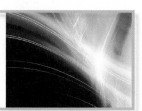

CAUTION: The **ammeter**, unlike the voltmeter, is a very low-impedance device. The ammeter is used to measure current and must be connected in series with the load to permit the load to limit the current flow *(Figure 10–13).* ■

An ammeter has a typical impedance of less than 0.1 ohm. If this meter is connected in parallel with the power supply, the impedance of the ammeter is the only thing to limit the amount of current flow in the circuit. Assume that an ammeter with a resistance of 0.1 ohm is connected across a 240-volt AC line. The current flow in this circuit would be 2400 amperes (240 V/0.1 Ω = 2400 A). A blinding flash of light would be followed by the destruction of the ammeter.

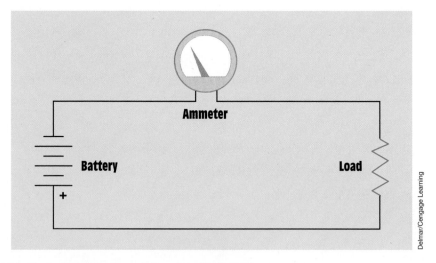

Delmar/Cengage Learning

FIGURE 10–13   An ammeter connects in series with the load.

FIGURE 10–14 In-line ammeter.

Ammeters connected directly into the circuit as shown in *Figure 10–13* are referred to as in-line ammeters. *Figure 10–14* shows an ammeter of this type.

10–6 Ammeter Shunts

DC ammeters are constructed by connecting a common moving-coil type of meter across a shunt. An **ammeter shunt** is a low-resistance device used to conduct most of the circuit current away from the meter movement. Because the meter movement is connected in parallel with the shunt, the voltage drop across the shunt is the voltage applied to the meter. Most ammeter shunts are manufactured to have a voltage drop of 50 millivolts (mv). If a 50-millivolt meter movement is connected across the shunt as shown in *Figure 10–15,* the pointer will move to the full-scale value when the rated current of the shunt is flowing. In the example shown, the ammeter shunt is rated to have a 50-millivolt drop when a 10-ampere current is flowing in the circuit. Because the meter movement has a full-scale voltage of 50 millivolts, it will indicate the full-scale value when 10 amperes of current are flowing through the shunt. An ammeter shunt is shown in *Figure 10–16.*

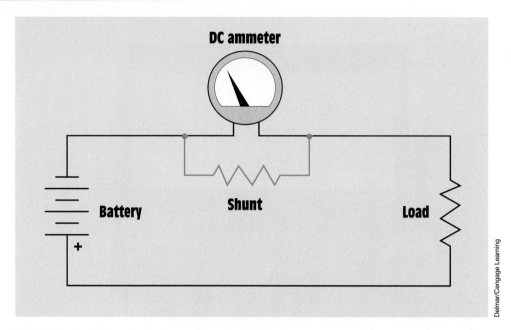

FIGURE 10–15 A shunt is used to set the value of the ammeter.

FIGURE 10–16 Ammeter shunt.

Ammeter shunts can be purchased to indicate different values. If the same 50-millivolt movement is connected across a shunt designed to drop 50 millivolts when 100 amperes of current flow through it, the meter will have a full-scale value of 100 amperes.

The resistance of an ammeter shunt can be calculated using Ohm's law. The resistance of a shunt designed to have a voltage drop of 50 millivolts when 100 amperes of current flow through it is

$$R = \frac{E}{I}$$

$$R = \frac{0.050 \text{ V}}{100 \text{ A}}$$

$$R = 0.0005 \; \Omega, \text{ or } 0.5 \text{ m}\Omega$$

In the preceding problem, no consideration was given to the electrical values of the meter movement. The reason is that the amount of current needed to operate the meter movement is so small compared with the 100-ampere circuit current it could have no meaningful effect on the resistance value of the shunt. When calculating the value for a low-current shunt, however, the meter values must be taken into consideration. For example, assume the meter has a voltage drop of 50 millivolts (0.050 V) and requires a current of 1 milliampere (0.001 A) to deflect the meter full scale. Using Ohm's law, it can be found that the meter has an internal resistance of 50 ohms (0.050 V/0.001 A = 50 Ω). Now assume that a shunt is to be constructed that will permit the meter to have a full-scale value of 10 milliamperes. If a total of 10 milliamperes is to flow through the circuit and 1 milliampere must flow through the meter, then 9 milliamperes must flow through the shunt *(Figure 10–17)*. Because the shunt must have a voltage drop of 50 millivolts when 9 milliamperes of current are flowing through it, its resistance must be 5.555 ohms (0.050 V/0.009 A = 5.555 Ω).

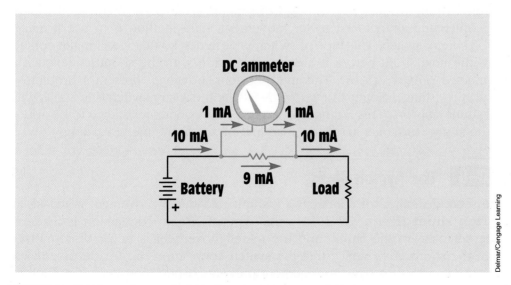

FIGURE 10–17   The total current is divided between the meter and the shunt.

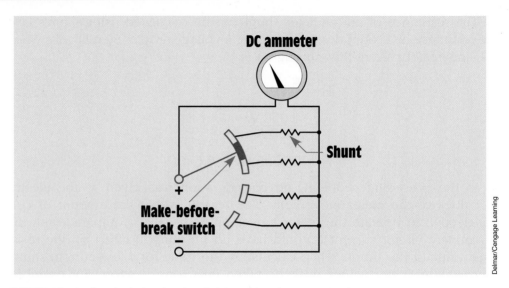

FIGURE 10–18   A make-before-break switch is used to change meter shunts.

10–7   Multirange Ammeters

Many ammeters, called **multirange ammeters,** are designed to operate on more than one range. This is done by connecting the meter movement to different shunts. ***When a multirange meter is used, care must be taken that the shunt is never disconnected from the meter.*** Disconnection would cause the meter movement to be inserted in series with the circuit, and full-circuit current would flow through the meter. Two basic methods are used for connecting shunts to a meter movement. One method is to use a make-before-break switch. This type of switch is designed so that it will make contact with the next shunt before it breaks connection with the shunt to which it is connected *(Figure 10–18)*. This method does, however, present a problem—contact resistance. Notice in *Figure 10–18* that the rotary switch is in series with the shunt resistors. This arrangement causes the contact resistance to be added to the shunt resistance and can cause inaccuracy in the meter reading.

10–8   The Ayrton Shunt

The second method of connecting a shunt to a meter movement is to use an **Ayrton shunt** *(Figure 10–19)*. In this type of circuit, connection is made to different parts of the shunt, and the meter movement is never disconnected from the shunt. Also notice that the switch connections are made external to the shunt and meter. This arrangement prevents contact resistance from affecting the accuracy of the meter.

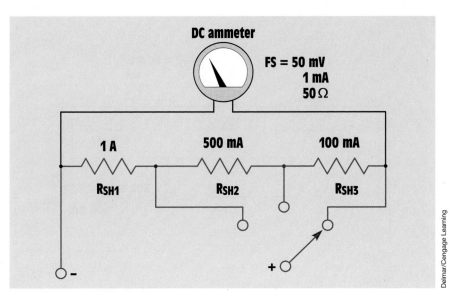

FIGURE 10–19 An Ayrton shunt.

Calculating the Resistor Values for an Ayrton Shunt

When an Ayrton shunt is used, the resistors are connected in parallel with the meter on some ranges and in series with the meter for other ranges. In this example, the meter movement has full-scale values of 50 millivolts, 1 milliampere, and 50 ohms of resistance. The shunt will permit the meter to have full-scale current values of 100 milliamperes, 500 milliamperes, and 1 ampere.

To find the resistor values, first calculate the resistance of the shunt when the range switch is set to permit a full-scale current of 100 milliampere *(Figure 10–20)*. When the range switch is set in this position, all three shunt resistors are connected in series across the meter movement. The formula for finding this resistance is

$$R_s = \frac{I_m \times R_m}{I_T}$$

where

R_s = resistance of the shunt

I_m = current of the meter movement

R_m = resistance of the meter movement

I_T = total circuit current

$$R_s = \frac{0.001 \text{ A} \times 50 \text{ }\Omega}{0.100 \text{ A}}$$

$R_s = 0.5 \text{ }\Omega$

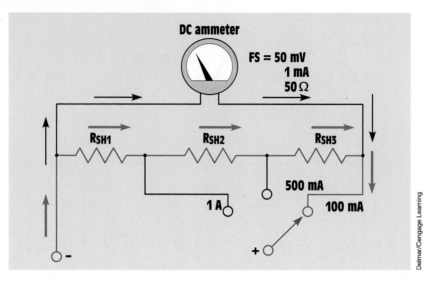

DC ammeter

FS = 50 mV
1 mA
50 Ω

R_{SH1} R_{SH2} R_{SH3}

500 mA

1 A

100 mA

+

–

FIGURE 10–20   The meter is in parallel with all shunt resistors.

Next, find the resistance of R_{SH1}, which is the shunt resistor used to produce a full-scale current of 1 ampere. When the selector switch is set in this position, R_{SH1} is connected in parallel with the meter and with R_{SH2} and R_{SH3}. R_{SH2} and R_{SH3}, however, are connected in series with the meter movement *(Figure 10–21)*. To calculate the value of this resistor, a variation of the previous formula is used. The new formula is

$$R_{SH1} = \frac{I_m \times R_{SUM}}{I_T}$$

where

R_{SH1} = the resistance of shunt 1

I_m = current of the meter movement

R_{SUM} = the sum of all the resistance in the circuit. Note that this is not the sum of the series-parallel combination. It is the sum of all the resistance. In this instance it will be 50.5 Ω (50 Ω [meter] + 0.5 Ω [shunt]).

I_T = total circuit current

$$R_{SH1} = \frac{0.001 \text{ A} \times 50.5 \text{ } \Omega}{1 \text{ A}}$$

$$R_{SH1} = 0.0505 \text{ } \Omega$$

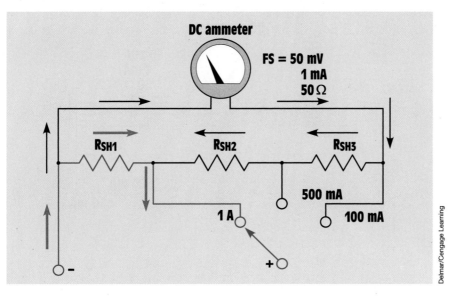

DC ammeter

FS = 50 mV
1 mA
50 Ω

R$_{SH1}$ R$_{SH2}$ R$_{SH3}$

500 mA

1 A

100 mA

+

−

Delmar/Cengage Learning

FIGURE 10–21   Current path through shunt and meter for a full-scale value of 1 ampere.

When the selector switch is changed to the 500-milliampere position, R$_{SH1}$ and R$_{SH2}$ are connected in series with each other and in parallel with the meter movement and R$_{SH3}$ *(Figure 10–22)*. The combined resistance value for R$_{SH1}$ and R$_{SH2}$ can be found using the formula

$$R_{SH1} \text{ and } R_{SH2} = \frac{I_m \times R_{SUM}}{I_T}$$

$$R_{SH1} \text{ and } R_{SH2} = \frac{0.001 \text{ A} \times 50.5 \text{ } \Omega}{0.5 \text{ A}}$$

$$R_{SH1} \text{ and } R_{SH2} = 0.101 \text{ } \Omega$$

Now that the total resistance for the sum of R$_{SH1}$ and R$_{SH2}$ is known, the value of R$_{SH2}$ can be found by subtracting it from the value of R$_{SH1}$.

$$R_{SH2} = 0.101 \text{ } \Omega - 0.0505 \text{ } \Omega$$

$$R_{SH2} = 0.0505 \text{ } \Omega$$

The value of R$_{SH3}$ can be found by subtracting the total shunt resistance from the values of R$_{SH1}$ and R$_{SH2}$.

$$R_{SH3} = 0.5 \text{ } \Omega - 0.0505 \text{ } \Omega - 0.0505 \text{ } \Omega$$

$$R_{SH3} = 0.399 \text{ } \Omega$$

The preceding procedure can be used to find the value of any number of shunt resistors for any value of current desired. Note, however, that this type of

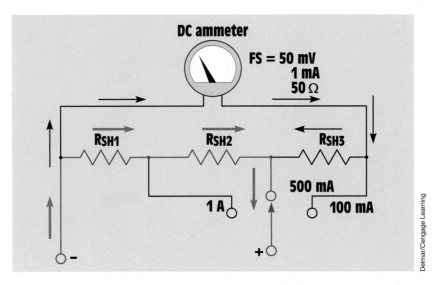

FIGURE 10–22 Current path through the meter and shunt for a full-scale value of 0.5 ampere.

FIGURE 10–23 DC ammeter with an Ayrton shunt.

shunt is not used for large current values because of the problem of switching contacts and contact size. The Ayrton shunt is seldom used for currents above 10 amperes. An ammeter with an Ayrton shunt is shown in *Figure 10–23*. The Ayrton shunt with all resistor values is shown in *Figure 10–24*.

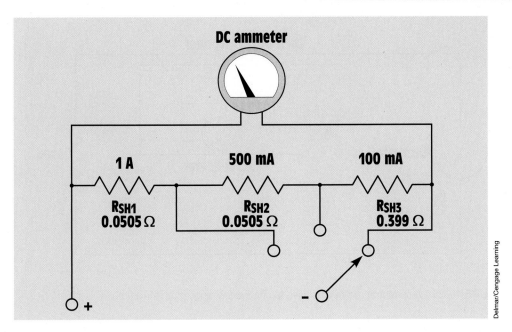

FIGURE 10–24 The Ayrton shunt with resistor values.

10–9 AC Ammeters

Shunts can be used with AC ammeters to increase their range but cannot be used to decrease their range. Most AC ammeters use a **current transformer** instead of shunts to change scale values. This type of ammeter is shown in *Figure 10–25*. The primary of the transformer is connected in series with the load, and the ammeter is connected to the secondary of the transformer. Notice that the range of the meter is changed by selecting different taps on the secondary of the current transformer. The different taps on the transformer provide different turns ratios between the primary and secondary of the transformer. The turns ratio is the ratio of the number of turns of wire in the primary as compared to the number of turns of wire in the secondary.

Calculating the Turns Ratio

In this example, it is assumed that an AC meter movement requires a current flow of 100 milliamperes to deflect the meter full scale. It is also assumed that the primary of the current transformer contains five turns of wire. A transformer will be designed to provide full-scale current readings of 1 ampere, 5 amperes,

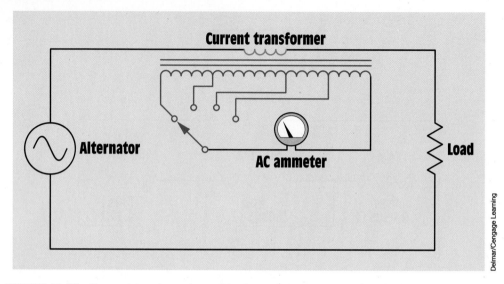

FIGURE 10–25 A current transformer is used to change the range of an AC ammeter.

and 10 amperes. To find the number of turns required in the secondary winding, the following formula can be used:

$$\frac{N_p}{N_s} = \frac{I_s}{I_p}$$

where

N_p = number of turns of wire in the primary

N_s = number of turns of wire in the secondary

I_p = current of the primary

I_s = current of the secondary

The number of turns of wire in the secondary to produce a full-scale current reading of 1 ampere can be calculated as follows:

$$\frac{5}{N_s} = \frac{0.1\ A}{1\ A}$$

Cross-multiplication is used to solve the problem. Cross-multiplication is accomplished by multiplying the bottom half of the equation on one side of the equal sign by the top half of the equation on the other side of the equal sign.

$$0.1\ A\ N_s = 5A\ \text{–turns}$$
$$N_s = 50\ \text{turns}$$

The transformer secondary must contain 50 turns of wire if the ammeter is to indicate a full-scale reading when 1 ampere of current flows through the primary winding.

The number of secondary turns can be found for the other values of primary current in the same way:

$$\frac{5}{N_s} = \frac{0.1 \text{ A}}{5 \text{ A}}$$

$$0.1 \text{ A } N_s = 25 \text{ A}$$

$$N_s = 250 \text{ turns}$$

$$\frac{5}{N_s} = \frac{0.1 \text{ A}}{10 \text{ A}}$$

$$0.1 \text{ A } N_s = 50 \text{ A}$$

$$N_s = 500 \text{ turns}$$

Current Transformers (CTs)

When a large amount of AC must be measured, a different type of current transformer is connected in the power line. These transformers have ratios that start at 200:5 and can have ratios of several thousand to five. These current transformers, generally referred to in industry as CTs, have a standard secondary current rating of 5 ampere AC. They are designed to be operated with a 5-ampere AC ammeter connected directly to their secondary winding, which produces a short circuit. CTs are designed to operate with the secondary winding shorted.

 CAUTION: The secondary winding of a CT should never be opened when there is power applied to the primary. This will cause the transformer to produce a step-up in voltage that could be high enough to kill anyone who comes in contact with it. ■

A current transformer is basically a toroid transformer. A toroid transformer is constructed with a hollow core similar to a doughnut *(Figure 10–26)*. When current transformers are used, the main power line is inserted through the opening in the transformer *(Figure 10–27)*. The power line acts as the primary of the transformer and is considered to be one turn.

Courtesy of Square D Company

FIGURE 10–26 A toroid current transformer.

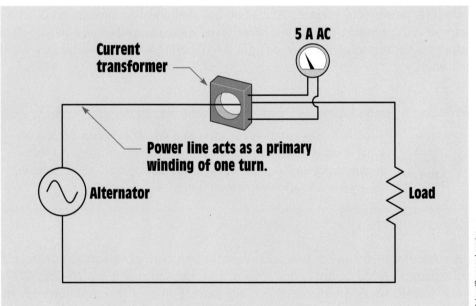

Delmar/Cengage Learning

FIGURE 10–27 Toroid transformer used to change the scale factor of an AC ammeter.

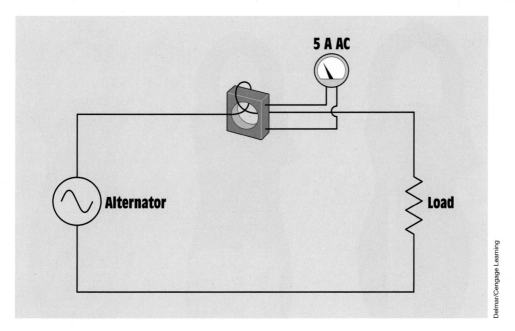

FIGURE 10–28   The primary conductor loops through the CT to produce a second turn, which changes the ratio.

The turns ratio of the transformer can be changed by looping the power wire through the opening in the transformer to produce a primary winding of more than one turn. For example, assume a current transformer has a ratio of 600:5. If the primary power wire is inserted through the opening, it will require a current of 600 amperes to deflect the meter full scale. If the primary power conductor is looped around and inserted through the window a second time, the primary now contains two turns of wire instead of one *(Figure 10–28)*. It now requires 300 amperes of current flow in the primary to deflect the meter full scale. If the primary conductor is looped through the opening a third time, it will require only 200 amperes of current flow to deflect the meter full scale.

10–10   Clamp-On Ammeters

Many electricians use the **clamp-on ammeter** *(Figure 10–29)*. The jaw of this type of meter is clamped around one of the conductors supplying power to the load *(Figure 10–30)*. The meter is clamped around only one of the lines. If the meter is clamped around more than one line, the magnetic fields of the wires cancel each other and the meter indicates zero.

Courtesy of Advanced Test Products

FIGURE 10–29 **(A)** Analog type clamp-on ammeter with vertical scale. **(B)** Analog type clamp-on ammeter with flat scale. **(C)** Clamp-on ammeter with digital scale.

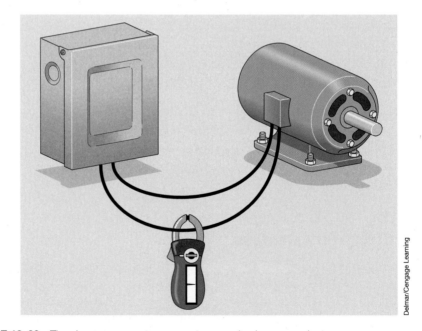

Delmar/Cengage Learning

FIGURE 10–30 The clamp-on ammeter connects around only one conductor.

The clamp-on meter also uses a current transformer to operate. The jaw of the meter is part of the core material of the transformer. When the meter is connected around the current-carrying wire, the changing magnetic field produced by the AC induces a voltage into the current transformer. The strength and frequency of the magnetic field determine the amount of voltage induced in the current transformer. Because 60 hertz is a standard frequency throughout the country, the amount of induced voltage is proportional to the strength of the magnetic field.

The clamp-on type ammeter can be given different range settings by changing the turns ratio of the secondary of the transformer just as is done on the in-line ammeter. The primary of the transformer is the conductor around which the movable jaw is connected. If the ammeter is connected around one wire, the primary has one turn of wire compared with the turns of the secondary. The turns ratio can be changed in the same manner that the ratio of the CT is changed. If two turns of wire are wrapped around the jaw of the ammeter *(Figure 10–31)*, the primary winding now contains two turns instead of one, and the turns ratio of the transformer is changed. The ammeter will now indicate double the amount of current in the circuit. The reading on the scale of the meter would have to be divided by 2 to get the correct reading. The ability to change the turns ratio of a clamp-on ammeter can be useful for measuring low currents. Changing the turns ratio is not limited to wrapping two turns of wire around the jaw of the ammeter. Any number of turns can be wrapped around the jaw of the ammeter, and the reading will be divided by that number.

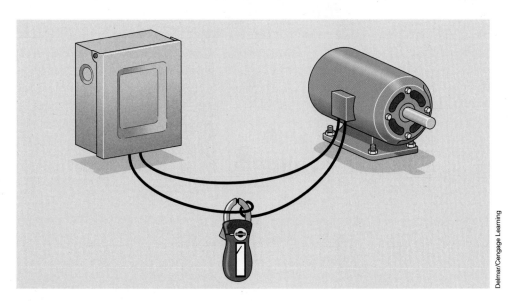

Delmar/Cengage Learning

FIGURE 10–31 Looping the conductor around the jaw of the ammeter changes the ratio.

10-11   DC–AC Clamp-On Ammeters

Most clamp-on ammeters that have the ability to measure both DC and AC do not operate on the principle of the current transformer. Current transformers depend on induction, which means that the current in the line must change direction periodically to provide a change of magnetic field polarity. It is the continuous change of field strength and direction that permits the current transformer to operate. The current in a DC circuit is unidirectional and does not change polarity, which would not permit the current transformer to operate.

DC–AC clamp-on ammeters *(Figure 10–32)* use the *Hall effect* as the basic principle of operation. The Hall effect was discovered by Edward H. Hall at

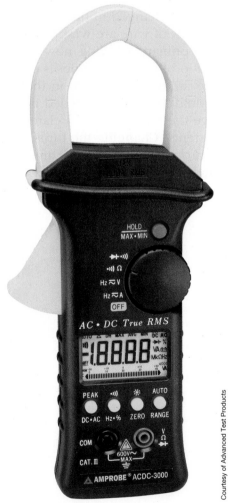

FIGURE 10–32   DC–AC clamp-on ammeter.

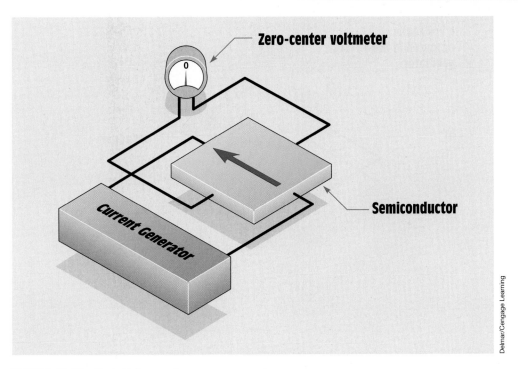

FIGURE 10-33   Basic Hall generator.

Johns Hopkins University in 1879. Hall originally used a piece of pure gold to produce the Hall effect, but today a semiconductor material is used because it has better operating characteristics and is less expensive. The device is often referred to as a *Hall generator. Figure 10–33* illustrates the operating principle of the Hall generator. A constant-current generator is used to supply a continuous current to the semiconductor chip. The leads of a zero-center voltmeter are connected across the opposite sides of the chip. As long as the current flows through the center of the semiconductor chip, no potential difference or voltage develops across the chip.

If a magnetic field comes near the chip *(Figure 10–34),* the electron path is distorted and the current no longer flows through the center of the chip. A voltage across the sides of the chip is produced. The voltage is proportional to the amount of current flow and the amount of current distortion. Because the current remains constant and the amount of distortion is proportional to the strength of the magnetic field, the voltage produced across the chip is proportional to the strength of the magnetic field.

If the polarity of the magnetic field were reversed *(Figure 10–35),* the current path would be distorted in the opposite direction, producing a voltage of the opposite polarity. Notice that the Hall generator produces a voltage

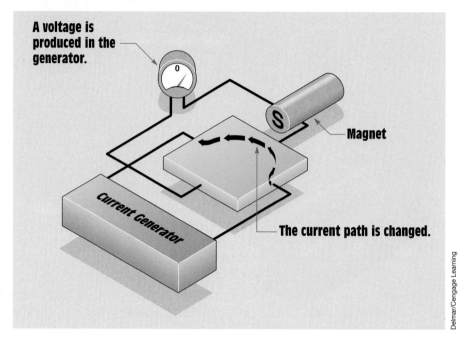

FIGURE 10–34 The presence of a magnetic field causes the Hall generator to produce a voltage.

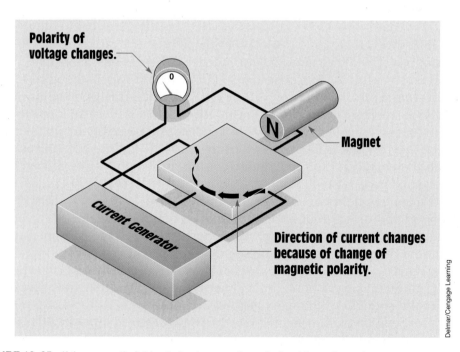

FIGURE 10–35 If the magnetic field polarity changes, the polarity of the voltage changes.

in the presence of a magnetic field. It makes no difference whether the field is moving or stationary. The Hall effect can therefore be used to measure DC or AC.

10–12   The Ohmmeter

The **ohmmeter** is used to measure resistance. The common volt-ohm-milliammeter (VOM) contains an ohmmeter. The ohmmeter has the only scale on a VOM that is nonlinear. The scale numbers increase in value as they progress from right to left. There are two basic types of analog ohmmeters—the series and the shunt. The series ohmmeter is used to measure high values of resistance, and the shunt type is used to measure low values of resistance. Regardless of the type used, the meter must provide its own power source to measure resistance. The power is provided by batteries located inside the instrument.

The Series Ohmmeter

A schematic for a basic series ohmmeter is shown in *Figure 10–36*. It is assumed that the meter movement has a resistance of 1000 ohms and requires a current of 50 microamperes to deflect the meter full scale. The power source will be a 3-volt battery. R_1, a fixed resistor with a value of 54 kilohms, is connected in series with the meter movement, and R_2, a variable resistor with a value of 10 kilohms, is connected in series with the meter and R_1.

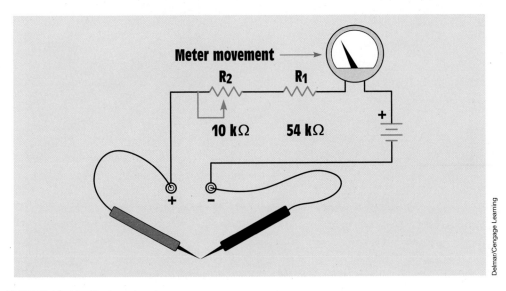

FIGURE 10–36   Basic series ohmmeter.

These resistance values were chosen to ensure there would be enough resistance in the circuit to limit the current flow through the meter movement to 50 microamperes. If Ohm's law is used to calculate the resistance needed (3 V/0.000050 A = 60,000 Ω), it will be seen that a value of 60 kilohms is needed. This circuit contains a total of 65,000 ohms (1000 Ω [meter] + 54,000 Ω + 10,000 Ω). The circuit resistance can be changed by adjusting the variable resistor to a value as low as 55,000 ohms, however, to compensate for the battery as it ages and becomes weaker.

When resistance is to be measured, the meter must first be zeroed. This is done with the ohms-adjust control, the variable resistor located on the front of the meter. To zero the meter, connect the leads *(Figure 10–36)* and turn the ohms-adjust knob until the meter indicates zero at the far right end of the scale *(Figure 10–37)*. When the leads are separated, the meter will again indicate infinity resistance at the left side of the scale. When the leads are connected across a resistance, the meter will again go up the scale. Because resistance has been added to the circuit, less than 50 microamperes of current will flow and the meter will indicate some value other than zero. *Figure 10–38* shows a meter indicating a resistance of 150 ohms, assuming the range setting is R×1.

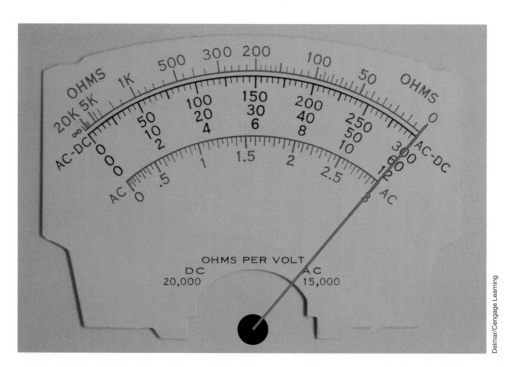

FIGURE 10–37   Adjusting the ohmmeter to zero.

Delmar/Cengage Learning

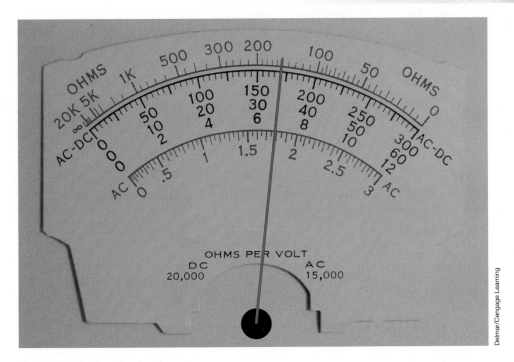

FIGURE 10–38 Reading the ohmmeter.

Ohmmeters can have different range settings such as R×1, R×100, R×1000, or R×10,000. These different scales can be obtained by adding different values of resistance in the meter circuit and resetting the meter to zero. ***An ohmmeter should always be readjusted to zero when the scale is changed.*** On the R×1 setting, the resistance is measured straight off the resistance scale located at the top of the meter. If the range is set for R×1000, however, the reading must be multiplied by 1000. The ohmmeter reading shown in *Figure 10–38* would be indicating a resistance of 150,000 ohms if the range had been set for R×1000. Notice that the ohmmeter scale is read backward from the other scales. Zero ohms is located on the far right side of the scale, and maximum ohms is located at the far left side. It generally takes a little time and practice to read the ohmmeter properly.

10–13 Shunt-Type Ohmmeters

The shunt-type ohmmeter is used for measuring low values of resistance. It operates on the same basic principle as an ammeter shunt. When using a shunt-type ohmmeter, place the unknown value of resistance in parallel with the meter movement. This placement causes part of the circuit current to bypass the meter *(Figure 10–39)*.

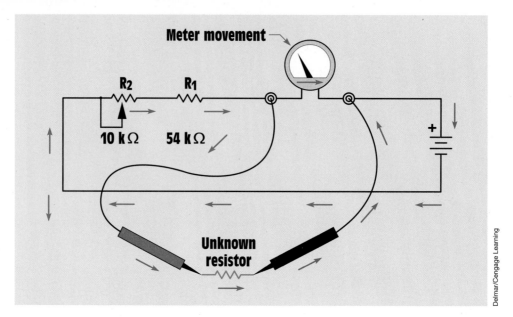

FIGURE 10–39 Shunt-type ohmmeter.

10–14 Digital Meters

Digital Ohmmeters

Digital ohmmeters display the resistance in figures instead of using a meter movement. When using a digital ohmmeter, care must be taken to notice the scale indication on the meter. For example, most digital meters will display a *K* on the scale to indicate kilohms or an M to indicate megohms (kilo means 1000 and mega means 1,000,000). If the meter is showing a resistance of 0.200 kilohms, it means 0.200 × 1000, or 200 ohms. If the meter indicates 1.65 megohms, it means 1.65 × 1,000,000, or 1,650,000 ohms.

Appearance is not the only difference between analog and digital ohmmeters. Their operating principle is different also. Analog meters operate by measuring the amount of current change in the circuit when an unknown value of resistance is added. Digital ohmmeters measure resistance by measuring the amount of voltage drop across an unknown resistance. In the circuit shown in *Figure 10–40,* a constant-current generator is used to supply a known amount of current to a resistor, R_x. It will be assumed that the amount of current supplied is 1 milliampere. The voltage dropped across the resistor is proportional to the resistance of the resistor and the amount of current flow. For example, assume the value of the unknown resistor is 4700 ohms. The voltmeter would indicate a drop of 4.7 volts when 1 milliampere of current flowed through

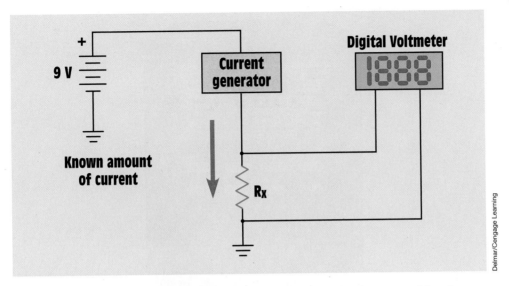

Delmar/Cengage Learning

FIGURE 10–40 Digital ohmmeters operate by measuring the voltage drop across a resistor when a known amount of current flows through it.

the resistor. The scale factor of the ohmmeter can be changed by changing the amount of current flow through the resistor. Digital ohmmeters generally exhibit an accuracy of about 1%.

The ohmmeter, whether digital or analog, must never be connected to a circuit when the power is turned on. Because the ohmmeter uses its own internal power supply, it has a very low operating voltage. Connecting a meter to power when it is set in the ohms position will probably damage or destroy the meter.

Digital Multimeters

Digital multimeters have become increasingly popular in the past few years. The most apparent difference between digital meters and analog meters is that digital meters display their reading in discrete digits instead of with a pointer and scale. A digital multimeter is shown in *Figure 10–41.* Some digital meters have a range switch similar to the range switch used with analog meters. This switch sets the full-range value of the meter. Many digital meters have voltage range settings from 200 millivolts to 2000 volts. The lower ranges are used for accuracy. For example, assume it is necessary to measure a voltage of 16 volts. The meter will be able to make a more accurate measurement when set on the 20-volt range than when set on the 2000-volt range.

Some digital meters do not contain a range setting control. These meters are known as autoranging meters. They contain a function control switch that permits selection of the electrical quantity to be measured, such as AC volts,

FIGURE 10–41   Digital multimeter.

Courtesy of Advanced Test Products

DC volts, ohms, and so on. When the meter probes are connected to the object to be tested, the meter automatically selects the proper range and displays the value.

Analog meters change scale value by inserting or removing resistance from the meter circuit *(Figure 10–7)*. The typical resistance of an analog meter is 20,000 ohms per volt for DC and 5000 ohms per volt for AC. If the meter is set for a full-scale value of 60 volts, there will be 1.2 megohms of resistance connected in series with the meter if it is being used to measure DC (60 V $\times$ 20,000 Ω/V = 1,200,000 Ω) and 300 kilohms if it is being used to measure AC (60 V $\times$ 5000 Ω/V = 300,000 Ω). The impedance of the meter is of little concern if it is used to measure circuits that are connected to a high-current source. For example, assume the voltage of a 480-volt panel is to be measured with a multimeter that has a resistance of 5000 ohms per volt. If the meter is set on the 600-volt range, the resistance connected in series with the meter is 3 megohms (600 V × 5000 Ω/V = 3,000,000 Ω). This resistance will permit a current of 160 microamperes to flow in the meter circuit (480 V/3,000,000 Ω = 0.000160 A). This 160 microamperes of current is not enough to affect the circuit being tested.

Now assume that this meter is to be used to test a 24-volt circuit that has a current flow of 100 microamperes. If the 60-volt range is used, the meter circuit contains a resistance of 300 kilohms (60 V × 5000 Ω/V = 300,000 Ω). Therefore, a current of 80 microamperes will flow when the meter is connected to the circuit (24 V/300,000 Ω = 0.000080 A). The connection of the meter to the circuit has changed the entire circuit operation. This phenomenon is known as the *loading effect*.

Digital meters do not have a loading effect. Most digital meters have an input impedance of about 10 megohms on all ranges. The input impedance is the ohmic value used to limit the flow of current through the meter. This impedance is accomplished by using field-effect transistors (FETs) and a voltage divider circuit. A simple schematic for such a circuit is shown in *Figure 10–42*. Notice that the meter input is connected across 10 megohms of resistance regardless of the range setting of the meter. If this meter is used to measure the voltage of the 24-volt circuit, a current of 2.4 microamperes will flow through the meter. This is not enough current to upset the rest of the circuit, and voltage measurements can be made accurately.

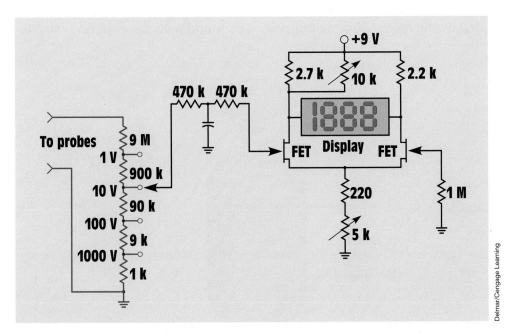

FIGURE 10–42 Digital voltmeter.

10–15 The Low-Impedance Voltage Tester

Another device used to test voltage is often referred to as a voltage tester. This device does measure voltage, but it does not contain a meter movement or digital display. It contains a coil and a plunger. The coil produces a magnetic field that is proportional to the amount of voltage to which the coil of the tester is connected. The higher the voltage to which the tester is connected, the stronger the magnetic field becomes. The plunger must overcome the force of a spring as it is drawn into the coil *(Figure 10–43)*. The plunger acts as a pointer to indicate the amount of voltage to which the tester is connected. The tester has an impedance of approximately 5000 ohms and can generally be used to measure voltages as high as 600 volts. ***The low-impedance voltage tester has a very large current draw compared with other types of voltmeters and should never be used to test low-power circuits.***

The relatively high current draw of the voltage tester can be an advantage when testing certain types of circuits, however, because it is not susceptible to giving the misleading voltage readings caused by high-impedance ground paths or feedback voltages that affect other types of voltmeters. An example of this advantage is shown in *Figure 10–44*. A transformer is used to supply power to a load. Notice that neither the output side of the transformer nor the load is connected to ground. If a high-impedance voltmeter is used to measure between one side of the transformer and a grounded point, it will most likely indicate some amount of voltage. That is because ground can act as a large capacitor and can permit a small amount of current to flow through the circuit created by the meter. This high-impedance ground path can support only a few

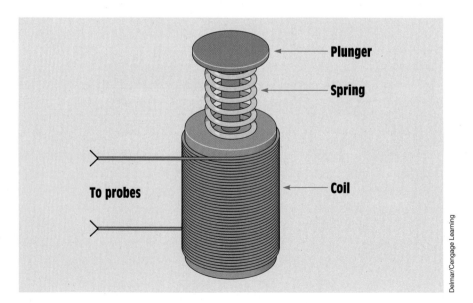

FIGURE 10–43 Low-impedance voltage tester.

Delmar/Cengage Learning

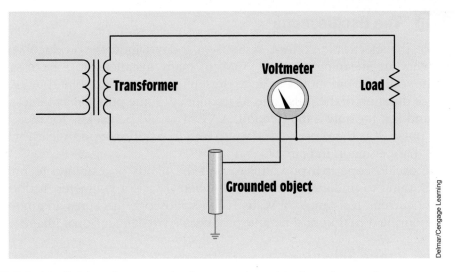

FIGURE 10–44   High-impedance ground paths can produce misleading voltage readings.

microamperes of current flow, but it is enough to operate the meter movement. If a voltage tester is used to make the same measurement, it will not show a voltage because there cannot be enough current flow to attract the plunger. A voltage tester is shown in *Figure 10–45*.

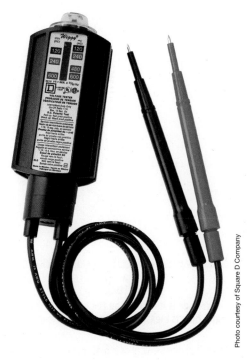

FIGURE 10–45   Wiggy voltage tester.

10–16   The Oscilloscope

Many of the electronic control systems in today's industry produce voltage pulses that are meaningless to a VOM. In many instances, it is necessary to know not only the amount of voltage present at a particular point, but also the length or duration of the pulse and its frequency. Some pulses may be less than 1 volt and last for only a millisecond. A VOM would be useless for measuring such a pulse. It is therefore necessary to use an **oscilloscope** to learn what is actually happening in the circuit.

The oscilloscope is a powerful tool in the hands of a trained technician. The first thing to understand is that an *oscilloscope* is a voltmeter. It does not measure current, resistance, or watts. The oscilloscope measures an amount of voltage during a period of time and produces a two-dimensional image.

Voltage Range Selection

The oscilloscope is divided into two main sections. One section is the voltage section, and the other is the time base. The display of the oscilloscope is divided by vertical and horizontal lines *(Figure 10–46)*. Voltage is measured on the vertical, or *Y*, axis of the display, and time is measured on the horizontal, or *X*, axis. When using a VOM, a range-selection switch is used to determine the full-scale value of the meter. Ranges of 600 volts, 300 volts, 60 volts, and 12 volts are common. The ability to change ranges permits more-accurate measurements to be made.

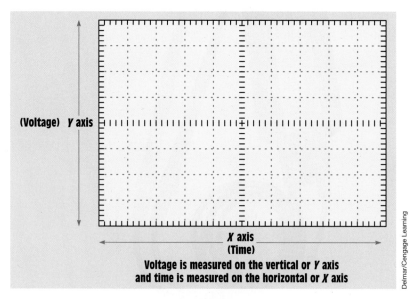

FIGURE 10–46   Voltage is measured on the vertical or *Y* axis and time is measured on the horizontal or *X* axis.

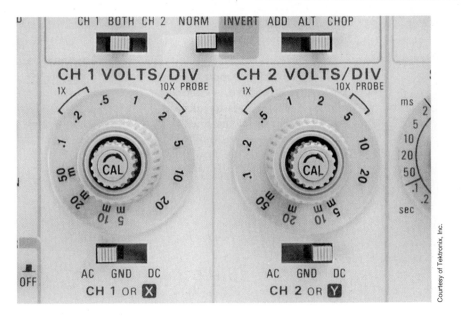

FIGURE 10–47   Voltage control of an analog oscilloscope.

Oscilloscopes can be divided into two main types: analog and digital. Analog oscilloscopes have been used for years and many are still in use; however, digital oscilloscopes are rapidly taking their place. Analog scopes generally employ some type of control knob to change their range of operation, *Figure 10–47*. The setting indicates the volts per division instead of volts full scale. The settings in *Figure 10–47* indicate that Channel 1 is set for 0.2 volts per division and Channel 2 is set for 0.5 volts per division.

Digital oscilloscopes often indicate their setting on the display instead of marking them on the face of the oscilloscope. Voltage control knobs for a four-channel digital oscilloscope are shown in *Figure 10–48*. The display of

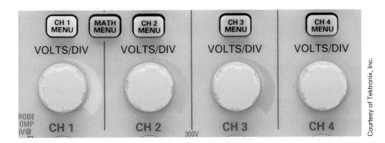

FIGURE 10–48   Voltage control knobs of a four channel digital oscilloscope.

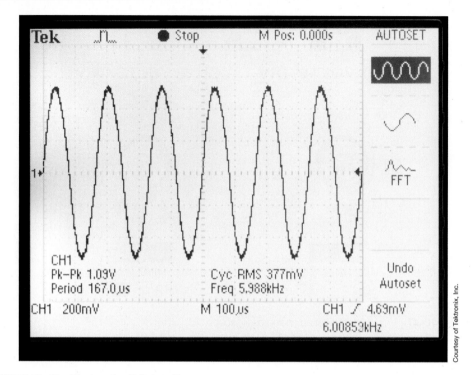

FIGURE 10–49 Display of a digital oscilloscope.

a typical digital oscilloscope is shown in *Figure 10–49*. In the lower left-hand corner of the display the notation CH1 200mV can be seen. This indicates that the voltage range has been set for 200 millivolts per division.

Oscilloscopes can display both positive and negative voltages. In the display shown in *Figure 10–50*, assume that a value of 0 volts has been set at the center line. The voltage shown at position A is positive with respect to 0 and the voltage at position B is negative with respect to 0. If the oscilloscope were set for a value of 2 volts per division, the value at point A would be 6 volts positive with respect to 0 and the value at point B would be 6 volts negative with respect to 0.

Another example of the oscilloscope's ability to display both positive and negative voltage values is in *Figure 10–51*. The waveform is basically an AC square wave. Notice that the voltage peaks at the leading edge of both the negative and positive waves. This could never be detected with a common voltmeter.

Many oscilloscopes have the ability to display more than one voltage at a time. Each voltage is generally referred to as a trace or channel. The oscilloscope shown in *Figure 10–52* has the ability to display four voltages or four traces. Many digital-type oscilloscopes have the ability to display a different color for each trace.

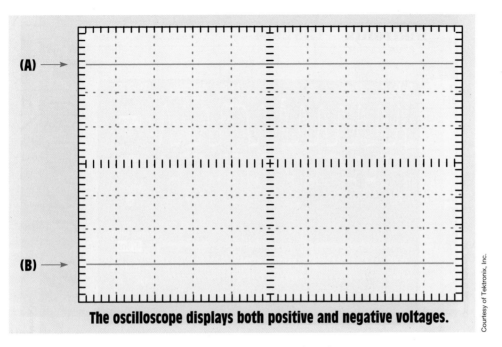

The oscilloscope displays both positive and negative voltages.

Courtesy of Tektronix, Inc.

FIGURE 10–50   The oscilloscope displays both positive and negative voltages.

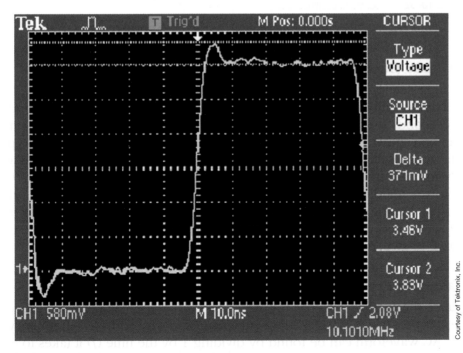

Courtesy of Tektronix, Inc.

FIGURE 10–51   AC waveform.

at 100 microseconds per division. The AC sine wave being displayed completes one complete cycle in 167 microseconds. The frequency is 5988 hertz, or 5.988 kilohertz. (1/0.000167). The oscilloscope display in *Figure 10–49* displays these values. Many oscilloscopes have the ability to measure frequency automatically and display the value for you.

Attenuated Probes

Most oscilloscopes use a probe that acts as an attenuator. An attenuator is a device that divides or makes smaller the input signal *(Figure 10–55)*. An attenuated probe is used to permit higher voltage readings than are normally possible. For example, most attenuated probes are 10 to 1. This means that if the voltage range switch is set for 5 volts per division, the display would actually indicate 50 volts per division. If the voltage range switch is set for 2 volts per division, each division on the display actually has a value of 20 volts per division.

Probe attenuators are made in different styles by different manufacturers. On some probes, the attenuator is located in the probe head itself, whereas on

Courtesy of Tektronix, Inc.

FIGURE 10–55 Oscilloscope attenuated probe.

others the attenuator is located at the scope input. Regardless of the type of attenuated probe used, it may have to be compensated or adjusted. In fact, probe compensation should be checked frequently. Different manufacturers use different methods for compensating their probes, so it is generally necessary to follow the procedures given in the operator's manual for the probe being used.

Oscilloscope Controls

The following is a list of common controls found on the oscilloscope. Refer to the oscilloscope shown in *Figure 10–56*.

1. **POWER**. The power switch is used to turn the oscilloscope ON or OFF.

2. **BEAM FINDER.** This control is used to locate the position of the trace if it is off the display. The BEAM FINDER button will indicate the approximate location of the trace. The position controls are then used to move the trace back on the display.

3. **PROBE ADJUST** (sometimes called calibrate). This is a reference voltage point used when compensating the probe. Most probe adjust points produce a square wave signal of about 0.5 volts.

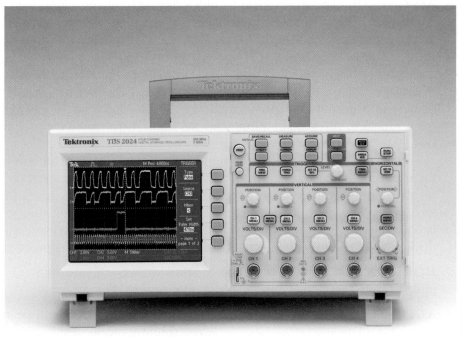

FIGURE 10–56   An oscilloscope.

4. **INTENSITY** and **FOCUS.** The INTENSITY control adjusts the brightness of the trace. A bright spot should never be left on the display because it will burn a spot on the face of the cathode ray tube (CRT). This burned spot results in permanent damage to the CRT. The FOCUS control sharpens the image of the trace.

5. **VERTICAL POSITION.** This is used to adjust the trace up or down on the display. If a dual-trace oscilloscope is being used, there will be two vertical POSITION controls. (A dual-trace oscilloscope contains two separate traces that can be used separately or together.)

6. **CH 1–BOTH–CH 2.** This control determines which channel of a dual-trace oscilloscope is to be used, or whether they are both to be used at the same time.

7. **ADD–ALT.–CHOP.** This control is active only when both traces are being displayed at the same time. The ADD adds the two waves together. ALT. stands for alternate. This alternates the sweep between Channel 1 and Channel 2. The CHOP mode alternates several times during one sweep. This generally makes the display appear more stable. The CHOP mode is generally used when displaying two traces at the same time.

8. **AC–GND–DC.** The AC is used to block any DC voltage when only the AC portion of the voltage is to be seen. For instance, assume an AC voltage of a few millivolts is riding on a DC voltage of several hundred volts. If the voltage range is set high enough so that 100 VDC can be seen on the display, the AC voltage cannot be seen. The AC section of this switch inserts a capacitor in series with the probe. The capacitor blocks the DC voltage and permits the AC voltage to pass. Because the 100 VDC has been blocked, the voltage range can be adjusted for millivolts per division, which will permit the AC signal to be seen.

    The GND section of the switch stands for ground. This section grounds the input so the sweep can be adjusted for 0 volt at any position on the display. The ground switch grounds at the scope and does not ground the probe. This permits the ground switch to be used when the probe is connected to a live circuit. The DC section permits the oscilloscope to display all of the voltage, both AC and DC, connected to the probe.

9. **HORIZONTAL POSITION.** This control adjusts the position of the trace from left to right.

10. **AUTO–NORMAL.** This determines whether the time base will be triggered automatically or operated in a free-running mode. If this control is operated in the NORM setting, the trigger signal is taken from the line to which the probe is connected. The scope is generally operated with the trigger set in the AUTO position.

11. **LEVEL.** The LEVEL control determines the amplitude the signal must be before the scope triggers.

12. **SLOPE.** The SLOPE permits selection as to whether the trace is triggered by a negative or positive waveform.

13. **INT.–LINE–EXT.** The *INT.* stands for internal. The scope is generally operated in this mode. In this setting, the trigger signal is provided by the scope. In the LINE mode, the trigger signal is provided from a sample of the line. The *EXT,* or external, mode permits the trigger pulse to be applied from an external source.

These are not all the controls shown on the oscilloscope in *Figure 10–56,* but they are the major controls. Most oscilloscopes contain these controls.

Interpreting Waveforms

The ability to interpret the waveforms on the display of the oscilloscope takes time and practice. When using the oscilloscope, one must keep in mind that the display shows the voltage with respect to time.

In *Figure 10–57,* it is assumed that the voltage range has been set for 0.5 volts per division, and the time base is set for 2 milliseconds per division. It is also assumed that 0 volt has been set on the center line of the display. The waveform shown is a square wave. The display shows that the voltage

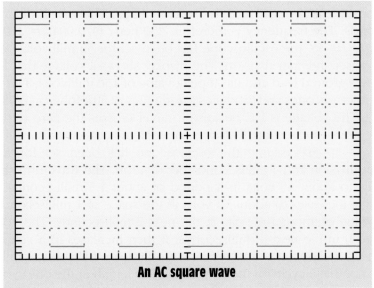

An AC square wave

FIGURE 10–57 AC square wave.

Delmar/Cengage Learning

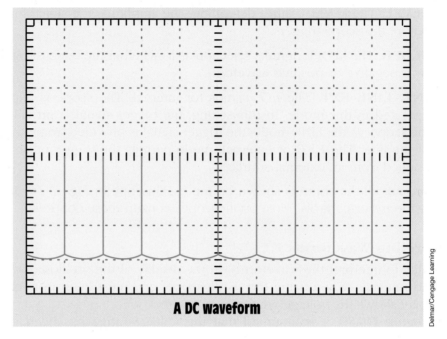

A DC waveform

Delmar/Cengage Learning

FIGURE 10–58   A DC waveform.

rises in the positive direction to a value of 1.4 volts and remains there for 2 milliseconds. The voltage then drops to 1.4 volts negative and remains there for 2 milliseconds before going back to positive. Because the voltage changes between positive and negative, it is an AC voltage. The length of one cycle is 4 milliseconds. The frequency is therefore 250 hertz (1/0.004s = 250 Hz).

In *Figure 10–58*, the oscilloscope has been set for 50 millivolts per division and 20 microseconds per division. The display shows a voltage that is negative to the probe's ground lead and has a peak value of 150 millivolts. The waveform lasts for 20 microseconds and produces a frequency of 50 kilohertz (1/0.000020s = 50,000 Hz). The voltage is DC because it never crosses the zero reference and goes in the positive direction. This type of voltage is called *pulsating DC*.

In *Figure 10–59*, assume the oscilloscope has been set for a value of 50 volts per division and 4 milliseconds per division. The waveform shown rises from 0 volts to about 45 volts in a period of about 1.5 milliseconds. The voltage gradually increases to about 50 volts in the space of 1 millisecond and then rises to a value of about 100 volts in the next 2 milliseconds. The voltage then decreases to 0 in the next 4 milliseconds. It then increases to a value of about 10 volts in 0.5 milliseconds and remains at that level for about 8 milliseconds. This is one complete cycle for the waveform. The length of the one cycle is about 16.6 milliseconds, which is a frequency of 60.2 hertz. (1/0.0166). The voltage is DC because it remains positive and never drops below the 0 line.

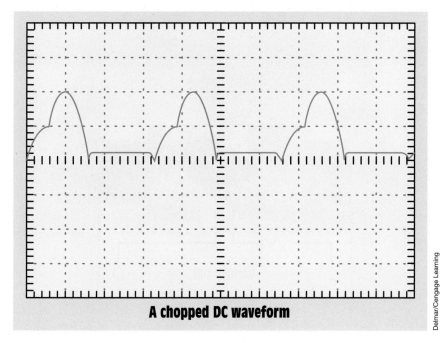

A chopped DC waveform

Delmar/Cengage Learning

FIGURE 10–59 A chopped DC waveform.

Learning to interpret the waveforms seen on the display of an oscilloscope will take time and practice, but it is well worth the effort. The oscilloscope is the only means by which many of the waveforms and voltages found in electronic circuits can be understood. Consequently, the oscilloscope is the single most valuable piece of equipment a technician can use.

10–17 The Wattmeter

The **wattmeter** is used to measure true power in a circuit. There are two basic types of wattmeters, dynamic and electronic. Dynamic wattmeters differ from d'Arsonval-type meters in that they do not contain a permanent magnet. They contain an electromagnet and a moving coil *(Figure 10–60)*. The electromagnets are connected in series with the load in the same manner that an ammeter is connected. The moving coil has resistance connected in series with it and is connected directly across the power source in the same manner as a voltmeter *(Figure 10–61)*.

Because the electromagnet is connected in series with the load, the current flow through the load determines the magnetic field strength of the stationary magnet. The magnetic field strength of the moving coil is determined by the amount of line voltage. The turning force of the coil is proportional to the strength of these two magnetic fields. The deflection of the meter against the spring is proportional to the amount of current flow and voltage.

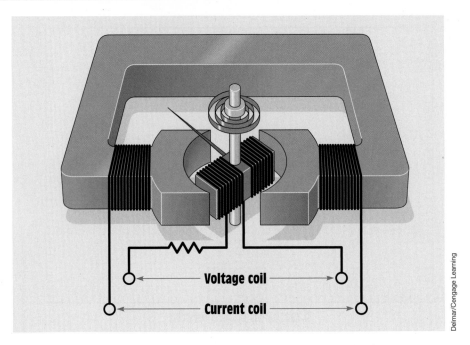

FIGURE 10–60 The wattmeter contains two coils—one for voltage and the other for current.

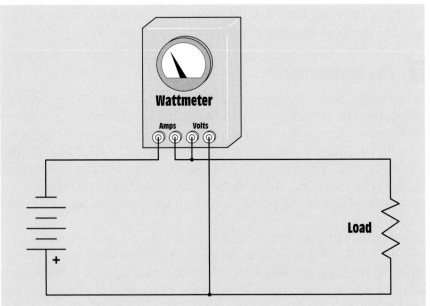

FIGURE 10–61 The current section of the wattmeter is connected in series with the load, and the voltage section is connected in parallel with the load.

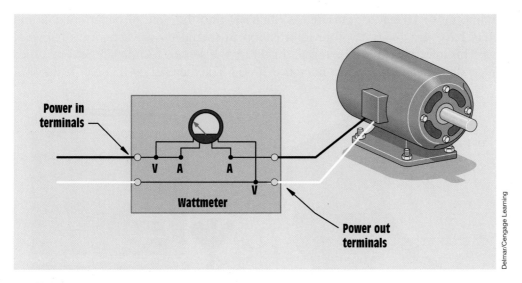

Delmar/Cengage Learning

FIGURE 10–62 Portable wattmeters often make connection to the voltage and current terminals inside the meter.

Because the wattmeter contains an electromagnet instead of a permanent magnet, the polarity of the magnetic field is determined by the direction of current flow. The same is true of the polarity of the moving coil connected across the source of voltage. If the wattmeter is connected into an AC circuit, the polarity of the two coils will reverse at the same time, producing a continuous torque. For this reason, the wattmeter can be used to measure power in either a DC or an AC circuit. However, if the connection of the stationary coil or the moving coil is reversed, the meter will attempt to read backward.

Dynamic-type wattmeters are being replaced by wattmeters that contain electronic circuitry to determine true power. They are less expensive and generally more accurate than the dynamic type. Like dynamic wattmeters, electronic-type meters contain amperage terminals that connect in series with the load and voltage terminals that connect in parallel with the load. Portable-type wattmeters often have terminals labeled "power in" and "power out." Connection to the current and voltage section of the meter is made inside the meter *(Figure 10–62)*. Analog-type electronic wattmeters use a standard d'Arsonval-type movement to indicate watts. The electronic circuit determines the true power of the circuit and then supplies the appropriate power to the meter movement. Wattmeters with digital displays are also available.

10–18 Recording Meters

On occasion, it becomes necessary to make a recording of an electrical value over a long period of time. Recording meters produce a graph of metered values during a certain length of time. They are used to detect spike voltages, or currents of short duration, or sudden drops in voltage, current, or power. Recording meters can show the amount of voltage or current, its duration, and the time of occurrence. Some meters have the ability to store information in memory over a period of several days. This information can be recalled later by the service technician. Several types of recording meters are shown in *Figure 10–63*. The meter shown in *Figure 10–63A*

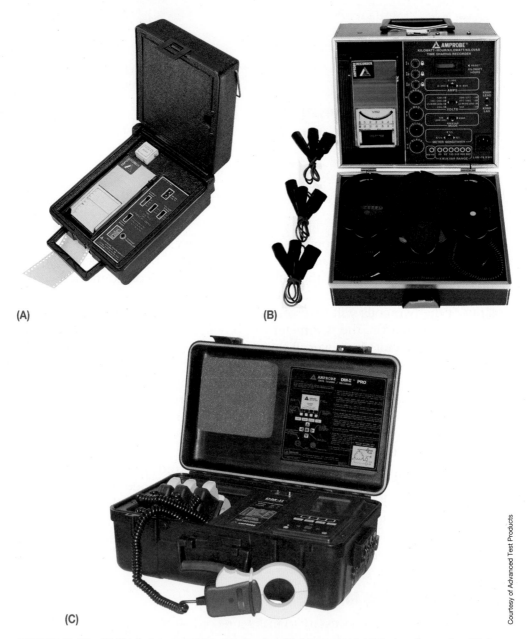

(A)   (B)

(C)

FIGURE 10–63   (A) Single-line recording volt-ammeter. **(B)** A kilowatt-kiloVAR recording meter. **(C)** A single-phase or three-phase voltage and current recording meter.

Courtesy of Advanced Test Products

is a single-line-recording volt-ammeter. It will record voltage or current or both for a single phase. A kilowatt-kiloVARs recording meter is shown in *Figure 10–63B*. This meter will record true power (kilowatts) or reactive power (kiloVARs) on a time-share basis. It can be used on single- or three-phase circuits and can also be used to determine circuit power factors. A chartless recorder is shown in *Figure 10–63C*. This instrument can record voltages and currents on single- or three-phase lines. The readings can be stored in memory for as long as 41 days.

10–19 Bridge Circuits

One of the most common devices used to measure values of resistance, inductance, and capacitance accurately is a **bridge circuit.** A bridge is constructed by connecting four components to form a parallel-series circuit. All four components are of the same type, such as four resistors, four inductors, or four capacitors. The bridge used to measure resistance is called a **Wheatstone bridge.** The basic circuit for a Wheatstone bridge is shown in *Figure 10–64*. The bridge operates on the principle that the sum of the voltage drops in a series circuit must equal the applied voltage. A galvanometer is used to measure

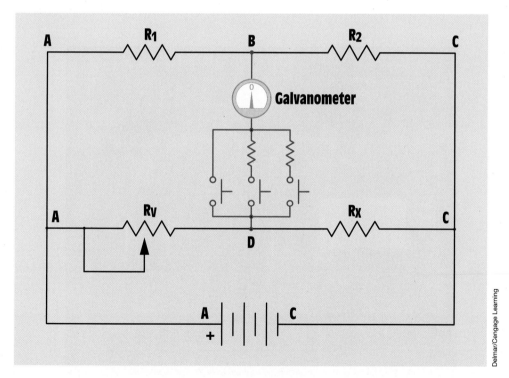

FIGURE 10–64 The Wheatstone bridge circuit is used to make accurate measurements of resistance and operates on the principle that the sum of the voltage drops in a series circuit must equal the applied voltage.

the voltage between points B and D. The galvanometer can be connected to different values of resistance or directly between points B and D. Values of resistance are used to determine the sensitivity of the meter circuit. When the meter is connected directly across the two points, its sensitivity is maximum.

In *Figure 10–64,* assume the battery has a voltage of 12 volts and that R_1 and R_2 are precision resistors and have the same value of resistance. Because R_1 and R_2 are connected in series and have the same value, each will have a voltage drop equal to one-half of the applied voltage, or 6 volts. This means that point B is 6 volts more negative than point A and 6 volts more positive than point C.

R_V (variable) and R_X (unknown) are connected in series with each other. R_X represents the unknown value of resistance to be measured. R_V can be adjusted for different resistive values. If the value of R_V is greater than the value of R_X, the voltage at point D will be more negative than the voltage at point B. This will cause the pointer of the zero-center galvanometer to move in one direction. If the value of R_V is less than R_X, the voltage at point D will be more positive than the voltage at point B, causing the pointer to move in the opposite direction. When the value of R_V becomes equal to that of R_X, the voltage at point D will become equal to the voltage at point B. When this occurs, the galvanometer will indicate zero. A Wheatstone bridge is shown in *Figure 10–65.*

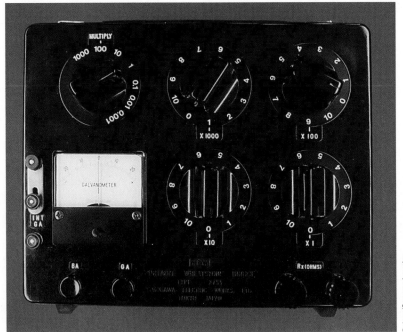

FIGURE 10–65 Wheatstone bridge.

Delmar/Cengage Learning

Summary

- The d'Arsonval type of meter movement is based on the principle that like magnetic fields repel.

- The d'Arsonval movement operates only on DC.

- Voltmeters have a high resistance and are designed to be connected directly across the power line.

- The steps to reading a meter are as follows:

  a. Determine what quantity the meter is set to measure.

  b. Determine the full-range value of the meter.

  c. Read the meter.

- Ammeters have a low resistance and must be connected in series with a load to limit the flow of current.

- Shunts are used to change the value of DC ammeters.

- AC ammeters use a current transformer to change the range setting.

- Clamp-on ammeters measure the flow of current by measuring the strength of the magnetic field around a conductor.

- Ohmmeters are used to measure the resistance in a circuit.

- Ohmmeters contain an internal power source, generally batteries.

- Ohmmeters must never be connected to a circuit that has power applied to it.

- Digital multimeters display their value in digits instead of using a meter movement.

- Digital multimeters generally have an input impedance of 10 megohms on all ranges.

- The oscilloscope measures the amplitude of voltage with respect to time.

- The frequency of a waveform can be determined by dividing 1 by the time of one cycle ($f = 1/t$).

- Wattmeters contain a stationary coil and a movable coil.

- The stationary coil of a wattmeter is connected in series with the load, and the moving coil is connected to the line voltage.

- The turning force of the dynamic wattmeter is proportional to the strength of the magnetic field of the stationary coil and the strength of the magnetic field of the moving coil.

- Digital ohmmeters measure resistance by measuring the voltage drop across an unknown resistor when a known amount of current flows through it.

- Low-impedance voltage testers are not susceptible to indicating a voltage caused by a high-impedance ground or a feedback.

- A bridge circuit can be used to accurately measure values of resistance, inductance, and capacitance.

Review Questions

1. To what is the turning force of a d'Arsonval meter movement proportional?

2. What type of voltage must be connected to a d'Arsonval meter movement?

3. A DC voltmeter has a resistance of 20,000 Ω per volt. What is the resistance of the meter if the range selection switch is set on the 250-V range?

4. What is the purpose of an ammeter shunt?

5. Name two methods used to make a DC multirange ammeter.

6. How is an ammeter connected into a circuit?

7. How is a voltmeter connected into a circuit?

8. An ammeter shunt has a voltage drop of 50 mV when 50 A of current flow through it. What is the resistance of the shunt?

9. What type of meter contains its own separate power source?

10. What electrical quantity does the oscilloscope measure?

11. What is measured on the Y axis of an oscilloscope?

12. What is measured on the X axis of an oscilloscope?

13. A waveform shown on the display of an oscilloscope completes one cycle in 50 μs. What is the frequency of the waveform?

14. What is the major difference between a wattmeter and a d'Arsonval meter?

15. What two factors determine the turning force of a wattmeter?

Practical Applications

*Y*ou are an electrician on the job. You have been give a multimeter that has the following AC voltage ranges: 30, 60, and 150. The meter states that it has a resistance of 5000 Ω/V. You need to be able to measure a voltage of 277 volts. How much resistance should be inserted in series with the meter to make the 30-volt range indicate a full scale value of 300 volts? ■

Practice Problems

Measuring Instruments

1. A d'Arsonval meter movement has a full-scale current value of 100 μA (0.000100 A) and a resistance of 5 kΩ (5000 Ω). What size resistor must be placed in series with this meter to permit it to indicate 10 V full scale?

2. The meter movement described in Question 1 is to be used to construct a multirange voltmeter. The meter is to have voltage ranges of 15 V, 60 V, 150 V, and 300 V *(Figure 10–66)*. Find the values of R_1, R_2, R_3, and R_4.

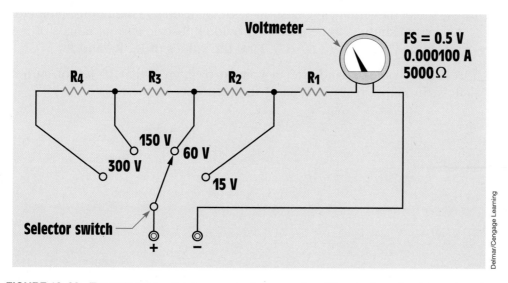

FIGURE 10–66 The multirange voltmeter operates by connecting different values of resistance in series with the meter movement.

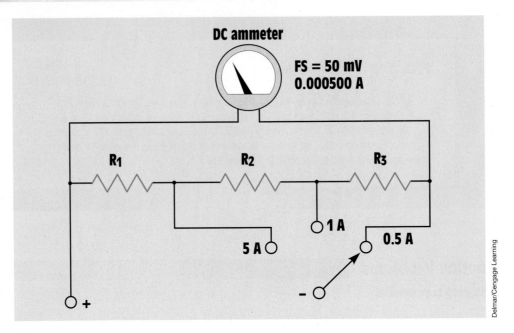

FIGURE 10–67 Ayrton shunt.

3. A meter movement has a full-scale value of 500 μA (0.000500 A) and 50 mV (0.050 V). A shunt is to be connected to the meter that permits it to have a full-scale current value of 2 A. What is the resistance of the shunt?

4. The meter movement in Question 3 is to be used as a multirange ammeter. An Ayrton shunt is to be used to provide full-scale current ranges of 5 A, 1 A, and 0.5 A *(Figure 10–67)*. Find the values of R_1, R_2, and R_3.

5. A digital voltmeter indicates a voltage of 2.5 V when 10 μA of current flow through a resistor. What is the resistance of the resistor?

Unit 11

Using Wire Tables and Determining Conductor Sizes

KEY TERMS

Ambient air temperature
American Wire Gauge (AWG)
Ampacity (current-carrying ability)
Circular mil
Correction factor
Damp locations
Dry locations
Insulation
Maximum operating temperature
MEGGER
Mil-foot
National Electrical Code (NEC)
Parallel conductors
Wet locations

Why You Need to Know

*B*eing able to determine the amount of current a conductor is permitted to carry or the size wire need for an installation is essential to any electrician, whether he or she works as an installation electrician or as a maintenance electrician. This unit

- explains how the amount of current a conductor is permitted to carry is not the same as selecting the proper wire for an installation and describes the differences.
- differentiates the different types of wire insulation and the appropriate use of each based on ambient temperatures.
- explains the method for using tools such as a MEGGER when determining the resistance of wire insulation.
- discusses how conductor length and size impact resistance and determine the required conductor size.
- provides the tools for determining ampacity rating of conductors when applying correction factors for wiring in a raceway.
- explains that, as a general rule, electricians select wire sizes from the *NEC*. However, there are instances where the wire run is too long or some special type of wire is being employed. In those instances, wire size and type are chosen by determining the maximum voltage drop and calculating the resistance of the wire. This unit explains how to determine wire size using the *NEC* and calculating wire resistance.

Courtesy of Niagara Mohawk Power Corporation.

Objectives

After studying this unit, you should be able to

■ select a conductor from the proper wire table.

■ discuss the different types of wire insulation.

■ determine insulation characteristics.

■ use correction factors to determine the proper ampacity rating of conductors.

■ determine the resistance of long lengths of conductors.

■ determine the proper wire size for loads located long distances from the power source.

■ list the requirements for using parallel conductors.

■ discuss the use of a MEGGER for testing insulation.

Preview

The size of the conductor needed for a particular application can be determined by several methods. The **National Electrical Code (NEC)** is used throughout industry to determine the conductor size for most applications. It is imperative that an electrician be familiar with Code tables and correction factors. In some circumstances, however, wire tables cannot be used, as in the case of extremely long wire runs or for windings of a transformer or motor. In these instances, the electrician should know how to determine the conductor size needed by calculating maximum voltage drop and resistance of the conductor. ■

11–1   The American Wire Gauge (AWG)

The American Wire Gauge was standardized in 1857 and is used mainly in the United States for the diameters of round, solid, nonferrous electrical wire. The gauge size is important for determining the current-carrying capacity of a conductor. Gauge sizes are determined by the number of draws necessary to produce a given diameter or wire. Electrical wire is made by drawing it through a succession of dies, *Figure 11–1*.

Each time a wire passes through a die, it is wrapped around a draw block several times. The draw block provides the pulling force necessary to draw the wire through the die. A 24 AWG wire would be drawn through 24 dies, each having a smaller diameter. In the field, wire size can be determined with a wire gauge, *Figure 11–2*. One side of the wire gauge lists the AWG size of the wire, *Figure 11–3*. The opposite side of the wire gauge indicates the diameter of the wire in thousandths of an inch, *Figure 11–4*. When determining wire size, first remove the insulation from around the conductor. The slots in the wire gauge, not the holes behind the slots, are used to determine the size, *Figure 11–5*.

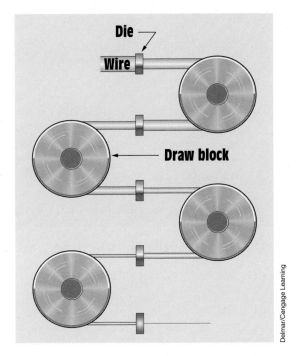

FIGURE 11–1   Wire is drawn through a succession of dies to produce the desired diameter.

FIGURE 11–2   Wire gauge

The largest AWG size is 4/0, which has an area of 211,600 circular mills (CM). Conductors with a larger area are measured in thousand circular mills. The next largest conductor past 4/0 is 250 thousand circular mills (250 kcmil). Conductors can be obtained up to 2000 kcmil. In practice, large conductors

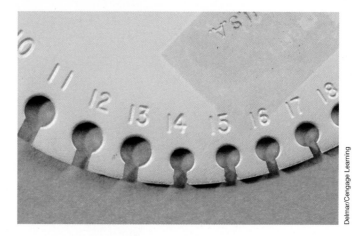

FIGURE 11–3 One side of the wire gauge is marked with the AWG size.

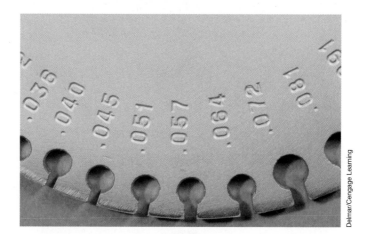

FIGURE 11–4 The other side of the wire gauge lists the diameter of the wire in thousandths of an inch.

FIGURE 11–5 The slot, not the hole, determines the wire size.

are difficult to pull through conduit. It is sometimes desirable to use parallel conductors instead of extremely large conductors.

11–2 Using the *NEC* Charts

NEC 310 deals with conductors for general wiring. *Table 310.15(B)(16)* through *Table 310.15(B)(19)* are generally used to select a wire size according to the requirements of the circuit. Each of these tables lists different conditions. The table used is determined by the wiring conditions. *Table 310.15(B)(16) (Figure 11–6)*

Table 310.15(B)16 Allowable Ampacities of Insulated Conductors Rated Up to and Including 2000 Volts, 60°C Through 90°C (140°F Through 194°F) Not More Than Three Current-Carrying Conductors in Raceway, Cable, or Earth (Direct Buried), on Ambient Temperature of 30°C (86°F)

Size AWG or kcmil	Temperature Rating of Conductors [See Table 310.104(A)]						Size AWG or kcmil
	60°C (140°F)	75°C (167°F)	90°C (194°F)	60°C (140°F)	75°C (167°F)	90°C (194°F)	
	Types TW,UF	Types RHW, THHW,THW, THWN,XHHW, USE,ZW	Types TBS,SA,SIS, FEP,FEPB,MI,RHH, RHW-2,THHN, THHW,THW-2, THWN-2,USE-2, XHH, XHHW, XHHW-2, ZW-2	Types TW,UF	Types RHW, THHW,THW, THWN,XHHW, USE,ZW	Types TBS,SA,SIS, FEP,FEPB,MI,RHH, RHW-2,THHN, THHW,THW-2, THWN-2,USE-2, XHH, XHHW, XHHW-2, ZW-2	
	COPPER			ALUMINUM OR COPPER-CLAD ALUMINUM			
18	-	-	14	-	-	-	-
16	-	-	18	-	-	-	-
14**	15	20	25	-	-	-	12**
12**	20	25	30	15	20	25	10**
10**	30	35	40	25	30	35	8
8	40	50	55	35	40	45	
6	55	65	75	40	50	60	6
4	70	85	95	55	65	75	4
3	85	100	115	65	75	85	3
2	95	115	130	75	90	100	2
1	110	130	145	85	100	115	1
1/0	125	150	170	100	120	135	1/0
2/0	145	175	195	115	135	150	2/0
3/0	165	200	225	130	155	175	3/0
4/0	195	230	260	150	180	205	4/0
250	215	255	290	170	205	230	250
300	240	285	320	190	230	260	300
350	260	310	350	210	250	280	350
400	280	335	380	225	270	305	400
500	320	380	430	260	310	350	500
600	350	420	475	285	340	385	600
700	385	460	520	310	375	425	700
750	400	475	535	320	385	435	750
800	410	490	555	330	395	445	800
900	435	520	585	355	425	480	900
1000	455	545	615	375	445	500	1000
1250	495	590	665	405	485	545	1250
1500	525	625	705	435	520	585	1500
1750	545	650	735	455	545	615	1750
2000	555	665	750	470	560	630	2000

*Refer to 310.15(B)(2) for the ampacity correction factors where the ambient temperature is other than 30°C (86°F)
** Refer to 240.4(D) for conductor overcurrent protection limitations.

FIGURE 11–6 *NEC Table 310.16. (Reprinted with permission from NFPA 70-2011, National Electrical Code, Copyright © 2011, National Fire Protection Association, Quincy, MA 02269. This reprinted material is not the complete and official position of the NFPA on the referenced subject, which is represented only by the standard in its entirety.)*

Table 310.15(B)17 Allowable Ampacities of Insulated Conductors Rated Up to and Including 2000 Volts in Free Air, Based on Ambient Air temperature of 30°C (86°F)

Size AWG or kcmil	Temperature Rating of Conductors [See Table 310.104(A)]						Size AWG or kcmil
	60°C (140°F)	75°C (167°F)	90°C (194°F)	60°C (140°F)	75°C (167°F)	90°C (194°F)	
	Types TW,UF	Types RHW, THHW,THW, THWN,XHHW, ZW	Types TBS,SA,SIS, FEP,FEPB,MI,RHH, RHW-2,THHN, THHW,THW-2, THWN-2,USE-2, XHH, XHHW, XHHW-2, ZW-2	Types TW,UF	Types RHW, THHW,THW, THWN,XHHW, ZW	Types TBS,SA,SIS, FEP,FEPB,MI,RHH, RHW-2,THHN, THHW,THW-2, THWN-2,USE-2, XHH, XHHW, XHHW-2, ZW-2	
	COPPER			ALUMINUM OR COPPER-CLAD ALUMINUM			
18	-	-	18	-	-	-	-
16	-	-	24	-	-	-	-
14 **	25	30	35	-	-	-	-
12 **	30	35	40	25	30	35	12 **
10 **	40	50	55	35	40	45	10 **
8	60	70	80	45	55	60	8
6	80	95	105	60	75	85	6
4	105	125	140	80	100	115	4
3	120	145	165	95	115	130	3
2	140	170	190	110	135	150	2
1	165	195	220	130	155	175	1
1/0	195	230	260	150	180	205	1/0
2/0	225	265	300	175	210	235	2/0
3/0	260	310	350	200	240	270	3/0
4/0	300	360	405	235	280	315	4/0
250	340	405	455	265	315	355	250
300	375	445	500	290	350	395	300
350	420	505	570	330	395	445	350
400	455	545	615	355	425	480	400
500	515	620	700	405	485	545	500
600	575	690	780	445	540	615	600
700	630	755	850	500	595	670	700
750	655	785	885	515	620	700	750
800	680	815	920	535	645	725	800
900	730	870	980	580	700	790	900
1000	780	935	1055	625	750	845	1000
1250	890	1065	1200	710	855	965	1250
1500	980	1175	1325	795	950	1070	1500
1750	1070	1280	1445	875	1050	1185	1750
2000	1155	1385	1560	960	1150	1295	2000

* Refer to 310.15(B)(2) for the ampacity correction factors where the ambient termperature is other than 30 °C (86 °F)

** Refer to 240.4(D) for conductor overcurrent protection limitation.

FIGURE 11–7   *NEC Table 310.17. (Reprinted with permission from NFPA 70-2011, National Electrical Code, Copyright © 2011, National Fire Protection Association, Quincy, MA 02269. This reprinted material is not the complete and official position of the NFPA on the referenced subject, which is represented only by the standard in its entirety.)*

lists **ampacities (current-carrying ability)** of not more than three single insulated conductors in raceway or cable or buried in the earth based on an **ambient** (surrounding) **air temperature** of 30°C (86°F). *Table 310.15(B)(17) (Figure 11–7)* lists ampacities of single insulated conductors in free air based on an ambient temperature of 30°C. *Table 310.15(B)(18) (Figure 11–8)* lists the ampacities of three single insulated conductors in raceway or cable based on an ambient temperature of 40°C (104°F). The conductors listed in *Table 310.15(B)(18)*

Table 310.15(B)18 Allowable Ampacities of Insulated Conductors Rated Up to and Including 2000 Volts, 150°C Through 250°C (302°F Through 482°F) Not More Than Three Current-Carrying Conductors in Raceway or Cable Based on Ambient Air Temperature of 40°C (104°F)

| | Temperature Rating of Conductors [See Table 310.104(A)] | | | | |
Size AWG or kcmil	150°C (302°F) Type Z COPPER	200°C (392°F) Types FEP, FEPB, PFA, SA COPPER	250°C (482°F) Types PFAH, TFE NICKEL OR NICKEL-COATED COPPER	150°C (302°F) Type Z ALUMINUM OR COPPER-CLAD ALUMINUM	Size AWG or kcmil
14	34	36	39	-	-
12	43	45	54	30	12
10	55	60	73	44	10
8	76	83	93	57	8
6	96	110	117	75	6
4	120	125	148	94	4
3	143	152	166	109	3
2	160	171	191	124	2
1	186	197	215	145	1
1/0	215	229	244	169	1/0
2/0	251	260	273	198	2/0
3/0	288	297	308	227	3/0
4/0	332	346	361	260	4/0

*Refer to 310.15(B)(2)(b) for the ampacity correction factors where the ambient temperature is other than 40 °C (104 °F)

FIGURE 11–8   *NEC Table 310.18. (Reprinted with permission from NFPA 70-2011, National Electrical Code, Copyright © 2011, National Fire Protection Association, Quincy, MA 02269. This reprinted material is not the complete and official position of the NFPA on the referenced subject, which is represented only by the standard in its entirety).*

and Table 310.15(B)(19) are generally used for high-temperature locations. The heading at the top of each table lists a different set of conditions.

11–3   Factors That Determine Ampacity

Conductor Material

One of the factors that determines the resistivity of wire is the material from which the wire is made. The wire tables list the current-carrying capacity of both copper and aluminum or copper-clad aluminum conductors. The currents listed in the left-hand half of *Table 310.15(B)(16),* for example, are for copper wire. The currents listed in the right-hand half of the table are for aluminum or

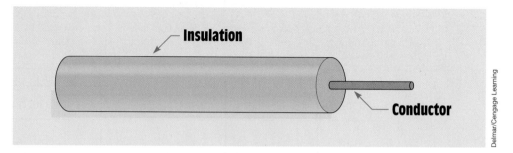

FIGURE 11–9 Insulation around conductor.

copper-clad aluminum. The table indicates that a copper conductor is permitted to carry more current than an aluminum conductor of the same size and insulation type. An 8 **American Wire Gauge (AWG)** copper conductor with Type TW insulation is rated to carry a maximum of 40 amperes. An 8 AWG aluminum conductor with Type TW insulation is rated to carry only 35 amperes. One of the columns of *Table 310.15(B)(18)* and *Table 310.15(B)(19)* gives the ampacity rating of nickel or nickel-coated copper conductors.

Insulation Type

Another factor that determines the amount of current a conductor is permitted to carry is the type of insulation used. This is due to the fact that different types of insulation can withstand more heat than others. The **insulation** is the non-conductive covering around the wire *(Figure 11–9)*. The voltage rating of the conductor is also determined by the type of insulation. The amount of voltage a particular type of insulation can withstand without breaking down is determined by the type of material it is made of and its thickness. *NEC Table 310.104(A)* (not shown due to space limitations) lists information concerning different types of insulation. The table is divided into columns that list the trade name; identification letters; **maximum operating temperature;** whether the insulation can be used in a wet, damp, or dry location; material; thickness; and outer covering.

■ **EXAMPLE 11–1**

Find the maximum operating temperature of Type RHW insulation. (Note: Refer to the *NEC.*)

Solution

Find Type RHW in the second column of *Table 310.104(A)*. The third column lists a maximum operating temperature of 75°C, or 167°F.

■ **EXAMPLE 11-2**

Can Type THHN insulation be used in wet locations?

Solution

Locate Type THHN insulation in the second column. The fourth column indicates that this insulation can be used in **dry** and **damp locations.** This type of insulation cannot be used in **wet locations.** For an explanation of the difference between damp and wet locations, consult "locations" in *Article 100* of the *NEC*.

A good thing to remember is that insulation materials that contain the letter *W*, such as RHW, THW, THWN, and so on may be used in wet locations.

11–4   Correction Factors

One of the main factors that determines the amount of current a conductor is permitted to carry is the ambient, or surrounding, air temperature. *Table 310.15(B)(16)*, for example, lists the ampacity of not more than three conductors in a raceway in free air. These ampacities are based on an ambient air temperature of 30°C, or 86°F. If these conductors are to be used in a location that has a higher ambient temperature, the ampacity of the conductor must be reduced because the resistance of copper or aluminum increases with an increase of temperature. Temperature correction factors can be found in *Table 310.15(B)(2)(a)* and *310.15(B)(2)(b)*. *Table 310.15(B)(2)(a)* is for conductors rated at 30°C or 86°F. The ampacity of conductors in *Table 310.15(B)(16)* and *Table 310.15(B)(17)* are based on an ambient air temperature of 30°C. The correction factors for conductors rated at 40°C are found in *Table 310.15(B)(2)(b)*. The ampacity of conductors in *Table 310.15(B)(18)* and *Table 310.15(B)(19)* are based on an ambient temperature of 40°C. The correction factors found in Table *310.15(B)(2)(a)* are shown in *Figure 11-10*.

More Than Three Conductors in a Raceway

Table 310.15(B)(16) and *Table 310.15(B)(18)* list three conductors in a raceway. If a raceway is to contain more than three conductors, the ampacity of the conductors must be derated because the heat from each conductor combines with the heat dissipated by the other conductors to produce a higher temperature inside the raceway. *NEC Table 310.15(B)(3)(a) (Figure 11–11)* lists these correction factors. If the raceway is used in an area with a greater ambient

Table 310.15(B)(2)(a) Ambient Temperature Correction Factors Based on 30 °C (86 °F)

For ambient temperatures other than 30 °C (86 °F) multiply the allowable ampacities specified in the ampacity tables by the appropriate correction factor shown below.

Ambient Temperature (°C)	Temperature Rating of Conductor			Ambient Temperature (°F)
	60°C	75°C	90°C	
10 or less	1.29	1.20	1.15	50 or less
11-15	1.22	1.15	1.12	51-59
16-20	1.15	1.11	1.08	60-68
21-25	1.08	1.05	1.04	69-77
26-30	1.00	1.00	1.00	78-86
31-35	0.91	0.94	0.96	87-95
36-40	0.82	0.88	0.91	96-104
41-45	0.71	0.82	0.87	105-113
46-50	0.58	0.75	0.82	114-122
51-55	0.41	0.67	0.76	123-131
56-60	-	0.58	0.71	132-140
61-65	-	0.47	0.65	141-149
66-70	-	0.33	0.58	150-158
71-75	-	-	0.50	159-167
76-80	-	-	0.41	168-176
80-85	-	-	0.29	177-185

FIGURE 11–10 *Ambient temperature correction factors. (Reprinted with permission from NFPA 70-2011, National Electrical Code, Copyright © 2011, National Fire Protection Association, Quincy, MA 02269. This reprinted material is not the complete and official position of the NFPA on the referenced subject, which is represented only by the standard in its entirety).*

Table 310.15(B)(3)(a) Adjustment Factors for More Than Three Current-Carrying conductors in a Raceway or Cable

Number of Conductors	Percent of Values in Table 310.15(B)(16) through Table 310.15(B)(19) as Adjusted for Ambient Temperature if Necessary
4-6	80
7-9	70
10-20	50
21-30	45
31-40	40
41 and above	35

FIGURE 11–11 *NEC Table 310.15(B)(3)(a). (Reprinted with permission from NFPA 70-2011, National Electrical Code, Copyright © 2011, National Fire Protection Association, Quincy, MA 02269. This reprinted material is not the complete and official position of the NFPA on the referenced subject, which is represented only by the standard in its entirety.)*

temperature than that listed in the appropriate wire table, the temperature correction factor must also be applied.

Determining Conductor Size Using the *NEC*

Using the *NEC* to determine the amount of current a conductor is permitted to carry and the proper size conductor to use for a particular application are not

■ **EXAMPLE 11-3**

What is the maximum ampacity of a 4 AWG copper conductor with Type THWN insulation used in an area with an ambient temperature of 43°C?

Solution

Determine the ampacity of a 4 AWG copper conductor with Type THWN insulation from the wire table. Type THWN insulation is located in the second column of *Table 310.15(B)(16)*. The table lists an ampacity of 85 A for this conductor. Locate 43°C in the far left-hand column of the correction factor chart shown in *Figure 11-10*; 43°C falls between 41°C and 45°C. Follow across to the 75°C column. The chart lists a correction factor of 0.82. The ampacity of the conductor in the above wire table is to be multiplied by the correction factor:

$$85 \text{ A} \times 0.82 = 69.7 \text{ A}$$

■ EXAMPLE 11-4

What is the maximum ampacity of a 1/0 AWG copper-clad aluminum conductor with Type RHH insulation if the conductor is to be used in an area with an ambient air temperature of 100°F?

Solution

Locate the column that contains Type RHH insulation in the copper-clad aluminum section of *Table 310.15(B)(16)*. The table indicates a maximum ampacity of 135 A for this conductor. The chart shown in *Figure 11-10* is used to determine the correction factor for this temperature. Fahrenheit degrees are located in the far right-hand column of the chart; 100°F falls between 97°F and 104°F. The correction factor for this temperature is 0.91. Multiply the ampacity of the conductor by this factor:

$$135 \text{ A} \times 0.91 = 122.85 \text{ A}$$

■ EXAMPLE 11-5

Twelve 14 AWG copper conductors with Type RHW insulation are to be run in a conduit. The conduit is used in an area that has an ambient temperature of 110°F. What is the maximum ampacity of these conductors?

Solution

Find the ampacity of a 14 AWG copper conductor with Type RHW insulation. Type RHW insulation is located in the second column of *Table 310.15(B)(16)*. A 14 AWG copper conductor has an ampacity of 20 A. Next, use the correction factor for an ambient temperature of 110°F in *Table 310.15(B)(2)(a)* shown in *Figure 11-10*. A correction factor of 0.82 will be used:

$$20 \text{ A} \times 0.82 = 16.4 \text{ A}$$

The correction factor located in *Table 310.15(B)(3)(a)* must now be used. The table indicates a correction factor of 50% when 10 through 20 conductors are run in a raceway:

$$16.4 \text{ A} \times 0.50 = 8.2 \text{ A}$$
$$16.4 \text{ A} \times 0.50 = 8.2 \text{ A}$$

Each 14 AWG conductor has a maximum current rating of 8.2 A.

the same thing. The several factors that must be considered when selecting a conductor for a specific job are of no consequence when determining the ampacity of a conductor. One of these factors is whether the load is a continuous or noncontinuous load. The *NEC* defines a continuous load as one where the maximum current is expected to continue for three hours or more. Most industrial motor and lighting loads, for example, would be considered continuous loads. *NEC 210.19(A)(1)* states that conductors must have an ampacity not less than the noncontinuous load plus 125% of the continuous load. Basically, the ampacity of a conductor must be 125% greater than the current rating of a continuous load.

Another factor that affects the selection of a conductor is termination temperature limitations. *NEC 110.14(C)* states that the temperature rating of a conductor must be selected so as not to exceed the lowest temperature rating of any connected conductor, termination, or device. Because the termination temperature rating of most devices is not generally known, *NEC 110.14(C)(1)(a)* states that conductors for circuits rated at 100 amperes or less are to be selected from the 60°C column. This does not mean that conductors with a higher temperature rating cannot be used, but their size must be selected from the 60°C column. The only exception to this is motors that are marked with *NEMA* (National Electrical Manufacturers Association) code letters B, C, or D. *NEC 110.14(C)(1)(a)(4)* permits conductors for motors with these code letters to be selected from the 75°C column.

■ EXAMPLE 11-6

Assume that a motor with a full-load current rating of 28 A is to be connected with copper conductors that have Type THW insulation. The motor is located in an area with an ambient temperature of 30°C. The motor is not marked with *NEMA* code letters, and the termination temperature is not known. What size conductors should be used?

Solution

Because a motor load is continuous, multiply the full-load current rating by 125%:

$$28 \text{ A} \times 1.25 = 35 \text{ A}$$

Refer to *NEC Table 310.15(B)(16)*. Type THW insulation is located in the 75°C column. Although this conductor is located in the 75°C column, the wire size must be chosen from the 60°C column because the termination temperature is not known and the motor does not contain *NEMA* code letters. The nearest wire size without going under 35 A is an 8 AWG.

■ EXAMPLE 11-7

Assume that a bank of heating resistors is rated at 28 kW and is connected to 240 V. The resistors operate for more than three hours at a time and are located in an area with an ambient temperature of 86°F. Aluminum conductors with Type THWN-2 insulation are to be used to connect the heaters. What size conductors should be used?

Solution

Determine the amperage of the load using Ohm's law:

$$I = \frac{P}{E}$$

$$I = \frac{28,000 \text{ W}}{240 \text{ V}}$$

$$I = 116.667 \text{ A}$$

Because the load is continuous, the conductors must have an ampacity 125% greater than the load current.

$$I = 116.667 \text{ A} \times 1.25$$

$$I = 145.834 \text{ A or } 146 \text{ A}$$

Refer to *NEC Table 310.15(B)(16)*. Type THWN-2 insulation is located in the 90°C column. The conductor size, however, must be chosen from the 75°C column. The nearest size aluminum conductor without going less than 146 A is 3/0 AWG.

The requirements of *NEC 110.14(C)* also require that conductors be selected from *Table 310.15(B)(16)* as a general rule. *NEC Table 310.15(B)(17)*, for example, lists the current-carrying capacity of conductors located in free air. Because the conductors are generally terminated inside an enclosure, however, they must be chosen from a table that lists the ampacity of conductors inside an enclosure.

For circuits with a current of 100 A or greater, *NEC 110.14(C)(1)(b)* permits conductors to be selected from the 75°C column.

Another factor to be considered when selecting a conductor for a particular application is the ambient temperature. *Table 310.15(B)(2)(a)* states: *"For ambient temperatures other than 30°C (86°F), multiply the allowable ampacities shown above by the appropriate factor shown below."* The wire tables are used to determine the maximum current-carrying capacity of conductors based on the type of material from which they are made, the type of insulation, and the surrounding air temperature (ambient temperature). The correction factors listed have a value less than 1 for any temperature greater than 30°C (86°F) because

conductors become more resistive as temperature increases. This reduces the amount of current they can safely carry.

When determining the conductor size needed for a particular application, you are not determining the maximum current a conductor can carry. You are determining the size conductor needed to carry the amount of current for the particular application. Therefore, when determining conductor size in an area where ambient temperature is a concern, you must divide the needed current by the correction factor instead of multiplying by it. Dividing by the correction factor will

---

■ EXAMPLE 11-8

An electric annealing oven is located in an area with an ambient temperature of 125°F. The oven contains a 50-kW electric heating element and is connected to 480 volts. The conductors are to be copper with type THHN insulation. The termination temperature is not known. The furnace is expected to operate more than three hours continuously. What size conductor should be employed to make this connection?

Solution

Determine the amount of current needed to operate the furnace.

$$I = \frac{P}{E}$$

$$I = \frac{50{,}000}{480}$$

$$I = 104.167 \text{ amperes}$$

Because the load is continuous, it must be increased by 125%:

$$104.167 \times 1.25 = 130.2 \text{ amperes}$$

The next step is to apply the correction factor for temperature. Type THHN insulation is located in the 90°C column. The correction factor for a temperature of 125°F is 0.76, as shown in *Figure 11-10*. To determine the current rating of the conductor at 125°F, divide the current by the correction factor:

$$I = \frac{130.2}{0.76}$$

$$I = 171.3 \text{ amperes}$$

Because the termination temperature is not known and the current is over 100 amperes, the conductor size will be selected from the 75°C column of *NEC Table 310.15(B)(16)*. A 2/0 AWG conductor will be used.

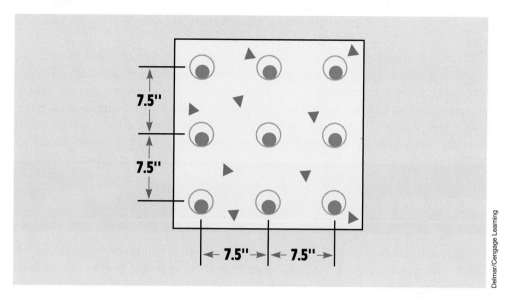

Delmar/Cengage Learning

FIGURE 11–12 Electric duct banks.

result in an increase of the amperage used in selecting the conductor size. This increase is needed to offset the effects of higher temperature on the conductor.

Duct Banks

Duct banks are often used when it becomes necessary to bury cables in the ground. An electric duct can be a single metallic or nonmetallic conduit. An electric duct bank is a group of electric ducts buried together as shown in *Figure 11–12*. When a duct bank is used, the center points of individual ducts should be separated by a distance of not less than 7.5 inches.

11–5 Calculating Conductor Sizes and Resistance

Although the wire tables in the *NEC* are used to determine the proper size wire for most installations, there are instances in which these tables are not used. The formula in *310.15(c)* of the *NEC* is used for ampacities not listed in the wire tables:

$$I = \sqrt{\frac{TC - (TA + \Delta TD)}{RDC(1 + YC)RCA}}$$

where

TC = conductor temperature in °C

TA = ambient temperature in °C

ΔTD = dielectric loss temperature rise

RDC = DC resistance of conductor at temperature TC

YC = component AC resistance resulting from skin
effect and proximity effect

RCA = effective thermal resistance between
conductor and surrounding ambient

Although this formula is seldom used by electricians, the *NEC* does permit its use under the supervision of an electrical engineer.

Long Wire Lengths

Another situation in which it becomes necessary to calculate wire sizes instead of using the tables in the *Code* is when the conductor becomes excessively long. The listed ampacities in the *Code* tables assume that the length of the conductor will not increase the resistance of the circuit by a significant amount. When the wire becomes extremely long, however, it is necessary to calculate the size of wire needed.

All wire contains resistance. As wire is added to a circuit, it has the effect of adding resistance in series with the load *(Figure 11–13)*. Four factors determine the resistance of a length of wire:

1. *The type material from which the wire is made.* Different types of material have different wire resistances. A copper conductor will have less resistance than an aluminum conductor of the same size and length. An aluminum conductor will have less resistance than a piece of iron wire the same size and length.

2. *The diameter of the conductor.* The larger the diameter, the less resistance it will have. A large-diameter pipe, for example, will have less resistance to the flow of water than will a small-diameter pipe *(Figure 11–14)*. The cross-sectional area of round wire is measured in circular mils (CM). One mil equals 0.001 inch. A **circular mil** is the cross-sectional area of the wire in mils squared. For example, assume a wire has a diameter of 0.064 inch. Sixty-four thousandths should be written as a whole number, not as a decimal or a fraction (64^2 [64 mil × 64 mil] = 4096 CM).

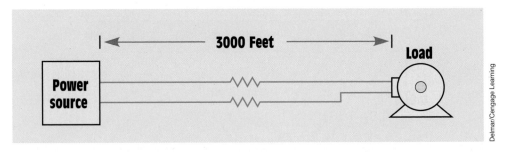

FIGURE 11–13 Long wire runs have the effect of adding resistance in series with the load.

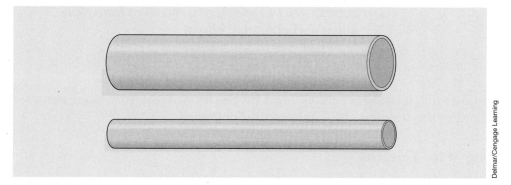

Delmar/Cengage Learning

FIGURE 11–14   A large pipe has less resistance to the flow of water than a small pipe.

3. *The length of the conductor.* The longer the conductor, the more resistance it will have. Adding length to a conductor has the same effect as connecting resistors in series.

4. *The temperature of the conductor.* As a general rule, most conductive materials will increase their resistance with an increase of temperature. Some exceptions to this rule are carbon, silicon, and germanium. If the coefficient of temperature for a particular material is known, its resistance at different temperatures can be calculated. Materials that increase their resistance with an increase of temperature have a positive coefficient of temperature. Materials that decrease their resistance with an increase of temperature have a negative coefficient of temperature.

In the English system of measure, a standard value of resistance called the **mil-foot** is used to determine the resistance of different lengths and sizes of wire. A mil-foot is a piece of wire 1 foot long and 1 mil in diameter *(Figure 11–15)*. A chart showing the resistance of a mil-foot of wire at 20°C

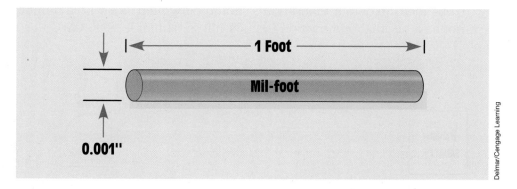

Delmar/Cengage Learning

FIGURE 11–15   A mil-foot is equal to a piece of wire one foot long and one thousandth of an inch in diameter.

Resistivity (K) of Materials

Material	Ω = CM ft at 20°C	Temp. coeff. (Ω per °C)
Aluminum	17	0.004
Carbon	22,000 aprx.	−0.0004
Constantan	295	0.000002
Copper	10.4	0.0039
Gold	14	0.004
Iron	60	0.0055
Lead	126	0.0043
Manganin	265	0.000000
Mercury	590	0.00088
Nichrome	675	0.0002
Nickel	52	0.005
Platinum	66	0.0036
Silver	9.6	0.0038
Tungsten	33.8	0.005

Delmar/Cengage Learning

FIGURE 11–16   Resistivity of materials.

is shown in *Figure 11–16*. Notice the wide range of resistances for different materials. The temperature coefficient of the different types of conductors is listed also.

Calculating Resistance

Now that a standard measure of resistance for different types of materials is known, the resistance of different lengths and sizes of these materials can be calculated. The formula for calculating resistance of a certain length, size, and type of wire is

$$R = \frac{K \times L}{CM}$$

where

R = resistance of the wire

K = ohms-CM per foot

$$L = \text{length of wire in feet}$$

$$CM = \text{circular mil (area of the wire)}$$

This formula can be converted to calculate other values in the formula such as, to find the SIZE of wire to use

$$CM = \frac{K \times L}{R}$$

to find the LENGTH of wire to use

$$L = \frac{R \times CM}{K}$$

to find the TYPE of wire to use

$$K = \frac{R \times CM}{L}$$

■ EXAMPLE 11-9

Find the resistance of a piece of 6 AWG copper wire 550 ft long. Assume a temperature of 20°C. The formula to be used is

$$R = \frac{K \times L}{CM}$$

Solution

The value for K can be found in the table in *Figure 11–16*. The table indicates a value of 10.4 Ω-CM per foot for a copper conductor. The length, L, was given as 550 ft, and the circular mil area of 6 AWG wire is listed as 26,250 in the table shown in *Figure 11–17*.

$$R = \frac{10.4\ \Omega\text{-CM/ft} \times 550\text{ft}}{26{,}250\ CM}$$

$$R = \frac{5720\ \Omega\text{-CM}}{26{,}250\ CM}$$

$$R = 0.218\ \Omega$$

B & S Gauge No.	Diam. in Mils	Area in Circular Mils	Ohms per 1000 Ft. (ohms per 100 meters)						Pounds per 1000 Ft. (kg per 100 meters)			
			Copper*		Copper*		Aluminum		Copper		Aluminum	
			68°F	(20°C)	167°F	(75°C)	68°F	(20°C)				
0000	460	211,600	0.049	(0.016)	0.0596	(0.0195)	0.0804	(0.0263)	640	(95.2)	195	(29.0)
000	410	167,800	0.0618	(0.020)	0.0752	(0.0246)	0.101	(0.033)	508	(75.5)	154	(22.9)
00	365	133,100	0.078	(0.026)	0.0948	(0.031)	0.128	(0.042)	403	(59.9)	122	(18.1)
0	325	105,500	0.0983	(0.032)	0.1195	(0.0392)	0.161	(0.053)	320	(47.6)	97	(14.4)
1	289	83,690	0.1239	(0.0406)	0.151	(0.049)	0.203	(0.066)	253	(37.6)	76.9	(11.4)
2	258	66,370	0.1563	(0.0512)	0.191	(0.062)	0.256	(0.084)	201	(29.9)	61.0	(9.07)
3	229	52,640	0.1970	(0.0646)	0.240	(0.079)	0.323	(0.106)	159	(23.6)	48.4	(7.20)
4	204	41,740	0.2485	(0.0815)	0.302	(0.099)	0.408	(0.134)	126	(18.7)	38.4	(5.71)
5	182	33,100	0.3133	(0.1027)	0.381	(0.125)	0.514	(0.168)	100	(14.9)	30.4	(4.52)
6	162	26,250	0.395	(0.129)	0.481	(0.158)	0.648	(0.212)	79.5	(11.8)	24.1	(3.58)
7	144	20,820	0.498	(0.163)	0.606	(0.199)	0.817	(0.268)	63.0	(9.37)	19.1	(2.84)
8	128	16,510	0.628	(0.206)	0.764	(0.250)	1.03	(0.338)	50.0	(7.43)	15.2	(2.26)
9	114	13,090	0.792	(0.260)	0.963	(0.316)	1.30	(0.426)	39.6	(5.89)	12.0	(1.78)
10	102	10,380	0.999	(0.327)	1.215	(0.398)	1.64	(0.538)	31.4	(4.67)	9.55	(1.42)
11	91	8,234	1.260	(0.413)	1.532	(0.502)	2.07	(0.678)	24.9	(3.70)	7.57	(1.13)
12	81	6,530	1.588	(0.520)	1.931	(0.633)	2.61	(0.856)	19.8	(2.94)	6.00	(0.89)
13	72	5,178	2.003	(0.657)	2.44	(0.80)	3.29	(1.08)	15.7	(2.33)	4.80	(0.71)
14	64	4,107	2.525	(0.828)	3.07	(1.01)	4.14	(1.36)	12.4	(1.84)	3.80	(0.56)
15	57	3,257	3.184	(1.044)	3.98	(1.27)	5.22	(1.71)	9.86	(1.47)	3.00	(0.45)
16	51	2,583	4.016	(1.317)	4.88	(1.60)	6.59	(2.16)	7.82	(1.16)	2.40	(0.36)
17	45.3	2,048	5.06	(1.66)	6.16	(2.02)	8.31	(2.72)	6.20	(0.922)	1.90	(0.28)
18	40.3	1,624	6.39	(2.09)	7.77	(2.55)	10.5	(3.44)	4.92	(0.713)	1.50	(0.22)
19	35.9	1,288	8.05	(2.64)	9.79	(3.21)	13.2	(4.33)	3.90	(0.580)	1.20	(0.18)
20	32	1,022	10.15	(3.33)	12.35	(4.05)	16.7	(5.47)	3.09	(0.459)	0.94	(0.14)
21	28.5	810	12.8	(4.2)	15.6	(5.11)	21.0	(6.88)	2.45	(0.364)	0.745	(0.11)
22	25.4	642	16.1	(5.3)	19.6	(6.42)	26.5	(8.69)	1.95	(0.290)	0.591	(0.09)
23	22.6	510	20.4	(6.7)	24.8	(8.13)	33.4	(10.9)	1.54	(0.229)	0.468	(0.07)
24	20.1	404	25.7	(8.4)	31.2	(10.2)	42.1	(13.8)	1.22	(0.181)	0.371	(0.05)
25	17.9	320	32.4	(10.6)	39.4	(12.9)	53.1	(17.4)	0.97	(0.14)	0.295	(0.04)
26	15.9	254	40.8	(13.4)	49.6	(16.3)	67.0	(22.0)	0.77	(0.11)	0.234	(0.03)
27	14.2	202	51.5	(16.9)	62.6	(20.5)	84.4	(27.7)	0.61	(0.09)	0.185	(0.03)
28	12.6	160	64.9	(21.3)	78.9	(25.9)	106	(34.7)	0.48	(0.07)	0.147	(0.02)
29	11.3	126.7	81.8	(26.8)	99.5	(32.6)	134	(43.9)	0.384	(0.06)	0.117	(0.02)
30	10	100.5	103.2	(33.8)	125.5	(41.1)	169	(55.4)	0.304	(0.04)	0.092	(0.01)
31	8.93	79.7	130.1	(42.6)	158.2	(51.9)	213	(69.8)	0.241	(0.04)	0.073	(0.01)
32	7.95	63.2	164.1	(53.8)	199.5	(65.4)	269	(88.2)	0.191	(0.03)	0.058	(0.01)
33	7.08	50.1	207	(68)	252	(82.6)	339	(111)	0.152	(0.02)	0.046	(0.01)
34	6.31	39.8	261	(86)	317	(104)	428	(140)	0.120	(0.02)	0.037	(0.01)
35	5.62	31.5	329	(108)	400	(131)	540	(177)	0.095	(0.01)	0.029	
36	5	25	415	(136)	505	(165)	681	(223)	0.076	(0.01)	0.023	
37	4.45	19.8	523	(171)	636	(208)	858	(281)	0.0600	(0.01)	0.0182	
38	3.96	15.7	660	(216)	802	(263)	1080	(354)	0.0476	(0.01)	0.0145	
39	3.53	12.5	832	(273)	1012	(332)	1360	(446)	0.0377	(0.01)	0.0115	
40	3.15	9.9	1049	(344)	1276	(418)	1720	(564)	0.0299	(0.01)	0.0091	
41	-											
42	2.5											
43	-											
44	1.97											

American Wire Gauge Table

* Resistance figures are given for standard annealed copper. For hard-drawn copper add 2%

FIGURE 11–17 American Wire Gauge table.

Delmar/Cengage Learning

■ EXAMPLE 11-10

An aluminum wire 2250 ft long cannot have a resistance greater than 0.2 Ω. What size aluminum wire must be used?

Solution

To find the size of wire, use

$$CM = \frac{K \times L}{R}$$

$$CM = \frac{17 \ \Omega\text{-CM/ft} \times 2250 \ \text{ft}}{0.2 \ \Omega}$$

$$CM = \frac{38,250 \ \Omega\text{-CM}}{0.2 \ \Omega}$$

$$CM = 191,250$$

The nearest standard size conductor for this installation can be found in the American Wire Gauge table. Because the resistance cannot be greater than 0.2 Ω, the conductor cannot be smaller than 191,250 CM. The nearest standard conductor size is 0000 AWG.

Good examples of when it becomes necessary to calculate the wire size for a particular installation can be seen in the following problems.

■ EXAMPLE 11-11

A manufacturing plant has a cooling pond located 4000 ft from the plant. Six pumps are used to circulate water between the pond and the plant. In cold weather, however, the pumps can freeze and fail to supply water. The plant owner decides to connect electric-resistance heaters to each pump to prevent this problem. The six heaters are connected to a two-conductor cable from the plant at a junction box *(Figure 11–18)*. The heaters operate on 480 V and have a total current draw of 50 A. What size copper conductors should be used to supply power to the heaters if the voltage drop at the junction box is to be kept to 3% of the applied voltage? Assume an average ambient temperature of 20°C.

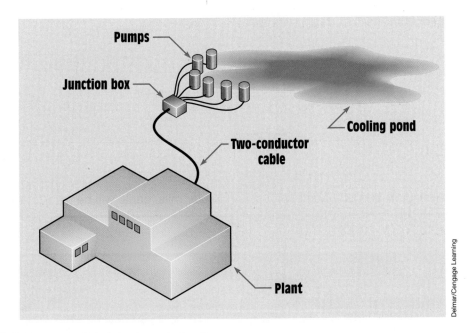

FIGURE 11–18   Calculating long wire lengths.

Solution

The first step in the solution of this problem is to determine the maximum amount of resistance the conductors can have without producing a voltage drop greater than 3% of the applied voltage. The maximum amount of voltage drop can be calculated by multiplying the applied voltage by 3%:

$$480 \text{ V} \times 0.03 = 14.4 \text{ V}$$

The maximum amount of resistance can now be calculated using Ohm's law:

$$R = \frac{E}{I}$$

$$R = \frac{14.4 \text{ V}}{50 \text{ A}}$$

$$R = 0.288 \ \Omega$$

The distance to the pond is 4000 ft. Because two conductors are used, the resistance of both conductors must be considered. Two conductors 4000 ft long will have the same resistance as one conductor 8000 ft long. For this reason, a length of 8000 ft is used in the formula:

$$CM = \frac{K \times L}{R}$$

$$CM = \frac{10.4 \ \Omega\text{-CM/ft} \times 8000 \ \text{ft}}{0.288 \ \Omega}$$

$$CM = \frac{83,200 \ \Omega\text{-CM}}{0.288 \ \Omega}$$

$$CM = 288,888.9$$

For this installation 300-kcmil (thousand circular mils) cable will be used.

■ EXAMPLE 11-12

The next problem concerns conductors used in a three-phase system. Assume that a motor is located 2500 ft from its power source and operates on 560 V. When the motor starts, it has a current draw of 168 A. The voltage drop at the motor terminals cannot be permitted to be greater than 5% of the source voltage during starting. What size aluminum conductors should be used for this installation?

Solution

The solution to this problem is very similar to the solution in the previous example. First, find the maximum voltage drop that can be permitted at the load by multiplying the source voltage by 5%:

$$E = 560 \ \text{V} \times 0.05$$
$$E = 28 \ \text{V}$$

The second step is to determine the maximum amount of resistance of the conductors. To calculate this value, the maximum voltage drop is divided by the starting current of the motor:

$$R = \frac{E}{I}$$

$$R = \frac{28 \ \text{V}}{168 \ \text{A}}$$

$$R = 0.167 \ \Omega$$

The next step is to calculate the length of the conductors. In the previous example, the lengths of the two conductors were added to find the total amount of wire resistance. In a single-phase system, each conductor must carry the same amount of current. During any period of time, one conductor is supplying current from the source to the load, and the other conductor completes the circuit by permitting the same amount of current to flow from the load to the source.

In a balanced three-phase circuit, three currents are 120° out of phase with each other *(Figure 11–19)*. These three conductors share the flow of current between source and load. In *Figure 11–19*, two lines labeled A and B have been drawn through the three current waveforms. Notice that at position A the current flow in phase 1 is maximum and in a positive direction. The current flow in phases 2 and 3 is less than maximum and in a negative direction. This condition corresponds to the example shown in *Figure 11–20*. Notice that maximum current is flowing in only one conductor. Less than maximum current is flowing in the other two conductors.

Observe the line marking position B in *Figure 11–19*. The current flow in phase 1 is zero, and the currents flowing in phases 2 and 3 are in opposite directions and less than maximum. This condition of current flow is illustrated in *Figure 11–21*. Notice that only two of the three phase lines are conducting current and that the current in each line is less than maximum.

Because the currents flowing in a three-phase system are never maximum at the same time, and at other times the current is divided between two phases, the total conductor resistance will not be the sum of two conductors. To calculate

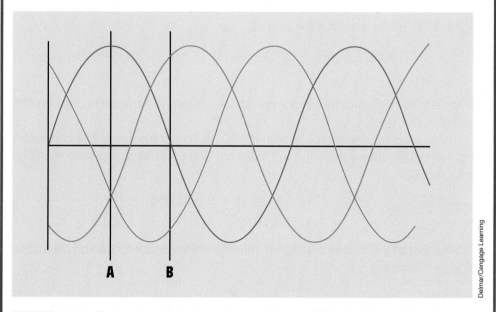

Delmar/Cengage Learning

FIGURE 11–19   The line currents in a three-phase system are 120° out of phase with each other.

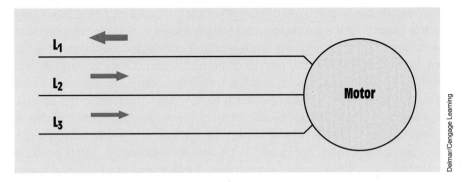

FIGURE 11–20 Current flows from lines 2 and 3 to line 1.

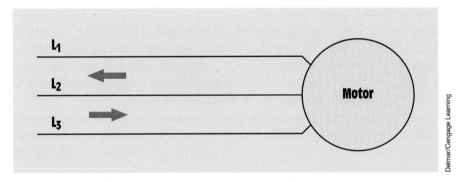

FIGURE 11–21 Current flows from line 3 to line 2.

the resistance of conductors in a three-phase system, a demand factor of 0.866 is used.

In this problem, the motor is located 2500 ft from the source. The conductor length is calculated by doubling the length of one conductor and then multiplying by 0.866:

$$L = 2500 \text{ ft} \times 2 \times 0.866$$
$$L = 4330 \text{ ft}$$

Now that all the factors are known, the size of the conductor can be calculated using the formula

$$CM = \frac{K \times L}{R}$$

where

$$K = \frac{17 \ \Omega\text{-CM}}{ft}$$

$$L = 4{,}330 \text{ ft}$$

$$R = 0.167 \ \Omega$$

$$CM = \frac{17 \ \Omega\text{-CM/ft} \times 4330 \text{ ft}}{0.167 \ \Omega}$$

$$CM = \frac{73{,}601 \ \Omega\text{-CM}}{0.167 \ \Omega}$$

$$CM = 440{,}778.443$$

Three 500-kcmil conductors will be used.

11–6   Calculating Voltage Drop

Sometimes it is necessary to calculate the voltage drop of an installation when the length, size of wire, and current are known. The following formula can be used to find the voltage drop of conductors used on a single-phase system:

$$E_D = \frac{2 \text{ KIL}}{CM}$$

where

$$E_D = \text{voltage drop}$$
$$K = \text{ohms per mil ft}$$
$$I = \text{current}$$
$$L = \text{length of conductor in ft}$$
$$CM = \text{circular mil area of the conductor}$$

■ EXAMPLE 11-13

A single-phase motor is located 250 ft from its power source. The conductors supplying power to the motor are 10 AWG copper. The motor has a full-load current draw of 24 A. What is the voltage drop across the conductors when the motor is in operation?

Solution

$$E_D = \frac{2 \times 10.4 \ \Omega\text{-CM/ft} \times 24 \ A \times 250 \ \text{ft}}{10,380 \ \text{CM}}$$

$$E_D = 12.023 \ V$$

A slightly different formula can be used to calculate the voltage drop on a three-phase system. Instead of multiplying KIL by 2, multiply KIL by the square root of 3:

$$E_D = \frac{\sqrt{3} \ \text{KIL}}{\text{CM}}$$

■ EXAMPLE 11–14

A three-phase motor is located 175 ft from its source of power. The conductors supplying power to the motor are 1/0 AWG aluminum. The motor has a full-load current draw of 88 A. What is the voltage drop across the conductors when the motor is operating at full load?

Solution

$$E_D = \frac{1.732 \times 17 \ \Omega\text{-CM/ft} \times 98 \ A \times 175}{105.500 \ \text{CM}}$$

$$E_D = 4.298 \ V$$

Coefficient of Temperature

The temperature of a conductor can greatly affect its resistance. Figure 11–16 lists the ohms-per-mil foot (K) at 20°C for various materials. The resistance of a material is generally given at 20°C because it is the standard used in the *American Engineers Handbook* and is considered a standard throughout the United States. The temperature coefficient can be used to determine the resistance of a material at different temperatures. Most conductors will increase their resistance with an increase of temperature. Semiconductor materials such as silicon, germanium, and carbon will exhibit a decrease of resistance with an increase of temperature. These materials have a negative coefficient of temperature.

■ EXAMPLE 11-15

Determine the ohms-per-mil foot at 75°C for a copper conductor?

Solution

Use the formula

$$R = R_{ref}[1 + \alpha(T - T_{ref})]$$

where

   R = Conductor resistance at temperature "T"

   R_{ref} = Conductor resistance at reference temperature (20°C in this example)

   α = Coefficient of resistance for the conductor material

   T = Conductor temperature in °C

   T_{ref} = Reference temperature that α is specified at for the conductor material.

$$R = 10.4[1 + 0.0039(75 - 20)]$$
$$R = 10.4[1 + 0.0039(55)]$$
$$R = 10.4[1 + 0.2145]$$
$$R = 10.4[1.2145]$$
$$R = 12.63$$

At a temperature of 75°C, copper would have a resisance of 12.63 ohms-per-mil foot.

11-7 Parallel Conductors

Under certain conditions, it may become necessary or advantageous to connect conductors in parallel. One such condition for **parallel conductors** is when the conductor is very large as in the earlier example, where it was calculated that the conductors supplying a motor 2500 feet from its source would have to be 500 kcmil. A 500-kcmil conductor is very large and difficult to handle. Therefore, it may be preferable to use parallel conductors for this installation. The *NEC* lists five conditions that must be met when conductors are connected in parallel *(310.10 (H))*. These conditions are listed here:

1. The conductors must be the same length.

2. The conductors must be made of the same material. For example, all parallel conductors must be either copper or aluminum. It is not permissible to use copper for one of the conductors and aluminum for the other.

3. The conductors must have the same circular mil area.

4. The conductors must use the same type of insulation.

5. The conductors must be terminated or connected in the same manner.

In the example, the actual conductor size needed was calculated to be 440,778.443 CM. This circular mil area could be obtained by connecting two 250-kcmil conductors in parallel for each phase, or three 000 (3/0) conductors in parallel for each phase. [Note: Each 000 (3/0) conductor has an area of 167,800 CM. This is a total of 503,400 CM.]

Another example of when it may be necessary to connect wires in parallel is when conductors of a large size must be run in a conduit. Conductors of a single phase are not permitted to be run in metallic conduits as shown in *Figure 11–22* [*NEC 300.5(I), NEC 300.20(A),* and *NEC 300.20(B)*], because when current flows through a conductor, a magnetic field is produced around the conductor. In an AC circuit, the current continuously changes direction and magnitude, which causes the magnetic field to cut through the wall of the metal conduit *(Figure 11–23)*. This cutting action of the magnetic field induces a current, called an *eddy current,* into the metal of the conduit. Eddy currents are currents that are induced into metals. They tend to move in a circular fashion similar to the eddies of a river, hence the name eddy currents *(Figure 11–24)*. Eddy currents can produce enough heat in high-current circuits to melt the insulation surrounding the conductors. All metal conduits can

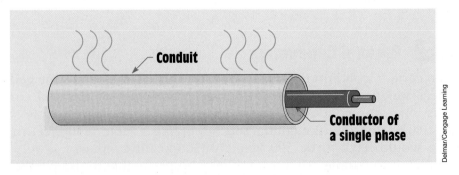

FIGURE 11–22 A single-phase conductor causes heat to be produced in the conduit.

Delmar/Cengage Learning

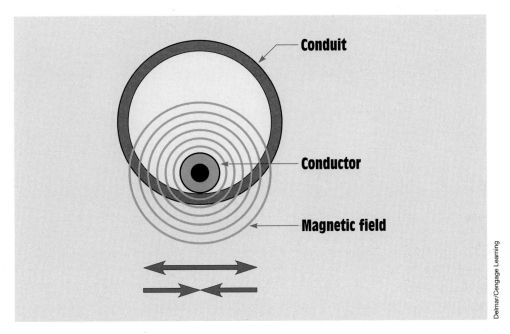

FIGURE 11–23 The magnetic field expands and contracts.

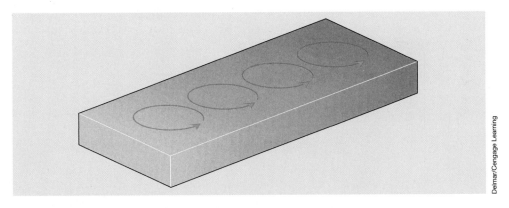

FIGURE 11–24 Eddy currents are currents induced in metals.

have eddy current induction, but conduits made of magnetic materials such as steel have an added problem with hysteresis loss. Hysteresis loss is caused by molecular friction *(Figure 11–25)*. As the direction of the magnetic field reverses, the molecules of the metal are magnetized with the opposite polarity and swing to realign themselves. This continuous aligning and realigning of the molecules produces heat caused by friction. Hysteresis losses become greater with an increase in frequency.

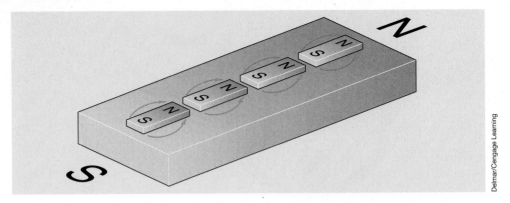

FIGURE 11-25 The molecules reverse direction each time the magnetic field changes direction.

To correct this problem, a conductor of each phase must be run in each conduit *(Figure 11–26)*. When all three phases are contained in a single conduit, the magnetic fields of the separate conductors cancel each other resulting in no current being induced in the walls of the conduit.

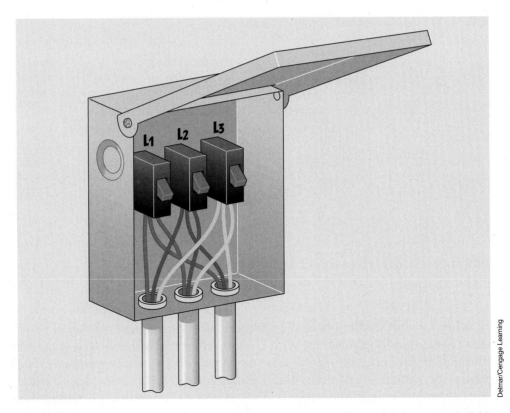

FIGURE 11-26 Each conduit contains a conductor from each phase. This permits the magnetic fields to cancel each other.

11–8   Testing Wire Installations

After the conductors have been installed in conduits or raceways, it is accepted practice to test the installation for grounds and shorts. This test requires an ohmmeter, which not only measures resistance in millions of ohms but also provides a high enough voltage to ensure that the insulation will not break down when rated line voltage is applied to the conductors. Most ohmmeters operate with a maximum voltage that ranges from 1.5 volts to about 9 volts depending on the type of ohmmeter and the setting of the range scale. To test wire insulation, a special type of ohmmeter, called a **MEGGER,** is used. The MEGGER is a megohmmeter that can produce voltages that range from about 250 to 5000 volts depending on the model of the meter and the range setting. One model of a MEGGER is shown in *Figure 11–27*. This instrument contains a hand crank that is connected to the rotor of a brushless AC generator. The advantage of this particular instrument is that it does not require the use of batteries. A range-selector switch permits the meter to be used as a standard ohmmeter or as a megohmmeter. When it is used as a megohmmeter, the selector switch permits the test voltage to be selected. Test voltages of 100 volts, 250 volts, 500 volts, and 1000 volts can be obtained.

MEGGER can also be obtained in battery-operated models *(Figure 11–28)*. These models are small, lightweight, and particularly useful when it becomes necessary to test the dielectric of a capacitor.

Wire installations are generally tested for two conditions, shorts and grounds. Shorts are current paths that exist between conductors. To test an installation for shorts, the MEGGER is connected across two conductors at a time *(Figure 11–29)*. The circuit is tested at rated voltage or slightly higher. The MEGGER indicates the resistance between the two conductors. Because both conductors are insulated, the resistance between them should

FIGURE 11–27   A hand-crank MEGGER.

Courtesy of Megger

FIGURE 11–28 Battery-operated MEGGER.

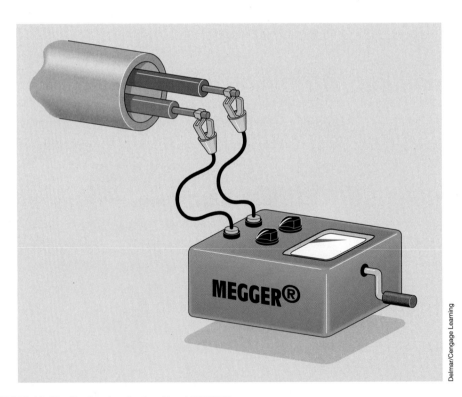

Delmar/Cengage Learning

FIGURE 11–29 Testing for shorts with a MEGGER.

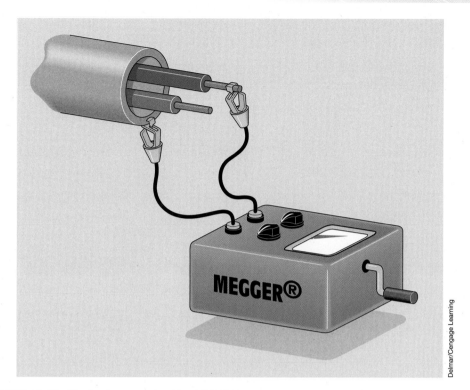

FIGURE 11–30   Testing for grounds with a MEGGER.

be extremely high. Each conductor should be tested against every other conductor in the installation.

To test the installation for grounds, one lead of the MEGGER is connected to the conduit or raceway *(Figure 11–30)*. The other meter lead is connected to one of the conductors. The conductor should be tested at rated voltage or slightly higher. Each conductor should be tested.

Summary

- The *NEC* is used to determine the wire size for most installations.
- The resistance of a wire is determined by four factors:

  a. the type of material from which the conductor is made

  b. the length of the conductor

  c. the area of the conductor

  d. the temperature of the conductor

- When conductors are used in high-temperature locations, their current capacity must be reduced.

- When there are more than three conductors in a raceway, their current-carrying capacity must be reduced.

- The amount of current a conductor can carry is affected by the type of insulation around the wire.

- In the English system, the mil foot is used as a standard for determining the resistance of different types of wire.

- After wires have been installed, they should be checked for shorts or grounds with a MEGGER.

Review Questions

Use of the *NEC* will be required to answer some of the following questions.

1. What is the maximum temperature rating of Type XHHW insulation when used in a wet location?

2. Name two types of conductor insulation designed to be used underground.

3. A 10 AWG copper conductor with Type THW insulation is to be run in free air. What is the maximum ampacity of this conductor if the ambient air temperature is 40°C?

4. Six 1/0 aluminum conductors are to be run in a conduit. Each conductor has Type THWN insulation, and the ambient air temperature is 30°C. What is the ampacity of each conductor?

5. Name five conditions that must be met for running conductors in parallel.

6. What is the largest solid (nonstranded) conductor listed in the wire tables?

7. Can Type TW cable be used in an area that has an ambient temperature of 65°C?

8. How is the grounded conductor in a flat multiconductor cable 4 or larger identified?

9. What three colors are ungrounded conductors not permitted to be?

10. Twenty-five 12 AWG copper conductors are run in conduit. Each conductor has Type THHN insulation. The conduit is located in an area that has an ambient temperature of 95°F. What is the ampacity of each conductor?

11. A single-phase load is located 2800 ft from its source. The load draws a current of 86 A and operates on 480 V. The maximum voltage drop at the load cannot be greater than 3%. What size aluminum conductors should be installed to operate this load?

12. It is decided to use parallel 0000 conductors to supply the load in Question 11. How many 0000 conductors will be needed?

13. A three-phase motor operates on 480 V and is located 1800 ft from the power source. The starting current is 235 A. What size copper conductors will be needed to supply this load if the voltage must not be permitted to drop below 6% of the terminal voltage during starting?

Practical Applications

You have been hired by a company to connect an outside lighting system to a panel box located 55 ft from the point of attachment. The conductors are to have Type THWN-2 insulation and the wire is copper. The termination temperature of the connections is not known. The lighting system is rated at 20 kW. The voltage is 240 V. Because the lights are expected to operate more than three hours at a time, they are considered continuous duty. Using the *NEC*, determine the wire size needed to make this connection. ∎

Practical Applications

You are a journeyman electrician in an industrial plant. You have been assigned the task of connecting a 25-kW electric heater to a 240-V, single-phase panel. The heater is located in an area with an ambient temperature of 90°F. The conductors are to be copper with Type TW insulation. The heater is expected to operate for more than three hours at a time. What size conductors would you use to connect this heater? ∎

Practical Applications

*Y*ou are an electrician working in an industrial plant. The company is constructing a maintenance shop located 225 feet from the main plant electrical system. You are to install a 400-ampere three-phase wire system in the new maintenance shop. The service is supplied by the existing plant service. The voltage is 480 volts and the voltage drop is to be kept at 3% or below at full load. The conductors are to be aluminum. What size conductors should be installed for the new service? Express your answer to the nearest standard wire size. (Note: Determine the wire size by calculating length, resistance, and voltage drop as opposed to using the *National Electrical Code*.) ■

Practical Applications

*U*sing the above example, determine the wire size needed for the 400-ampere service in accordance with the *National Electrical Code*. Assume that the conductors have type THWH insulation. (Note: As a general rule, for each 100 feet of conductor length the wire size is increased one standard size because of voltage drop.) ■

Practice Problems

Using Wire Tables and Determining Conductor Sizes

1. A 2 AWG copper conductor is 450 ft long. What is the resistance of this wire? Assume the ambient temperature to be 20°C.

2. An 8 AWG conductor is 500 ft long and has a resistance of 1.817 Ω. The ambient temperature is 20°C. Of what material is the wire made?

3. Three 500-kcmil copper conductors with Type RHH insulation are to be used in an area that has an ambient temperature of 58°C. What is the maximum current-carrying capacity of these conductors?

4. Eight 10 AWG aluminum conductors with Type THWN insulation are installed in a single conduit. What is the maximum current-carrying capacity of these conductors? Assume an ambient temperature of 30°C.

5. A three-phase motor is connected to 480 V and has a starting current of 522 A. The motor is located 300 ft from the power source. The voltage

drop to the motor cannot be greater than 5% during starting. What size copper conductors should be connected to the motor?

6. A 50-hp DC motor is connected to 250 volts. The motor does not contain a NEMA code. The motor is expected to operate for more than three hours at a time. The conductors are to be copper with type RHW-2 insulation. The motor is located in an area with an ambient temperature of 112°F. The two conductors are to be run in conduit. What size conductors should be used to connect this motor? (Note: 1 hp = 746 watts.)

7. A bank of electric heaters has a power rating of 25 kW. The heaters are connected to 480 volts. The heaters will operate for more than three hours at a time. The conductors supplying the heaters are to be aluminum with type THWN-2 insulation. The heaters are located in an area with an ambient temperature of 48°C. There will be six conductors in the conduit. What size conductors should be used for this installation?

8. A 15-hp squirrel-cage induction motor is connected to 240 volts. The motor has a NEMA code B. The motor will operate for more than three hours at a time. The motor is in an area with an ambient temperature of 86°F. There are to be three conductors in the raceway. The conductors are copper with THW-2 insulation. What size conductors should be used for this installation? (Note: 1 hp = 746 watts.)

9. Determine the resistance of a 16 AWG copper conductor located in an area with an ambient temperature of 188.6°F (87°C) with a length of 250 feet (76.2 m).

10. Determine the ohms-per-mill foot of an aluminum conductor located in an area with a temperature of 104°F (40°C).

Use the *NEC* to determine the ampacity of the following conductors.

Size	Material	Insulation	Ambient Temperature	Conductors in Raceway	Amps
#10 AWG	Copper	RHW	44°C	3	
350 kcmil	Copper	XHH	128°F	6	
#2 AWG	Aluminum	TW	86°F	2	
3/0 AWG	Aluminum	XHHW-2	38°C	9	
500 kcmil	Copper	THWN	48°C	6	
#6 AWG	Copper	THW-2	150°F	3	
2/0 AWG	Aluminum	UF	86°F	12	
750 kcmil	Aluminum	RHW-2	34°F	6	

IV Small Sources of Electricity

Unit 12

Conduction in Liquids and Gases

KEY TERMS

Acids	Electrolytes
Alkalies	Electron impact
Anode	Electroplating
Arc	Ion
Cathode	Ionization potential
Copper sulfate	Metallic salt
Cuprous cyanide	Sulfuric acid
Electrolysis	X-rays

Why You Need to Know

If you want to understand how batteries produce electricity, how fluorescent lights work, or how it is possible to electroplate objects, you need to have an understanding of how electricity is conducted in gases and liquids. This unit.

- explains how electricity is conducted in gases and liquids and how it is different from conduction through metals such as silver, copper, or aluminum. Whereas metals rely on the conduction of electrons, gases and liquids rely on the movement of ions.
- explains the process of separating elements electrically.
- describes what X-rays are and how they are used.
- defines the ionization potential.

Objectives

After studying this unit, you should be able to

- define positive and negative ions.

- discuss electrical conduction in a gas.

- discuss electrical conduction in a liquid.

- discuss several processes that occur as a result of ionization.

Preview

The conduction of electric current is generally thought of as electrons moving through a wire. Many processes, however, depend on electric current flowing through a gas or liquid. Batteries, for example, would not work if conduction could not take place through a liquid, and fluorescent lighting operates on the principle of conduction through a gas. Conduction through a gas or liquid does not depend on the flow of individual electrons as is the case with metallic conductors. Conduction in gases and liquids depends on the movement of ions. ∎

12–1   The Ionization Process: Magnesium and Chlorine

An **ion** is a charged atom. Atoms that have a deficiency of electrons are known as positive ions. Atoms that have gained extra electrons are known as negative ions.

A good example of how ionization occurs can be seen by the combination of two atoms, magnesium and chlorine. The magnesium atom contains two valence electrons and is considered to be a metal. Chlorine contains seven valence electrons and is considered a nonmetal *(Figure 12–1)*.

When magnesium is heated in the presence of chlorine gas, a magnesium atom combines with two chlorine atoms to form a **metallic salt** called magnesium chloride *(Figure 12–2)*. When this process occurs, the magnesium atom gives up its two valence electrons to the two chlorine atoms. These atoms are no longer called atoms; they are now called ions. Because the magnesium atom gave up two electrons, it has a positive charge and has become a positive ion. The two chlorine atoms have each gained an electron and are now negative ions.

Magnesium chloride is similar to table salt, which is sodium chloride. Both are formed by combining a metal, sodium or magnesium, with chlorine. If either of these salts is mixed with a nonconducting liquid, such as distilled water, it will become a conductor. A simple experiment can be performed to demonstrate conduction through a liquid. In *Figure 12–3*, copper electrodes have been placed in a glass. One electrode is connected to a lamp. The other is connected to one side of a 120-volt circuit. The other side of the lamp is connected to the other side of the circuit. The lamp acts as a load to prevent the circuit from drawing excessive current. If pure water is poured into the glass,

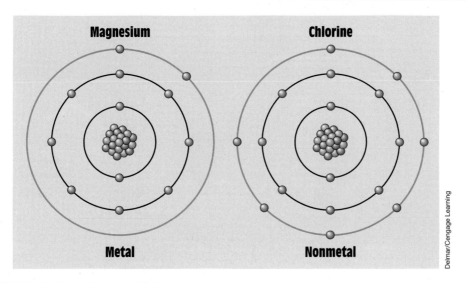

FIGURE 12–1 Magnesium and chlorine atoms.

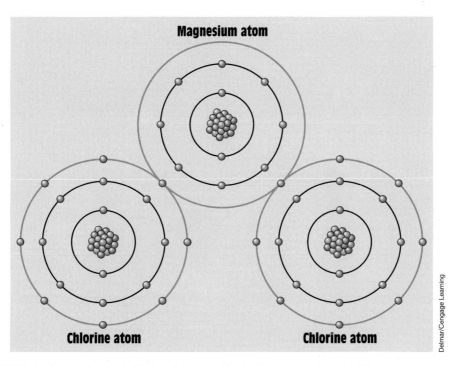

FIGURE 12–2 Magnesium and chlorine atoms combine to form magnesium chloride.

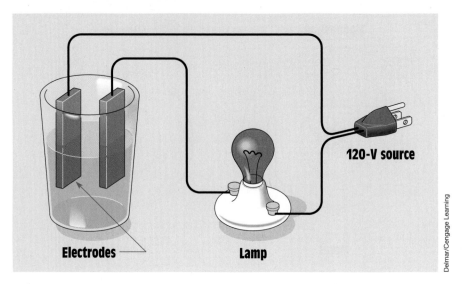

FIGURE 12–3 Conduction in a liquid.

nothing seems to happen because pure water is an insulator. If salt is added to the water, the lamp begins to glow *(Figure 12–4)*. The lamp glows dimly at first and increases in brightness as salt is added to the water. The glow continues to brighten until the solution becomes saturated with salt and current is limited by the resistance of the filament in the lamp.

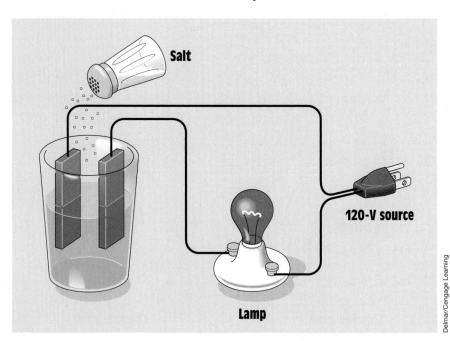

FIGURE 12–4 Adding salt to the water causes the lamp to glow.

Salt is not the only compound that can be used to promote conduction in a liquid. **Acids, alkalies,** and other types of metallic salts can be used. These solutions are often referred to as **electrolytes.**

12–2 Other Types of Ions

The ions discussed so far are formed by charging a single atom. There are other types of ions, however, that are formed from groups of atoms. These ions are extremely useful to the electric industry.

Sulfuric Acid

One of the most useful ion compounds is **sulfuric acid.** It is the electrolyte for the lead-acid batteries used in automobiles. The chemical formula for sulfuric acid, H_2SO_4, indicates that one molecule of sulfuric acid contains two atoms of hydrogen, one atom of sulfur, and four atoms of oxygen. Sulfuric acid in its pure form is not useful as an electrolyte. When it is combined with water, H_2O, the molecules separate into ions of H^+, H^+, and SO_4^{--}. The plus sign indicates that each of two hydrogen atoms has lost its electrons and is now a positive ion. The SO_4^{--} is called a sulfate ion and indicates that the two electrons lost by the hydrogen atoms are connected to the sulfur atom and four oxygen atoms. The two extra electrons actually hold the sulfate ion together.

Copper Sulfate

Another useful compound is **copper sulfate,** $CuSO_4$. In its natural form, copper sulfate is in the form of blue crystals. When it is mixed with water, it dissolves and forms two separate ions, Cu^{++} and SO_4^{--}. The Cu^{++} is called a cupric ion and indicates that a copper atom has lost two of its electrons. The SO_4^{--} is the same sulfate ion formed when sulfuric acid is mixed with water. This copper sulfate solution is used in copper electroplating processes.

Cuprous Cyanide

Cuprous cyanide is used to electroplate copper to iron. It is a very poisonous solid that becomes Cu^+ and CN^- when mixed in solution. The Cu^+ is a cuprous ion, and CN^- is a cyanide ion.

12–3 Electroplating

Electroplating is the process of depositing atoms of one type of metal on another. Several factors are always true in an electroplating process:

1. *The electrolyte solution must contain ions of the metal to be plated.*

2. *Metal ions are always positively charged.*

3. *The object to be plated must be connected to the negative power terminal.* The negative terminal is called the **cathode** and refers to the terminal where electrons enter the circuit.

4. *Direct current is used as the power source.*

5. *The positive terminal is made of the same metal that is to form the coating.* The positive terminal is referred to as the **anode** and refers to the terminal where electrons leave the circuit.

An example of the electroplating process is shown in *Figure 12–5*. Note that the object to be plated has been connected to the negative battery terminal, or cathode, and the copper bar has been connected to the positive terminal, or anode. Both objects are submerged in a cuprous cyanide solution. When power is applied to the circuit, positively charged copper ions in the solution move toward the object to be plated, and the negatively charged cyanide ions move toward the copper bar. When the copper ion contacts the object, it receives an electron and becomes a neutral copper atom. These neutral copper atoms form a copper covering over the object to be plated. The thickness of the coating is determined by the length of time the process is permitted to continue.

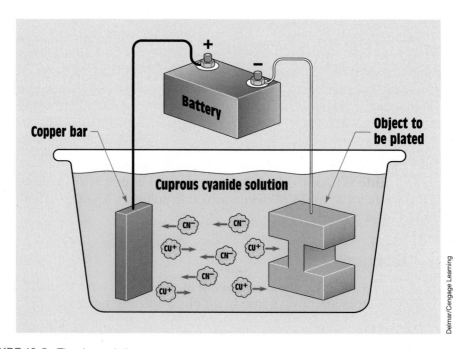

FIGURE 12–5 The electroplating process.

When the negatively charged cyanide ions contact the positive copper bar, copper atoms on the surface of the bar lose an electron and become positive ions. The copper ion is attracted into the solution and flows toward the object to be plated. Just as many copper ions are being formed as are being attracted to the object to be plated. The solution therefore remains at a constant strength.

One of the most common uses for electroplating is the production of pure metals. Impure, or raw, copper is connected to the positive terminal, and pure copper is connected to the negative terminal. Because only pure atoms of copper become ions, impurities remain in solution or never leave the positive plate. Impure copper has a high resistance and is not suitable for use as an electrical conductor. Wire manufacturers require electrolytically purified copper because of its low-resistance characteristics.

12–4 Electrolysis

The term **electrolysis** refers to the process of separating elements electrically. Elements are often chemically combined with other elements and must be separated. Although aluminum oxide is an abundant compound, aluminum was once very rare because extracting aluminum from aluminum oxide was difficult and expensive. Aluminum became inexpensive when a process was discovered for separating the aluminum and oxygen atoms electrically. In this process, electrons are removed from oxygen ions and returned to aluminum ions. The aluminum ions then become aluminum atoms.

12–5 Conduction in Gases

At atmospheric pressure, air is an excellent insulator. In *Figure 12–6,* two electrodes are separated by an air gap. The electrodes are connected to a variable-voltage power supply. Assume that the output voltage of the power supply is zero when the switch is turned on. As the output voltage is increased, the ammeter will remain at zero until the voltage reaches a certain potential. Once this potential has been reached, an **arc** will be established across the electrodes and current will begin to flow *(Figure 12–7)*. The amount of current is limited by the impedance of the rest of the circuit, because once an arc has been established, it has a very low resistance. For example, once the arc has been established, the voltage can be reduced and the arc will continue because molecules of air become ionized and conduction is maintained by the ionized gas. One factor that determines the voltage required is the distance between the two conductors. The farther apart they are, the higher the voltage must be. When using an electric arc welder, the arc is first established by touching the electrode to the object to be welded and then withdrawing the electrode once conduction begins. Recall that once an arc has been established, the amount of voltage required to maintain it is reduced.

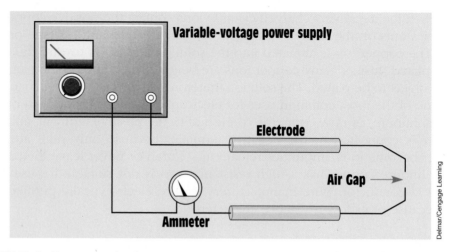

FIGURE 12–6 Air acts as an insulator.

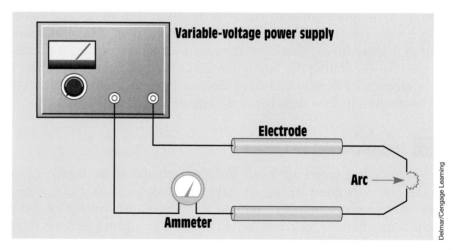

FIGURE 12–7 When the voltage becomes high enough, an arc is established.

Ionization Potential in a Gas

The amount of voltage, or potential, an electron must possess to cause ionization is called the **ionization potential.** The following factors determine the amount of voltage required to cause conduction in a gas-filled envelope.

1. *Atmospheric pressure.* The amount of air pressure greatly influences the voltage required to reach ionization potential. Atmospheric pressure is 14.7 pounds per square inch (PSI) at sea level. If the pressure is increased, such as in an automobile engine, the amount of voltage required to reach

ionization potential increases greatly. The ignition system of most automobiles produces voltages that range from 20,000 volts to 70,000 volts.

If the atmospheric pressure is reduced, the amount of voltage required to reach ionization potential is reduced. This is the principle of operation of the gas-filled tubes that have been used in industry for many years. Gas-filled tubes operate with lower than normal atmospheric pressure inside the tube.

2. *The type of gas in the surrounding atmosphere.* The type of gas in the atmosphere can greatly influence the amount of voltage required to cause ionization. Sodium vapor, for example, requires a potential of approximately 5 volts. Mercury vapor requires 10.4 volts; neon, 21.5 volts; and helium, 24.5 volts. Before an electron can ionize an atom of mercury, it must possess a potential difference of 10.4 volts.

Electron Impact

Conduction in a gas is different from conduction in a liquid or metal. The most important factor in the ionization of a gas is **electron impact.** The process begins when an electron is freed from the negative terminal inside a gas-filled tube or envelope and begins to travel toward the positive terminal *(Figure 12–8)*. The electron is repelled from the negative terminal and attracted to the positive terminal. The speed the electron attains is proportional to the applied voltage and the distance it travels before striking a gas molecule. If the electron's speed (potential) is great enough, it will liberate other electrons from the gas molecule, which in turn flow toward the positive terminal and strike other molecules *(Figure 12–9)*. These electrons are very energetic and are the main source of current flow in a gas. When electrons are removed from the molecules, the molecules become positive ions similar to those in a liquid.

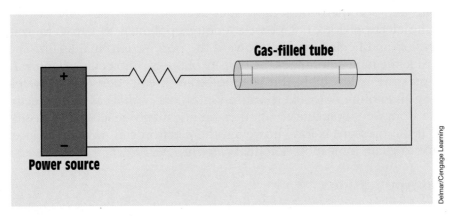

FIGURE 12–8   Conduction in a gas.

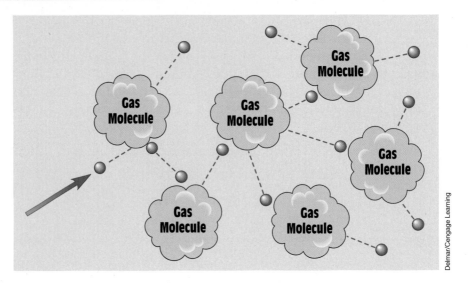

FIGURE 12–9   Electron impact frees other electrons from gas molecules.

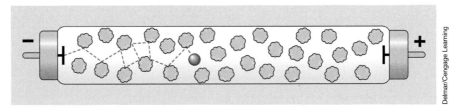

FIGURE 12–10   Under high pressure, the electron can travel only a short distance between impacts with gas molecules.

These positive ions are attracted to the negative terminal, where they receive electrons and become neutral molecules again.

If the electron does not possess enough potential to cause ionization, it bounces off and begins traveling toward the positive terminal again. The pressure inside the tube determines the density of the gas molecules *(Figure 12–10)*. If the pressure is too high, the gas molecules are so dense that the electron cannot gain enough potential to cause ionization. In this case, the electron will bounce from one molecule to another until it finally reaches the positive terminal. If the pressure is low, however, the electron can travel a great enough distance to attain ionization potential *(Figure 12–11)*.

Useful Applications

Ionization in a gas can be very useful. One use is the production of light. Often, when an electron strikes a gas molecule that does not have enough energy to

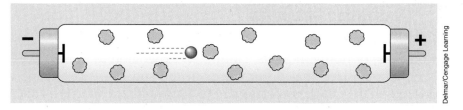

FIGURE 12–11 Under low pressure, the electron can travel a greater distance before striking a gas atom.

release other electrons, the molecule will emit light to rid itself of the excess energy. The color of the light is characteristic of the type of gas. Some examples of this type of lighting are sodium vapor, mercury vapor, and neon.

Another useful device that operates on the principle of conduction in a gas was invented in 1895 by Wilhelm Roentgen. During experimentation in his laboratory, he discovered that when cathode rays struck metal or glass, a new kind of radiation was produced. These new rays passed through wood, paper, and air without effect. They passed through thin metal better than through thick metal and through flesh easier than through bone. Roentgen named these new rays **X-rays.** Today, X-rays are used by doctors throughout the world.

Probably one of the most familiar devices that operates on the principle of conduction in a gas is the cathode ray tube. In 1897 J. J. Thomson measured the weight of the negative particle of the electric charge. He also found that a beam of negative particles could be bent by magnetic fields *(Figure 12–12).* Today the cathode ray tube is used in television sets, computer display terminals, and oscilloscopes.

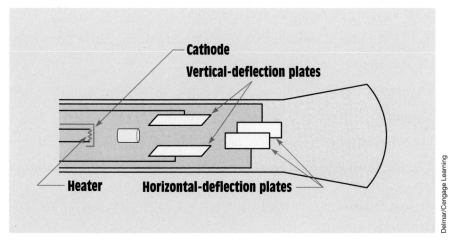

FIGURE 12–12 Cathode ray tube.

12–6   Ionization in Nature

One of the most familiar phenomena caused by ionization of gas is the aurora borealis, or northern lights. During an expedition to the arctic, Kristian Birkeland hypothesized that disturbances in the sun cause large amounts of electrons and high-speed protons to be given off and blown into space. When these highly energetic particles reach the earth, they strike the very thin layers of the upper atmosphere and cause ionization to occur. The different colors are caused by different layers of the atmosphere being ionized.

A buildup in the atmosphere of charged particles that are not strong enough to cause lightning will often produce a continuous discharge of light around objects such as the masts of ships or church steeples. This glowing discharge is often referred to as Saint Elmo's fire by sailors.

Summary

- Electrical conduction in a liquid is caused by a movement of charged atoms called ions.

- Pure water is an insulator and becomes a conductor if an acid, an alkali, or a metallic salt is added.

- Solutions of acids, alkalies, and metallic salts are referred to as electrolytes.

- Electroplating is the process of depositing the atoms of one type of metal onto another.

- Several factors that affect the electroplating of metals are

  a. the electrolyte solution must contain ions of the metal to be plated.

  b. metal ions are always positively charged.

  c. the object to be plated must be connected to the negative power terminal.

  d. direct current is used as the power source.

  e. the positive terminal is made of the same metal that forms the coating.

- The negative terminal of a device is called the cathode.

- The positive terminal of a device is called the anode.

- One of the most common uses for electroplating is the production of pure metals.

- The term *electrolysis* refers to the process of separating elements electrically.

- Conduction through a gas depends on the ionization of gas molecules.

- The amount of voltage required to cause ionization of a gas is called the ionization potential.

- Gas can be ionized more easily when it is at a low pressure than when it is at a high pressure.

- The ionization potential is different for different gases.

- The most important factor in the ionization of a gas is electron impact.

Review Questions

1. Conduction in a liquid depends on the movement of _____.

2. What is a negative ion?

3. What is a positive ion?

4. Name three basic substances that can be used to produce ionization in a liquid.

5. What is an electrolyte?

6. What is used to hold a sulfate ion together?

7. What is the negative terminal of a power source called?

8. What is the positive terminal of a power source called?

9. What determines the thickness of the coating during an electroplating process?

10. What is electrolysis?

11. What is ionization potential?

12. What is the most important factor in the ionization of a gas?

13. Name two factors that determine the speed an electron attains inside a gas environment.

14. What determines the density of the molecules in a gas environment?

15. What determines the color of light emitted by a gas-filled tube?

Unit 13 📀 VIDEO

Batteries and Other Sources of Electricity

Why You Need to Know

Methods other than conductors cutting through magnetic fields can be employed to produce electricity. Although magnetic devices such as generators and alternators are certainly the greatest producers of electricity, they are not the only ones. Batteries are common in everyday life, and an understanding of how they operate is essential. This unit

- discusses different types of batteries and how to identify their current capacity for correct use. Some batteries can be recharged and others cannot, and you must understand those differences.
- explains why different types of batteries produce different voltages and what determines the amount of current a battery can produce.
- illustrates how to calculate the voltage and current capacity of batteries by connecting them in series and/or parallel.
- discusses solar cells, thermocouples, and piezoelectricity.

KEY TERMS

Battery
Cell
Current capacity
Electromotive series of metals
Hydrometer
Internal resistance
Load test
Nickel-cadmium (nicad) cell
Piezoelectricity
Primary cell
Secondary cell
Specific gravity
Thermocouple
Voltaic cell
Voltaic pile

Objectives

After studying this unit, you should be able to

- discuss the differences between primary and secondary cells.
- list voltages for different types of cells.
- discuss different types of primary cells.
- construct a cell from simple materials.
- discuss different types of secondary cells.
- connect batteries in series and parallel to obtain desired voltage and ampere hour (A-hr) ratings.
- discuss the operation of solar cells.
- connect solar cells in series or parallel to produce the desired output voltage and current capacity.
- discuss the operation of thermocouples.
- discuss the piezoelectric effect.

Courtesy of Power-Sonic Corporation.

Preview

Most of the electric power in the world is produced by large rotating machines called alternators. There are other sources of electricity, however, that are smaller and are used for emergency situations or for the operation of portable electric devices. The most common of these small power sources is the battery. Batteries are used to start automobiles and to operate toys, flashlights, portable communications equipment, computers, watches, calculators, and hundreds of other devices. They range in size from small enough to fit into a hearing aid to large enough to operate electric forklifts and start diesel trucks. ■

13–1   History of the Battery

In 1791, Luigi Galvani was conducting experiments in anatomy using dissected frog legs preserved in a salt solution. Galvani suspended the frog legs by means of a copper wire. He noticed that when he touched the leg with an iron scalpel the leg would twitch. Galvani realized that the twitch was caused by electricity, but he thought that the electricity was produced by the muscular contraction of the frog's leg. This was the first recorded incident of electricity being produced by chemical action.

   In 1800, Alessandro Volta repeated Galvani's experiment. Volta, however, concluded that the electricity was produced by the chemical action of the copper wire, the iron scalpel, and the salt solution. Further experiments led Volta

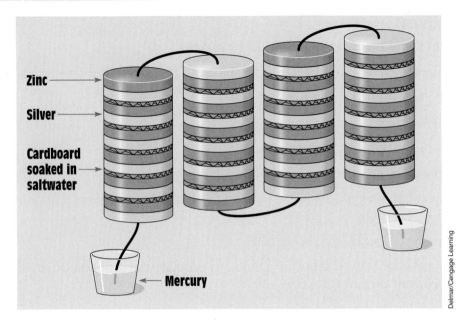

Zinc

Silver

Cardboard soaked in saltwater

Mercury

FIGURE 13–1   A voltaic pile.

to produce the first practical battery *(Figure 13–1)*. The battery was constructed using zinc and silver discs separated by a piece of cardboard soaked in brine, or saltwater. Volta called his battery a **voltaic pile** because it was a series of individual cells connected together. Each cell produced a certain amount of voltage depending on the materials used to make the cell. A **battery** is actually several cells connected together, although the word *battery* is often used in reference to a single cell. The schematic symbols for an individual **cell** and for a battery are shown in *Figure 13–2*.

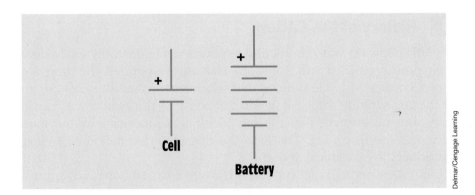

Cell

Battery

FIGURE 13–2   Schematic symbols used to represent an individual cell and a battery.

13-2 Cells

A **voltaic cell** can be constructed using virtually any two unlike metals and an acid, alkaline, or salt solution. A very simple cell can be constructed as shown in *Figure 13–3*. In this example, a copper wire is inserted in one end of a potato and an aluminum wire is inserted in the other. The acid in the potato acts as the electrolyte. If a high-impedance voltmeter is connected to the wires, a small voltage can be measured.

Another example of a simple voltaic cell is shown in *Figure 13–4*. In this example, two coins of different metal, a nickel and a penny, are separated by a

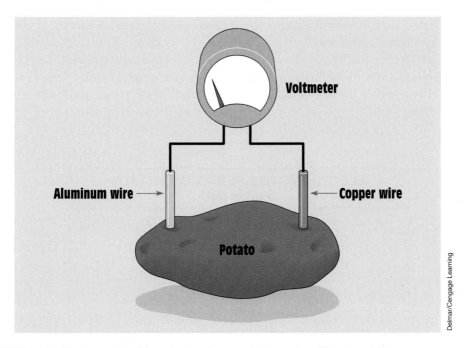

FIGURE 13–3 Simple voltaic cell constructed from a potato and two different metals.

FIGURE 13–4 A voltaic cell constructed from two coins and a piece of paper.

piece of paper. The paper has been wetted with saliva from a person's mouth. The saliva contains acid or alkali that acts as the electrolyte. When a high-impedance voltmeter is connected across the two coins, a small voltage is produced.

13-3   Cell Voltage

The amount of voltage produced by an individual cell is determined by the materials from which it is made. When a voltaic cell is constructed, the plate metals are chosen on the basis of how easily one metal will give up electrons compared with the other. A special list of metals called the **electromotive series of metals** is shown in *Figure 13–5*. This table lists metals in the order of their ability to accept or receive electrons. The metals at the top accept electrons more easily than those at the bottom. The farther apart the metals are on the list, the higher the voltage developed by the cell. One of the first practical cells to be constructed was the zinc-copper cell. This cell uses zinc and copper as the active metals and a solution of water and hydrochloric acid as the electrolyte. Notice that zinc is located closer to the top of the list

ELECTROMOTIVE SERIES OF METALS
(Partial list)
Lithium
Potassium
Calcium
Sodium
Magnesium
Aluminum
Manganese
Zinc
Chromium
Iron
Cadmium
Cobalt
Nickel
Tin
Lead
Antimony
Copper
Silver
Mercury
Platinum
Gold

Delmar/Cengage Learning

FIGURE 13–5   A partial list of the electromotive series of metals.

CELL	NEGATIVE PLATE	POSITIVE PLATE	ELECTROLYTE	VOLTS PER CELL
Primary Cells				
Carbon-zinc (Leclanche)	Zinc	Carbon, manganese dioxide	Ammonium chloride	1.5
Alkaline	Zinc	Manganese dioxide	Potassium hydroxide	1.5
Mercury	Zinc	Mercuric oxide	Potassium hydroxide	1.35
Silver-zinc	Zinc	Silver Oxide	Potassium hydroxide	1.6
Zinc-air	Zinc	Oxygen	Potassium hydroxide	1.4
Edison-Lalande	Zinc	Copper oxide	Sodium hydroxide	0.8
Secondary Cells				
Lead-acid	Lead	Lead dioxide	Dilute sulfuric acid	2.2
Nickel-iron (Edison)	Iron	Nickel oxide	Potassium hydroxide	1.4
Nickel-cadmium	Cadmium	Nickel hydroxide	Potassium hydroxide	1.2
Silver-zinc	Zinc	Silver oxide	Potassium hydroxide	1.5
Silver-cadium	Cadmium	Silver oxide	Potassium hydroxide	1.1

Delmar/Cengage Learning

FIGURE 13–6 Voltaic cells.

than copper. Because zinc accepts electrons more readily than copper, zinc is the negative electrode and copper is the positive.

Although it is possible to construct a cell from virtually any two unlike metals and an electrolyte solution, not all combinations are practical. Some metals corrode rapidly when placed in an electrolyte solution, and some produce chemical reactions that cause a buildup of resistance. In practice, relatively few metals can be used to produce a practical cell.

The table in *Figure 13–6* lists common cells. The table is divided into two sections. One section lists primary cells and the other lists secondary cells. A **primary cell** is a cell that *cannot be recharged*. The chemical reaction of a primary cell causes one of the electrodes to be eaten away as power is produced. When a primary cell becomes discharged, it should be replaced with a new cell. A **secondary cell** *can be recharged*. The recharging process is covered later in this unit.

13–4 Primary Cells

An example of a primary cell is the zinc-copper cell shown in *Figure 13–7*. This cell consists of two electrodes (terminals that conduct electricity into or away from a conducting substance), one zinc and one copper, suspended in a solution of water and hydrochloric acid. This cell will produce approximately 1.08 volts. The electrolyte contains positive H^+ ions, which attract electrons from the zinc

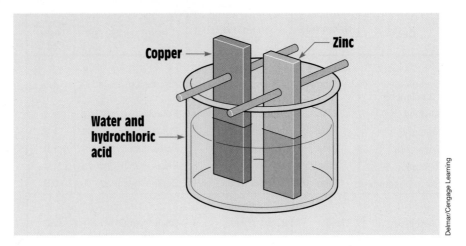

FIGURE 13–7   Zinc-copper cell.

atoms, which have two valence electrons. When an electron bonds with the H^+ ion, it becomes a neutral H atom. The neutral H atoms combine to form H_2 molecules of hydrogen gas. This hydrogen can be seen bubbling away in the electrolyte solution. The zinc ions, Zn^{++}, are attracted to negative Cl^- ions in the electrolyte solution.

The copper electrode provides another source of electrons that can be attracted by the H^+ ions. When a circuit is completed between the zinc and copper electrodes *(Figure 13–8)*, electrons are attracted from the zinc electrode to replace the electrons in the copper electrode that are attracted into the solution. After some period of time, the zinc electrode dissolves as a result of the zinc ions being distributed in the solution.

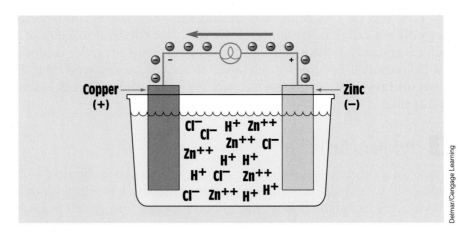

FIGURE 13–8   Electrons flow from the zinc to the copper electrode.

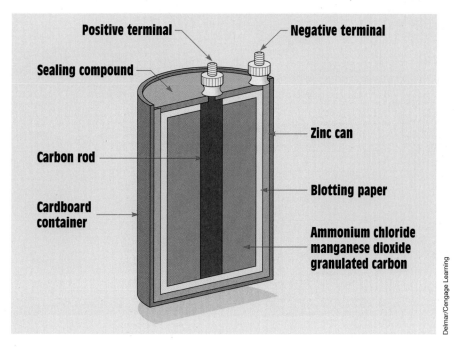

FIGURE 13–9 Carbon-zinc cell.

The Carbon-Zinc Cell

One of the first practical primary cells was invented by Leclanché. It is known as the carbon-zinc cell, or the Leclanché cell. Leclanché used a carbon rod as an electrode, instead of copper, and a mixture of ammonium chloride, manganese dioxide, and granulated carbon as the electrolyte. The mixture was packed inside a zinc container that acted as the negative electrode *(Figure 13–9)*. This cell is often referred to as a dry cell because the electrolyte is actually a paste instead of a liquid. The use of a paste permits the cell to be used in any position without spilling the electrolyte. As the cell is discharged, the zinc container is eventually dissolved. When the container has dissolved, the cell should be discarded immediately and replaced. Failure to discard the cell can result in damage to equipment.

Alkaline Cells

Another primary cell very similar to the carbon-zinc cell is the alkaline cell. This cell uses a zinc can as the negative electrode and manganese dioxide, MnO_2, as the positive electrode. The major electrolytic ingredient is potassium hydroxide. Like the carbon-zinc cell, the alkaline cell contains a paste electrolyte and is considered a dry cell also. The voltage developed is the same as in the carbon-zinc cell, 1.5 volts. The major advantage of the alkaline cell is longer life. The average alkaline cell can supply from three to five

Courtesy of GE Industrial Systems

FIGURE 13–10 Alkaline-manganese cell.

times the power of a carbon-zinc cell, depending on the discharge rate when the cell is being used. The major disadvantage is cost. Although some special alkaline cells can be recharged, the number of charge and discharge cycles is limited. In general, the alkaline cell is considered to be a nonrechargeable cell. A D-size alkaline-manganese cell is shown in *Figure 13–10*. Several different sizes and case styles of alkaline and mercury cells are shown in *Figure 13–11*.

One of the devices shown in *Figure 13–11* is a true battery and not a cell. The battery is rectangular and has two snap-on type terminal connections at the top. This battery actually contains six individual cells rated at 1.5 volts each. The cells are connected in series and provide a terminal voltage of 9 volts. The internal makeup of the battery is shown in *Figure 13–12*.

Button Cells

Another common type of primary cell is the button cell. The button cell is so named because it resembles a button. Button cells are commonly used in cameras, watches, hearing aids, and handheld calculators. Most button cells are constructed using mercuric oxide as the anode, zinc as the cathode, and potassium hydroxide as the electrolyte *(Figure 13–13)*. Although these cells are more expensive than other cells, they have a high-energy density and a long life. The mercury-zinc cell has a voltage of 1.35 volts.

Courtesy of GE Industrial Systems

FIGURE 13–11   Different sizes and case styles of primary cells.

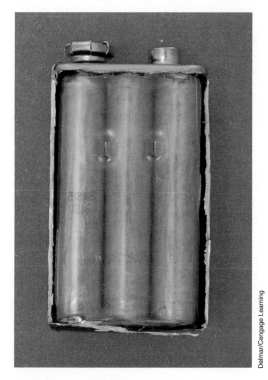

Delmar/Cengage Learning

FIGURE 13–12   Internal makeup of a 9-V battery.

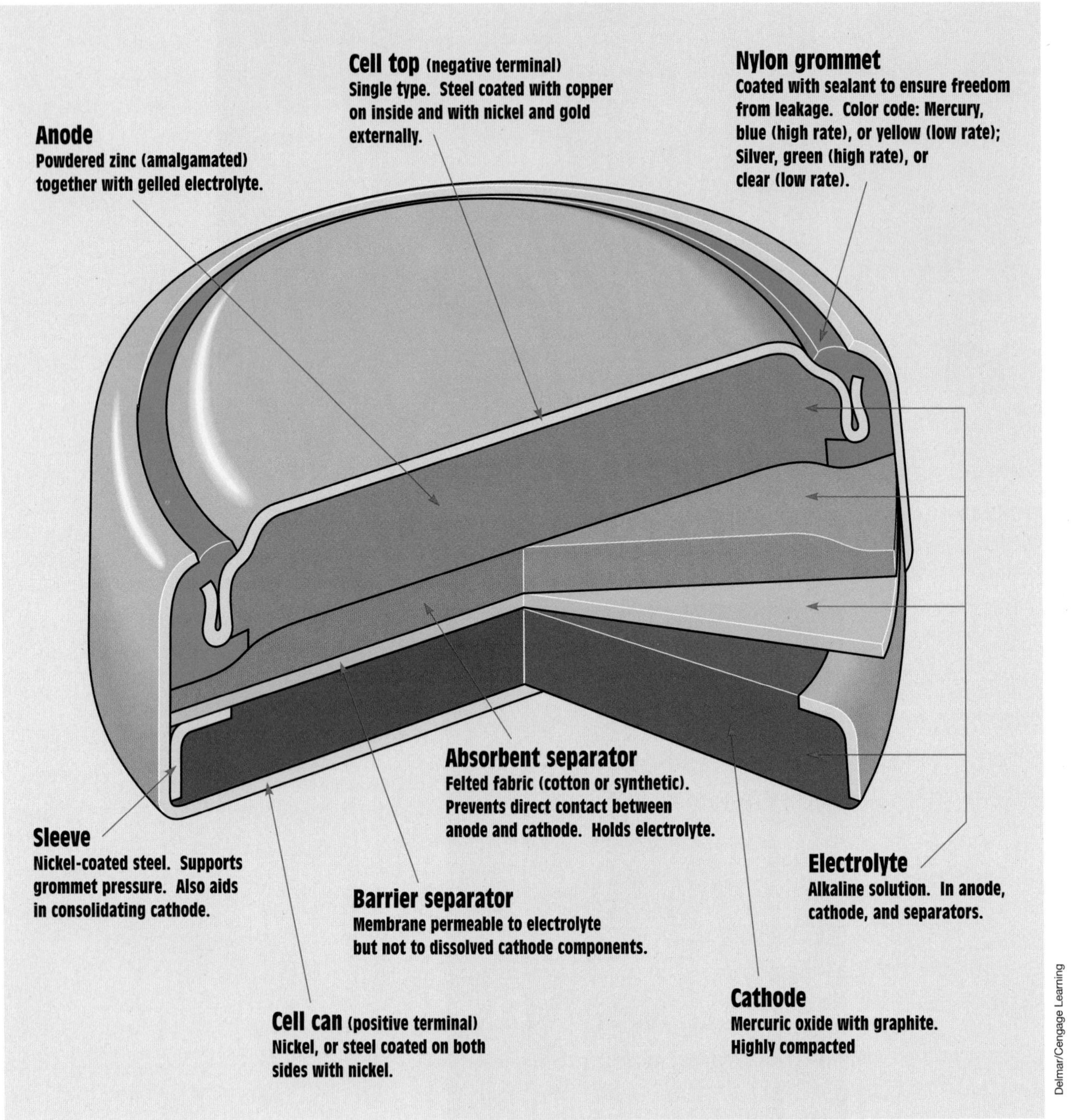

Anode
Powdered zinc (amalgamated) together with gelled electrolyte.

Cell top (negative terminal)
Single type. Steel coated with copper on inside and with nickel and gold externally.

Nylon grommet
Coated with sealant to ensure freedom from leakage. Color code: Mercury, blue (high rate), or yellow (low rate); Silver, green (high rate), or clear (low rate).

Sleeve
Nickel-coated steel. Supports grommet pressure. Also aids in consolidating cathode.

Absorbent separator
Felted fabric (cotton or synthetic). Prevents direct contact between anode and cathode. Holds electrolyte.

Barrier separator
Membrane permeable to electrolyte but not to dissolved cathode components.

Electrolyte
Alkaline solution. In anode, cathode, and separators.

Cell can (positive terminal)
Nickel, or steel coated on both sides with nickel.

Cathode
Mercuric oxide with graphite. Highly compacted

Delmar/Cengage Learning

FIGURE 13–13 Mercury button cell.

Another type of button cell that is less common because of its higher cost is the silver-zinc cell. This cell is the same as the mercury-zinc cell except that it uses silver oxide as the cathode material instead of mercuric oxide. The silver-zinc cell does have one distinct advantage over the mercury-zinc cell. The silver-zinc cell develops a voltage of 1.6 volts as compared with the 1.35 volts developed by the mercury-zinc cell. This higher voltage can be of major importance in some electronic circuits.

A variation of the silver-zinc cell uses divalent silver oxide as the cathode material instead of silver oxide. The chemical formula for silver oxide is Ag_2O. The chemical formula for divalent silver oxide is Ag_2O_2. One of the factors that determines the energy density of a cell is the amount of oxygen in the cathode material. Although divalent silver oxide contains twice as much oxygen as silver oxide, it does not produce twice the energy density. The increased energy density is actually about 10% to 15%. Using divalent silver oxide does have one disadvantage, however. The compound is less stable, which can cause the cell to have a shorter shelf life.

Lithium Cells

Lithium cells should probably be referred to as the lithium system because there are several different types of lithium cells. Lithium cells can have voltages that range from 1.9 volts to 3.6 volts, depending on the material used to construct the cell. Lithium is used as the anode material because of its high affinity for oxygen. In fact, for many years lithium had to be handled in an airless and moisture-free environment because of its extreme reactivity with oxygen. Several different cathode materials can be used, depending on the desired application. One type of lithium cell uses a solid electrolyte of lithium hydroxide. The use of a solid electrolyte produces a highly stable compound, resulting in a shelf life that is measured in decades. This cell produces a voltage of 1.9 volts but has an extremely low current capacity. The output current of the lithium cell is measured in microamperes (millionths of an ampere) and is used to power watches with liquid crystal displays and to maintain memory circuits in computers.

Other lithium cells use liquid electrolytes and can provide current outputs comparable to alkaline-manganese cells. Another lithium system uses a combination electrolyte-cathode material. One of these is sulfur dioxide, SO_2. This combination produces a voltage of 2.9 volts. In another electrolyte-cathode system, sodium chloride, $NaCl$, is used. This combination provides a terminal voltage of 3.6 volts.

Although lithium cells are generally considered to be primary cells, some types are rechargeable. The amount of charging current, however, is critical for these cells. An incorrect amount of charging current can cause the cell to explode.

Current Capacity and Cell Ratings

The amount of current a particular type of cell can deliver is determined by the surface area of its plates. A D cell can deliver more current than a C cell, and a C cell can deliver more current than an AA cell. The amount of power a cell can deliver is called its **current capacity.** To determine a cell's current capacity, several factors must be included, such as the type of cell, the rate of current flow, the voltage, and the length of time involved. Primary cells are generally limited by size and weight and therefore do not contain a large amount of power.

One of the common ratings for primary cells is the milliampere-hour (mA-hr). A milliampere (mA) is 1/1000 of an ampere. Therefore, if a cell can provide a current of 1 milliampere for 1 hour it will have a rating of 1 milliampere-hour. An average D-size alkaline cell has a capacity of approximately 10,000 milliampere-hours. Some simple calculations would reveal that this cell should be able to supply 100 milliamperes of current for a period of 100 hours, or 200 milliamperes of current for 50 hours.

Another common measure of a primary cell's current capacity is watt-hours (W-hr). Watt-hours are determined by multiplying the cell's milliampere-hour rating by its terminal voltage. If the alkaline-manganese cell just discussed has a voltage of 1.5 volts, its watt-hour capacity would be 15 watt-hours (10,000 mA-hr $\times$ 1.5 V = 15,000 mW-hr, or 15 W-hr). The chart in *Figure 13–14* shows the watt-hour capacity for several sizes of alkaline-manganese cells. The chart lists the cell size, the volume of the cell, and the watt-hours per cubic inch.

The amount of power a cell contains depends not only on the volume of the cell but also on the type of cell being used. The chart in *Figure 13–15* compares the watt-hours per cubic inch for different types of cells.

Internal Resistance

Batteries actually have two voltage ratings, one at no load and the other at normal load. The cell's rated voltage at normal load is the one used. The no-load voltage of a cell is greater because of the **internal resistance** of the cell. All cells have some amount of internal resistance. For example, the alkaline cell discussed previously had a rating of 10,000 milliampere-hours, or 10 ampere-hours (A-hr).

ALKALINE-MANGANESE CELL SIZE	VOLUME IN CUBIC INCHES	WATT-HOURS PER CUBIC INCH
D cell	3.17	4.0
C cell	1.52	4.2
AA cell	0.44	4.9
AAA cell	0.20	5.1

Delmar/Cengage Learning

FIGURE 13–14   Watt-hours per cubic inch for different size cells.

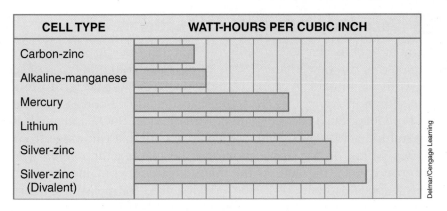

FIGURE 13–15 Watt-hours per cubic inch for different types of cells.

Theoretically, the cell should be able to deliver 10 amperes of current for one hour or 20 amperes of current for a half-hour. In practice, the cell could not deliver 10 amperes of current even under a short-circuit condition. *Figure 13–16* illustrates what happens when a DC ammeter is connected directly across the terminals of a D-size alkaline cell. Assume that the ammeter indicates a current flow of 4.5 amperes and that the terminal voltage of the cell has dropped to 0.5 volt. By applying Ohm's law, it can be determined that the cell has an internal resistance of 0.111 ohm (0.5 V/4.5 A = 0.111 Ω).

As the cell ages and power is used, the electrodes and electrolyte begin to deteriorate. This causes them to become less conductive, which results in an increase of internal resistance. As the internal resistance increases, the terminal voltage decreases.

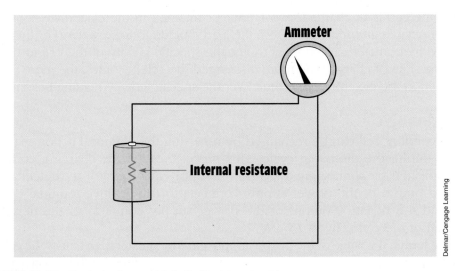

FIGURE 13–16 Short-circuit current is limited by internal resistance.

13–5   Secondary Cells: Lead-Acid Batteries

The secondary cell is characterized by the fact that once its stored energy has been depleted, it can be recharged. One of the most common types of secondary cells is the lead acid. A string of individual cells connected to form one battery is shown in *Figure 13–17*. A single lead-acid cell consists of one plate made of pure lead, Pb; a second plate of lead dioxide, PbO_2; and an electrolyte of dilute sulfuric acid, H_2SO_4, with a specific gravity that can range from 1.215 to 1.28, depending on the application and the manufacturer *(Figure 13–18)*. **Specific gravity** is a measure of the amount of acid contained in the water. Water has a specific gravity of 1.000. A device used for measuring the specific gravity of a cell is called a **hydrometer** *(Figure 13–19)*.

Discharge Cycle

When a load is connected between the positive and negative terminals, the battery begins to discharge its stored energy through the load *(Figure 13–20)*. The process is called the discharge cycle. Each of the lead atoms on the surface of the negative plate loses two electrons to become a Pb^{++} ion. These positive ions attract SO_4^{--} ions from the electrolyte. As a result, a layer of lead sulfate, $PbSO_4$, forms on the negative lead plate.

The positive plate is composed of lead dioxide, PbO_2. Each of the Pb atoms lacks four electrons that were given to the oxygen atom when the compound was formed and the Pb became a Pb^{++++} ion. Each of the Pb^{++++} ions takes two electrons from the load circuit to become a Pb^{++} ion. The Pb^{++} ions cannot hold the oxygen atoms that are released into the electrolyte solution and combine with hydrogen atoms to form molecules of water, H_2O. The remaining Pb^{++} ions combine with SO_4^{--} ions and form a layer of lead sulfate around the positive plate.

As the cell is discharged, two H^+ ions contained in the electrolyte combine with an oxygen atom liberated from the lead dioxide to form water. For this reason, the hydrometer can be used to test the specific gravity of the cell to determine the state of charge. The more discharged the cell becomes, the more water is formed in the electrolyte and the lower the specific gravity reading becomes.

Charging Cycle

The secondary cell can be recharged by reversing the chemical action that occurred during the discharge cycle. This process, called the charging cycle, is accomplished by connecting a DC power supply or generator to the cell. The positive output of the power supply connects to the positive terminal of the cell, and the negative output of the power supply connects to the negative terminal of the cell *(Figure 13–21)*.

The terminal voltage of the power supply must be greater than that of the cell or battery. As current flows through the cell, hydrogen is produced at the negative plate

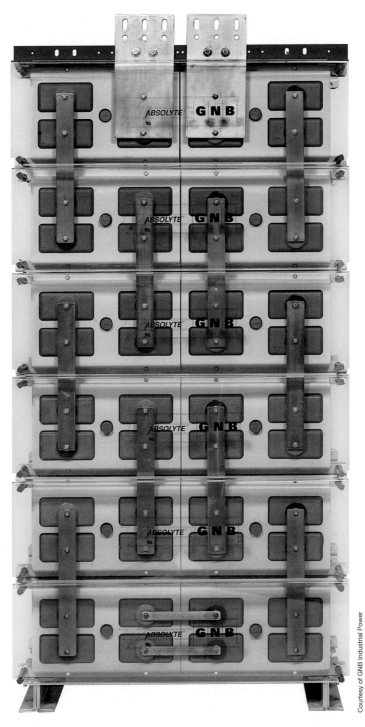

Courtesy of GNB Industrial Power

FIGURE 13–17   Lead-acid storage batteries.

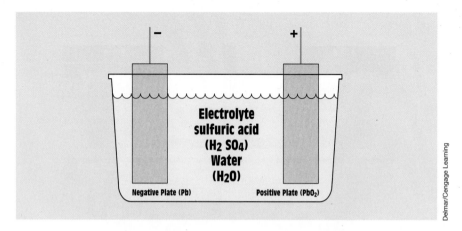

FIGURE 13–18   Basic lead-acid cell.

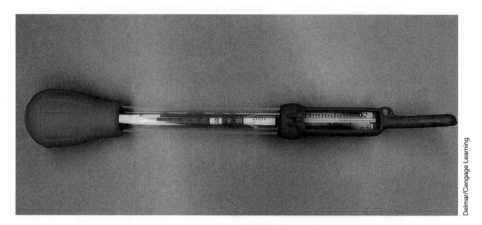

FIGURE 13–19   Hydrometer.

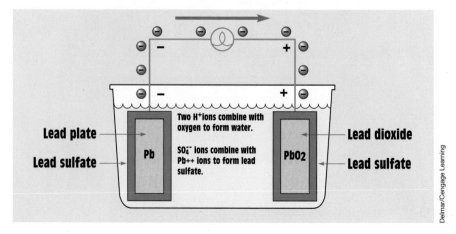

FIGURE 13–20   Discharge cycle of a lead-acid cell.

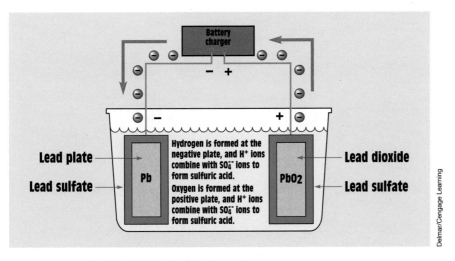

FIGURE 13–21 A lead-acid cell during the charging cycle.

and oxygen is produced at the positive plate. If the cell is in a state of discharge, both plates are covered with a layer of lead sulfate. The H^+ ions move toward the negative plate and combine with SO_4^{--} ions to form new molecules of sulfuric acid. As a result, Pb^{++} ions are left at the plate. These ions combine with electrons being supplied by the power supply and again become neutral lead atoms.

At the same time, water molecules break down at the positive plate. The hydrogen atoms combine with SO_4^{--} ions in the electrolytic solution to form sulfuric acid. The oxygen atom recombines with the lead dioxide to form a Pb^{++} ion. As electrons are removed from the lead dioxide by the power supply, the Pb^{++} ions become Pb^{++++} ions.

As the cell is charged, sulfuric acid is again formed in the electrolyte. The hydrometer can again be used to determine the state of charge. When the cell is fully charged, the electrolyte should be back to its original strength.

Most lead-acid cells contain multiple plates *(Figure 13–22)*. One section of plates is connected together to form a single positive plate, while the other section forms a single negative plate. This arrangement increases the surface area and thus the current capacity of the cell.

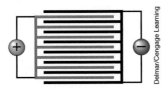

FIGURE 13–22 Multiple plates increase the surface area and the amount of current the cell can produce.

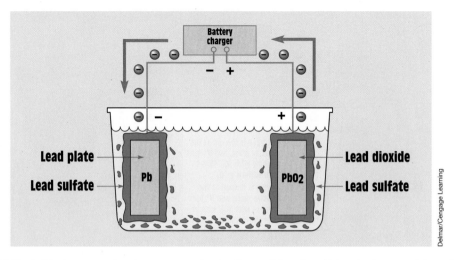

FIGURE 13–23 Improper charging can cause lead sulfate to flake off and fall to the bottom of the cell.

Cautions Concerning Charging

Theoretically, a secondary cell should be able to be discharged and recharged indefinitely. In practice, however, it cannot. If the cell is overcharged, that is, the amount of charge current is too great, the lead sulfate does not have a chance to dissolve back into the electrolyte and become acid. High charging current or mechanical shock can cause large flakes of lead sulfate to break away from the plates and fall to the bottom of the cell *(Figure 13–23)*. These flakes can no longer recombine with the H$^+$ ions to become sulfuric acid. Therefore, the electrolyte is permanently weakened. If the flakes build up to a point that they touch the plates, they cause a short circuit and the cell can no longer operate.

CAUTION: ***Overcharging can also cause hydrogen gas to form:*** Because hydrogen is the most explosive element known, sparks or open flames should be kept away from batteries or cells, especially during the charging process. Overcharging also causes excess heat that can permanently damage the cell. The accepted temperature limit for most lead-acid cells is 110°F (43°C).

The proper amount of charging current can vary from one type of battery to another, and manufacturers' specifications should be followed when possible. A general rule concerning charging current is that the current should not be greater than 1/10 the ampere-hour capacity. An 80-ampere-hour battery, for instance, should not be charged with a current greater than 8 amperes. ■

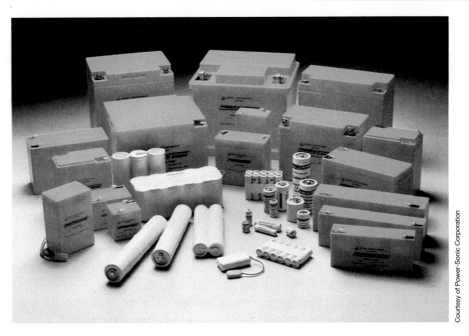

Courtesy of Power-Sonic Corporation

FIGURE 13–24   Sealed lead-acid and nicad cells and batteries.

Sealed Lead-Acid Batteries

Sealed lead-acid batteries have become increasingly popular in the past several years. They come in different sizes, voltage ratings, ampere-hour ratings, and case styles *(Figure 13–24)*. These batteries are often referred to as gel cells because the sulfuric acid electrolyte is suspended in an immobilized gelatin state. This treatment prevents spillage and permits the battery to be used in any position. Gel cells use a cast grid constructed of lead-calcium, which is free of antimony. The calcium is used to add strength to the grid. The negative plate is actually a lead paste material, and the positive plate is made of lead-dioxide paste. A one-way pressure-relief valve set to open at 2–6 PSI is used to vent any gas buildup during charging.

Ratings for Lead-Acid Batteries

One of the most common ratings for lead-acid batteries is the ampere-hour rating. The ampere-hour rating for lead-acid batteries is determined by measuring the battery's ability to produce current for a 20-hour period at 80°F. A battery with the ability to produce a current of 4 amperes for 20 hours would have a rating of 80 ampere-hours.

Another common battery rating, especially for automotive batteries, is cold-cranking amperes. This rating has nothing to do with the ampere-hour rating of the battery. Cold-cranking amperes are the maximum amount of initial current the battery can supply at 68°F (20°C).

Testing Lead-Acid Batteries

It is sometimes necessary to test the state of charge or condition of a lead-acid battery. The state of charge can often be tested with a hydrometer, as previously described. As batteries age, however, the specific gravity remains low even after the battery has been charged. When this happens, it is an indication that the battery has lost part of its materials because of lead sulfate flaking off the plates and falling to the bottom of the battery. When this occurs there is no way to recover the material.

Another standard test for lead-acid batteries is the **load test.** This test will probably reveal more information concerning the condition of the battery than any other test. To perform a load test, the amount of test current should be three times the ampere-hour capacity. The voltage should not drop below 80% of the terminal voltage for a period of three minutes. For example, an 80-ampere-hour, 12-volt battery is to be load tested. The test current will be 240 amperes (80 A $\times$ 3), and the voltage should not drop below 9.6 volts (12 V $\times$ 0.80) for a period of three minutes.

13–6 Other Secondary Cells

Nickel-Iron Batteries (Edison Battery)

The nickel-iron cell is often referred to as the Edison cell, or Edison battery. The nickel-iron battery was developed in 1899 for use in electric cars being built by the Edison Company. The negative plate is a nickeled steel grid that contains powdered iron. The positive plates are nickel tubes that contain nickel oxides and nickel hydroxides. The electrolyte is a solution of 21% potassium hydroxide.

The nickel-iron cell weighs less than lead-acid cells but has a lower energy density. The greatest advantage of the nickel-iron cell is its ability to withstand deep discharges and to recover without harm to the cell. The nickel-iron cell can also be left in a state of discharge for long periods of time without harm. Because these batteries need little maintenance, they are sometimes found in portable and emergency lighting equipment. They are also used to power electric mine locomotives and electric forklifts.

The nickel-iron battery does have two major disadvantages. One is high cost. These batteries cost several times more than comparable lead-acid batteries. The second disadvantage is high internal resistance. Nickel-iron batteries do not have the ability to supply the large initial currents needed to start gasoline or diesel engines.

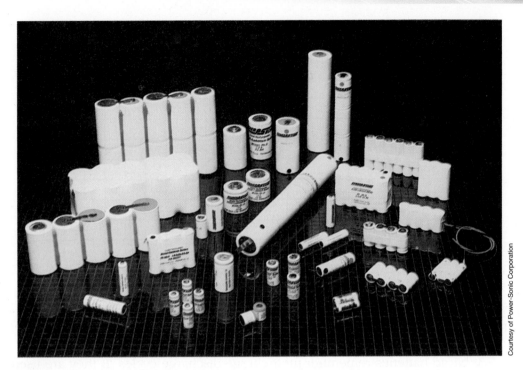

Courtesy of Power-Sonic Corporation

FIGURE 13–25 Nickel-cadmium batteries.

Nickel-Cadmium Batteries

The **nickel-cadmium (nicad) cell** was first developed in Sweden by Junger and Berg in 1898. The positive plate is constructed of nickel hydroxide mixed with graphite. The graphite is used to increase conductivity. The negative plate is constructed of cadmium oxide, and the electrolyte is potassium hydroxide with a specific gravity of approximately 1.2. Nicad batteries have extremely long life spans. On the average, they can be charged and discharged about 2000 times. Nicad batteries can be purchased in a variety of case styles *(Figure 13–25)*.

Nickel-cadmium batteries have the ability to produce large amounts of current, similar to the lead-acid battery, but do not experience the voltage drop associated with the lead-acid battery *(Figure 13–26)*.

Nickel-cadmium batteries do have some disadvantages:

1. The nicad battery develops only 1.2 volts per cell as compared with 1.5 volts for carbon-zinc and alkaline primary cells, or 2 volts for lead-acid cells.

2. Nicad batteries cost more initially than lead-acid batteries.

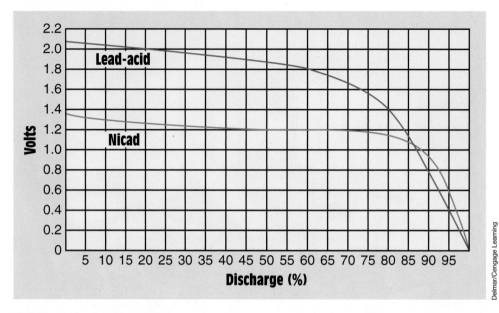

FIGURE 13–26 Typical discharge curves for nicad and lead-acid cells.

3. Nicad batteries remember their charge-discharge cycles. If they are used at only low currents and are permitted to discharge through only part of a cycle and are then recharged, over a long period of time they will develop a characteristic curve to match this cycle.

Nickel-Metal Hydride Cells

Nickel-metal hydride cells (Ni-MH) are similar to nickel-cadmium cells in many respects. Both exhibit a voltage of 1.2 volts per cell, and both have very similar charge and discharge curves. Nickel-metal hydride cells exhibit some improved characteristics over nickel-cadmium cells, however. They have about a 40% higher energy density, and they do not exhibit as great a problem of memory accumulation. Nickel-metal hydride cells are also more environmentally friendly than nickel-cadmium cells. The positive electrode is made of nickel oxyhydroxide (Ni00H), and the negative electrode is made of metal hydride. The electrolyte is an aqueous (watery) potassium-hydroxide solution. Nickel-metal hydride cells are replacing nickel-cadmium cells in many applications because they have a greater energy density and very little problem with memory accumulation.

Lithium-Ion Cells

Lithium-ion cells are very popular for portable equipment such as notebook computers, video cameras, cell phones, and many others. Lithium-ion cells can be recharged and offer a very high energy density for their size and weight.

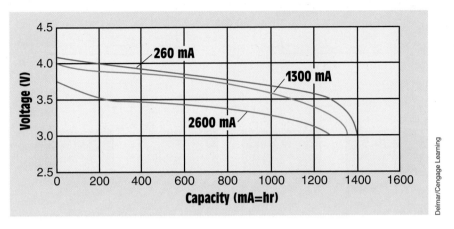

FIGURE 13–27   Typical discharge curves for lithium-ion cells.

They exhibit a voltage of 3.6 volts per cell, which is the same voltage that can be obtained by connecting three nickel-cadmium (Ni-Cd) or three nickel-metal hydride (Ni-MH) cells in series. Lithium-ion cells also exhibit a weight-energy density that is about three times greater than nickel-cadmium cells. Under proper charging conditions, these cells can be recharged about 500 times. They also exhibit a rather flat discharge curve *(Figure 13–27)*, which makes them an ideal choice for electronic devices that require a constant voltage. Unlike nickel-cadmium cells and to some extent nickel-metal hydride cells, lithium-ion cells do not have the problem of memory accumulation. They can be recharged to their full capacity each time they are charged.

Lithium-ion cells can safely be recharged because they do not contain metallic lithium. The positive electrode or anode is made of lithium metallic oxide, and the negative electrode or cathode is made of carbon. These cells work by transferring lithium ions between the cathode and anode during discharging and charging. Although lithium-ion cells are safe to recharge, they do require a special charger. Chargers for these cells generally produce a constant voltage and constant current. Overcurrent or overvoltage during charging can cause early deterioration of the cell.

13–7   Series and Parallel Battery Connections

When batteries or cells are connected in series, their voltages add and their current capacities remain the same. In *Figure 13–28*, four batteries, each having a voltage of 12 volts and 60 ampere-hours, are connected in series. Connecting them in series has the effect of maintaining the surface area of the plates and increasing the number of cells. The connection shown in *Figure 13–28* will have an output voltage of 48 volts and a current capacity of 60 ampere-hours.

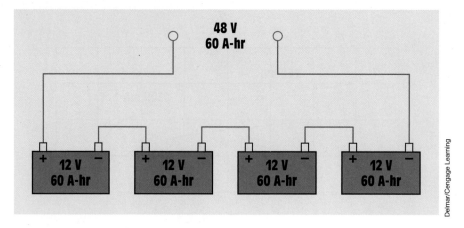

FIGURE 13–28 When batteries are connected in series, their voltages add and their ampere-hour capacities remain the same.

Connecting batteries or cells in parallel *(Figure 13–29)* has the effect of increasing the area of the plates. In this example, the same four batteries are connected in parallel. The output voltage remains 12 volts, but the ampere-hour capacity has increased to 240 ampere-hours.

CAUTION: *Batteries of different voltages should never be connected in parallel. The batteries could be damaged or one of the batteries could explode.* ■

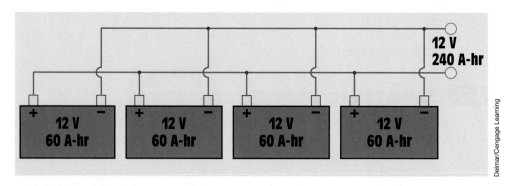

FIGURE 13–29 When batteries are connected in parallel, their voltages remain the same and their ampere-hour capacities add.

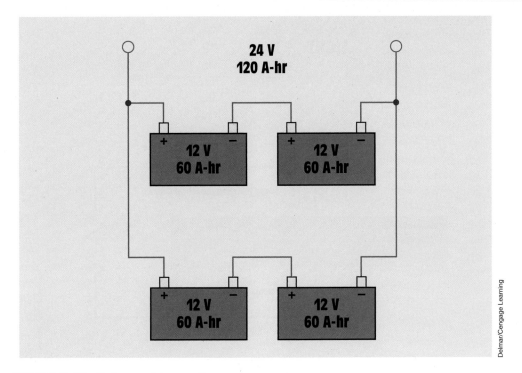

FIGURE 13–30   Series-parallel connection.

Batteries can also be connected in a series-parallel combination. In *Figure 13–30*, the four batteries have been connected in such a manner that the output has a value of 24 volts and 120 ampere-hours. To make this connection, the four batteries were divided into two groups of two batteries each. The batteries of each group were connected in series to produce an output of 24 volts at 60 ampere-hours. These two groups were then connected in parallel to provide an output of 24 volts at 120 ampere-hours.

13–8   Other Small Sources of Electricity

Solar Cells

Although batteries are the largest source of electricity after alternators and generators, they are not the only source. One source of electricity is the photovoltaic cell or solar cell. Solar cells are constructed from a combination of P- and N-type semiconductor material. Semiconductors are made from materials that contain four valence electrons. The most common semiconductor materials are silicon and germanium. Impurities must be added to the pure semiconductor material to form P- and N-type materials. If a material containing three valence electrons is added to a pure semiconductor material, P-type material is formed. P-type material has a lack of electrons. N-type material is formed when a

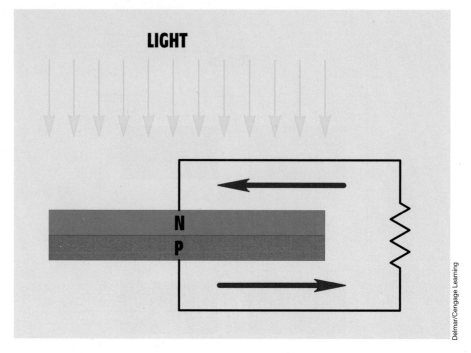

FIGURE 13–31   Solar cells are formed by bonding P- and N-type semiconductor materials.

material containing five valence electrons is added to a pure semiconductor material. N-type material has an excess of electrons.

Light is composed of particles called photons. A photon is a small package of pure energy that contains no mass. When photons strike the surface of the photocell, the energy contained in the photon is given to a free electron. This additional energy causes the electron to cross the junction between the two types of semiconductor material and produce a voltage *(Figure 13–31).*

 GREEN TIP: Solar cells can produce electricity without the use of any fossil fuels. Solar cell arrays are often used to charge batteries that in turn supply power to rural locations where access to power lines is not available. ■

The amount of voltage produced by a solar cell is determined by the material from which it is made. Silicon solar cells produce an open-circuit voltage of 0.5 volts per cell in direct sunlight. The amount of current a cell can deliver is determined by the surface area of the cell. Because the solar cell produces

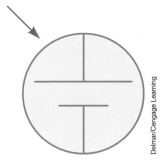

FIGURE 13–32 Schematic symbol for a solar cell.

a voltage in the presence of light, the schematic symbol for a solar cell is the same as that used to represent a single voltaic cell with the addition of an arrow to indicate it is receiving light *(Figure 13–32)*.

 GREEN TIP: Solar collectors are a very efficient method of heating water. Many countries throughout the world depend almost entirely on solar energy for heating residential water. ■

It is often necessary to connect solar cells in series or parallel or both to obtain desired amounts of voltage and current. For example, assume that an array of photovoltaic cells is to be used to charge a 12-volt lead-acid battery. The charging voltage is to be 14 volts, and the charging current should be 0.5 ampere. Now assume that each cell produces 0.5 volt with a short-circuit current of 0.25 ampere. In order to produce 14 volts, it will be necessary to connect 28 solar cells in series. An output of 14 volts with a current capacity of 0.25 ampere will be produced. To produce an output of 14 volts with a current capacity of 0.5 ampere, it will be necessary to connect a second set of 28 cells in series and parallel this set with the first set *(Figure 13–33)*.

Thermocouples

In 1822, a German scientist named Seebeck discovered that when two dissimilar metals are joined at one end and that junction is heated, a voltage is produced *(Figure 13–34)*. This is known as the Seebeck effect. The device produced by the joining of two dissimilar metals for the purpose of producing electricity with heat is called a **thermocouple.** The amount of voltage produced by a thermocouple is determined by

1. the type of materials used to produce the thermocouple.

2. the temperature difference between the junction and the open ends.

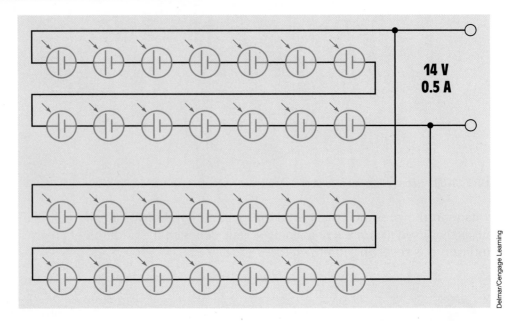

FIGURE 13–33   Series-parallel connection of solar cells.

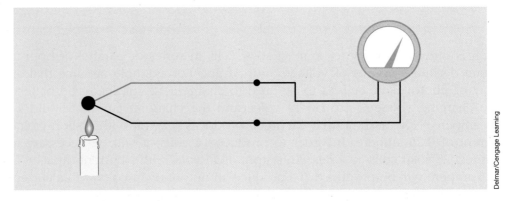

FIGURE 13–34   The thermocouple is made by forming a junction of two different types of metal.

The chart in *Figure 13–35* shows common types of thermocouples. The different metals used in the construction of thermocouples are shown, as well as the normal temperature ranges of the thermocouples.

The amount of voltage produced by a thermocouple is small, generally on the order of millivolts (1 mV = 0.001 V). The polarity of the voltage of some thermocouples is determined by the temperature. For example, a type J thermocouple produces 0 volt at about 32°F. At temperatures above 32°F, the iron wire is positive and the constantan wire is negative. At temperatures below 32°F, the iron wire becomes negative and the constantan wire becomes positive. At a temperature of 300°F, a type J thermocouple will produce a voltage

TYPE	MATERIAL		Degrees F	Degrees C
J	Iron	Constantan	−328 to +32 +32 to +1432	−200 to 0 0 to 778
K	Chromel	Alumel	−328 to +32 +32 to +2472	−200 to 0 0 to 1356
T	Copper	Constantan	−328 to +32 +32 to 752	−200 to 0 0 to 400
E	Chromel	Constantan	−328 to +32 +32 to 1832	−200 to 0 0 to 1000
R	Platinum 13% rhodium	Platinum	−32 to +3232	0 to 1778
S	Platinum 10% rhodium	Platinum	−32 to +3232	0 to 1778
B	Platinum 6% rhodium	Platinum	−32 to +3092	0 to 1700

Delmar/Cengage Learning

FIGURE 13–35 Thermocouple chart.

of about 7.9 millivolts. At a temperature of −300°F, it will produce a voltage of about −7.9 millivolts.

Because thermocouples produce such low voltages, they are often connected in series, as shown in *Figure 13–36*. This connection is referred to as a thermopile. Thermocouples and thermopiles are generally used for measuring temperature and are sometimes used to detect the presence of a pilot light in appliances that operate with natural gas. The thermocouple is heated by the pilot light. The current produced by the thermocouple is used to produce a magnetic field, which holds a gas valve open to permit gas to flow to the main burner. If the pilot light goes out, the thermocouple ceases to produce current and the valve closes *(Figure 13–37)*.

Piezoelectricity

The word *piezo*, pronounced "pee-ay-zo," is derived from the Greek word for pressure. **Piezoelectricity** is produced by some materials when they are

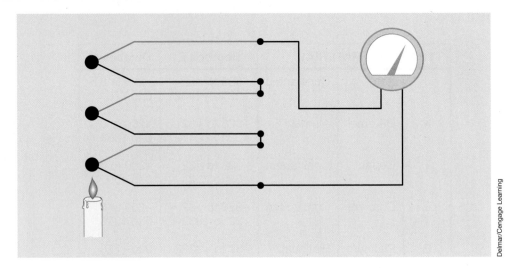

FIGURE 13–36 A thermopile is a series connection of thermocouples.

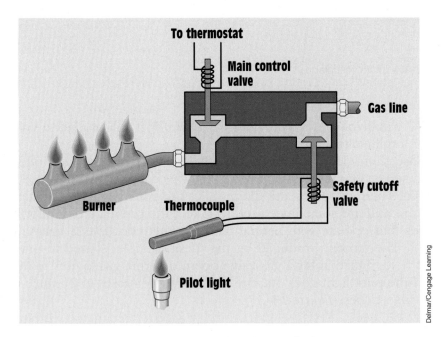

FIGURE 13–37 A thermocouple provides power to the safety cutoff valve.

placed under pressure. The pressure can be caused by compression, twisting, bending, or stretching. Rochelle salt (sodium potassium tartrate) is often used as the needle, or stylus, of a phonograph. Grooves in the record cause the crystal to be twisted back and forth, producing an alternating voltage. This voltage is amplified and is heard as music or speech. Rochelle salt crystals are also used as the pickup for microphones. The vibrations caused by sound waves produce stress on the crystals. This stress causes the crystals to produce an alternating voltage that can be amplified.

Another crystal used to produce the piezoelectric effect is barium titanate. Barium titanate can actually produce enough voltage and current to flash a small neon lamp when struck by a heavy object *(Figure 13–38)*. Industry uses the crystal's ability to produce voltage in transducers for sensing pressure and mechanical vibration of machine parts.

Quartz crystal has been used for many years as the basis for crystal oscillators. In this application, an AC voltage close to the natural mechanical vibration frequency of a slice of quartz is applied to opposite surfaces of the crystal. This causes the quartz to vibrate at its natural frequency. The frequency is extremely constant for a particular slice of quartz. Quartz crystals have been used to change the frequency range of two-way radios for many years.

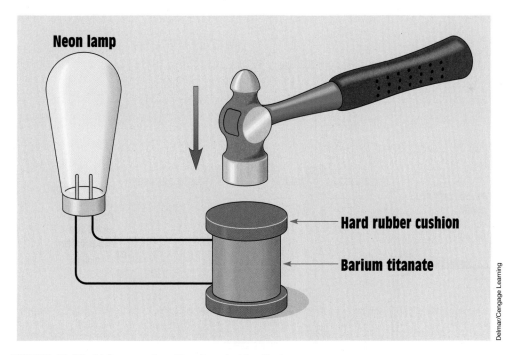

FIGURE 13–38 Voltage produced by piezoelectric effect.

Fuel Cells

Another device that employs chemical processes to produce electricity is the fuel cell. Similar to voltaic cells, the fuel cell transfers electrons from one material to another. An important difference between the voltaic cell and the fuel cell, however, is that the material giving the electrons and the material accepting the electrons are not consumed in the process. Fuel cells require high-purity fuel and a catalytic surface for the reaction to take place. They also require auxiliary equipment such as gas containers and pressure controls. Research is under way to develop fuel cells for powering automobiles and as a source of power for homes.

A common type of fuel cell is the hydrogen-oxygen cell (*Figure 13–39.*)

The cell contains hollow, porous, carbon electrodes immersed in a potassium hydroxide solution. Hydrogen gas is supplied to one electrode and oxygen gas is supplied to the other. The carbon electrodes contain metal or metal oxide that acts as a catalyst. A catalyst is a substance that aids the chemical reaction. In this case, the catalyst helps hydrogen molecules, which are pairs of hydrogen atoms, to separate into single atoms that combine with the negatively charged hydroxide ions of the electrolyte. The combination of the hydrogen

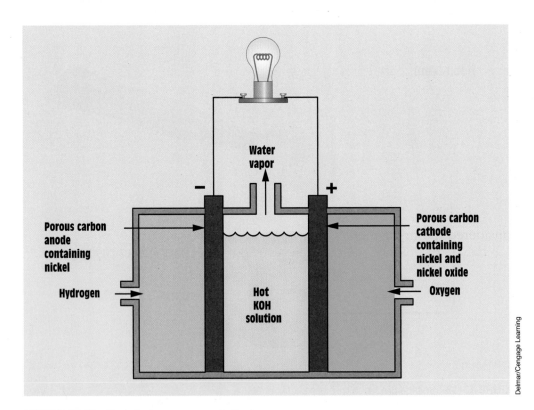

FIGURE 13–39 Basic hydrogen-oxygen fuel cell.

and hydroxide ion forms a molecule of water, H_2O, with one electron left over. The excess electrons are attracted to the carbon electrode that is supplied with oxygen. The hydroxide ions are re-formed at the cathode at the same rate as they are used up at the anode. The result is that hydrogen and oxygen are converted into water.

The alkaline-hydrogen-oxygen fuel cell just discussed was one the very early types introduced in the 1960s. There are other types of fuel cells today that offer different characteristics and advantages.

The Polymer Electrolyte Membrane (PEM) Fuel Cell

Polymer electrolyte membrane fuel cells are often called proton exchange membrane fuel cells. These cells offer high-power density combined with low weight and volume compared with other types of fuel cells. PEM fuel cells employ a solid polymer as the electrolyte. The electrodes are porous carbon combined with a platinum catalyst. They require only hydrogen, oxygen from the air, and water to operate. They do not need corrosive fluids like some other fuel cells. PEM fuel cells operate at a relatively low temperature, around 176°F (80°C).

Direct Methanol Fuel Cells

Direct methanol fuel cells are power by pure methanol mixed with steam. The methanol-steam mixture is fed directly to the fuel cell anode. These fuel cells have an advantage in that they do not have the fuel storage problems of cells that rely on pure hydrogen. Because methanol is a liquid, it is much easier to store. Methanol also has a higher energy density than hydrogen, although it does not contain as much energy as gasoline or diesel fuel.

Phosphoric Acid Fuel Cells

Phosphoric acid fuel cells use liquid phosphoric acid as an electrolyte. The acid is contained in a Teflon-bonded silicon carbide matrix. The electrodes are composed of porous carbon containing a platinum catalyst. This cell is considered the first generation of modern fuel cells and has been used commercially for stationary power generation and to power large vehicles such as city buses.

Molten Carbonate Fuel Cells

Molten carbonate fuel cells are being developed for natural gas and coal-fired power plants. These fuel cells operate at very high temperatures, typically 1200°F (650°C) and above. The high operating temperature has an advantage in that nonprecious metals can be used as a catalyst in the anode and cathode electrodes, resulting in much lower cost. Molten carbonate fuel cells can reach efficiencies as high as 60%.

Solid Oxide Fuel Cells

Solid oxide fuel cells use a hard nonporous ceramic compound as the electrolyte. Because the electrolyte is a solid, the fuel cell does not have to be constructed in the plate-like configuration typical of other types of fuel cells. These fuel cells operate at very high temperatures, 1,830°F (1000°C). The high operating temperature has an advantage in that nonprecious metals can be used as a catalyst in the electrodes. Also, the high operating temperature of the solid oxide fuel cell permits re-forming of fuels internally, enabling the use of a variety of fuels, which helps reduce the cost of adding a re-former to the system. Solid oxide fuel cells have an efficiency of 50% to 60%.

GREEN TIP: Fuel cells can be used to produce electricity with very little harmful effect to the environment. ■

Summary

- The first practical battery was invented by Alessandro Volta in 1800.
- A voltaic cell converts chemical energy into electrical energy.
- A battery is a group of cells connected together.
- Primary cells cannot be recharged.
- Secondary cells can be recharged.
- The amount of voltage produced by a cell is determined by the materials from which it is made.
- A voltaic cell can be constructed from almost any two unlike metals and an electrolyte of acid, alkali, or salt.
- Voltaic cells depend on the movement of ions in solution to produce electricity.
- The amount of current a cell can provide is determined by the area of the cell plates.
- Primary cells are often rated in milliampere-hours or watt-hours.
- The amount of energy a battery can contain is called its capacity or energy density.
- Secondary cells can be recharged by reversing the current flow through them.

- A hydrometer is a device for measuring the specific gravity of the electrolyte.
- When lead-acid batteries are charged, hydrogen gas is produced.
- Nickel-cadmium batteries have a life span of about 2000 charge-discharge cycles.
- When cells are connected in series, their voltages add and their ampere-hour capacities remain the same.
- When cells are connected in parallel, their voltages remain the same and their ampere-hour capacities add.
- Cells or batteries of different voltages should never be connected in parallel.
- Solar cells produce electricity in the presence of light.
- The amount of voltage produced by a solar cell is determined by the materials from which it is made.
- The amount of current produced by a solar cell is determined by the surface area.
- Thermocouples produce a voltage when the junction of two unlike metals is heated.
- The amount of voltage produced by a thermocouple is determined by the type of materials used and the temperature difference between the ends of the junction.
- The voltage polarity of some thermocouples is determined by the temperature.
- Thermocouples connected in series to produce a higher voltage are called a thermopile.
- Some crystals can produce a voltage when placed under pressure.
- The production of voltage by application of pressure is called the piezoelectric effect.

Review Questions

1. What is a voltaic cell?
2. What factors determine the amount of voltage produced by a cell?
3. What determines the amount of current a cell can provide?
4. What is a battery?
5. What is a primary cell?
6. What is a secondary cell?

7. What material is used as the positive electrode in a zinc-mercury cell?

8. What is another name for the Leclanché cell?

9. What is used as the electrolyte in a carbon-zinc cell?

10. What is the advantage of the alkaline cell as compared with the carbon-zinc cell?

11. What material is used as the positive electrode in an alkaline cell?

12. How is the A-hr capacity of a lead-acid battery determined?

13. What device is used to test the specific gravity of a cell?

14. A 6-V lead-acid battery has an A-hr rating of 180 A-hr. The battery is to be load tested. What should be the test current, and what are the maximum permissible amount and duration of the voltage drop?

15. Three 12-V, 100-A-hr batteries are connected in series. What are the output voltage and A-hr capacity of this connection?

16. What is the voltage produced by a silicon solar cell?

17. What determines the current capacity of a solar cell?

18. A solar cell can produce a voltage of 0.5 V and has a current capacity of 0.1 A. How many cells should be connected in series and parallel to produce an output of 6 V at 0.3 A?

19. What determines the amount of voltage produced by a thermocouple?

20. A thermocouple is to be used to measure a temperature of 2800°F. Which type or types of thermocouples can be used to measure this temperature?

21. What materials are used in the construction of a type J thermocouple?

22. What does the word *piezo* mean?

Practical Applications

Your job is to order and connect lead-acid cells used to supply an uninterruptible power supply (UPS). The battery output voltage must be 126 V and have a current capacity of not less than 250 A-hr. Each cell has a rating of 2 V and 100 A-hr. How many cells must be ordered and how would you connect them? ∎

Practical Applications

A bank of nickel-cadmium cells is used as the emergency lighting supply for a hospital. There are 100 cells connected in series and each has an A-hr rating of 120 A-hr. The bank has to be replaced, and the manufacturer is no longer supplying nickel-cadmium cells. It is decided to replace the cells with lead-acid cells. How many lead-acid cells will be required if each has an A-hr rating of 60 A-hr, and how should they be connected? ■

Practical Applications

A n office building uses a bank of 63 lead-acid cells connected in series with a capacity of 80 amp-hours each to provide battery backup for their computers. The lead-acid cells are to be replaced with nickel-metal hydride cells with a capacity of 40 amp-hours each. How many nickel-metal hydride cells will be required to replace the lead-acid cells and how should they be connected? ■

Unit 14

Magnetic Induction

Why You Need to Know

Magnetic induction is one of the most important concepts in the study of electricity. Devices such as generators, alternators, motors, and transformers operate on this principle. This unit

- explains why current cannot rise in an inductor instantly.
- describes what determines the amount of inducted voltage and its polarity.
- describes what a voltage spike is, when it occurs, and how to control it.
- illustrates how an induced current produces a magnetic field around a coil.
- explains how the amount of voltage induced in a conductor relates to magnetic induction and lines of flux.
- discusses time constants and how the rise time of current in an inductor can be determined.
- introduces the exponential curve. Understanding the exponential curve is important because so many things in both science and nature operate on this principle. Not only does the current in an inductor rise at an exponential rate, but also clothes hung on a line dry at an exponential rate.

KEY TERMS

Eddy current

Electromagnetic induction

Exponential curve

Henry (H)

Hysteresis loss

Left-hand generator rule

Lenz's law

Metal oxide varistor (MOV)

R-L time constant

Speed

Strength of magnetic field

Turns of wire

Voltage spike

Weber (Wb)

Objectives

After studying this unit, you should be able to

- discuss electromagnetic induction.

- list factors that determine the amount and polarity of an induced voltage.

- discuss Lenz's law.

- discuss an exponential curve.

- list devices used to help prevent induced voltage spikes.

Preview

Electromagnetic induction is one of the most important concepts in the electrical field. It is the basic operating principle underlying alternators, transformers, and most AC motors. It is imperative that anyone desiring to work in the electrical field have an understanding of the principles involved. ∎

14–1   Electromagnetic Induction

In Unit 4, it was stated that one of the basic laws of electricity is that whenever current flows through a conductor, a magnetic field is created around the conductor *(Figure 14–1)*. The direction of the current flow determines the polarity of the magnetic field, and the amount of current determines the strength of the magnetic field.

That basic law in reverse is the principle of **electromagnetic induction,** which states that ***whenever a conductor cuts through magnetic lines of flux, a voltage is induced into the conductor.*** The conductor in *Figure 14–2* is connected to a zero-center microammeter, creating a complete circuit. When the conductor is moved downward through the magnetic lines of flux, the induced voltage causes electrons to flow in the direction indicated by the arrows. This flow of electrons causes the pointer of the meter to be deflected from the zero-center position.

If the conductor is moved upward, the polarity of induced voltage is reversed and the current flows in the opposite direction *(Figure 14–3)*. Consequently, the pointer is deflected in the opposite direction.

The polarity of the induced voltage can also be changed by reversing the polarity of the magnetic field *(Figure 14–4)*. In this example, the conductor is again moved downward through the lines of flux, but the polarity of the magnetic field has been reversed. Therefore, the polarity of the induced voltage is the opposite of that in *Figure 14–2,* and the pointer of the meter is deflected in

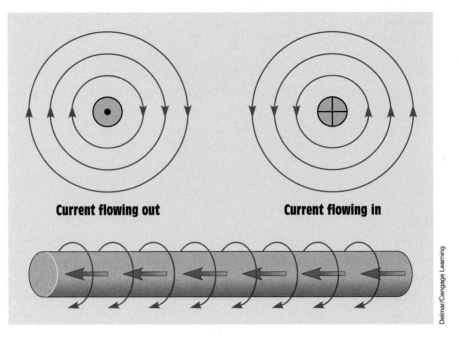

Current flowing out   **Current flowing in**

Delmar/Cengage Learning

FIGURE 14–1   Current flowing through a conductor produces a magnetic field around the conductor.

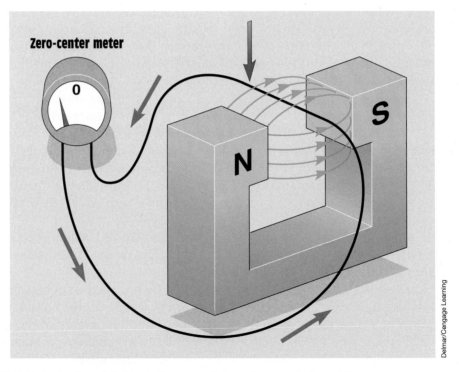

Zero-center meter

N

S

Delmar/Cengage Learning

FIGURE 14–2   A voltage is induced when a conductor cuts magnetic lines of flux.

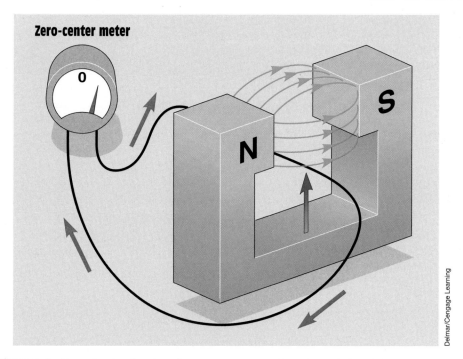

FIGURE 14–3   Reversing the direction of movement reverses the polarity of the voltage.

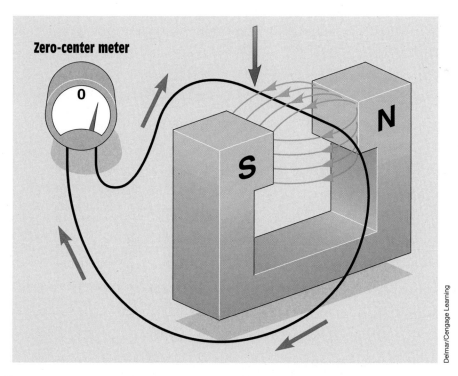

FIGURE 14–4   Reversing the polarity of the magnetic field reverses the polarity of the voltage.

the opposite direction. It can be concluded that *the polarity of the induced voltage is determined by the polarity of the magnetic field in relation to the direction of movement.*

14–2   Fleming's Left-Hand Generator Rule

Fleming's **left-hand generator rule** can be used to determine the relationship of the motion of the conductor in a magnetic field to the direction of the induced current. To use the left-hand rule, place the thumb, forefinger, and center finger at right angles to each other, as shown in *Figure 14–5. The forefinger points in the direction of the field flux,* assuming that magnetic lines of force are in a direction of north to south. *The thumb points in the direction of thrust,* or movement of the conductor, *and the center finger shows the direction of the current induced into the armature.* An easy method of remembering which finger represents which quantity follows:

THumb = **TH**rust

Forefinger = **F**lux

Center finger = **C**urrent

The left-hand rule can be used to clearly illustrate that if the polarity of the magnetic field is changed or if the direction of armature rotation is changed, the direction of induced current also changes.

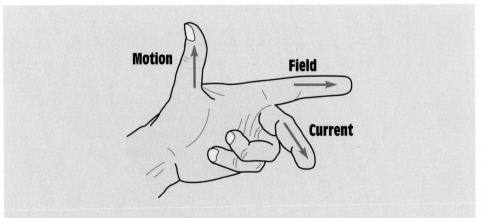

FIGURE 14–5   Left-hand generator rule.

Delmar/Cengage Learning

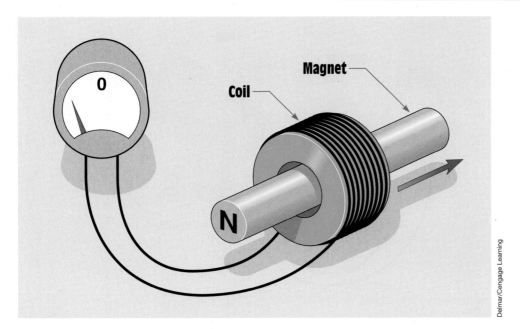

FIGURE 14–6 Voltage is induced by a moving magnetic field.

14–3 Moving Magnetic Fields

The important factors concerning electromagnetic induction are a conductor, a magnetic field, and relative motion. In practice, it is often desirable to move the magnet instead of the conductor. Most AC generators or alternators operate on this principle. In *Figure 14–6,* a coil of wire is held stationary while a magnet is moved through the coil. As the magnet is moved, the lines of flux cut through the windings of the coil and induce a voltage into them.

14–4 Determining the Amount of Induced Voltage

Three factors determine the amount of voltage that will be induced in a conductor:

1. the number of **turns of wire**

2. the **strength of the magnetic field** (flux density)

3. the **speed** of the cutting action

In order to induce 1 volt in a conductor, the conductor must cut 100,000,000 lines of magnetic flux in 1 second. In magnetic measurement, 100,000,000 lines of flux are equal to 1 **weber (Wb).** Therefore, if a conductor cuts magnetic lines of flux at a rate of 1 weber per second, a voltage of 1 volt is induced.

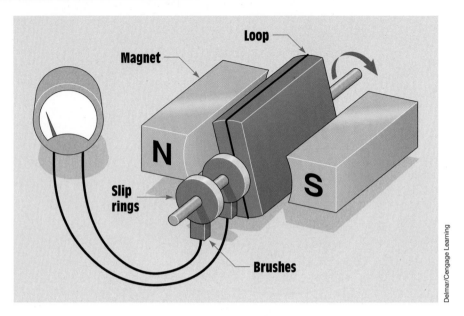

FIGURE 14–7   A single-loop generator.

A simple one-loop generator is shown in *Figure 14–7*. The loop is attached to a rod that is free to rotate. This assembly is suspended between the poles of two stationary magnets. If the loop is turned, the conductor cuts through magnetic lines of flux and a voltage is induced into the conductor.

If the speed of rotation is increased, the conductor cuts more lines of flux per second and the amount of induced voltage increases. If the speed of rotation remains constant and the strength of the magnetic field is increased, there will be more lines of flux per square inch. When there are more lines of flux, the number of lines cut per second increases and the induced voltage increases. If more turns of wire are added to the loop *(Figure 14–8)*, more flux lines are cut per second and the amount of induced voltage increases again. Adding more turns has the effect of connecting single conductors in series, and the amount of induced voltage in each conductor adds.

14–5   Lenz's Law

When a voltage is induced in a coil and there is a complete circuit, current flows through the coil *(Figure 14–9)*. When current flows through the coil, a magnetic field is created around the coil. This magnetic field develops a polarity opposite that of the moving magnet. The magnetic field developed by the induced current acts to attract the moving magnet and pull it back inside the coil.

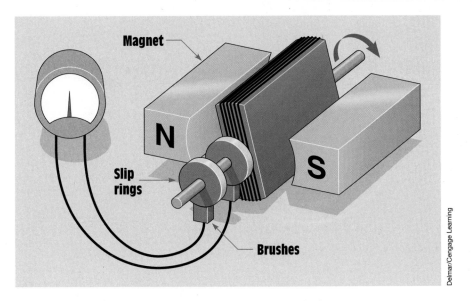

FIGURE 14–8   Increasing the number of turns increases the induced voltage.

FIGURE 14–9   An induced current produces a magnetic field around the coil.

If the direction of motion is reversed, the polarity of the induced current is reversed, and the magnetic field created by the induced current again opposes the motion of the magnet. This principle was first noticed by Heinrich Lenz many years ago and is summarized in **Lenz's law,** which states that ***an induced voltage or current opposes the motion that causes it.*** From this basic principle, other laws concerning inductors have been developed. One is that

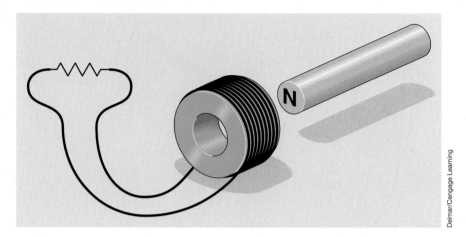

FIGURE 14–10 No current flows through the coil.

inductors always oppose a change of current. The coil in *Figure 14–10,* for example, has no induced voltage and therefore no induced current. If the magnet is moved toward the coil, however, magnetic lines of flux begin to cut the conductors of the coil and a current is induced in the coil. The induced current causes magnetic lines of flux to expand outward around the coil *(Figure 14–11)*. As this expanding magnetic field cuts through the conductors of the coil, a voltage is induced in the coil. The polarity of the voltage is such that it opposes the induced current caused by the moving magnet.

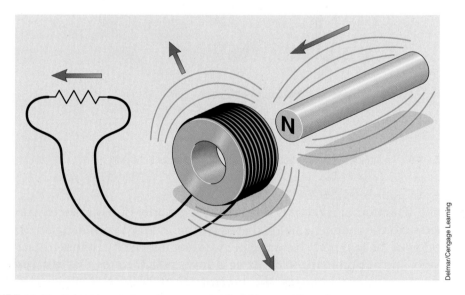

FIGURE 14–11 Induced current produces a magnetic field around the coil.

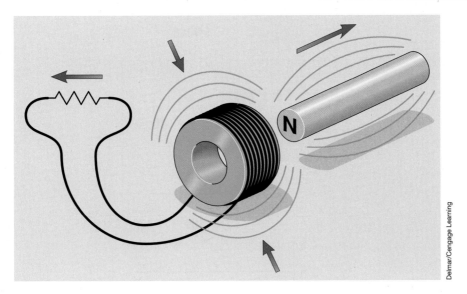

FIGURE 14–12   The induced voltage forces current to flow in the same direction.

If the magnet is moved away, the magnetic field around the coil collapses and induces a voltage in the coil *(Figure 14–12)*. Because the direction of movement of the collapsing field has been reversed, the induced voltage is opposite in polarity, forcing the current to flow in the same direction.

14–6   Rise Time of Current in an Inductor

When a resistive load is suddenly connected to a source of DC *(Figure 14–13)*, the current instantly rises to its maximum value. The resistor shown in *Figure 14–13* has a value of 10 ohms and is connected to a 20-volt source. When the switch is closed, the current instantly rises to a value of 2 amperes (20 V/10 Ω = 2 A).

If the resistor is replaced with an inductor that has a wire resistance of 10 ohms and the switch is closed, the current cannot instantly rise to its maximum value of 2 amperes *(Figure 14–14)*. As current begins to flow through an in-ductor, the expanding magnetic field cuts through the conductors, inducing a voltage into them. In accord with Lenz's law, the induced voltage is opposite in polarity to the applied voltage. The induced voltage therefore acts like a resistance to hinder the flow of current through the inductor *(Figure 14–15)*.

The induced voltage is proportional to the rate of change of current (speed of the cutting action). When the switch is first closed, current flow through the coil tries to rise instantly. This extremely fast rate of current change

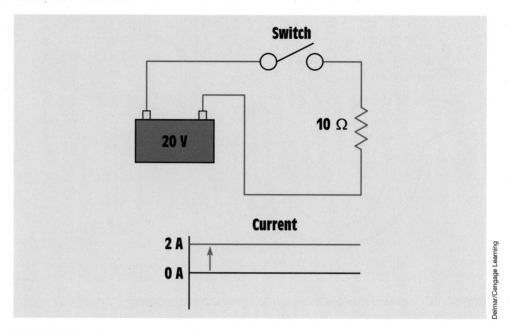

FIGURE 14–13 The current rises instantly in a resistive circuit.

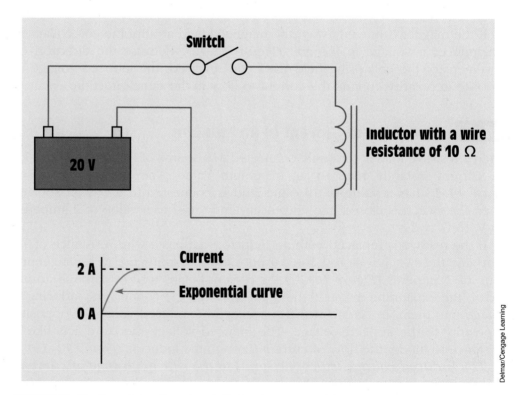

FIGURE 14–14 Current rises through an indicator at an exponential rate.

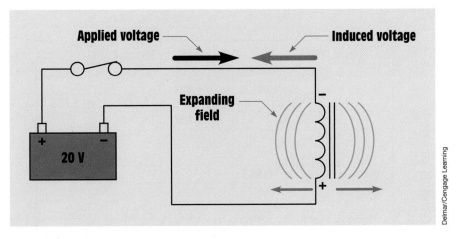

FIGURE 14–15 The applied voltage is opposite in polarity to the induced voltage.

induces maximum voltage in the coil. As the current flow approaches its maximum Ohm's law value—in this example, 2 amperes—the rate of change becomes less and the amount of induced voltage decreases.

14–7 The Exponential Curve

The **exponential curve** describes the rate of certain occurrences. The curve is divided into five time constants. During each time constant, the current rises an amount equal to 63.2% of some value. An exponential curve is shown in *Figure 14–16*. In this example, current must rise from zero to a value of 1.5 amperes at an exponential rate. In this example, 100 milliseconds are required for the current to rise to its full value. Because the current requires a total of 100 milliseconds to rise to its full value, each time constant is 20 milliseconds (100 ms/5 time constants = 20 ms per time constant). During the first time constant, the current rises from 0 to 63.2% of its total value, or 0.984 ampere (1.5 A × 0.632 = 0.948 A). The remaining current is now 0.552 ampere (1.5 A − 0.948 A = 0.552 A).

During the second time constant, the current rises 63.2% of the remaining value or 0.349 ampere (0.552 A × 0.632 = 0.349 A). At the end of the second time constant, the current has reached a total value of 1.297 amperes (0.948 A + 0.349 A = 1.297 A). The remaining current is now 0.203 ampere (1.5 A − 1.297 A = 0.203 A). During the third time constant, the current rises 63.2% of the remaining 0.203 ampere or 0.128 ampere (0.203 A × 0.632 = 0.128 A). The total current at the end of the third time constant is 1.425 amperes (1.297 A + 0.128 A = 1.425 A).

Because the current increases at a rate of 63.2% during each time constant, it is theoretically impossible to reach the total value of 1.5 amperes. After five

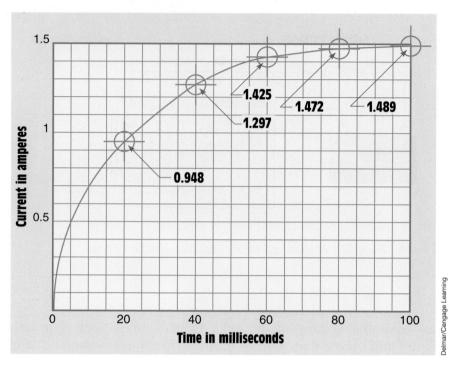

FIGURE 14–16 An exponential curve.

time constants, however, the current has reached approximately 99.3% of the maximum value and for all practical purposes is considered to be complete.

The exponential curve can often be found in nature. If clothes are hung on a line to dry, they will dry at an exponential rate. Another example of the exponential curve can be seen in *Figure 14–17*. In this example, a bucket has been

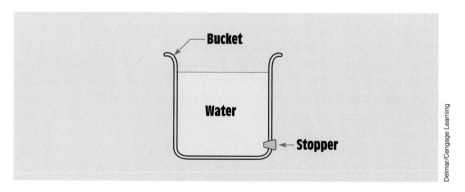

FIGURE 14–17 Exponential curves can be found in nature.

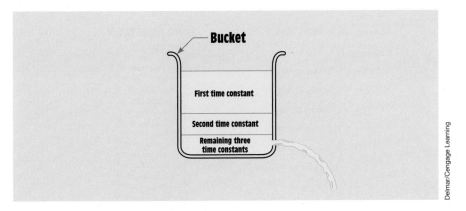

FIGURE 14–18   Water flows from a bucket at an exponential rate.

filled to a certain mark with water. A hole has been cut at the bottom of the bucket and a stopper placed in the hole. When the stopper is removed from the bucket, water flows out at an exponential rate. Assume, for example, it takes five minutes for the water to flow out of the bucket. Exponential curves are always divided into five time constants, so in this case each time constant has a value of one minute. In *Figure 14–18,* if the stopper is removed and water is permitted to drain from the bucket for a period of one minute before the stopper is replaced, during that first time constant 63.2% of the water in the bucket will drain out. If the stopper is again removed for a period of one minute, 63.2% of the water remaining in the bucket will drain out. Each time the stopper is removed for a period of one time constant, the bucket will lose 63.2% of its remaining water.

14–8   Inductance

The unit of measurement for inductance is the **henry (H),** and inductance is represented by the letter *L*. *A coil has an inductance of 1 henry when a current change of 1 ampere per second results in an induced voltage of 1 volt.*

The amount of inductance a coil has is determined by its physical properties and construction. A coil wound on a nonmagnetic core material such as wood or plastic is referred to as an *air-core* inductor. If the coil is wound on a core made of magnetic material such as silicon steel or soft iron, it is referred to as an *iron-core* inductor. Iron-core inductors produce more inductance with fewer turns than air-core inductors because of the good magnetic path provided by the core material. Iron-core inductors cannot be used for high-frequency applications, however, because of **eddy current** loss and **hysteresis loss** in the core material.

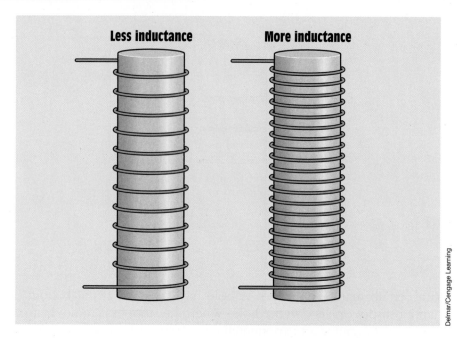

FIGURE 14–19 Inductance is determined by the physical construction of the coil.

Another factor that determines inductance is how far the windings are separated from each other. If the turns of wire are far apart, they will have less inductance than turns wound closer together *(Figure 14–19).*

The inductance of a coil can be determined using the formula

$$L = \frac{0.4\pi N^2 \mu A}{l}$$

where

L = inductance in henrys

π = 3.1416

N = number of turns of wire

μ = permeability of the core material

A = cross-sectional area of the core

l = length of the core

The formula indicates that the inductance is proportional to the number of turns of wire, the type of core material used, and the cross-sectional area of the core, but inversely proportional to core length. An inductor is basically an electromagnet that changes its polarity at regular intervals. Because the inductor is an electromagnet, the same factors that affect magnets affect inductors. The permeability of the core material is just as important to an inductor as it is to any other electromagnet. Flux lines pass through a core material with a high permeability (such as silicon, steel, or soft iron) better than through a material with a low permeability (such as brass, copper, or aluminum). Once the core material has become saturated, however, the permeability value becomes approximately 1 and an increase in turns of wire has only a small effect on the value of inductance.

14–9 R-L Time Constants

The time necessary for current in an inductor to reach its full Ohm's law value, called the **R-L time constant,** can be computed using the formula

$$T = \frac{L}{R}$$

where

T = time in seconds

L = inductance in henrys

R = resistance in ohms

This formula computes the time of one time constant.

■ EXAMPLE 14–1

A coil has an inductance of 1.5 H and a wire resistance of 6 Ω. If the coil is connected to a battery of 3 V, how long will it take the current to reach its full Ohm's law value of 0.5 A (3 V/6 Ω = 0.5 A)?

Solution

To find the time of one time constant, use the formula

$$T = \frac{L}{R}$$

$$T = \frac{1.5 \text{ H}}{6 \text{ Ω}}$$

$$T = 0.25 \text{ s}$$

The time for one time constant is 0.25 s. Because five time constants are required for the current to reach its full value of 0.5 A, 0.25 s will be multiplied by 5:

$$0.25 \text{ s} \times 5 = 1.25 \text{ s}$$

14–10 Induced Voltage Spikes

A **voltage spike** may occur when the current flow through an inductor stops, and the current also decreases at an exponential rate *(Figure 14–20)*. As long as a complete circuit exists when the power is interrupted, there is little or no problem. In the circuit shown in *Figure 14–21*, a resistor and inductor are connected in parallel. When the switch is closed, the battery supplies current

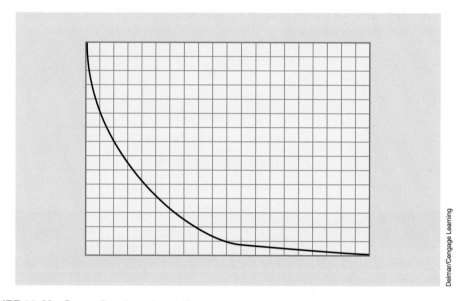

FIGURE 14–20 Current flow through an inductor decreases at an exponential rate.

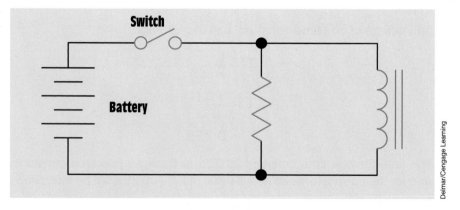

FIGURE 14–21 The resistor helps prevent voltage spikes caused by the inductor.

to both. When the switch is opened, the magnetic field surrounding the inductor collapses and induces a voltage into the inductor. The induced voltage attempts to keep current flowing in the same direction. Recall that inductors oppose a change of current. The amount of current flow and the time necessary for the flow to stop is determined by the resistor and the properties of the inductor. The amount of voltage produced by the collapsing magnetic field is determined by the maximum current in the circuit and the total resistance in the circuit. In the circuit shown in *Figure 14–21,* assume that the inductor has a wire resistance of 6 ohms and the resistor has a resistance of 100 ohms. Also assume that when the switch is closed, a current of 2 amperes will flow through the inductor. These spikes can be on the order of hundreds or even thousands of volts and can damage circuit components, especially in circuits containing solid-state devices such as diodes, transistors, intergrated circuits, and so on. The amount of induced voltage can be determined if the inductance of the coil, the amount of current change, and the amount of time change are known, by using the formula

$$EMF = -L\left(\frac{\Delta I}{\Delta t}\right)$$

where

L = inductance in henrys (A negative sign is placed in front of the L because the induced voltage is always opposite in polarity to the voltage that produces it.)

ΔI = change of current

Δt = change of time

Assume that a 1.5-henry inductor has a current flow of 2.5 amperes. When a switch is opened, the current changes from 2.5 amperes to 0.5 amperes in 0.005 second. How much voltage is induced into the inductor?

$$EMF = -L\left(\frac{\Delta I}{\Delta t}\right)$$

$$EMF = -1.5\,H\left(\frac{2\,A}{0.005\,s}\right)$$

$$EMF = -1.5\,H \times 400$$

$$EMF = -600\,V$$

When the switch is opened, a series circuit exists composed of the resistor and inductor *(Figure 14–22).* The maximum voltage developed in this circuit would be 212 volts (2 A × 106 Ω = 212 V). If the circuit resistance were increased, the induced voltage would become greater. If the circuit resistance were decreased, the induced voltage would become less.

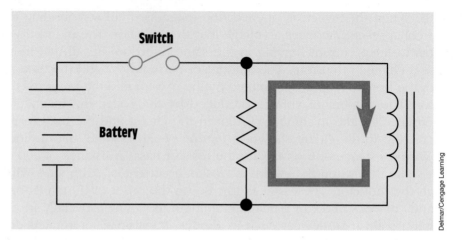

Delmar/Cengage Learning

FIGURE 14–22 When the switch is opened, a series path is formed by the resistor and inductor.

Another device often used to prevent induced voltage spikes when the current flow through an inductor is stopped is the diode *(Figure 14–23)*. The diode is an electronic component that operates like an electric check valve. The diode permits current to flow through it in only one direction. The diode is connected in parallel with the inductor in such a manner that when voltage is applied to the circuit, the diode is reverse biased and acts like an open switch. When the diode is reverse biased, no current flows through it.

When the switch is opened, the induced voltage produced by the collapsing magnetic field is opposite in polarity to the applied voltage. The diode

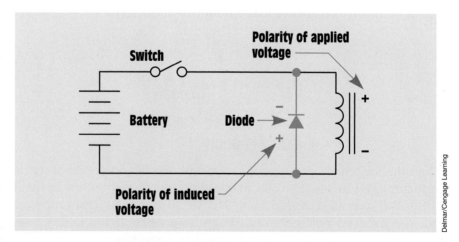

Delmar/Cengage Learning

FIGURE 14–23 A diode is used to prevent induced voltage spikes.

then becomes forward-biased and acts like a closed switch. Current can now flow through the diode and complete a circuit back to the inductor. A silicon diode has a forward voltage drop of approximately 0.7 volts regardless of the current flowing through it. When the switch opens, the diode becomes the load, and the induced voltage caused by the collapsing magnetic field of the inductor becomes the source. The diode and inductor are now series connected. In a series circuit, the applied or source voltage must equal the sum of the voltage drops in the circuit. In this case, the diode is the total con-nected load. Because the diode will not permit a voltage drop greater than about 0.7 volt, the source voltage is limited to 0.7 volt also. The circuit energy is dissipated as heat by the diode. The diode can be used for this purpose in DC circuits only; it cannot be used for this purpose in AC circuits.

A device that can be used for spike suppression in either DC or AC circuits is the **metal oxide varistor (MOV).** The MOV is a bidirectional device, which means that it conducts current in either direction and can therefore be used in AC circuits. The MOV is an extremely fast-acting solid-state component that ex-hibits a change of resistance when the voltage reaches a certain point. Assume that the MOV shown in *Figure 14–24* has a voltage rating of 140 volts and that the voltage applied to the circuit is 120 volts. When the switch is closed and current flows through the circuit, a magnetic field is established around the in-ductor *(Figure 14–25)*. As long as the voltage applied to the MOV is less than 140 volts, it will exhibit an extremely high resistance, in the range of several hundred thousand ohms.

When the switch is opened, current flow through the coil suddenly stops and the magnetic field collapses. This sudden collapse of the magnetic field causes an extremely high voltage to be induced in the coil. When this induced voltage reaches 140 volts, however, the MOV suddenly changes from a high resistance to a low resistance, preventing the voltage from becoming greater than 140 volts *(Figure 14–26)*.

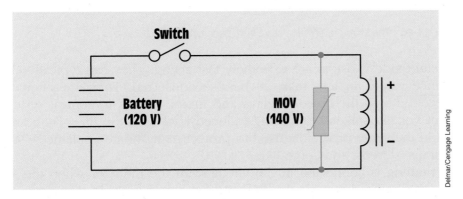

FIGURE 14–24   Metal oxide varistor used to suppress a voltage spike.

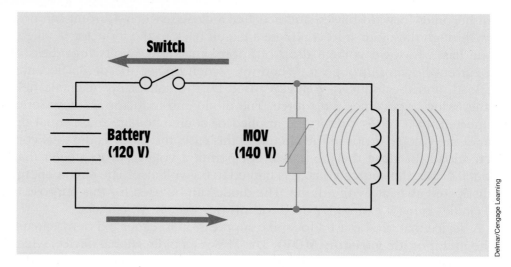

FIGURE 14–25 When the switch is closed, a magnetic field is established around the inductor.

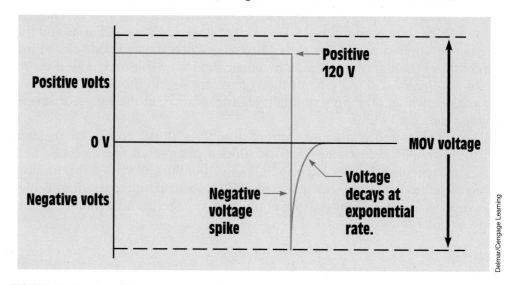

FIGURE 14–26 The MOV prevents the spike from becoming too high.

Metal oxide varistors are extremely fast acting. They can typically change resistance values in less than 20 nanoseconds (ns). They are often found connected across the coils of relays and motor starters in control systems to prevent voltage spikes from being induced back into the line. They are also found in the surge protectors used to protect many home appliances, such as televisions, stereos, and computers.

If nothing is connected in the circuit with the inductor when the switch opens, the induced voltage can become extremely high. In this instance, the resistance of the circuit is the air gap of the switch contacts, which is practically

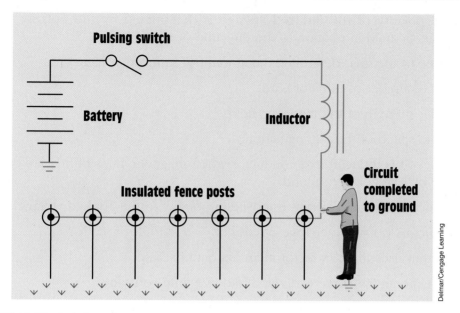

FIGURE 14–27 An inductor is used to produce a high voltage for an electric fence.

infinite. The inductor will attempt to produce any voltage necessary to prevent a change of current. Inductive voltage spikes can reach thousands of volts. This is the principle of operation of many high-voltage devices, such as the ignition systems of many automobiles.

Another device that uses the collapsing magnetic field of an inductor to produce a high voltage is the electric fence charger *(Figure 14–27)*. The switch is constructed in such a manner that it pulses on and off. When the switch closes, current flows through the inductor and a magnetic field is produced around the inductor. When the switch opens, the magnetic field collapses and induces a high voltage across the inductor. If anything or anyone standing on the ground touches the fence, a circuit is completed through the object or person and the ground. The coil is generally constructed of many turns of very small wire. This construction provides the coil with a high resistance and limits current flow when the field collapses.

Summary

- When current flows through a conductor, a magnetic field is created around the conductor.

- When a conductor is cut by a magnetic field, a voltage is induced in the conductor.

- The polarity of the induced voltage is determined by the polarity of the magnetic field in relation to the direction of motion.
- Three factors that determine the amount of induced voltage are

  a. the number of turns of wire.

  b. the strength of the magnetic field.

  c. the speed of the cutting action.

- One volt is induced in a conductor when magnetic lines of flux are cut at a rate of 1 weber per second.
- Induced voltage is always opposite in polarity to the applied voltage.
- Inductors oppose a change of current.
- Current rises in an inductor at an exponential rate.
- An exponential curve is divided into five time constants.
- Each time constant is equal to 63.2% of some value.
- Inductance is measured in units called henrys (H).
- A coil has an inductance of 1 henry when a current change of 1 ampere per second results in an induced voltage of 1 volt.
- Air-core inductors are inductors wound on cores of nonmagnetic material.
- Iron-core inductors are wound on cores of magnetic material.
- The amount of inductance an inductor will have is determined by the number of turns of wire and the physical construction of the coil.
- Inductors can produce extremely high voltages when the current flowing through them is stopped.
- Two devices used to help prevent large spike voltages are the resistor and the diode.

Review Questions

1. What determines the polarity of magnetism when current flows through a conductor?

2. What determines the strength of the magnetic field when current flows through a conductor?

3. Name three factors that determine the amount of induced voltage in a coil.

4. How many lines of magnetic flux must be cut in 1 s to induce a voltage of 1 V?

5. What is the effect on induced voltage of adding more turns of wire to a coil?

6. Into how many time constants is an exponential curve divided?

7. Each time constant of an exponential curve is equal to what percentage of the maximum amount of charge?

8. An inductor has an inductance of 0.025 H and a wire resistance of 3 Ω. How long will it take the current to reach its full Ohm's law value?

9. Refer to the circuit shown in *Figure 14–21*. Assume that the inductor has a wire resistance of 0.2 Ω and the resistor has a value of 250 Ω. If a current of 3 A is flowing through the inductor, what will be the maximum induced voltage when the switch is opened?

10. What electronic component is often used to prevent large voltage spikes from being produced when the current flow through an inductor is suddenly terminated?

Practical Applications

*Y*ou are an electrician working in a plant that uses programmable controllers to perform much of the logic for the motor controls. The plant produces highly flammable chemicals, so the programmable controllers use a 24-volt VDC output that is intrinsically safe in a hazardous area. You are having trouble with the output modules of the programmable controllers that are connected to DC control relays. These outputs are going bad at an unusually high rate, and you suspect that spike voltages produced by the relay coils are responsible. To test your theory, you connect an oscilloscope to an output that is operating a DC relay coil and watch the display when the programmable controller turns the relay off. You discover that there is a high voltage spike produced by the relay. What device would you use to correct this problem, and how would you install it? ■

V Basics of Alternating Current

Unit 15

Basic Trigonometry and Vectors

KEY TERMS

Adjacent

Cosine

Hypotenuse

Opposite

Oscar Had A Heap
   Of Apples

Parallelogram
   method

Pythagoras

Pythagorean
   theorem

Right angle

Right triangle

Sine

Tangent

Vector

Vector addition

Why You Need to Know

*I*n DC circuits, the product of the current and voltage always equals the power or watts. You will find that this is not true in AC circuits, however, because the voltage and current can become out of phase with each other. This can be illustrated through the use of basic trigonometry. This unit

- explains how the phase difference between voltage and current is determined by the type of load or loads connected to the circuit.
- explains how voltage and current relationships in AC circuits are based on right triangles.
- describes several different methods, all of them involving vectors, that can be employed to determine these phase relationships.
- illustrates how the Pythagorean theorem provides the formula needed for electricians to understand 90° angles and how to calculate voltages and current in AC circuits.

Objectives

After studying this unit, you should be able to

- define a right triangle.

- discuss the Pythagorean theorem.

- solve problems concerning right triangles using the Pythagorean theorem.

- solve problems using sines, cosines, and tangents.

Preview

Before beginning the study of AC, a brief discussion of right triangles and the mathematical functions involving them is appropriate. Many AC formulas are based on right triangles, because, depending on the type of load, the voltage and current in an AC circuit can be out of phase with each other by approximately 90°. The exact amount of the out-of-phase condition is determined by different factors, which are covered in later units. ∎

15–1 Right Triangles

A **right triangle** is a triangle that contains a **right** or (90°), **angle** *(Figure 15–1)*. The **hypotenuse** is the longest side of a right triangle and is always opposite the right angle. Several thousand years ago, a Greek mathematician named **Pythagoras** made some interesting discoveries concerning triangles that contain right angles. One of these discoveries was that if the two sides of a

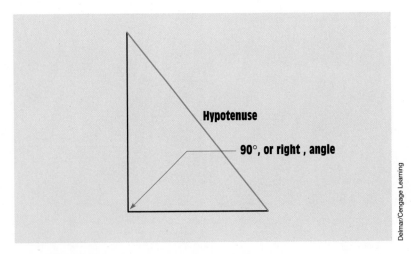

Hypotenuse

90°, or right , angle

FIGURE 15–1 A right triangle.

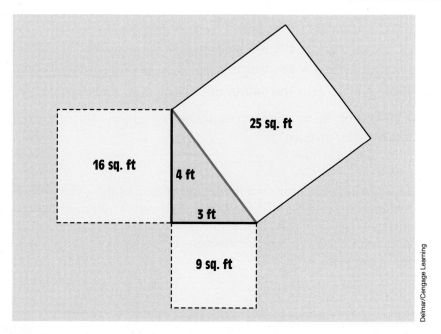

FIGURE 15–2 The Pythagorean theorem states that the sum of the squares of the sides of a right triangle is equal to the square of the hypotenuse.

right triangle are squared, their sum equals the square of the hypotenuse. For example, assume a right triangle has one side 3 feet long and the other side is 4 feet long *(Figure 15–2)*. If the side that is 3 feet long is squared, it produces an area of 9 square feet (3 ft × 3 ft = 9 sq. ft). If the side that is 4 feet long is squared, it produces an area of 16 square feet (4 ft × 4 ft = 16 sq. ft). The sum of the areas of these two sides equals the square area formed by the hypotenuse. In this instance, the hypotenuse has an area of 25 square feet (9 sq. ft + 16 sq. ft = 25 sq. ft). Now that the area, or square, of the hypotenuse is known, its length can be determined by finding the square root of its area. The length of the hypotenuse is 5 feet, because the square root of 25 is 5.

15–2 The Pythagorean Theorem

From this knowledge concerning the relationship of the length of the sides to the length of the hypotenuse, Pythagoras derived a formula known as the **Pythagorean theorem:**

$$c^2 = a^2 + b^2$$

where

$$c = \text{the length of the hypotenuse}$$
$$a = \text{the length of one side}$$
$$b = \text{the length of the other side}$$

If the lengths of the two sides are known, the length of the hypotenuse can be found using the formula

$$c = \sqrt{a^2 + b^2}$$

It is also possible to determine the length of one of the sides if the length of the other side and the length of the hypotenuse are known. Because the sum of the squares of the two sides equals the square of the hypotenuse, the square of one side equals the difference between the square of the hypotenuse and the square of the other side.

■ EXAMPLE 15–1

The triangle shown in *Figure 15–3* has a hypotenuse (C) 18 in. long. One side (B) is 7 in. long. What is the length of the second side (A)?

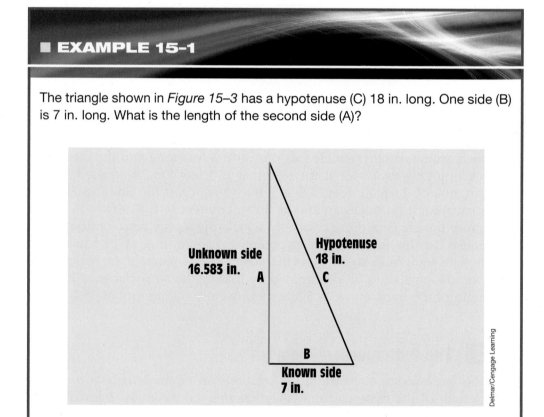

FIGURE 15–3   Using the Pythagorean theorem to find the length of one side.

Solution

Transpose the formula $c^2 = a^2 + b^2$ to find the value of a. This can be done by subtracting b^2 from both sides of the equation. The result is the formula:

$$a^2 = c^2 - b^2$$

Now take the square root of each side of the equation so that the answer is equal to a, not a^2:

$$a = \sqrt{c^2 - b^2}$$

The formula can now be used to find the length of the unknown side:

$$a = \sqrt{(18 \text{ in.})^2 - (7 \text{ in.})^2}$$

$$a = \sqrt{324 \text{ in.}^2 - 49 \text{ in.}^2}$$

$$a = \sqrt{275 \text{ in.}^2}$$

$$a = 16.583 \text{ in.}$$

15-3   Sines, Cosines, and Tangents

A second important concept concerning right triangles is the relationship of the length of the sides to the angles. Because one angle is always 90°, only the two angles that are not right angles are of concern. Another important fact concerning triangles is that the sum of the angles of any triangle equals 180°; because one angle of a right triangle is 90°, the sum of the two other angles must equal 90°.

It was discovered long ago that the number of degrees in each of these two angles is proportional to the lengths of the three sides. This relationship is expressed as the *sine, cosine,* or *tangent* of a particular angle. The function used is determined by which sides are known and which of the two angles is to be found.

When using sines, cosines, or tangents, the sides are designated the *hypotenuse,* the **opposite,** and the **adjacent.** The hypotenuse is always the longest side of a right triangle, but the opposite and adjacent sides are determined by which of the two angles is to be found. In *Figure 15–4,* a right triangle has its sides labeled A, B, and HYPOTENUSE. The two unknown angles are labeled X and Y. To determine which side is opposite an angle, draw a line bisecting the angle. This bisect line would intersect the opposite side. In *Figure 15–5,* side A is opposite angle X, and side B is opposite angle Y. If side A is opposite angle X,

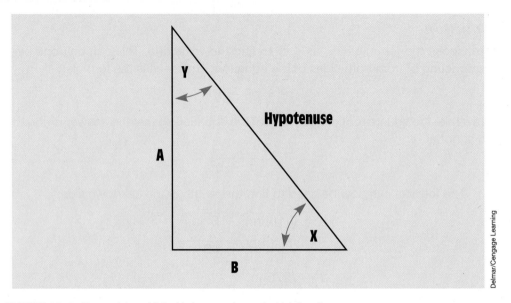

Delmar/Cengage Learning

FIGURE 15–4 Determining which side is opposite and which is adjacent.

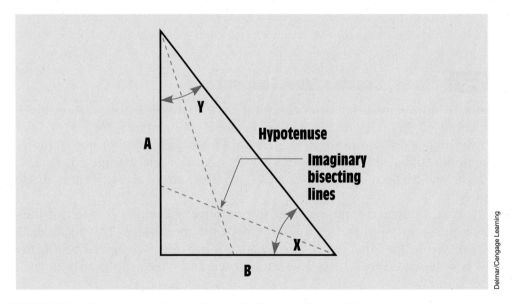

Delmar/Cengage Learning

FIGURE 15–5 Side A is opposite angle X, and side B is opposite angle Y.

then side B is adjacent to angle X; and if side B is opposite angle Y, then side A is adjacent to angle Y *(Figure 15–6)*.

The **sine** function is equal to the opposite side divided by the hypotenuse:

$$\text{sine } \angle X = \frac{\text{opposite}}{\text{hypotenuse}}$$

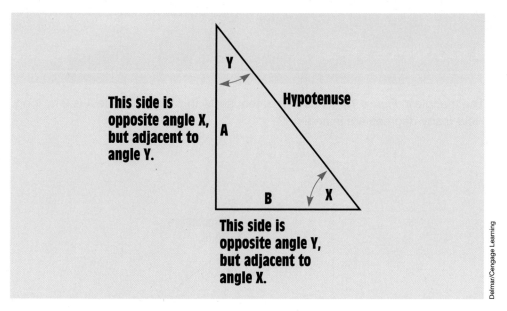

FIGURE 15–6 Determining opposite and adjacent sides.

The **cosine** function is equal to the adjacent side divided by the hypotenuse:

$$\text{cosine} \angle X = \frac{\text{adjacent}}{\text{hypotenuse}}$$

The **tangent** function is equal to the opposite side divided by the adjacent side:

$$\text{tangent} \angle X = \frac{\text{opposite}}{\text{adjacent}}$$

A simple saying is often used to help remember the relationship of the trigonometric function and the side of the triangle. This saying is **Oscar Had A Heap Of Apples.** To use this simple saying, write down *sine, cosine, and tangent.* The first letter of each word becomes the first letter of each of the sides. *O* stands for opposite, *H* stands for hypotenuse, and *A* stands for adjacent:

$$\text{sine} \angle X = \frac{\text{Oscar}}{\text{Had}} \qquad \text{(Opposite)}{\qquad}\text{(Hypotenuse)}$$

$$\text{cosine} \angle X = \frac{\text{A}}{\text{Heap}} \qquad \text{(Adjacent)}{\qquad}\text{(Hypotenuse)}$$

$$\text{tangent} \angle X = \frac{\text{Of}}{\text{Apples}} \qquad \text{(Opposite)}{\qquad}\text{(Adjacent)}$$

Once the sine, cosine, or tangent of the angle has been determined, the angle can be found by using the trigonometric functions on a scientific calculator or from the trigonometric tables located in Appendices A and B.

■ EXAMPLE 15-2

The triangle in *Figure 15–7* has a hypotenuse 14 in. long, and side A is 9 in. long. How many degrees are in angle X?

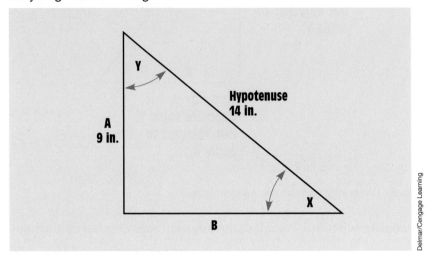

FIGURE 15–7 Finding the angles using trigonometric functions.

Solution

Because the lengths of the hypotenuse and the opposite side are known, the sine function is used to find the angle:

$$\text{sine} \angle \text{X} = \frac{\text{opposite}}{\text{hypotenuse}}$$

$$\text{sine} \angle \text{X} = \frac{9}{14}$$

$$\text{sine} \angle \text{X} = 0.643$$

Note that 0.643 is the sine of the angle, not the angle. To find the angle, use the SIN function on a scientific calculator. If you are using a scientific calculator, it will be necessary to use the ARC SIN function or INV SIN function to find the answer. If the number 0.643 is entered and the SIN key is pressed, an answer of 0.0112 results. This is the sine of a 0.643° angle. If the sine, cosine, or tangent of an angle is known, the ARC key or INV key (depending on the manufacturer of the calculator) must be pressed before the SIN, COS, or TAN key is pressed:

$$\text{SIN}^{-1} \text{ or ARCSIN } 0.643 = 40°$$

■ EXAMPLE 15-3

Using the same triangle *(Figure 15–7)*, determine the number of degrees in angle Y.

Solution

In this example, the lengths of the hypotenuse and the adjacent side are known. The cosine function can be used to find the angle:

$$\text{cosine} \angle Y = \frac{\text{adjacent}}{\text{hypotenuse}}$$

$$\text{cosine} \angle Y = \frac{9}{14}$$

$$\text{cosine} \angle Y = 0.643$$

To find what angle corresponds to the cosine of 0.643, use the trigonometric tables in Appendices A and B or the COS function of a scientific calculator:

$$\text{COS}^{-1} \text{ or ARC COS } 0.643 = 50°$$

15-4   Formulas

Some formulas that can be used to find the angles and lengths of different sides follow:

$$\sin \angle \theta = \frac{O}{H} \qquad\qquad \cos \angle \theta = \frac{A}{H} \qquad\qquad \tan \angle \theta = \frac{O}{A}$$

$$\text{Adj.} = \cos \angle \theta \times \text{Hyp.} \qquad\qquad \text{Adj.} = \frac{O}{\tan \angle \theta}$$

$$\text{Opp} = \sin \angle \theta \times \text{Hyp.} \qquad\qquad \text{Opp.} = \text{Adj.} \times \tan \angle \theta$$

$$\text{Hyp.} = \frac{O}{\sin \angle \theta} \qquad\qquad \text{Hyp.} = \frac{A}{\cos \angle \theta}$$

15-5   Practical Application

Although the purpose of this unit is to provide preparation for the study of AC circuits, basic trigonometry can provide answers to other problems that may be encountered on the job. Assume that it is necessary to know the height of a

FIGURE 15–8   Using trigonometry to measure the height of a tall building.

tall building *(Figure 15–8)*. Now assume that the only tools available to make this measurement are a 1-foot ruler, a tape measure, and a scientific calculator. To make the measurement, find a relatively flat area in the open sunlight. Hold the ruler upright and measure the shadow cast by the sun *(Figure 15–9)*. Assume the length of the shadow to be 7.5 inches. Using the length of the

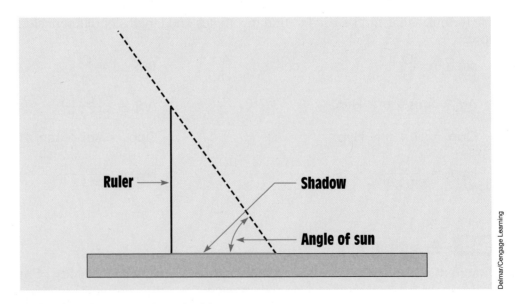

FIGURE 15–9   Determining the angle of the sun.

shadow as one side of a right triangle and the ruler as the other side, the angle of the sun can now be determined. The two known sides are the opposite and the adjacent. The tangent function corresponds to these two sides. The angle of the sun is

$$\tan \angle \theta = \frac{O}{A}$$

$$\tan \angle \theta = \frac{12}{7.5}$$

$$\tan \angle \theta = 1.6$$

$$= 57.99°$$

The tape measure can now be used to measure the shadow cast by the building on the ground. Assume the length of the shadow to be 35 feet. If the height of the building is used as one side of a right triangle and the shadow is used as the other side, the height can be found because the angle of the sun is known *(Figure 15–10).*

Opp. = Adj. × tan ∠θ

Opp. = 35 ft × tan 57.99°

Opp. = 35 ft × 1.6

Opp. = 56 ft

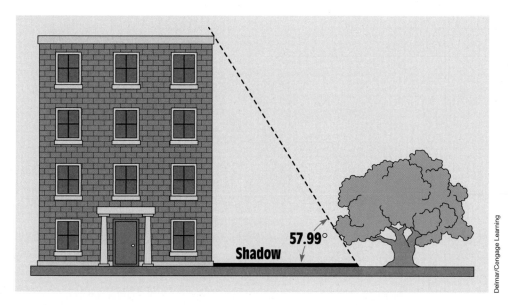

FIGURE 15–10   Measuring the building's shadow.

Vectors

Using the right triangle is just one method of graphically showing how angular quantities can be added. Another method of illustrating this concept is with the use of vectors. A **vector** is a line that indicates both magnitude and direction. The magnitude is indicated by its length, and the direction is indicated by its angle of rotation from 0°. Vectors should not be confused with *scalars,* which are used to represent magnitude only and do not take direction into consideration. Imagine, for example, that you are in a strange city and you ask someone for directions to a certain building. If the person said, "Walk three blocks," that would be a scalar because it contains only the magnitude, three blocks. If the person said, "Walk three blocks south," it would be a vector because it contains both the magnitude, three blocks, and the direction, south.

Zero degrees is indicated by a horizontal line. An arrow is placed at one end of the line to indicate direction. The magnitude can represent any quantity such as inches, meters, miles, volts, amperes, ohms, power, and so on. A vector with a magnitude of 5 at an angle of 0° is shown in *Figure 15–11*. Vectors rotate in a counterclockwise direction. Assume that a vector with a magnitude of 3 is to be drawn at a 45° angle from the first vector *(Figure 15–12)*. Now assume that a third vector with a magnitude of 4 is to be drawn in a direction of 120° *(Figure 15–13)*. Notice that the direction of the third vector is referenced from the horizontal 0° line and not from the second vector line, which was drawn at an angle of 45°.

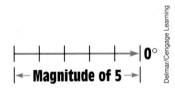

FIGURE 15–11   A vector with a magnitude of 5 and an angle of 0°.

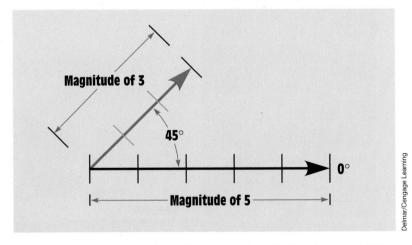

FIGURE 15–12   A second vector with a magnitude of 3 and a direction of 45°.

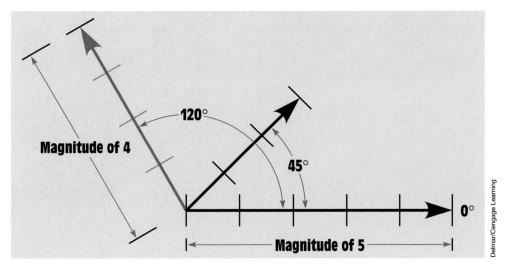

FIGURE 15–13 A third vector with a magnitude of 4 and a direction of 120° is added.

Adding Vectors

Because vectors are used to represent quantities such as volts, amperes, ohms, power, and so on, they can be added, subtracted, multiplied, and divided. In electrical work, however, addition is the only function needed, so it is the only one discussed. Several methods can be used to add vectors. Regardless of the method used, because vectors contain both magnitude and direction, they must be added with a combination of geometric and algebraic addition. This method is often referred to as **vector addition.**

One method is to connect the starting point of one vector to the end point of another. This method works especially well when all vectors are in the same direction. Consider the circuit shown in *Figure 15–14*. In this circuit, two batteries, one rated at 6 volts and the other rated at 4 volts, are connected in such a manner that their voltages add. If vector addition is used, the starting point of one vector is placed at the end point of the other. Notice that the sum of the two vector quantities is equal to the sum of the two voltages, 10 volts. Another example of this type of vector addition is shown in *Figure 15–15*. In this example, three resistors are connected in series. The first resistor has a resistance of 80 ohms; the second, a resistance of 50 ohms; and the third, a resistance of 30 ohms. Because there is no phase angle shift of voltage or current, the impedance is 160 ohms, which is the sum of the resistances of the three resistors.

Adding Vectors with Opposite Directions

To add vectors that are exactly opposite in direction (180° apart), subtract the magnitude of the larger vector from the magnitude of the smaller. The resultant

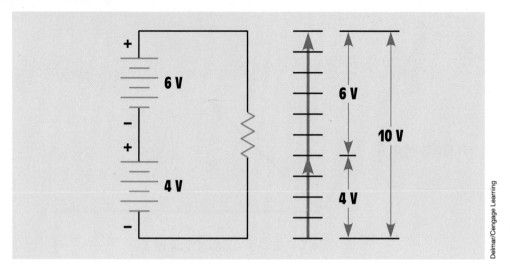

FIGURE 15–14   Adding vectors in the same direction.

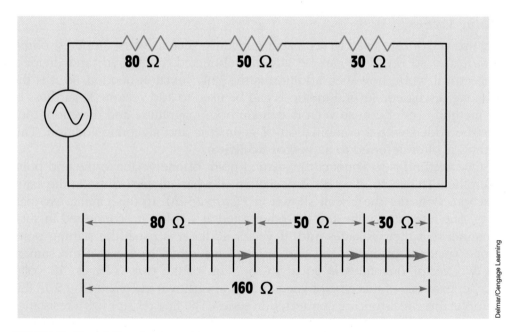

FIGURE 15–15   Addition of series resistors.

is a vector with the same direction as the vector with the larger magnitude. If one of the batteries in *Figure 15–14* were reversed, the two voltages would oppose each other *(Figure 15–16)*. This means that 4 volts of the 6-volt battery A would have to be used to overcome the voltage of battery B. The resultant would be a vector with a magnitude of 2 volts in the same direction as the

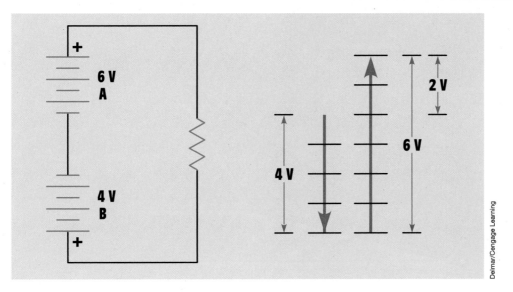

FIGURE 15–16   Adding vectors with opposite directions.

6-volt battery. In algebra, this is the same operation as adding a positive number and a negative number ($+6 + [-4] = +2$). When the -4 is brought out of brackets, the equation becomes $6 - 4 = 2$.

Adding Vectors of Different Directions

Vectors that have directions other than 180° from each other can also be added. *Figure 15–17* illustrates the addition of a vector with a magnitude of 4 and a direction of 15° to a vector with a magnitude of 3 and a direction of 60°. The addition is made by connecting the starting point of the second vector to the ending point of the first vector. The resultant is drawn from the starting point of the first vector to the ending point of the second. It is possible to add several different vectors using this method. *Figure 15–18* illustrates the addition of several different vectors and the resultant.

The Parallelogram Method of Vector Addition

The **parallelogram method** can be used to find the resultant of two vectors that originate at the same point. A parallelogram is a four-sided figure whose opposite sides form parallel lines. A rectangle, for example, is a parallelogram with 90° angles. Assume that a vector with a magnitude of 24 and a direction of 26° is to be added to a vector with a magnitude of 18 and a direction of 58°. Also assume that the two vectors originate from the same point

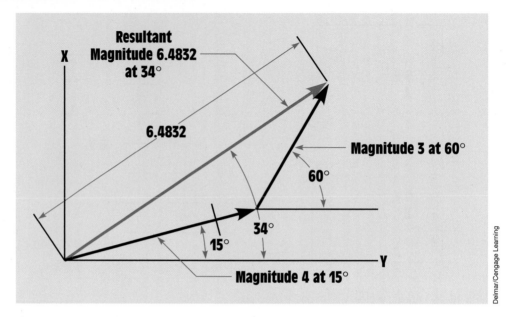

FIGURE 15–17 Adding two vectors with different directions.

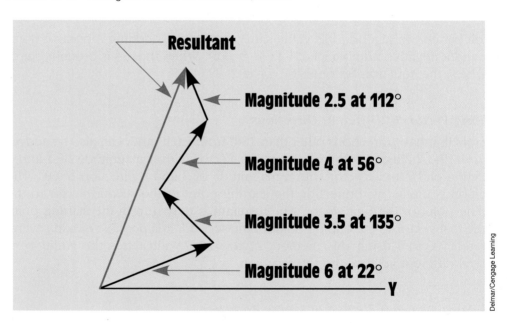

FIGURE 15–18 Adding vectors with different magnitudes and directions.

(Figure 15–19). To find the resultant of these two vectors, form a parallelogram using the vectors as two of the sides. The resultant is drawn from the corner of the parallelogram where the two vectors intersect to the opposite corner *(Figure 15–20).*

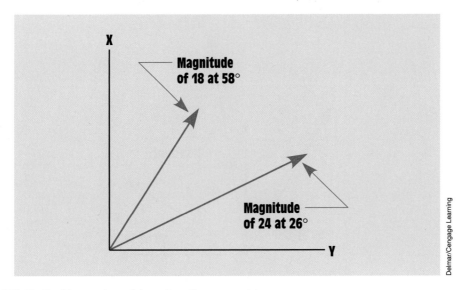

FIGURE 15–19 Vectors that originate from the same point.

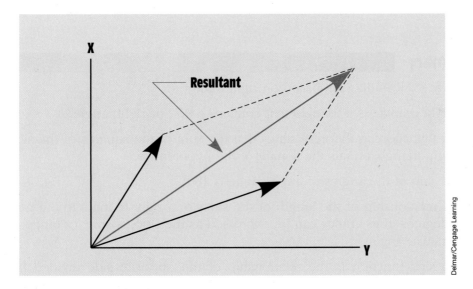

FIGURE 15–20 The parallelogram method of vector addition.

Another example of the parallelogram method of vector addition is shown in *Figure 15–21*. In this example, one vector has a magnitude of 50 and a direction of 32°. The second vector has a magnitude of 60 and a direction of 120°. The resultant is found by using the two vectors as two of the sides of a parallelogram.

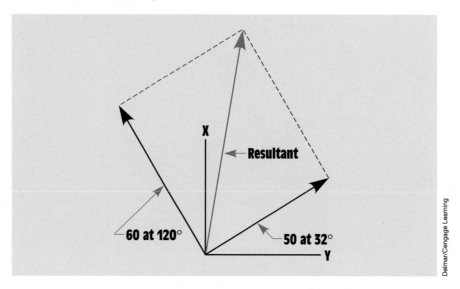

FIGURE 15–21 The parallelogram method of vector addition, second example.

Summary

- Many AC formulas are based on right triangles.

- A right triangle is a triangle that contains a 90°, or right, angle.

- The Pythagorean theorem states that the sum of the squares of the sides of a right triangle equals the square of the hypotenuse.

- The sum of the angles of any triangle is 180°.

- The relationship of the length of the sides of a right triangle to the number of degrees in its angles can be expressed as the sine, cosine, or tangent of a particular angle.

- The sine function is the relationship of the opposite side divided by the hypotenuse.

- The cosine function is the relationship of the adjacent side divided by the hypotenuse.

- The tangent function is the relationship of the opposite side divided by the adjacent side.

- The hypotenuse is always the longest side of a right triangle.

- A simple saying that can be used to help remember the relationship of the trigonometric functions to the sides of a right triangle is *Oscar Had A Heap Of Apples.*

- Vectors are lines that indicate both magnitude and direction.

- Scalars indicate magnitude only.

Review Questions

1. Which trigonometric function is used to find the angle if the length of the hypotenuse and of the adjacent side are known?

Refer to *Figure 15–22* to answer the following questions.

2. If side A has a length of 18.5 ft and side B has a length of 28 ft, what is the length of the hypotenuse?

3. Side A has a length of 12 m, and angle Y is 12°. What is the length of side B?

4. Side A has a length of 6 in. and angle Y is 45°. What is the length of the hypotenuse?

5. The hypotenuse has a length of 65 in., and side A has a length of 31 in. What is angle X?

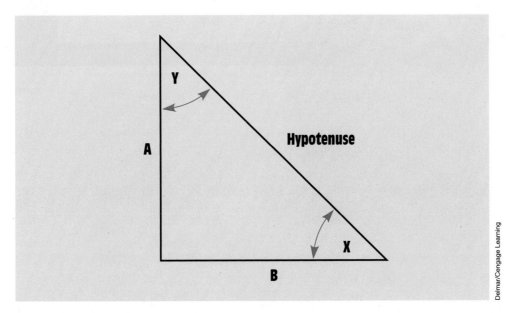

FIGURE 15–22   Finding the values of a right triangle.

6. The hypotenuse has a length of 83 ft and side B has a length of 22 ft. What is the length of side A?

7. Side A has a length of 1.25 in., and side B has a length of 2 in. What is angle Y?

8. Side A has a length of 14 ft, and angle X is 61°. What is the length of the hypotenuse?

9. Using the dimensions in Question 8, what is the length of side B?

10. Angle Y is 36°. What is angle X?

Practical Applications

A parking lot is 275 ft by 200 ft *(Figure 15–23)*. A trade size 4 conduit has been buried 2 ft beneath the concrete. The conduit runs at an angle from one corner of the parking lot to the other. Two pull boxes have been placed in the conduit run. The pull boxes are equally spaced along the length of conduit. Your job is to pull conductors through the conduit from one end of the run to the other.

(A) Allowing 12 ft of extra conductor, what is the length of each conductor? The conductors are not to be spliced.

(B) How much cable will be required to pull these conductors through the conduit, allowing an extra 5 ft at one end that is to be connected to the pulling device and 3 ft extra that is to be used to connect the conductors to the cable? ∎

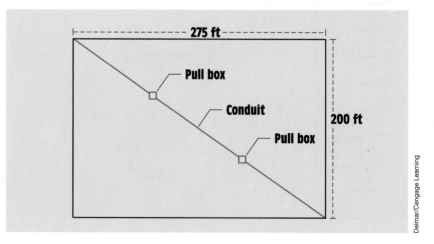

FIGURE 15–23   Determining length.

Practical Applications

You desire to add a solar heating system to your swimming pool to extend the length of time you can use the pool. For maximum efficiency, the collector should be placed on an angle that will permit the sun's rays to strike the collector at a 90° angle *(Figure 15–24)*. When the sun is at its highest point of the day, you place a yardstick perpendicular to a flat concrete surface. The stick casts a shadow 3.5 ft long. At what angle with respect to the ground should the solar collector be placed? ■

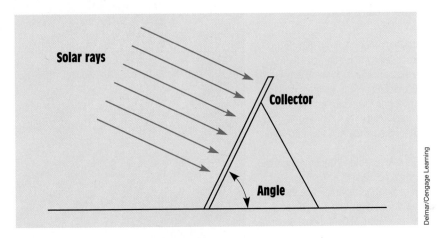

FIGURE 15–24   The sun's rays should strike the face of the solar collector at a 90° angle.

Practice Problems

Basic Trigonometry

Refer to *Figure 15–22* to find the missing values in the following chart.

∠ X	∠ Y	Side A	Side B	Hyp.
40°				8
	33°	72		
		38		63
52°			14	
	42°		156	

Unit 16 📀 DVD VIDEO

Alternating Current

Why You Need to Know

Most of the electric power in the world is AC. Electricians work with AC about 99% of the time. In this unit, you will learn about different types of waveforms, different methods employed to measure AC values, and the relationship of voltage and current in a circuit that contains pure resistance. This unit

- introduces the sine wave. The most common of all AC waveforms is the sine wave, because it is produced by a rotating machine. All rotating machines operate on the principle of the sine wave. The pistons in an internal combustion engine, for example, travel up and down inside the cylinder in a sine wave pattern because they are connected to a rotating crankshaft.
- explains that the main advantage for the use of AC is that AC voltage can be transformed and DC voltage cannot.
- presents the concepts of eddy currents and skin effect that act as resistance to the flow of current in an AC circuit.
- explains the differences between peak, RMS, and average voltage values. These are very important concepts and should be understood by anyone working in the electrical field. Some electric components such as capacitors and solid-state devices are very sensitive to voltage value. Many of these devices list a peak rating of voltage instead of the RMS value. It then becomes imperative that you know how to determine the peak value from the RMS value.

KEY TERMS

Amplitude	Peak
Average	Resistive loads
Cycle	Ripple
Effective	RMS (root-mean-
Frequency	square) value
Hertz (Hz)	Sine wave
In phase	Skin effect
Linear wave	Triangle wave
Oscillators	True power

Objectives

After studying this unit, you should be able to

■ discuss differences between DC and AC.

■ compute instantaneous values of voltage and current for a sine wave.

■ compute peak, RMS, and average values of voltage and current.

■ discuss the phase relationship of voltage and current in a pure resistive circuit.

Courtesy of Niagara Mohawk Power Corporation.

Preview

Most of the electric power produced in the world is AC. It is used to operate everything from home appliances, such as television sets, computers, microwave ovens, and electric toasters, to the largest motors found in industry. AC has several advantages over DC that make it a better choice for the large-scale production of electric power. ■

16–1   Advantages of AC

Probably the single greatest advantage of AC is the fact that AC can be transformed and DC cannot. A transformer permits voltage to be stepped up or down. Voltage can be stepped up for the purpose of transmission and then stepped back down when it is to be used by some device. Transmission voltages of 69 kilovolts, 138 kilovolts, and 345 kilovolts are common. The advantage of high-voltage transmission is that less current is required to produce the same amount of power. The reduction of current permits smaller wires to be used, which results in a savings of material.

In the very early days of electric power generation, Thomas Edison, an American inventor, proposed powering the country with low-voltage DC. He reasoned that low-voltage DC was safer for people to use than higher-voltage AC. A Serbian immigrant named Nikola Tesla, however, argued that DC was impractical for large-scale applications. The disagreement was finally settled at the 1904 World's Fair held in St. Louis, Missouri. The 1904 World's Fair not only introduced the first ice cream cone and the first iced tea but was also the first World's Fair to be lighted with "electric candles." At that time, the only two companies capable of providing electric lighting for the World's Fair were the Edison Company, headed by Thomas Edison, and the Westinghouse Company, headed by George Westinghouse, a close friend of Nikola Tesla. The Edison Company submitted a bid of over $1 per lamp to light the fair with low-voltage DC. The Westinghouse Company submitted a bid of less than 25 cents per lamp to light the fair using higher-voltage AC. This set the precedent for how electric power would be supplied throughout the world.

16–2 AC Waveforms

Square Waves

AC differs from DC in that AC reverses its direction of flow at periodic intervals *(Figure 16–1)*. AC waveforms can vary depending on how the current is produced. One waveform frequently encountered is the square wave *(Figure 16–2)*. It is assumed that the oscilloscope in *Figure 16–2* has been adjusted so that 0 volts is represented by the center horizontal line. The waveform shows that the voltage is in the positive direction for some length of time and then changes polarity. The voltage remains negative for some length of time and then changes back to positive again. Each time the voltage reverses polarity, the current flow through the circuit changes direction. A square wave could be produced by a simple single-pole double-throw switch connected to two

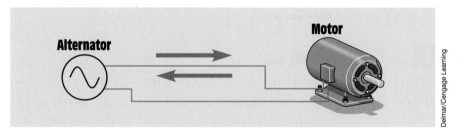

FIGURE 16–1 AC flows first in one direction and then in the other.

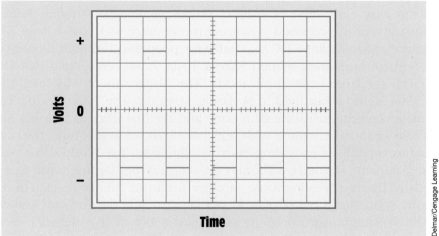

FIGURE 16–2 Square wave AC.

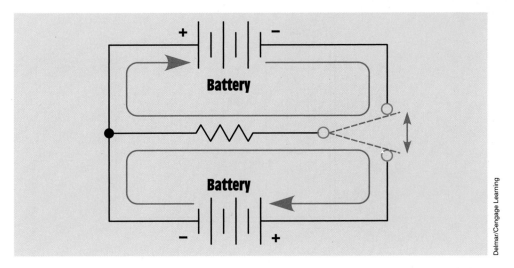

FIGURE 16–3   Square wave AC produced with a switch and two batteries.

batteries as shown in *Figure 16–3.* Each time the switch position is changed, current flows through the resistor in a different direction. Although this circuit will produce a square wave AC, it is not practical. Square waves are generally produced by electronic devices called **oscillators.** The schematic diagram of a simple square wave oscillator is shown in *Figure 16–4.* In this circuit, two bipolar transistors are used as switches to reverse the direction of current flow through the windings of the transformer.

Triangle Waves

Another common AC waveform is the **triangle wave** shown in *Figure 16–5.* The triangle wave is a **linear wave,** one in which the voltage rises at a constant rate with respect to time. Linear waves form straight lines when plotted on a graph. For example, assume that the waveform shown in *Figure 16–5* reaches a maximum

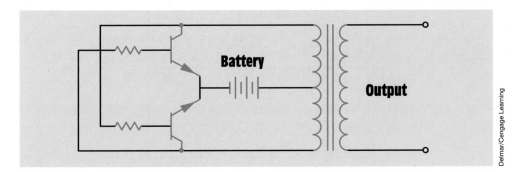

FIGURE 16–4   Square wave oscillator.

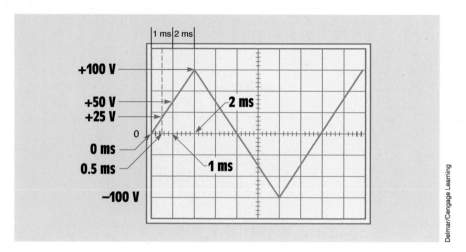

FIGURE 16–5 Triangle wave.

positive value of 100 volts after 2 milliseconds. The voltage will be 25 volts after 0.5 milliseconds, 50 volts after 1 milliseconds, and 75 volts after 1.5 milliseconds.

Sine Waves

The most common of all AC waveforms is the **sine wave** *(Figure 16–6)*. Sine waves are produced by all rotating machines. The sine wave contains a total of 360 electric degrees. It reaches its peak positive voltage at 90°, returns to a value of 0 volts at 180°, increases to its maximum negative voltage at 270°, and returns to 0 volts at 360°. Each complete waveform of 360° is called a **cycle.** The number

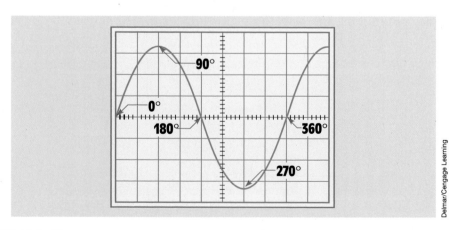

FIGURE 16–6 Sine wave.

of complete cycles that occur in one second is called the **frequency.** Frequency is measured in **hertz (Hz).** The most common frequency in the United States and Canada is 60 hertz. This means that the voltage increases from zero to its maximum value in the positive direction, returns to zero, increases to its maximum value in the negative direction, and returns to zero 60 times each second.

Sine waves are so named because the voltage at any point along the waveform is equal to the maximum, or peak, value times the sine of the angle of rotation. Figure 16–7 illustrates one half of a loop of wire cutting through lines of magnetic flux. The flux lines are shown with equal spacing between each line, and the arrow denotes the arc of the loop as it cuts through the lines of flux. Notice the number of flux lines that are cut by the loop during the first 30° of rotation. Now notice the number of flux lines that are cut during the second and third 30° of rotation. Because the loop is cutting the flux lines at an angle, it must travel a greater distance between flux lines during the first degrees of rotation. Consequently, fewer flux lines are cut per second, which results in a lower induced voltage. Recall that 1 volt is induced in a conductor when it cuts lines of magnetic flux at a rate of 1 weber per second (Wb/s). One weber is equal to 100,000,000 lines of flux.

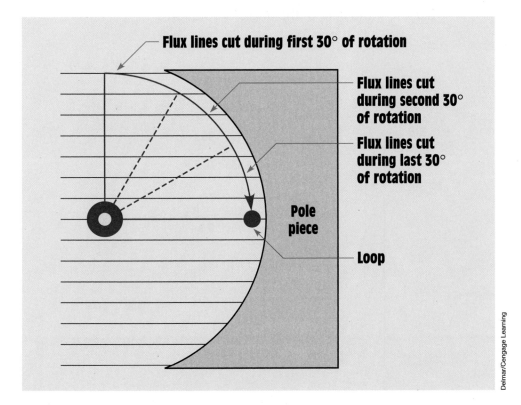

FIGURE 16–7   As the loop approaches 90° of rotation, the flux lines are cut at a faster rate.

When the loop has rotated 90°, its direction of motion is perpendicular to the flux lines and is cutting them at the maximum rate, which results in the highest, or peak, voltage being induced in the loop. The voltage at any point during the rotation is equal to the maximum induced voltage times the sine of the angle of rotation. For example, if the induced voltage after 90° of rotation is 100 volts, the voltage after 30° of rotation will be 50 volts because the sine of a 30° angle is 0.5 (100 × 0.5 = 50 V). The induced voltage after 45° of rotation is 70.7 volts because the sine of a 45° angle is 0.707 (100 × 0.707 = 70.7 V). A sine wave showing the instantaneous voltage values after different degrees of rotation is shown in *Figure 16–8*. The instantaneous voltage value is the value of voltage at any instant on the waveform.

The following formula can be used to determine the instantaneous value at any point along the sine wave:

$$E_{(INST)} = E_{(MAX)} \times \sin \angle \theta$$

where

$$E_{(INST)} = \text{the voltage at any point on the waveform}$$
$$E_{(MAX)} = \text{the maximum, or peak, voltage}$$
$$\sin \angle \theta = \text{the sine of the angle of rotation}$$

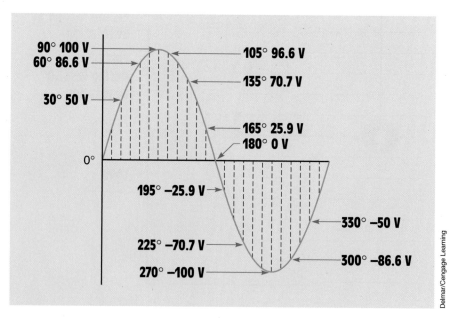

FIGURE 16–8   Instantaneous values of voltage along a sine wave.

■ EXAMPLE 16-1

A sine wave has a maximum voltage of 138 V. What is the voltage after 78° of rotation?

Solution

$$E_{(INST)} = E_{(MAX)} \sin \angle\theta$$

$$E_{(INST)} = 138 \times 0.978 \text{ (sin of 78°)}$$

$$E_{(INST)} = 134.964 \text{ V}$$

The formula can be changed to find the maximum value if the instantaneous value and the angle of rotation are known or to find the angle if the maximum and instantaneous values are known:

$$E_{(MAX)} = \frac{E_{(INST)}}{\sin \angle\theta}$$

$$\sin \angle\theta = \frac{E_{(INST)}}{E_{(MAX)}}$$

■ EXAMPLE 16-2

A sine wave has an instantaneous voltage of 246 V after 53° of rotation. What is the maximum value the waveform will reach?

Solution

$$E_{(MAX)} = \frac{E_{(INST)}}{\sin \angle\theta}$$

$$E_{(MAX)} = \frac{246 \text{ V}}{0.799}$$

$$E_{(MAX)} = 307.885 \text{ V}$$

■ EXAMPLE 16-3

A sine wave has a maximum voltage of 350 V. At what angle of rotation will the voltage reach 53 V?

Solution

$$\sin \angle\theta = \frac{E_{(INST)}}{E_{(MAX)}}$$

$$\sin \angle\theta = \frac{53\ V}{350\ V}$$

$$\sin \angle\theta = 0.151$$

Note: 0.151 is the *sine* of the angle, not the angle.

$$\angle\theta = 8.685°$$

16–3   Sine Wave Values

Several measurements of voltage and current are associated with sine waves. These measurements are peak to peak, peak, RMS, and average. A sine wave showing peak to peak, peak, and RMS measurements is shown in *Figure 16–9*.

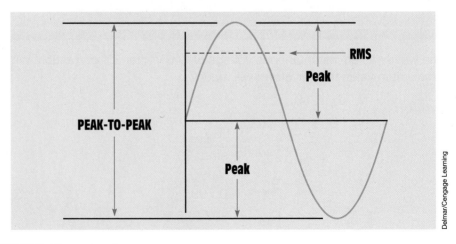

FIGURE 16–9   Sine wave values.

Peak-to-Peak and Peak Values

The peak-to-peak value is measured from the maximum value in the positive direction to the maximum value in the negative direction. The peak-to-peak value is often the simplest measurement to make when using an oscilloscope.

The **peak** value, or **amplitude,** is measured from zero to the highest value obtained in either the positive or negative direction. The peak value is one-half of the peak-to-peak value.

RMS Values

In *Figure 16–10,* a 100-volt battery is connected to a 100-ohm resistor. This connection will produce 1 ampere of current flow, and the resistor will dissipate 100 watts of power in the form of heat. An AC alternator that produces a peak voltage of 100 volts is also shown connected to a 100-ohm resistor. A peak current of 1 ampere will flow in the circuit, but the resistor will dissipate only 50 watts in the form of heat. The reason is that the voltage produced by a pure source of DC, such as a battery, is one continuous value *(Figure 16–11).* The AC sine wave, however, begins at zero, increases to the maximum value, and decreases back to zero during an equal period of time. Because the sine wave has a value of 100 volts for only a short period of time and is less than 100 volts during the rest of the half-cycle, it cannot produce as much power as 100 volts of DC.

The solution to this problem is to use a value of AC voltage that will produce the same amount of power as a like value of DC voltage. This AC value

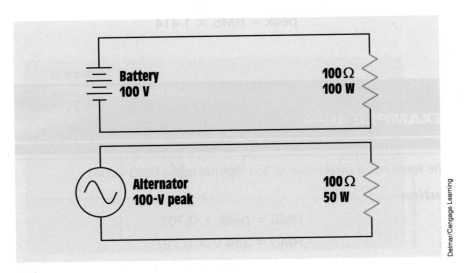

FIGURE 16–10   DC compared with a sine wave AC.

of wide copper tape or a wide, flat, braided cable. Braided cable is less affected by skin effect because it contains many small conductors, which provide a large amount of surface area.

Summary

- Most of the electric power generated in the world is AC.

- AC can be transformed and DC cannot.

- AC reverses its direction of flow at periodic intervals.

- The most common AC waveform is the sine wave.

- There are 360° in one complete sine wave.

- One complete waveform is called a cycle.

- The number of complete cycles that occur in 1 second is called the frequency.

- Sine waves are produced by rotating machines.

- Frequency is measured in hertz (Hz).

- The instantaneous voltage at any point on a sine wave is equal to the peak, or maximum, voltage times the sine of the angle of rotation.

- The peak-to-peak voltage is the amount of voltage measured from the positive-most peak to the negative-most peak.

- The peak value is the maximum amount of voltage attained by the waveform.

- The RMS value of voltage will produce as much power as a like amount of DC voltage.

- The average value of voltage is used when an AC sine wave is changed into DC.

- The current and voltage in a pure resistive circuit are in phase with each other.

- True power, or watts, can be produced only when current and voltage are both positive or both negative.

- Resistance in AC circuits is characterized by the fact that the resistive part will produce heat.

- There are three basic types of AC loads: resistive, inductive, and capacitive.

- The electrons in an AC circuit are forced toward the outside of the conductor by eddy current induction in the conductor itself.

- Skin effect is proportional to frequency.

- Skin effect can be reduced by using conductors with a large surface area.

Review Questions

1. What is the most common type of AC waveform?

2. How many degrees are there in one complete sine wave?

3. At what angle does the voltage reach its maximum negative value on a sine wave?

4. What is frequency?

5. A sine wave has a maximum value of 230 V. What is the voltage after 38° of rotation?

6. A sine wave has a voltage of 63 V after 22° of rotation. What is the maximum voltage reached by this waveform?

7. A sine wave has a maximum value of 560 V. At what angle of rotation will the voltage reach a value of 123 V?

8. A sine wave has a peak value of 433 V. What is the RMS value?

9. A sine wave has a peak-to-peak value of 88 V. What is the average value?

10. A DC voltage has an average value of 68 V. What is the RMS value?

Practical Applications

*Y*ou are an electrician working on an overhead crane. The crane uses a large electromagnet to pick up large metal pipes. The magnet must have a minimum of 200 VDC to operate properly. The crane has an AC source of 240 V. You are given four diodes that have a peak voltage rating of 400 V each. These diodes are to be used to form a bridge rectifier to convert the AC voltage into DC voltage. Is the voltage rating of the diodes sufficient? To the nearest volt, what will be the DC output voltage of the bridge rectifier? ■

Practical Applications

*Y*ou are a journeyman electrician working in a large office building. The fluorescent lighting system is operated at 277 V. You have been instructed to replace the existing light ballasts with a new electronic type that is more efficient. The manufacturer of the ballast states that the maximum peak operating voltage for the ballast is 350 V. Will the new electronic ballast operate on the building's lighting system without harm? ■

Practice Problems

Refer to the AC formulas in Appendix D to answer the following questions.

Sine Wave Values

Fill in all the missing values.

Peak Volts	Inst. Volts	Degrees
347	208	
780		43.5
	24.3	17.6
224	5.65	
48.7		64.6
	240	45
87.2	23.7	
156.9		82.3
	62.7	34.6
1256	400	
15,720		12
	72.4	34.8

$$E_{(INST)} = E_{(MAX)} \times \sin \angle \theta$$

$$E_{(MAX)} = \frac{E_{(INST)}}{\sin \angle \theta}$$

$$\sin \angle \theta = \frac{E_{(INST)}}{E_{(MAX)}}$$

Peak, RMS, and Average Values

Fill in all the missing values.

Peak	RMS	Full-Wave Rectified Average
12.7		
	53.8	
		164.2
1235		
	240	
		16.6
339.7		
	12.6	
		9
123.7		
	74.8	
		108

VI

Alternating Current (AC) Circuits Containing Inductance

Unit 17

Inductance in AC Circuits

KEY TERMS

Current lags voltage

Induced voltage

Inductance (L)

Inductive reactance, (X_L)

Quality (Q)

Reactance

Reactive power (VARs$_L$)

Why You Need to Know

*I*nductance is one the three major types of loads found in alternating current circuits. Electricians need to understand the impact on a pure inductive circuit and how current lags voltage when the effect is applied in an AC circuit. This unit

- explains how properties other than resistance can limit the flow of current.
- introduces another measurement called impedance. Impedance is the total current-limiting effect in an AC circuit and can be comprised of more than one element, such as resistance and inductance. Without an understanding of inductance, you will never be able to understand many of the concepts to follow in later units.

Objectives

After studying this unit, you should be able to

- discuss the properties of inductance in an AC circuit.
- discuss inductive reactance.
- calculate values of inductive reactance and inductance.
- discuss the relationship of voltage and current in a pure inductive circuit.
- be able to calculate values for inductors connected in series or parallel.
- discuss reactive power (VARs).
- determine the Q of a coil.

Preview

This unit discusses the effects of inductance on AC circuits. The unit explains how current is limited in an inductive circuit as well as the effect inductance has on the relationship of voltage and current. ∎

17–1 Inductance

Inductance (L) is one of the primary types of loads in AC circuits. Some amount of inductance is present in all AC circuits because of the continually changing magnetic field (*Figure 17–1*). The amount of inductance of a single conductor is extremely small, and, in most instances, it is not considered in circuit calculations. Circuits are generally considered to contain inductance when any type of

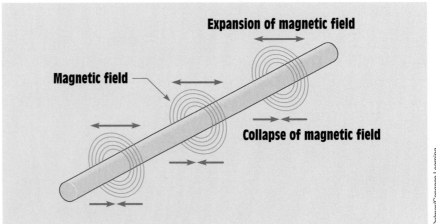

FIGURE 17–1 A continually changing magnetic field induces a voltage into any conductor.

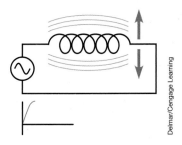

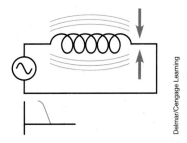

FIGURE 17–2   As current flows through a coil, a magnetic field is created around the coil.

FIGURE 17–3   As current flow decreases, the magnetic field collapses.

load that contains a coil is used. For circuits that contain a coil, inductance *is* considered in circuit calculations. Loads such as motors, transformers, lighting ballast, and chokes all contain coils of wire.

In Unit 14, it was discussed that whenever current flows through a coil of wire, a magnetic field is created around the wire *(Figure 17–2)*. If the amount of current decreases, the magnetic field collapses *(Figure 17–3)*. Recall from Unit 14 several facts concerning inductance:

1. When magnetic lines of flux cut through a coil, a voltage is induced in the coil.

2. An induced voltage is always opposite in polarity to the applied voltage. This is often referred to as counter-electromotive force (CEMF).

3. The amount of induced voltage is proportional to the rate of change of current.

4. An inductor opposes a change of current.

The inductors in *Figure 17–2* and *Figure 17–3* are connected to an alternating voltage. Therefore, the magnetic field continually increases, decreases, and reverses polarity. Because the magnetic field continually changes magnitude and direction, a voltage is continually being induced in the coil. This **induced voltage** is 180° out of phase with the applied voltage and is always in opposition to the applied voltage *(Figure 17–4)*. Because the induced voltage is always in opposition to the applied voltage, the effective applied voltage is reduced by the induced voltage. For example, assume an inductor is connected to a 120-volt AC line. Now assume that the inductor has an induced voltage of 116 volts. Because the induced voltage subtracts from the applied voltage, there are only 4 volts to push current through the wire resistance of the coil (120 V − 116 V = 4).

Calculating the Induced Voltage

The amount of induced voltage in an inductor can be calculated if the resistance of the wire in the coil and the amount of circuit current are known. For example, assume that an ohmmeter is used to measure the actual amount of

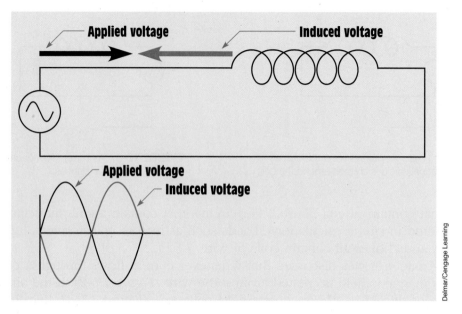

FIGURE 17–4 The applied voltage and induced voltage are 180 degrees out of phase with each other.

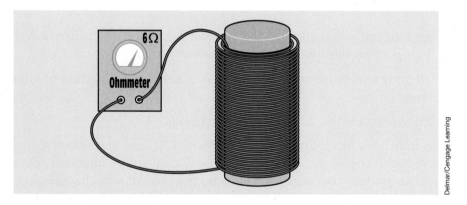

FIGURE 17–5 Measuring the resistance of a coil.

resistance in a coil, and the coil is found to contain 6 ohms of wire resistance *(Figure 17–5)*. Now assume that the coil is connected to a 120-volt AC circuit and an ammeter measures a current flow of 0.8 ampere *(Figure 17–6)*. Ohm's law can now be used to determine the amount of voltage necessary to push 0.8 ampere of current through 6 ohms of resistance:

$$E = I \times R$$
$$E = 0.8 \, A \times 6$$
$$E = 4.8 \, V$$

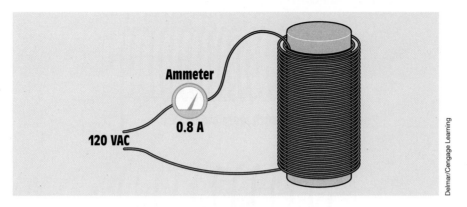

Ammeter

0.8 A

120 VAC

Delmar/Cengage Learning

FIGURE 17–6   Measuring circuit current with an ammeter.

Because only 4.8 volts are needed to push the current through the wire resistance of the inductor, the remainder of the 120 volts is used to overcome the coil's induced voltage of 119.904 volts ($\sqrt{(120\ V)^2 - (4.8V)^2} = 119.904$ volts).

17–2   Inductive Reactance

Notice that the induced voltage is able to limit the flow of current through the circuit in a manner similar to resistance. This induced voltage is *not* resistance, but it can limit the flow of current just as resistance does. This current-limiting property of the inductor is called **reactance** and is symbolized by the letter X. This reactance is caused by inductance, so it is called **inductive reactance** and is symbolized by X_L, pronounced "X sub L." Inductive reactance is measured in ohms just as resistance is and can be calculated when the values of inductance and frequency are known. The following formula can be used to find inductive reactance:

$$X_L = 2 \pi fL$$

where

X_L = inductive reactance

2 = a constant

π = 3.1416

f = frequency in hertz (Hz)

L = inductance in henrys (H)

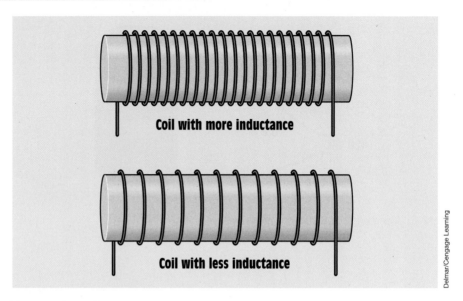

Delmar/Cengage Learning

FIGURE 17–7   Coils with turns closer together produce more inductance than coils with turns farther apart.

The inductive reactance, in ohms, is caused by the induced voltage and is therefore proportional to the three factors that determine induced voltage:

1. the *number* of turns of wire;

2. the *strength* of the magnetic field;

3. the *speed* of the cutting action (relative motion between the inductor and the magnetic lines of flux).

The number of turns of wire and the strength of the magnetic field are determined by the physical construction of the inductor. Factors such as the size of wire used, the number of turns, how close the turns are to each other, and the type of core material determine the amount of inductance (in henrys, H) of the coil *(Figure 17–7)*. The speed of the cutting action is proportional to the frequency (hertz). An increase of frequency causes the magnetic lines of flux to cut the conductors at a faster rate and thus produces a higher induced voltage or more inductive reactance.

■ EXAMPLE 17–1

The inductor shown in *Figure 17–8* has an inductance of 0.8 H and is connected to a 120-V, 60-Hz line. How much current will flow in this circuit if the wire resistance of the inductor is negligible?

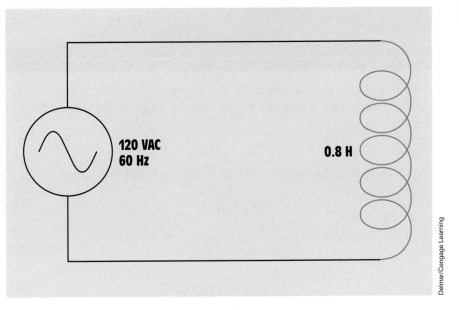

Delmar/Cengage Learning

FIGURE 17–8   Circuit current is limited by inductive reactance.

Solution

The first step is to determine the amount of inductive reactance of the inductor:

$$X_L = 2\pi fL$$
$$X_L = 2 \times 3.1416 \times 60 \text{ Hz} \times 0.8$$
$$X_L = 301.594 \ \Omega$$

Because inductive reactance is the current-limiting property of this circuit, it can be substituted for the value of R in an Ohm's law formula:

$$I = \frac{E}{X_L}$$

$$I = \frac{120 \text{ V}}{301.594 \ \Omega}$$

$$I = 0.398 \text{ A}$$

If the amount of inductive reactance is known, the inductance of the coil can be determined using the formula

$$L = \frac{X_L}{2\pi f}$$

■ EXAMPLE 17-2

Assume an inductor with a negligible resistance is connected to a 36-V, 400-Hz line. If the circuit has a current flow of 0.2 A, what is the inductance of the inductor?

Solution

The first step is to determine the inductive reactance of the circuit:

$$X_L = \frac{E}{I}$$

$$X_L = \frac{36\ V}{0.2\ A}$$

$$X_L = 180\ \Omega$$

Now that the inductive reactance of the inductor is known, the inductance can be determined:

$$L = \frac{X_L}{2\pi f}$$

$$L = \frac{180\ \Omega}{2 \times 3.1416 \times 400\ Hz}$$

$$L = 0.0716\ H$$

■ EXAMPLE 17-3

An inductor with negligible resistance is connected to a 480-V, 60-Hz line. An ammeter indicates a current flow of 24 A. How much current will flow in this circuit if the frequency is increased to 400 Hz?

Solution

The first step in solving this problem is to determine the amount of inductance of the coil. Because the resistance of the wire used to make the inductor is negligible, the current is limited by inductive reactance. The inductive reactance can be found by substituting X_L for R in an Ohm's law formula:

$$X_L = \frac{E}{I}$$

$$X_L = \frac{480 \text{ V}}{24 \text{ A}}$$

$$X_L = 20 \text{ } \Omega$$

Now that the inductive reactance is known, the inductance of the coil can be found using the formula

$$L = \frac{X_L}{2\pi f}$$

Note: When using a frequency of 60 hertz, $2 \times \pi \times 60 = 376.992$. To simplify calculations, this value is generally rounded to 377. Because 60 hertz is the major frequency used throughout the United States, 377 should be memorized because it is used in many calculations:

$$L = \frac{20 \text{ } \Omega}{377 \text{ Hz}}$$

$$L = 0.053 \text{ H}$$

Because the inductance of the coil is determined by its physical construction, it does not change when connected to a different frequency. Now that the inductance of the coil is known, the inductive reactance at 400 hertz can be calculated:

$$X_L = 2\pi f L$$

$$X_L = 2 \times 3.1416 \times 400 \text{ Hz} \times 0.053$$

$$X_L = 133.204 \text{ } \Omega$$

The amount of current flow can now be found by substituting the value of inductive reactance for resistance in an Ohm's law formula:

$$I = \frac{E}{X_L}$$

$$I = \frac{480 \text{ V}}{133.204 \text{ } \Omega}$$

$$I = 3.603 \text{ A}$$

17–3 Schematic Symbols

The schematic symbol used to represent an inductor depicts a coil of wire. Several symbols for inductors are shown in *Figure 17–9*. The symbols shown with the two parallel lines represent iron-core inductors, and the symbols without the parallel lines represent air-core inductors.

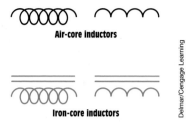

Air-core inductors

Iron-core inductors

FIGURE 17–9   Schematic symbols for inductors.

17-4   Inductors Connected in Series

When inductors are connected in series *(Figure 17–10)*, the total inductance of the circuit (L_T) equals the sum of the inductances of all the inductors:

$$L_T = L_1 + L_2 + L_3$$

The total inductive reactance (X_{LT}) of inductors connected in series equals the sum of the inductive reactances for all the inductors:

$$X_{LT} = X_{L1} + X_{L2} + X_{L3}$$

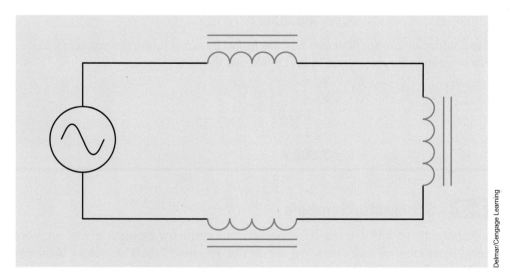

FIGURE 17–10   Inductors connected in series.

■ EXAMPLE 17-4

Three inductors are connected in series. Inductor 1 has an inductance of 0.6 H, Inductor 2 has an inductance of 0.4 H, and Inductor 3 has an inductance of 0.5 H. What is the total inductance of the circuit?

Solution

$$L_T = 0.6\ H + 0.4\ H + 0.5\ H$$
$$L_T = 1.5\ H$$

■ EXAMPLE 17-5

Three inductors are connected in series. Inductor 1 has an inductive reactance of 180 Ω, Inductor 2 has an inductive reactance of 240 Ω, and Inductor 3 has an inductive reactance of 320 Ω. What is the total inductive reactance of the circuit?

Solution

$$X_{LT} = 180\ \Omega + 240\ \Omega + 320\ \Omega$$
$$X_{LT} = 740\ \Omega$$

17-5   Inductors Connected in Parallel

When inductors are connected in parallel *(Figure 17–11)*, the total inductance can be found in a manner similar to finding the total resistance of a parallel circuit. The reciprocal of the total inductance is equal to the sum of the reciprocals of all the inductors:

$$\frac{1}{L_T} = \frac{1}{L_1} + \frac{1}{L_2} + \frac{1}{L_3}$$

or

$$L_T = \frac{1}{\frac{1}{L_1} + \frac{1}{L_2} + \frac{1}{L_3}}$$

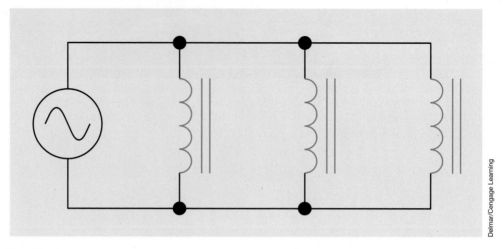

FIGURE 17–11   Inductors connected in parallel.

Another formula that can be used to find the total inductance of parallel inductors is the product-over-sum formula:

$$L_T = \frac{L_1 \times L_2}{L_1 + L_2}$$

If the values of all the inductors are the same, total inductance can be found by dividing the inductance of one inductor by the total number of inductors:

$$L_T = \frac{L}{N}$$

Similar formulas can be used to find the total inductive reactance of inductors connected in parallel:

$$\frac{1}{X_{LT}} = \frac{1}{X_{L1}} + \frac{1}{X_{L2}} + \frac{1}{X_{L3}}$$

or

$$X_{LT} = \frac{1}{\dfrac{1}{X_{L1}} + \dfrac{1}{X_{L2}} + \dfrac{1}{X_{L3}}}$$

or

$$X_{LT} = \frac{X_{L1} \times X_{L2}}{X_{L1} + X_{L2}}$$

or

$$X_{LT} = \frac{X_L}{N}$$

■ EXAMPLE 17-6

Three inductors are connected in parallel. Inductor 1 has an inductance of 2.5 H, Inductor 2 has an inductance of 1.8 H, and Inductor 3 has an inductance of 1.2 H. What is the total inductance of this circuit?

Solution

$$L_T = \frac{1}{\dfrac{1}{2.5\ H} + \dfrac{1}{1.8\ H} + \dfrac{1}{1.2\ H}}$$

$$L_T = \frac{1}{(1.788)}$$

$$L_T = 0.559\ H$$

17–6   Voltage and Current Relationships in an Inductive Circuit

In Unit 16, it was discussed that when current flows through a pure resistive circuit, the current and voltage are in phase with each other. ***In a pure inductive circuit, the current lags the voltage by 90°.*** At first this may seem to be an impossible condition until the relationship of applied voltage and induced voltage is considered. How the current and applied voltage can become 90° out of phase with each other can best be explained by comparing the relationship of the current and induced voltage *(Figure 17–12)*. Recall that the induced voltage is proportional to the rate of change of the current (speed of cutting action). At the beginning of the waveform, the current is shown at its maximum value in the negative direction. At this time, the current is not changing, so induced voltage is zero. As the current begins to decrease in value, the magnetic field produced by the flow of current decreases or collapses and begins to induce a voltage into the coil as it cuts through the conductors *(Figure 17–3)*.

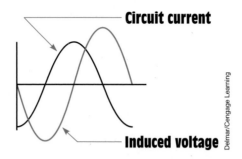

FIGURE 17–12   Induced voltage is proportional to the rate of change of current.

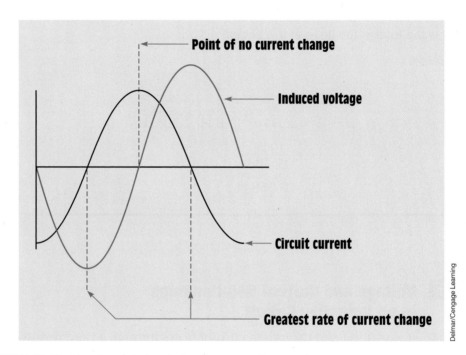

FIGURE 17–13   No voltage is induced when the current does not change.

The greatest rate of current change occurs when the current passes from negative, through zero and begins to increase in the positive direction *(Figure 17–13)*. Because the current is changing at the greatest rate, the induced voltage is maximum. As current approaches its peak value in the positive direction, the rate of change decreases, causing a decrease in the induced voltage. The induced voltage will again be zero when the current reaches its peak value and the magnetic field stops expanding.

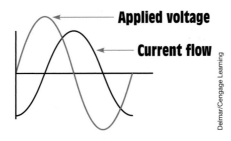

FIGURE 17–14   The current lags the applied voltage by 90°.

It can be seen that the current flowing through the inductor is leading the induced voltage by 90°. Because the induced voltage is 180° out of phase with the applied voltage, the current lags the applied voltage by 90° *(Figure 17–14)*.

17–7   Power in an Inductive Circuit

In a pure resistive circuit, the true power, or watts, is equal to the product of the voltage and current. In a pure inductive circuit, however, no true power, or watts, is produced. Recall that voltage and current must both be either positive or negative before true power can be produced. Because the voltage and current are 90° out of phase with each other in a pure inductive circuit, the current and voltage will be at different polarities 50% of the time and at the same polarity 50% of the time. During the period of time that the current and voltage have the same polarity, power is being given to the circuit in the form of creating a magnetic field. When the current and voltage are opposite in polarity, power is being given back to the circuit as the magnetic field collapses and induces a voltage back into the circuit. Because power is stored in the form of a magnetic field and then given back, no power is used by the inductor. Any power used in an inductor is caused by losses such as the resistance of the wire used to construct the inductor, generally referred to as I^2R losses, eddy current losses, and hysteresis losses.

The current and voltage waveform in *Figure 17–15* has been divided into four sections: A, B, C, and D. During the first time period, indicated by A, the current is negative and the voltage is positive. During this period, energy is being given to the circuit as the magnetic field collapses. During the second time period, B, both the voltage and current are positive. Power is being used to produce the magnetic field. In the third time period, C, the current is positive and the voltage is negative. Power is again being given back to the circuit as the field collapses. During the fourth time period, D, both the voltage and current are negative. Power is again being used to produce the magnetic field. If the amount of power used to produce the magnetic field is subtracted from the power given back, the result will be zero.

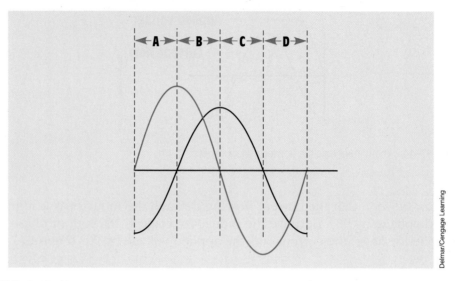

Delmar/Cengage Learning

FIGURE 17–15 Voltage and current relationships during different parts of a cycle.

17–8 Reactive Power

Although essentially no true power is being used, except by previously mentioned losses, an electrical measurement called volt-amperes-reactive **(VARs)** is used to measure the **reactive power** in a pure inductive circuit. VARs can be calculated in the same way as watts except that inductive values are substituted for resistive values in the formulas. VARs is equal to the amount of current flowing through an inductive circuit times the voltage applied to the inductive part of the circuit. Several formulas for calculating VARs are

$$\text{VARs} = E_L \times I_L$$

$$\text{VARs} = \frac{E_L^2}{X_L}$$

$$\text{VARs} = I_L^2 \times X_L$$

where

E_L = voltage applied to an inductor

I_L = current flow through an inductor

X_L = inductive reactance

17-9   Q of an Inductor

So far in this unit, it has been generally assumed that an inductor has no resistance and that inductive reactance is the only current-limiting factor. In reality, that is not true. Because inductors are actually coils of wire, they all contain some amount of internal resistance. Inductors actually appear to be a coil connected in series with some amount of resistance *(Figure 17–16)*. The amount of resistance compared with the inductive reactance determines the **quality (Q)** of the coil. Inductors that have a higher ratio of inductive reactance to resistance are considered to be inductors of higher quality. An inductor constructed with a large wire will have a low wire resistance and therefore a higher Q *(Figure 17–17)*. Inductors constructed with many turns of small wire have a much higher resistance and therefore a lower Q. To determine the Q of an inductor, divide the inductive reactance by the resistance:

$$Q = \frac{X_L}{R}$$

Although inductors have some amount of resistance, inductors that have a Q of 10 or greater are generally considered to be pure inductors. Once the ratio of inductive reactance becomes 10 times as great as resistance, the amount of resistance is considered negligible. For example, assume an inductor has an inductive reactance of 100 ohms and a wire resistance of 10 ohms. The

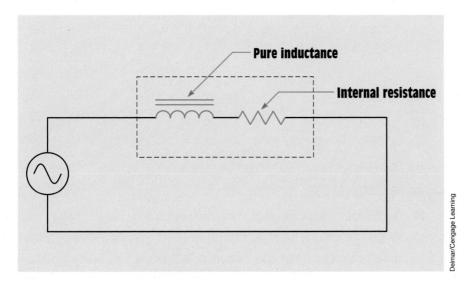

FIGURE 17–16   Inductors contain internal resistance.

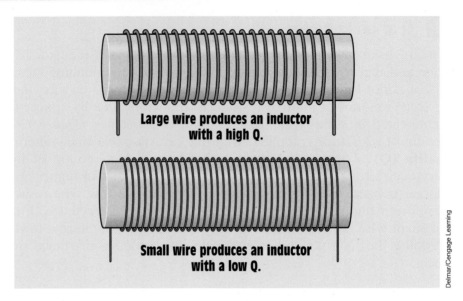

FIGURE 17–17   The Q of an inductor is a ratio of inductive reactance as compared to resistance. The letter Q stands for quality.

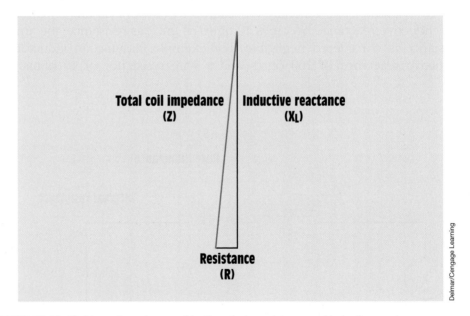

FIGURE 17–18   Coil impedance is a combination of wire resistance and inductive reactance.

inductive reactive component in the circuit is 90° out of phase with the resistive component. This relationship produces a right triangle *(Figure 17–18)*. The total current-limiting effect of the inductor is a combination of the inductive reactance and resistance. This total current-limiting effect is called impedance and is symbolized by the letter Z. The impedance of the circuit is represented by the hypotenuse of the right triangle formed by the inductive reactance and the resistance. To calculate the value of impedance for the coil, the inductive reactance and the resistance must be added. Because these two components form the legs of a right triangle and the impedance forms the hypotenuse, the Pythagorean theorem discussed in Unit 15 can be used to calculate the value of impedance:

$$Z = \sqrt{R^2 + X_L{}^2}$$
$$Z = \sqrt{10^2 + 100^2}$$
$$Z = \sqrt{10,100}$$
$$Z = 100.499 \ \Omega$$

Notice that the value of total impedance for the inductor is only 0.5 ohm greater than the value of inductive reactance.

If it should become necessary to determine the true inductance of an inductor, the resistance of the wire must be taken into consideration. Assume that an inductor is connected to a 480-volt, 60-Hz power source and that an ammeter indicates a current flow of 0.6 ampere. Now assume that an ohmmeter measures 150 Ω of wire resistance in the inductor. What is the inductance of the inductor?

To determine the inductance, it will be necessary to first determine the amount of inductive reactance as compared to the wire resistance. The total current-limiting value (impedance) can be found with Ohm's law:

$$Z = \frac{E}{I}$$
$$Z = \frac{480}{0.6}$$
$$Z = 800 \ \Omega$$

The total current-limiting effect of the inductor is 800 Ω. This value is a combination of both the inductive reactance of the inductor and the wire resistance. Because the resistive part and the inductive reactance part of the inductor are 90° out of phase with each other, they form the legs of a right triangle with the impedance forming the hypotenuse of the triangle *(Figure 17-19)*. To determine the amount of inductive reactance, use the following formula:

$$X_L = \sqrt{Z^2 - R^2}$$
$$X_L = \sqrt{800^2 - 150^2}$$
$$X_L = 785.812 \ \Omega$$

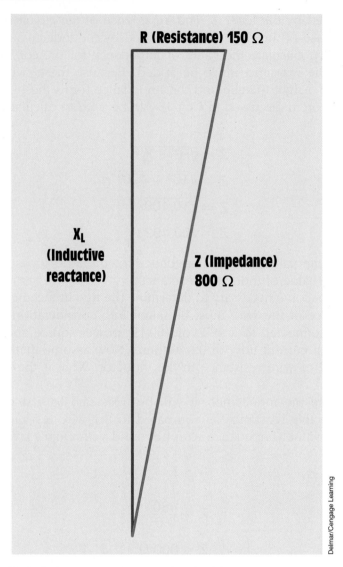

FIGURE 17–19   The inductive reactance forms one leg of a right triangle.

Now that the amount of inductive reactance has been determined, the inductance can be calculated using the formula

$$L = \frac{X_L}{2\pi f}$$

$$L = \frac{785.812}{2 \times 3.1416 \times 60}$$

$$L = 2.084 \text{ henrys}$$

Summary

- Induced voltage is proportional to the rate of change of current.

- Induced voltage is always opposite in polarity to the applied voltage.

- Inductive reactance is a countervoltage that limits the flow of current, as does resistance.

- Inductive reactance is measured in ohms.

- Inductive reactance is proportional to the inductance of the coil and the frequency of the line.

- Inductive reactance is symbolized by X_L.

- Inductance is measured in henrys (H) and is symbolized by the letter L.

- When inductors are connected in series, the total inductance is equal to the sum of all the inductors.

- When inductors are connected in parallel, the reciprocal of the total inductance is equal to the sum of the reciprocals of all the inductors.

- The current lags the applied voltage by 90° in a pure inductive circuit.

- All inductors contain some amount of resistance.

- The Q of an inductor is the ratio of the inductive reactance to the resistance.

- Inductors with a Q of 10 are generally considered to be "pure" inductors.

- Pure inductive circuits contain no true power or watts.

- Reactive power is measured in VARs.

- VARs is an abbreviation for volt-amperes-reactive.

Review Questions

1. How many degrees are the current and voltage out of phase with each other in a pure resistive circuit?

2. How many degrees are the current and voltage out of phase with each other in a pure inductive circuit?

3. To what is inductive reactance proportional?

4. Four inductors, each having an inductance of 0.6 H, are connected in series. What is the total inductance of the circuit?

5. Three inductors are connected in parallel. Inductor 1 has an inductance of 0.06 H; Inductor 2 has an inductance of 0.05 H; and Inductor 3 has an inductance of 0.1 H. What is the total inductance of this circuit?

6. If the three inductors in Question 5 were connected in series, what would be the inductive reactance of the circuit? Assume the inductors are connected to a 60-Hz line.

7. An inductor is connected to a 240-V, 1000-Hz line. The circuit current is 0.6 A. What is the inductance of the inductor?

8. An inductor with an inductance of 3.6 H is connected to a 480-V, 60-Hz line. How much current will flow in this circuit?

9. If the frequency in Question 8 is reduced to 50 Hz, how much current will flow in the circuit?

10. An inductor has an inductive reactance of 250 Ω when connected to a 60-Hz line. What will be the inductive reactance if the inductor is connected to a 400-Hz line?

Practical Applications

You are working as an electrician installing fluorescent lights. You notice that the lights were made in Europe and that the ballasts are rated for operation on a 50-Hz system. Will these ballasts be harmed by overcurrent if they are connected to 60 Hz? If there is a problem with these lights, what will be the most likely cause of the trouble? ■

Practical Applications

You have the task of ordering a replacement inductor for one that has become defective. The information on the nameplate has been painted over and cannot be read. The machine that contains the inductor operates on 480 V at a frequency of 60 Hz. Another machine has an identical inductor in it, but its nameplate has been painted over also. A clamp-on ammeter indicates a current of 18 A, and a voltmeter indicates a voltage drop across the inductor of 324 V in the machine that is still in operation. After turning off the power and locking out the panel, you disconnect the inductor in the operating machine and measure a wire resistance of 1.2 Ω with an ohmmeter. Using the identical inductor in the operating machine as an example, what inductance value should you order and what would be the minimum VAR rating of the inductor? Should you be concerned with the amount of wire resistance in the inductor when ordering? Explain your answers. ■

Practice Problems

Inductive Circuits

1. Fill in all the missing values. Refer to the following formulas:

$$X_L = 2\pi fL$$

$$L = \frac{X_L}{2\pi f}$$

$$f = \frac{X_L}{2\pi L}$$

Inductance (H)	Frequency (Hz)	Inductive Reactance (Ω)
1.2	60	
0.085		213.628
	1000	4712.389
0.65	600	
3.6		678.584
	25	411.459
0.5	60	
0.85		6408.849
	20	201.062
0.45	400	
4.8		2412.743
	1000	40.841

2. What frequency must be applied to a 33-mH inductor to produce an inductive reactance of 99.526 Ω?

3. An inductor is connected to a 120-volt, 60-Hz line and has a current flow of 4 amperes. An ohmmeter indicates that the inductor has a wire resistance of 12 Ω. What is the inductance of the inductor?

4. A 0.75-henry inductor has a wire resistance of 90 Ω. When connected to a 60-Hz power line, what is the total current-limiting effect of the inductor?

5. An inductor has a current flow of 3 amperes when connected to a 240-volt, 60-Hz power line. The inductor has a wire resistance of 15 Ω. What is the Q of the inductor?

Unit **18**

Resistive-Inductive Series Circuits

Why You Need to Know

*I*n previous units, you learned that current and voltage are in phase with each other in a pure resistive circuit and that current and voltage are 90° out of phase with each other in a pure inductive circuit. This unit

- describes what happens when resistive and inductive elements are combined in the same circuit. Although there are some applications for connecting resistors and inductors in series with each other, more often you will encounter devices that appear to have both elements connected in series with each other. A very good example of this is AC motors and transformers. Motors and transformers are inductive because they contain wound coils. At no load, these devices appear to be very inductive and current is limited by inductive reactance. As load is added, electrical energy is converted to some other form and they appear to become more resistive.
- explains what power factor is and how to correct for its impact. Without a knowledge of what happens when inductance and resistance are connected in the same circuit, you will never be able to understand such concepts as power factor and power factor correction.

KEY TERMS

Angle theta ($\angle \theta$)
Apparent power (VA)
Power factor (PF)
Quadrature power
Total current (I_T)
Wattless power

Objectives

After studying this unit, you should be able to

- discuss the relationship of resistance and inductance in an AC series circuit.

- define power factor.

- calculate values of voltage, current, apparent power, true power, reactive power, impedance, resistance, inductive reactance, and power factor in an RL series circuit.

- calculate the phase angle for current and voltage in an RL series circuit.

- connect an RL series circuit and make measurements using test instruments.

- discuss vectors and be able to plot electrical quantities using vectors.

Preview

This unit covers the relationship of resistance and inductance used in the same circuit. The resistors and inductors are connected in series. Concepts such as circuit impedance, power factor, and vector addition are introduced. Although it is true that some circuits are basically purely resistive or purely inductive, many circuits contain a combination of both resistive and inductive elements. ∎

18–1 R-L Series Circuits

When a pure resistive load is connected to an AC circuit, the voltage and current are in phase with each other. When a pure inductive load is connected to an AC circuit, the voltage and current are 90° out of phase with each other *(Figure 18–1)*. When a circuit containing both resistance, R, and inductance, L, is connected to an AC circuit, the voltage and current will be out of phase with each other by some amount between 0° and 90°. *The exact amount of phase angle difference is determined by the ratio of resistance as compared to inductance.* In the following example, a series circuit containing 30 ohms of resistance (R) and 40 ohms of inductive reactance (X_L) is connected to a 240-volt, 60-hertz line *(Figure 18–2)*. It is assumed that the inductor has negligible resistance. The following unknown values will be calculated:

Z—total circuit impedance

I—current flow

E_R—voltage drop across the resistor

P—watts (true power)

L—inductance of the inductor

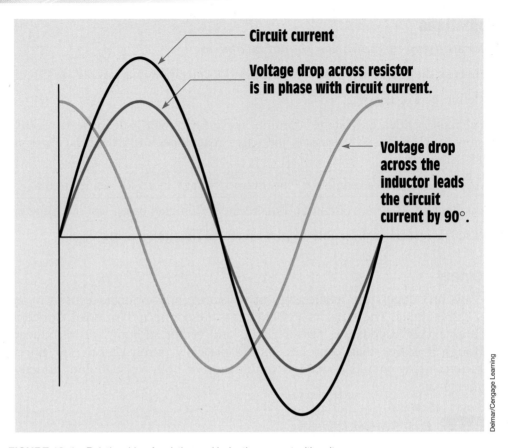

Circuit current

Voltage drop across resistor is in phase with circuit current.

Voltage drop across the inductor leads the circuit current by 90°.

Delmar/Cengage Learning

FIGURE 18–1 Relationship of resistive and inductive current with voltage.

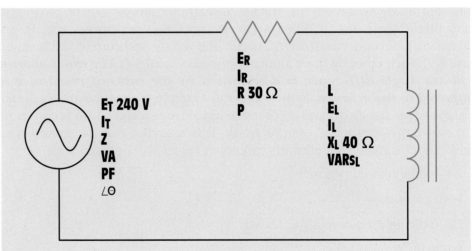

E_R
I_R
R 30 Ω
P

E_T 240 V
I_T
Z
VA
PF
∠Θ

L
E_L
I_L
X_L 40 Ω
VARS$_L$

Delmar/Cengage Learning

FIGURE 18–2 R-L series circuit.

E_L—voltage drop across the inductor

VARs—reactive power

VA—apparent power

PF—power factor

$\angle\theta$—the angle the voltage and current are out of phase with each other

18–2   Impedance

In Unit 17, impedance was defined as a measure of the part of the circuit that impedes, or hinders, the flow of current. It is measured in ohms and symbolized by the letter Z. In this circuit, impedance is a combination of resistance and inductive reactance.

In a series circuit, the total resistance is equal to the sum of the individual resistors. In this instance, however, the total impedance (Z) is the sum of the resistance and the inductive reactance. It would first appear that the sum of these two quantities should be 70 ohms (30 Ω + 40 Ω = 70 Ω). In practice, however, the resistive part of the circuit and the reactive part of the circuit are out of phase with each other by 90°. To find the sum of these two quantities, vector addition must be used. Because these two quantities are 90° out of phase with each other, the resistive and inductive reactance form the two legs of a right triangle, and the impedance is the hypotenuse *(Figure 18–3)*.

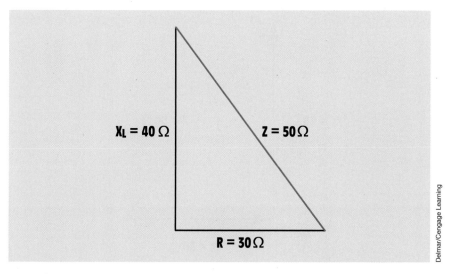

FIGURE 18–3   Impedance is the combination of resistance and inductive reactance.

The total impedance (Z) can be calculated using the formula

$$Z = \sqrt{R^2 + X_L^2}$$

$$Z = \sqrt{(30\ \Omega)^2 + (40\ \Omega)^2}$$

$$Z = \sqrt{900\ \Omega^2 + 1600\ \Omega^2}$$

$$Z = \sqrt{2500\ \Omega^2}$$

$$Z = 50\ \Omega$$

To find the impedance of the circuit in *Figure 18–3* using vector addition, connect the starting point of one vector to the ending point of the other. Because resistance and inductive reactance are 90° out of phase with each other, the two vectors must be placed at a 90° angle. If the resistive vector has a magnitude of 30 ohms and the inductive vector has a magnitude of 40 ohms, the resultant (impedance) will have a magnitude of 50 ohms *(Figure 18–4)*. Notice that the result is the same as that found using the right triangle.

The parallelogram method of vector addition can also be used to find the total impedance of the circuit shown in *Figure 18–2*. The resistance forms a vector with a magnitude of 30 ohms and a direction of 0°. The inductive reactance forms a vector with a magnitude of 40 ohms and a direction of 90°. When lines are extended to form a parallelogram and a resultant is drawn, the resultant will have a magnitude of 50 ohms, which is the impedance of the

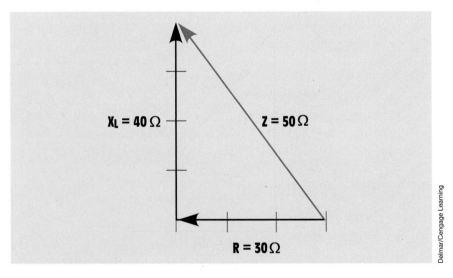

FIGURE 18–4   Determining impedance using vector addition.

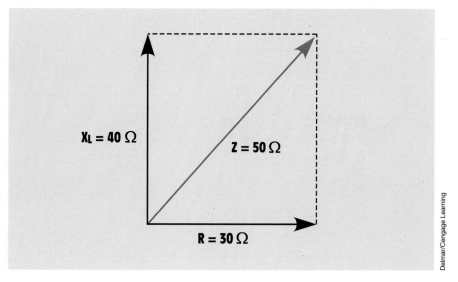

FIGURE 18–5   Finding the impedance of the circuit using the parallelogram method of vector addition.

circuit *(Figure 18–5)*. Some students of electricity find the right triangle concept easier to understand, and others find vectors more helpful. For this reason, this text uses both methods to help explain the relationship of voltage, current, and power in AC circuits.

18–3   Total Current

One of the primary laws for series circuits is that the current must be the same in any part of the circuit. This law holds true of R-L series circuits also. Because the impedance is the total current-limiting component of the circuit, it can be used to replace R in an Ohm's law formula. The **total current (I$_T$)** flow through the circuit can be calculated by dividing the total applied voltage by the total current-limiting factor. Total current can be found by using the formula

$$I_T = \frac{E_T}{Z}$$

The total current for the circuit in *Figure 18–2,* then, is

$$I_T = \frac{240 \text{ V}}{50 \text{ }\Omega}$$

$$I_T = 4.8 \text{ A}$$

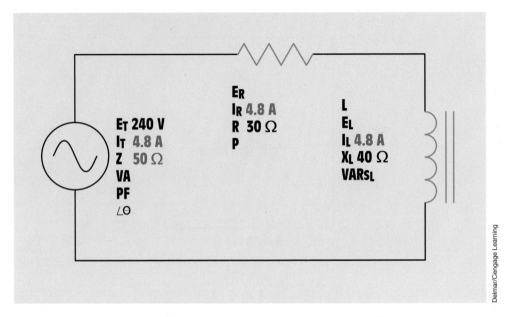

Delmar/Cengage Learning

FIGURE 18–6 Total voltage divided by impedance equals total current.

In a series circuit, the current is the same at any point in the circuit. Therefore, 4.8 amperes of current flow through both the resistor and the inductor. These values can be added to the circuit, as shown in *Figure 18–6*.

18–4 Voltage Drop across the Resistor

Now that the amount of current flow through the resistor is known, the voltage drop across the resistor (E_R) can be calculated using the formula:

$$E_R = I_R \times R$$

The voltage drop across the resistor in our circuit is

$$E_R = 4.8\ A \times 30\ \Omega$$
$$E_R = 144\ V$$

Notice that the amount of voltage dropped across the resistor was found using quantities that pertained only to the resistive part of the circuit. The amount of voltage dropped across the resistor could not be found using a formula such as

$$E_R = I_R \times X_L$$

or

$$E_R = I_R \times Z$$

Inductive reactance (X_L) is an inductive quantity, and impedance (Z) is a circuit total quantity. These quantities cannot be used with Ohm's law to find resistive quantities. They can, however, be used with vector addition to find like resistive quantities. For example, both inductive reactance and impedance are measured in ohms. The resistive quantity that is measured in ohms is resistance (R). If the impedance and inductive reactance of a circuit were known, they could be used with the following formula to find the circuit resistance:

$$R = \sqrt{Z^2 - X_L{}^2}$$

Note: Refer to the Resistive-Inductive Series Circuits section of the AC formulas listed in Appendix B.

18–5   Watts

True power (P) for the circuit can be calculated by using any of the watts formulas with pure resistive parts of the circuit. Watts (W) can be calculated, for example, by multiplying the voltage dropped across the resistor (E_R) by the current flow through the resistor (I_R), or by squaring the voltage dropped across the resistor and dividing by the resistance of the resistor, or by squaring the current flow through the resistor and multiplying by the resistance of the resistor. Watts cannot be calculated by multiplying the total voltage (E_T) by the current flow through the resistor or by multiplying the square of the current by the inductive reactance. Recall that true power, or watts, can be produced only during periods of time when the voltage and current are both positive or both negative.

In an R-L series circuit, the current is the same through both the resistor and the inductor. The voltage dropped across the resistor, however, is in phase with the current, and the voltage dropped across the inductor is 90° out of phase with the current (*Figure 18–7*). Because true power, or watts, can be produced only when the current and voltage are both positive or both negative, only resistive parts of the circuit can produce watts.

The formula used in this example is

$$P = E_R \times I_R$$
$$P = 144 \text{ V} \times 4.8 \text{ A}$$
$$P = 691.2 \text{ W}$$

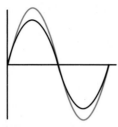

Voltage dropped across the resistor is in phase with the current.

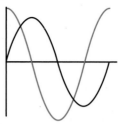

Voltage dropped across the Inductor is 90° out of phase with the current.

Delmar/Cengage Learning

FIGURE 18–7   Relationship of current and voltage in an R-L series circuit.

18–6   Calculating the Inductance

The amount of inductance can be calculated using the formula

$$L = \frac{X_L}{2\pi f}$$

$$L = \frac{40 \ \Omega}{377}$$

$$L = 0.106 \ H$$

18–7   Voltage Drop across the Inductor

The voltage drop across the inductor (E_L) can be calculated using the formula

$$E_L = I_L \times X_L$$
$$E_L = 4.8 \ A \times 40 \ \Omega$$
$$E_L = 192 \ V$$

Notice that only inductive quantities were used to find the voltage drop across the inductor.

18–8   Total Voltage

Although the total applied voltage in this circuit is known (240 volts), the total voltage is also equal to the sum of the voltage drops, just as it is in any other series circuit. Because the voltage dropped across the resistor is in phase with the current and the voltage dropped across the inductor is 90° out of phase with the current, vector addition must be used. The total voltage will be the hypotenuse of a right triangle, and the resistive and inductive voltage drops will form the legs of the triangle *(Figure 18–8)*. This relationship of voltage drops can also be represented using the parallelogram method of vector addition as shown in *Figure 18–9*.

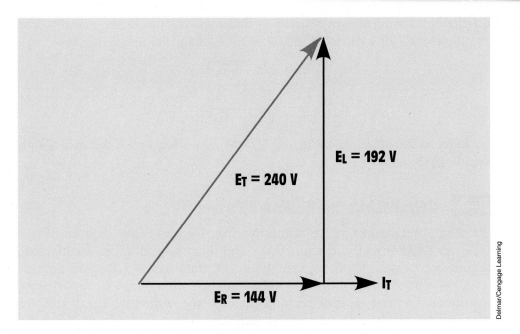

FIGURE 18–8 Relationship of resistive and inductive voltage drops in an R-L series circuit.

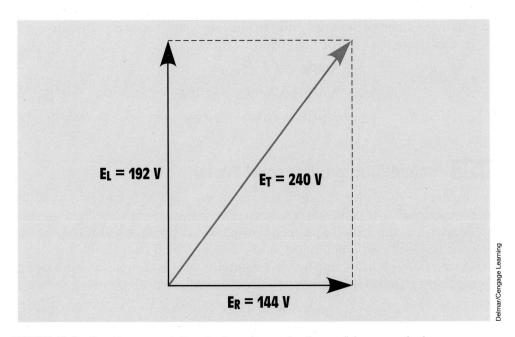

FIGURE 18–9 Graphic representation of voltage drops using the parallelogram method of vector addition.

The following formulas can be used to find total voltage or the voltage drops across the resistor or inductor if the other two voltage values are known:

$$E_T = \sqrt{E_R{}^2 + E_L{}^2}$$
$$E_R = \sqrt{E_T{}^2 - E_L{}^2}$$
$$E_L = \sqrt{E_T{}^2 - E_R{}^2}$$

Note: Refer to the Resistive-Inductive Series Circuits section of the AC formulas listed in Appendix B.

18–9   Calculating the Reactive Power

VARs is an abbreviation for volt-amperes-reactive and is the amount of reactive power (VARs) in the circuit. VARs should not be confused with watts, which is true power. VARs represents the product of the volts and amperes that are 90° out of phase with each other, such as the voltage dropped across the inductor and the current flowing through the inductor. Recall that true power can be produced only during periods of time when the voltage and current are both positive or both negative *(Figure 18–10)*. During these periods, the power is being stored in the form of a magnetic field. During the periods that voltage and current have opposite signs, the power is returned to the circuit. For this reason, VARs is often referred to as **quadrature power, or wattless power.** It can be calculated in a manner similar to watts except that reactive values of voltage and current are used instead of resistive values. In this example, the formula used is

$$VARs = I_L{}^2 \times X_L$$
$$VARs = (4.8\ A)^2 \times 40\ \Omega$$
$$VARs = 921.6$$

18–10   Calculating the Apparent Power

Volt-amperes (VA) is the apparent power of the circuit. It can be calculated in a manner similar to watts or VARs, except that total values of voltage and current are used. It is called **apparent power (VA)** because it is the value that would be found if a voltmeter and ammeter were used to measure the circuit voltage and current and then these measured values were multiplied together *(Figure 18–11)*. In this example, the formula used is

$$VA = E_T \times I_T$$
$$VA = 240\ V \times 4.8\ A$$
$$VA = 1152$$

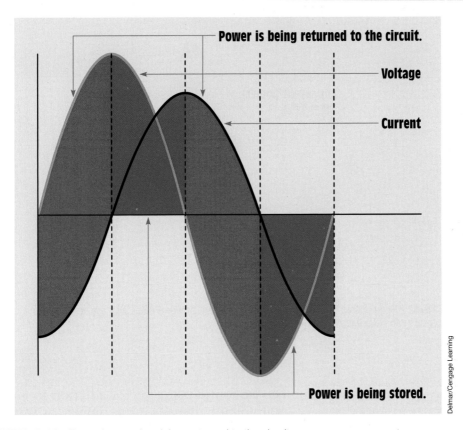

FIGURE 18–10   Power is stored and then returned to the circuit.

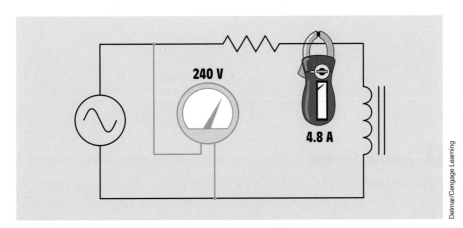

FIGURE 18–11   Apparent power is the product of measured values (240 V × 4.8 A = 1152 VA).

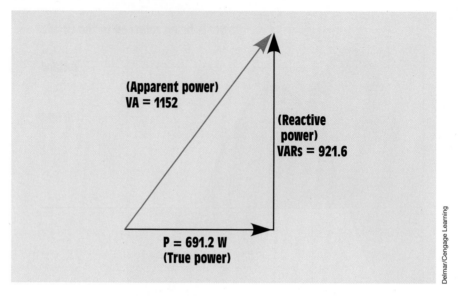

FIGURE 18–12   Relationship of true power (watts), reactive power (VARs), and apparent power (volt-amperes) in an R-L series circuit.

The apparent power can also be found using vector addition in a manner similar to impedance or total voltage. Because true power, or watts, is a pure resistive component and VARs is a pure reactive component, they form the legs of a right triangle. The apparent power is the hypotenuse of this triangle *(Figure 18–12)*. This relationship of the three power components can also be plotted using the parallelogram method *(Figure 18–13)*. The following formulas can be used to calculate the values of apparent power, true power, and reactive power when the other two values are known:

$$VA = \sqrt{P^2 + VARs^2}$$
$$P = \sqrt{VA^2 - VARs^2}$$
$$VARs = \sqrt{VA^2 - P^2}$$

18–11   Power Factor

Power factor (PF) is a ratio of the true power to the apparent power. It can be calculated by dividing any resistive value by its like total value. For example, power factor can be calculated by dividing the voltage drop across the resistor

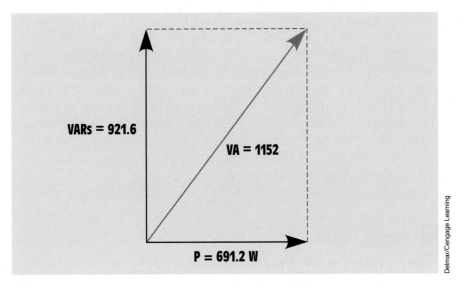

FIGURE 18–13 Using the parallelogram method to plot the relationship of volt-amperes, watts, and VARs.

by the total circuit voltage; or by dividing resistance by impedance; or by dividing watts by volt-amperes:

$$PF = \frac{E_R}{E_T}$$

$$PF = \frac{R}{Z}$$

$$PF = \frac{P}{VA}$$

Power factor is generally expressed as a percentage. The decimal fraction calculated from the division will therefore be changed to a percent by multiplying it by 100. In this circuit, the formula used is

$$PF = \frac{W}{VA}$$

$$PF = \frac{691.2 \ W}{1152 \ VA}$$

$$PF = 0.6 \times 100, \text{ or } 60\%$$

Note that in a series circuit, the power factor cannot be calculated using current because current is the same in all parts of the circuit.

Power factor can become very important in an industrial application. Most power companies charge a substantial surcharge when the power factor drops below a certain percent. The reason for this is that electric power is sold on the basis of true power, or watt-hours, consumed. The power company, however, must supply the apparent power. Assume that an industrial plant has a power factor of 60% and is consuming 5 megawatts of power. At a power factor of 60%, the power company must actually supply 8.333 megavolt-amperes (5 MW/0.6 = 8.333 MVA). If the power factor were to be corrected to 95%, the power company would have to supply only 5.263 megavolt-amperes to furnish the same amount of power to the plant.

18–12   Angle Theta

The angular displacement by which the voltage and current are out of phase with each other is called **angle theta ($\angle\theta$).** Because the power factor is the ratio of true power to apparent power, the phase angle of voltage and current is formed between the resistive leg of the right triangle and the hypotenuse *(Figure 18–14)*. The resistive leg of the triangle is adjacent to the angle, and the total leg is the hypotenuse. The trigonometric function that corresponds to the adjacent side and the hypotenuse is the cosine. Angle theta is the cosine of watts divided by volt-amperes. Watts divided by volt-amperes is

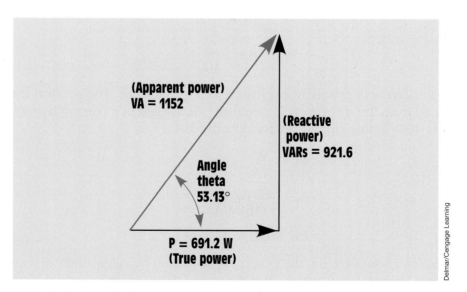

FIGURE 18–14   The angle theta is the relationship of true power to apparent power.

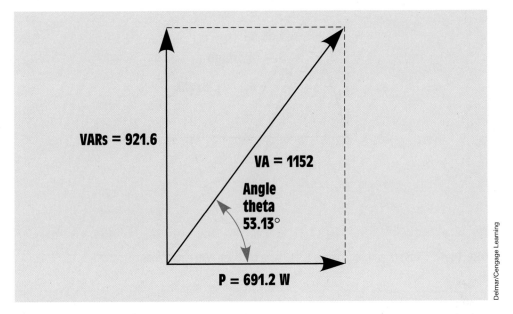

VARs = 921.6

VA = 1152

Angle
theta
53.13°

P = 691.2 W

Delmar/Cengage Learning

FIGURE 18–15 The angle theta can be found using vectors provided by the parallelogram method.

also the power factor. Therefore, the cosine of angle theta ($\angle\theta$) is the power factor (PF):

$$\cos \angle\theta = PF$$
$$\cos \angle\theta = 0.6$$
$$\angle\theta = 53.13°$$

The vectors formed using the parallelogram method of vector addition can also be used to find angle theta as shown in *Figure 18–15*. Notice that the total quantity, volt-amperes, and the resistive quantity, watts, are again used to determine angle theta.

Because this circuit contains both resistance and inductance, the current is lagging the voltage by 53.13° *(Figure 18–16)*. Angle theta can also be determined by using any of the other trigonometric functions:

$$\sin \angle\theta = \frac{VARs}{VA}$$

$$\tan \angle\theta = \frac{VARs}{P}$$

Now that all the unknown values have been calculated, they can be filled in as shown in *Figure 18–17*.

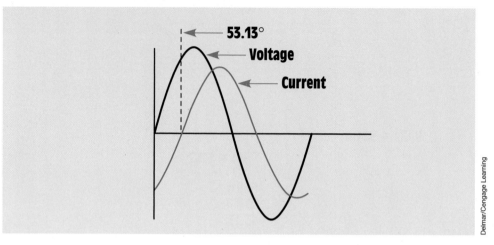

FIGURE 18–16 Current and voltage are 53.13° out of phase with each other.

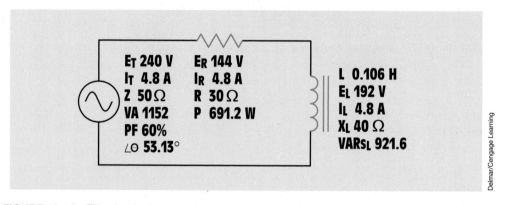

FIGURE 18–17 Filling in all unknown values.

■ EXAMPLE 18–1

Two resistors and two inductors are connected in series *(Figure 18–18)*. The circuit is connected to a 130-V, 60-Hz line. The first resistor has a power dissipation of 56 W, and the second resistor has a power dissipation of 44 W. One inductor has a reactive power of 152 VARs and the second a reactive power of 88 VARs. Find the unknown values in this circuit.

Solution

The first step is to find the total amount of true power and the total amount of reactive power. The total amount of true power can be calculated by adding the

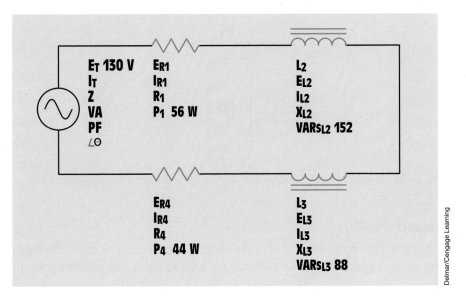

FIGURE 18–18   An R-L series circuit containing two resistors and two inductors.

values of the resistors together. A vector diagram of these two values would reveal that they both have a direction of 0° *(Figure 18–19)*.

$$P_T = P_1 + P_4$$
$$P_T = 56 \text{ W} + 44 \text{ W}$$
$$P_T = 100 \text{ W}$$

The total reactive power in the circuit can be found in the same manner. Like watts, the VARs are both inductive and are therefore in the same direction *(Figure 18–20)*. The total reactive power will be the sum of the two reactive power ratings:

$$VARs_T = VARs_{L2} + VARs_{L3}$$
$$VARs_T = 152 + 88$$
$$VARs_T = 240$$

FIGURE 18–19   The two true power (watts) vectors are in the same direction.

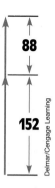

FIGURE 18–20 The two reactive power (VARs) vectors are in the same direction.

Apparent Power

Now that the total amount of true power and the total amount of reactive power are known, the apparent power (VA) can be calculated using vector addition:

$$VA = \sqrt{P_T^2 + VARs_T^2}$$
$$VA = \sqrt{(100 \text{ W})^2 + (240 \text{ VARs})^2}$$
$$VA = 260$$

The parallelogram method of vector addition is shown in *Figure 18–21* for this calculation.

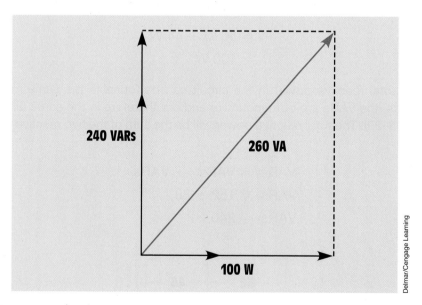

FIGURE 18–21 Power vector for the circuit.

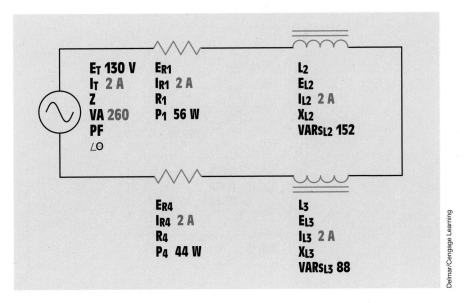

FIGURE 18–22 Adding circuit values.

Total Circuit Current

Now that the apparent power is known, the total circuit current can be found using the applied voltage and Ohm's law:

$$I_T = \frac{VA}{E_T}$$

$$I_T = \frac{260\ VA}{130\ V}$$

$$I_T = 2\ A$$

Because this is a series circuit, the current must be the same at all points in the circuit. The known values added to the circuit are shown in *Figure 18–22*.

Other Circuit Values

Now that the total circuit current has been found, other values can be calculated using Ohm's law.

Impedance

$$Z = \frac{E_T}{I_T}$$

$$Z = \frac{130\ V}{2\ A}$$

$$Z = 65\ \Omega$$

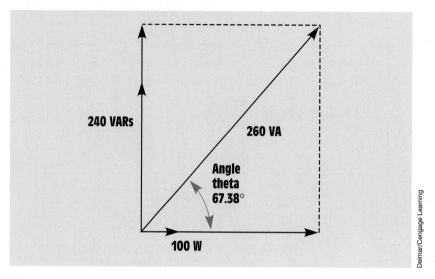

FIGURE 18–23   Angle theta for the circuit.

Power Factor

$$PF = \frac{W_T}{VA}$$

$$PF = \frac{100 \text{ W}}{260 \text{ VA}}$$

$$PF = 38.462\%$$

Angle Theta

Angle theta is the cosine of the power factor. A vector diagram showing this relationship is shown in *Figure 18–23*.

$$\cos \angle\theta = PF$$

$$\cos \angle\theta = 0.38462$$

$$\angle\theta = 67.38°$$

The relationship of voltage and current for this circuit is shown in *Figure 18–24*.

E_{R1}

$$E_{R1} = \frac{P_1}{I_{R1}}$$

$$E_{R1} = \frac{56 \text{ W}}{2 \text{ A}}$$

$$E_{R1} = 28 \text{ V}$$

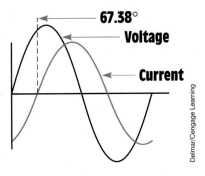

67.38°

Voltage

Current

Delmar/Cengage Learning

FIGURE 18–24   Voltage and current are 67.38° out of phase with each other.

R₁

$$R_1 = \frac{E_{R1}}{I_{R1}}$$

$$R_1 = \frac{28\ V}{2\ A}$$

$$R_1 = 14\ \Omega$$

E₍L2₎

$$E_{L2} = \frac{VARs_{L2}}{I_{L2}}$$

$$E_{L2} = \frac{152\ VARs}{2\ A}$$

$$E_{L2} = 76\ V$$

X₍L2₎

$$X_{L2} = \frac{E_{L2}}{I_{L2}}$$

$$X_{L2} = \frac{76\ V}{2\ A}$$

$$X_{L2} = 38\ \Omega$$

L₂

$$L_2 = \frac{X_{L2}}{2\pi f}$$

$$L_2 = \frac{38\ \Omega}{377}$$

$$L_2 = 0.101\ H$$

E$_{L3}$

$$E_{L3} = \frac{VARs_{L3}}{I_{L3}}$$

$$E_{L3} = \frac{88 \text{ VARs}}{2 \text{ A}}$$

$$E_{L3} = 44 \text{ V}$$

X$_{L3}$

$$X_{L3} = \frac{E_{L3}}{I_{L3}}$$

$$X_{L3} = \frac{44 \text{ V}}{2 \text{ A}}$$

$$X_{L3} = 22 \ \Omega$$

L$_3$

$$L_3 = \frac{X_{L3}}{2\pi f}$$

$$L_3 = \frac{22 \ \Omega}{377}$$

$$L_3 = 0.0584$$

E$_{R4}$

$$E_{R4} = \frac{P_4}{I_{R4}}$$

$$E_{R4} = \frac{44 \text{ W}}{2 \text{ A}}$$

$$E_{R4} = 22 \text{ V}$$

R$_4$

$$R_4 = \frac{E_{R4}}{I_{R4}}$$

$$R_4 = \frac{22 \text{ V}}{2 \text{ A}}$$

$$R_4 = 11 \ \Omega$$

The complete circuit with all values is shown in *Figure 18–25*.

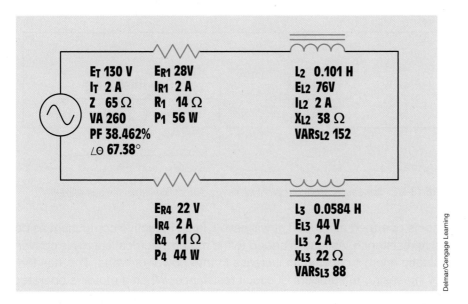

FIGURE 18–25   All values for the circuit.

Many people working the electrical field encounter R-L series circuits almost every day in ways that they do not realize. A good example of this is an electric motor. As far as the circuit is concerned, a motor is actually an R-L series circuit. The motor winding is basically an inductor. It is a coil of wire surrounded by an iron core. The wire does have resistance, however. Because there is only one path for current flow through the circuit, it is a series circuit *(Figure 18–26)*. When

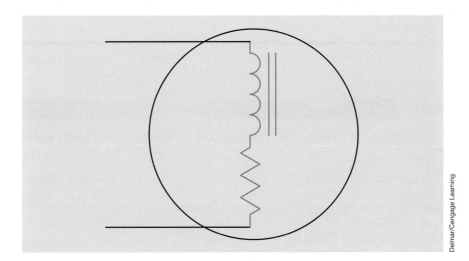

FIGURE 18–26   A motor winding appears to be an R-L series circuit.

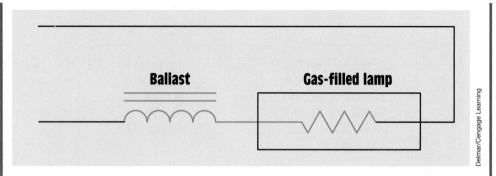

FIGURE 18–27 A ballast is used to prevent excessive current flow when the gas ionizes.

a motor is operated at no load, it will have a large inductive component as compared to resistance. As load is added to the motor, electrical energy is converted into kinetic energy, causing an increase in true power or watts. This has the effect of increasing the amount of circuit resistance. When a motor is operated at full load, the circuit has become an R-L series circuit with more resistance than inductance.

Another example of an RL series circuit is a luminaire (lighting fixture) that depends on the ionization of gas such as fluorescent, mercury vapor, sodium vapor, and so on. When a gas is ionized, it basically becomes a short circuit. A ballast is used to prevent excessive current flow when the gas ionizes. The ballast is connected in series with the lamp *(Figure 18–27)*. When the gas in the lamp ionizes, the inductive reactance of the ballast limits current. The lamp appears as a resistive load after ionization takes place. Without an understanding of the relationship of inductance and resistance in a series circuit, it would be impossible to understand how this everyday electric device actually works.

Summary

- In a pure resistive circuit, the voltage and current are in phase with each other.

- In a pure inductive circuit, the voltage and current are 90° out of phase with each other.

- In an R-L series circuit, the voltage and current will be out of phase with each other by some value between 0° and 90°.

- The amount the voltage and current are out of phase with each other is determined by the ratio of resistance to inductance.

■ Total circuit values include total voltage, E_T; total current, I_T; volt-amperes, VA; and impedance, Z.

■ Pure resistive values include voltage drop across the resistor, E_R; current flow through the resistor, I_R; resistance, R; and watts, P.

■ Pure inductive values include inductance of the inductor, L; voltage drop across the inductor, E_L; current through the inductor, I_L; inductive reactance, X_L; and inductive VARs, $VARs_L$.

■ Angle theta measures the phase angle difference between the applied voltage and total circuit current.

■ The cosine of angle theta is equal to the power factor.

■ Power factor is a ratio of true power to apparent power.

Review Questions

1. What is the relationship of voltage and current (concerning phase angle) in a pure resistive circuit?

2. What is the relationship of voltage and current (concerning phase angle) in a pure inductive circuit?

3. What is power factor?

4. A circuit contains a 20-Ω resistor and an inductor with an inductance of 0.093 H. If the circuit has a frequency of 60 Hz, what is the total impedance of the circuit?

5. An R-L series circuit has a power factor of 86%. How many degrees are the voltage and current out of phase with each other?

6. An R-L series circuit has an apparent power of 230 VA and a true power of 180 W. What is the reactive power?

7. The resistor in an R-L series circuit has a voltage drop of 53 V, and the inductor has a voltage drop of 28 V. What is the applied voltage of the circuit?

8. An R-L series circuit has a reactive power of 1234 VARs and an apparent power of 4329 VA. How many degrees are voltage and current out of phase with each other?

9. An R-L series circuit contains a resistor and an inductor. The resistor has a value of 6.5 Ω. The circuit is connected to 120 V and has a current flow of 12 A. What is the inductive reactance of this circuit?

10. What is the voltage drop across the resistor in the circuit in Question 9?

Practical Applications

*A*n AC electric motor is connected to a 240-V, 60-Hz source. A clamp-on am-meter with a peak hold function reveals that the motor has an inrush current of 34 A when the motor is first started. Your job is to reduce the inrush current to a value of 20 A by connecting a resistor in series with the motor. The resistor will be shunted out of the circuit after the motor is started. Using an ohmmeter, you find that the motor has a wire resistance of 3 Ω. How much resistance should be connected in series with the motor to reduce the starting current to 20 A? ■

Practical Applications

*Y*ou are a journeyman electrician working in an industrial plant. Your task is to connect an inductor to a 480-V, 60-Hz line. To determine the proper conduc-tor and fuse size for this installation, you need to know the amount of current the inductor will draw from the line. The nameplate on the inductor indicates that it has an inductance of 0.1 H. An ohmmeter reveals that it has a wire resistance of 10 Ω. How much current should this inductor draw when connected to the line? ■

Practice Problems

Refer to the circuit shown in *Figure 18–2* and the Resistive-Inductive Series Circuits section of the AC formulas listed in Appendix B.

1. Assume that the circuit shown in *Figure 18–2* is connected to a 480-V, 60-Hz line. The inductor has an inductance of 0.053 H, and the resistor has a resistance of 12 Ω.

E_T 480V	E_R _____	E_L _____
I_T _____	I_R _____	I_L _____
Z _____	R 12 Ω	X_L _____
VA _____	P _____	$VARS_L$ _____
PF _____	$\angle \theta$ _____	L 0.053 H

2. Assume that the voltage drop across the resistor, E_R, is 78 V, that the voltage drop across the inductor, E_L, is 104 V, and the circuit has a total impedance, Z, of 20 Ω. The frequency of the AC voltage is 60 Hz.

E_T _____	E_R 78 V	E_L 104 V
I_T _____	I_R _____	I_L _____
Z 20 Ω	R _____	X_L _____
VA _____	P _____	$VARS_L$ _____
PF _____	$\angle\theta$ _____	L _____

3. Assume the circuit shown in *Figure 18–2* has an apparent power of 144 VA and a true power of 115.2 W. The inductor has an inductance of 0.15915 H, and the frequency is 60 Hz.

E_T _____	E_R _____	E_L _____
I_T _____	I_R _____	I_L _____
Z _____	R _____	X_L _____
VA 144	P 115.2 W	$VARS_L$ _____
PF _____	$\angle\theta$ _____	L 0.15915 H

4. Assume the circuit shown in *Figure 18–2* has a power factor of 78%, an apparent power of 374.817 VA, and a frequency of 400 Hz. The inductor has an inductance of 0.0382 H.

E_T _____	E_R _____	E_L _____
I_T _____	I_R _____	I_L _____
Z _____	R _____	X_L _____
VA 374.817	P _____	$VARS_L$ _____
PF 78%	$\angle\theta$ _____	L 0.0382 H

5. In an RL series circuit, E_T = 240 volts, R = 60 Ω, and X_L = 75 Ω. Find E_L.

6. In an RL series circuit, E_T = 208 volts, R = 2.4 kΩ, and X_L = 1.5 kΩ. Find PF.

7. In an RL series circuit, E_T = 120 volts, R = 35 Ω, and X_L = 48 Ω. Find reactive power.

8. In an RL series circuit, the apparent power is 560 VA, PF = 62%. Find reactive power.

9. An RL series circuit is connected to a 60-Hz, 208-volt power line. A clamp-on ammeter indicates a current of 18 amperes. The inductor has a reactive power of 1860 VARs. What is the value of R?

10. In an RL series circuit, $Z = 88\ \Omega$, $R = 32\ \Omega$. Find X_L.

11. In an RL series circuit, apparent power = 450 VA, reactive power = 224 VARs. Find true power.

12. In an RL series circuit, $\angle\theta = 22°$, true power = 94 watts. Find reactive power.

13. An RL series circuit contains two resistors and two inductors. The resistors have values of 120 Ω and 300 Ω. The inductors have reactive values of 220 Ω and 470 Ω. Find impedance.

14. An RL series circuit contains two resistors and two inductors. The resistors dissipate powers of 96 watts and 125 watts. The inductors have reactive powers of 100 VARs and 78 VARs. What is the power factor?

15. An RL series circuit contains two resistors and two inductors. The resistors are 68 Ω and 124 Ω. The inductors have inductive reactances of 44 Ω and 225 Ω. The total voltage is 240 volts. Find the voltage drop across the 124-Ω resistor.

16. An RL series circuit contains two resistors and two inductors. The resistors are 86 kΩ and 68 kΩ. The inductors have inductive reactances of 24 kΩ and 56 kΩ. The total voltage is 480 volts. Find the voltage drop across the 56-kΩ inductor.

Unit 19

Resistive-Inductive Parallel Circuits

KEY TERMS

Current flow through
the inductor (I_L)
Current flow through
the resistor (I_R)

Why You Need to Know

*M*any times, resistive and inductive loads are connected in parallel with each other. This can be seen in every kind of electrical environment. Incandescent lights in a home are resistive loads, but they are often connected in parallel with a motor or transformer. A common heat-pump system in a home is a perfect example of this type circuit. The compressor and blower motors are inductive loads, but the electric heat strips are resistive loads. To understand how these different loads affect the circuit, you must understand resistive-inductive parallel circuits. This unit

- illustrates how to calculate the effect on voltage, current, and impedance when a parallel circuit contains both resistance and inductance.
- explains how to apply the relationship of power factor to true power in a parallel circuit containing resistive-inductive factors.

Objectives

After studying this unit, you should be able to

- discuss the operation of a parallel circuit containing resistance and inductance.

- calculate circuit values of an RL parallel circuit.

- connect an RL parallel circuit and measure circuit values with test instruments.

Preview

This unit discusses circuits that contain resistance and inductance connected in parallel with each other. Mathematical calculations are used to show the relationship of current and voltage on the entire circuit and the relationship of current through different branches of the circuit. ■

19–1   Resistive-Inductive Parallel Circuits

A circuit containing a resistor and an inductor connected in parallel is shown in *Figure 19–1*. Because the voltage applied to any device in parallel must be the same, the voltage applied to the resistor and inductor must be in phase and have the same value. The current flow through the inductor will be 90° out of phase with the voltage, and the current flow through the resistor will be in phase with the voltage *(Figure 19–2)*. This configuration produces a phase angle difference of 90° between the current flow through a pure inductive load and a pure resistive load *(Figure 19–3)*.

The amount of phase angle shift between the total circuit current and voltage is determined by the ratio of the amount of resistance to the amount of inductance. The circuit power factor is still determined by the ratio of true power to apparent power.

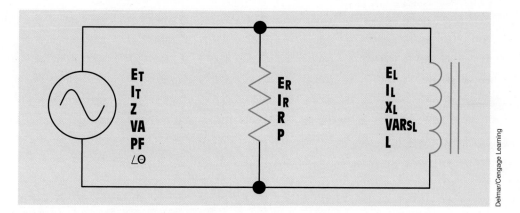

E_T
I_T
Z
VA
PF
$\angle\Theta$

E_R
I_R
R
P

E_L
I_L
X_L
$VARS_L$
L

Delmar/Cengage Learning

FIGURE 19–1   A resistive-inductive parallel circuit.

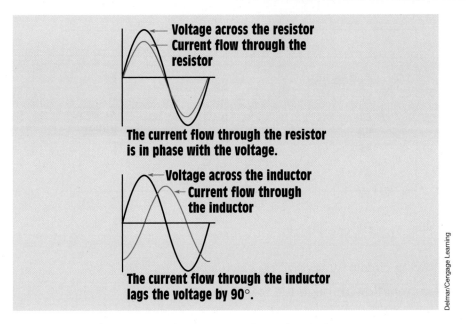

FIGURE 19–2 Relationship of voltage and current in an RL parallel circuit.

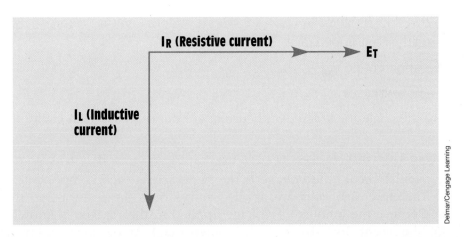

FIGURE 19–3 Resistive and inductive currents are 90° out of phase with each other in an RL parallel circuit.

19–2 Calculating Circuit Values

In the circuit shown in *Figure 19–4*, a resistance of 15 ohms is connected in parallel with an inductive reactance of 20 ohms. The circuit is connected to a voltage of 240 volts AC and a frequency of 60 hertz. In this example problem,

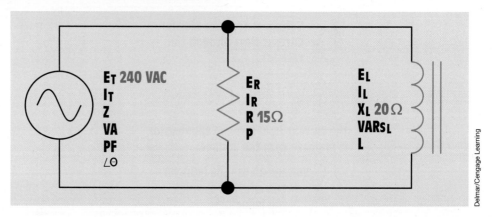

FIGURE 19–4 Typical RL parallel circuit.

the following circuit values will be calculated:

I_R—current flow through the resistor

P—watts (true power)

I_L—current flow through the inductor

VARs—reactive power

I_T—total circuit current

Z—total circuit impedance

VA—apparent power

PF—power factor

$\angle\theta$—the angle the voltage and current are out of phase with each other

Resistive Current

In any parallel circuit, the voltage is the same across each component in the circuit. Therefore, 240 volts are applied across both the resistor and the inductor. Because the amount of voltage applied to the resistor is known, the amount of **current flow through the resistor (I_R)** can be calculated by using the formula:

$$I_R = \frac{E}{R}$$

$$I_R = \frac{240\ V}{15\ \Omega}$$

$$I_R = 16\ A$$

Watts

True power (P), or watts (W), can be calculated using any of the watts formulas and pure resistive values. The amount of true power in this circuit is calculated using the formula:

$$P = E_R \times I_R$$
$$P = 240 \text{ V} \times 16 \text{ A}$$
$$P = 3840 \text{ W}$$

Inductive Current

Because the voltage applied to the inductor is known, the current flow can be found by dividing the voltage by the inductive reactance. The amount of **current flow through the inductor (I_L)** is calculated using the formula:

$$I_L = \frac{E}{X_L}$$
$$I_L = \frac{240 \text{ V}}{20 \text{ }\Omega}$$
$$I_L = 12 \text{ A}$$

VARs

The amount of reactive power (VARs) is calculated using the formula:

$$\text{VARs} = E_L \times I_L$$
$$\text{VARs} = 240 \text{ V} \times 12 \text{ A}$$
$$\text{VARs} = 2880$$

Inductance

Because the frequency and the inductive reactance are known, the inductance of the coil can be found using the formula:

$$L = \frac{X_L}{2\pi f}$$
$$L = \frac{20 \text{ }\Omega}{377}$$
$$L = 0.0531 \text{ H}$$

Total Current

The total current (I_T) flow through the circuit can be calculated by adding the current flow through the resistor and the inductor. Because these two currents

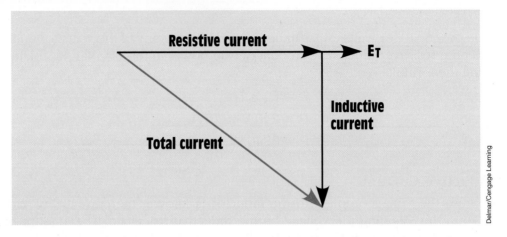

Delmar/Cengage Learning

FIGURE 19–5 Relationship of resistive, inductive, and total current in an RL parallel circuit.

are 90° out of phase with each other, vector addition is used. If these current values were plotted, they would form a right triangle similar to the one shown in *Figure 19–5*. Notice that the current flow through the resistor and inductor forms the legs of a right triangle and the total current is the hypotenuse Because the resistive and inductive currents form the legs of a right triangle and the total current forms the hypotenuse, the Pythagorean theorem can be used to add these currents together:

$$I_T = \sqrt{I_R{}^2 + I_L{}^2}$$
$$I_T = \sqrt{(16\ A)^2 + (12\ A)^2}$$
$$I_T = \sqrt{256\ A^2 + 144\ A^2}$$
$$I_T = \sqrt{400\ A^2}$$
$$I_T = 20\ A$$

The parallelogram method for plotting the total current is shown in *Figure 19–6*.

Impedance

Now that the total current and total voltage are known, the total impedance (Z) can be calculated by substituting Z for R in an Ohm's law formula. The total impedance of the circuit can be calculated using the formula:

$$Z = \frac{E}{I_T}$$
$$Z = \frac{240\ V}{20\ A}$$
$$Z = 12\ \Omega$$

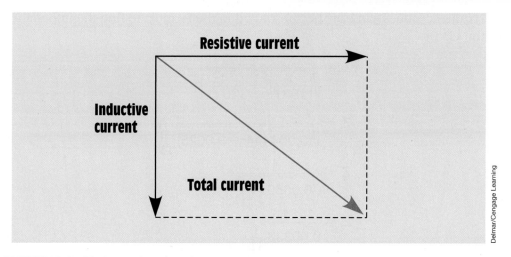

FIGURE 19–6   Plotting total current using the parallelogram method.

The value of impedance can also be found if total current and voltage are not known. In a parallel circuit, the reciprocal of the total resistance is equal to the sum of the reciprocals of each resistor. This same rule can be amended to permit a similar formula to be used in an RL parallel circuit. Because resistance and inductive reactance are 90° out of phase with each other, vector addition must be used when the reciprocals are added. The initial formula is:

$$\left(\frac{1}{Z}\right)^2 = \left(\frac{1}{R}\right)^2 + \left(\frac{1}{X_L}\right)^2$$

This formula states that the square of the reciprocal of the impedance is equal to the sum of the squares of the reciprocals of resistance and inductive reactance. To remove the square from the reciprocal of the impedance, take the square root of both sides of the equation:

$$\frac{1}{Z} = \sqrt{\left(\frac{1}{R}\right)^2 + \left(\frac{1}{X_L}\right)^2}$$

Notice that the formula can now be used to find the reciprocal of the impedance, not the impedance. To change the formula so that it is equal to the impedance, take the reciprocal of both sides of the equation:

$$Z = \frac{1}{\sqrt{\left(\frac{1}{R}\right)^2 + \left(\frac{1}{X_L}\right)^2}}$$

Numeric values can now be substituted in the formula to find the impedance of the circuit:

$$Z = \frac{1}{\sqrt{\left(\frac{1}{15\ \Omega}\right)^2 + \left(\frac{1}{20\ \Omega}\right)^2}}$$

$$Z = \frac{1}{\sqrt{(0.004444 + 0.0025)\ \frac{1}{\Omega^2}}}$$

$$Z = \frac{1}{\sqrt{0.006944\ \frac{1}{\Omega^2}}}$$

$$Z = \frac{1}{0.08333\ \frac{1}{\Omega}}$$

$$Z = 12\ \Omega$$

Another formula that can be used to determine the impedance of resistance and inductive reactance connected in parallel is:

$$Z = \frac{R \times X_L}{\sqrt{R^2 + X_L{}^2}}$$

Substituting the same values for resistance and inductive reactance in this formula will result in the same answer:

$$Z = \frac{15\ \Omega \times 20\ \Omega}{\sqrt{(15\ \Omega)^2 + (20\ \Omega)^2}}$$

$$Z = \frac{300\ \Omega}{\sqrt{625\ \Omega^2}}$$

$$Z = \frac{300\ \Omega}{25\ \Omega}$$

$$Z = 12\ \Omega$$

Apparent Power

The apparent power (VA) can be calculated by multiplying the circuit voltage by the total current flow. The relationship of volt-amperes, watts, and VARs is the same for an RL parallel circuit as it is for an RL series circuit. The reason is that power adds in any type of circuit. Because the true power and reactive power are 90° out of phase with each other, they form a right triangle with apparent power as the hypotenuse *(Figure 19–7)*.

$$VA = E_T \times I_T$$

$$VA = 240\ V \times 20\ A$$

$$VA = 4800$$

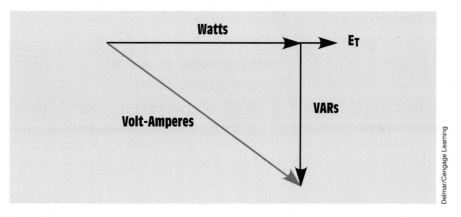

Delmar/Cengage Learning

FIGURE 19–7   Relationship of apparent power (volt-amperes), true power (watts), and reactive power (VARs) in an RL parallel circuit.

Power Factor

Power factor (PF) in an RL parallel circuit is the relationship of apparent power to the true power just as it was in the RL series circuit. There are some differences in the formulas used to calculate power factor in a parallel circuit, however. In an RL series circuit, power factor could be calculated by dividing the voltage dropped across the resistor by the total, or applied, voltage. In a parallel circuit, the voltage is the same, but the currents are different. Therefore, power factor can be calculated by dividing the current flow through the resistive parts of the circuit by the total circuit current:

$$PF = \frac{I_R}{I_T}$$

Another formula that changes involves resistance and impedance. In a parallel circuit, the total circuit impedance will be less than the resistance. Therefore, if power factor is to be calculated using impedance and resistance, the impedance must be divided by the resistance:

$$PF = \frac{Z}{R}$$

The circuit power factor in this example will be calculated using the formula:

$$PF = \frac{P}{VA} \times 100$$

$$PF = \frac{3840 \text{ W}}{4800 \text{ VA}} \times 100$$

$$PF = 0.80, \quad \text{or} \quad 80\%$$

Angle Theta

The cosine of angle theta ($\angle\theta$) is equal to the power factor:

$$\cos \angle\theta = 0.80$$
$$\angle\theta = 36.87°$$

A vector diagram using apparent power, true power, and reactive power is shown in *Figure 19–8*. Notice that angle theta is the angle produced by the apparent power and the true power. The relationship of current and voltage for this circuit is shown in *Figure 19–9*. The circuit with all values is shown in Figure *19–10*.

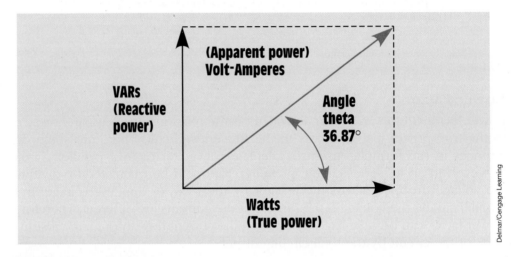

FIGURE 19–8 Angle theta.

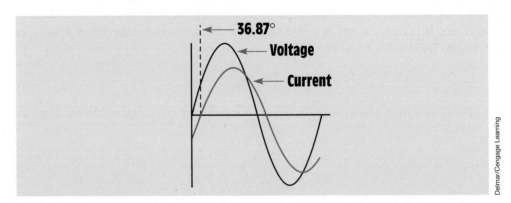

FIGURE 19–9 The current is 36.87° out of phase with the voltage.

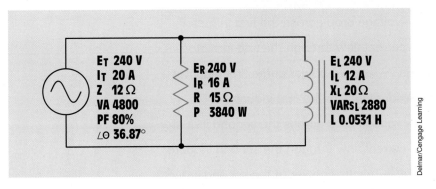

FIGURE 19–10 All values have been found.

■ EXAMPLE 19–1

In this circuit, one resistor is connected in parallel with two inductors *(Figure 19–11)*. The frequency is 60 Hz. The circuit has an apparent power of 6120 VA, the resistor has a resistance of 45 Ω, the first inductor has an inductive reactance of 40 Ω, and the second inductor has an inductive reactance of 60 Ω. It is assumed that both inductors have a Q greater than 10 and their resistance is negligible. Find the following missing values:

Z—total circuit impedance

I_T—total circuit current

E_T—applied voltage

E_R—voltage drop across the resistor

I_R—current flow through the resistor

P—watts (true power)

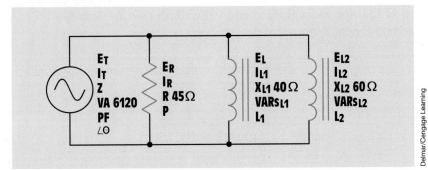

FIGURE 19–11 Example Circuit 2.

E_{L1}—voltage drop across the first inductor

I_{L1}—current flow through the first inductor

$VARS_{L1}$—reactive power of the first inductor

L_1—inductance of the first inductor

E_{L2}—voltage drop across the second inductor

I_{L2}—current flow through the second inductor

$VARS_{L2}$—reactive power of the second inductor

L_2—inductance of the second inductor

$\angle\theta$—the angle that voltage and current are out of phase with each other

Solution

Impedance

Before it is possible to calculate the impedance of the circuit, the total amount of inductive reactance for the circuit must be found. Because these two inductors are connected in parallel, the reciprocal of their inductive reactances must be added. This will give the reciprocal of the total inductive reactance:

$$\frac{1}{X_{LT}} = \frac{1}{X_{L1}} + \frac{1}{X_{L2}}$$

To find the total inductive reactance, take the reciprocal of both sides of the equation:

$$X_{LT} = \frac{1}{\dfrac{1}{X_{L1}} + \dfrac{1}{X_{L2}}}$$

Refer to the formulas for pure inductive circuits shown in the AC formulas section of the Appendix B. Numeric values can now be substituted in the formula to find the total inductive reactance:

$$X_{LT} = \frac{1}{\dfrac{1}{40\ \Omega} + \dfrac{1}{60\ \Omega}}$$

$$X_{LT} = \frac{1}{(0.025 + 0.01667)\dfrac{1}{\Omega}}$$

$$X_{LT} = 24\ \Omega$$

Now that the total amount of inductive reactance for the circuit is known, the impedance can be calculated using the formula:

$$Z = \frac{1}{\sqrt{\left(\frac{1}{R}\right)^2 + \left(\frac{1}{X_{LT}}\right)^2}}$$

$$Z = \frac{1}{\sqrt{\left(\frac{1}{45\ \Omega}\right)^2 + \left(\frac{1}{24\ \Omega}\right)^2}}$$

$$Z = 21.176\ \Omega$$

A diagram showing the relationship of resistance, inductive reactance, and impedance is shown in *Figure 19–12*.

$\mathbf{E_T}$

Now that the circuit impedance and the apparent power are known, the applied voltage can be calculated using the formula:

$$E_T = \sqrt{VA \times Z}$$

$$E_T = \sqrt{6120\ VA \times 21.176\ \Omega}$$

$$E_T = 359.996\ V$$

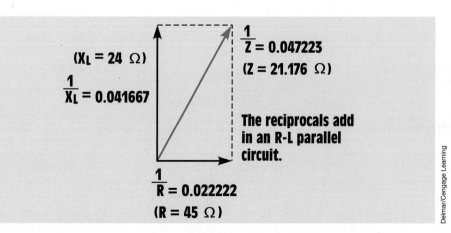

FIGURE 19–12   Relationship of resistance, inductive reactance, and impedance for Circuit 2.

E_R, E_{L1}, E_{L2}

In a parallel circuit, the voltage must be the same across any leg or branch. Therefore, 360 V is dropped across the resistor, the first inductor, and the second inductor:

$$E_R = 359.996 \text{ V}$$
$$E_{L1} = 359.996 \text{ V}$$
$$E_{L2} = 359.996 \text{ V}$$

I_T

The total current of the circuit can now be calculated using the formula:

$$I_T = \frac{E_T}{Z}$$

$$I_T = \frac{359.996 \text{ V}}{21.176 \text{ }\Omega}$$

$$I_T = 17 \text{ A}$$

The remaining values of the circuit can be found using Ohm's law. Refer to the Resistive-Inductive Parallel Circuits listed in the AC formula section in Appendix B.

I_R

$$I_R = \frac{E_R}{R}$$

$$I_R = \frac{359.996 \text{ V}}{45 \text{ }\Omega}$$

$$I_R = 8 \text{ A}$$

P

$$P = E_R \times I_R$$
$$P = 359.996 \text{ V} \times 8 \text{ }\Omega$$
$$P = 2879.968 \text{ W}$$

I_{L1}

$$I_{L1} = \frac{E_{L1}}{X_{L1}}$$

$$I_{L1} = \frac{359.996 \text{ V}}{40 \text{ }\Omega}$$

$$I_{L1} = 9 \text{ A}$$

VARs$_{L1}$

$$VARs_{L1} = E_{L1} \times I_{L1}$$
$$VARs_{L1} = 359.996 \text{ V} \times 9 \text{ A}$$
$$VARs_{L1} = 3239.964$$

L$_1$

$$L_1 = \frac{X_{L1}}{2\pi f}$$
$$L_1 = \frac{40 \ \Omega}{377}$$
$$L_1 = 0.106 \text{ H}$$

I$_{L2}$

$$I_{L2} = \frac{E_{L2}}{X_{L2}}$$
$$I_{L2} = \frac{359.996 \text{ V}}{60 \ \Omega}$$
$$I_{L2} = 6 \text{ A}$$

VARs$_{L2}$

$$VARs_{L2} = E_{L2} \times I_{L2}$$
$$VARs_{L2} = 359.996 \times 6$$
$$VARs_{L2} = 2159.976$$

L$_2$

$$L_2 = \frac{X_{L2}}{2\pi f}$$
$$L_2 = \frac{60 \ \Omega}{377}$$
$$L_2 = 0.159 \text{ H}$$

PF

$$PF = \frac{W}{VA}$$

$$PF = \frac{2879.968 \text{ W}}{6120 \text{ VA}}$$

$$PF = 47.06\%$$

∠θ

$$\cos \angle\theta = PF$$

$$\cos \angle\theta = 0.4706$$

$$\angle\theta = 61.93°$$

A vector diagram showing angle theta is shown in *Figure 19–13*. The vectors used are those for apparent power, true power, and reactive power. The phase relationship of voltage and current for this circuit is shown in *Figure 19–14*, and the circuit with all completed values is shown in *Figure 19–15*.

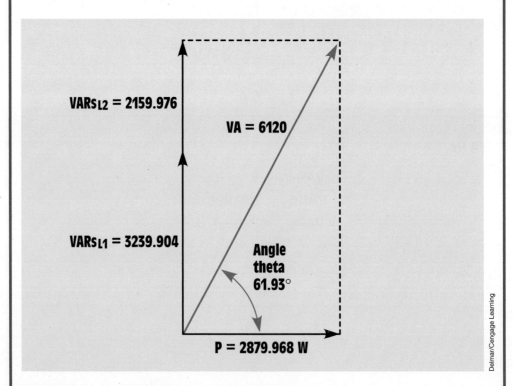

VARS$_{L2}$ = 2159.976

VA = 6120

VARS$_{L1}$ = 3239.904

Angle theta 61.93°

P = 2879.968 W

FIGURE 19–13   Angle theta determined by power vectors.

Delmar/Cengage Learning

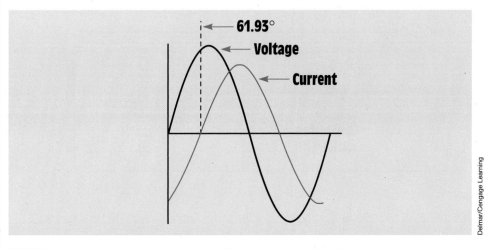

FIGURE 19–14   Voltage and current are 61.93° out of phase with each other.

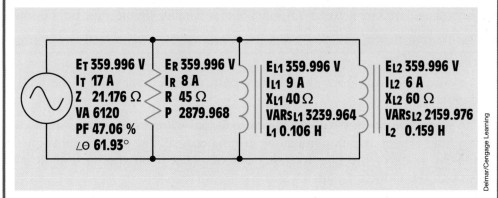

FIGURE 19–15   Completed values for Circuit 2.

Practical Applications

RL parallel circuits exist throughout the electrical field. Many circuits contain both inductive and resistive loads connected to the same branch circuit *(Figure 19–16)*. A ceiling fan, for example, often has a light kit attached to the fan. The fan is an inductive load, and the incandescent lamps are a resistive load. Branch circuits can also contain incandescent lamps, fluorescent lights, and motors connected on the same circuit. All these loads are connected in parallel. ■

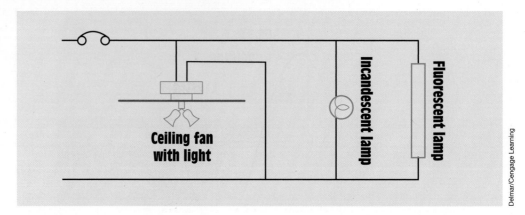

FIGURE 19–16 Branch circuits often contain resistive and inductive loads connected in parallel.

Summary

- The voltage applied across components in a parallel circuit must be the same.

- The current flowing through resistive parts of the circuit will be in phase with the voltage.

- The current flowing through inductive parts of the circuit will lag the voltage by 90°.

- The total current in a parallel circuit is equal to the sum of the individual currents. Vector addition must be used because the current through the resistive parts of the circuit is 90° out of phase with the current flowing through the inductive parts.

- The impedance of an RL parallel circuit can be calculated by using vector addition to add the reciprocals of the resistance and inductive reactance.

- Apparent power, true power, and reactive power add in any kind of a circuit. Vector addition must be used, however, because true power and reactive power are 90° out of phase with each other.

Review Questions

1. When an inductor and a resistor are connected in parallel, how many degrees out of phase are the current flow through the resistor and the current flow through the inductor?

2. An inductor and resistor are connected in parallel to a 120-V, 60-Hz line. The resistor has a resistance of 50 ohms, and the inductor has an inductance of 0.2 H. What is the total current flow through the circuit?

3. What is the impedance of the circuit in Question 2?

4. What is the power factor of the circuit in Question 2?

5. How many degrees out of phase are the current and voltage in Question 2?

6. In the circuit shown in *Figure 19–1*, the resistor has a current flow of 6.5 A and the inductor has a current flow of 8 A. What is the total current in this circuit?

7. A resistor and an inductor are connected in parallel. The resistor has a resistance of 24 ohms, and the inductor has an inductive reactance of 20 ohms. What is the impedance of this circuit?

8. The RL parallel circuit shown in *Figure 19–1* has an apparent power of 325 VA. The circuit power factor is 66%. What is the true power in this circuit?

9. The RL parallel circuit shown in *Figure 19–1* has an apparent power of 465 VA and a true power of 320 W. What is the reactive power?

10. How many degrees out of phase are the total current and voltage in Question 9?

Practical Applications

*I*ncandescent lighting of 500 W is connected in parallel with an inductive load. A clamp-on ammeter reveals a total circuit current of 7 A. What is the inductance of the load connected in parallel with the incandescent lights? Assume a voltage of 120 V and a frequency of 60 Hz. ∎

Practical Applications

*Y*ou are working on a residential heat pump. The heat pump is connected to a 240-V, 60-Hz powerline. The compressor has a current draw of 34 amperes when operating. The compressor has a power factor of 70%. The back-up strip heat is rated at 10 kW. You need to know the amount of total current draw that will occur if the strip heat comes on while the compressor is operating. ∎

Practice Problems

Refer to the circuit shown in *Figure 19–1*. Use the AC formulas in the Resistive-Inductive Parallel Circuits section of Appendix B.

1. Assume that the circuit shown in *Figure 19–1* is connected to a 60-Hz line and has a total current flow of 34.553 A. The inductor has an inductance of 0.02122 H, and the resistor has a resistance of 14 Ω.

E_T _____	E_R _____	E_L _____
I_T 34.553A _____	I_R _____	I_L _____
Z _____	R 14 Ω _____	X_L _____
VA _____	P _____	VARS$_L$ _____
PF _____	$\angle\theta$ _____	L 0.02122 H

2. Assume that the current flow through the resistor, I_R, is 15 A; the current flow through the inductor, I_L, is 36 A; and the circuit has an apparent power of 10,803 VA. The frequency of the AC voltage is 60 Hz.

E_T _____	E_R _____	E_L _____
I_T _____	I_R 15 A _____	I_L 36 A
Z _____	R _____	X_L _____
VA 10,803	P _____	VARS$_L$ _____
PF _____	$\angle\theta$ _____	L _____

3. Assume that the circuit in *Figure 19–1* has an apparent power of 144 VA and a true power of 115.2 W. The inductor has an inductance of 0.15915 H, and the frequency is 60 Hz.

E_T _____	E_R _____	E_L _____
I_T _____	I_R _____	I_L _____
Z _____	R _____	X_L _____
VA 144 _____	P 115.2 _____	VARS$_L$ _____
PF _____	$\angle\theta$ _____	L 0.15915 H _____

4. Assume that the circuit in *Figure 19–1* has a power factor of 78%, an apparent power of 374.817 VA, and a frequency of 400 Hz. The inductor has an inductance of 0.0382 H.

E_T _____	E_R _____	E_L _____
I_T _____	I_R _____	I_L _____

Z _____   R _____ X_L _____

VA 374.817   P _____ $VARS_L$ _____

PF 78% $\angle\theta$ _____   L 0.0382 H

5. In an RL parallel circuit, R = 240 Ω and X_L = 360 Ω. Find impedance.

6. In an RL parallel circuit, I_T = 0.25 amps, I_R = 0.125 amps. The inductor has a reactive power of 75 VARs. What is E_T?

7. In an RL parallel circuit, E_T = 120 volts, R = 120 Ω, X_L = 150 Ω. Find I_L.

8. In an RL parallel circuit, E_T = 48 volts, I_T = 0.25 amps, R = 320 Ω. Find X_L.

9. In an RL parallel circuit, E_T = 240 volts, R = 560 Ω, and X_L = 330 Ω. Find reactive power.

10. In an RL parallel circuit, E_T = 240 volts, R = 560 Ω, and X_L = 330 Ω. Find apparent power.

11. In an RL parallel circuit, E_T = 208 volts, R = 2.4 kΩ, and X_L = 1.8 kΩ. Find I_T.

12. In an RL parallel circuit, E_T = 480 volts, R = 16 Ω, and X_L = 24 Ω. Find PF.

13. In an RL parallel circuit, I_T = 1.25 amps, R = 1.2 kΩ, and X_L = 1 kΩ. Find I_R.

14. In an RL parallel circuit, true power = 4.6 watts and reactive power = 5.4 VARs. What is the apparent power?

15. An RL parallel circuit is connected to 240 volts at 60 Hz. The resistor has a resistance of 68 Ω, and the inductor has an inductive reactance of 48 Ω. What is the reactance of the inductor?

16. An RL parallel circuit has an applied voltage of 208 volts and a total current of 2 amperes. The resistor has a value 180 Ω. Find I_L.

VII

AC Circuits Containing Capacitors

KEY TERMS

Capacitor
Dielectric
Dielectric constant
Dielectric stress
Electrolytic
Exponential
Farad
HIPOT
JAN (Joint Army-
  Navy) standard
Leakage current
Nonpolarized
  capacitors
Plates
Polarized capacitors
RC time constant
Surface area
Variable capacitors

Unit 20
Capacitors

Why You Need to Know

Capacitors are one of the major electric devices. This unit

■ explains how capacitors are constructed, how they are charged and discharged, the differences between different types of capacitors, and their markings.
■ describes how capacitance is measured and the importance of the voltage rating.
■ explains how capacitors store energy in an electrostatic field and how current can flow only when the capacitors are charging or discharging.

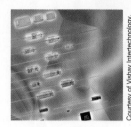

Objectives

After studying this unit, you should be able to

■ list the three factors that determine the capacitance of a capacitor.

■ discuss the electrostatic charge.

■ discuss the differences between nonpolarized and polarized capacitors.

■ calculate values for series and parallel connections of capacitors.

■ calculate an RC time constant.

Preview

Capacitors perform a variety of jobs such as power factor correction, storing an electric charge to produce a large current pulse, timing circuits, and electronic filters. Capacitors can be nonpolarized or polarized depending on the application. Nonpolarized capacitors can be used in both AC and DC circuits, whereas polarized capacitors can be used in DC circuits only. Both types are discussed in this unit. ■

20–1   Capacitors

CAUTION: *It is the habit of some people to charge a capacitor to high voltage and then hand it to another person. Although some people think this is comical, it is an extremely dangerous practice. Capacitors have the ability to supply an almost infinite amount of current. Under some conditions, a capacitor can have enough energy to cause a person's heart to go into fibrillation.* ■

This statement is not intended to strike fear into the heart of anyone working in the electrical field. It is intended to make you realize the danger that capacitors can pose under certain conditions.

Capacitors *are devices that oppose a change of voltage.* The simplest type of capacitor is constructed by separating two metal **plates** by some type of insulating material called the **dielectric** (Figure 20–1). Figure 20–2 shows the capacitor symbol. Three factors determine the capacitance of a capacitor:

1. the surface area of the plates

2. the distance between the plates

3. the type of dielectric used

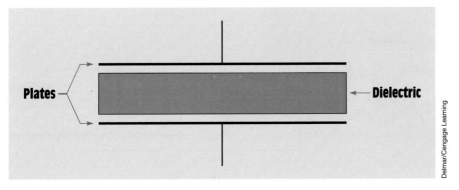

FIGURE 20–1 A capacitor is made by separating two metal plates with a dielectric.

FIGURE 20–2 Symbol generally used to indicate a capacitor.

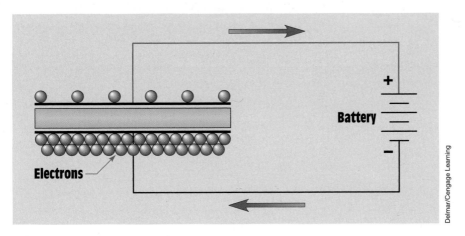

FIGURE 20–3 A capacitor can be charged by removing electrons from one plate and depositing electrons on the other plate.

The greater the **surface area** of the plates, the more capacitance a capacitor will have. If a capacitor is charged by connecting it to a DC source *(Figure 20–3)*, electrons are removed from the plate connected to the positive battery terminal and are deposited on the plate connected to the negative terminal.

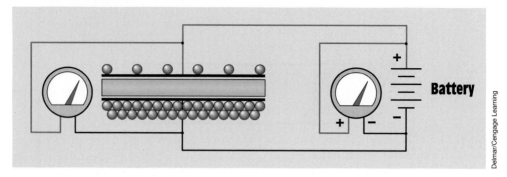

FIGURE 20–4 Current flows until the voltage across the capacitor is equal to the voltage of the battery.

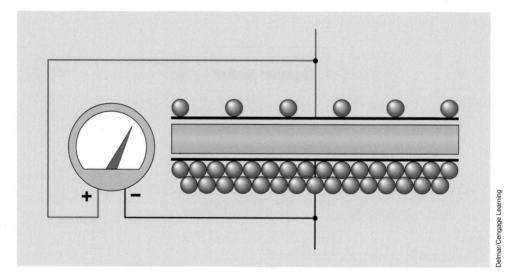

FIGURE 20–5 The capacitor remains charged after the battery is removed from the circuit.

This flow of current continues until a voltage equal to the battery voltage is established across the plates of the capacitor *(Figure 20–4)*. When these two voltages become equal, the flow of electrons stops. The capacitor is now charged. If the battery is disconnected from the capacitor, the capacitor will remain charged as long as there is no path by which the electrons can move from one plate to the other *(Figure 20–5)*. A good rule to remember concerning a capacitor and current flow is that ***current can flow only during the period of time that a capacitor is either charging or discharging.***

In theory, it should be possible for a capacitor to remain in a charged condition forever. In actual practice, however, it cannot. No dielectric is a perfect

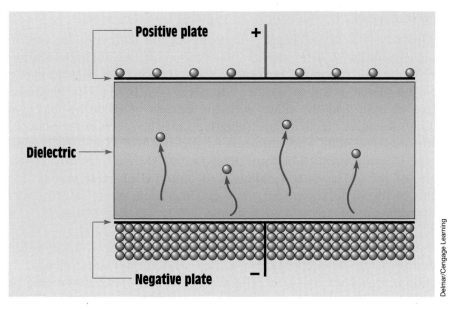

Positive plate +

Dielectric

Negative plate −

Delmar/Cengage Learning

FIGURE 20–6 Electrons eventually leak through the dielectric. This flow of electrons is known as leakage current.

insulator, and electrons eventually move through the dielectric from the negative plate to the positive, causing the capacitor to discharge *(Figure 20–6)*. This current flow through the dielectric is called **leakage current** and is proportional to the resistance of the dielectric and the charge across the plates. If the dielectric of a capacitor becomes weak, it will permit an excessive amount of leakage current to flow. A capacitor in this condition is often referred to as a *leaky capacitor.*

20–2 Electrostatic Charge

Two other factors that determine capacitance are the type of dielectric used and the distance between the plates. To understand these concepts, it is necessary to understand how a capacitor stores energy. In previous units, it was discussed that an inductor stores energy in the form of a magnetic field. A capacitor stores energy in an electrostatic field.

The term *electrostatic* refers to electric charges that are stationary, or not moving. They are very similar to the static electric charges that form on objects that are good insulators, as discussed in Unit 3. The electrostatic field is formed when electrons are removed from one plate and deposited on the other.

Dielectric Stress

When a capacitor is not charged, the atoms of the dielectric are uniform as shown in *Figure 20–7*. The valence electrons orbit the nucleus in a circular pattern. When the capacitor becomes charged, however, a potential exists between the plates of the capacitor. The plate with the lack of electrons has a positive charge, and the plate with the excess of electrons has a negative charge. Because electrons are negative particles, they are repelled away from the negative plate and attracted to the positive plate. This attraction causes the electron orbit to become stretched as shown in *Figure 20–8*. This stretching of the atoms of the dielectric is called **dielectric stress.** Placing the atoms of the dielectric under stress has the same effect as drawing back a bowstring with an arrow and holding it *(Figure 20–9);* that is, it stores energy.

The amount of dielectric stress is proportional to the voltage difference between the plates. The greater the voltage, the greater the dielectric stress. If the voltage becomes too great, the dielectric will break down and permit current to flow between the plates. At this point the capacitor becomes shorted. Capacitors have a voltage rating that should not be exceeded. The voltage rating indicates the maximum amount of voltage the dielectric is intended to withstand without breaking down. The amount of voltage

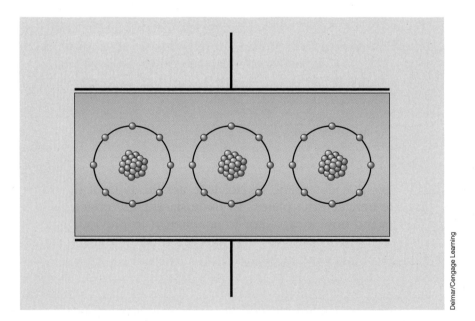

FIGURE 20–7 Atoms of the dielectric in an uncharged capacitor.

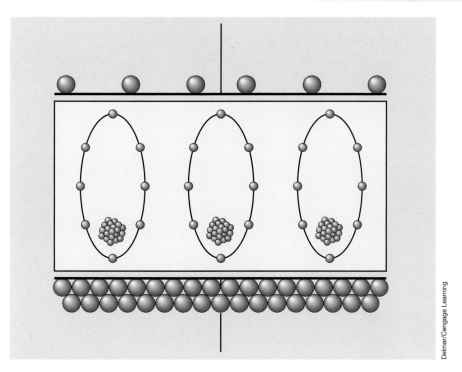

FIGURE 20–8   Atoms of the dielectric in a charged capacitor.

Delmar/Cengage Learning

applied to a capacitor is critical to its life span. Capacitors operated above their voltage rating will fail relatively quickly. Many years ago, the U.S. military made a study of the voltage rating of a capacitor relative to its life span. The results showed that a capacitor operated at one half its rated voltage will have a life span approximately eight times longer than a capacitor operated at the rated voltage.

The energy of the capacitor is stored in the dielectric in the form of an electrostatic charge. It is this electrostatic charge that permits the capacitor to produce extremely high currents under certain conditions. If the leads of a capacitor are shorted together, it has the effect of releasing the drawn-back bowstring *(Figure 20–9)*. When the bowstring is released, the arrow is propelled forward at a high rate of speed. The same is true for the capacitor. When the leads are shorted, the atoms of the dielectric snap back to their normal position. Shorting causes the electrons on the negative plate to be literally blown off and attracted to the positive plate. Capacitors can produce currents of thousands of amperes for short periods of time.

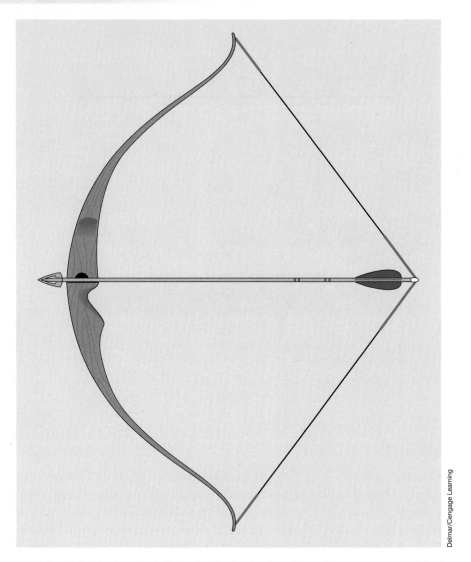

FIGURE 20–9 Dielectric stress is similar to drawing back a bowstring with an arrow and holding it.

This principle is used to operate the electronic flash of many cameras. Electronic flash attachments contain a small glass tube filled with a gas called xenon. Xenon produces a very bright white light similar to sunlight when the gas is ionized. A large amount of power is required, however, to produce a bright flash. A battery capable of directly ionizing the xenon would be very large and expensive and would have a potential of about 500 volts. The simple circuit shown in *Figure 20–10* can be used to overcome the problem. In this circuit, two small 1.5-volt batteries are connected to an oscillator. The oscillator

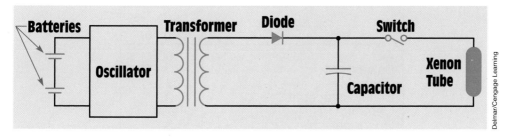

FIGURE 20–10   Energy is stored in a capacitor.

changes the DC of the batteries into square wave AC. The AC is then connected to a transformer, and the voltage is increased to about 500 volts peak.
A diode changes the AC voltage back into DC and charges the capacitor. The
capacitor charges to the peak value of the voltage waveform. When the switch
is closed, the capacitor suddenly discharges through the xenon tube and supplies the power needed to ionize the gas. It may take several seconds to store
enough energy in the capacitor to ionize the gas in the tube, but the capacitor
can release the stored energy in a fraction of a second.

To understand how the capacitor can supply the energy needed, consider
the amount of gunpowder contained in a 0.357 cartridge. If the powder were to
be removed from the cartridge and burned in the open air, it would be found
that the actual amount of energy contained in the powder is very small. This
amount of energy would not even be able to raise the temperature by a noticeable amount in a small enclosed room. If this same amount of energy is converted into heat in a fraction of a second, however, enough force is developed
to propel a heavy projectile with great force. This same principle is at work
when a capacitor is charged over some period of time and then discharged in
a fraction of a second.

20–3   Dielectric Constant

Because much of the capacitor's energy is stored in the dielectric, the type of
dielectric used is extremely important in determining the amount of capacitance a capacitor will have. Different materials are assigned a number called
the **dielectric constant.** Vacuum is assigned the number 1 and is used as a
reference. Air has a dielectric constant of approximately 1.0006. For example,
assume that a capacitor uses vacuum as the dielectric and is found to have a
capacitance of 1 microfarad (1 μF). Now assume that some material is placed
between the plates without changing the spacing and the capacitance value
becomes 5 microfarad. This material has a dielectric constant of 5. A chart
showing the dielectric constant of different materials is shown in *Figure 20–11*.

Material	Dielectric constant
Air	1.006
Bakelite	4.0 –10.0
Castor oil	4.3 – 4.7
Cellulose acetate	7.0
Ceramic	1200
Dry paper	3.5
Hard rubber	2.8
Insulating oils	2.2 – 4.6
Lucite	2.4 – 3.0
Mica	6.4 – 7.0
Mycalex	8.0
Paraffin	1.9 – 2.2
Porcelain	5.5
Pure water	81
Pyrex glass	4.1 – 4.9
Rubber compounds	3.0 – 7.0
Teflon	2
Titanium dioxide compounds	90 – 170

Delmar/Cengage Learning

FIGURE 20–11   Dielectric constant of different materials.

20–4   Capacitor Ratings

The basic unit of capacitance is the **farad** and is symbolized by the letter *F*. It receives its name from a famous scientist named Michael Faraday. *A capacitor has a capacitance of one farad when a change of 1 volt across its plates results in a movement of 1 coulomb:*

$$Q = C \times V$$

where

$$Q = \text{charge in coulombs}$$
$$C = \text{capacitance in farads}$$
$$V = \text{charging voltage}$$

Although the farad is the basic unit of capacitance, it is seldom used because it is an extremely large amount of capacitance. The following formula can be used to determine the capacitance of a capacitor when the area of the plates, the dielectric constant, and the distance between the plates are known:

$$C = \frac{K \times A}{4.45\ D}$$

where

$$C = \text{capacitance in pF (picofarads)}$$
$$K = \text{dielectric constant (F/in.)}$$
$$A = \text{area of one plate in sq. in.}$$
$$D = \text{distance between the plates in in.}$$

■ EXAMPLE 20-1

What would be the plate area of a 1-F (farad) capacitor if air is used as the dielectric and the plates are separated by a distance of 1 in.?

Solution

The first step is to convert the preceding formula to solve for area:

$$A = \frac{C \times 4.45 \times D}{K}$$

$$A = \frac{1{,}000{,}000{,}000{,}000 \times 4.45 \times 1}{1.006}$$

$$A = 4{,}447{,}000{,}000{,}000 \text{ sq. in.}$$
$$A = 1107.7 \text{ sq. miles}$$

Because the basic unit of capacitance is so large, other units such as the microfarad (μF), nanofarad (nF), and picofarad (pF) are generally used:

$$\mu F = \frac{1}{1{,}000{,}000} \ (1 \times 10^{-6}) \text{ of a farad}$$

$$nF = \frac{1}{1{,}000{,}000{,}000} \ (1 \times 10^{-9}) \text{ of a farad}$$

$$pF = \frac{1}{1{,}000{,}000{,}000{,}000} \ (1 \times 10^{-12}) \text{ of a farad}$$

The picofarad is sometimes referred to as a micro-microfarad and is symbolized by μμF.

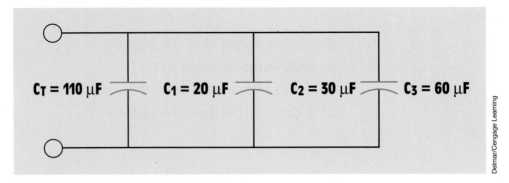

FIGURE 20–12 Capacitors connected in parallel.

20–5 Capacitors Connected in Parallel

Connecting capacitors in parallel *(Figure 20–12)* has the same effect as increasing the plate area of one capacitor. In the example shown, three capacitors having a capacitance of 20 microfarads, 30 microfarads, and 60 microfarads are connected in parallel. The total capacitance of this connection is

$$C_T = C_1 + C_2 + C_3$$
$$C_T = 20\ \mu F + 30\ \mu F + 60\ \mu F$$
$$C_T = 110\ \mu F$$

20–6 Capacitors Connected in Series

Connecting capacitors in series *(Figure 20–13)* has the effect of increasing the distance between the plates, thus reducing the total capacitance of the circuit. The total capacitance can be calculated in a manner similar to calculating

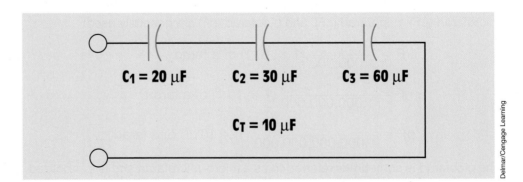

FIGURE 20–13 Capacitors connected in series.

parallel resistance. The following formulas can be used to find the total capacitance when capacitors are connected in series:

$$C_T = \cfrac{1}{\cfrac{1}{C_1} + \cfrac{1}{C_2} + \cfrac{1}{C_3}}$$

or

$$C_T = \frac{C_1 \times C_2}{C_1 + C_2}$$

or

$$C_T = \frac{C}{N}$$

where

$$C = \text{capacitance of one capacitor}$$
$$N = \text{number of capacitors connected in series}$$

Note: The last formula can be used only when all the capacitors connected in series are of the same value.

■ EXAMPLE 20-2

What is the total capacitance of three capacitors connected in series if C_1 has a capacitance of 20 μF, C_2 has a capacitance of 30 μF, and C_3 has a capacitance of 60 μF?

Solution

$$C_T = \cfrac{1}{\cfrac{1}{C_1} + \cfrac{1}{C_2} + \cfrac{1}{C_3}}$$

$$C_T = \cfrac{1}{\cfrac{1}{20 \ \mu F} + \cfrac{1}{30 \ \mu F} + \cfrac{1}{60 \ \mu F}}$$

$$C_T = 10 \ \mu F$$

20–7   Capacitive Charge and Discharge Rates

Capacitors charge and discharge at an **exponential** rate. A charge curve for a capacitor is shown in *Figure 20–14*. The curve is divided into five time constants, and during each time constant the voltage changes by an amount equal to 63.2% of the maximum amount that it can change. In *Figure 20–14,* it is assumed that a capacitor is to be charged to a total of 100 volts. At the end of the first time constant, the voltage has reached 63.2% of 100, or 63.2 volts. At the end of the second time constant, the voltage reaches 63.2% of the remaining voltage, or 86.4 volts. This pattern continues until the capacitor has been charged to 100 volts.

The capacitor discharges in the same manner *(Figure 20–15)*. At the end of the first time constant, the voltage will decrease by 63.2% of its charged value. In this example, the voltage decreases from 100 volts to 36.8 volts in the first time constant. At the end of the second time constant, the voltage will drop to 13.6 volts; and by the end of the third time constant, the voltage will drop to 5 volts. The voltage will continue to drop at this rate until it reaches approximately 0 after five time constants.

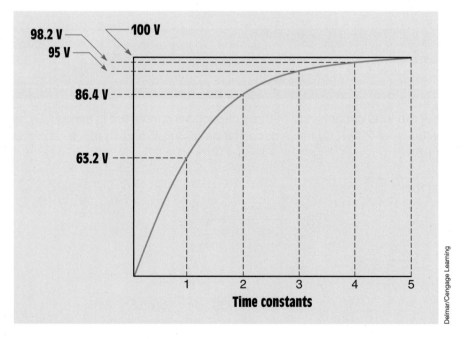

FIGURE 20–14   Capacitors charge at an exponential rate.

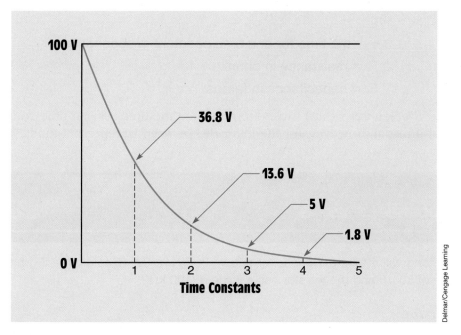

FIGURE 20–15 Capacitor discharge curve.

20–8 RC Time Constants

When a capacitor is connected in a circuit with a resistor, the amount of time needed to charge the capacitor—that is, the **RC time constant**—can be determined very accurately *(Figure 20–16)*. The formula for determining charge time is

$$\tau = R \times C$$

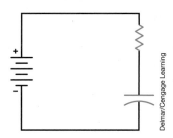

FIGURE 20–16 The charge time of the capacitor can be determined very accurately.

where

τ = the time for one time constant in seconds

R = resistance in ohms

C = capacitance in farads

The Greek letter τ (tau) is used to represent the time for one time constant. It is not unusual, however, for the letter t to be used to represent time.

■ EXAMPLE 20-3

How long will it take the capacitor shown in *Figure 20–15* to charge if it has a value of 50 μF and the resistor has a value of 100 kΩ?

Solution

$$\tau = R \times C$$
$$\tau = 0.000050 \text{ F} \times 100,000 \text{ }\Omega$$
$$\tau = 5 \text{ s}$$

The formula is used to find the time for one time constant. Five time constants are required to charge the capacitor:

$$\text{Total time} = 5 \text{ s} \times 5 \text{ time constants}$$
$$\text{Total time} = 25 \text{ s}$$

■ EXAMPLE 20-4

How much resistance should be connected in series with a 100-pF capacitor to give it a total charge time of 0.2 s?

Solution

Change the preceding formula to solve for resistance:

$$R = \frac{\tau}{C}$$

The total charge time is to be 0.2 s. The value of τ is therefore 0.2/5 = 0.04 s. Substitute these values in the formula:

$$R = \frac{0.04 \text{ s}}{100^{\times 10^{-12}} \text{ F}}$$

$$R = 400 \text{ M}\Omega$$

■ EXAMPLE 20-5

A 500-kΩ resistor is connected in series with a capacitor. The total charge time of the capacitor is 15 s. What is the capacitance of the capacitor?

Solution

Change the base formula to solve for the value of capacitance:

$$C = \frac{\tau}{R}$$

Because the total charge time is 15 s, the time of one time constant will be 3 s (15 s/5 = 3 s):

$$C = \frac{3 \text{ s}}{500,000 \text{ }\Omega}$$

$$C = 0.000006\text{F}$$

or

$$C = 6 \text{ }\mu\text{F}$$

20–9 Applications for Capacitors

Capacitors are among the most used of electric components. They are used for power factor correction in industrial applications; in the start windings of many single-phase AC motors; to produce phase shifts for SCR and Triac circuits; to filter pulsating DC; and in RC timing circuits. (SCRs and Triacs are solid-state electronic devices used throughout industry to control high-current circuits.) Capacitors are used extensively in electronic circuits for control of frequency and pulse generation. The type of capacitor used is dictated by the circuit application.

20–10 Nonpolarized Capacitors

Capacitors can be divided into two basic groups, nonpolarized and polarized. **Nonpolarized capacitors** are often referred to as *AC capacitors* because they are not sensitive to polarity connection. Nonpolarized capacitors can be connected to either DC or AC circuits without harm to the capacitor. Nonpolarized capacitors are constructed by separating metal plates by some type of dielectric *(Figure 20–1)*. These capacitors can be obtained in many different styles and case types *(Figure 20–17)*.

A common type of AC capacitor called the paper capacitor or oil-filled paper capacitor is often used in motor circuits and for power factor correction *(Figure 20–18)*. It derives its name from the type of dielectric used. This capacitor is constructed by separating plates made of metal foil with thin sheets

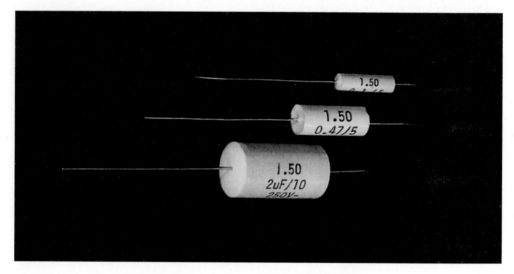

FIGURE 20–17 Nonpolarized capacitors. *(Courtesy of Vishay Intertechnology)*

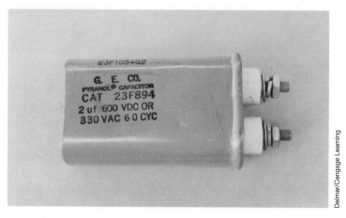

FIGURE 20–18 Oil-filled paper capacitor.

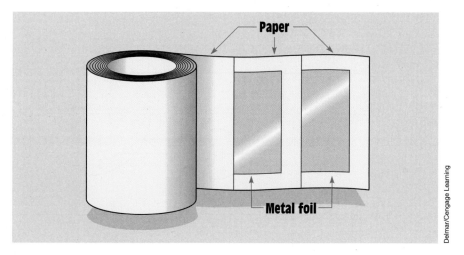

FIGURE 20–19 Oil-filled paper capacitor.

of paper soaked in a dielectric oil *(Figure 20–19)*. These capacitors are often used as the run or starting capacitors for single-phase motors. Many manufacturers of oil-filled capacitors identify one terminal with an arrow, a painted dot, or a stamped dash in the capacitor can *(Figure 20–20)*. This identified terminal marks the connection to the plate that is located nearer to the metal container or can. It has long been known that when a capacitor's dielectric breaks down and permits a short circuit to ground, the plate nearer to the outside case most

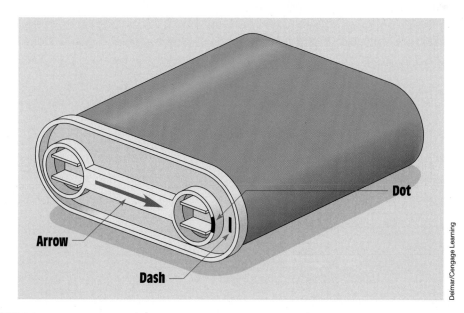

FIGURE 20–20 Marks indicate plate nearest capacitor case.

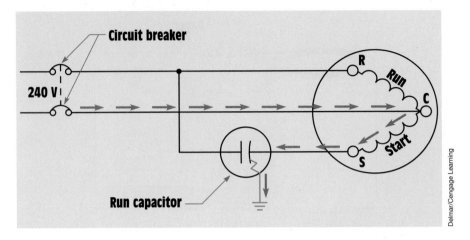

FIGURE 20–21   Identified capacitor terminal connected to motor start winding (incorrect connection).

often becomes grounded. For this reason, it is generally desirable to connect the identified capacitor terminal to the line side instead of to the motor start winding.

In *Figure 20–21,* the run capacitor has been connected in such a manner that the identified terminal is connected to the start winding of a single-phase motor. If the capacitor should become shorted to ground, a current path exists through the motor start winding. The start winding is an inductive-type load, and inductive reactance will limit the value of current flow to ground. Because the flow of current is limited, it will take the circuit breaker or fuse some time to open the circuit and disconnect the motor from the power line. This time delay can permit the start winding to overheat and become damaged.

In *Figure 20–22,* the run capacitor has been connected in the circuit in such a manner that the identified terminal is connected to the line side. If the

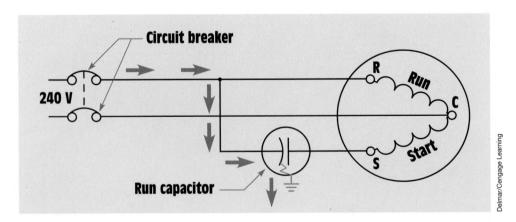

FIGURE 20–22   Identified terminal connected to the line (correct connections).

capacitor should become shorted to ground, a current path exists directly to ground, bypassing the motor start winding. When the capacitor is connected in this manner, the start winding does not limit current flow and permits the fuse or circuit breaker to open the circuit almost immediately.

20–11 Polarized Capacitors

Polarized capacitors are generally referred to as **electrolytic** capacitors. These capacitors are sensitive to the polarity they are connected to and have one terminal identified as positive or negative *(Figure 20–23)*. Polarized capacitors can be used in DC circuits only. If their polarity connection is reversed, the capacitor can be damaged and will sometimes explode. The advantage of electrolytic capacitors is that they can have very high capacitance in a small case.

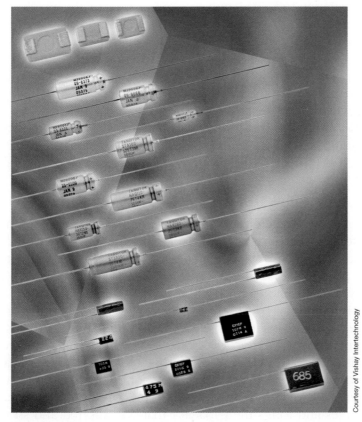

Courtesy of Vishay Intertechnology

FIGURE 20–23   Polarized capacitors.

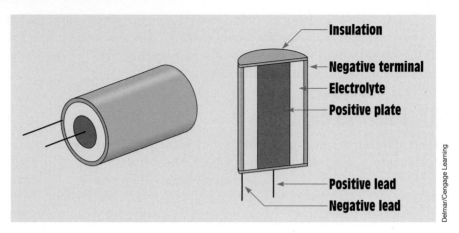

FIGURE 20–24   Wet-type electrolytic capacitor.

The two basic types of electrolytic capacitors are the wet type and the dry type. The wet-type electrolytic capacitor *(Figure 20–24)* has a positive plate made of aluminum foil. The negative plate is actually an electrolyte made from a borax solution. A second piece of aluminum foil is placed in contact with the electrolyte and becomes the negative terminal. When a source of DC is connected to the capacitor, the borax solution forms an insulating oxide film on the positive plate. This film is only a few molecules thick and acts as the insulator to separate the plates. The capacitance is very high because the distance between the plates is so small.

If the polarity of the wet-type electrolytic capacitor becomes reversed, the oxide insulating film dissolves and the capacitor becomes shorted. If the polarity connection is corrected, the film reforms and restores the capacitor.

AC Electrolytic Capacitors

This ability of the wet-type electrolytic capacitor to be shorted and then reformed is the basis for a special type of nonpolarized electrolytic capacitor called the AC electrolytic capacitor. This capacitor is used as the starting capacitor for many small single-phase motors, as the run capacitor in many ceiling fan motors, and for low-power electronic circuits when a nonpolarized capacitor with a high capacitance is required. The AC electrolytic capacitor is made by connecting two wet-type electrolytic capacitors together inside the same case *(Figure 20–25)*. In the example shown, the two wet-type electrolytic capacitors have their negative terminals connected. When AC is applied to the leads, one capacitor will be connected to reverse polarity and become shorted. The other capacitor will be connected to the correct polarity and will form. During the next half cycle, the polarity changes and forms the capacitor that was shorted and shorts the other capacitor. An AC electrolytic capacitor is shown in *Figure 20–26*.

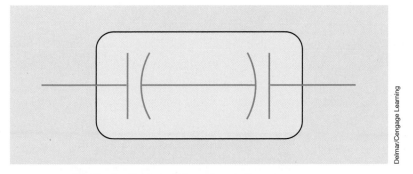

Delmar/Cengage Learning

FIGURE 20–25   Two wet-type electrolytic capacitors connect to form an AC electrolytic capacitor.

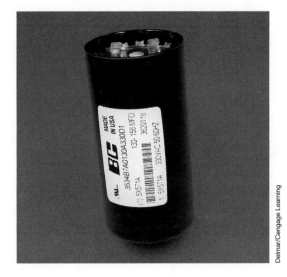

Delmar/Cengage Learning

FIGURE 20–26   An AC electrolytic capacitor.

Dry-Type Electrolytic Capacitors

The dry-type electrolytic capacitor is very similar to the wet type except that gauze is used to hold the borax solution. This prevents the capacitor from leaking. Although the dry-type electrolytic capacitor has the advantage of being relatively leak proof, it does have one disadvantage. If the polarity connection should become reversed and the oxide film is broken down, it will not reform when connected to the proper polarity. Reversing the polarity of a dry-type electrolytic capacitor permanently damages the capacitor.

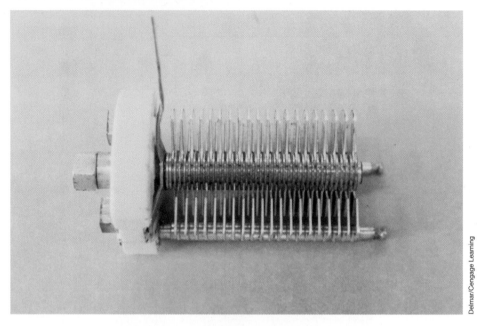

FIGURE 20–27   A variable capacitor.

20–12   Variable Capacitors

Variable capacitors are constructed in such a manner that their capacitance value can be changed over a certain range. They generally contain a set of movable plates, which are connected to a shaft, and a set of stationary plates *(Figure 20–27)*. The movable plates can be interleaved with the stationary plates to increase or decrease the capacitance value. Because air is used as the dielectric and the plate area is relatively small, variable capacitors are generally rated in picofarads. Another type of small variable capacitor is called the trimmer capacitor *(Figure 20–28)*. This capacitor has one movable plate and one stationary plate. The capacitance value is changed by turning

FIGURE 20–28   A trimmer capacitor.

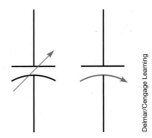

Delmar/Cengage Learning

FIGURE 20–29   Variable capacitor symbols.

an adjustment screw that moves the movable plate closer to or farther away from the stationary plate. *Figure 20–29* shows schematic symbols used to represent variable capacitors.

20–13   Capacitor Markings

Different types of capacitors are marked in different ways. Large AC oil-filled paper capacitors generally have their capacitance and voltage values written on the capacitor. The same is true for most electrolytic and small nonpolarized capacitors. Other types of capacitors, however, depend on color codes or code numbers and letters to indicate the capacitance value, tolerance, and voltage rating. Although color coding for capacitors has been abandoned in favor of direct marking by most manufacturers, it is still used by some. Also, many older capacitors with color codes are still in use. For this reason, we discuss color coding for several types of capacitors. Unfortunately, there is no actual set standard used by all manufacturers. The color codes presented are probably the most common. An identification chart for postage stamp (so called because of their size and shape) mica capacitors and tubular paper or tubular mica capacitors is shown in *Figure 20–30*. Note that most postage stamp mica capacitors use a five-dot color code. There are six-dot color codes, however. When a six-dot color code is used, the third dot represents a third digit and the rest of the code is the same as a five-dot code. The capacitance values given are in picofarads. Although these markings are typical, there is no actual standard and it may be necessary to use the manufacturer's literature to determine the true values.

A second method for color coding mica capacitors is called the EIA (Electronic Industries Association) standard, or the **JAN (Joint Army-Navy) standard.** The JAN standard is used for electronic components intended for military use. When the EIA standard is employed, the first dot is white. In some instances, the first dot may be silver instead of white. This indicates that the capacitor's dielectric is paper instead of mica. When the JAN standard is used, the first dot is black. The second and third dots represent digits, the fourth dot is the multiplier, the fifth dot is the tolerance, and the sixth dot indicates classes A to E of temperature and leakage coefficients.

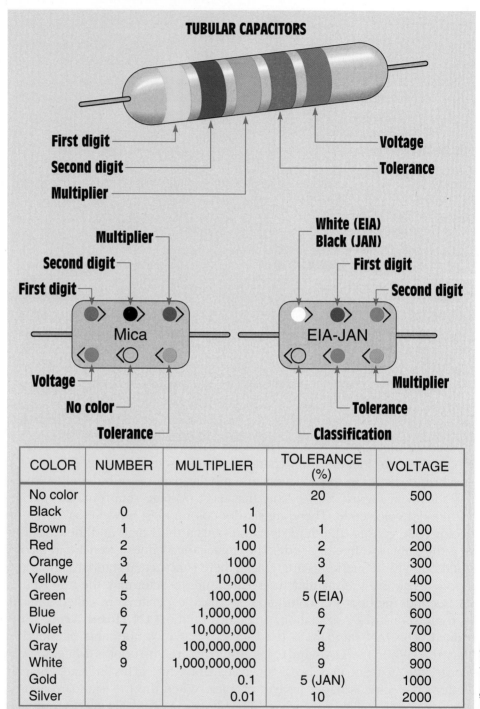

COLOR	NUMBER	MULTIPLIER	TOLERANCE (%)	VOLTAGE
No color			20	500
Black	0	1		
Brown	1	10	1	100
Red	2	100	2	200
Orange	3	1000	3	300
Yellow	4	10,000	4	400
Green	5	100,000	5 (EIA)	500
Blue	6	1,000,000	6	600
Violet	7	10,000,000	7	700
Gray	8	100,000,000	8	800
White	9	1,000,000,000	9	900
Gold		0.1	5 (JAN)	1000
Silver		0.01	10	2000

Delmar/Cengage Learning

FIGURE 20–30 Identification of mica and tubular capacitors.

20–14 Temperature Coefficients

The temperature coefficient indicates the amount of capacitance change with temperature. Temperature coefficients are listed in parts per million (ppm) per degree Celsius. A positive temperature coefficient indicates that the capacitor will increase its capacitance with an increase in temperature. A negative temperature coefficient indicates that the capacitance will decrease with an increase in temperature.

20–15 Ceramic Capacitors

Another capacitor that often uses color codes is the ceramic capacitor *(Figure 20–31)*. This capacitor generally has one band that is wider than the others. The wide band indicates the temperature coefficient, and the other bands are first and second digits, multiplier, and tolerance.

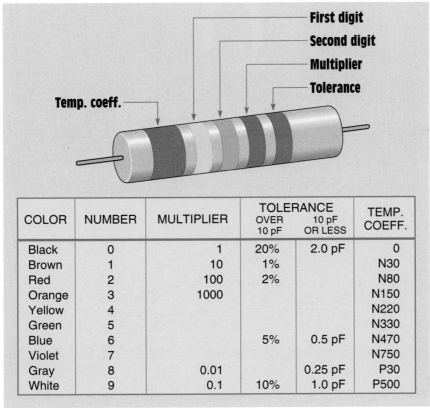

COLOR	NUMBER	MULTIPLIER	TOLERANCE OVER 10 pF	TOLERANCE 10 pF OR LESS	TEMP. COEFF.
Black	0	1	20%	2.0 pF	0
Brown	1	10	1%		N30
Red	2	100	2%		N80
Orange	3	1000			N150
Yellow	4				N220
Green	5				N330
Blue	6		5%	0.5 pF	N470
Violet	7				N750
Gray	8	0.01		0.25 pF	P30
White	9	0.1	10%	1.0 pF	P500

Delmar/Cengage Learning

FIGURE 20–31   Color codes for ceramic capacitors.

20–16 Dipped Tantalum Capacitors

A dipped tantalum capacitor is shown in *Figure 20–32*. This capacitor has the general shape of a match head but is somewhat larger. Color bands and dots determine the value, tolerance, and voltage. The capacitance value is given in picofarads.

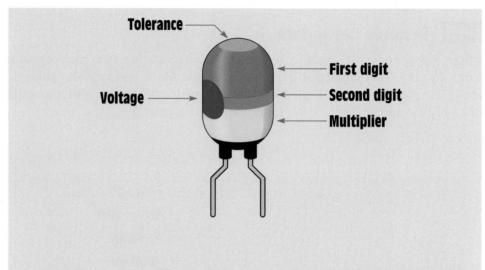

COLOR	NUMBER	MULTIPLIER	TOLERANCE (%)	VOLTAGE
			No dot 20	
Black	0			4
Brown	1			6
Red	2			10
Orange	3			15
Yellow	4	10,000		20
Green	5	100,000		25
Blue	6	1,000,000		35
Violet	7	10,000,000		50
Gray	8			
White	9			3
Gold			5	
Silver			10	

FIGURE 20–32 Dipped tantalum capacitors.

20–17   Film Capacitors

Not all capacitors use color codes to indicate values. Some capacitors use numbers and letters. A film-type capacitor is shown in *Figure 20–33*. This capacitor is marked 105 K. The value can be read as follows:

1. The first two numbers indicate the first two digits of the value.

2. The third number is the multiplier. Add the number of zeros to the first two numbers indicated by the multiplier. In this example, add five zeros to 10. The value is given in picofarads (pF). This capacitor has a value of 1,000,000 pF or 1 μF.

3. The K is the tolerance. In this example, K indicates a tolerance of ±10%.

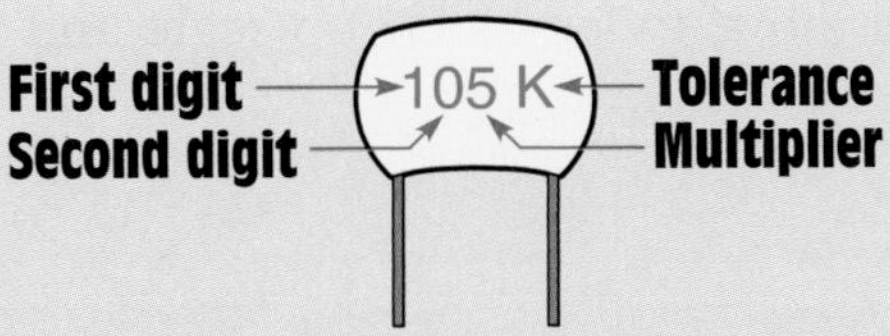

NUMBER	MULTIPLIER	TOLERANCE		
			10 pF or less	Over 10 pF
0	1	B	0.1 pF	
1	10	C	0.25 pF	
2	100	D	0.5 pF	
3	1000	F	1.0 pF	1%
4	10,000	G	2.0 pF	2%
5	100,000	H		3%
6		J		5%
7		K		10%
8	0.01	M		20%
9	0.1			

FIGURE 20–33   Film-type capacitors.

20–18 Testing Capacitors

Testing capacitors is difficult at best. Small electrolytic capacitors are generally tested for shorts with an ohmmeter. If the capacitor is not shorted, it should be tested for leakage using a variable DC power supply and a microammeter *(Figure 20–34)*. When rated voltage is applied to the capacitor, the microammeter should indicate zero current flow.

Large AC oil-filled capacitors can be tested in a similar manner. To test the capacitor accurately, two measurements must be made. One is to measure the capacitance value of the capacitor to determine if it is the same or approximately the same as the rated value. The other is to test the strength of the dielectric.

The first test should be made with an ohmmeter. With the power disconnected, connect the terminals of an ohmmeter directly across the capacitor terminals *(Figure 20–35)*. This test determines if the dielectric is shorted. When the ohmmeter is connected, the needle should swing up scale and return to infinity. The amount of needle swing is determined by the capacitance of the capacitor. Then reverse the ohmmeter connection, and the needle should move twice as far up scale and return to the infinity setting.

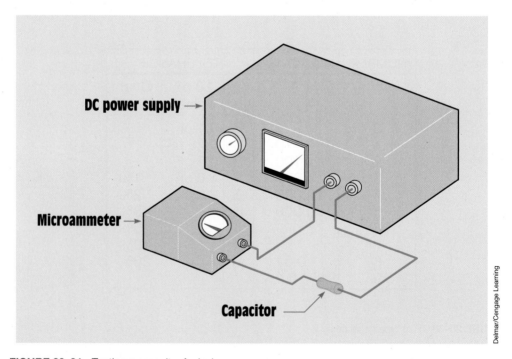

FIGURE 20–34 Testing a capacitor for leakage.

Delmar/Cengage Learning

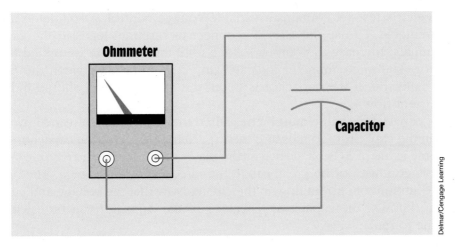

Delmar/Cengage Learning

FIGURE 20–35 Testing the capacitor with an ohmmeter.

If the ohmmeter test is successful, the dielectric must be tested at its rated voltage. This is called a dielectric strength test. To make this test, a dielectric test set must be used *(Figure 20–36)*. This device is often referred to as a **HIPOT** because of its ability to produce a high voltage or high potential.

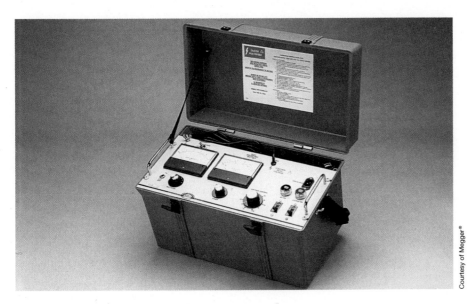

Courtesy of Megger®

FIGURE 20–36 A dielectric test set.

The dielectric test set contains a variable voltage control, a voltmeter, and a microammeter. To use the HIPOT, connect its terminal leads to the capacitor terminals. Increase the output voltage until rated voltage is applied to the capacitor. The microammeter indicates any current flow between the plates of the dielectric. If the capacitor is good, the microammeter should indicate zero current flow.

The capacitance value must be measured to determine if there are any open plates in the capacitor. To measure the capacitance value of the capacitor, connect some value of AC voltage across the plates of the capacitor *(Figure 20–37)*. This voltage must not be greater than the rated capacitor voltage. Then measure the amount of current flow in the circuit. Now that the voltage and current flow are known, the capacitive reactance of the capacitor can be calculated using the formula:

$$X_C = \frac{E}{I}$$

After the capacitive reactance has been determined, the capacitance can be calculated using the formula:

$$C = \frac{1}{2\pi f X_C}$$

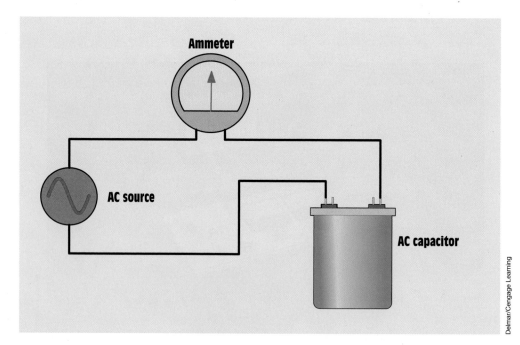

FIGURE 20–37 Determining the capacitance value.

Delmar/Cengage Learning

Note: Capacitive reactance is measured in ohms and limits current flow in a manner similar to inductive reactance. Capacitive reactance is covered fully in Unit 21.

Summary

- Capacitors are devices that oppose a change of voltage.
- Three factors that determine the capacitance of a capacitor are
  a. the surface area of the plates.
  b. the distance between the plates.
  c. the type of dielectric used.
- A capacitor stores energy in an electrostatic field.
- Current can flow only during the time a capacitor is charging or discharging.
- Capacitors charge and discharge at an exponential rate.
- The basic unit of capacitance is the farad (F).
- Capacitors are generally rated in microfarads (μF), nanofarads (nF), or pico-farads (pF).
- When capacitors are connected in parallel, their capacitance values add.
- When capacitors are connected in series, the reciprocal of the total capaci-tance is equal to the sum of the reciprocals of all the capacitors.
- The charge and discharge times of a capacitor are proportional to the amount of capacitance and resistance in the circuit.
- Five time constants are required to charge or discharge a capacitor.
- Nonpolarized capacitors are often called AC capacitors.
- Nonpolarized capacitors can be connected to DC or AC circuits.
- Polarized capacitors are often referred to as electrolytic capacitors.
- Polarized capacitors can be connected to DC circuits only.
- There are two basic types of electrolytic capacitors, the wet type and the dry type.
- Wet-type electrolytic capacitors can be re-formed if reconnected to the correct polarity.
- Dry-type electrolytic capacitors will be permanently damaged if connected to the incorrect polarity.

- Capacitors are often marked with color codes or with numbers and letters.
- To test a capacitor for leakage, a microammeter should be connected in series with the capacitor and rated voltage applied to the circuit.

Review Questions

1. What is the dielectric?

2. List three factors that determine the capacitance of a capacitor.

3. A capacitor uses air as a dielectric and has a capacitance of 3 μF. A dielectric material is inserted between the plates without changing the spacing, and the capacitance becomes 15 μF. What is the dielectric constant of this material?

4. In what form is the energy of a capacitor stored?

5. Four capacitors having values of 20 μF, 50 μF, 40 μF, and 60 μF are connected in parallel. What is the total capacitance of this circuit?

6. If the four capacitors in Question 5 were to be connected in series, what would be the total capacitance of the circuit?

7. A 22-μ capacitor is connected in series with a 90-kΩ resistor. How long will it take this capacitor to charge?

8. A 450-pF capacitor has a total charge time of 0.5 second. How much resistance is connected in series with the capacitor?

9. Can a nonpolarized capacitor be connected to a DC circuit?

10. Explain how an AC electrolytic capacitor is constructed.

11. What type of electrolytic capacitor will be permanently damaged if connected to the incorrect polarity?

12. A 500-nF capacitor is connected to a 300-kΩ resistor. What is the total charge time of this capacitor?

13. A film-type capacitor is marked 253 H. What are the capacitance value and tolerance of this capacitor?

14. A postage stamp mica capacitor has the following color marks starting at the upper left dot: yellow, violet, brown, green, no color, and blue. What are the capacitance value, tolerance, and voltage rating of this capacitor?

15. A postage stamp capacitor has the following color marks starting at the upper-left dot: black, orange, orange, black, silver, and white. What are the capacitance value and tolerance of this capacitor?

Practical Applications

You are changing the starting relay on a central air-conditioning unit when you notice that the identifying mark on the compressor-run capacitor is connected to the run winding of the compressor. Should you change the capacitor connection so that the identifying mark is facing the line side or is it correct as connected? Explain your answer. ■

Practical Applications

You are an electrician working in an industrial plant. You discover the problem with a certain machine is a defective capacitor. The capacitor is connected to a 240-volt AC circuit. The information on the capacitor reveals that it has a capacitance value of 10 μF and a voltage rating of 240 VAC. The only 10-μF AC capacitor in the storeroom is marked with a voltage rating of 350 WVDC. Can this capacitor be used to replace the defective capacitor? Explain your answer. ■

Practical Applications

You find that a 25-μF capacitor connected to 480 VAC is defective. The storeroom has no capacitors with a 480-VAC rating. However, you find two capacitors rated at 50 μF and 370 VAC. Can these two capacitors be connected in such a manner that they can replace the defective capacitor? If yes, explain how they are connected and why the capacitors will not be damaged by the lower voltage rating. If no, explain why they cannot be used without damage to the capacitor. ■

Practice Problems

RC Time Constants

1. Fill in all the missing values. Refer to the formulas that follow.

Resistance	Capacitance	Time constant	Total time
150 kΩ	100 μF		
350 kΩ			35 s
	350 pF	0.05 s	
	0.05 μF		10 s
1.2 MΩ	0.47 μF		
	12 μF	0.05 s	
86 kΩ			1.5 s
120 kΩ	470 pF		
	250 nF		100 ms
	8 μF		150 μs
100 kΩ		150 ms	
33 kΩ	4 μF		

$$\tau = RC$$

$$R = \frac{\tau}{C}$$

$$C = \frac{\tau}{R}$$

Total time = $\tau \times 5$

2. Two capacitors having values of 80 μF and 60 μF are connected in series. What is the total capacitance?

3. Three capacitors having values of 120 μF, 20 μF, and 60 μF are connected in parallel. What is the total capacitance?

4. Three capacitors having values of 2.2 μF, 280 nF, and 470 pF are connected in parallel. What is the total capacitance?

5. A 470-μF capacitor is connected in series with a 120-kΩ resistor. How long will it take the capacitor to charge completely?

Unit 21

Capacitance in AC Circuits

OUTLINE

KEY TERMS

Appears to flow

Capacitive reactance (X_C)

Out of phase

Voltage rating

Why You Need to Know

Capacitors are one of the three major types of AC loads and are the exact opposite of inductors in almost every respect. Although a capacitor is an open circuit, you will see how current can appear to flow through it. This unit

- illustrates how to calculate the current-limiting effect of a capacitor.
- discusses the different voltage ratings that are listed on nonpolarized capacitors.
- illustrates how to connect the capacitor into an AC circuit and calculate capacitance charge and discharge rates.
- discusses the effects of frequency in a capacitance circuit.

Objectives

After studying this unit, you should be able to

- explain why current appears to flow through a capacitor when it is connected to an AC circuit.

- discuss capacitive reactance.

- calculate the value of capacitive reactance in an AC circuit.

- calculate the value of capacitance in an AC circuit.

- discuss the relationship of voltage and current in a pure capacitive circuit.

Preview

In Unit 20, it was discussed that a capacitor is composed of two metal plates separated by an insulating material called the dielectric. Because there is no complete circuit between the plates, current cannot flow through the capacitor. The only time that current can flow is during the period of time that the capacitor is being charged or discharged. ■

21–1   Connecting the Capacitor into an AC Circuit

When a capacitor *(Figure 21–1)* is connected to an AC circuit, current **appears to flow** through the capacitor. The reason is that in an AC circuit, the current continually changes direction and polarity. To understand this concept, consider the hydraulic circuit shown in *Figure 21–2*. Two tanks are connected to a common pump. Assume Tank A to be full and Tank B to be empty. Now assume that the pump pumps water from Tank A to Tank B. When Tank B

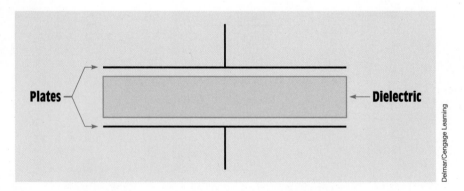

Plates —

 — **Dielectric**

Delmar/Cengage Learning

FIGURE 21–1   A basic capacitor.

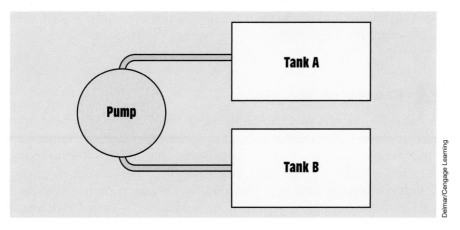

FIGURE 21–2   Water can flow continuously, but not between the two tanks.

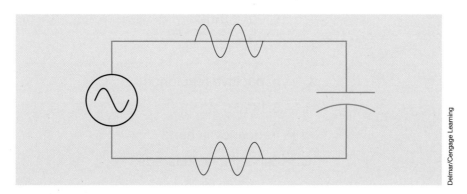

FIGURE 21–3   A capacitor connected to an AC circuit.

becomes full, the pump reverses and pumps the water from Tank B back into Tank A. Each time a tank becomes filled, the pump reverses and pumps water back into the other tank. Notice that water is continually flowing in this circuit, but there is no direct connection between the two tanks.

A similar action takes place when a capacitor is connected to an AC circuit *(Figure 21–3)*. In this circuit, the AC generator or alternator charges one plate of the capacitor positive and the other plate negative. During the next half cycle, the voltage changes polarity and the capacitor discharges and recharges to the opposite polarity also. As long as the voltage continues to increase, decrease, and change polarity, current flows from one plate of the capacitor to the other. If an ammeter were placed in the circuit, it would indicate a continuous flow of current, giving the appearance that current is flowing through the capacitor.

21–2   Capacitive Reactance

As the capacitor is charged, an impressed voltage is developed across its plates as an electrostatic charge is built up *(Figure 21–4)*. The impressed voltage is the voltage provided by the electrostatic charge. This impressed voltage opposes the applied voltage and limits the flow of current in the circuit. This countervoltage is similar to the countervoltage produced by an inductor. The countervoltage developed by the capacitor is called *reactance* also. Because this countervoltage is caused by capacitance, it is called **capacitive reactance (X_C)** and is measured in ohms. The formula for finding capacitive reactance is

$$X_C = \frac{1}{2\pi fC}$$

where

$$X_C = \text{capacitive reactance}$$

$$\pi = 3.1416$$

$$f = \text{frequency in hertz}$$

$$C = \text{capacitance in farads}$$

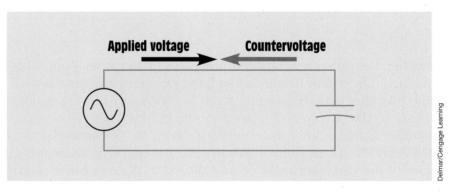

Applied voltage   **Countervoltage**

Delmar/Cengage Learning

FIGURE 21–4   Countervoltage limits the flow of current.

■ EXAMPLE 21-1

A 35-μF capacitor is connected to a 120-V, 60-Hz line. How much current will flow in this circuit?

Solution

The first step is to calculate the capacitive reactance. Recall that the value of C in the formula is given in F. This must be changed to the capacitive units being used—in this case, μF:

$$X_C = \frac{1}{2 \times 3.1416 \times 60 \times (35 \times 10^{-6})}$$

$$X_C = 75.788 \ \Omega$$

Now that the value of capacitive reactance is known, it can be used like resistance in an Ohm's law formula. Because capacitive reactance is the current-limiting factor, it will replace the value of R:

$$I = \frac{E}{X_C}$$

$$I = \frac{120 \ V}{75.788 \ \Omega}$$

$$I = 1.583 \ A$$

21-3   Calculating Capacitance

If the value of capacitive reactance is known, the capacitance of the capacitor can be found using the formula

$$C = \frac{1}{2\pi f X_C}$$

■ EXAMPLE 21-2

A capacitor is connected into a 480-V, 60-Hz circuit. An ammeter indicates a current flow of 2.6 A. What is the capacitance value of the capacitor?

Solution

The first step is to calculate the value of capacitive reactance. Because capacitive reactance, like resistance, limits current flow, it can be substituted for R in an Ohm's law formula:

$$X_C = \frac{E}{I}$$

$$X_C = \frac{480 \text{ V}}{2.6 \text{ A}}$$

$$X_C = 184.615 \ \Omega$$

Now that the capacitive reactance of the circuit is known, the value of capacitance can be found:

$$C = \frac{1}{2\pi f X_C}$$

$$C = \frac{1}{2 \times 3.1416 \times 60 \text{ Hz} \times 184.615 \ \Omega}$$

$$C = \frac{1}{69,598.378}$$

$$C = 0.000014368 \text{ F} = 14.368 \ \mu\text{F}$$

21-4 Voltage and Current Relationships in a Pure Capacitive Circuit

Earlier in this text, it was shown that the current in a pure resistive circuit is in phase with the applied voltage and that current in a pure inductive circuit lags the applied voltage by 90°. In this unit, it will be shown that in a pure capacitive circuit the current will *lead* the applied voltage by 90°.

When a capacitor is connected to an AC, the capacitor charges and discharges at the same rate and time as the applied voltage. The charge in

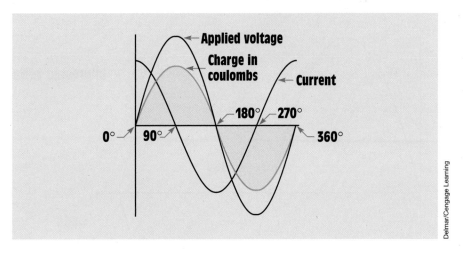

FIGURE 21–5 Capacitive current leads the applied voltage by 90°.

coulombs is equal to the capacitance of the capacitor times the applied voltage $(Q = C \times V)$. When the applied voltage is zero, the charge in coulombs and impressed voltage is also zero. When the applied voltage reaches its maximum value, positive or negative, the charge in coulombs and impressed voltage also reaches maximum *(Figure 21–5)*. The impressed voltage follows the same curve as the applied voltage.

In the waveform shown, voltage and charge are both zero at 0°. Because there is no charge on the capacitor, there is no opposition to current flow, which is shown to be maximum. As the applied voltage increases from zero toward its positive peak at 90°, the capacitor begins to charge at the same time. The charge produces an impressed voltage across the plates of the capacitor that opposes the flow of current. The impressed voltage is 180° **out of phase** with the applied voltage *(Figure 21–6)*. When the applied voltage reaches 90° in the positive direction, the charge reaches maximum, the impressed voltage reaches peak in the negative direction, and the current flow is zero.

As the applied voltage begins to decrease, the capacitor begins to discharge, causing the current to flow in the opposite or negative direction. When the applied voltage and charge reach zero at 180°, the impressed voltage is zero also and the current flow is maximum in the negative direction. As the applied voltage and charge increase in the negative direction, the increase of the impressed voltage across the capacitor again causes the current to decrease. The applied voltage and charge reach maximum negative after 270° of rotation. The impressed voltage reaches maximum positive and the current has decreased to zero *(Figure 21–7)*. As the applied voltage decreases

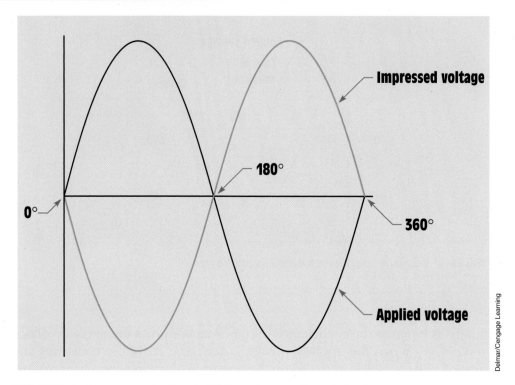

FIGURE 21–6   The impressed voltage is 180° out of phase with applied voltage.

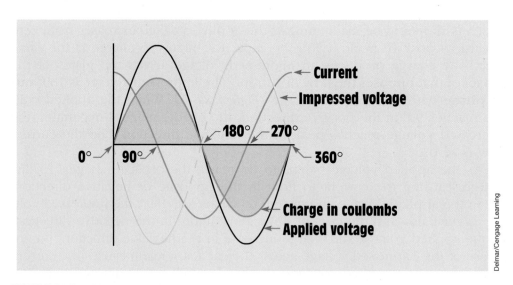

FIGURE 21–7   Voltage, current, and charge relationships for a capacitive circuit.

from its maximum negative value, the capacitor again begins to discharge. This causes the current to flow in the positive direction. The current again reaches its maximum positive value when the applied voltage and charge reach zero after 360° of rotation.

21–5 Power in a Pure Capacitive Circuit

Because the current flow in a pure capacitive circuit leads the applied voltage by 90°, the voltage and current have the same polarity for half the time during one cycle and have opposite polarities the other half of the time *(Figure 21–8)*. During the period of time that the voltage and current have the same polarity, energy is being stored in the capacitor in the form of an electrostatic field. When the voltage and current have opposite polarities, the capacitor is discharging and the energy is returned to the circuit. When the values of current and voltage for one full cycle are added, the sum equals zero just as it does with pure inductive circuits. Therefore, there is no true power, or watts, produced in a pure capacitive circuit.

The power value for a capacitor is reactive power and is measured in VARs, just as it is for an inductor. Inductive VARs and capacitive VARs are 180° out of phase with each other, however *(Figure 21–9)*. To distinguish between inductive and capacitive VARs, inductive VARs are shown as VARs$_L$ and capacitive VARs are shown as VARs$_C$.

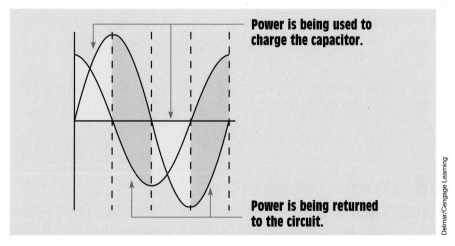

FIGURE 21–8 A pure capacitive circuit has no true power (watts). The power required to charge the capacitor is returned to the circuit when the capacitor discharges.

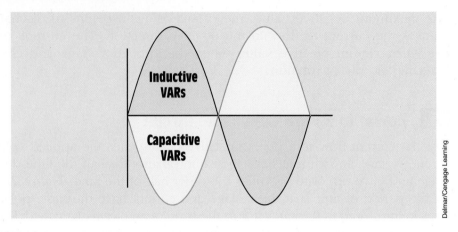

FIGURE 21–9   Inductive VARs and capacitive VARs are 180° out of phase with each other.

21–6   Quality of a Capacitor

The quality (Q) of a capacitor is generally very high. As with inductors, it is a ratio of the resistance to capacitive reactance:

$$Q = \frac{R}{X_C}$$

The R value for a capacitor is generally very high because it is the equivalent resistance of the dielectric between the plates of the capacitor. If a capacitor is leaky, however, the dielectric will appear to be a much lower resistance and the Q rating will decrease.

Q for a capacitor can also be found by using other formulas. One of these formulas follows:

$$Q = \frac{VARs_C}{W}$$

where the reactive power is represented by VARs. Another formula that can be used sets Q equal to the reciprocal of the power factor:

$$Q = \frac{1}{PF}$$

21–7   Capacitor Voltage Rating

The **voltage rating** of a capacitor is actually the voltage rating of the dielectric. ***Voltage rating is extremely important concerning the life of the capacitor and should never be exceeded.*** Unfortunately, there are no set standards concerning how voltage ratings are marked. It is not unusual to see capacitors marked VOLTS AC, VOLTS DC, PEAK VOLTS, and WVDC

(WORKING VOLTS DC). The voltage rating of electrolytic or polarized capacitors is always given in DC volts. The voltage rating of nonpolarized capacitors, however, can be given as AC or DC volts.

If a nonpolarized capacitor has a voltage rating given in AC volts, the voltage indicated is the RMS value. If the voltage rating is given as PEAK or as DC volts, it indicates the peak value of AC volts. If a capacitor is to be connected to an AC circuit, it is necessary to calculate the peak value if the voltage rating is given as DC volts.

■ **EXAMPLE 21-3**

An AC oil-filled capacitor has a voltage rating of 300 WVDC. Will the voltage rating of the capacitor be exceeded if the capacitor is connected to a 240-V, 60-Hz line?

Solution

The DC voltage rating of the capacitor indicates the peak value of voltage. To determine whether the voltage rating will be exceeded, find the peak value of 240 V by multiplying by 1.414:

$$\text{Peak} = 240 \text{ V} \times 1.414$$
$$\text{Peak} = 339.36 \text{ V}$$

The answer is that the capacitor voltage rating will be exceeded.

21–8 Effects of Frequency in a Capacitive Circuit

One of the factors that determines the capacitive reactance of a capacitor is the frequency. Capacitive reactance is inversely proportional to frequency. As the frequency increases, the capacitive reactance decreases. The chart in *Table 21–1* shows the capacitive reactance for different values of capacitance at different frequencies. Frequency has an effect on capacitive reactance because the capacitor charges and discharges faster at a higher frequency. Recall that current is a rate of electron flow. A current of 1 A is 1 coulomb per second:

$$I = \frac{C}{t}$$

where

$$I = \text{current}$$
$$C = \text{charge in coulombs}$$
$$t = \text{time in seconds}$$

Capacitance	Capacitive reactance			
	30 Hz	60 Hz	400 Hz	1000 Hz
10 pF	530.515 MΩ	265.258 MΩ	39.789 MΩ	15.915 MΩ
350 pF	15.156 MΩ	7.579 MΩ	1.137 MΩ	454.727 kΩ
470 nF	11.286 kΩ	5.644 kΩ	846.567 kΩ	338.627 kΩ
750 nF	7.074 kΩ	3.537 kΩ	530.515 kΩ	212.206 Ω
1 μF	5.305 kΩ	2.653 kΩ	397.886 Ω	159.155 Ω
25 μF	212.206 Ω	106.103 Ω	15.915 Ω	6.366 Ω

TABLE 21–1 Capacitive Reactance Is Inversely Proportional to Frequency

Assume that a capacitor is connected to a 30-hertz line, and 1 coulomb of charge flows each second. If the frequency is doubled to 60 hertz, 1 coulomb of charge will flow in 0.5 second because the capacitor is being charged and discharged twice as fast *(Figure 21–10)*. This means that in a period of 1 second, 2 coulombs of charge will flow. Because the capacitor is being charged and discharged at a faster rate, the opposition to current flow is decreased.

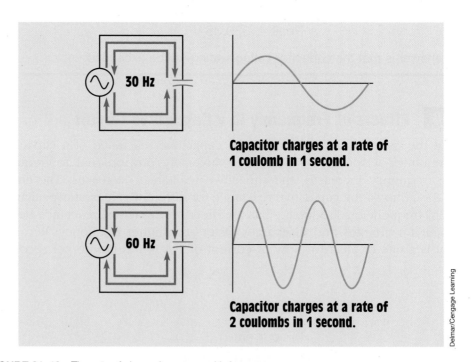

FIGURE 21–10 The rate of charge increases with frequency.

21–9   Series Capacitors

■ EXAMPLE 21-4

Three capacitors with values of 10 μF, 30 μF, and 15 μF are connected in series to a 480-V, 60-Hz line *(Figure 21–11)*. Find the following circuit values:

X_{C1}—capacitive reactance of the first capacitor

X_{C2}—capacitive reactance of the second capacitor

X_{C3}—capacitive reactance of the third capacitor

X_{CT}—total capacitive reactance for the circuit

C_T—total capacitance for the circuit

I_T—total circuit current

E_{C1}—voltage drop across the first capacitor

$VARs_{C1}$—reactive power of the first capacitor

E_{C2}—voltage drop across the second capacitor

$VARs_{C2}$—reactive power of the second capacitor

E_{C3}—voltage drop across the third capacitor

$VARs_{C3}$—reactive power of the third capacitor

$VARs_{CT}$—total reactive power for the circuit

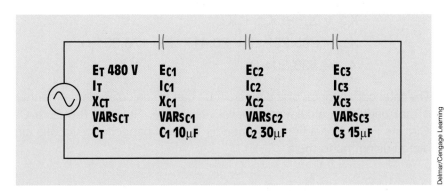

E_T 480 V	E_{C1}	E_{C2}	E_{C3}
I_T	I_{C1}	I_{C2}	I_{C3}
X_{CT}	X_{C1}	X_{C2}	X_{C3}
$VARs_{CT}$	$VARs_{C1}$	$VARs_{C2}$	$VARs_{C3}$
C_T	C_1 10 μF	C_2 30 μF	C_3 15 μF

Delmar/Cengage Learning

FIGURE 21–11   Capacitors connected in series.

Solution

Because the frequency and the capacitance of each capacitor are known, the capacitive reactance for each capacitor can be found using the formula

$$X_C = \frac{1}{2\pi fC}$$

Recall that the value for C in the formula is in farads and the capacitors in this problem are rated in microfarads.

$$X_{C1} = \frac{1}{2\pi fC}$$

$$X_{C1} = \frac{1}{377 \times 0.000010}$$

$$X_{C1} = 265.252 \ \Omega$$

$$X_{C2} = \frac{1}{2\pi fC}$$

$$X_{C2} = \frac{1}{377 \times 0.000030}$$

$$X_{C2} = 88.417 \ \Omega$$

$$X_{C3} = \frac{1}{2\pi fC}$$

$$X_{C3} = \frac{1}{377 \times 0.000015}$$

$$X_{C3} = 176.835 \ \Omega$$

Because there is no phase angle shift among the three capacitive reactances, the total capacitive reactance is the sum of the three reactances (*Figure 21–12*):

$$X_{CT} = X_{C1} + X_{C2} + X_{C3}$$
$$X_{CT} = 265.252 \ \Omega + 88.417 \ \Omega + 176.835 \ \Omega$$
$$X_{CT} = 530.504 \ \Omega$$

The total capacitance of a series circuit can be calculated in a manner similar to that used for calculating parallel resistance. Refer to the Pure Capacitive Circuits Formula section of Appendix B. Total capacitance in this circuit is calculated using the formula

$$C_T = \frac{1}{\dfrac{1}{C_1} + \dfrac{1}{C_2} + \dfrac{1}{C_3}}$$

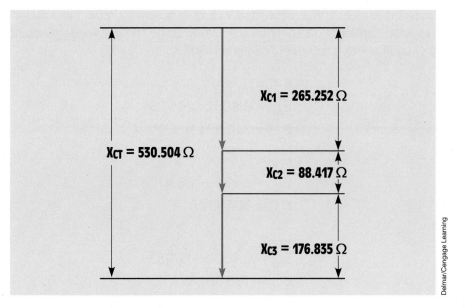

FIGURE 21–12   Vector sum for capacitive reactance.

$$C_T = \cfrac{1}{\cfrac{1}{10\ \mu F} + \cfrac{1}{30\ \mu f} + \cfrac{1}{15\ \mu F}}$$

$$C_T = \cfrac{1}{0.2\ \cfrac{1}{\mu F}}$$

$$C_T = 5\ \mu F$$

The total current can be found by using the total capacitive reactance to substitute for R in an Ohm's law formula:

$$I_T = \frac{E_{CT}}{X_{CT}}$$

$$I_T = \frac{480\ V}{530.504\ \Omega}$$

$$I_T = 0.905\ A$$

Because the current is the same at any point in a series circuit, the voltage drop across each capacitor can now be calculated using the capacitive reactance of each capacitor and the current flowing through it.

$$E_{C1} = I_{C1} \times X_{C1}$$
$$E_{C1} = 0.905 \times 265.25$$
$$E_{C1} = 240.051 \text{ V}$$

$$E_{C2} = I_{C2} \times X_{C2}$$
$$E_{C2} = 0.905 \times 88.417$$
$$E_{C2} = 80.017 \text{ V}$$

$$E_{C3} = I_{C3} \times X_{C3}$$
$$E_{C3} = 0.905 \times 176.83$$
$$E_{C3} = 160.031 \text{ V}$$

Now that the voltage drops of the capacitors are known, the reactive power of each capacitor can be found.

$$VARS_{C1} = E_{C1} \times I_{C1}$$
$$VARS_{C1} = 240.051 \times 0.905$$
$$VARS_{C1} = 217.246$$

$$VARS_{C2} = E_{C2} \times I_{C2}$$
$$VARS_{C2} = 80.017 \times 0.905$$
$$VARS_{C2} = 72.415$$

$$VARS_{C3} = E_{C3} \times I_{C3}$$
$$VARS_{C3} = 160.031 \times 0.905$$
$$VARS_{C3} = 144.828$$

Power, whether true power, apparent power, or reactive, will add in any type of circuit. The total reactive power in this circuit can be found by taking the sum of all the VARs for the capacitors or by using total values of voltage and current and Ohm's law:

$$VARS_{CT} = VARS_{C1} + VARS_{C2} + VARS_{C3}$$
$$VARS_{CT} = 217.248 + 72.415 + 144.828$$
$$VARS_{CT} = 434.491$$

The circuit with all calculated values is shown in *Figure 21–13*.

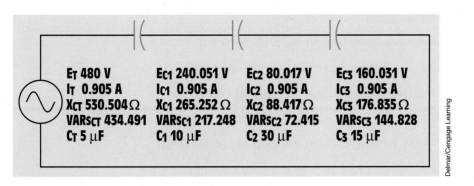

FIGURE 21–13 Series circuit 1 with all values.

21–10 Parallel Capacitors

■ EXAMPLE 21–5

Three capacitors having values of 50 μF, 75 μF, and 20 μF are connected in parallel to a 60-Hz line. The circuit has a total reactive power of 787.08 VARs *(Figure 21–14)*. Find the following unknown values:

X_{C1}—capacitive reactance of the first capacitor

X_{C2}—capacitive reactance of the second capacitor

X_{C3}—capacitive reactance of the third capacitor

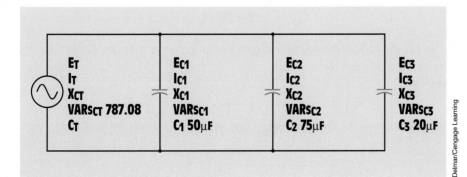

FIGURE 21–14 Capacitors connected in parallel.

X_{CT}—total capacitive reactance for the circuit

E_T—total applied voltage

I_{C1}—current flow through the first capacitor

$VARS_{C1}$—reactive power of the first capacitor

I_{C2}—current flow through the second capacitor

$VARS_{C2}$—reactive power of the second capacitor

I_{C3}—current flow through the third capacitor

$VARS_{C3}$—reactive power of the third capacitor

I_T—total circuit current

    Because the frequency of the circuit and the capacitance of each capacitor are known, the capacitive reactance of each capacitor can be calculated using the formula:

$$X_C = \frac{1}{2\pi fC}$$

Note: Refer to the Pure Capacitive Circuits Formula section of Appendix B.

$$X_{C1} = \frac{1}{377 \times 0.000050\ F}$$

$$X_{C1} = 53.05\ \Omega$$

$$X_{C2} = \frac{1}{377 \times 0.000075\ F}$$

$$X_{C2} = 35.367\ \Omega$$

$$X_{C3} = \frac{1}{377 \times 0.000020\ F}$$

$$X_{C3} = 132.626\ \Omega$$

    The total capacitive reactance can be found in a manner similar to finding the resistance of parallel resistors:

$$X_{CT} = \frac{1}{\dfrac{1}{X_{C1}} + \dfrac{1}{X_{C2}} + \dfrac{1}{X_{C3}}}$$

$$X_{CT} = \frac{1}{\dfrac{1}{53.05\ \Omega} + \dfrac{1}{35.367\ \Omega} + \dfrac{1}{132.626\ \Omega}}$$

$$X_{CT} = \frac{1}{0.05467 \, \frac{1}{\Omega}}$$

$$X_{CT} = 18.292 \ \Omega$$

Now that the total capacitive reactance of the circuit is known and the total reactive power is known, the voltage applied to the circuit can be found using the formula

$$E_T = \sqrt{VARS_{CT} \times X_{CT}}$$

$$E_T = \sqrt{787.08 \ VARs \times 18.292 \ \Omega}$$

$$E_T = 119.989 \ V$$

In a parallel circuit, the voltage must be the same across each branch of the circuit. Therefore, 120 V is applied across each capacitor.

Now that the circuit voltage is known, the amount of total current for the circuit and the amount of current in each branch can be found using Ohm's law:

$$I_{CT} = \frac{E_{CT}}{X_{CT}}$$

$$I_{CT} = \frac{119.989 \ V}{18.292 \ \Omega}$$

$$I_{CT} = 6.56 \ A$$

$$I_{C1} = \frac{E_{C1}}{X_{C1}}$$

$$I_{C1} = \frac{119.989 \ V}{53.05 \ \Omega}$$

$$I_{C1} = 2.262 \ A$$

$$I_{C2} = \frac{E_{C2}}{X_{C2}}$$

$$I_{C2} = \frac{119.989 \ V}{35.367 \ \Omega}$$

$$I_{C2} = 3.393 \ A$$

$$I_{C3} = \frac{E_{C3}}{X_{C3}}$$

$$I_{C3} = \frac{119.989 \ V}{132.626 \ \Omega}$$

$$I_{C3} = 0.905 \ A$$

The amount of reactive power for each capacitor can now be calculated using Ohm's law:

$$\text{VARS}_{C1} = E_{C1} \times I_{C1}$$
$$\text{VARS}_{C1} = 119.989 \text{ V} \times 2.262 \text{ A}$$
$$\text{VARS}_{C1} = 271.415$$

$$\text{VARS}_{C2} = E_{C2} \times I_{C2}$$
$$\text{VARS}_{C2} = 119.989 \text{ V} \times 3.393 \text{ A}$$
$$\text{VARS}_{C2} = 407.123$$

$$\text{VARS}_{C3} = E_{C3} \times I_{C3}$$
$$\text{VARS}_{C3} = 119.987 \text{ V} \times 0.905 \text{ A}$$
$$\text{VARS}_{C3} = 108.590$$

To make a quick check of the circuit values, add the VARs for all the capacitors and see if they equal the total circuit VARs:

$$\text{VARS}_{CT} = \text{VARS}_{C1} + \text{VARS}_{C2} + \text{VARS}_{C3}$$
$$\text{VARS}_{CT} = 271.415 + 407.123 + 108.590$$
$$\text{VARS}_{CT} = 787.128$$

The slight difference in answers is caused by the rounding off of values. The circuit with all values is shown in *Figure 21–15.*

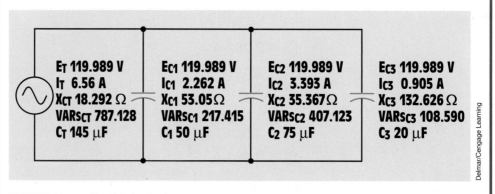

FIGURE 21–15 Parallel circuit with completed values.

Summary

- When a capacitor is connected to an AC circuit, current appears to flow through the capacitor.

- Current appears to flow through a capacitor because of the continuous increase and decrease of voltage and because of the continuous change of polarity in an AC circuit.

- The current flow in a pure capacitive circuit is limited by capacitive reactance.

- Capacitive reactance is inversely proportional to the capacitance of the capacitor and the frequency of the AC line.

- Capacitive reactance is measured in ohms.

- In a pure capacitive circuit, the current leads the applied voltage by 90°.

- There is no true power, or watts, in a pure capacitive circuit.

- Capacitive power is reactive and is measured in VARs, as is inductance.

- Capacitive and inductive VARs are 180° out of phase with each other.

- The Q of a capacitor is the ratio of the resistance to the capacitive reactance.

- Capacitor voltage ratings are given as volts AC, peak volts, and volts DC.

- A DC voltage rating for an AC capacitor indicates the peak value of voltage.

Review Questions

1. Can current flow through a capacitor?

2. What two factors determine the capacitive reactance of a capacitor?

3. How many degrees are the current and voltage out of phase in a pure capacitive circuit?

4. Does the current in a pure capacitive circuit lead or lag the applied voltage?

5. A 30-μF capacitor is connected into a 240-V, 60-Hz circuit. What is the current flow in this circuit?

6. A capacitor is connected into a 1250-V, 1000-Hz circuit. The current flow is 80 A. What is the capacitance of the capacitor?

7. A capacitor is to be connected into a 480-V, 60-Hz line. If the capacitor has a voltage rating of 600 VDC, will the voltage rating of the capacitor be exceeded?

8. On the average, by what factor is the life expectancy of a capacitor increased if the capacitor is operated at half its voltage rating?

9. A capacitor is connected into a 277-V, 400-Hz circuit. The circuit current is 12 A. What is the capacitance of the capacitor?

10. A capacitor has a voltage rating of 350 VAC. Can this capacitor be connected into a 450-VDC circuit without exceeding the voltage rating of the capacitor?

Practical Applications

You are working as an electrician in an industrial plant. You are given an AC oil-filled capacitor to install on a 480-V, 60-Hz AC line. The capacitor has the following marking: (15 μF 600 VDC). Will this capacitor be damaged if it is installed? Explain your answer. ■

Practical Applications

You are working in an industrial plant. You have been instructed to double the capacitance connected to a machine. The markings on the capacitor, however, are not visible. The capacitor is connected to 560 volts and an ammeter indicates a current of 6 amperes flowing to the capacitor. What size capacitor should be connected in parallel with the existing capacitor? What is the minimum AC voltage rating of the new capacitor? What is the minimum DC voltage rating of the new capacitor? What is the minimum kVAR size that can be used in this installation? ■

Practice Problems

Capacitive Circuits

1. Fill in all the missing values. Refer to the formulas that follow.

$$X_C = \frac{1}{2\pi f C}$$

$$C = \frac{1}{2\pi f X_C}$$

$$f = \frac{1}{2\pi C X_C}$$

Capacitance	X_C	Frequency
38 μF		60 Hz
	78.8 Ω	400 Hz
250 pF	4.5 kΩ	
234 μF		10 kHz
	240 Ω	50 Hz
10 μF	36.8 Ω	
560 nF		2 MHz
	15 kΩ	60 Hz
75 μF	560 Ω	
470 pF		200 kHz
	6.8 kΩ	400 Hz
34 μF	450 Ω	

2. A 4.7-μF capacitor is connected to a 60-Hz power source. What is the capacitive reactance of the capacitor?

3. A capacitor is connected to a 208-volt, 60-Hz power source. An ammeter indicates a current flow of 0.28 amperes. What is the capacitance of the capacitor?

4. A 0.47-μF capacitor is connected to a 240-volt power source. An ammeter indicates a current of 0.2 ampere. What is the frequency of the power source?

5. Three capacitors having capacitance values of 20 μF, 40 μF, and 50 μF are connected in parallel to a 60-Hz power line. An ammeter indicates a circuit current of 8.6 amperes. How much current is flowing through the 40-μF capacitor?

6. A capacitor has a capacitive reactance of 300 Ω when connected to a 60-Hz power line. What is the capacitive reactance if the frequency is increased to 100 Hz?

7. A pure capacitive circuit is connected to a 480-volt, 60-Hz power source. An ammeter indicates a current flow of 24 amperes. The circuit current must be reduced to 16 amperes by connecting a second capacitor in series with the first. What is the value of the existing capacitor? What value capacitor should be connected in series with the original capacitor to limit the circuit current to 16 amperes?

Unit **22**

Resistive-Capacitive Series Circuits

OUTLINE

KEY TERMS

Capacitance (C)

Total voltage (E_T)

Voltage drop across the capacitor (E_C)

Voltage drop across the resistor (E_R)

Why You Need to Know

*T*he relationship of voltage, current, impedance, and power when resistance and capacitance are connected in series with each other is not unlike that of resistive-inductive series circuits. This unit

- describes the effect of voltage, current, power, and impedance in series circuits that contain both resistance and capacitance.
- illustrates how to calculate voltage drop across a resistor and how voltage and current are out of phase in an amount less than 90°.

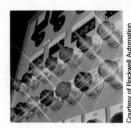

Objectives

After studying this unit, you should be able to

■ discuss the relationship of resistance and capacitance in an AC series circuit.

■ calculate values of voltage, current, apparent power, true power, reactive power, impedance, resistance, inductive reactance, and power factor in an RC series circuit.

■ calculate the phase angle for current and voltage in an RC series circuit.

■ connect an RC series circuit and make measurements using test instruments.

Preview

In this unit, the relationship of voltage, current, impedance, and power in a resistive-capacitive series circuit is discussed. As with any other type of series circuit, the current flow must be the same through all parts of the circuit. This unit explores the effect of voltage drop across each component; the relationship of resistance, reactance, and impedance; and the differences between true power, reactive power, and apparent power. ■

22–1 Resistive-Capacitive Series Circuits

When a pure capacitive load is connected to an AC circuit, the voltage and current are 90° out of phase with each other. In a capacitive circuit, the current leads the voltage by 90 electric degrees. When a circuit containing both resistance and capacitance is connected to an AC circuit, the voltage and current will be out of phase with each other by some amount between 0° and 90°. The exact amount of phase angle difference is determined by the ratio of resistance to capacitance. Resistive-capacitive series circuits are similar to resistive-inductive series circuits, covered in Unit 18. Other than changing a few formulas, the procedure for solving circuit values is the same.

In the following example, a series circuit containing 12 ohms of resistance and 16 ohms of capacitive reactance is connected to a 240-volts, 60-hertz line *(Figure 22–1)*. The following unknown values are calculated:

Z—total circuit impedance

I—total current

E_R—voltage drop across the resistor

P—watts (true power)

C—capacitance

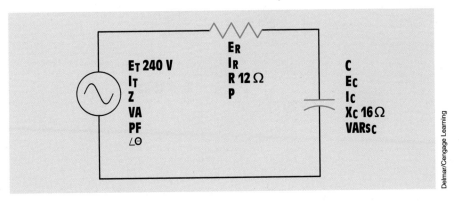

FIGURE 22–1 Resistive-capacitive series circuit.

E_C—voltage drop across the capacitor

$VARs_C$—volt-amperes-reactive (reactive power)

VA—volt-amperes (apparent power)

PF—power factor

$\angle\theta$—angle theta (the angle the voltage and current are out of phase with each other)

22-2 Impedance

The total impedance (Z) is the total current-limiting element in the circuit. It is a combination of both resistance and capacitive reactance. Because this is a series circuit, the current-limiting elements must be added. Resistance and capacitive reactance are 90° out of phase with each other, forming a right triangle with impedance being the hypotenuse *(Figure 22–2)*. A vector diagram illustrating this relationship is shown in *Figure 22–3*. Impedance can be calculated using the formula

$$Z = \sqrt{R^2 + X_C^2}$$
$$Z = \sqrt{(12\ \Omega)^2 + (16\ \Omega)^2}$$
$$Z = \sqrt{144\ \Omega^2 + 256\ \Omega^2}$$
$$Z = \sqrt{400\ \Omega^2}$$
$$Z = 20\ \Omega$$

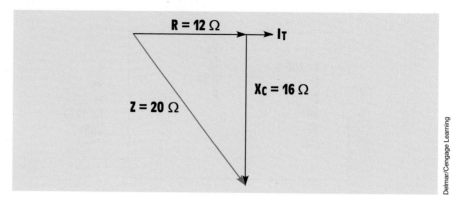

FIGURE 22–2   Resistance and capacitive reactance are 90° out of phase with each other.

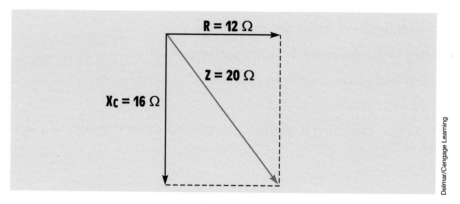

FIGURE 22–3   Impedance vector for example circuit 1.

22–3   Total Current

Now that the total impedance of the circuit is known, the total current flow (I_T) can be calculated using the formula

$$I_T = \frac{E}{Z}$$

$$I_T = \frac{240\ V}{20\ \Omega}$$

$$I_T = 12\ A$$

22-4   Voltage Drop Across the Resistor

In a series circuit, the current is the same at any point in the circuit. Therefore, 12 amperes of current flow through both the resistor and the capacitor. The **voltage drop across the resistor (E_R)** can be calculated by using the formula

$$E_R = I \times R$$
$$E_R = 12 \text{ A} \times 12 \text{ } \Omega$$
$$E_R = 144 \text{ V}$$

22-5   True Power

True power (P) for the circuit can be calculated by using any of the watts formulas as long as values that apply only to the resistive part of the circuit are used. Recall that current and voltage must be in phase with each other for true power to be produced. The formula used in this example is

$$P = E_R \times I$$
$$P = 144 \text{ V} \times 12 \text{ A}$$
$$P = 1728 \text{ W}$$

22-6   Capacitance

The amount of **capacitance (C)** can be calculated using the formula

$$C = \frac{1}{2\pi f X_C}$$

$$C = \frac{1}{377 \times 16 \text{ } \Omega}$$

$$C = \frac{1}{6032}$$

$$C = 0.0001658 \text{ F} = 165.8 \text{ } \mu\text{F}$$

22–7   Voltage Drop Across the Capacitor

The **voltage drop across the capacitor (E$_c$)** can be calculated using the formula

$$E_c = I \times X_c$$
$$E_c = 12 \text{ A} \times 16 \text{ }\Omega$$
$$E_c = 192 \text{ V}$$

22–8   Total Voltage

Although the amount of **total voltage (E$_T$)** applied to the circuit is given as 240 volt in this circuit, it is possible to calculate the total voltage if it is not known by adding together the voltage drop across the resistor and the voltage drop across the capacitor. In a series circuit, the voltage drops across the resistor and capacitor are 90° out of phase with each other and vector addition must be used. These two voltage drops form the legs of a right triangle, and the total voltage forms the hypotenuse *(Figure 22–4)*. The total voltage can be calculated using the following formula

$$E_T = \sqrt{E_R{}^2 + E_C{}^2}$$
$$E_T = \sqrt{(144 \text{ V})^2 + (192 \text{ V})^2}$$
$$E_T = 240 \text{ V}$$

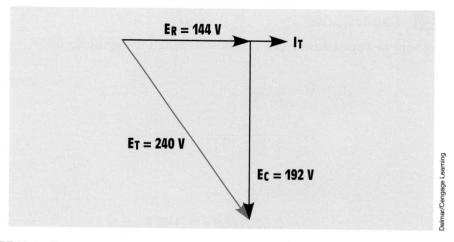

FIGURE 22–4   The voltage drops across the resistor and capacitor are 90° out of phase with each other.

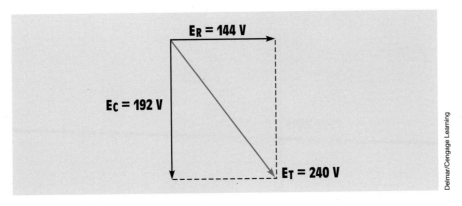

FIGURE 22–5 Voltage vector for example circuit.

A vector diagram illustrating the voltage relationships for this circuit is shown in *Figure 22–5*.

22–9 Reactive Power

The **reactive power (VARs$_c$)** in the circuit can be calculated in a manner similar to that used for watts except that reactive values of voltage and current are used instead of resistive values. In this example, the formula used is

$$VARs_c = E_c \times I$$
$$VARs_c = 192 \text{ V} \times 12 \text{ A}$$
$$VARs_c = 2304$$

22–10 Apparent Power

The apparent power (VA) of the circuit can be calculated in a manner similar to that used for watts or VARs$_c$, except that total values of voltage and current are used. In this example, the formula used is

$$VA = E_T \times I$$
$$VA = 240 \text{ V} \times 12 \text{ A}$$
$$VA = 2880$$

The apparent power can also be determined by vector addition of the true power and reactive power *(Figure 22–6)*.

$$VA = \sqrt{P^2 + VARs_c^2}$$

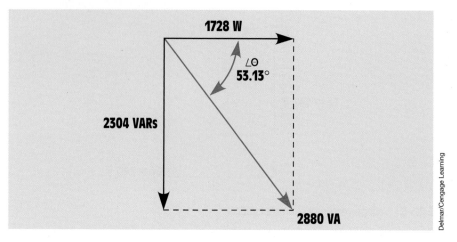

FIGURE 22–6   Apparent power vector for example circuit.

22–11   Power Factor

Power factor (PF) is a ratio of the true power to the apparent power. It can be calculated by dividing any resistive value by its like total value. In this circuit, the formula used is

$$PF = \frac{P}{VA}$$

$$PF = \frac{1728\ W}{2880\ VA}$$

$$PF = 0.6 \times 100,\ or\ 60\%$$

22–12   Angle Theta

The power factor of a circuit is the cosine of the phase angle. Because the power factor of this circuit is 0.6, angle theta ($\angle\theta$) is

$$\cos \angle\theta = PF$$

$$\cos \angle\theta = 0.6$$

$$\angle\theta = 53.13°$$

In this circuit, the current leads the applied voltage by 53.13° *(Figure 22–7)*.

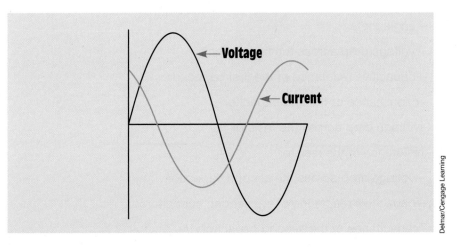

FIGURE 22–7   The current leads the voltage by 53.13°.

■ **EXAMPLE 22-1**

One resistor and two capacitors are connected in series to a 42.5-V, 60-Hz line *(Figure 22–8)*. Capacitor 1 has a reactive power of 3.75 VARs, the resistor has a true power of 5 W, and the second capacitor has a reactive power of 5.625 VARs. Find the following values:

VA—apparent power

I_T—total circuit current

Z—impedance of the circuit

PF—power factor

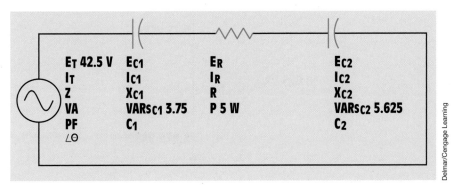

FIGURE 22–8   Example circuit.

$\angle\theta$—angle theta

E_{C1}—voltage drop across the first capacitor

X_{C1}—capacitive reactance of the first capacitor

C_1—capacitance of the first capacitor

E_R—voltage drop across the resistor

R—resistance of the resistor

E_{C2}—voltage drop across the second capacitor

X_{C2}—capacitive reactance of the second capacitor

C_2—capacitance of the second capacitor

Solution

Because the reactive power of the two capacitors is known and the true power of the resistor is known, the apparent power can be found using the formula

$$VA = \sqrt{P^2 + VARs_C^2}$$

In this circuit, $VARs_C$ is the sum of the VARs of the two capacitors. A power triangle for this circuit is shown in *Figure 22–9*.

$$VA = \sqrt{(5 \text{ W})^2 + (3.75 \text{ VARs} + 5.625 \text{ VARs})^2}$$

$$VA = \sqrt{25 \text{ W}^2 + 87.891 \text{ VARs}^2}$$

$$VA = 10.625$$

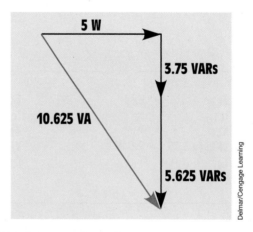

FIGURE 22–9   Power triangle for example circuit.

Now that the apparent power and applied voltage are known, the total circuit current can be calculated using the formula

$$I_T = \frac{VA}{E_T}$$

$$I_T = \frac{10.625 \text{ VA}}{42.5 \text{ V}}$$

$$I_T = 0.25 \text{ A}$$

In a series circuit, the current must be the same at any point in the circuit. Therefore, I_{C1}, I_R, and I_{C2} will all have a value of 0.25 A.

The impedance of the circuit can now be calculated using the formula

$$Z = \frac{E_T}{I_T}$$

$$Z = \frac{42.5 \text{ V}}{0.25 \text{ A}}$$

$$Z = 170 \text{ }\Omega$$

The power factor is calculated using the formula

$$PF = \frac{P}{VA}$$

$$PF = \frac{5 \text{ W}}{10.625 \text{ VA}}$$

$$PF = 0.4706, \text{ or } 47.06\%$$

The cosine of angle theta is the power factor:

$$\cos \angle\theta = 0.4706$$

$$\angle\theta = 61.93°$$

A vector diagram is shown in *Figure 22–10* illustrating the relationship of angle theta to the reactive power, true power, and apparent power.

Now that the current through each circuit element is known and the power of each element is known, the voltage drop across each element can be calculated:

$$E_{C1} = \frac{VARS_{C1}}{I_{C1}}$$

$$E_{C1} = \frac{3.75 \text{ VARs}}{0.25 \text{ A}}$$

$$E_{C1} = 15 \text{ V}$$

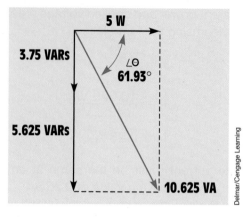

FIGURE 22–10 Vector relationship of reactive, true, and apparent power.

$$E_R = \frac{P}{I_R}$$

$$E_R = \frac{5\ W}{0.25\ A}$$

$$E_R = 20\ V$$

$$E_{C2} = \frac{VARS_{C2}}{I_{C2}}$$

$$E_{C2} = \frac{5.625\ VARS}{0.25\ A}$$

$$E_{C2} = 22.5\ V$$

The capacitive reactance of the first capacitor is

$$X_{C1} = \frac{E_{C1}}{I_{C1}}$$

$$X_{C1} = \frac{15\ V}{0.25\ A}$$

$$X_{C1} = 60\ \Omega$$

The capacitance of the first capacitor is

$$C_1 = \frac{1}{2\pi f X_{C1}}$$

$$C_1 = \frac{1}{377 \times 30 \ \Omega}$$

$$C_1 = 0.0000442 \ \text{F, or } 44.2 \ \mu\text{F}$$

The resistance of the resistor is

$$R = \frac{E_R}{I_R}$$

$$R = \frac{20 \ \text{V}}{0.25 \ \text{A}}$$

$$R = 80 \ \Omega$$

The capacitive reactance of the second capacitor is

$$X_{C2} = \frac{E_{C2}}{I_{C2}}$$

$$X_{C2} = \frac{22.5 \ \text{V}}{0.25 \ \text{A}}$$

$$X_{C2} = 90 \ \Omega$$

The capacitance of the second capacitor is

$$C_2 = \frac{1}{2\pi f X_{C2}}$$

$$C_2 = \frac{1}{377 \times 90 \ \Omega}$$

$$C_2 = 0.0000295 \ \text{F, or } 29.5 \ \mu\text{F}$$

The completed circuit with all values is shown in *Figure 22–11*.

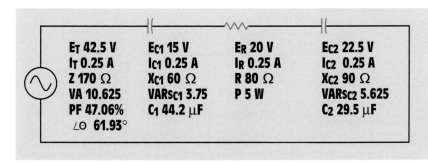

E_T 42.5 V	E_{C1} 15 V	E_R 20 V	E_{C2} 22.5 V
I_T 0.25 A	I_{C1} 0.25 A	I_R 0.25 A	I_{C2} 0.25 A
Z 170 Ω	X_{C1} 60 Ω	R 80 Ω	X_{C2} 90 Ω
VA 10.625	VARS$_{C1}$ 3.75	P 5 W	VARS$_{C2}$ 5.625
PF 47.06%	C_1 44.2 µF		C_2 29.5 µF
∠Θ 61.93°			

Delmar/Cengage Learning

FIGURE 22–11 Example circuit with completed values.

■ EXAMPLE 22-2

A small indicating lamp has a rating of 2 W when connected to 120 V. The lamp must be connected to a voltage of 480 V at 60 Hz. A capacitor will be connected in series with the lamp to reduce the circuit current to the proper value. What value of capacitor will be needed to perform this job?

Solution

The first step is to determine the amount of current the lamp will normally draw when connected to a 120-V line:

$$I_{LAMP} = \frac{P}{E}$$

$$I_{LAMP} = \frac{2\ W}{120\ V}$$

$$I_{LAMP} = 0.0167\ A$$

The next step is to determine the amount of voltage that must be dropped across the capacitor when a current of 0.01667 A flows through it. Because the voltage dropped across the resistor and the voltage dropped across the capacitor are 90° out of phase with each other, vectors must be used to determine the voltage drop across the capacitor *(Figure 22–12)*. The voltage drop across the capacitor can be calculated using the formula

$$E_C = \sqrt{E_T^2 - E_R^2}$$
$$E_C = \sqrt{(480\ V)^2 - (120\ V)^2}$$
$$E_C = \sqrt{216,000\ V^2}$$
$$E_C = 464.758\ V$$

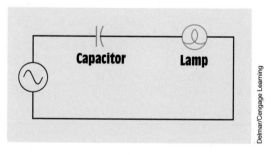

FIGURE 22–12 Determining voltage drop across the capacitor.

Delmar/Cengage Learning

Now that the voltage drop across the capacitor and the amount of current flow are known, the capacitive reactance can be calculated:

$$X_C = \frac{E_C}{I}$$

$$X_C = \frac{464.758 \text{ V}}{0.0167 \text{ A}}$$

$$X_C = 27{,}829.82 \ \Omega$$

The amount of capacitance needed to produce this capacitive reactance can now be calculated using the formula

$$C = \frac{1}{2\pi f X_C}$$

$$C = \frac{1}{377 \times 27{,}829.82 \ \Omega}$$

$$C = 0.0000000953 \text{ F, or } 95.3 \text{ nF}$$

The circuit containing the lamp and capacitor is shown in *Figure 22–13*.

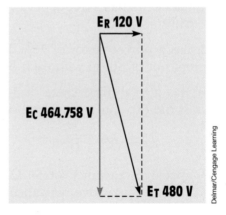

FIGURE 22–13   The capacitor reduces the current to the lamp.

Summary

- In a pure capacitive circuit, the voltage and current are 90° out of phase with each other.

- In a pure resistive circuit, the voltage and current are in phase with each other.

▪ In a circuit containing resistance and capacitance, the voltage and current will be out of phase with each other by some amount between 0° and 90°.

▪ The amount of phase angle difference between voltage and current in an RC series circuit is the ratio of resistance to capacitance.

▪ In a series circuit, the current flow through all components is the same. Therefore, the voltage drops across the resistive and capacitive parts become out of phase with each other.

▪ True power can be produced by resistive parts of the circuit only.

▪ Power factor is the ratio of true power to apparent power.

Review Questions

Refer to the formulas in the Resistive-Capacitive Series Circuits Formula section of Appendix B.

1. In a pure capacitive circuit, does the current lead or lag the voltage?

2. A series circuit contains a 20-Ω resistor and a capacitor with a capacitance of 110.5 μF. If the circuit has a frequency of 60 Hz, what is the total impedance of the circuit?

3. An RC series circuit has a power factor of 76%. How many degrees are the voltage and current out of phase with each other?

4. An RC series circuit has a total impedance of 84 Ω. The resistor has a value of 32 Ω. What is the capacitive reactance of the capacitor?

$$X_C = \sqrt{Z^2 - R^2}$$

5. A capacitor has a capacitive reactance of 50 Ω when connected to a 60-Hz line. What will be the capacitive reactance if the capacitor is connected to a 1000-Hz line?

Practice Problems

Refer to the formulas in the Resistive-Capacitive Series Circuits Formula section of Appendix B and to *Figure 22–1*.

1. Assume that the circuit shown in *Figure 22–1* is connected to a 480-V, 60-Hz line. The capacitor has a capacitance of 165.782 μF, and the resistor has a resistance of 12 Ω. Find the missing values.

E_T 480 V _____ E_R _____ E_C _____

I_T _____ I_R _____ I_C _____

Z _____ R 12 Ω _____ X_C _____

VA _____ P _____ $VARs_C$ _____

PF _____ $\angle\theta$ _____ C 165.782 μF _____

2. Assume that the voltage drop across the resistor, E_R, is 78 V; the voltage drop across the capacitor, E_C, is 104 V; and the circuit has a total impedance, Z, of 20 Ω. The frequency of the AC voltage is 60 Hz. Find the missing values.

E_T _____ E_R 78 V _____ E_C 104 V _____

I_T _____ I_R _____ I_C _____

Z 20 Ω _____ R _____ X_C _____

VA _____ P _____ $VARs_C$ _____

PF _____ $\angle\theta$ _____ C _____

3. Assume the circuit shown in *Figure 22–1* has an apparent power of 432 VA and a true power of 345.6 W. The capacitor has a capacitance of 15.8919 μF, and the frequency is 60 Hz. Find the missing values.

E_T _____ E_R _____ E_C _____

I_T _____ I_R _____ I_C _____

Z _____ R _____ X_C _____

VA 432 _____ P 345.6 W _____ $VARs_C$ _____

PF _____ $\angle\theta$ _____ C 15.8919 μF _____

4. Assume the circuit in *Figure 22–1* has a power factor of 68%, an apparent power of 300 VA, and a frequency of 400 Hz. The capacitor has a capacitance of 4.7125 μF. Find the missing values.

E_T _____ E_R _____ E_C _____

I_T _____ I_R _____ I_C _____

Z _____ R _____ X_C _____

VA 300 _____ P _____ $VARs_C$ _____

PF 68% _____ $\angle\theta$ _____ C 4.7125 μF _____

5. In a series RC circuit, E_T = 240 volts, R = 60 Ω, and X_C = 85 Ω. Find E_C.

6. In a series RC circuit, E_T = 120 volts, R = 124 Ω, and X_C = 64 Ω. Find reactive power.

7. In a series RC circuit, E_T = 208 volts, I_T = 2.4 amperes, and R = 45 Ω. Find the power factor.

8. In a series RC circuit, E_T = 460 volts and $\angle\theta$ = 44°. Find E_C.

9. In a series RC circuit, E_T = 240 volts at 60 Hz. An ammeter indicates a total current of 0.75 amperes. The resistor has a value of 140 Ω. What is the capacitance of the capacitor?

10. In a series RC circuit, the apparent power is 4,250 VA and the reactive power is 2125 VARs. What is the true power?

Unit 23

Resistive-Capacitive Parallel Circuits

Why You Need to Know

*T*he relationship of voltage, current, impedance, and power when resistance and capacitance are connected in parallel with each other is very similar to resistive-inductive parallel circuits. Although it may seem that the units on resistive-capacitive series circuits and resistive-capacitive parallel circuits are a repeat of the information covered previously, they are an important step toward understanding what happens when elements of resistance, inductance, and capacitance are combined into the same circuit. This unit

- discusses parallel circuits that contain both resistance and capacitance and the effect on voltage, current, power, and impedance.
- illustrates how to calculate circuit current and voltage and their relationship in a resistive-capacitive parallel circuit.

OUTLINE

KEY TERMS

Circuit impedance (Z)

Current flow through the capacitor (I_C)

Phase angle shift

Total circuit current (I_T)

Objectives

After studying this unit, you should be able to

■ discuss the operation of a parallel circuit containing resistance and capacitance.

■ calculate circuit values of an RC parallel circuit.

■ connect an RC parallel circuit and measure circuit values with test instruments.

Preview

This unit discusses the relationship of different electrical quantities such as voltage, current, impedance, and power in a circuit that contains both resistance and capacitance connected in parallel. Because all components connected in parallel must share the same voltage, the current flow through different components will be out of phase with each other. The effect this condition has on other circuit quantities is explored. ■

23–1   Operation of RC Parallel Circuits

When resistance and capacitance are connected in parallel, the voltage across all the devices will be in phase and will have the same value. The current flow through the capacitor, however, will be 90° out of phase with the current flow through the resistor *(Figure 23–1)*. The amount of **phase angle shift** between the total circuit current and voltage is determined by the ratio of the amount

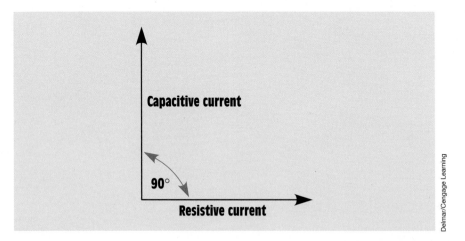

FIGURE 23–1   Current flow through the capacitor is 90° out of phase with current flow through the resistor.

of resistance to the amount of capacitance. The circuit power factor is still determined by the ratio of resistance and capacitance.

23–2 Calculating Circuit Values

■ EXAMPLE 23–1

In the RC parallel circuit shown in *Figure 23–2*, assume that a resistance of 30 Ω is connected in parallel with a capacitive reactance of 20 Ω. The circuit is connected to a voltage of 240 VAC and a frequency of 60 Hz. Calculate the following circuit values:

I_R—current flow through the resistor

P—watts (true power)

I_C—current flow through the capacitor

$VARs_C$—volt-amperes reactive (reactive power)

C—capacitance of the capacitor

I_T—total circuit current

Z—total impedance of the circuit

VA—volt-amperes (apparent power)

PF—power factor

$\angle\theta$—angle theta (phase angle of voltage and current)

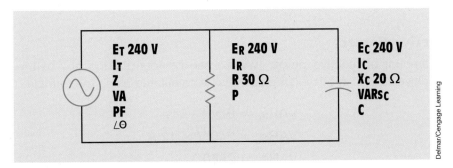

FIGURE 23–2 Resistive-capacitive parallel circuit.

Solution

Resistive Current

The amount of current flow through the resistor (I_R) can be calculated by using the formula

$$I_R = \frac{E}{R}$$

$$I_R = \frac{240 \text{ V}}{30 \text{ }\Omega}$$

$$I_R = 8 \text{ A}$$

True Power

The amount of total true power (P) in the circuit can be determined by using any of the values associated with the pure resistive part of the circuit. In this example, true power is found using the formula

$$P = E \times I_R$$

$$P = 240 \text{ V} \times 8 \text{ A}$$

$$P = 1920 \text{ W}$$

Capacitive Current

The amount of **current flow through the capacitor (I_C)** is calculated using the formula

$$I_C = \frac{E}{X_C}$$

$$I_C = \frac{240 \text{ V}}{20 \text{ }\Omega}$$

$$I_C = 12 \text{ A}$$

Reactive Power

The amount of reactive power ($VARs_C$) can be found using any of the total capacitive values. In this example, $VARs_C$ is calculated using the formula

$$VARs_C = E \times I_C$$

$$VARs_C = 240 \text{ V} \times 12 \text{ A}$$

$$VARs_C = 2880$$

Capacitance

The capacitance of the capacitor can be calculated using the formula

$$C = \frac{1}{2\pi f X_C}$$

$$C = \frac{1}{377 \times 20\ \Omega}$$

$$C = \frac{1}{7540}$$

$$C = 0.0001326\ F = 132.6\ \mu F$$

Total Current

The voltage is the same across all legs of a parallel circuit. The current flow through the resistor is in phase with the voltage, and the current flow through the capacitor is leading the voltage by 90° *(Figure 23–3)*. The 90° difference in capacitive and resistive current forms a right triangle as shown in *Figure 23–4*. Because these two currents are connected in parallel, vector addition can be used to find the total current flow in the circuit

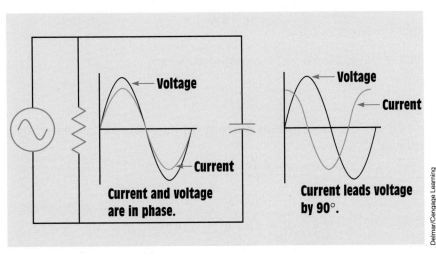

FIGURE 23–3 Phase relationship of current and voltage in an RC parallel circuit.

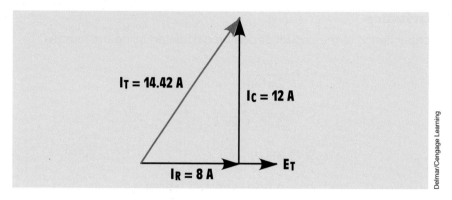

FIGURE 23–4 Resistance current and capacitive current are 90° out of phase with each other.

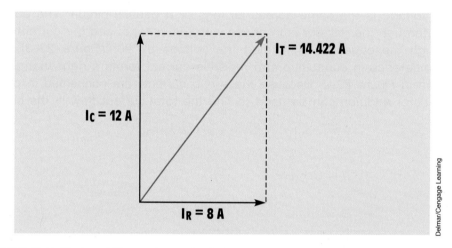

FIGURE 23–5 Vector addition can be used to find total current.

(Figure 23–5). The **total circuit current (I_T)** flow can be calculated by using the formula:

$$I_T = \sqrt{I_R^2 + I_C^2}$$

$$I_T = \sqrt{(8A)^2 + (12\ A)^2}$$

$$I_T = \sqrt{64\ A^2 + 144\ A^2}$$

$$I_T = \sqrt{208\ A^2}$$

$$I_T = 14.422\ A$$

Impedance

The total **circuit impedance (Z)** can be found by using any of the total values and substituting Z for R in an Ohm's law formula. The total impedance of this circuit is calculated using the formula

$$Z = \frac{E}{I_T}$$

$$Z = \frac{240 \text{ V}}{14.422 \text{ A}}$$

$$Z = 16.641 \ \Omega$$

The impedance can also be found by adding the reciprocals of the resistance and capacitive reactance. Because the resistance and capacitive reactance are 90° out of phase with each other, vector addition must be used:

$$Z = \frac{1}{\sqrt{\left(\frac{1}{R}\right)^2 + \left(\frac{1}{X_C}\right)^2}}$$

Another formula that can be used to determine the impedance in a circuit that contains both resistance and capacitive reactance is

$$Z = \frac{R \times X_C}{\sqrt{R^2 + X_C^2}}$$

Apparent Power

The apparent power (VA) can be calculated by multiplying the circuit voltage by the total current flow:

$$VA = E \times I_T$$

$$VA = 240 \text{ V} \times 14.422 \text{ A}$$

$$VA = 3461.28$$

Power Factor

The power factor (PF) is the ratio of true power to apparent power. The circuit power factor can be calculated using the formula

$$PF = \frac{W}{VA} \times 100$$

$$PF = \frac{1920}{3461.28}$$

$$PF = 0.5547, \text{ or } 55.47\%$$

Angle Theta

The cosine of angle theta ($\angle\theta$) is equal to the power factor:

$$\cos \angle\theta = 0.5547$$

$$\angle\theta = 56.31°$$

A vector diagram of apparent, true, and reactive power is shown in *Figure 23–6*. Angle theta is the angle developed between the apparent and true power. The complete circuit with all values is shown in *Figure 23–7*.

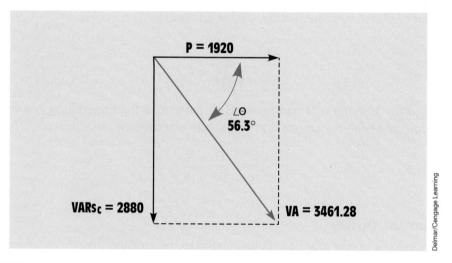

FIGURE 23–6 Vector relationship of apparent, true, and reactive power.

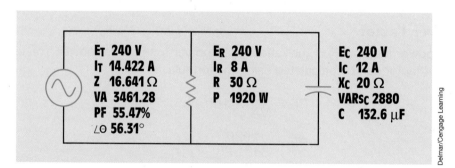

FIGURE 23–7 Example circuit with all calculated values.

■ EXAMPLE 23-2

In this circuit, a resistor and a capacitor are connected in parallel to a 400-Hz line. The power factor is 47.1%, the apparent power is 4086.13 VA, and the capacitance of the capacitor is 33.15 μF *(Figure 23–8)*. Find the following unknown values:

$\angle\theta$—angle theta

P—true power

$VARs_C$—capacitive VARs

X_C—capacitive reactance

E_C—voltage drop across the capacitor

I_C—capacitive current

E_R—voltage drop across the resistor

I_R—resistive current

R—resistance of the resistor

E_T—applied voltage

I_T—total circuit current

Z—impedance of the circuit

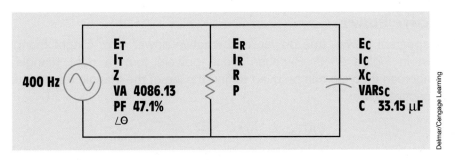

FIGURE 23–8   Example circuit.

Solution
Angle Theta

The power factor is the cosine of angle theta. To find angle theta, change the power factor from a percentage into a decimal fraction by dividing by 100:

$$PF = \frac{47.1}{100}$$

$$PF = 0.471$$

$$\cos \angle\theta = 0.471$$

$$\angle\theta = 61.9°$$

True Power

The power factor is determined by the ratio of true power to apparent power:

$$PF = \frac{P}{VA}$$

This formula can be changed to calculated the true power when the power factor and apparent power are known (refer to the Resistive-Capacitive Parallel Circuits Formula Section of Appendix B).

$$P = VA \times PF$$

$$P = 4086.13 \times 0.471$$

$$P = 1924.47 \text{ W}$$

Reactive Power

The apparent power, true power, and reactive power form a right triangle as shown in *Figure 23–9*. Because these powers form a right triangle, the Pythagorean theorem can be used to find the leg of the triangle represented by the reactive power:

$$VARs_C = \sqrt{VA^2 - P^2}$$

$$VARs_C = \sqrt{4086.13^2 - 1924.47^2}$$

$$VARs_C = \sqrt{12,992,873.6}$$

$$VARs_C = 3604.56$$

A vector diagram showing the relationship of apparent power, true power, reactive power, and angle theta is shown in *Figure 23–10*.

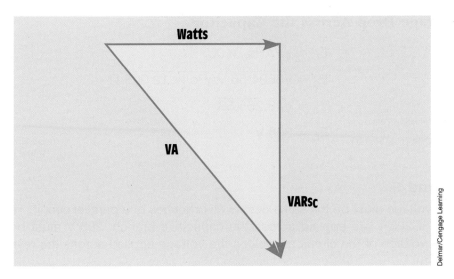

FIGURE 23–9 Right triangle formed by the apparent, true, and reactive powers.

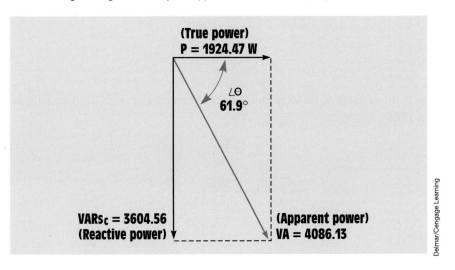

FIGURE 23–10 Vector diagram of apparent power, true power, reactive power, and angle theta.

Capacitive Reactance

Because the capacitance of the capacitor and the frequency are known, the capacitive reactance can be found using the formula

$$X_C = \frac{1}{2\pi fC}$$

$$X_C = \frac{1 \text{ Hz}}{2 \times 3.1416 \times 400 \text{ Hz} \times 0.00003315 \text{ F}}$$

$$X_C = 12.003 \ \Omega$$

Voltage Drop Across the Capacitor

$$E_C = \sqrt{VARs_C \times X_C}$$

$$E_C = \sqrt{3604.56 \text{ VARs} \times 12.003 \text{ } \Omega}$$

$$E_C = \sqrt{43{,}265.53}$$

$$E_C = 208 \text{ V}$$

E_R and E_T

The voltage must be the same across all branches of a parallel circuit. Therefore, if 208 V are applied across the capacitive branch, 208 V must be the total voltage of the circuit as well as the voltage applied across the resistive branch:

$$E_T = 208 \text{ V}$$

$$E_R = 208 \text{ V}$$

I_C

The amount of current flowing in the capacitive branch can be calculated using the formula

$$I_C = \frac{E_C}{X_C}$$

$$I_C = \frac{208 \text{ V}}{12 \text{ } \Omega}$$

$$I_C = 17.331$$

I_R

The amount of current flowing through the resistor can be calculated using the formula

$$I_R = \frac{P}{E_R}$$

$$I_R = \frac{1924.47 \text{ W}}{208}$$

$$I_R = 9.25 \text{ A}$$

Resistance

The amount of resistance can be calculated using the formula

$$R = \frac{E_R}{I_R}$$

$$R = \frac{208 \text{ V}}{9.25 \text{ A}}$$

$$R = 22.49 \ \Omega$$

Total Current

The total current can be calculated using Ohm's law or by vector addition because both the resistive and capacitive currents are known. Vector addition is used in this example:

$$I_T = \sqrt{I_R{}^2 + I_C{}^2}$$

$$I_T = \sqrt{(9.25 \text{ A})^2 + (17.33 \text{ A})^2}$$

$$I_T = \sqrt{385.89 \text{ A}^2}$$

$$I_T = 19.64 \text{ A}$$

Impedance

The impedance of the circuit is calculated using the formula

$$Z = \frac{R \times X_C}{\sqrt{R^2 + X_C{}^2}}$$

$$Z = \frac{22.479 \ \Omega \times 12.003 \ \Omega}{\sqrt{(22.479 \ \Omega)^2 + (12.003 \ \Omega)^2}}$$

$$Z = \frac{269.815 \ \Omega}{\sqrt{649.377 \ \Omega^2}}$$

$$Z = \frac{269.815 \ \Omega}{25.483 \ \Omega}$$

$$Z = 10.588 \ \Omega$$

The complete circuit with all values is shown in *Figure 23–11*.

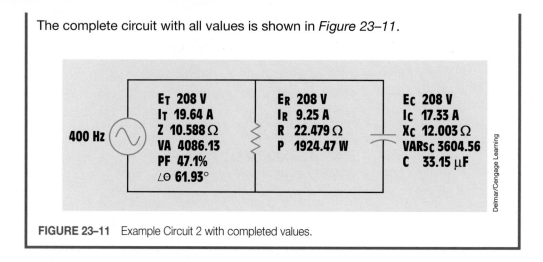

400 Hz

E_T 208 V	E_R 208 V	E_C 208 V
I_T 19.64 A	I_R 9.25 A	I_C 17.33 A
Z 10.588 Ω	R 22.479 Ω	X_C 12.003 Ω
VA 4086.13	P 1924.47 W	VARS$_C$ 3604.56
PF 47.1%		C 33.15 μF
∠θ 61.93°		

Delmar/Cengage Learning

FIGURE 23–11 Example Circuit 2 with completed values.

Summary

- The current flow in the resistive part of the circuit is in phase with the voltage.
- The current flow in the capacitive part of the circuit leads the voltage by 90°.
- The amount the current and voltage are out of phase with each other is determined by the ratio of resistance to capacitance.
- The voltage is the same across any leg of a parallel circuit.
- The circuit power factor is the ratio of true power to apparent power.

Review Questions

1. When a capacitor and a resistor are connected in parallel, how many degrees out of phase are the current flow through the resistor and the current flow through the capacitor?

2. A capacitor and a resistor are connected in parallel to a 120-V, 60-Hz line. The resistor has a resistance of 40 Ω, and the capacitor has a capacitance of 132.6 μF. What is the total current flow through the circuit?

3. What is the impedance of the circuit in Question 2?

4. What is the power factor of the circuit in Question 2?

5. How many degrees out of phase are the current and voltage in Question 2?

Practice Problems

Refer to the formulas in the Resistive-Capacitive Parallel Circuits section of Appendix B and to *Figure 23–2*.

1. Assume that the circuit shown in *Figure 23–2* is connected to a 60-Hz line and has a total current flow of 10.463 A. The capacitor has a capacitance of 123.626 μF, and the resistor has a resistance of 14 Ω. Find the missing values.

E_T _____	E_R _____	E_C _____
I_T 10.463 A	I_R _____	I_C _____
Z _____	R 14 Ω	X_C _____
VA _____	P _____	VARS$_C$ _____
PF _____	$\angle\theta$ _____	C 132.626 μF

2. Assume that the circuit is connected to a 400-Hz line and has a total impedance of 21.6 Ω. The resistor has a resistance of 36 Ω, and the capacitor has a current flow of 2 A through it. Find the missing values.

E_T _____	E_R _____	E_C _____
I_T _____	I_R _____	I_C 2 A
Z 21.6 Ω	R 36 Ω	X_C _____
VA _____	P _____	VARS$_C$ _____
PF _____	$\angle\theta$ _____	C _____

3. Assume that the circuit shown in *Figure 23–2* is connected to a 600-Hz line and has a current flow through the resistor of 65.6 A and a current flow through the capacitor of 124.8 A. The total impedance of the circuit is 2.17888 Ω. Find the missing values.

E_T _____	E_R _____	E_C _____
I_T _____	I_R 65.6 A	I_C 124.8 A
Z 2.17888 Ω	R _____	X_C _____
VA _____	P _____	VARS$_C$ _____
PF _____	$\angle\theta$ _____	C _____

4. Assume that the circuit shown in *Figure 23–2* is connected to a 1000-Hz line and has a true power of 486.75 W and a reactive power of

187.5 VARs. The total current flow in the circuit is 7.5 A. Find the missing values.

E_T _____	E_R _____	E_C _____
I_T 7.5 A	I_R _____	I_C _____
Z _____	R _____	X_C _____
VA _____	P 486.75 W	VARS$_C$ 187.5
PF _____	$\angle\theta$ _____	C _____

5. In an RC parallel circuit, R = 3.6 kΩ and X_C = 4.7 kΩ. Find Z.

6. In an RC parallel circuit, I_R = 0.6 amperes, R = 24 Ω, and X_C = 33 Ω. Find I_C.

7. In an RC parallel circuit, E_T = 120 volts, I_T = 1.2 amperes, and R = 240 Ω. Find X_C.

8. In an RC parallel circuit, the apparent power is 3400 VA and $\angle\theta$ = 58°. Find reactive power.

9. In an RC parallel circuit, the true power is 780 watts and the reactive power is 560 VARs. Find the power factor.

10. In an RC parallel circuit, E_T = 7.5 volts at 1 kHz. The circuit current is 2.214 amperes. R = 4 Ω. What is the value of the capacitor connected in the circuit?

Objectives

After studying this unit, you should be able to

- discuss AC circuits that contain resistance, inductance, and capacitance connected in series.

- connect an RLC series circuit.

- calculate values of impedance, inductance, capacitance, power, VARs, reactive power, voltage drop across individual components, power factor, and phase angle of voltage and current.

- discuss series resonant circuits.

Preview

Circuits containing resistance, inductance, and capacitance connected in series are presented in this unit. Electrical quantities for voltage drop, impedance, and power are calculated for the total circuit values and for individual components. Circuits that become resonant at a certain frequency are presented as well as the effect a resonant circuit has on electrical quantities such as voltage, current, and impedance. ■

24–1   RLC Series Circuits

When an AC circuit contains elements of resistance, inductance, and capacitance connected in series, the ***current is the same*** through all components, but the ***voltages dropped across the elements are out of phase*** with each other. The voltage dropped across the resistance is in phase with the current; the voltage dropped across the inductor leads the current by 90°; and the voltage dropped across the capacitor lags the current by 90° *(Figure 24–1)*. An RLC series circuit is shown in *Figure 24–2*. The ratio of resistance, inductance, and capacitance determines how much the applied voltage leads or lags the circuit current. If the circuit contains more inductive VARs than capacitive VARs, the current lags the applied voltage and the power factor is a **lagging power factor.** If there are more capacitive VARs than inductive VARs, the current leads the voltage and the power factor is a **leading power factor.**

Because inductive reactance and capacitive reactance are 180° out of phase with each other, they cancel each other in an AC circuit. This cancellation can permit the impedance of the circuit to become less than either or both of the reactances, producing a high amount of current flow through the circuit. When Ohm's law is applied to the circuit values, note that the voltage drops developed across these components can be higher than the applied voltage.

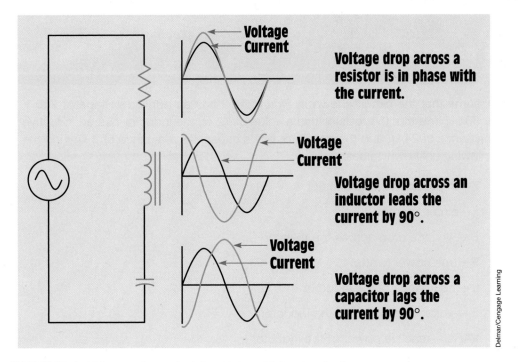

FIGURE 24–1   Voltage and current relationship in an RLC series circuit.

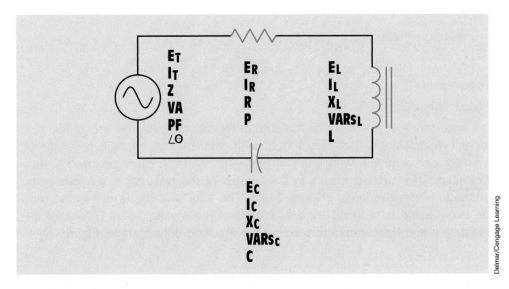

FIGURE 24–2   Resistive-inductive-capacitive series circuit.

■ EXAMPLE 24-1

Assume that the circuit shown in *Figure 24–2* has an applied voltage of 240 V at 60 Hz and that the resistor has a value of 12 Ω, the inductor has an inductive reactance of 24 Ω, and the capacitor has a capacitive reactance of 8 Ω. Find the following unknown values:

Z—impedance of the circuit

I_T—circuit current

E_R—voltage drop across the resistor

P—true power (watts)

L—inductance of the inductor

E_L—voltage drop across the inductor

$VARs_L$—reactive power of the inductor

C—capacitance

E_C—voltage drop across the capacitor

$VARs_C$—reactive power of the capacitor

VA—volt-amperes (apparent power)

PF—power factor

$\angle\theta$—angle theta

Solution
Total Impedance

The impedance of the circuit is the sum of resistance, inductive reactance, and capacitive reactance. Because inductive reactance and capacitive reactance are 180° out of phase with each other, vector addition must be used to find their sum. This method results in the smaller of the two reactive values being subtracted from the larger *(Figure 24–3)*. The smaller value is eliminated, and the larger value is reduced by the amount of the smaller value. The total impedance is the hypotenuse formed by the resulting right triangle *(Figure 24–4)*.

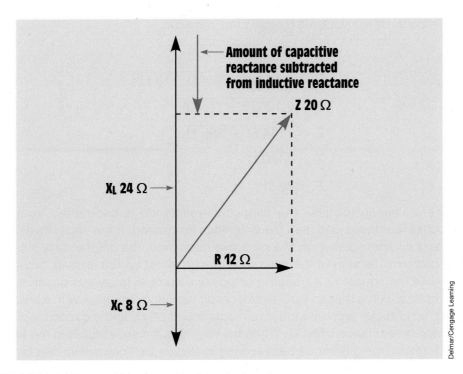

FIGURE 24–3 Vector addition is used to determine impedance.

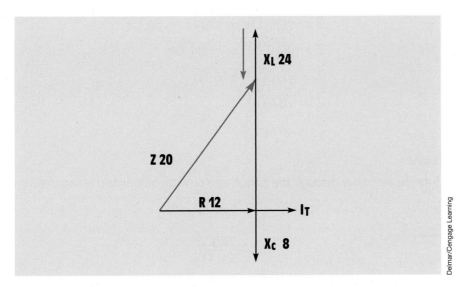

FIGURE 24–4 Right triangle formed by circuit impedance.

The impedance is calculated by using the formula

$$Z = \sqrt{R^2 + (X_L - X_C)^2}$$

$$Z = \sqrt{(12 \ \Omega)^2 + (24 \ \Omega - 8 \ \Omega)^2}$$

$$Z = \sqrt{(12 \ \Omega)^2 + (16 \ \Omega)^2}$$

$$Z = \sqrt{144 \ \Omega + 256 \ \Omega}$$

$$Z = \sqrt{400 \ \Omega}$$

$$Z = 20 \ \Omega$$

In the preceding formula, the capacitive reactance is subtracted from the inductive reactance and then the difference is squared. If the capacitive reactance is a larger value than the inductive reactance, the difference is a negative number. The sign of the difference has no effect on the answer, however, because the square of a negative or positive number is always positive. For example, assume that an RLC series circuit contains a resistor with a value of 10 Ω, an inductor with an inductive reactance of 30 Ω, and a capacitor with a capacitive reactance of 54 Ω. When these values are substituted in the previous formula, the difference between the inductive and capacitive reactances is a negative number:

$$Z = \sqrt{R^2 + (X_L - X_C)^2}$$

$$Z = \sqrt{(10 \ \Omega)^2 + (30 \ \Omega - 54 \ \Omega)^2}$$

$$Z = \sqrt{(10 \ \Omega)^2 + (-24 \ \Omega)^2}$$

$$Z = \sqrt{(100 + 576 \ \Omega)^2}$$

$$Z = \sqrt{676 \ \Omega}$$

$$Z = 26 \ \Omega$$

Current

The total current flow through the circuit can now be calculated using the formula

$$I_T = \frac{E_T}{Z}$$

$$I_T = \frac{240 \ V}{20 \ \Omega}$$

$$I_T = 12 \ A$$

In a series circuit, the current flow is the same at any point in the circuit. Therefore, 12 A flow through each of the circuit components.

Resistive Voltage Drop

The voltage drop across the resistor can be calculated using the formula

$$E_R = I_R \times R$$
$$E_R = 12 \text{ A} \times 12 \text{ } \Omega$$
$$E_R = 144 \text{ V}$$

Watts

The true power of the circuit can be calculated using any of the pure resistive values. In this example, true power is found using the formula

$$P = E_R \times I$$
$$P = 144 \text{ V} \times 12 \text{ A}$$
$$P = 1728 \text{ W}$$

Inductance

The amount of inductance in the circuit can be calculated using the formula

$$L = \frac{X_L}{2\pi f}$$
$$L = \frac{24 \text{ } \Omega}{377}$$
$$L = 0.0637 \text{H}$$

Voltage Drop Across the Inductor

The amount of voltage drop across the inductor can be calculated using the formula

$$E_L = I \times X_L$$
$$E_L = 12 \text{ A} \times 24 \text{ } \Omega$$
$$E_L = 288 \text{ V}$$

Notice that the voltage drop across the inductor is greater than the applied voltage.

Inductive VARs

The amount of reactive power of the inductor can be calculated by using inductive values:

$$\text{VARS}_L = E_L \times I$$
$$\text{VARS}_L = 288 \text{ V} \times 12 \text{ A}$$
$$\text{VARS}_L = 3456$$

Capacitance

The amount of capacitance in the circuit can be calculated by using the formula

$$C = \frac{1}{2\pi f X_C}$$

$$C = \frac{1}{377 \times 8\ \Omega}$$

$$C = \frac{1}{3016}$$

$$C = 0.000331565\ \text{F, or } 331.565\ \mu\text{F}$$

Voltage Drop Across the Capacitor

The voltage dropped across the capacitor can be calculated using the formula

$$E_C = I \times X_C$$

$$E_C = 12\ \text{A} \times 8\ \Omega$$

$$E_C = 96\ \text{V}$$

Capacitive VARs

The amount of capacitive VARs can be calculated using the formula

$$\text{VARs}_C = E_C \times I$$

$$\text{VARs}_C = 96\ \text{V} \times 12\ \text{A}$$

$$\text{VARs}_C = 1152$$

Apparent Power

The VAs (apparent power) can be calculated by multiplying the applied voltage and the circuit current:

$$VA = E_T \times I$$

$$VA = 240\ \text{V} \times 12\ \text{A}$$

$$VA = 2880$$

The apparent power can also be found by vector addition of true power, inductive VARs, and capacitive VARs *(Figure 24–5)*. As with the addition of resistance, inductive reactance, and capacitive reactance, inductive VARs, VARs_L, and capacitive VARs, VARs_C are 180° out of phase with each other. The result is the elimination

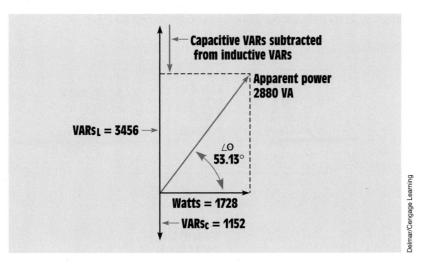

FIGURE 24–5 Vector addition of apparent power, true power, and reactive power.

of the smaller and a reduction of the larger. The following formula can be used to determine apparent power:

$$VA = \sqrt{P^2 + (VARs_L - VARs_C)^2}$$
$$VA = \sqrt{(1728 \text{ W})^2 + (3456 \text{ VARs}_L - 1152 \text{ VARs}_C)^2}$$
$$VA = \sqrt{(1728 \text{ W})^2 + (2304 \text{ VARs})^2}$$
$$VA = \sqrt{8{,}294{,}400}$$
$$VA = 2880$$

Power Factor

The power factor can be calculated by dividing the true power of the circuit by the apparent power. The answer is multiplied by 100 to change the decimal into a percent:

$$PF = \frac{W}{VA} \times 100$$
$$PF = \frac{1728 \text{ W}}{2880 \text{ VA}} \times 100$$
$$PF = 0.06 \times 100$$
$$PF = 60\%$$

Angle Theta

The power factor is the cosine of angle theta:

$$\cos \angle\theta = 0.60$$
$$\angle\theta = 53.13°$$

Delmar/Cengage Learning

The circuit with all calculated values is shown in *Figure 24–6.*

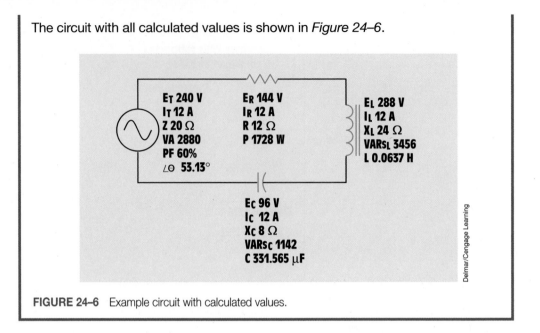

FIGURE 24–6   Example circuit with calculated values.

■ EXAMPLE 24-2

An RLC series circuit contains a capacitor with a capacitance of 66.3 μF, an inductor with an inductance of 0.0663 H, and a resistor with a value of 8 Ω connected to a 120-V, 60-Hz line *(Figure 24–7)*. How much current will flow in this circuit?

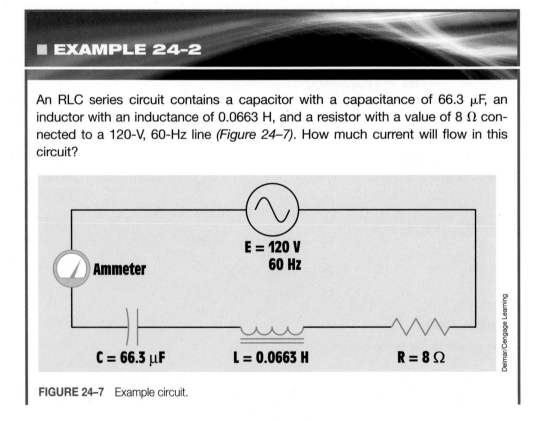

FIGURE 24–7   Example circuit.

Solution

The first step in solving this problem is to find the values of capacitive and inductive reactance:

$$X_C = \frac{1}{2\pi fC}$$

$$X_C = \frac{1}{377 \times 0.0000663 \text{ F}}$$

$$X_C = 40.008 \ \Omega$$

$$X_L = 2\pi fL$$

$$X_L = 377 \times 0.0663 \text{ H}$$

$$X_L = 24.995 \ \Omega$$

Now that the capacitive and inductive reactance values are known, the circuit impedance can be found using the formula

$$Z = \sqrt{R^2 + (X_L - X_C)^2}$$

$$Z = \sqrt{(8 \ \Omega)^2 + (24.995 \ \Omega - 40.008 \ \Omega)^2}$$

$$Z = \sqrt{64 \ \Omega + 225.39 \ \Omega}$$

$$Z = 17.011 \ \Omega$$

Now that the circuit impedance is known, the current flow can be found using Ohm's law

$$I_T = \frac{E_T}{Z}$$

$$I_T = \frac{120 \text{ V}}{17.011 \ \Omega}$$

$$I_T = 7.054 \text{ A}$$

■ EXAMPLE 24-3

The RLC series circuit shown in *Figure 24–8* contains an inductor with an inductive reactance of 62 Ω and a capacitor with a capacitive reactance of 38 Ω. The circuit is connected to a 208-V, 60-Hz line. How much resistance should be connected in the circuit to limit the circuit current to a value of 8 A?

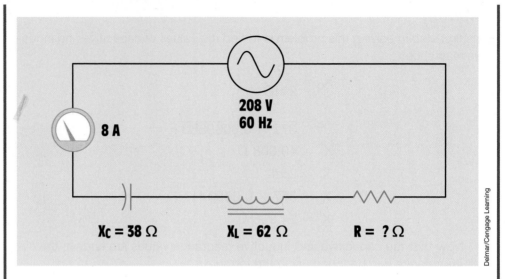

FIGURE 24–8 Example circuit.

Solution

The first step is to determine the total impedance necessary to limit the circuit current to a value of 8 A:

$$Z = \frac{E_T}{I_T}$$

$$Z = \frac{208 \text{ V}}{8 \text{ A}}$$

$$Z = 26 \ \Omega$$

The formula for finding impedance in an RLC series circuit can now be adjusted to find the missing resistance value (refer to the Resistive-Inductive-Capacitive Series Circuits section in Appendix B):

$$R = \sqrt{Z^2 - (X_L - X_C)^2}$$

$$R = \sqrt{(26 \ \Omega)^2 - (62 \ \Omega - 38 \ \Omega)^2}$$

$$R = \sqrt{(26 \ \Omega)^2 - (24 \ \Omega)^2}$$

$$R = \sqrt{100 \ \Omega^2}$$

$$R = 10 \ \Omega$$

24-2   Series Resonant Circuits

When an inductor and capacitor are connected in series *(Figure 24–9)*, there is one frequency at which the inductive reactance and capacitive reactance become equal. The reason for this is that, as frequency increases, inductive reactance increases and capacitive reactance decreases. The point at which the two reactances become equal is called **resonance.** Resonant circuits are used to provide great increases of current and voltage at the resonant frequency. The following formula can be used to determine the resonant frequency when the values of inductance (I) and capacitance (C) are known:

$$f_R = \frac{1}{2\pi\sqrt{LC}}$$

where

f_R = frequency at resonance

L = inductance in henrys

C = capacitance in farads

In the circuit shown in *Figure 24–9*, an inductor has an inductance of 0.0159 henry and a wire resistance in the coil of 5 ohms. The capacitor connected in series with the inductor has a capacitance of 1.59 microfarads. This circuit reaches resonance at 1000 hertz, when both the inductor and capacitor produce reactances of 100 ohms. At this point, the two reactances are equal and opposite in direction and the only current-limiting factor in the circuit is the 5 ohms of wire resistance in the coil *(Figure 24–10)*.

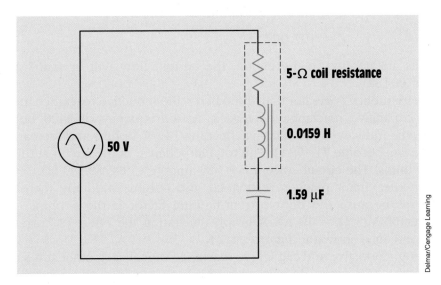

5-Ω **coil resistance**

0.0159 H

50 V

1.59 μF

Delmar/Cengage Learning

FIGURE 24–9   LC series circuit.

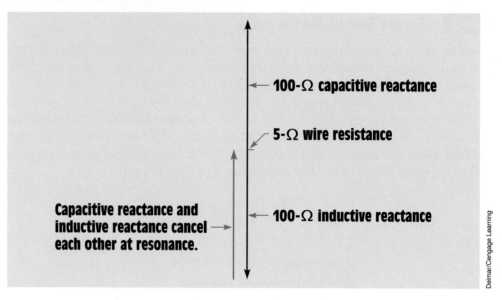

FIGURE 24–10 Inductive reactance and capacitive reactance become equal at resonance.

During the period of time that the circuit is not at resonance, current flow is limited by the combination of inductive reactance and capacitive reactance. At 600 hertz, the inductive reactance is 59.942 ohms and the capacitive reactance is 166.829 ohms. The total circuit impedance is

$$Z = \sqrt{R^2 + (X_L - X_C)^2}$$
$$Z = \sqrt{(5\ \Omega)^2 + (59.942\ \Omega - 166.829\ \Omega)^2}$$
$$Z = 107.004\ \Omega$$

If 50 volts are applied to the circuit, the current flow will be 0.467 ampere (50 V/107.004 Ω).

If the frequency is greater than 1000 hertz, the inductive reactance increases and the capacitive reactance decreases. At a frequency of 1400 hertz, for example, the inductive reactance has become 139.864 ohms and the capacitive reactance has become 71.498 Ω. The total impedance of the circuit at this point is 68.549 ohms. The circuit current is 0.729 ampere (50 V/68.549 Ω).

When the circuit reaches resonance, the current suddenly increases to 10 amperes because the only current-limiting factor is the 5 ohms of wire resistance (50 V/5 Ω = 10 A). A graph illustrating the effect of current in a resonant circuit is shown in *Figure 24–11*.

Although inductive and capacitive reactance cancel each other at resonance, each is still a real value. In this example, both the inductive reactance and capacitive reactance have an ohmic value of 100 ohms at the resonant frequency.

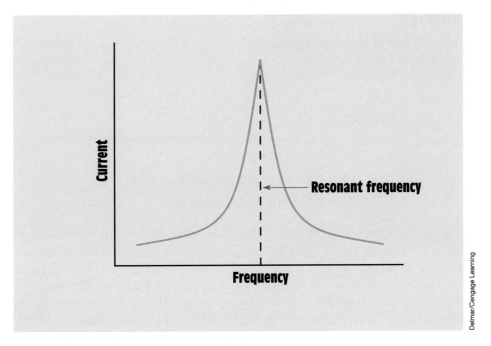

FIGURE 24–11 The current increases sharply at the resonant frequency.

The voltage drop across each component is proportional to the reactance and the amount of current flow. If voltmeters were connected across each component, a voltage of 1000 volts would be seen (10 A × 100 Ω = 1000 V) *(Figure 24–12).*

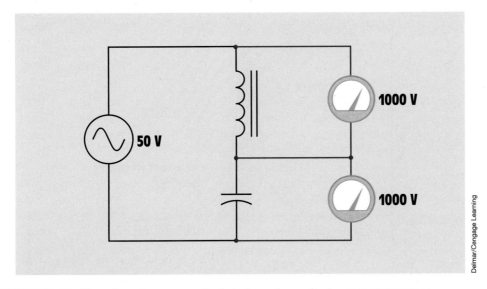

FIGURE 24–12 The voltage drops across the inductor and capacitor increase at resonance.

Bandwidth

The rate of current increase and decrease is proportional to the quality (Q) of the components in the circuit:

$$B = \frac{f_R}{Q}$$

where

B = bandwidth

Q = quality of circuit components

f_R = frequency at resonance

High-Q components result in a sharp increase of current as illustrated by the curve of *Figure 24–11*. Not all series resonant circuits produce as sharp an increase or decrease of current as illustrated in *Figure 24–11*. The term used to describe this rate of increase or decrease is **bandwidth.** Bandwidth is a frequency measurement. It is the difference between the two frequencies at which the current is at a value of 0.707 of the maximum current value *(Figure 24–13)*:

$$B = f_2 - f_1$$

Assume the circuit producing the curve in *Figure 24–13* reaches resonance at a frequency of 1000 hertz. Also assume that the circuit reaches a maximum value of 1 ampere at resonance. The bandwidth of this circuit can be determined by

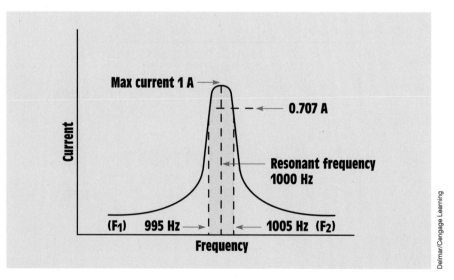

FIGURE 24–13 A narrow-band resonant circuit is produced by high-Q inductors and capacitors.

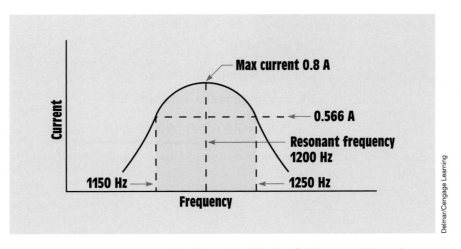

Max current 0.8 A

0.566 A

Resonant frequency
1200 Hz

1150 Hz

1250 Hz

Current

Frequency

Delmar/Cengage Learning

FIGURE 24–14   A wide-band resonant circuit is produced by low-Q inductors and capacitors.

finding the lower and upper frequencies on either side of 1000 hertz at which the current reaches a value of 0.707 of the maximum value. In this illustration that is 0.707 ampere (1 A × 0.707 B = 0.707 A). Assume the lower frequency value to be 995 hertz and the upper value to be 1005 hertz. This circuit has a bandwidth of 10 hertz (1005 Hz − 995 Hz = 10 Hz). When resonant circuits are constructed with components that have a relatively high Q, the difference between the two frequencies is small. These circuits are said to have a *narrow* bandwidth.

In a resonant circuit using components with a lower Q rating, the current does not increase as sharply, as shown in *Figure 24–14*. In this circuit, it is assumed that resonance is reached at a frequency of 1200 hertz and the maximum current flow at resonance is 0.8 ampere. The bandwidth is determined by the difference between the two frequencies at which the current is at a value of 0.566 ampere (0.8 A × 0.707 = 0.566 A). Assume the lower frequency to be 1150 hertz and the upper frequency to be 1250 hertz. This circuit has a bandwidth of 100 hertz (1250 Hz − 1150 Hz = 100 Hz). This circuit is said to have a *wide* bandwidth.

Determining Resonant Values of L and C

There are times when it is necessary to determine what value of inductance or capacitance is needed to resonate with an existing value for inductance or capacitance. The two formulas shown can be used to determine these values for a series or parallel resonant circuit:

$$L = \frac{1}{4\pi^2 f_R^2 C} \quad \text{or} \quad C = \frac{1}{4\pi^2 f_R^2 L}$$

where

$$\pi = 3.1416$$

$$f_R = \text{resonant frequency}$$

$$C = \text{capacitance in farads}$$

$$L = \text{inductance in henrys}$$

What value of capacitance would be needed to resonate at a frequency of 1200 Hz with a 0.2-henry inductor?

$$C = \frac{1}{4\pi^2 f_R^2 L}$$

$$C = \frac{1}{4 \times 3.1416^2 \times (1200 \text{ Hz})^2 \times 0.2 \text{ H}}$$

$$C = \frac{1}{4 \times 9.869 \times 1{,}440{,}000 \text{ Hz}^2 \times 0.2 \text{ H}}$$

$$C = \frac{1}{11{,}369{,}088}$$

$$C = 0.000{,}000{,}088 \text{ farads or } 0.088 \text{ } \mu\text{F or } 88 \text{ nF}$$

What value of inductance is needed to resonate with a 50-microfarad capacitor at 400 hertz?

$$L = \frac{1}{4\pi^2 f_R^2 C}$$

$$L = \frac{1}{4 \times 3.1416^2 \times (400 \text{ Hz})^2 \times 0.000{,}050 \text{ F}}$$

$$L = \frac{1}{4 \times 9.869 \times 160{,}000 \text{ Hz}^2 \times 0.000{,}050 \text{ F}}$$

$$L = \frac{1}{315.808}$$

$$L = 0.003166 \text{ henry, or } 3.166 \text{ mH}$$

Summary

- The voltage dropped across the resistor in an RLC series circuit will be in phase with the current.

- The voltage dropped across the inductor in an RLC series circuit will lead the current by 90°.

■ The voltage dropped across the capacitor in an RLC series circuit will lag the current by 90°.

■ Vector addition can be used in an RLC series circuit to find values of total voltage, impedance, and apparent power.

■ In an RLC circuit, inductive and capacitive values are 180° out of phase with each other. Adding them results in the elimination of the smaller value and a reduction of the larger value.

■ LC resonant circuits increase the current and voltage drop at the resonant frequency.

■ Resonance occurs when inductive reactance and capacitive reactance become equal.

■ The rate the current increases is proportional to the Q of the circuit components.

■ Bandwidth is determined by calculating the upper and lower frequencies at which the current reaches a value of 0.707 of the maximum value.

■ Bandwidth is inversely proportional to the Q of the components in the circuit.

Review Questions

1. What is the phase angle relationship of current and the voltage dropped across a pure resistance?

2. What is the phase angle relationship of current and the voltage dropped across an inductor?

3. What is the phase angle relationship of current and the voltage dropped across a capacitor?

4. An AC circuit has a frequency of 400 Hz. A 16-Ω resistor, a 0.0119-H inductor, and a 16.6-μF capacitor are connected in series. What is the total impedance of the circuit?

5. If 440 V are connected to the circuit, how much current will flow?

6. How much voltage would be dropped across the resistor, inductor, and capacitor in this circuit?

$E_R =$ _____ V

$E_L =$ _____ V

$E_C =$ _____ V

7. What is the true power of the circuit in Question 6?

8. What is the apparent power of the circuit in Question 6?

9. What is the power factor of the circuit in Question 6?

10. How many degrees are the voltage and current out of phase with each other in the circuit in Question 6?

Practical Applications

Y ou are an electrician working in a plant. A series resonant circuit is to be used to produce a high voltage at a frequency of 400 Hz. The inductor has an inductance of 15 mH and a wire resistance of 2 Ω. How much capacitance should be connected in series with the inductor to produce a resonant circuit? The voltage supplied to the circuit is 240 V at 400 Hz. What is the minimum voltage rating of the capacitor? ■

Practice Problems

Refer to the Resistive-Inductive-Capacitive Series Circuits Formula section of Appendix B and to *Figure 24–2*.
   Find all the missing values in the following problems.

1. The circuit shown in *Figure 24–2* is connected to a 120-V, 60-Hz line. The resistor has a resistance of 36 Ω, the inductor has an inductive reactance of 100 Ω, and the capacitor has a capacitive reactance of 52 Ω.

E_T 120 V	E_R ————	E_L ————	E_C ————
I_T ————	I_R ————	I_L ————	I_C ————
Z ————	R 36 Ω	X_L 100 Ω	X_C 52 Ω
VA ————	P ————	$VARS_L$ ————	$VARS_C$ ————
PF ————	∠θ ————	L ————	C ————

2. The circuit is connected to a 400-Hz line with an applied voltage of 35.678 V. The resistor has a true power of 14.4 W, and there are 12.96 inductive VARs and 28.8 capacitive VARs.

E_T 35.678 Ω	E_R ————	E_L ————	E_C ————
I_T ————	I_R ————	I_L ————	I_C ————
Z ————	R ————	X_L————	X_C————
VA ————	P 14.4 W	$VARS_L$ 12.96	$VARS_C$ 28.8
PF ————	∠θ ————	L ————	C ————

3. The circuit is connected to a 60-Hz line. The apparent power in the circuit is 29.985 VA, and the power factor is 62.5%. The resistor has a voltage drop of 14.993 V, the inductor has an inductive reactance of 60 Ω, and the capacitor has a capacitive reactance of 45 Ω.

E_T _____	E_R 14.993 V	E_L _____	E_C _____
I_T _____	I_R _____	I_L _____	I_C _____
Z _____	R _____	X_L 60 Ω	X_C 45 Ω
VA 29.985	P _____	VARS$_L$ _____	VARS$_C$ _____
PF 62.5%	$\angle\theta$ _____	L _____	C _____

4. This circuit is connected to a 1000-Hz line. The resistor has a voltage drop of 185 V, the inductor has a voltage drop of 740 V, and the capacitor has a voltage drop of 444 V. The circuit has an apparent power of 51.8 VA.

E_T _____	E_R 185 V	E_L 740 V	E_C 444 V
I_T _____	I_R _____	I_L _____	I_C _____
Z _____	R _____	X_L _____	X_C _____
VA 51.8	P _____	VARS$_L$ _____	VARS$_C$ _____
PF _____	$\angle\theta$ _____	L _____	C _____

5. A series RLC circuit contains a 4-kΩ resistor, an inductor with an inductive reactance of 3.5 kΩ, and a capacitor with a capacitive reactance of 2.4 kΩ. A 120-VAC, 60-Hz power source is connected to the circuit. How much voltage is dropped across the inductor?

6. A series RLC circuit contains a resistor with a true power of 18 watts, an inductor with a reactive power of 24 VARs, and a capacitor with a reactive power of 34 VARs. What is the circuit power factor?

7. Is the power factor in Question 6 a leading or lagging power factor? Explain your answer.

8. A series RLC circuit contains a resistor with a resistance of 8 Ω, an inductor with a reactance of 12 Ω, and a capacitor with a reactance of 16 Ω. E_R = 17.8 volts. Find E_C.

9. A series RLC circuit has an applied voltage of 240 volts and an apparent power of 600 VA. The circuit power factor is 62%. Find the value of the resistor.

10. A series RLC circuit is connected to a 60-Hz power line. The resistor has a value of 240 Ω. The inductor has an inductance of 0.796 henrys and the capacitor has a capacitance of 5.89 μF. Find Z.

Unit 25

Resistive-Inductive-Capacitive Parallel Circuits

Why You Need to Know

*R*LC parallel circuits are used throughout the electrical field. They are employed in electronic equipment as filters to separate one frequency from another. They are used in industry to produce large increases in current flow for induction heating applications and to correct the power factor of inductive loads in an effort to reduce the amperage supplied to a particular load. RLC parallel resonant circuits are often referred to as "tank" circuits. Tank circuits are the principle used in many motor-control applications as proximity sensors that detect the presence or absence of metal. This unit

- discusses the values and how to calculate voltage, current, power, and impedance in parallel circuits that contain resistance, capacitance, and inductance.
- gives a step-by-step procedure for determining and correcting power factor to the desired percentage.

KEY TERMS

Power factor correction
Tank circuits
Unity

Objectives

After studying this unit, you should be able to

- discuss parallel circuits that contain resistance, inductance, and capacitance.

- calculate the values of an RLC parallel circuit.

- calculate values of impedance, inductance, capacitance, power, reactive power, current flow through individual components, power factor, and phase angle from measurements taken.

- discuss the operation of a parallel resonant circuit.

- calculate the power factor correction for an AC motor.

Preview

Circuits containing elements of resistance, inductance, and capacitance connected in parallel are discussed in this unit. Electrical quantities of current, impedance, and power are calculated for the entire circuit as well as for individual components. Parallel resonant circuits, or tank circuits, and their effect on voltage, current, and impedance are also presented. ∎

25–1 RLC Parallel Circuits

When an AC circuit contains elements of resistance, inductance, and capacitance connected in parallel, the ***voltage dropped across each element is the same. The currents flowing through each branch, however, will be out of phase with each other*** *(Figure 25–1)*. The current flowing through a pure

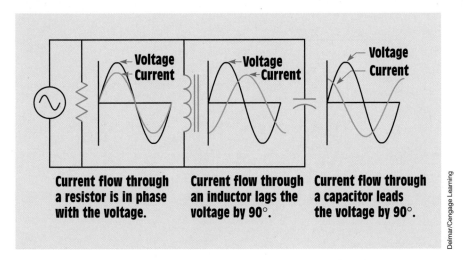

Current flow through a resistor is in phase with the voltage.

Current flow through an inductor lags the voltage by 90°.

Current flow through a capacitor leads the voltage by 90°.

FIGURE 25–1 Voltage and current relationship in an RLC parallel circuit. The voltage is the same across each branch, but the currents are out of phase.

resistive element will be in phase with the applied voltage. The current flowing through a pure inductive element lags the applied voltage by 90 electrical degrees, and the current flowing through a pure capacitive element will lead the voltage by 90 electrical degrees. The phase angle difference between the applied voltage and the total current is determined by the ratio of resistance, inductance, and capacitance connected in parallel. As with an RLC series circuit, if the inductive VARs is greater than the capacitive VARs, the current will lag the voltage and the power factor will be lagging. If the capacitive VARs is greater, the current will lead the voltage and the power factor will be leading.

■ EXAMPLE 25-1

Assume that the RLC parallel circuit shown in *Figure 25–2* is connected to a 240-V, 60-Hz line. The resistor has a resistance of 12 Ω, the inductor has an inductive reactance of 8 Ω, and the capacitor has a capacitive reactance of 16 Ω. Complete the following unknown values:

Z—impedance of the circuit

I_T—total circuit current

I_R—current flow through the resistor

P—true power (watts)

L—inductance of the inductor

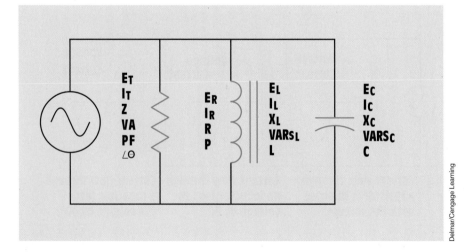

FIGURE 25–2   RLC parallel circuit.

I_L—current flow through the inductor

$VARS_L$—reactive power of the inductor

C—capacitance

I_C—current flow through the capacitor

$VARS_C$—reactive power of the capacitor

VA—volt-amperes (apparent power)

PF—power factor

$\angle\theta$—angle theta

Solution

Impedance

The impedance of the circuit is the reciprocal of the sum of the reciprocals of the legs. Because these values are out of phase with each other, vector addition must be used:

$$Z = \frac{1}{\sqrt{\left(\frac{1}{R}\right)^2 + \left(\frac{1}{X_L} - \frac{1}{X_C}\right)^2}}$$

$$Z = \frac{1}{\sqrt{\left(\frac{1}{12\Omega}\right)^2 + \left(\frac{1}{8\Omega} - \frac{1}{16\Omega}\right)^2}}$$

$$Z = \frac{1}{\sqrt{(0.006944 + 0.003906)\frac{1}{\Omega}}}$$

$$Z = \frac{1}{\sqrt{(0.01085)\frac{1}{\Omega}}}$$

$$Z = \frac{1}{(0.10416)\frac{1}{\Omega}}$$

$$Z = 9.601\Omega$$

To find the total impedance of the previous example using a scientific calculator, press the following keys. Note that the calculator automatically carries each answer to the maximum number of decimal places. This increases the accuracy of the answer.

Note that this is intended to illustrate how total parallel resistance can be determined using many scientific calculators. Some calculators may require a different key entry or pressing the equal key at the end.

Another formula that can be used to determine the total impedance of a circuit containing resistance, inductive reactance, and capacitive reactance is

$$Z = \frac{R \times X}{\sqrt{R^2 + X^2}}$$

where

$$X = \frac{X_L \times X_C}{X_L + X_C}$$

In this formula, X_L is a positive number and X_C is a negative number. Therefore, Z will be either positive or negative depending on whether the circuit is more inductive (positive) or capacitive (negative). To find the total impedance of this circuit using this formula, first determine the value of X:

$$X = \frac{X_L \times X_C}{X_L + X_C}$$

$$X = \frac{8\ \Omega \times (-16\ \Omega)}{8\ \Omega + (-16\ \Omega)}$$

$$X = \frac{-128\ \Omega^2}{-8\Omega}$$

$$X = 16\ \Omega$$

Now that the value of X has been determined, the impedance can be calculated using the formula

$$Z = \frac{R \times X}{\sqrt{R^2 + X^2}}$$

$$Z = \frac{12\ \Omega \times 16\ \Omega}{\sqrt{(12\ \Omega)^2 + (16\ \Omega)^2}}$$

$$Z = \frac{192\ \Omega}{\sqrt{400\ \Omega}}$$

$$Z = \frac{192\ \Omega}{20\ \Omega}$$

$$Z = 9.6\ \Omega$$

Resistive Current

The next unknown value to be found is the current flow through the resistor. This can be calculated by using the formula

$$I_R = \frac{E}{R}$$

$$I_R = \frac{240\ V}{12\ \Omega}$$

$$I_R = 20\ A$$

True Power

The true power, or watts (W), can be calculated using the formula

$$P = E \times I_R$$
$$P = 240\ V \times 20\ A$$
$$P = 4800\ W$$

Inductive Current

The amount of current flow through the inductor can be calculated using the formula

$$I_L = \frac{E}{R}$$

$$I_L = \frac{240\ V}{12\ \Omega}$$

$$I_L = 30\ A$$

Inductive VARs

The amount of reactive power, or VARs, produced by the inductor can be calculated using the formula

$$VARs_L = E \times I_L$$
$$VARs_L = 240\ V \times 30\ A$$
$$VARs_L = 7200$$

Inductance

The amount of inductance in the circuit can be calculated using the formula

$$L = \frac{X_L}{2\pi f}$$

$$L = \frac{8\ \Omega}{377}$$

$$L = 0.0212\ \text{H}$$

Capacitive Current

The current flow through the capacitor can be calculated using the formula

$$I_C = \frac{E}{X_C}$$

$$I_C = \frac{240\ \text{V}}{16\ \Omega}$$

$$I_C = 15\ \text{A}$$

Capacitance

The amount of circuit capacitance can be calculated using the formula

$$C = \frac{1}{2\pi f X_C}$$

$$C = \frac{1}{377 \times 16\ \Omega}$$

$$C = 0.000165782\ \text{F} = 165.782\ \mu\text{F}$$

Capacitive VARs

The capacitive VARs can be calculated using the formula

$$\text{VARs}_C = E \times I_C$$
$$\text{VARs}_C = 240 \times 15$$
$$\text{VARs}_C = 3600$$

Total Circuit Current

The amount of total current flow in the circuit can be calculated by vector addition of the current flowing through each leg of the circuit *(Figure 25–3)*. The inductive current is 180° out of phase with the capacitive current. These two currents tend to cancel each other, resulting in the elimination of the smaller and reduction

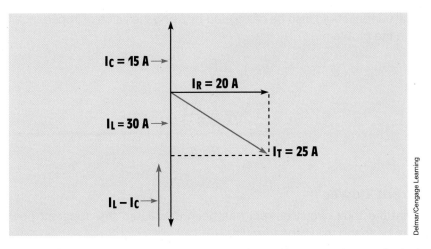

FIGURE 25–3   Vector diagram of resistive, inductive, and capacitive currents in example circuit.

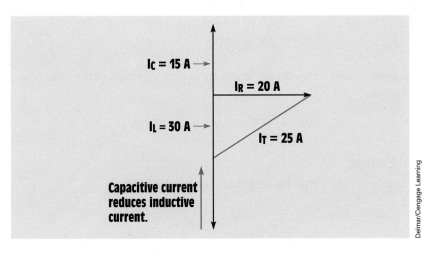

FIGURE 25–4   Inductive and capacitive currents cancel each other.

of the larger. The total circuit current is the hypotenuse of the resulting right triangle *(Figure 25–4)*. The following formula can be used to find total circuit current:

$$I_T = \sqrt{I_R^2 + (I_L - I_C)^2}$$
$$I_T = \sqrt{(20\ A)^2 + (30\ A - 15\ A)^2}$$
$$I_T = \sqrt{(20\ A)^2 + (15\ A)^2}$$
$$I_T = \sqrt{400\ A + 225\ A}$$
$$I_T = \sqrt{625\ A}$$
$$I_T = 25\ A$$

The total current could also be calculated by using the value of impedance found earlier in the problem:

$$I_T = \frac{E}{Z}$$

$$I_T = \frac{240 \text{ V}}{9.6 \text{ }\Omega}$$

$$I_T = 25 \text{ A}$$

Apparent Power

Now that the total circuit current has been calculated the apparent power, or VAs, can be found using the formula

$$VA = E \times I_T$$
$$VA = 240 \text{ V} \times 25 \text{ A}$$
$$VA = 6000$$

The apparent power can also be found by vector addition of the true power and reactive power:

$$VA = \sqrt{P^2 + (VARs_L - VARs_C)^2}$$

Power Factor

The power factor can now be calculated using the formula

$$PF = \frac{W}{VA} \times 100$$

$$PF = \frac{4800 \text{ W}}{6000 \text{ VA}} \times 100$$

$$PF = 0.80 \times 100$$

$$PF = 80\%$$

Angle Theta

The power factor is the cosine of angle theta. Angle theta is therefore

$$\cos \angle\theta = 0.80$$
$$\angle\theta = 36.87°$$

The circuit with all calculated values is shown in *Figure 25–5.*

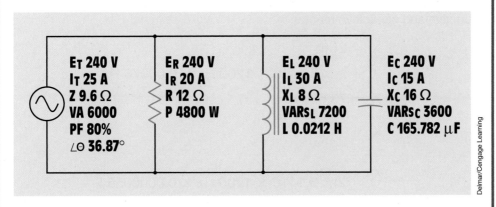

FIGURE 25–5   Example circuit with all calculated values.

■ EXAMPLE 25-2

In the circuit shown in *Figure 25–6,* a resistor, an inductor, and a capacitor are connected to a 1200-Hz power source. The resistor has a resistance of 18 Ω, the inductor has an inductance of 9.76 mH (0.00976 H), and the capacitor has a capacitance of 5.5 μF. What is the impedance of this circuit?

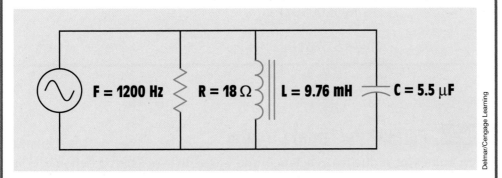

FIGURE 25–6   Example circuit.

Solution

The first step in finding the impedance of this circuit is to find the values of inductive and capacitive reactance:

$$X_L = 2\pi fL$$

$$X_L = 2 \times 3.1416 \times 1200 \text{ Hz} \times 0.00976 \text{ H}$$

$$X_L = 73.589 \ \Omega$$

$$X_C = \frac{1}{2\pi fC}$$

$$X_C = \frac{1}{2 \times 3.1416 \times 1200 \text{ Hz} \times 0.0000055 \text{ F}}$$

$$X_C = 24.114 \ \Omega$$

The impedance of the circuit can now be calculated using the formula

$$Z = \frac{1}{\sqrt{\left(\frac{1}{R}\right)^2 + \left(\frac{1}{X_L} - \frac{1}{X_C}\right)^2}}$$

$$Z = \frac{1}{\sqrt{\left(\frac{1}{18 \ \Omega}\right)^2 + \left(\frac{1}{74.589 \ \Omega} - \frac{1}{24.114 \ \Omega}\right)^2}}$$

$$Z = \frac{1}{\sqrt{(0.0555^2 + (0.0134 - 0.0415)^2)\frac{1}{\Omega^2}}}$$

$$Z = \frac{1}{\sqrt{(0.00387)}\ \frac{1}{\Omega}}$$

$$Z = 16.075 \ \Omega$$

25–2   Parallel Resonant Circuits

When values of inductive reactance and capacitive reactance become equal, they are said to be resonant. In a parallel circuit, inductive current and capacitive current cancel each other because they are 180° out of phase with each other. This produces minimum line current at the point of resonance. An LC

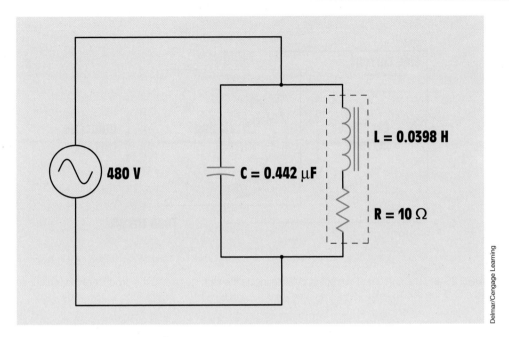

Delmar/Cengage Learning

FIGURE 25–7 Parallel resonant circuit.

parallel circuit is shown in *Figure 25–7*. LC parallel circuits are often referred to as **tank circuits.** In the example circuit, the inductor has an inductance of 0.0398 henrys and a wire resistance of 10 ohms. The capacitor has a capacitance of 0.442 microfarad. This circuit will reach resonance at 1200 hertz, when both the capacitor and inductor exhibit reactances of 300 ohms each.

Calculating the values for a parallel resonant circuit is a bit more involved than calculating the values for a series resonant circuit. In theory, when a parallel circuit reaches resonance, the total circuit current should reach zero and total circuit impedance should become infinite because the capacitive current and inductive currents cancel each other. In practice, the quality (Q) of the circuit components determines total circuit current and therefore total circuit impedance. Because capacitors generally have an extremely high Q by their very nature, the Q of the inductor is the determining factor *(Figure 25–8):*

$$Q = \frac{I_{TANK}}{I_{LINE}}$$

or

$$Q = \frac{X_L}{R}$$

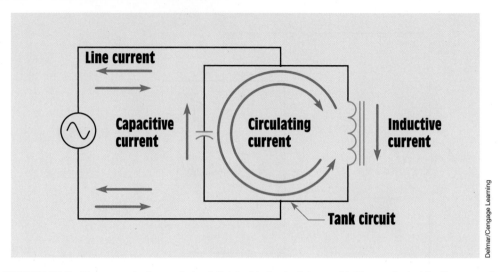

Delmar/Cengage Learning

FIGURE 25–8   The amount of current circulating inside the tank is equal to the product of the line current and the Q of the circuit.

In the example shown in *Figure 25–7*, the inductor has an inductive reactance of 300 ohms at the resonant frequency and a wire resistance of 10 ohms. The Q of this inductor is 30 at the resonant frequency ($Q = X_L/R$). To determine the total circuit current at resonance, it is first necessary to determine the amount of current flow through each of the components at the resonant frequency. Because this is a parallel circuit, the inductor and capacitor will have the alternator voltage of 480 volts applied to them. At the resonant frequency of 1200 hertz, both the inductor and capacitor will have a current flow of 1.6 ampere (480 V/300 Ω = 1.6 A). The total current flow in the circuit will be the in-phase current caused by the wire resistance of the coil *(Figure 25–9)*. This value can be calculated by dividing the circulating current inside the LC parallel loop by the Q of the circuit. The total current in this circuit will be 0.0533 ampere (1.6 A/30 = 0.0533 A). Now that the total circuit current is known, the total impedance at resonance can be found using Ohm's law:

$$Z = \frac{E}{I_T}$$

$$Z = \frac{480 \text{ V}}{0.0533 \text{ A}}$$

$$Z = 9006.63 \ \Omega$$

Another method of calculating total current for a tank circuit is to determine the true power in the circuit caused by the resistance of the coil. The resistive part of a coil is considered to be in series with the reactive part *(Figure 25–7)*.

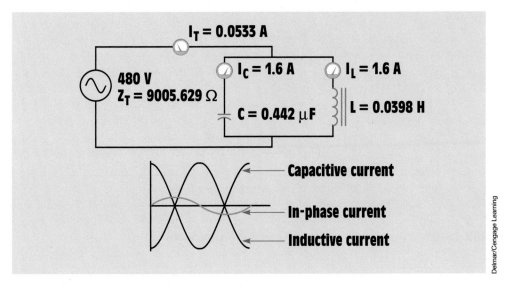

FIGURE 25–9   Total current is equal to the in-phase current.

Because this is a series connection, the current flow through the inductor is the same for both the reactive and resistive elements. The coil has a resistance of 10 ohms. The true power produced by the coil can be calculated by using the formula

$$P = I^2R$$
$$P = (1.6 \text{ A})^2 \times (10 \text{ }\Omega)$$
$$P = 25.6 \text{ W}$$

Now that the true power is known, the total circuit current can be found using Ohm's law:

$$I_T = \frac{P}{E}$$

$$I_T = \frac{25.6 \text{ W}}{480 \text{ V}}$$

$$I_T = 0.0533 \text{ A}$$

Graphs illustrating the decrease of current and increase of impedance in a parallel resonant circuit are shown in *Figure 25–10*.

Bandwidth

The bandwidth for a parallel resonant circuit is determined in a manner similar to that used for a series resonant circuit. The bandwidth of a parallel circuit is

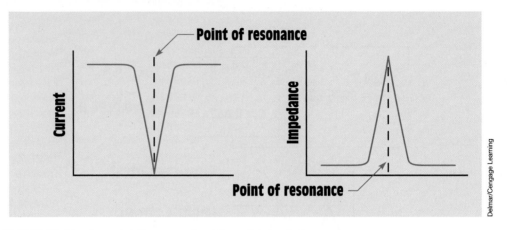

FIGURE 25–10 Characteristic curves of an LC parallel circuit at resonance.

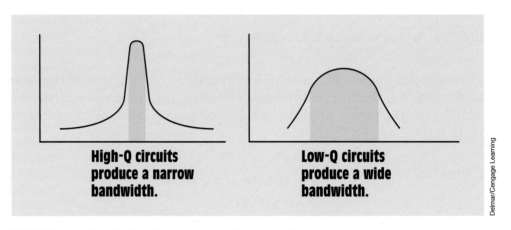

FIGURE 25–11 The Q of the circuit determines the bandwidth.

determined by calculating the frequency on either side of resonance at which the impedance is 0.707 of maximum. As in series resonant circuits, the Q of the parallel circuit determines the bandwidth. Circuits that have a high Q will have a narrow bandwidth, and circuits with a low Q will have a wide bandwidth *(Figure 25–11)*.

Induction Heating

The tank circuit is often used when a large amount of current flow is needed. Recall that the formula for Q of a parallel resonant circuit is

$$Q = \frac{I_{TANK}}{I_{LINE}}$$

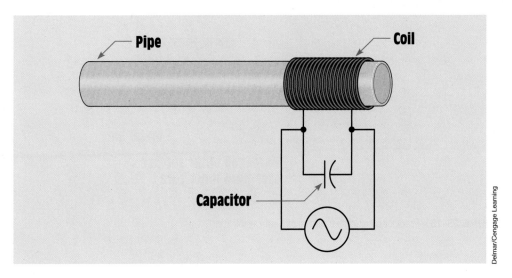

Pipe

Coil

Capacitor

Delmar/Cengage Learning

FIGURE 25–12   Induction heating system.

If this formula is changed, it can be seen that the current circulating inside the tank is equal to the line current times the Q of the circuit:

$$I_{TANK} = I_{LINE} \times Q$$

A high-Q circuit can produce an extremely high current inside the tank with very little line current. A good example of this is an induction heater used to heat pipe for tempering *(Figure 25–12)*. In this example the coil is the inductor and the pipe acts as the core of the inductor. The capacitor is connected in parallel with the coil to produce resonance at a desired frequency. The pipe is heated by eddy current induction. Assume that the coil has a Q of 10. If this circuit has a total current of 100 amperes, then 1000 amperes of current flow in the tank. This 1000 amperes is used to heat the pipe. Because the pipe acts as a core for the inductor, the inductance of the coil changes when the pipe is not in the coil. Therefore, the circuit is resonant only during the times that the pipe is in the coil.

Induction heaters of this type have another advantage over other methods that heat pipe with flames produced by oil- or gas-fired furnaces. When induction heating is used, the resonant frequency can be changed by adding or subtracting capacitance in the tank circuit. This ability to control the frequency greatly affects the tempering of metal. If the frequency is relatively low, 400 hertz or less, the metal is heated evenly. If the frequency is increased to 1000 hertz or greater, skin effect causes most of the heating effect to localize at the surface of the metal. This localization at the surface permits a hard coating to develop at the outer surface of the metal without greatly changing the temper of the inside of the metal *(Figure 25–13)*.

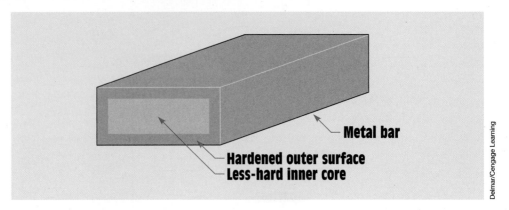

FIGURE 25–13   Frequency controls depth of heat penetration.

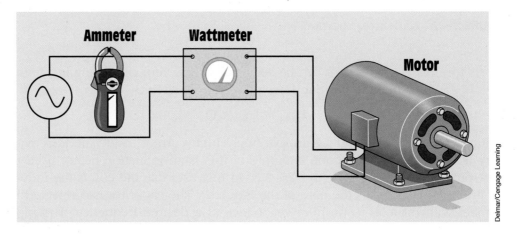

FIGURE 25–14   Determining motor power factor.

Power Factor Correction

Another very common application for LC parallel circuits is the **correction of power factor.** Assume that a motor is connected to a 240-volt single-phase line with a frequency of 60 Hz *(Figure 25–14)*. An ammeter indicates a current flow of 10 amperes and a wattmeter indicates a true power of 1630 watts when the motor is at full load. In this problem, the existing power factor will be determined and then the amount of capacitance needed to correct the power factor will be calculated.

Although an AC motor is an inductive device, when it is loaded it must produce true power to overcome the load connected to it. For this reason, the motor appears to be a resistance connected in series with an inductance *(Figure 25–15)*. Also, the inductance of the motor remains constant

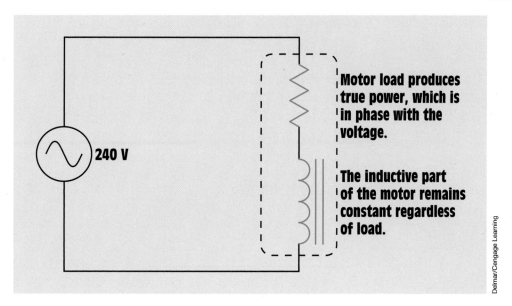

Motor load produces true power, which is in phase with the voltage.

The inductive part of the motor remains constant regardless of load.

240 V

Delmar/Cengage Learning

FIGURE 25–15   Equivalent motor circuit.

regardless of the load connected to it. Recall that true power, or watts, is produced only when electrical energy is converted into some other form. A resistor produces true power because it converts electrical energy into thermal (heat) energy. In the case of a motor, electrical energy is being converted into both thermal energy and kinetic (mechanical) energy. When a motor is operated at a no-load condition, the current is relatively small in comparison with the full-load current. At no load, most of the current is used to magnetize the iron core of the stator and rotor. This current is inductive and is 90° out of phase with the voltage. The only true power produced at no load is caused by motor losses, such as eddy currents being induced into the iron core, the heating effect caused by the resistance of the wire in the windings, hysteresis losses, and the small amount of mechanical energy required to overcome the losses of bearing friction and windage. At no load, the motor would appear to be a circuit containing a large amount of inductance and a small amount of resistance *(Figure 25–16)*.

As load is added to the motor, more electrical energy is converted into kinetic energy to drive the load. The increased current used to produce the mechanical energy is in-phase with the voltage. This causes the circuit to appear to be more resistive. By the time the motor reaches full load, the circuit appears to be more resistive than inductive *(Figure 25–17)*. Notice that as load is added or removed, only the resistive value of the motor changes, which means that once the power factor has been corrected, it will remain constant regardless of the motor load.

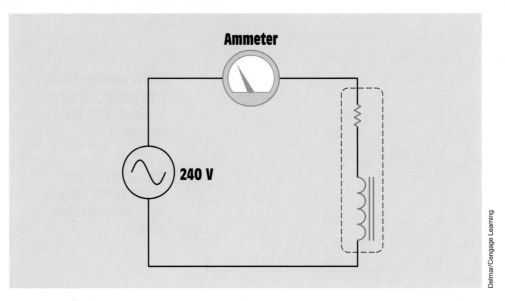

FIGURE 25–16   At no load, the motor appears to be an RL series circuit with a large amount of inductance and a small amount of resistance.

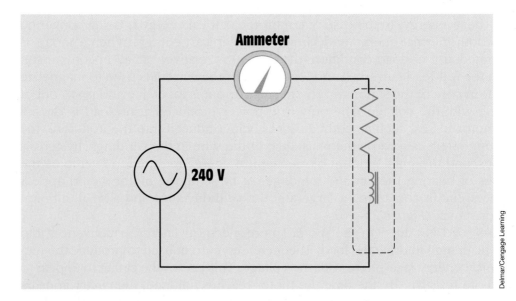

FIGURE 25–17   At full load, the motor appears to be an RL series circuit with a large amount of resistance and a smaller amount of inductance.

When determining the amount of capacitance needed to correct power factor, it is helpful to use a step-by-step procedure. The first step in this procedure is to determine the apparent power of the circuit:

$$VA = E \times I$$
$$VA = 240 \times 10$$
$$VA = 2400$$

Now that the apparent power is known, the power factor can be calculated using the formula

$$PF = \frac{P}{VA}$$
$$PF = \frac{1630}{2400}$$
$$PF = 0679 \ or \ 67.9\%$$

The second step is to determine the amount of reactive power in the circuit. The reactive part of the circuit can be determined by finding the reactive power produced by the inductance. Inductive VARs can be calculated using the formula

$$VARs_L = \sqrt{VA^2 - P^2}$$
$$VARs_L = \sqrt{2400^2 - 1630^2}$$
$$VARs_L = 1761.562$$

To correct the power factor to 100% or **unity,** an equal amount of capacitive VARs would be connected in parallel with the motor. In actual practice, however, it is generally not considered practical to correct the power factor to unity or 100%. It is common practice to correct motor power factor to a value of about 95%. Step three in this example is to determine the amount of apparent power necessary to produce a power factor of 95%:

$$VA = \frac{P}{PF}$$
$$VA = \frac{1630}{0.95}$$
$$VA = 1715.789$$

The fourth step in this example is to determine the amount of inductive VARs that would result in an apparent power of 1715.789 VA. The inductive VARs needed to produce this amount of apparent power can now be determined using the formula

$$VARS_L = \sqrt{VA^2 - P^2}$$

$$VARS_L = \sqrt{1715.789^2 - 1630^2}$$

$$VARS_L = 535.754$$

The fifth step in this example is to determine the amount of capacitive VARs needed to reduce the inductive VARs from 1715.789 to 535.754. To find the capacitive VARs needed to produce a total reactive power of 535.754 in the circuit, subtract the amount of reactive power needed from the present amount:

$$VARS_C = VARS_{L(present)} - VARS_{L(needed)}$$

$$VARS_C = 1761.562 - 535.754$$

$$VARS_C = 1225.808$$

Step six is to determine the amount of capacitive reactance that would produce 1225.808 capacitive VARs. To determine the capacitive reactance needed to produce the required reactive power at 240 volts, the following formula can be used:

$$X_C = \frac{E^2}{VARS_C}$$

$$X_C = \frac{240^2}{1225.808}$$

$$X_C = 46.989 \ \Omega$$

The last step in this example is to determine the amount of capacitance needed to produce a capacitive reactance of 46.989 Ω. The amount of capacitance needed to produce the required capacitive reactance at 60 Hz can be calculated using the formula

$$C = \frac{1}{2\pi f X_C}$$

$$C = \frac{1}{377 \times 46.989}$$

$$C = 56.45 \mu F$$

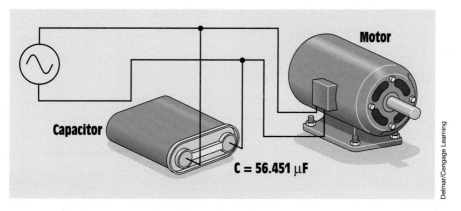

Motor

Capacitor

$C = 56.451 \, \mu F$

Delmar/Cengage Learning

FIGURE 25–18 The capacitor corrects the power factor to 95%.

The power factor will be corrected to 95% when a capacitor with a capacitance of 56.45 μF is connected in parallel with the motor *(Figure 25–18)*.

GREEN TIPS: Power factor correction results in less power loss on conductors by lowering the amount of current flow through the conductor. Reduced current flow results in less power loss due to heating the conductor. ▪

Summary

- The voltage applied to all legs of an RLC parallel circuit is the same.
- The current flow in the resistive leg will be in phase with the voltage.
- The current flow in the inductive leg will lag the voltage by 90°.
- The current flow in the capacitive leg will lead the voltage by 90°.
- Angle theta for the circuit is determined by the amounts of inductance and capacitance.
- An LC resonant circuit is often referred to as a tank circuit.
- When an LC parallel circuit reaches resonance, the line current drops and the total impedance increases.

■ When an LC parallel circuit becomes resonant, the total circuit current is determined by the amount of pure resistance in the circuit.

■ Total circuit current and total impedance in a resonant tank circuit are proportional to the Q of the circuit.

■ Motor power factor can be corrected by connecting capacitance in parallel with the motor. The same amount of capacitive VARs must be connected as inductive VARs.

Review Questions

1. An AC circuit contains a 24-Ω resistor, a 15.9-mH inductor, and a 13.3-μF capacitor connected in parallel. The circuit is connected to a 240-V, 400-Hz power supply. Find the following values.

   X_L = _____ Ω

   X_C = _____ Ω

   I_R = _____ A

   I_L = _____ A

   I_C = _____ A

   P = _____ W

   $VARS_L$ = _____

   $VARS_C$ = _____

   I_T = _____ A

   VA = _____

   PF = _____ %

   $\angle \theta$ = _____ °

2. An RLC parallel circuit contains a resistor with a resistance of 16 Ω, an inductor with an inductive reactance of 8 Ω, and a capacitor with a capacitive reactance of 20 Ω. What is the total impedance of this circuit?

3. The circuit shown in *Figure 25–2* has a current of 38 A flowing through the resistor, 22 A flowing through the inductor, and 7 A flowing through the capacitor. What is the total circuit current?

4. A tank circuit contains a capacitor and an inductor that produce 30 Ω of reactance at the resonant frequency. The inductor has a Q of 15. The voltage of 277 V is connected to the circuit. What is the total circuit current at the resonant frequency?

5. A 0.796-mH inductor produces an inductive reactance of 50 Ω at 10 kHz. What value of capacitance will be needed to produce a resonant circuit at this frequency?

6. An AC motor is connected to a 560-V, 60-Hz line. The motor has a current draw at full load of 53 A. A wattmeter indicates a true power of 18,700 W. Find the power factor of the motor and the amount of capacitance that should be connected in parallel with the motor to correct the power factor to 100%, or unity.

Practical Applications

A single-phase AC motor is connected to a 240-V, 60-Hz supply. A clamp-on ammeter indicates the motor has a current draw of 15 A at full load. A watt-meter connected to the motor indicates a true power of 2.2 kW. What is the power factor of the motor, and how much capacitance is needed to correct the power factor to 95%? ∎

Practice Problems

Refer to the Resistive-Inductive-Capacitive Parallel Circuits Formula section of Appendix B and to *Figure 25–2*.

Find all the missing values in the following problems.

1. The circuit in *Figure 25–2* is connected to a 120-V, 60-Hz line. The resistor has a resistance of 36 Ω, the inductor has an inductive reactance of 40 Ω, and the capacitor has a capacitive reactance of 50 Ω.

E_T 120 V	E_R _____	E_L _____	E_C _____
I_T _____	I_R _____	I_L _____	I_C _____
Z _____	R 36 Ω	X_L 40 Ω	X_C 50 Ω
VA _____	P _____	$VARS_L$ _____	$VARS_C$ _____
PF _____	∠θ _____	L _____	C _____

2. The circuit in *Figure 25–2* is connected to a 400-Hz line with a total current flow of 22.267 A. There is a true power of 3840 W, and the inductor

has a reactive power of 1920 VARs. The capacitor has a reactive power of 5760 VARs.

E_T _____	E_R _____	E_L _____	E_C _____
I_T 22.267 A	I_R _____	I_L _____	I_C _____
Z _____	R _____	X_L _____	X_C _____
VA _____	P 3840 W	VARs$_L$ 1920	VARs$_C$ 5760
PF _____	$\angle\theta$ _____	L _____	C _____

3. The circuit in *Figure 25–2* is connected to a 60-Hz line. The apparent power in the circuit is 48.106 VA. The resistor has a resistance of 12 Ω. The inductor has an inductive reactance of 60 Ω, and the capacitor has a capacitive reactance of 45 Ω.

E_T _____	E_R _____	E_L _____	E_C _____
I_T _____	I_R _____	I_L _____	I_C _____
Z _____	R 12 Ω	X_L 60 Ω	X_C 45 Ω
VA 48.106	P _____	VARs$_L$ _____	VARs$_C$ _____
PF _____	$\angle\theta$ _____	L _____	C _____

4. The circuit in *Figure 25–2* is connected to a 1000-Hz line. The resistor has a current flow of 60 A, the inductor has a current flow of 150 A, and the capacitor has a current flow of 70 A. The circuit has a total impedance of 4.8 Ω.

E_T _____	E_R _____	E_L _____	E_C _____
I_T _____	I_R 60 A	I_L 150 A	I_C 70 A
Z 4.8 Ω	R _____	X_L _____	X_C _____
VA _____	P _____	VARs$_L$ _____	VARs$_C$ _____
PF _____	$\angle\theta$ _____	L _____	C _____

5. In an RLC parallel circuit, the resistor has resistance of 24 kΩ, the inductor has a reactance of 36 kΩ, and the capacitor has a reactance of 14 kΩ. Find Z.

6. In an RLC parallel circuit, the resistor has a resistance of 60 Ω, the inductor has a reactance of 180 Ω, and the capacitor has a reactance of 80 Ω. Find the circuit power factor.

7. In an RLC parallel circuit, the true power is 260 watts. The inductor has a reactive power of 360 VARs, and the capacitor has a reactive power

of 760 VARs. By how many degrees are the voltage and current out of phase with each other?

8. An RLC parallel circuit has an apparent power of 400 VA. The inductor has a reactive power of 450 VARs, and the capacitor has a reactive power of 200 VARs. Find the true power or watts.

9. In an RLC parallel circuit, the resistor has a value of 12 Ω and a current flow of 40 amperes. What is the voltage drop across the capacitor (E_C)?

10. An RLC parallel circuit is connected to 240 volts. The resistor has a power dissipation of 600 watts. If the circuit power factor is 75%, what is the total circuit current (I_T)?

Unit 26

Filters

Why You Need to Know

*F*ilters are employed throughout the electrical field to separate different frequencies. They are the underlying principle behind radio and television. Without filters, it would be impossible to separate the different stations. Some filters are designed to pass particular frequencies, and others are designed to block particular frequencies. This unit introduces the different kinds of filters and the principles on which they work.

KEY TERMS

Filter circuits   Q of the inductor
Notch filters   Reject
One-tenth   Trimmer capacitor
Pass   Trimmer inductor
PI (π) filter

Objectives

After studying this unit, you should be able to

- discuss the necessity of filter circuits.
- discuss the operation of low-pass filters.
- discuss the operation of high-pass filters.
- discuss the operation of bandpass filters.
- discuss the operation of band-reject filters.

Preview

Filter circuits are used to either **pass** (offer little opposition to) or **reject** (offer much opposition to) different frequencies. Many common electronic devices must use filters to operate. Radios and televisions are prime examples of these devices. Literally thousands of different radio and television stations are broadcasting at the same time. Filters are used to select one particular station from the thousands available.

In previous units, it was discussed that inductors and capacitors can be used to produce a circuit that is resonant at a particular frequency. Resonant circuits are one type of filter. In previous examples, fixed values of inductance and capacitance were used to produce circuit resonance at one particular frequency. In actual practice, most filter circuits are constructed with variable components so that they can be tuned for different frequencies or fine-tuned to adjust for the tolerance in inductor or capacitor values. Some filter circuits use a variable capacitor, some use a variable inductor, and some use both. Examples of tank circuits with a **trimmer inductor** and a **trimmer capacitor** are shown in *Figure 26–1*. ■

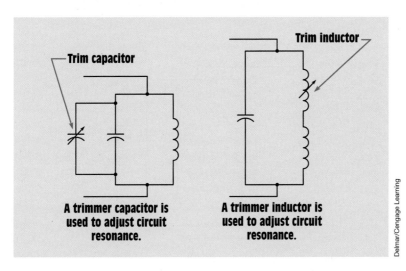

FIGURE 26–1 Trimmer capacitors and trimmer inductors are used to overcome tolerance differences in capacitors and inductors.

26–1   Broadband Tuning

Some filter circuits must be capable of tuning over a broad band of frequencies. When this is the case, it is common practice to switch from one set of inductors or capacitors to another and then use a variable capacitor or inductor to adjust for a particular frequency. An example of a circuit that switches between different inductors and then uses a variable capacitor for tuning is shown in *Figure 26–2*. A circuit that functions by switching between different capacitors with an inductor for tuning is shown in *Figure 26–3*. Although either method can be used, tuning with a variable capacitor is the most common. Variable capacitors generally provide a wider range of tuning than variable inductors.

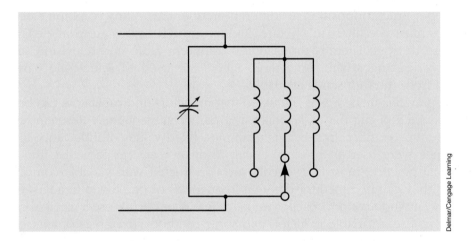

FIGURE 26–2   Broadband filter using multiple inductors and a tuning capacitor.

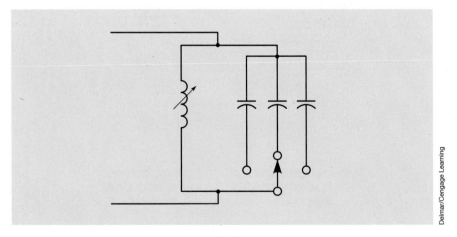

FIGURE 26–3   Broadband filter using multiple capacitors and a trimmer inductor.

26–2   Low-Pass Filters

Filter circuits can be divided into different types depending on the frequencies they either pass or reject. Low-pass filters offer little opposition to current flow when the frequency is low but increase their opposition dramatically after the frequency reaches a certain point *(Figure 26–4)*. Low-pass filters can be constructed in different ways. One of the most common types of low-pass filters is an inductor connected in series with a load resistor *(Figure 26–5)*. At low

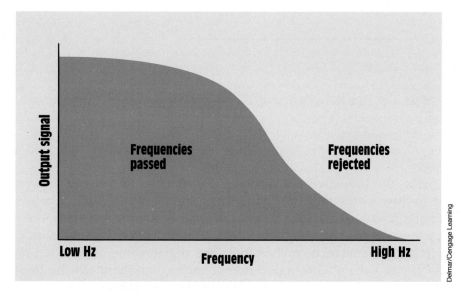

FIGURE 26–4   Low-pass filters pass low frequencies and reject high frequencies.

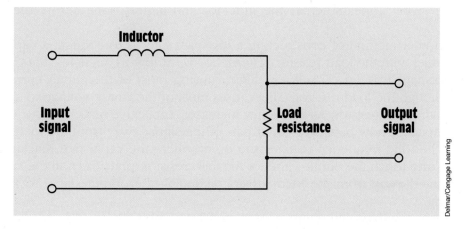

FIGURE 26–5   A low-pass filter can be constructed by connecting an inductor in series with the load resistance.

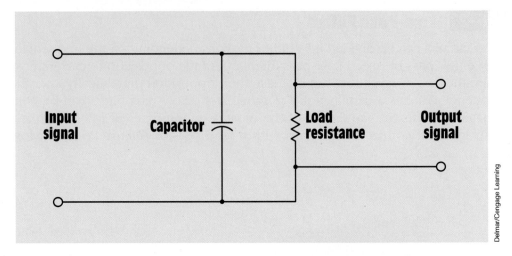

FIGURE 26–6   A low-pass filter using a capacitor connected in parallel with the load resistance.

frequencies, the inductive reactance of the inductor is very low, permitting current to flow with little opposition. This permits most of the circuit voltage drop to appear across the load resistor. As the frequency increases, the inductive reactance increases, permitting less current to flow through the circuit. When inductive reactance reaches the same ohmic value as the load resistor, half the circuit voltage is dropped across each component. As the frequency continues to increase, more voltage drops across the inductor and less voltage drops across the load resistor. When the frequency becomes high enough that the inductive reactance is 10 times greater than the load resistance, practically all the signal appears across the inductor and almost none across the load resistor.

Another method of constructing a low-pass filter is to connect a capacitor in parallel with the load resistor *(Figure 26–6)*. At low frequencies, the capacitive reactance of the capacitor is high, causing most of the circuit current to flow though the load resistor. This causes most of the circuit voltage to appear across the load resistor. As frequency increases, capacitive reactance decreases. The capacitor now begins to shunt part of the signal away from the load resistor. When the frequency reaches the point where the capacitive reactance is about **one-tenth** the ohmic value of the load resistor, practically all the current is shunted away from the load resistor, reducing the output signal to almost nothing.

Some low-pass filters are constructed by combining an inductor and capacitor *(Figure 26–7)*. When this is done, a sharper rejection curve is produced

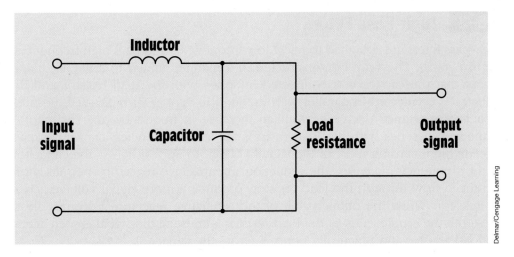

FIGURE 26–7   Low-pass filter combining the effects of an inductor and a capacitor.

(Figure 26–8). One characteristic of this type of filter is that the output signal will peak at the resonant frequency when the values of inductive reactance and capacitive reactance are equal. This will be followed by a sharp drop in the output signal.

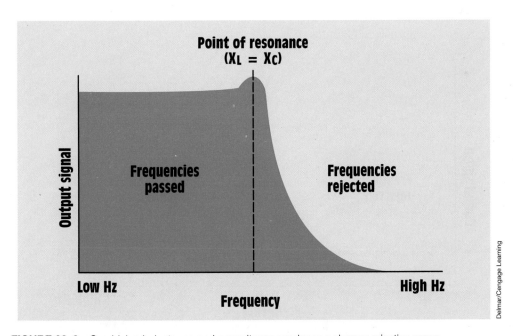

FIGURE 26–8   Combining inductance and capacitance produces a sharper rejection curve.

26–3 High-Pass Filters

High-pass filters are designed to reject low frequencies and pass high frequencies *(Figure 26–9)*. They can be constructed in a manner similar to low-pass filters, except that the capacitor is connected in series with the load resistor and the inductor is connected in parallel with the load resistor *(Figure 26–10)*. Because capacitive reactance decreases with an increase in frequency, the capacitor is connected in series with the load resistor. At low frequencies, capacitive reactance is high, causing most of the signal voltage to appear across the capacitor. As the frequency increases, the reduction in capacitive reactance permits more current to flow through the load resistor, producing more signal voltage across the resistor. When the ohmic value of the capacitive reactance becomes about one-tenth the ohmic value of the load resistor, almost all the signal is seen across the load resistor and practically none is across the capacitor.

The parallel inductor operates in the opposite way. At low frequencies, inductive reactance is low, shunting most of the current away from the load resistor. As frequency increases, inductive reactance becomes high, permitting more current to flow through the load resistor, producing an increase in output signal strength. When the ohmic value of the inductive reactance becomes about 10 times greater than the ohmic value of the load resistor, practically all the signal appears across the load resistor.

The effects of the capacitor and inductor can also be combined in high-pass filters *(Figure 26–11)*. The combination of inductance and capacitance in the

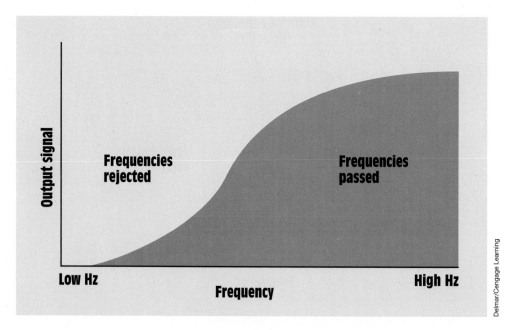

FIGURE 26–9   High-pass filters reject low frequencies and pass high frequencies.

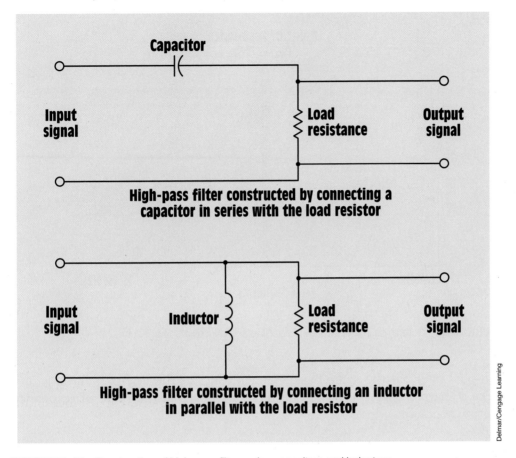

**High-pass filter constructed by connecting a
capacitor in series with the load resistor**

**High-pass filter constructed by connecting an inductor
in parallel with the load resistor**

Delmar/Cengage Learning

FIGURE 26–10 Construction of high-pass filters using capacitors and inductors.

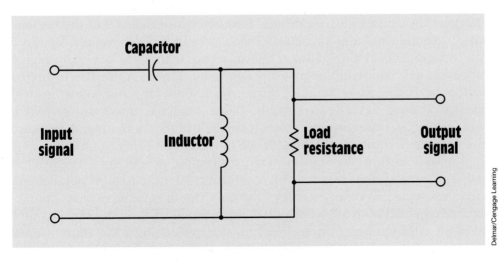

Delmar/Cengage Learning

FIGURE 26–11 High-pass filters combining the effects of an inductor and a capacitor.

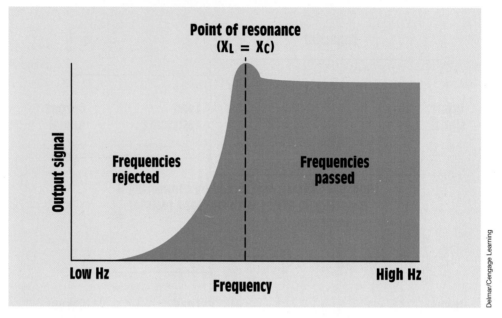

FIGURE 26–12 Combining inductance and capacitance produces a sharper curve.

high-pass filter has the same basic characteristics as the low-pass filter. It produces a sharper curve, and the output signal peaks at the point of resonance *(Figure 26–12)*.

26–4 Bandpass Filters

Bandpass filters permit a certain range of frequencies to pass while rejecting frequencies on either side. The desired frequency to pass is set at the resonant point of the inductor and capacitor. The bandwidth is determined by the Q of the components *(Figure 26–13)*. Because capacitors are generally high-Q components, the bandwidth is basically determined by the **Q of the inductor.** Inductors with a high Q produce a narrow bandwidth, and inductors with a low Q produce a wide bandwidth. Basic bandpass filters are shown in *Figure 26–14*. One filter uses the series resonance of the inductor and capacitor. At the resonant frequency, the impedance of the circuit is low, permitting most of the current to flow through the load resistance, producing a high output signal. The second filter operates by connecting a tank circuit in parallel with the load resistance. The impedance of the tank circuit is relatively low until the resonant frequency is reached. The low impedance of the tank circuit shunts most of the current away from the load resistor, producing a low output signal. The impedance of the tank circuit becomes high at the resonant frequency, permitting most of the circuit current to flow through the load resistor.

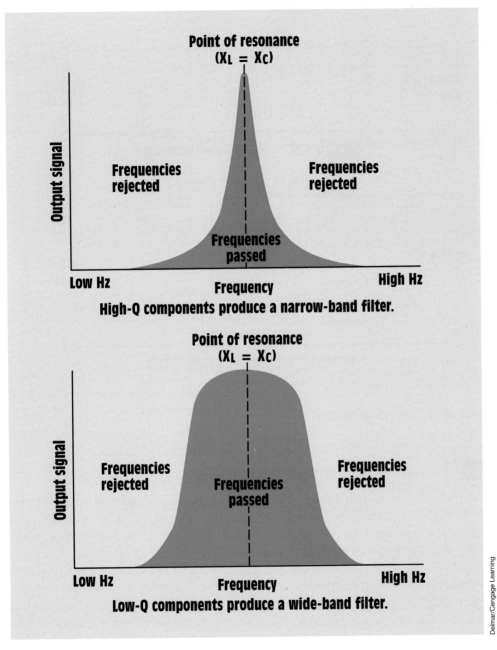

FIGURE 26–13 The bandwidth is controlled by the Q of the components.

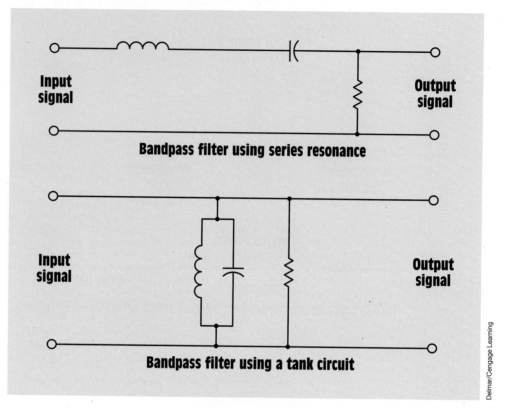

Delmar/Cengage Learning

FIGURE 26–14   Basic bandpass filters.

26–5   Band-Rejection (Notch) Filters

Band-rejection or **notch filters** function in a manner opposite that of band-pass filters. Band-rejection filters are designed to exhibit low impedance to frequencies on either side of their resonant frequency. Band-rejection filters can be narrow band or wide band depending on the Q of the components *(Figure 26–15)*. Two common connections for band-rejection filters are shown in *Figure 26–16*. The parallel tank circuit connected in series with the load resistor exhibits a low impedance until the resonant frequency is reached. At that point the impedance increases and dramatically reduces the current flow to the load resistor.

The second band-rejection filter employs a series resonant circuit connected in parallel with the load resistance. The impedance of the series resonant circuit remains high at frequencies other than the resonant frequency, permitting most of the circuit current to flow through the load resistance. The circuit impedance dramatically drops when the frequency reaches resonance, shunting most of the circuit current away from the load resistance.

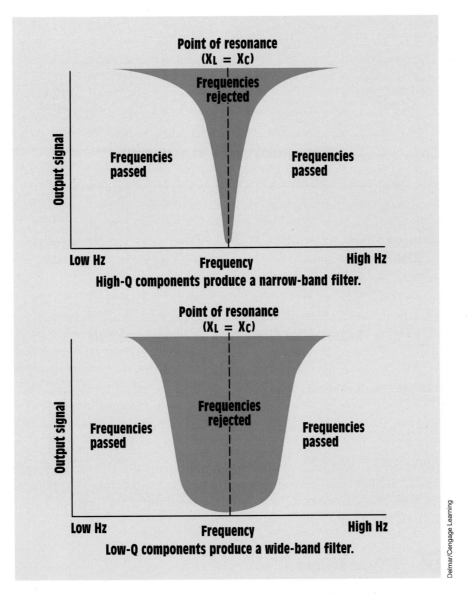

FIGURE 26–15   Band-rejection filters pass most frequencies and reject only a certain range.

26–6   T Filters

The filters discussed thus far have been simple connections involving inductors, capacitors, or a combination of the two. Filter circuits are often much more complex in nature. Filter circuits often fall into one of two classifications, T or

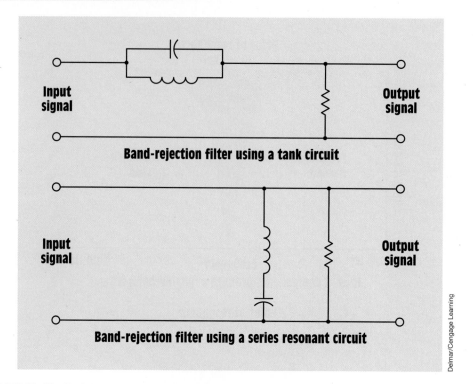

Band-rejection filter using a tank circuit

Band-rejection filter using a series resonant circuit

Delmar/Cengage Learning

FIGURE 26–16   Basic band-rejection filters.

PI. *T filters* are so named because when schematically drawn they resemble the letter *T*. T filters are basically constructed by connecting two components in series with one component in parallel. Several different types of T filters are shown in *Figure 26–17*.

26–7   PI-Type Filters

PI-type filters are so named because they resemble the Greek letter PI (π) when connected. **PI (π) filters** are constructed with two parallel components and one series component. Several examples of PI-type filters are shown in *Figure 26–18*.

26–8   Crossover Networks

A good example of where and how filters are used can be seen in the typical crossover network for a set of speakers. Speakers are designed to produce sound from a particular range of frequencies. Tweeter speakers produce sounds

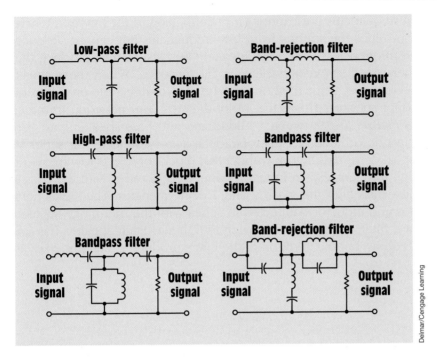

FIGURE 26–17 Basic T-type filters.

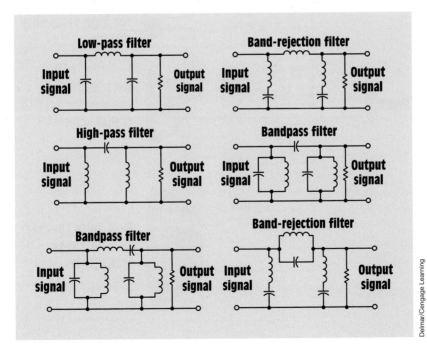

FIGURE 26–18 Basic PI-type filters.

Delmar/Cengage Learning

from high frequencies but cannot produce the sound for low frequencies. Bass speakers produce the sounds from low frequencies but cannot produce sound from high frequencies. Mid-range speakers are designed to produce the sound that neither bass nor tweeter speakers can produce. When these speakers are connected together, some method must be used to direct the proper frequencies to the proper speaker. This is the job of the crossover network. In the example shown in *Figure 26–19*, tweeter, mid-range, and bass speakers are connected together in one cabinet. A 4-microfarad capacitor connected in series with the tweeter speaker forms a high-pass filter that blocks or rejects low frequencies but passes high frequencies. An inductor connected in series with the bass speaker forms a low-pass filter that passes only low frequencies. The series network of a capacitor and inductor connected in series with the mid-range speaker forms a bandpass filter set for approximately 707 hertz. Each speaker has an ohmic value of 8 ohms, but the addition of the filters produce a total impedance for the

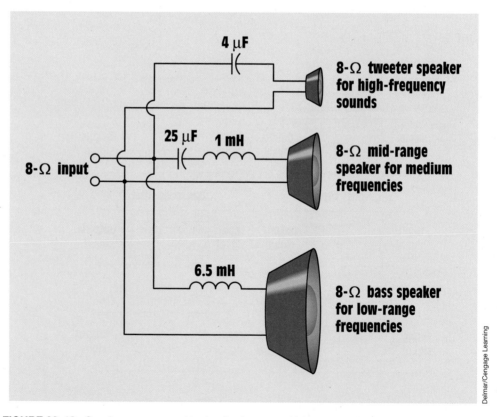

FIGURE 26–19   Speaker crossover network using low-pass, high-pass, and bandpass filters.

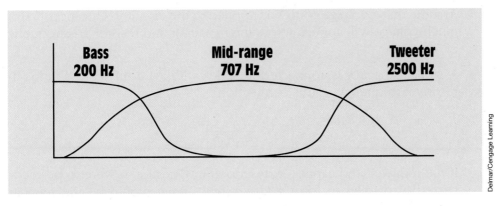

Delmar/Cengage Learning

FIGURE 26–20   Filters direct the proper frequency to each speaker.

entire connection of approximately 8 ohms. A chart illustrating the frequency response of the crossover network is shown in *Figure 26–20*.

Summary

- Low-pass filters pass low frequencies and reject high frequencies.
- High-pass filters pass high frequencies and reject low frequencies.
- Bandpass filters pass a certain range of frequencies and reject lower and higher frequencies on either side of the resonant value.
- Band-rejection filters are sometime called notch filters.
- Band-rejection filters pass lower and higher frequencies on either side of their resonant value.
- More complex filter arrangements are generally divided into two classifications: T and PI.
- T filters are characterized by using two series components and one parallel component.
- PI filters are characterized by using two parallel components and one series component.

Review Questions

1. A 47-Ω load resistor is connected in series with a 10-μF capacitor. At what frequency will the capacitor reach the same value as the load resistor?

2. At what frequency will the capacitor in Question 1 have a capacitive reactance 10 times greater than the load resistance?

3. Assume that a voltage of 16 V is applied to the circuit in Question 1. How much voltage will appear across the capacitor and resistor at a frequency of 500 Hz?

4. A 20-μF capacitor is connected in series with a 1.2-mH coil. What is the resonant frequency of this connection?

$$\left(f_R = \frac{1}{2\pi\sqrt{LC}} \right)$$

5. If the inductor and capacitor described in Question 4 were to be reconnected in parallel to form a tank circuit, what would be the resonant frequency of the tank circuit?

6. A 6-mH choke coil is connected in series with a 10-Ω resistor. At what frequency will the inductive reactance of the choke coil be equal to the ohmic value of the resistor?

7. A 3-mH choke coil is connected in series with a capacitor. What size capacitor is needed to resonate this circuit at a frequency of 1000 Hz?

8. A 2.2-μF capacitor is connected in parallel with an inductor to form a tank circuit. The circuit has a resonant frequency of 1400 Hz. What is the size of the inductor?

Three-Phase Power

Courtesy of Getty Images Inc.

Unit 27

Three-Phase Circuits

KEY TERMS

Delta connection   Star connection
Line current   Three-phase VARs
Line voltage   Three-phase watts
Phase current   Wye connection
Phase voltage

Why You Need to Know

Most of the power produced in the world is three phase. The importance of understanding the principles concerning three-phase power cannot be overstated. Three-phase power is used to run almost all industry throughout the United States and Canada. Even residential power is derived from a three-phase power system. This unit

- discusses the basics of three-phase power generation as the most common electric power source in the world.
- explains how three-phase power is produced and the basic connections used by devices intended to operate on three-phase power.
- illustrates how to calculate voltage, current, and power for different types of three-phase loads.
- explains why electricians must be able to understand and calculate the different voltages in a three-phase circuit.
- determines the power factor and how to correct it in a three-phase system.

Objectives

After studying this unit, you should be able to

- discuss the differences between three-phase and single-phase voltages.

- discuss the characteristics of delta and wye connections.

- calculate voltage and current values for delta and wye circuits.

- connect delta and wye circuits and make measurements with measuring instruments.

- calculate the amount of capacitance needed to correct the power factor of a three-phase motor.

Preview

Most of the electric power generated in the world today is three phase. Three-phase power was first conceived by Nikola Tesla. In the early days of electric power generation, Tesla not only led the battle concerning whether the nation should be powered with low-voltage DC or high-voltage AC, but he also proved that three-phase power was the most efficient way that electricity could be produced, transmitted, and consumed. ∎

27–1 Three-Phase Circuits

There are several reasons why three-phase power is superior to single-phase power:

1. The horsepower rating of three-phase motors and the kilovolt-ampere rating of three-phase transformers are about 150% greater than for single-phase motors or transformers with a similar frame size.

2. The power delivered by a single-phase system pulsates *(Figure 27–1)*. The power falls to zero three times during each cycle. The power delivered by a three-phase circuit pulsates also, but it never falls to zero *(Figure 27–2)*. In a three-phase system, the power delivered to the load is the same at any instant. This produces superior operating characteristics for three-phase motors.

3. In a balanced three-phase system, the conductors need be only about 75% the size of conductors for a single-phase, two-wire system of the same kilovolt-ampere (kVA) rating. This savings helps offset the cost of supplying the third conductor required by three-phase systems.

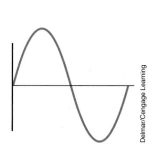

FIGURE 27–1   Single-phase power falls to zero two times each cycle.

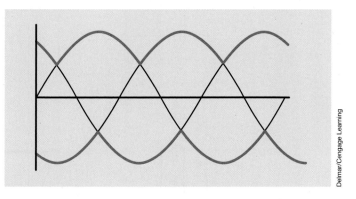

FIGURE 27–2   Three-phase power never falls to zero.

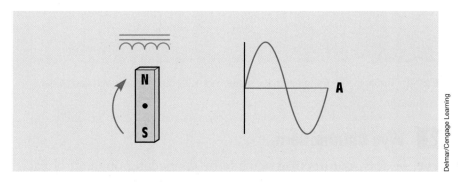

FIGURE 27–3   Producing a single-phase voltage.

A single-phase alternating voltage can be produced by rotating a magnetic field through the conductors of a stationary coil, as shown in *Figure 27–3*.

Because alternate polarities of the magnetic field cut through the conductors of the stationary coil, the induced voltage changes polarity at the same speed as the rotation of the magnetic field. The alternator shown in *Figure 27–3* is single phase because it produces only one AC voltage.

If three separate coils are spaced 120° apart, as shown in *Figure 27–4*, three voltages 120° out of phase with each other are produced when the magnetic field cuts through the coils. This is the manner in which a three-phase voltage is produced. There are two basic three-phase connections: the wye, or star, and the delta.

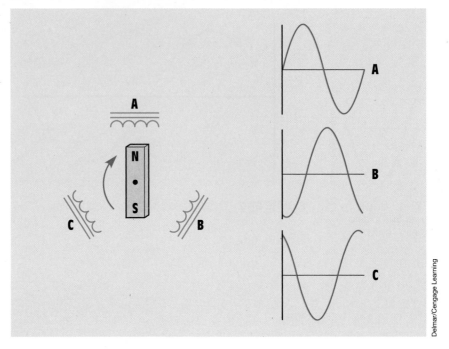

FIGURE 27–4 The voltages of a three-phase system are 120° out of phase with each other.

27–2 Wye Connections

The **wye,** or **star, connection** is made by connecting one end of each of the three-phase windings together *(Figure 27–5)*. The voltage measured across a single winding, or phase, is known as the **phase voltage** *(Figure 27–6)*. The voltage measured between the lines is known as the line-to-line voltage, or simply as the **line voltage.**

FIGURE 27–5 A wye connection is formed by joining one end of each of the windings together.

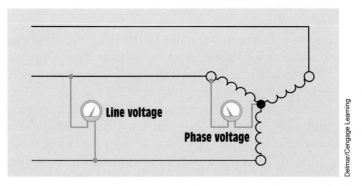

FIGURE 27–6 Line and phase voltages are different in a wye connection.

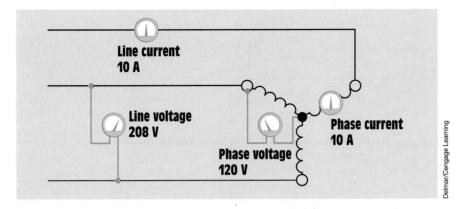

FIGURE 27–7 Line current and phase current are the same in a wye connection.

In *Figure 27–7*, ammeters have been placed in the phase winding of a wye-connected load and in the line that supplies power to the load. Voltmeters have been connected across the input to the load and across the phase. A line voltage of 208 volts has been applied to the load. Notice that the voltmeter connected across the lines indicates a value of 208 volts, but the voltmeter connected across the phase indicates a value of 120 volts.

In a wye-connected system, the line voltage is higher than the phase voltage by a factor of the square root of 3 (1.732). Two formulas used to calculate the voltage in a wye-connected system are

$$E_{Line} = E_{Phase} \times 1.732$$

and

$$E_{Phase} = \frac{E_{Line}}{1.732}$$

Notice in *Figure 27–7* that 10 amperes of current flow in both the phase and the line. *In a wye-connected system,* **phase current** *and* **line current** *are the same:*

$$I_{Line} = I_{Phase}$$

Voltage Relationships in a Wye Connection

Many students of electricity have difficulty at first understanding why the line voltage of the wye connection used in this illustration is 208 volts instead of 240 volts. Because line voltage is measured across two phases that have a voltage of 120 volts each, it would appear that the sum of the two voltages should be 240 volts. One cause of this misconception is that many students are familiar with the 240/120-volt connection supplied to most homes. If voltage is measured across

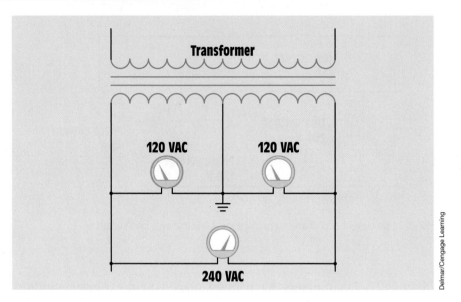

FIGURE 27–8   Single-phase transformer with grounded center tap.

the two incoming lines, a voltage of 240 volts will be seen. If voltage is measured from either of the two lines to the neutral, a voltage of 120 volts will be seen. The reason for this is that this is a single-phase connection derived from the center tap of a transformer *(Figure 27–8)*. The center tap is the midpoint of two out-of-phase voltages *(Figure 27–9)*. The vector sum of these two voltages is 240 volts.

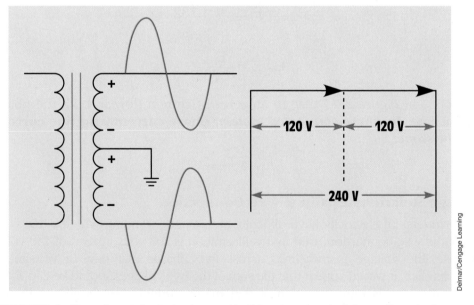

FIGURE 27–9   The voltages of a single-phase residential system are out of phase with each other.

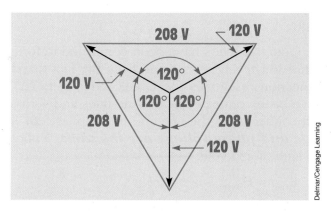

FIGURE 27–10   Vector sum of the voltages in a three-phase wye connection.

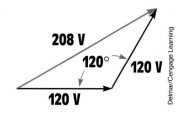

FIGURE 27–11   Adding voltage vectors of two-phase voltage values.

Three-phase voltages are 120° out of phase with each other, not in phase with each other. If the three voltages are drawn 120° apart, it will be seen that the vector sum of these voltages is 208 volts *(Figure 27–10)*. Another illustration of vector addition is shown in *Figure 27–11*. In this illustration, two-phase voltage vectors are added, and the resultant is drawn from the starting point of one vector to the end point of the other. The parallelogram method of vector addition for the voltages in a wye-connected three-phase system is shown in *Figure 27–12*.

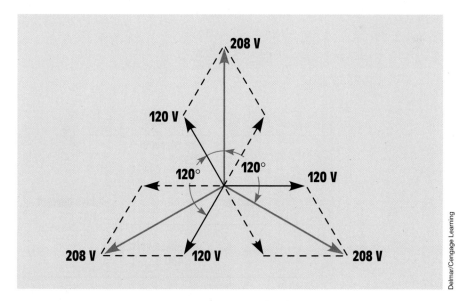

FIGURE 27–12   The parallelogram method of adding three-phase vectors.

27-3 Delta Connections

In *Figure 27–13*, three separate inductive loads have been connected to form a **delta connection.** This connection receives its name from the fact that a schematic diagram of this connection resembles the Greek letter delta (Δ). In *Figure 27–14*, voltmeters have been connected across the lines and across the phase. Ammeters have been connected in the line and in the phase. *In a delta connection, line voltage and phase voltage are the same.* Notice that both voltmeters indicate a value of 480 volts.

$$E_{Line} = E_{Phase}$$

The line current and phase current, however, are different. *The line current of a delta connection is higher than the phase current by a factor of the square root of 3 (1.732).* In the example shown, it is assumed that each of the phase windings has a current flow of 10 amperes. The current in each of the lines, however, is 17.32 amperes. The reason for this difference in current is that current flows through different windings at different times in a three-phase circuit. During some periods of time, current will flow between two lines only. At other times, current will flow from two lines to the third *(Figure 27–15)*. The delta connection is similar to a parallel connection because there is always more than one path for current flow. Because these currents are 120° out of phase with each other, vector addition must be used when finding the sum of the currents *(Figure 27–16)*. Formulas for determining the current in a delta connection are

$$I_{Line} = I_{Phase} \times 1.732$$

and

$$I_{Phase} = \frac{I_{Line}}{1.732}$$

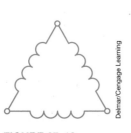

FIGURE 27–13
Three-phase delta
connection.

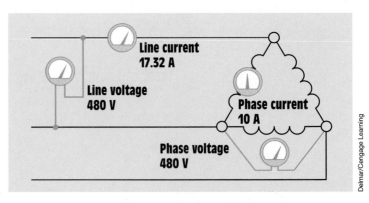

FIGURE 27–14 Voltage and current relationships in a delta connection.

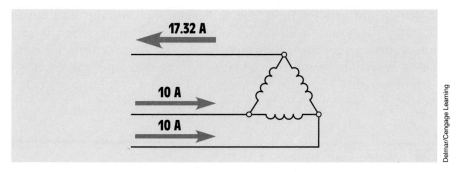

FIGURE 27–15 Division of currents in a delta connection.

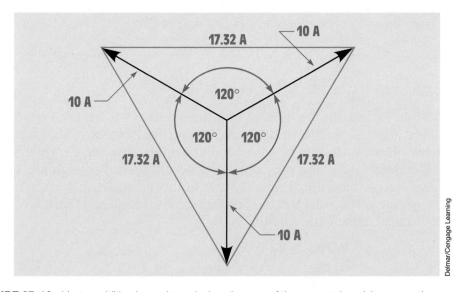

FIGURE 27–16 Vector addition is used to calculate the sum of the currents in a delta connection.

27–4 Three-Phase Power

Students sometimes become confused when calculating values of power in three-phase circuits. One reason for this confusion is that there are actually two formulas that can be used. If *line* values of voltage and current are known, the apparent power of the circuit can be calculated using the formula

$$VA = \sqrt{3} \times E_{Line} \times I_{Line}$$

If the *phase* values of voltage and current are known, the apparent power can be calculated using the formula

$$VA = 3 \times E_{Phase} \times I_{Phase}$$

Notice that in the first formula, the line values of voltage and current are multiplied by the square root of 3. In the second formula, the phase values of voltage and current are multiplied by 3. The first formula is used more because it is generally more convenient to obtain line values of voltage and current because they can be measured with a voltmeter and clamp-on ammeter.

27–5 Watts and VARs

Watts and VARs can be calculated in a similar manner. **Three-phase watts** can be calculated by multiplying the apparent power by the power factor:

$$P = \sqrt{3} \times E_{Line} \times I_{Line} \times PF$$

or

$$P = 3 \times E_{Phase} \times I_{Phase} \times PF$$

Note: When calculating the power of a pure resistive load, the voltage and current are in phase with each other and the power factor is 1.

Three-phase VARs can be calculated in a similar manner, except that voltage and current values of a pure reactive load are used. For example, a pure capacitive load is shown in *Figure 27–17*. In this example, it is assumed that the line voltage is 560 volts and the line current is 30 amperes. Capacitive VARs can be calculated using the formula

$$VARs_C = \sqrt{3} \times E_{Line\ (Capacitive)} \times I_{Line\ (Capacitive)}$$
$$VARs_C = 1.732 \times 560\ V \times 30\ A$$
$$VARs_C = 29{,}097.6$$

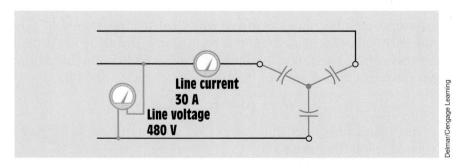

FIGURE 27–17 Pure capacitive three-phase load.

27–6   Three-Phase Circuit Calculations

In the following examples, values of line and phase voltage, line and phase current, and power are calculated for different types of three-phase connections.

■ EXAMPLE 27-1

A wye-connected three-phase alternator supplies power to a delta-connected resistive load *(Figure 27–18)*. The alternator has a line voltage of 480 volts. Each resistor of the delta load has 8 ohms of resistance. Find the following values:

$E_{L(Load)}$—line voltage of the load

$E_{P(Load)}$—phase voltage of the load

$I_{P(Load)}$—phase current of the load

$I_{L(Load)}$—line current to the load

$I_{L(Alt)}$—line current delivered by the alternator

$I_{P(Alt)}$—phase current of the alternator

$E_{P(Alt)}$—phase voltage of the alternator

P—true power

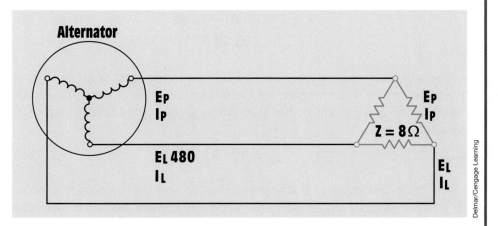

FIGURE 27–18   Calculating three-phase values using a wye-connected power source and a delta-connected load (example circuit).

Solution

The load is connected directly to the alternator. Therefore, the line voltage supplied by the alternator is the line voltage of the load:

$$E_{L(Load)} = 480 \text{ V}$$

The three resistors of the load are connected in a delta connection. In a delta connection, the phase voltage is the same as the line voltage:

$$E_{P(Load)} = E_{L(Load)}$$

$$E_{P(Load)} = 480 \text{ V}$$

Each of the three resistors in the load is one phase of the load. Now that the phase voltage is known (480 V), the amount of phase current can be calculated using Ohm's law:

$$I_{P(Load)} = \frac{E_{P(Load)}}{Z}$$

$$I_{P(Load)} = \frac{480 \text{ V}}{8 \text{ } \Omega}$$

$$I_{P(Load)} = 60 \text{ V}$$

The three load resistors are connected as a delta with 60 A of current flow in each phase. The line current supplying a delta connection must be 1.732 times greater than the phase current:

$$I_{L(Load)} = I_{P(Load)} \times 1.732$$

$$I_{L(Load)} = 60 \text{ A} \times 1.732$$

$$I_{L(Load)} = 103.92 \text{ A}$$

The alternator must supply the line current to the load or loads to which it is connected. In this example, only one load is connected to the alternator. Therefore, the line current of the load is the same as the line current of the alternator:

$$I_{L(Alt)} = 103.92 \text{ A}$$

The phase windings of the alternator are connected in a wye connection. In a wye connection, the phase current and line current are equal. The phase current of the alternator is therefore the same as the alternator line current:

$$I_{P(Alt)} = 103.92 \text{ A}$$

The phase voltage of a wye connection is less than the line voltage by a factor of the square root of 3. The phase voltage of the alternator is

$$E_{P(Alt)} = \frac{E_{L(Alt)}}{1.732}$$

$$E_{P(Alt)} = \frac{480\ V}{1.732}$$

$$E_{P(Alt)} = 277.136\ V$$

In this circuit, the load is pure resistive. The voltage and current are in phase with each other, which produces a unity power factor of 1. The true power in this circuit is calculated using the formula

$$P = 1.732 \times E_{L(Alt)} \times I_{L(Alt)} \times PF$$

$$P = 1.732 \times 480\ V \times 103.92\ A \times 1$$

$$P = 86{,}394.931\ W$$

■ EXAMPLE 27-2

A delta-connected alternator is connected to a wye-connected resistive load *(Figure 27–19)*. The alternator produces a line voltage of 240 V and the resistors have a value of 6 Ω each. Find the following values:

$E_{L(Load)}$—line voltage of the load

$E_{P(Load)}$—phase voltage of the load

$I_{P(Load)}$—phase current of the load

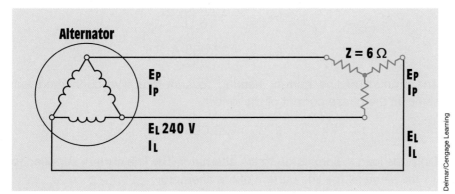

FIGURE 27–19   Calculating three-phase values using a delta-connected source and a wye-connected load (example circuit).

$I_{L(Load)}$ — line current to the load

$I_{L(Alt)}$ — line current delivered by the alternator

$I_{P(Alt)}$ — phase current of the alternator

$E_{P(Alt)}$ — phase voltage of the alternator

P — true power

Solution

As was the case in the previous example, the load is connected directly to the output of the alternator. The line voltage of the load must therefore be the same as the line voltage of the alternator:

$$E_{L(Load)} = 240 \text{ V}$$

The phase voltage of a wye connection is less than the line voltage by a factor of 1.732:

$$E_{P(Load)} = \frac{240 \text{ V}}{1.732}$$

$$E_{P(Load)} = 138.568 \text{ V}$$

Each of the three 6-Ω resistors is one phase of the wye-connected load. Because the phase voltage is 138.568 V, this voltage is applied to each of the three resistors. The amount of phase current can now be determined using Ohm's law:

$$I_{P(Load)} = \frac{E_{P(Load)}}{Z}$$

$$I_{P(Load)} = \frac{138.568 \text{ V}}{6 \text{ }\Omega}$$

$$I_{P(Load)} = 23.095 \text{ A}$$

The amount of line current needed to supply a wye-connected load is the same as the phase current of the load:

$$I_{L(Load)} = 23.095 \text{ A}$$

Only one load is connected to the alternator. The line current supplied to the load is the same as the line current of the alternator:

$$I_{L(Alt)} = 23.095 \text{ A}$$

The phase windings of the alternator are connected in delta. In a delta connection, the phase current is less than the line current by a factor of 1.732:

$$I_{P(Alt)} = \frac{I_{L(Alt)}}{1.732}$$

$$I_{P(Alt)} = \frac{23.095 \text{ A}}{1.732}$$

$$I_{P(Alt)} = 13.334 \text{ A}$$

The phase voltage of a delta is the same as the line voltage:

$$E_{P(Alt)} = 240 \text{ V}$$

Because the load in this example is pure resistive, the power factor has a value of unity, or 1. Power is calculated by using the line values of voltage and current:

$$P = 1.732 \times E_L \times I_L \times PF$$
$$P = 1.732 \times 240 \text{ V} \times 23.095 \text{ A} \times 1$$
$$P = 9600.13 \text{ W}$$

■ EXAMPLE 27-3

The phase windings of an alternator are connected in wye. The alternator produces a line voltage of 440 V and supplies power to two resistive loads. One load contains resistors with a value of 4 Ω each, connected in wye. The second load contains resistors with a value of 6 Ω each, connected in delta *(Figure 27–20)*. Find the following circuit values:

$E_{L(Load\ 2)}$—line voltage of Load 2

$E_{P(Load\ 2)}$—phase voltage of Load 2

$I_{P(Load\ 2)}$—phase current of Load 2

$I_{L(Load\ 2)}$—line current to Load 2

$E_{P(Load\ 1)}$—phase voltage of Load 1

$I_{P(Load\ 1)}$—phase current of Load 1

$I_{L(Load\ 1)}$—line current to Load 1

$I_{L(Alt)}$—line current delivered by the alternator

$I_{P(Alt)}$—phase current of the alternator

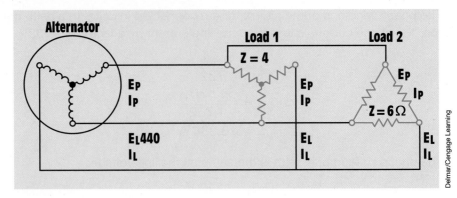

FIGURE 27–20 Calculating three-phase values using a wye-connected source and two three-phase loads (example circuit).

$E_{P(Alt)}$—phase voltage of the alternator

P—true power

Solution

Both loads are connected directly to the output of the alternator. The line voltage for both Load 1 and Load 2 is the same as the line voltage of the alternator:

$$E_{L(Load\ 2)} = 440\ V$$

$$E_{L(Load\ 1)} = 440\ V$$

Load 2 is connected as a delta. The phase voltage is the same as the line voltage:

$$E_{P(Load\ 2)} = 440\ V$$

Each of the resistors that constitutes a phase of Load 2 has a value of 6 Ω. The amount of phase current can be found using Ohm's law:

$$I_{P(Load\ 2)} = \frac{E_{P(Load\ 2)}}{Z}$$

$$I_{P(Load\ 2)} = \frac{440\ V}{6\ \Omega}$$

$$I_{P(Load\ 2)} = 73.333\ A$$

The line current supplying a delta-connected load is 1.732 times greater than the phase current. The amount of line current needed for Load 2 can be calculated by increasing the phase current value by 1.732:

$$I_{L(Load\ 2)} = I_{P(Load\ 2)} \times 1.732$$

$$I_{L(Load\ 2)} = 73.333\ A \times 1.732$$

$$I_{L(Load\ 2)} = 127.013\ A$$

The resistors of Load 1 are connected to form a wye. The phase voltage of a wye connection is less than the line voltage by a factor of 1.732:

$$E_{P(Load\ 1)} = \frac{E_{L(Load\ 1)}}{1.732}$$

$$E_{P(Load\ 1)} = \frac{440\ V}{1.732}$$

$$E_{P(Load\ 1)} = 254.042\ V$$

Now that the voltage applied to each of the 4-Ω resistors is known, the phase current can be calculated using Ohm's law:

$$I_{P(Load\ 1)} = \frac{E_{P(Load\ 1)}}{Z}$$

$$I_{P(Load\ 1)} = \frac{254.042\ V}{4\ \Omega}$$

$$I_{P(Load\ 1)} = 63.511\ A$$

The line current supplying a wye-connected load is the same as the phase current. Therefore, the amount of line current needed to supply Load 1 is

$$I_{L(Load\ 1)} = 63.511\ A$$

The alternator must supply the line current needed to operate both loads. In this example, both loads are resistive. The total line current supplied by the alternator is the sum of the line currents of the two loads:

$$I_{L(Alt)} = I_{L(Load\ 1)} + I_{L(Load\ 2)}$$

$$I_{L(Alt)} = 63.511\ A + 127.013\ A$$

$$I_{L(Alt)} = 190.524\ A$$

Because the phase windings of the alternator in this example are connected in a wye, the phase current is the same as the line current:

$$I_{P(Alt)} = 190.524\ A$$

The phase voltage of the alternator is less than the line voltage by a factor of 1.732:

$$E_{P(Alt)} = \frac{440\ V}{1.732}$$

$$E_{P(Alt)} = 254.042\ V$$

Both of the loads in this example are resistive and have a unity power factor of 1. The total power in this circuit can be found by using the line voltage and total line current supplied by the alternator:

$$P = 1.732 \times E_L \times I_L \times PF$$
$$P = 1.732 \times 440 \text{ V} \times 190.524 \text{ A} \times 1$$
$$P = 145{,}194.53 \text{ W}$$

■ EXAMPLE 27-4

A wye-connected three-phase alternator with a line voltage of 560 V supplies power to three different loads *(Figure 27–21)*. The first load is formed by three resistors with a value of 6 Ω each, connected in a wye; the second load comprises three inductors with an inductive reactance of 10 Ω each, connected in delta; and the third load comprises three capacitors with a capacitive reactance of 8 Ω each, connected in wye. Find the following circuit values:

$E_{L(Load\ 3)}$ — line voltage of Load 3 (capacitive)

$E_{P(Load\ 3)}$ — phase voltage of Load 3 (capacitive)

$I_{P(Load\ 3)}$ — phase current of Load 3 (capacitive)

$I_{L(Load\ 3)}$ — line current to Load 3 (capacitive)

$E_{L(Load\ 2)}$ — line voltage of Load 2 (inductive)

$E_{P(Load\ 2)}$ — phase voltage of Load 2 (inductive)

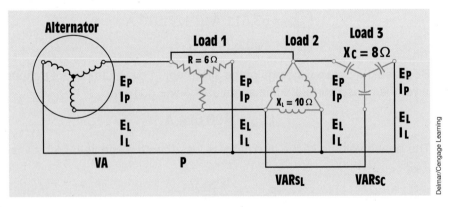

FIGURE 27–21 Calculating three-phase values with a wye-connected source supplying power to a resistive, inductive, and capacitive load (example circuit).

$I_{P(Load\ 2)}$—phase current of Load 2 (inductive)

$I_{L(Load\ 2)}$—line current to Load 2 (inductive)

$E_{L(Load\ 1)}$—line voltage of Load 1 (resistive)

$E_{P(Load\ 1)}$—phase voltage of Load 1 (resistive)

$I_{P(Load\ 1)}$—phase current of Load 1 (resistive)

$I_{L(Load\ 1)}$—line current to Load 1 (resistive)

$I_{L(Alt)}$—line current delivered by the alternator

$E_{P(Alt)}$—phase voltage of the alternator

P—true power

$VARs_L$—reactive power of the inductive load

$VARs_C$—reactive power of the capacitive load

VA—apparent power

PF—power factor

Solution

All three loads are connected to the output of the alternator. The line voltage connected to each load is the same as the line voltage of the alternator:

$$E_{L(Load\ 3)} = 560\ V$$
$$E_{L(Load\ 2)} = 560\ V$$
$$E_{L(Load\ 1)} = 560\ V$$

27-7 Load 3 Calculations

Load 3 is formed from three capacitors with a capacitive reactance of 8 ohms each, connected in a wye. Because this load is wye connected, the phase voltage is less than the line voltage by a factor of 1.732:

$$E_{P(Load\ 3)} = \frac{E_{L(Load\ 3)}}{1.732}$$

$$E_{P(Load\ 3)} = \frac{560\ V}{1.732}$$

$$E_{P(Load\ 3)} = 323.326\ V$$

Now that the voltage applied to each capacitor is known, the phase current can be calculated using Ohm's law:

$$I_{P(Load\ 3)} = \frac{E_{P(Load\ 3)}}{X_C}$$

$$I_{P(Load\ 3)} = \frac{323.326\ V}{8\ \Omega}$$

$$I_{P(Load\ 3)} = 40.416\ A$$

The line current required to supply a wye-connected load is the same as the phase current:

$$I_{L(Load\ 3)} = 40.416\ A$$

The reactive power of Load 3 can be found using a formula similar to the formula for calculating apparent power. Because Load 3 is pure capacitive, the current and voltage are 90° out of phase with each other and the power factor is zero:

$$VARs_C = 1.732 \times E_{L(Load\ 3)} \times I_{L(Load\ 3)}$$

$$VARs_C = 1.732 \times 560\ V \times 40.416\ A$$

$$VARs_C = 39,200.287$$

27–8   Load 2 Calculations

Load 2 comprises three inductors connected in a delta with an inductive reactance of 10 ohms each. Because the load is connected in delta, the phase voltage is the same as the line voltage:

$$E_{L(Load\ 2)} = 560\ V$$

The phase current can be calculated by using Ohm's law:

$$I_{P(Load\ 2)} = \frac{E_{P(Load\ 2)}}{X_L}$$

$$I_{P(Load\ 2)} = \frac{560\ V}{10\ \Omega}$$

$$I_{P(Load\ 2)} = 56\ A$$

The amount of line current needed to supply a delta-connected load is 1.732 times greater than the phase current of the load:

$$I_{L(Load\ 2)} = I_{P(Load\ 2)} \times 1.732$$

$$I_{L(Load\ 2)} = 56\ A \times 1.732$$

$$I_{L(Load\ 2)} = 96.992\ A$$

Because Load 2 is made up of inductors, the reactive power can be calculated using the line values of voltage and current supplied to the load:

$$VARs_L = 1.732 \times E_{L(Load\ 2)} \times I_{L(Load\ 2)}$$
$$VARs_L = 1.732 \times 560\ V \times 96.992\ A$$
$$VARs_L = 94,074.481$$

27–9   Load 1 Calculations

Load 1 consists of three resistors with a resistance of 6 ohms each, connected in wye. In a wye connection, the phase voltage is less than the line voltage by a factor of 1.732. The phase voltage for Load 1 is the same as the phase voltage for Load 3:

$$E_{P(Load\ 1)} = 323.326\ V$$

The amount of phase current can now be calculated using the phase voltage and the resistance of each phase:

$$I_{P(Load\ 1)} = \frac{E_{P(Load\ 1)}}{R}$$
$$I_{P(Load\ 1)} = \frac{323.326\ V}{6\ \Omega}$$
$$I_{P(Load\ 1)} = 53.888\ A$$

Because the resistors of Load 1 are connected in a wye, the line current is the same as the phase current:

$$I_{L(Load\ 1)} = 53.888\ A$$

Because Load 1 is pure resistive, true power can be calculated using the line and phase current values:

$$P = 1.732 \times E_{L(Load\ 1)} \times I_{L(Load\ 1)}$$
$$P = 1.732 \times 560\ V = 53.888\ A$$
$$P = 52,267.049\ W$$

27–10   Alternator Calculations

The alternator must supply the line current for each of the loads. In this problem, however, the line currents are out of phase with each other. To find the total line current delivered by the alternator, vector addition must be used. The current flow in Load 1 is resistive and in phase with the line voltage. The current flow in Load 2 is inductive and lags the line voltage by 90°. The current flow in Load 3 is capacitive and leads the line voltage by 90°. A formula similar to the formula used to find total current flow in an RLC parallel circuit can be employed to find the total current delivered by the alternator:

$$I_{L(Alt)} = \sqrt{I_{L(Load\ 1)}^2 + \left(I_{L(Load\ 2)} - I_{L(Load\ 3)}\right)^2}$$

$$I_{L(Alt)} = \sqrt{(53.888\ A)^2 + (96.992\ A - 40.416\ A)^2}$$

$$I_{L(Alt)} = 78.133\ A$$

The apparent power can now be found using the line voltage and current values of the alternator:

$$VA = 1.732 \times E_{L(Alt)} \times I_{L(Alt)}$$

$$VA = 1.732 \times 560V \times 78.133\ A$$

$$VA = 75,782.759$$

The circuit power factor is the ratio of apparent power and true power:

$$PF = \frac{W}{VA}$$

$$PF = \frac{52,267.049\ W}{75,782.759\ VA}$$

$$PF = 69\%$$

27–11 Power Factor Correction

Correcting the power factor of a three-phase circuit is similar to the procedure used to correct the power factor of a single-phase circuit.

■ EXAMPLE 27-5

A three-phase motor is connected to a 480-V, 60-Hz line *(Figure 27–22)*. A clamp-on ammeter indicates a running current of 68 A at full load, and a three-phase wattmeter indicates a true power of 40,277 W. Calculate the motor power factor

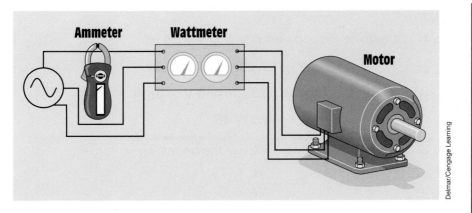

Delmar/Cengage Learning

FIGURE 27–22   Determining apparent and true power for a three-phase motor.

first. Then find the amount of capacitance needed to correct the power factor to 95%. Assume that the capacitors used for power factor correction are to be connected in wye and the capacitor bank is then to be connected in parallel with the motor.

When determining the amount of capacitance needed to correct power factor, it is helpful to follow a procedure. The procedure in this example will consist of nine steps.

Step 1:   Determine the apparent power of the circuit.

Step 2:   Determine the power factor of the circuit.

Step 3:   Determine the reactive power of the circuit.

Step 4:   Determine the amount of apparent power that would produce the desired power factor.

Step 5:   Determine the amount of reactive power that would produce the desired amount of apparent power.

Step 6:   Determine the capacitive VARs necessary to produce the desired reactive power.

Step 7:   Determine the amount of capacitive current necessary to produce the capacitive VARs needed.

Step 8:   Determine the capacitive reactance of each capacitor.

Step 9:   Determine the capacitance value of each capacitor.

Solution

Step 1:   Determine the apparent power of the circuit.

$$VA = 1.732 \times E_L \times I_L$$

$$VA = 1.732 \times 480 \text{ V} \times 68 \text{ A}$$

$$VA = 56,532.48$$

Step 2:   Determine the power factor of the circuit.

$$PF = \frac{P}{VA}$$

$$PF = \frac{40,277 \text{ W}}{56,532.48 \text{ V}}$$

$$PF = 0.7124 \text{ or } 71.24\%$$

Step 3:   Determine the reactive power of the circuit.

$$VARs_L = \sqrt{VA^2 - P^2}$$

$$VARs_L = \sqrt{56,532.46^2 - 40,277^2}$$

$$VARs_L = 39,669.693$$

Step 4:   Determine the amount of apparent power that would produce the desired power factor.

$$VA = \frac{P}{PF}$$

$$VA = \frac{40,277}{0.95}$$

$$VA = 42,396.842$$

Step 5:   Determine the amount of reactive power that would produce the desired amount of apparent power.

$$VARs_L = \sqrt{VA^2 - P^2}$$

$$VARs_L = \sqrt{42,396.842^2 - 40,277^2}$$

$$VARs_L = 13,238.409$$

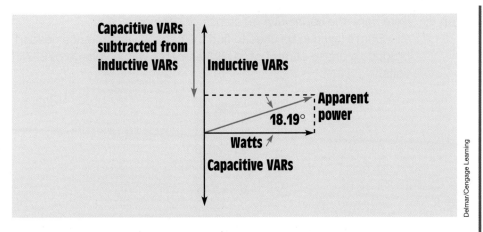

Capacitive VARs subtracted from inductive VARs

Inductive VARs

18.19°

Apparent power

Watts

Capacitive VARs

Delmar/Cengage Learning

FIGURE 27–23   Vector relationship of powers to correct motor power factor.

Step 6:   *Determine the capacitive VARs necessary to produce the desired reactive power.*

$$VARS_C = VARS_{present\ time} - VARS_{desired}$$

$$VARS_C = 39{,}669.693 - 13{,}238.409$$

$$VARS_C = 26{,}431.284$$

To correct the power factor to 95%, the inductive VARs must be reduced from 39,669.693 to 13,238.409. This can be done by connecting a bank of capacitors in the circuit that will produce a total of 26,431.284 capacitive VARs. This amount of capacitive VARs will reduce the inductive VARs to the desired amount *(Figure 27–23)*.

Step 7:   *Determine the amount of capacitive current necessary to produce the capacitive VARs needed.*

$$I_C = \frac{VARS_C}{E_L \times 1.732}$$

$$I_C = \frac{26{,}431.284\ VARS_C}{480 \times 1.732}$$

$$I_C = 31.793\ A$$

The capacitive load bank is to be connected in a wye. Therefore, the phase current is the same as the line current. The phase voltage, however, is less than the line voltage by a factor of 1.732. The phase voltage is 277.136 volts.

Step 8: *Determine the capacitive reactance of each capacitor.*

Ohm's law can be used to find the capacitive reactance needed to produce a phase current of 31.793 amperes with a voltage of 277.136 volts:

$$X_C = \frac{E_{Phase}}{I_{Phase}}$$

$$X_C = \frac{277.136 \text{ V}}{31.793 \text{ A}}$$

$$X_C = 8.717 \ \Omega$$

Step 9: *Determine the capacitance value of each capacitor:*

$$C = \frac{1}{2\pi f X_C}$$

$$C = \frac{1}{377 \times 8.717}$$

$$C = 304.293 \ \mu F$$

When a bank of wye-connected capacitors with a value of 304.293 μF is each connected in parallel with the motor, the power factor is corrected to 95% *(Figure 27–24)*.

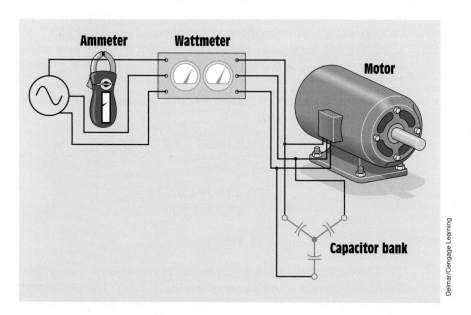

FIGURE 27–24 A wye-connected bank of capacitors is used to correct motor power factor.

Delmar/Cengage Learning

Summary

- The voltages of a three-phase system are 120° out of phase with each other.
- The two types of three-phase connections are wye and delta.
- Wye connections are characterized by the fact that one terminal of each of the devices is connected together.
- In a wye connection, the phase voltage is less than the line voltage by a factor of 1.732. The phase current and line current are the same.
- In a delta connection, the phase voltage is the same as the line voltage. The phase current is less than the line current by a factor of 1.732.

Review Questions

1. How many degrees out of phase with each other are the voltages of a three-phase system?

2. What are the two main types of three-phase connections?

3. A wye-connected load has a voltage of 480 V applied to it. What is the voltage dropped across each phase?

4. A wye-connected load has a phase current of 25 A. How much current is flowing through the lines supplying the load?

5. A delta connection has a voltage of 560 V connected to it. How much voltage is dropped across each phase?

6. A delta connection has 30 A of current flowing through each phase winding. How much current is flowing through each of the lines supplying power to the load?

7. A three-phase load has a phase voltage of 240 V and a phase current of 18 A. What is the apparent power of this load?

8. If the load in Question 7 is connected in a wye, what would be the line voltage and line current supplying the load?

9. An alternator with a line voltage of 2400 V supplies a delta-connected load. The line current supplied to the load is 40 A. Assuming the load is a balanced three-phase load, what is the impedance of each phase?

10. What is the apparent power of the circuit in Question 9?

Practical Applications

*Y*ou are working in an industrial plant. A bank of capacitors is to be used to correct the power factor of a 480-V, three-phase motor. The capacitor bank is connected in a wye. Each of the three capacitors is rated at 25 μF and 600 VDC. You have been told to reconnect these capacitors in delta. Can these capacitors be changed from wye to delta without harm to the capacitors? ■

Practical Applications

*Y*ou are a journeyman electrician working in an industrial plant. A 480-V, three-phase, 60-Hz power panel has a current draw of 216 A. A three-phase wattmeter indicates a true power of 86 kW. You have been instructed to reduce the current draw on the panel by adding three 400-μF capacitors to the system. The capacitors are to be connected in wye. Each capacitor will be connected to one of the three-phase lines. What should be the current flow on the system after the bank of capacitors is added to the circuit? ■

Practical Applications

*Y*ou are an electrician working in an industrial plant. A 30-hp three-phase induction motor has a current draw of 36 amperes at full load. The motor is connected to a 480-volt line. A three-phase wattmeter indicates a true power of 22 kW. Determine the power factor of the motor and the amount of capacitance needed to correct the power factor to 95%. Also determine the minimum voltage rating of the capacitors. The capacitors are to be connected in wye. ■

Practice Problems

1. Refer to the circuit shown in *Figure 27–18* to answer the following questions, but assume that the alternator has a line voltage of 240 V and the load has an impedance of 12 Ω per phase. Find all the missing values.

$E_{P(A)}$ _____ $E_{P(L)}$ _____

$I_{P(A)}$ _____ $I_{P(L)}$ _____

$E_{L(A)}$ 240 V $E_{L(L)}$ _____

$I_{L(A)}$ _____ $I_{L(L)}$ _____

P _____ $Z_{(PHASE)}$ 12 Ω

2. Refer to the circuit shown in *Figure 27–19* to answer the following questions, but assume that the alternator has a line voltage of 4160 V and the load has a resistance of 60 Ω per phase. Find all the missing values.

$E_{P(A)}$ _____ $E_{P(L)}$ _____

$I_{P(A)}$ _____ $I_{P(L)}$ _____

$E_{L(A)}$ 4160 V $E_{L(L)}$ _____

$I_{L(A)}$ _____ $I_{L(L)}$ _____

P _____ $Z_{(PHASE)}$ 60 Ω

3. Refer to the circuit shown in *Figure 27–20* to answer the following questions, but assume that the alternator has a line voltage of 560 V. Load 1 has an resistance of 5 Ω per phase, and Load 2 has a resistance of 8 Ω per phase. Find all the missing values.

$E_{P(A)}$ _____ $E_{P(L1)}$ _____ $E_{P(L2)}$ _____

$I_{P(A)}$ _____ $I_{P(L1)}$ _____ $I_{P(L2)}$ _____

$E_{L(A)}$ 560 V $E_{L(L1)}$ _____ $E_{L(L2)}$_____

$I_{L(A)}$ _____ $I_{L(L1)}$ _____ $I_{L(L2)}$ _____

P _____ $Z_{(PHASE)}$ 5 Ω $Z_{(PHASE)}$ 8 Ω

4. Refer to the circuit shown in *Figure 27–21* to answer the following questions, but assume that the alternator has a line voltage of 480 V. Load 1 has a resistance of 12 Ω per phase. Load 2 has an inductive reactance of 16 Ω per phase, and Load 3 has a capacitive reactance of 10 Ω per phase. Find all the missing values.

$E_{P(A)}$ _____ $E_{P(L1)}$ _____ $E_{P(L2)}$ _____ $E_{P(L3)}$ _____

$I_{P(A)}$ _____ $I_{P(L1)}$ _____ $I_{P(L2)}$ _____ $I_{P(L3)}$ _____

$E_{L(A)}$ 480 V $E_{L(L1)}$ _____ $E_{L(L2)}$ _____ $E_{L(L3)}$ _____

$I_{L(A)}$ _____ $I_{L(L1)}$ _____ $I_{L(L2)}$ _____ $I_{L(L3)}$ _____

VA _____ $R_{(PHASE)}$ 12 Ω $X_{L(PHASE)}$ 16 Ω $X_{C(PHASE)}$ 10 Ω

                      P _____ $VARs_L$ _____ $VARs_C$ _____

X Transformers

Unit 28

Single-Phase Transformers

KEY TERMS

Autotransformers
Constant-current transformer
Control transformer
Distribution transformer
Excitation current
Flux leakage
Inrush current
Isolation transformers
Laminated
Primary winding
Secondary winding
Step-down transformer
Step-up transformer
Tape-wound core
Toroid core
Transformer
Turns ratio
Volts-per-turn ratio

Why You Need to Know

*U*nderstanding how transformers change values of voltage and current is important to the information presented in later units. Many AC motors, for example, operate on these principles. This unit

- discusses transformers and how they are divided into three major types.
- determines values of voltage and current and illustrates different methods that can be employed to determine these values.
- describes how windings determine the primary and secondary voltage and how the nameplate rating provides key information when providing protection.
- discusses installation and testing.

Objectives

After studying this unit, you should be able to

- discuss the different types of transformers.

- calculate values of voltage, current, and turns for single-phase transformers using formulas.

- calculate values of voltage, current, and turns for single-phase transformers using the turns ratio.

- connect a transformer and test the voltage output of different windings.

- discuss polarity markings on a schematic diagram.

- test a transformer to determine the proper polarity marks.

Preview

Transformers are among the most common devices found in the electrical field. They range in size from less than one cubic inch to the size of rail cars. Their ratings can range from milli-volt-amperes (mVA) to giga-volt-amperes (GVA). It is imperative that anyone working in the electrical field have an understanding of transformer types and connections. This unit presents transformers intended for use in single-phase installations. The two main types of voltage transformers, isolation transformers and autotransformers, are discussed. ∎

28–1 Single-Phase Transformers

A **transformer** is a magnetically operated machine that can change values of voltage, current, and impedance without a change of frequency. Transformers are the most efficient machines known. Their efficiencies commonly range from 90% to 99% at full load. Transformers can be divided into three classifications:

1. Isolation transformer

2. Autotransformer

3. Current transformer (current transformers were discussed in Unit 10)

All values of a transformer are proportional to its turns ratio. This does not mean that the exact number of turns of wire on each winding must be known to determine different values of voltage and current for a transformer. What must be known is the *ratio* of turns. For example, assume a transformer has two windings. One winding, the primary, has 1000 turns of wire; and the other, the secondary, has 250 turns of wire *(Figure 28–1)*. The **turns ratio** of this transformer is 4 to 1, or 4:1 (1000 turns/250 turns = 4). This indicates

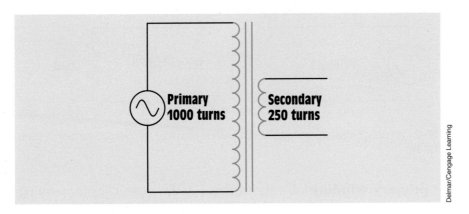

Delmar/Cengage Learning

FIGURE 28–1   All values of a transformer are proportional to its turns ratio.

there are four turns of wire on the primary for every one turn of wire on the secondary.

Transformer Formulas

Different formulas can be used to find the values of voltage and current for a transformer. The following is a list of standard formulas, where

$$N_P = \text{number of turns in the primary}$$
$$N_S = \text{number of turns in the secondary}$$
$$E_P = \text{voltage of the primary}$$
$$E_S = \text{voltage of the secondary}$$
$$I_P = \text{current in the primary}$$
$$I_S = \text{current in the secondary}$$

$$\frac{E_P}{E_S} = \frac{N_P}{N_S}$$

$$\frac{E_P}{E_S} = \frac{I_S}{I_P}$$

$$\frac{N_P}{N_S} = \frac{I_S}{I_P}$$

or

$$E_P \times N_S = E_S \times N_P$$
$$E_P \times I_P = E_S \times I_S$$
$$N_P \times I_P = N_S \times I_S$$

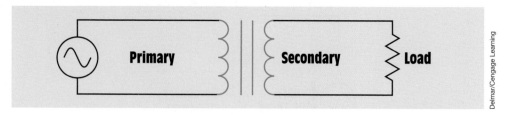

FIGURE 28–2 An isolation transformer has its primary and secondary windings electrically separated from each other.

The **primary winding** of a transformer is the power input winding. It is the winding that is connected to the incoming power supply. The **secondary winding** is the load winding, or output winding. It is the side of the transformer that is connected to the driven load *(Figure 28–2).*

28–2 Isolation Transformers

The transformers shown in *Figure 28–1* and *Figure 28–2* are **isolation transformers.** This means that the secondary winding is physically and electrically isolated from the primary winding. There is no electric connection between the primary and secondary winding. This transformer is magnetically coupled, not electrically coupled. This line isolation is often a very desirable characteristic. The isolation transformer greatly reduces any voltage spikes that originate on the supply side before they are transferred to the load side. Some isolation transformers are built with a turns ratio of 1:1. A transformer of this type has the same input and output voltages and is used for the purpose of isolation only.

The reason that the isolation transformer can greatly reduce any voltage spikes before they reach the secondary is because of the rise time of current through an inductor. Recall from Unit 14 that DC in an inductor rises at an exponential rate *(Figure 28–3).* As the current increases in value, the expanding magnetic field cuts through the conductors of the coil and induces a voltage that is opposed to the applied voltage. The amount of induced voltage is proportional to the rate of change of current. This simply means that the faster current attempts to increase, the greater the opposition to that increase is. Spike voltages and currents are generally of very short duration, which means that they increase in value very rapidly *(Figure 28–4).* This rapid change of value causes the opposition to the change to increase just as rapidly. By the time the spike has been transferred to the secondary winding of the transformer, it has been eliminated or greatly reduced *(Figure 28–5).*

The basic construction of an isolation transformer is shown in *Figure 28–6.* A metal core is used to provide good magnetic coupling between the two windings. The core is generally made of laminations stacked together. Laminating the core helps reduce power losses caused by eddy current induction.

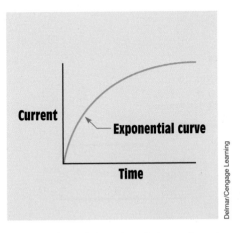

FIGURE 28–3   DC through an inductor rises at an exponential rate.

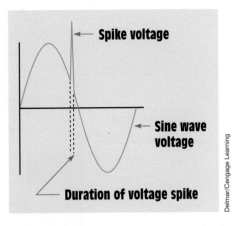

FIGURE 28–4   Voltage spikes are generally of very short duration.

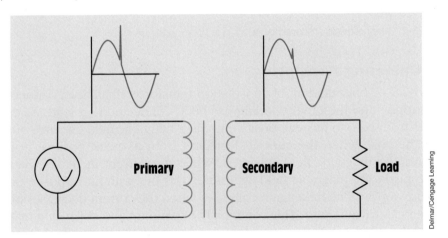

FIGURE 28–5   The isolation transformer greatly reduces the voltage spike.

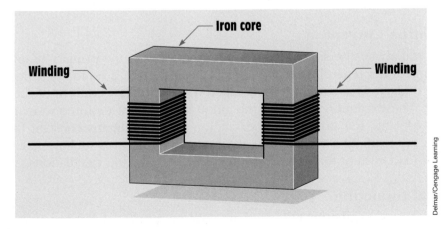

FIGURE 28–6   Basic construction of an isolation transformer.

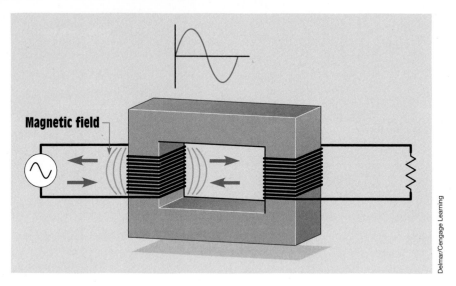

Delmar/Cengage Learning

FIGURE 28–7   Magnetic field produced by AC.

Basic Operating Principles

In *Figure 28–7,* one winding of an isolation transformer has been connected to an AC supply, and the other winding has been connected to a load. As current increases from zero to its peak positive point, a magnetic field expands outward around the coil. When the current decreases from its peak positive point toward zero, the magnetic field collapses. When the current increases toward its negative peak, the magnetic field again expands but with an opposite polarity of that previously. The field again collapses when the current decreases from its negative peak toward zero. This continually expanding and collapsing magnetic field cuts the windings of the primary and induces a voltage into it. This induced voltage opposes the applied voltage and limits the current flow of the primary. When a coil induces a voltage into itself, it is known as *self-induction.*

Excitation Current

There will always be some amount of current flow in the primary of any voltage transformer regardless of type or size even if there is no load connected to the secondary. This current flow is called the **excitation current** of the transformer. The excitation current is the amount of current required to magnetize the core of the transformer. The excitation current remains constant from no load to full load. As a general rule, the excitation current is such a small part of the full-load current that it is often omitted when making calculations.

Mutual Induction

Because the secondary windings of an isolation transformer are wound on the same core as the primary, the magnetic field produced by the primary winding

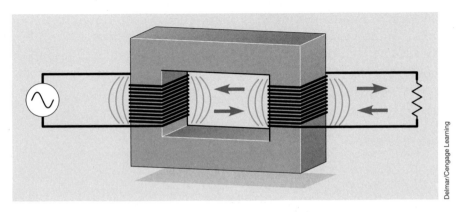

FIGURE 28–8 The magnetic field of the primary induces a voltage into the secondary.

also cuts the windings of the secondary *(Figure 28–8)*. This continually changing magnetic field induces a voltage into the secondary winding. The ability of one coil to induce a voltage into another coil is called *mutual induction*. The amount of voltage induced in the secondary is determined by the ratio of the number of turns of wire in the secondary to those in the primary. For example, assume the primary has 240 turns of wire and is connected to 120 VAC. This gives the transformer a **volts-per-turn ratio** of 0.5 (120 V/240 turns = 0.5 V per turn). Now assume the secondary winding contains 100 turns of wire. Because the transformer has a volts-per-turn ratio of 0.5 volt per turn, the secondary voltage is 50 volts (100 turns × 0.5 V/turn = 50 V per turn).

Transformer Calculations

In the following examples, values of voltage, current, and turns for different transformers are calculated.

Assume that the isolation transformer shown in *Figure 28–2* has 240 turns of wire on the primary and 60 turns of wire on the secondary. This is a ratio of 4:1 (240 turns/60 turns = 4). Now assume that 120 volts are connected to the primary winding. What is the voltage of the secondary winding?

$$\frac{E_P}{E_S} = \frac{N_P}{N_S}$$

$$\frac{120 \text{ V}}{E_S} = \frac{240 \text{ turns}}{60 \text{ turns}}$$

$$240 \text{ turns } E_S = 7200 \text{ V-turns}$$

$$E_S = \frac{7200 \text{ V-turns}}{240 \text{ turns}}$$

$$E_S = 30 \text{ V}$$

The transformer in this example is known as a **step-down transformer** because it has a lower secondary voltage than primary voltage.

Now assume that the load connected to the secondary winding has an impedance of 5 ohms. The next problem is to calculate the current flow in the secondary and primary windings. The current flow of the secondary can be calculated using Ohm's law because the voltage and impedance are known:

$$I = \frac{E}{Z}$$

$$I = \frac{30\ V}{5\ \Omega}$$

$$I = 6\ A$$

Now that the amount of current flow in the secondary is known, the primary current can be calculated using the formula

$$\frac{E_P}{E_S} = \frac{I_S}{I_P}$$

$$\frac{120\ V}{30\ V} = \frac{6\ A}{I_P}$$

$$120\ V\ I_P = 180\ VA$$

$$I_P = \frac{180\ VA}{120\ V}$$

$$I_P = 1.5\ A$$

Notice that the primary voltage is higher than the secondary voltage but the primary current is much less than the secondary current. *A good rule for any type of transformer is that power in must equal power out.* If the primary voltage and current are multiplied together, the product should equal the product of the voltage and current of the secondary:

Primary	Secondary
$120 \times 1.5 = 180\ VA$	$30 \times 6 = 180\ VA$

In this example, assume that the primary winding contains 240 turns of wire and the secondary contains 1200 turns of wire. This is a turns ratio of 1:5 (1200 turns/240 turns = 5). Now assume that 120 volts are connected to the primary winding. Calculate the voltage output of the secondary winding:

$$\frac{E_P}{E_S} = \frac{N_P}{N_S}$$

$$\frac{120 \text{ V}}{E_S} = \frac{240 \text{ turns}}{1200 \text{ turns}}$$

$$240 \text{ turns } E_S = 144{,}000 \text{ V-turns}$$

$$E_S = \frac{144{,}000 \text{ V-turns}}{240 \text{ turns}}$$

$$E_S = 600 \text{ V}$$

Notice that the secondary voltage of this transformer is higher than the primary voltage. This type of transformer is known as a **step-up transformer.**

Now assume that the load connected to the secondary has an impedance of 2400 ohms. Find the amount of current flow in the primary and secondary windings. The current flow in the secondary winding can be calculated using Ohm's law:

$$I = \frac{E}{Z}$$

$$I = \frac{600 \text{ V}}{2400 \text{ } \Omega}$$

$$I = 0.25 \text{ A}$$

Now that the amount of current flow in the secondary is known, the primary current can be calculated using the formula

$$\frac{E_P}{E_S} = \frac{I_S}{I_P}$$

$$\frac{120 \text{ V}}{600 \text{ V}} = \frac{0.25 \text{ A}}{I_P}$$

$$120 \text{ V } I_P = 150 \text{ VA}$$

$$I_P = 1.25 \text{ A}$$

Notice that the amount of power input equals the amount of power output:

Primary	Secondary
120 V × 1.25 A = 150 VA	600 V × 0.25 A = 150 VA

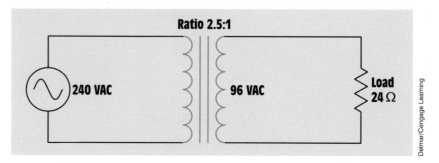

FIGURE 28–9 Calculating transformer values using the turns ratio.

Calculating Isolation Transformer Values Using the Turns Ratio

As illustrated in the previous examples, transformer values of voltage, current, and turns can be calculated using formulas. It is also possible to calculate these same values using the turns ratio. To make calculations using the turns ratio, a ratio is established that compares some number to 1, or 1 to some number. For example, assume a transformer has a primary rated at 240 volts and a secondary rated at 96 volts *(Figure 28–9)*. The turns ratio can be calculated by dividing the higher voltage by the lower voltage:

$$\text{Ratio} = \frac{240 \text{ V}}{96 \text{ V}}$$

$$\text{Ratio} = 2.5{:}1$$

This ratio indicates that there are 2.5 turns of wire in the primary winding for every 1 turn of wire in the secondary. The side of the transformer with the lowest voltage will always have the lowest number (1) of the ratio.

Now assume that a resistance of 24 ohms is connected to the secondary winding. The amount of secondary current can be found using Ohm's law:

$$I_S = \frac{96}{24}$$

$$I_S = 4 \text{ A}$$

The primary current can be found using the turns ratio. Recall that the volt-amperes of the primary must equal the volt-amperes of the secondary. Because the primary voltage is greater, the primary current will have to be less than the secondary current:

$$I_P = \frac{I_S}{\text{turns ratio}}$$

$$I_P = \frac{4 \text{ A}}{2.5}$$

$$I_P = 1.6 \text{ A}$$

To check the answer, find the volt-amperes of the primary and secondary:

Primary	Secondary
240 V × 1.6 A = 384 VA	96 V × 4 A = 384 VA

Now assume that the secondary winding contains 150 turns of wire. The primary turns can be found by using the turns ratio also. Because the primary voltage is higher than the secondary voltage, the primary must have more turns of wire:

$$N_P = N_S \times \text{turns ratio}$$
$$N_P = 150 \text{ turns} \times 2.5$$
$$N_P = 375 \text{ turns}$$

In the next example, assume an isolation transformer has a primary voltage of 120 volts and a secondary voltage of 500 volts. The secondary has a load impedance of 1200 ohms. The secondary contains 800 turns of wire *(Figure 28–10)*.

The turns ratio can be found by dividing the higher voltage by the lower voltage:

$$\text{Ratio} = \frac{500 \text{ V}}{120 \text{ V}}$$
$$\text{Ratio} = 1.4{:}17$$

The secondary current can be found using Ohm's law:

$$I_S = \frac{500 \text{ V}}{1200 \ \Omega}$$
$$I_S = 0.417 \text{ A}$$

In this example, the primary voltage is lower than the secondary voltage. Therefore, the primary current must be higher:

$$I_P = I_S \times \text{turns ratio}$$
$$I_P = 0.417 \text{ A} \times 4.17$$
$$I_P = 1.739 \text{ A}$$

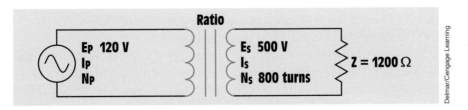

Ratio

Ep 120 V
Ip
Np

Es 500 V
Is
Ns 800 turns

Z = 1200 Ω

Delmar/Cengage Learning

FIGURE 28–10 Calculating transformer values.

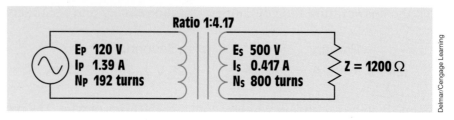

Delmar/Cengage Learning

FIGURE 28–11   Transformer with calculated values.

To check this answer, calculate the volt-amperes of both windings:

Primary	Secondary
120 V × 1.739 A = 208.68 VA	500 V × 0.417 A = 208.5 VA

The slight difference in answers is caused by rounding off values.

Because the primary voltage is less than the secondary voltage, the turns of wire in the primary is less also:

$$N_P = \frac{N_S}{\text{turns ratio}}$$

$$N_P = \frac{800 \text{ turns}}{4.17}$$

$$N_P = 192 \text{ turns}$$

Figure 28–11 shows the transformer with all calculated values.

Multiple-Tapped Windings

It is not uncommon for isolation transformers to be designed with windings that have more than one set of lead wires connected to the primary or secondary. These are called *multiple-tapped windings*. The transformer shown in *Figure 28–12* contains a secondary winding rated at 24 volts. The primary winding contains several taps, however. One of the primary lead wires is labeled C and is the common for the other leads. The other leads are labeled 120 volts, 208 volts, and 240 volts. This transformer is designed in such a manner that it can be connected to different primary voltages without changing the value of the secondary voltage. In this example, it is assumed that the secondary winding has a total of 120 turns of wire. To maintain the proper turns ratio, the primary would have 600 turns of wire between C and 120 volts, 1040 turns between C and 208 volts, and 1200 turns between C and 240 volts.

The isolation transformer shown in *Figure 28–13* contains a single primary winding. The secondary winding, however, has been tapped at several points. One of the secondary lead wires is labeled C and is common to the other lead wires. When rated voltage is applied to the primary, voltages of 12 volts, 24 volts,

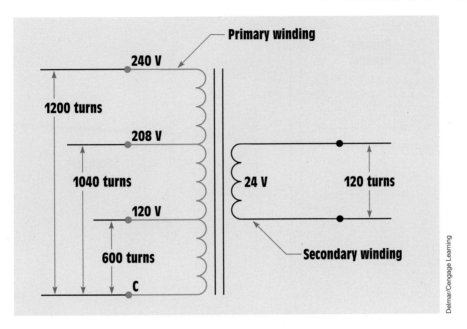

FIGURE 28–12 Transformer with multiple-tapped primary winding.

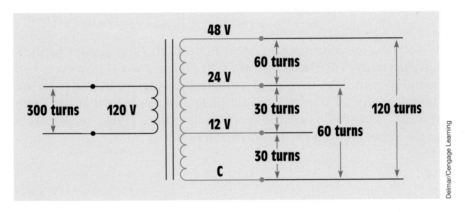

FIGURE 28–13 Transformer secondary with multiple taps.

and 48 volts can be obtained at the secondary. It should also be noted that this arrangement of taps permits the transformer to be used as a center-tapped transformer for two of the voltages. If a load is placed across the lead wires labeled C and 24, the lead wire labeled 12 volts becomes a center tap. If a load is placed across the C and 48 lead wires, the 24 volts lead wire becomes a center tap.

In this example, it is assumed that the primary winding has 300 turns of wire. To produce the proper turns ratio would require 30 turns of wire between C and 12 volts, 60 turns of wire between C and 24 volts, and 120 turns of wire between C and 48 volts.

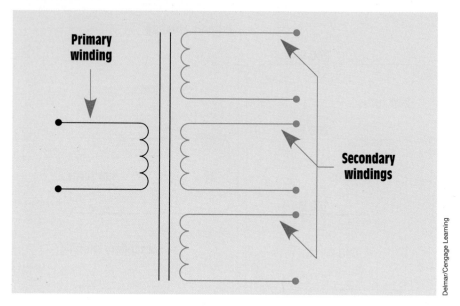

Primary winding

Secondary windings

Delmar/Cengage Learning

FIGURE 28–14 Transformer with multiple secondary windings.

The isolation transformer shown in *Figure 28–14* is similar to the trans-former in *Figure 28–13*. The transformer in *Figure 28–14,* however, has mul-tiple secondary windings instead of a single secondary winding with multiple taps. The advantage of the transformer in *Figure 28–14* is that the secondary windings are electrically isolated from each other. These secondary wind-ings can be either step-up or step-down depending on the application of the transformer.

Calculating Values for Isolation Transformers with Multiple Secondaries

When calculating the values of an isolation transformer with multiple second-ary windings, each secondary must be treated as a different transformer. For example, the transformer in *Figure 28–15* contains one primary winding and three secondary windings. The primary is connected to 120 VAC and contains 300 turns of wire. One secondary has an output voltage of 560 volts and a load impedance of 1000 ohms; the second secondary has an output voltage of 208 volts and a load impedance of 400 ohms; and the third secondary has an output voltage of 24 volts and a load impedance of 6 ohms. The current, turns of wire, and ratio for each secondary and the current of the primary will be found.

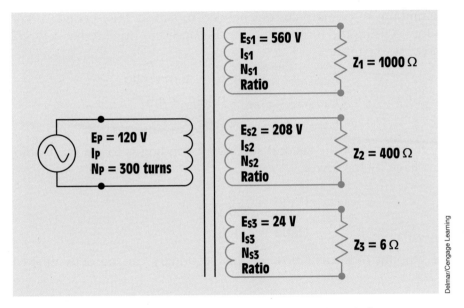

FIGURE 28–15   Calculating values for a transformer with multiple secondary windings.

The first step is to calculate the turns ratio of the first secondary. The turns ratio can be found by dividing the smaller voltage into the larger:

$$\text{Ratio} = \frac{E_{S1}}{E_P}$$

$$\text{Ratio} = \frac{560 \text{ V}}{120 \text{ V}}$$

$$\text{Ratio} = 1{:}4.67$$

The current flow in the first secondary can be calculated using Ohm's law:

$$I_{S1} = \frac{560 \text{ V}}{1000 \text{ V}}$$

$$I_{S1} = 0.56 \text{ A}$$

The number of turns of wire in the first secondary winding is found using the turns ratio. Because this secondary has a higher voltage than the primary, it must have more turns of wire:

$$N_{S1} = N_P \times \text{turns ratio}$$

$$N_{S1} = 300 \text{ turns} \times 4.67$$

$$N_{S1} = 1401 \text{ turns}$$

The amount of primary current needed to supply this secondary winding can be found using the turns ratio also. Because the primary has less voltage, it requires more current:

$$I_{P(FIRST\ SECONDARY)} = I_{S1} \times \text{turns ratio}$$
$$I_{P(FIRST\ SECONDARY)} = 0.56A \times 4.67$$
$$I_{P(FIRST\ SECONDARY)} = 2.61\ A$$

The turns ratio of the second secondary winding is found by dividing the higher voltage by the lower:

$$\text{Ratio} = \frac{208\ V}{120\ V}$$
$$\text{Ratio} = 1:1.73$$

The amount of current flow in this secondary can be determined using Ohm's law:

$$I_{S2} = \frac{208\ V}{400\ \Omega}$$
$$I_{S2} = 0.52\ A$$

Because the voltage of this secondary is greater than the primary, it has more turns of wire than the primary. The number of turns of this secondary is found using the turns ratio:

$$N_{S2} = N_P \times \text{turns ratio}$$
$$N_{S2} = 300\ \text{turns} \times 1.73$$
$$N_{S2} = 519\ \text{turns}$$

The voltage of the primary is less than this secondary. The primary therefore requires a greater amount of current. The amount of current required to operate this secondary is calculated using the turns ratio:

$$I_{P(SECOND\ SECONDARY)} = I_{S2} \times \text{turns ratio}$$
$$I_{P(SECOND\ SECONDARY)} = 0.52A \times 1.732$$
$$I_{P(SECOND\ SECONDARY)} = 0.9\ A$$

The turns ratio of the third secondary winding is calculated in the same way as the other two. The larger voltage is divided by the smaller:

$$\text{Ratio} = \frac{120\ V}{24\ V}$$
$$\text{Ratio} = 5:1$$

The primary current is found using Ohm's law:

$$I_{S3} = \frac{24 \text{ V}}{6 \text{ }\Omega}$$

$$I_{S3} = 4 \text{ A}$$

The output voltage of the third secondary is less than the primary. The number of turns of wire is therefore less than the primary turns:

$$N_{S3} = \frac{N_P}{\text{turns ratio}}$$

$$N_{S3} = \frac{300 \text{ turns}}{5}$$

$$N_{S3} = 60 \text{ turns}$$

The primary has a higher voltage than this secondary. The primary current is therefore less by the amount of the turns ratio:

$$I_{P \text{ (THIRD SECONDARY)}} = \frac{I_{S3}}{\text{turns ratio}}$$

$$I_{P \text{ (THIRD SECONDARY)}} = \frac{4 \text{ A}}{5}$$

$$I_{P \text{ (THIRD SECONDARY)}} = 0.8 \text{ A}$$

The primary must supply current to each of the three secondary windings. Therefore, the total amount of primary current is the sum of the currents required to supply each secondary:

$$I_{P(TOTAL)} = I_{P1} + I_{P2} + I_{P3}$$

$$I_{P(TOTAL)} = 2.61\text{A} + 0.9\text{A} + 0.8\text{A}$$

$$I_{P(TOTAL)} = 4.31 \text{ A}$$

The transformer with all calculated values is shown in *Figure 28–16.*

Distribution Transformers

A common type of isolation transformer is the **distribution transformer** *(Figure 28–17)*. This type of transformer changes the high voltage of power company distribution lines to the common 240/120 volts used to supply power to most homes and many businesses. In this example, it is assumed that the primary is connected to a 7200-volt line. The secondary is 240 volts with a center tap. The center tap is grounded and becomes the **neutral conductor** or common conductor. If voltage is measured across the entire secondary, a voltage of 240 volts is seen. If voltage is measured from either line to the center tap, half of the secondary voltage, or 120 volts, is seen *(Figure 28–18)*. The

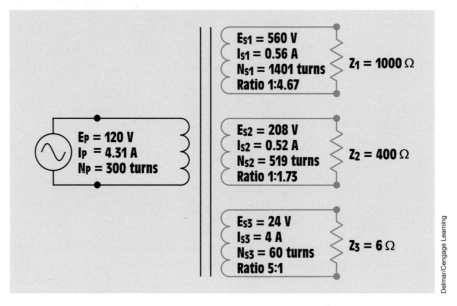

FIGURE 28–16 The transformer with all calculated values.

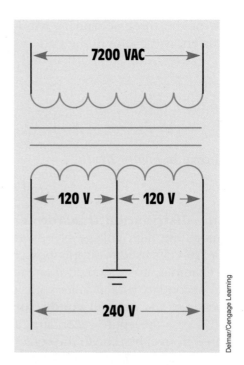

FIGURE 28–17 Distribution transformer.

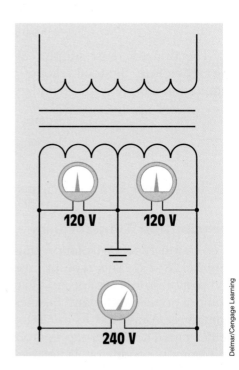

FIGURE 28–18 The voltage from either line to neutral is 120 volts. The voltage across the entire secondary winding is 240 volts.

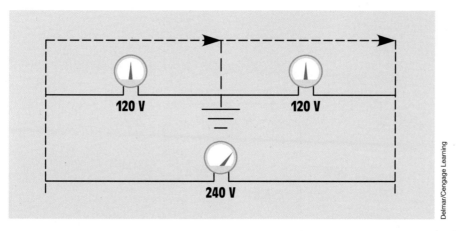

FIGURE 28–19 The voltages across the secondary are out of phase with each other.

reason this occurs is that the grounded neutral conductor becomes the center point of two out-of-phase voltages. If a vector diagram is drawn to illustrate this condition, you will see that the grounded neutral conductor is connected to the center point of the two out-of-phase voltages *(Figure 28–19)*. Loads that are intended to operate on 240 volts, such as water heaters, electric-resistance heating units, and central air conditioners are connected directly across the lines of the secondary *(Figure 28–20)*.

Loads that are intended to operate on 120 volts connect from the center tap, or neutral, to one of the secondary lines. The function of the neutral is to carry the difference in current between the two secondary lines and maintain

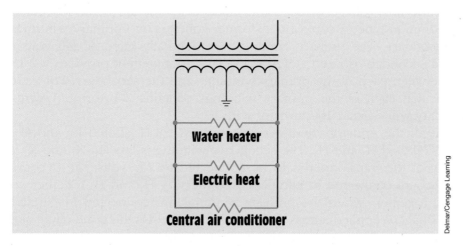

FIGURE 28–20 Loads of 240 volts connect directly across the secondary winding.

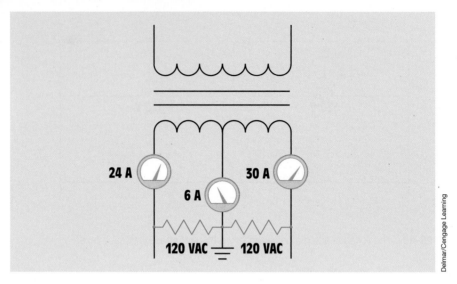

FIGURE 28–21   The neutral carries the sum of the unbalanced load.

a balanced voltage. In *Figure 28–21,* one of the secondary lines has a current flow of 30 amperes and the other has a current flow of 24 amperes. The neutral conducts the sum of the unbalanced load. In this example, the neutral current is 6 amperes (30A − 24A = 6A).

Control Transformers

Another common type of isolation transformer found throughout industry is the **control transformer** *(Figure 28–22).* The control transformer is used to reduce the line voltage to the value needed to operate control circuits. The most common type of control transformer contains two primary windings and one secondary. The primary windings are generally rated at 240 volts each, and the secondary is rated at 120 volts. This arrangement provides a 2:1 turns ratio between each of the primary windings and the secondary. For example, assume that each of the primary windings contains 200 turns of wire. The secondary will contain 100 turns of wire.

One of the primary windings in *Figure 28–23* is labeled H_1 and H_2. The other is labeled H_3 and H_4. The secondary winding is labeled X_1 and X_2. If the primary of the transformer is to be connected to 240 volts, the two primary windings are connected in parallel by connecting H_1 and H_3 together and H_2 and H_4 together. When the primary windings are connected in parallel, the same voltage is applied across both windings. This has the same effect as using one primary winding with a total of 200 turns of wire. A turns ratio of 2:1 is maintained, and the secondary voltage is 120 volts.

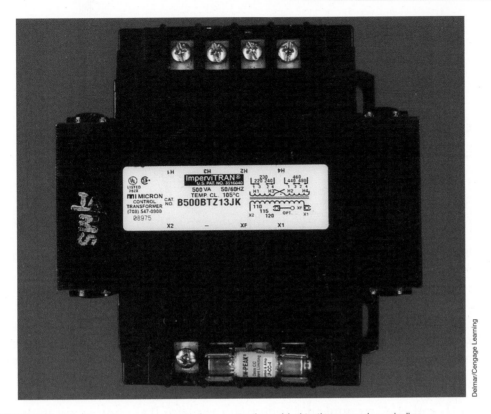

FIGURE 28–22   Control transformer with fuse protection added to the secondary winding.

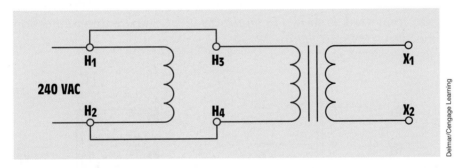

FIGURE 28–23   Control transformer connected for 240-volt operation.

If the transformer is to be connected to 480 volts, the two primary windings are connected in series by connecting H_2 and H_3 together *(Figure 28–24)*. The incoming power is connected to H_1 and H_4. Series-connecting the primary windings has the effect of increasing the number of turns in the primary to 400. This produces a turns ratio of 4:1. When 480 volts are connected to the primary, the secondary voltage will remain at 120.

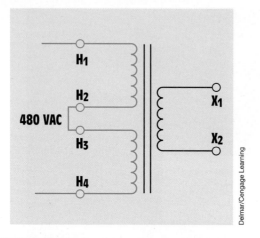

FIGURE 28–24 Control transformer connected for 480-volt operation.

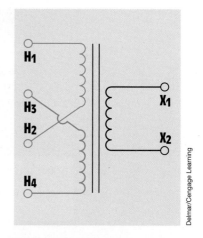

FIGURE 28–25 The primary windings of a control transformer are crossed.

The primary leads of a control transformer are generally cross-connected *(Figure 28–25)*. This is done so that metal links can be used to connect the primary for 240- or 480-volt operation. If the primary is to be connected for 240-volt operation, the metal links will be connected under screws *(Figure 28–26)*. Notice that leads H_1 and H_3 are connected together and leads H_2 and H_4 are connected together. Compare this connection with the connection shown in *Figure 28–23*.

If the transformer is to be connected for 480-volt operation, terminals H_2 and H_3 are connected as shown in *Figure 28–27*. Compare this connection with the connection shown in *Figure 28–24*.

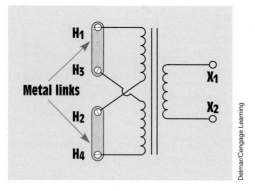

FIGURE 28–26 Metal links connect transformer for 240-volt operation.

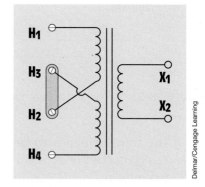

FIGURE 28–27 Control transformer connected for 480-volt operation.

Transformer Core Types

Several types of cores are used in the construction of transformers. Most cores are made from thin steel punchings **laminated** together to form a solid metal core. The core for a 600-mega-volt-ampere (MVA) three-phase transformer is shown in *Figure 28–28*. Laminated cores are preferred because a thin layer of oxide forms on the surface of each lamination and acts as an insulator to reduce the formation of eddy currents inside the core material. The amount of core material needed for a particular transformer is determined by the power rating of the transformer. The amount of core material must be sufficient to prevent saturation at full load. The type and shape of the core generally determine the amount of magnetic coupling between the windings and to some extent the efficiency of the transformer.

The transformer illustrated in *Figure 28–29* is known as a core-type transformer. The windings are placed around each end of the core material. As a general rule, the low-voltage winding is placed closest to the core and the high-voltage winding is placed over the low-voltage winding.

Courtesy of Houston Lighting and Power

FIGURE 28–28 Core of a 600-MVA three-phase transformer.

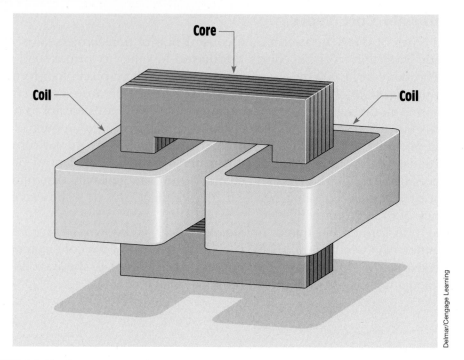

FIGURE 28–29 A core-type transformer.

Delmar/Cengage Learning

The shell-type transformer is constructed in a similar manner to the core type, except that the shell type has a metal core piece through the middle of the window *(Figure 28–30)*. The primary and secondary windings are wound around the center core piece with the low-voltage winding being closest to the metal core. This arrangement permits the transformer to be surrounded by the core and provides excellent magnetic coupling. When the transformer is in operation, all the magnetic flux must pass through the center core piece. It then divides through the two outer core pieces.

The H-type core shown in *Figure 28–31* is similar to the shell-type core in that it has an iron core through its center around which the primary and secondary windings are wound. The H core, however, surrounds the windings on four sides instead of two. This extra metal helps reduce stray leakage flux and improves the efficiency of the transformer. The H-type core is often found on high-voltage distribution transformers.

The **tape-wound core** or **toroid core** *(Figure 28–32)* is constructed by tightly winding one long continuous silicon steel tape into a spiral. The tape may or may not be housed in a plastic container, depending on the application. This type of core does not require steel punchings laminated together. Because the core is one continuous length of metal, **flux leakage** is kept to a

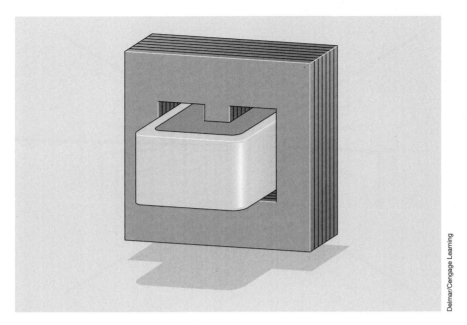

Delmar/Cengage Learning

FIGURE 28–30   A shell-type transformer.

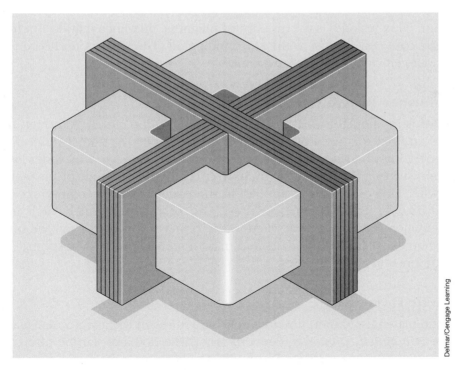

Delmar/Cengage Learning

FIGURE 28–31   A transformer with an H-type core.

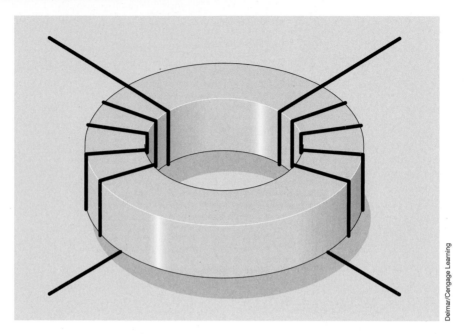

Delmar/Cengage Learning

FIGURE 28–32   A toroid transformer.

minimum. Flux leakage is the amount of magnetic flux lines that do not follow the metal core and are lost to the surrounding air. The tape-wound core is one of the most efficient core designs available.

Transformer Inrush Current

A reactor is an inductor used to add inductance to the circuit. Although transformers and reactors are both inductive devices, there is a great difference in their operating characteristics. Reactors are often connected in series with a low-impedance load to prevent **inrush current** (the amount of current that flows when power is initially applied to the circuit) from becoming excessive *(Figure 28–33).* Transformers, however, can produce extremely high inrush currents when power is first applied to the primary winding. The type of core used when constructing inductors and transformers is primarily responsible for this difference in characteristics.

Magnetic Domains

Magnetic materials contain tiny magnetic structures in their molecular material known as *magnetic domains* (See Unit 4). These domains can be affected by outside sources of magnetism. *Figure 28–34* illustrates a magnetic domain that has not been polarized by an outside magnetic source.

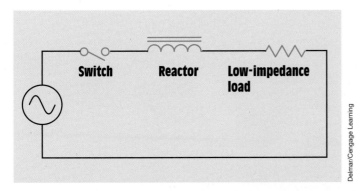
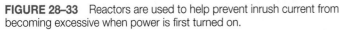

FIGURE 28–33 Reactors are used to help prevent inrush current from becoming excessive when power is first turned on.

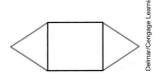

FIGURE 28–34 Magnetic domain in neutral position.

Now assume that the north pole of a magnet is placed toward the top of the material that contains the magnetic domains *(Figure 28–35)*. Notice that the structure of the domain has changed to realign the molecules in the direction of the outside magnetic field. If the polarity of the magnetic pole is changed *(Figure 28–36)*, the molecular structure of the domain changes to realign itself with the new magnetic lines of flux. This external influence can be produced by an electromagnet as well as a permanent magnet.

In certain types of cores, the molecular structure of the domain snaps back to its neutral position when the magnetizing force is removed. This type of core is used in the construction of reactors or chokes *(Figure 28–37)*. A core of this type is constructed by separating sections of the steel laminations with

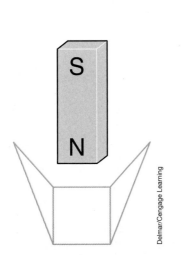

FIGURE 28–35 Domain influenced by a north magnetic field.

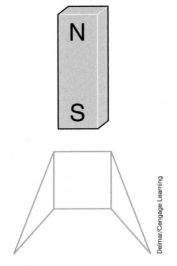

FIGURE 28–36 Domain influenced by a south magnetic field.

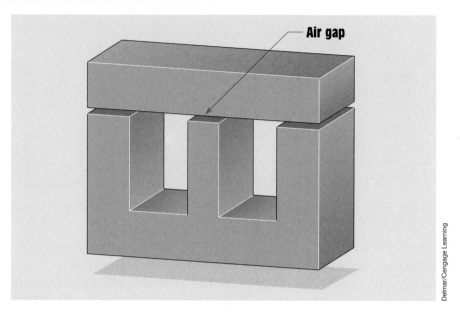

Air gap

Delmar/Cengage Learning

FIGURE 28–37   The core of an inductor contains an air gap.

an air gap. This air gap breaks the magnetic path through the core material and is responsible for the domains returning to their neutral position once the magnetizing force is removed.

The core construction of a transformer, however, does not contain an air gap. The steel laminations are connected together in such a manner as to produce a very low reluctance path for the magnetic lines of flux. In this type of core, the domains remain in their set position once the magnetizing force has been removed. This type of core "remembers" where it was last set. This was the principle of operation of the core memory of early computers. It is also the reason that transformers can have extremely high inrush currents when they are first connected to the powerline.

The amount of inrush current in the primary of a transformer is limited by three factors:

1. the amount of applied voltage,

2. the resistance of the wire in the primary winding, and

3. the flux change of the magnetic field in the core. The amount of flux change determines the amount of inductive reactance produced in the primary winding when power is applied.

Figure 28–38 illustrates a simple isolation-type transformer. The AC applied to the primary winding produces a magnetic field around the winding. As the current changes in magnitude and direction, the magnetic lines of flux change also. Because the lines of flux in the core are continually changing polarity,

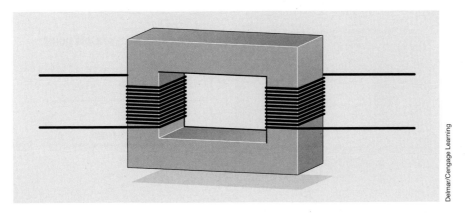

FIGURE 28–38   Isolation transformer.

the magnetic domains in the core material are changing also. As stated previously, the magnetic domains in the core of a transformer remember their last set position. For this reason, the point on the waveform at which current is disconnected from the primary winding can have a great bearing on the amount of inrush current when the transformer is reconnected to power. For example, assume the power supplying the primary winding is disconnected at the zero crossing point *(Figure 28–39)*. In this instance, the magnetic domains would be set at the neutral point. When power is restored to the primary winding, the core material can be magnetized by either magnetic polarity. This permits a change of flux, which is the dominant current-limiting factor. In this instance, the amount of inrush current would be relatively low.

If the power supplying current to the primary winding is interrupted at the peak point of the positive or negative half cycle, however, the domains in the core material will be set at that position. *Figure 28–40* illustrates this condition. It is assumed that the current was stopped as it reached its peak positive point. If

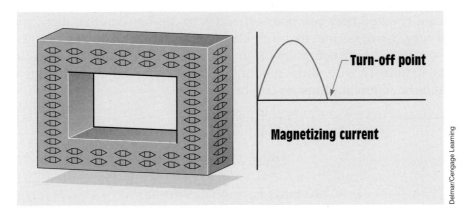

FIGURE 28–39   Magnetic domains are left in the neutral position.

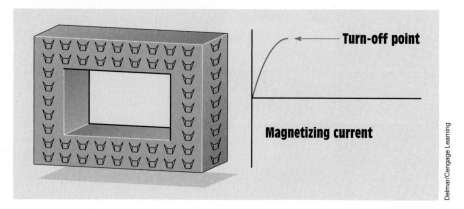

Delmar/Cengage Learning

FIGURE 28–40   Domains are set at one end of magnetic polarity.

the power is reconnected to the primary winding during the positive half cycle, only a very small amount of flux change can take place. Because the core material is saturated in the positive direction, the primary winding of the transformer is essentially an air-core inductor, which greatly decreases the inductive characteristics of the winding. The inrush current in this situation would be limited by the resistance of the winding and a very small amount of inductive reactance.

This characteristic of transformers can be demonstrated with a clamp-on ammeter that has a "peak-hold" capability. If the ammeter is connected to one of the primary leads and power is switched on and off several times, the amount of inrush current varies over a wide range.

28–3   Autotransformers

Autotransformers are one-winding transformers. They use the same winding for both the primary and secondary. The primary winding in *Figure 28–41* is between points B and N and has a voltage of 120 volts applied to it. If the turns of wire are counted between points B and N, it can be seen that there are 120 turns of wire. Now assume that the selector switch is set to point D. The load is now connected between points D and N. The secondary of this transformer contains 40 turns of wire. If the amount of voltage applied to the load is to be calculated the following formula can be used:

$$\frac{E_P}{E_S} = \frac{N_P}{N_S}$$

$$\frac{120\ V}{E_S} = \frac{120\ turns}{40\ turns}$$

$$120\ turns\ E_S = 4800\ V\text{-}turns$$

$$E_S = 40\ V$$

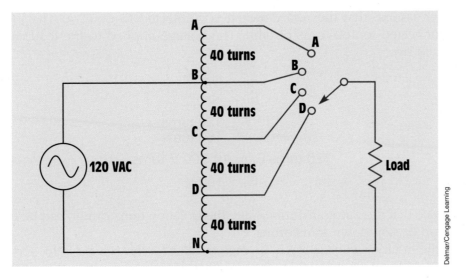

FIGURE 28–41 An autotransformer has only one winding used for both the primary and secondary.

Assume that the load connected to the secondary has an impedance of 10 ohms. The amount of current flow in the secondary circuit can be calculated using the formula

$$I = \frac{E}{Z}$$

$$I = \frac{40\ V}{10\ \Omega}$$

$$I = 4\ A$$

The primary current can be calculated by using the same formula that was used to calculate primary current for an isolation type of transformer:

$$\frac{E_P}{E_S} = \frac{I_S}{I_P}$$

$$\frac{120\ V}{40\ V} = \frac{4\ A}{I_P}$$

$$120\ V\ I_P = 160\ VA$$

$$I_P = 1.333\ A$$

The amount of power input and output for the autotransformer must be the same, just as they are in an isolation transformer:

Primary	Secondary
120 V × 1.333 A = 160 VA	40 V × 4 A = 160 VA

Now assume that the rotary switch is connected to point A. The load is now connected to 160 turns of wire. The voltage applied to the load can be calculated by

$$\frac{E_P}{E_S} = \frac{N_P}{N_S}$$

$$\frac{120 \text{ V}}{E_S} = \frac{120 \text{ turns}}{160 \text{ turns}}$$

$$120 \text{ turns } E_S = 19{,}200 \text{ V-turns}$$

$$E_S = 160 \text{ V}$$

Notice that the autotransformer, like the isolation transformer, can be either a step-up or step-down transformer.

If the rotary switch shown in *Figure 28–41* were to be removed and replaced with a sliding tap that made contact directly to the transformer winding, the turns ratio could be adjusted continuously. This type of transformer is commonly referred to as a Variac or Powerstat depending on the manufacturer. A cutaway view of a variable autotransformer is shown in *Figure 28–42*. The windings are wrapped around a tape-wound toroid core inside a plastic case. The tops of the windings have been milled flat to provide a commutator. A carbon brush makes contact with the windings.

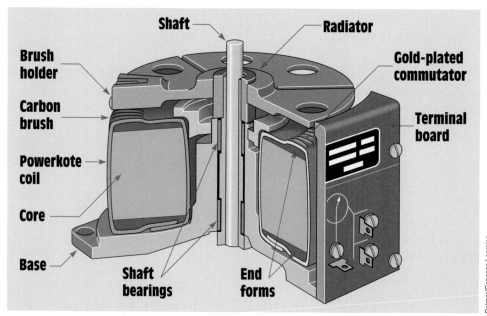

FIGURE 28–42   Cutaway view of a Powerstat.

FIGURE 28–43   Three-phase autotransformer.

Autotransformers are often used by power companies to provide a small increase or decrease to the line voltage. They help provide voltage regulation to large powerlines. A three-phase autotransformer is shown in *Figure 28–43*. This transformer is contained in a housing filled with transformer oil, which acts as a coolant and prevents moisture from forming in the windings.

The autotransformer does have one disadvantage. Because the load is connected to one side of the powerline, there is no line isolation between the incoming power and the load. This can cause problems with certain types of equipment and must be a consideration when designing a power system.

28–4   Transformer Polarities

To understand what is meant by transformer polarity, the voltage produced across a winding must be considered during some point in time. In a 60-herz AC circuit, the voltage changes polarity 60 times per second. When discussing transformer polarity, it is necessary to consider the relationship between the different windings at the same point in time. It is therefore assumed that this point in time is when the peak positive voltage is being produced across the winding.

Polarity Markings on Schematics

When a transformer is shown on a schematic diagram, it is common practice to indicate the polarity of the transformer windings by placing a dot beside one end of each winding as shown in *Figure 28–44*. These dots signify that the polarity is the same at that point in time for each winding. For example, assume the voltage applied to the primary winding is at its peak positive value at the terminal indicated by the dot. The voltage at the dotted lead of the secondary will be at its peak positive value at the same time.

This same type of polarity notation is used for transformers that have more than one primary or secondary winding. An example of a transformer with a multisecondary is shown in *Figure 28–45*.

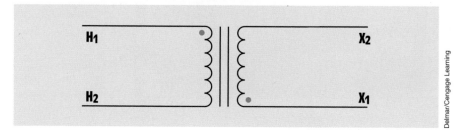

FIGURE 28–44 Transformer polarity dots.

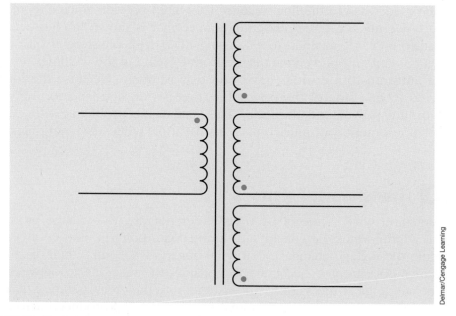

FIGURE 28–45 Polarity marks for multiple secondaries.

Additive and Subtractive Polarities

The polarity of transformer windings can be determined by connecting them as an autotransformer and testing for additive or subtractive polarity, often referred to as a boost or buck connection. This is done by connecting one lead of the secondary to one lead of the primary and measuring the voltage across both windings *(Figure 28–46)*. The transformer shown in the example has a primary voltage rating of 120 volts and a secondary voltage rating of 24 volts. This same circuit has been redrawn in *Figure 28–47* to show the connection more clearly. Notice that the secondary winding has been connected in series with the primary winding. The transformer now contains only one

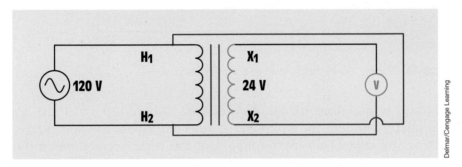

FIGURE 28–46   Connecting the secondary and primary windings forms an autotransformer.

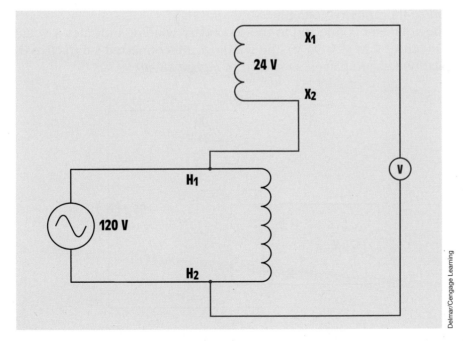

FIGURE 28–47   Redrawing the connection.

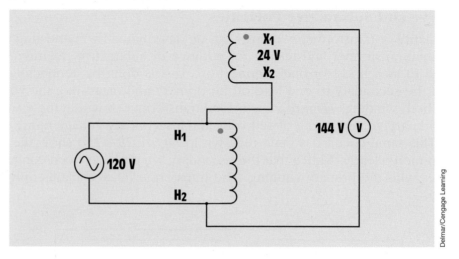

FIGURE 28–48 Placing polarity dots to indicate additive polarity.

winding and is therefore an autotransformer. When 120 volts are applied to the primary winding, the voltmeter connected across the secondary indicates either the *sum* of the two voltages or the *difference* between the two voltages. If this voltmeter indicates 144 volts (120 V + 24 V = 144 V), the windings are connected additive (boost) and polarity dots can be placed as shown in *Figure 28–48*. Notice in this connection that the secondary voltage is added to the primary voltage.

If the voltmeter connected to the secondary winding indicates a voltage of 96 volts (120 V − 24 V = 96 V), the windings are connected subtractive (buck) and polarity dots are placed as shown in *Figure 28–49*.

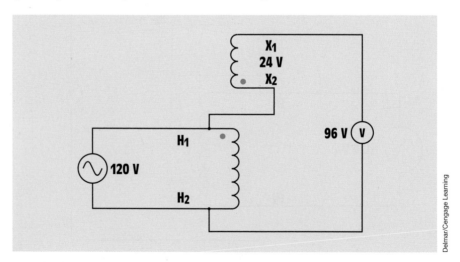

FIGURE 28–49 Polarity dots indicate subtractive polarity.

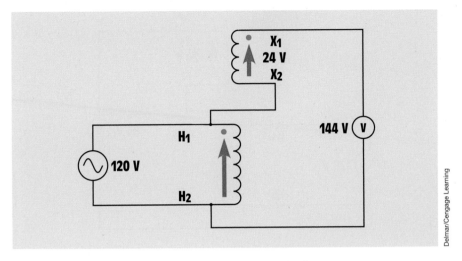

FIGURE 28–50 Arrows help indicate the placement of the polarity dots.

Using Arrows to Place Dots

To help in the understanding of additive and subtractive polarity, arrows can be used to indicate a direction of greater-than or less-than values. In *Figure 28–50,* arrows have been added to indicate the direction in which the dot is to be placed. In this example, the transformer is connected additive, or boost, and both arrows point in the same direction. Notice that the arrow points to the dot. In *Figure 28–51,* it is seen that values of the two arrows add to produce 144 volts.

In *Figure 28–52,* arrows have been added to a subtractive, or buck, connection. In this instance, the arrows point in opposite directions and the voltage of one tries to cancel the voltage of the other. The result is that the smaller value is eliminated and the larger value is reduced as shown in *Figure 28–53.*

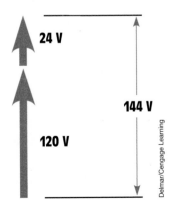

FIGURE 28–51 The values of the arrows add to indicate additive polarity (boost connection).

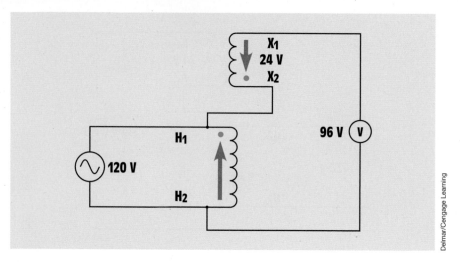

FIGURE 28–52 The arrows help indicate subtractive polarity.

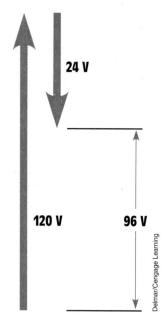

FIGURE 28–53 The values of the arrows subtract (buck connection).

28–5 Voltage and Current Relationships in a Transformer

When the primary of a transformer is connected to power but there is no load connected to the secondary, current is limited by the inductive reactance of the primary. At this time, the transformer is essentially an inductor and the excitation current is lagging the applied voltage by 90° *(Figure 28–54)*.

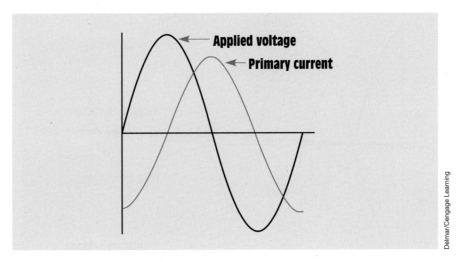

FIGURE 28–54 At no load, the primary current lags the voltage by 90°.

The primary current induces a voltage in the secondary. This induced voltage is proportional to the rate of change of current. The secondary voltage is maximum during the periods that the primary current is changing the most (0°, 180°, and 360°), and it will be zero when the primary current is not changing (90° and 270°). A plot of the primary current and secondary voltage shows that the secondary voltage lags the primary current by 90° *(Figure 28–55)*. Because the secondary voltage lags the primary current by 90° and the applied voltage leads the primary current by 90°, the secondary voltage is 180° out of phase with the applied voltage and in phase with the induced voltage in the primary.

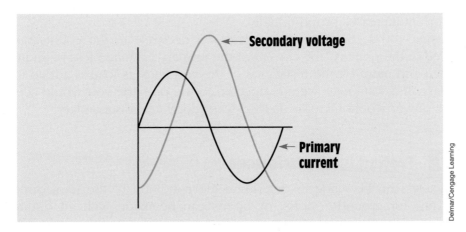

FIGURE 28–55 The secondary voltage lags the primary current by 90°.

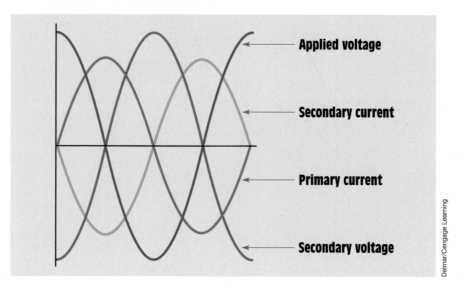

FIGURE 28–56   Voltage and current relationships of the primary and secondary windings.

Adding Load to the Secondary

When a load is connected to the secondary, current begins to flow. Because the transformer is an inductive device, the secondary current lags the secondary voltage by 90°. Because the secondary voltage lags the primary current by 90°, the secondary current is 180° out of phase with the primary current *(Figure 28–56)*.

The current of the secondary induces a countervoltage in the secondary windings that is in opposition to the countervoltage induced in the primary. The countervoltage of the secondary weakens the countervoltage of the primary and permits more primary current to flow. As secondary current increases, primary current increases proportionally.

Because the secondary current causes a decrease in the countervoltage produced in the primary, the current of the primary is limited less by inductive reactance and more by the resistance of the windings as load is added to the secondary. If a wattmeter were connected to the primary, you would see that the true power would increase as load was added to the secondary.

28–6 Testing the Transformer

Several tests can be made to determine the condition of the transformer. A simple test for grounds, shorts, or opens can be made with an ohmmeter *(Figure 28–57)*. Ohmmeter A is connected to one lead of the primary and one lead of the secondary. This test checks for shorted windings between the

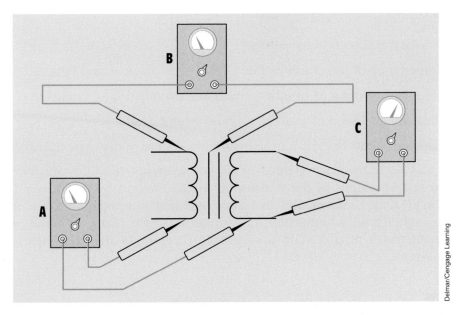

Delmar/Cengage Learning

FIGURE 28–57   Testing a transformer with an ohmmeter.

primary and secondary. The ohmmeter should indicate infinity. If there is more than one primary or secondary winding, all isolated windings should be tested for shorts. Ohmmeter B illustrates testing the windings for grounds. One lead of the ohmmeter is connected to the case of the transformer, and the other is connected to the winding. All windings should be tested for grounds, and the ohmmeter should indicate infinity for each winding. Ohmmeter C illustrates testing the windings for continuity. The wire resistance of the winding should be indicated by the ohmmeter.

If the transformer appears to be in good condition after the ohmmeter test, it should then be tested for shorts and grounds with a megohmmeter. A MEGGER® will reveal problems of insulation breakdown that an ohmmeter will not. Large oil-filled transformers should have the condition of the dielectric oil tested at periodic intervals. This test involves taking a sample of the oil and performing certain tests for dielectric strength and contamination.

28–7   Transformer Nameplates

Most transformers contain a nameplate that lists information concerning the transformer. *NEC* ® *450.11* requires the following information:

1. Name of manufacturer

2. Rated kilovolt-ampere

3. Frequency

4. Primary and secondary voltage

5. Impedance of transformers rated 25 kilovolt-ampere and larger

6. Required clearances of transformers with ventilating openings

7. Amount and kind of insulating liquid where used

8. The temperature class for the insulating system of dry-type transformers

Notice that the transformer is rated in kilovolt-amperes, not kilowatts, because the true power output of the transformer is determined by the power factor of the load. Other information that may be listed is temperature rise in °C, model number, and whether the transformer is single phase or three phase. Many nameplates also contain a schematic diagram of the windings to aid in connection.

28–8   Determining Maximum Current

The nameplate does not list the current rating of the windings. Because power input must equal power output, the current rating for a winding can be determined by dividing the kilovolt-ampere rating by the winding voltage. For example, assume a transformer has a kilovolt-ampere rating of 0.5 kilovolt-ampere, a primary voltage of 480 volts, and a secondary voltage of 120 volts. To determine the maximum current that can be supplied by the secondary, divide the kVA rating by the secondary voltage:

$$I_S = \frac{kVA}{E_S}$$

$$I_S = \frac{500 \text{ VA}}{120 \text{ V}}$$

$$I_S = 4.167 \text{ A}$$

The primary current can be calculated in the same way:

$$I_P = \frac{kVA}{E_P}$$

$$I_P = \frac{500 \text{ VA}}{480 \text{ V}}$$

$$I_P = 1.042 \text{ A}$$

Transformers with multiple secondary windings will generally have the current rating listed with the voltage rating.

28–9   Transformer Impedance

Transformer impedance is determined by the physical construction of the transformer. Factors such as the amount and type of core material, wire size used to construct the windings, the number of turns, and the degree of magnetic coupling between the windings greatly affect the transformer's impedance. Impedance is expressed as a percent (%Z or %IZ) and is measured by connecting a short circuit across the low-voltage winding of the transformer and then connecting a variable voltage source to the high-voltage winding *(Figure 28–58)*. The variable voltage is then increased until rated current flows in the low-voltage winding. The transformer impedance is determined by calculating the percentage of variable voltage as compared to the rated voltage of the high-voltage winding.

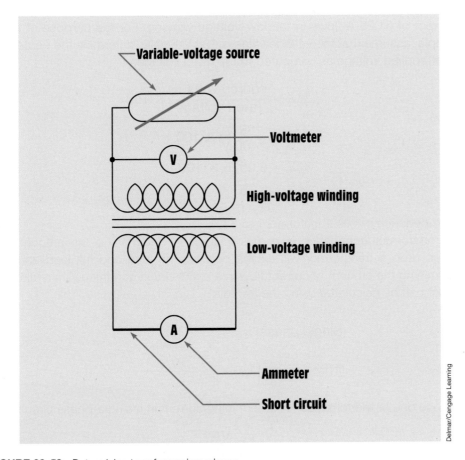

FIGURE 28–58   Determining transformer impedance.

■ EXAMPLE 28-1

Assume that the transformer shown in *Figure 28–58* is a 2400/480-V, 15-kVA transformer. To determine the impedance of the transformer, first calculate the full-load current rating of the secondary winding:

$$I = \frac{VA}{E}$$

$$I = \frac{15,000 \text{ VA}}{480 \text{ V}}$$

$$I = 31.25 \text{ A}$$

Next, increase the source voltage connected to the high-voltage winding until a current of 31.25 A flows in the low-voltage winding. For the purpose of this example, assume that the voltage value is 138 V. Finally, determine the percentage of applied voltage as compared to the rated voltage:

$$\%Z = \frac{\text{source voltage}}{\text{rated voltage}} \times 100$$

$$\%Z = \frac{138 \text{ V}}{2400 \text{ V}} \times 100$$

$$\%Z = 0.0575 \times 100$$

$$\%Z = 5.75$$

The impedance of this transformer is 5.75%.

Transformer impedance is a major factor in determining the amount of voltage drop a transformer will exhibit between no load and full load and in determining the amount of current flow in a short-circuit condition. Short-circuit current can be calculated using the formula

$$\text{(Single phase) } I_{SC} = \frac{VA}{E \times \%Z}$$

$$\text{(Three phase) } I_{SC} = \frac{VA}{E \times \sqrt{3} \times \%Z}$$

because one of the formulas for determining current in a single-phase circuit is

$$I = \frac{VA}{E}$$

and one of the formulas for determining current in a three-phase circuit is

$$I = \frac{VA}{E \times \sqrt{3}}$$

The preceding formulas for determining short-circuit current can be modified to show that the short-circuit current can be calculated by dividing the rated secondary current by the %Z:

$$I_{SC} = \frac{I_{Rated}}{\%Z}$$

■ EXAMPLE 28-2

A single-phase transformer is rated at 50 kVA and has a secondary voltage of 240 V. The nameplate reveals that the transformer has an internal impedance (%Z) of 2.5%. What is the short-circuit current for this transformer?

$$I_{secondary} = \frac{50,000 \text{ VA}}{240 \text{ V}}$$

$$I_{secondary} = 208.3 \text{ A}$$

$$I_{short\ circuit} = \frac{208.3 \text{ A}}{\%Z}$$

$$I_{short\ circuit} = \frac{208.3 \text{ A}}{0.025}$$

$$I_{short\ circuit} = 8,333.3 \text{ A}$$

It is sometimes necessary to calculate the amount of short-circuit current when determining the correct fuse rating for a circuit. The fuse must have a high enough "interrupt" rating to clear the fault in the event of a short circuit.

Constant-Current Transformers

A very special type of isolation transformer is the **constant-current transformer,** often referred to as a **current regulator.** Constant-current transformers are designed to deliver a constant output current, generally 6.6 amperes, under varying load conditions. They are most often employed to provide power to series-connected street lamps. Street lamps are often connected in series instead of parallel because of the savings in wire. Series-connected lamps require a single conductor to be connected from lamp to lamp instead of two conductors, *Figure 28–59.*

When lamps are connected in series, some device must be used to continue the circuit in the event that one or more lamps should fail. Some lights use a reactor coil connected in parallel with the lamp, *Figure 28–60.* Another method uses a film cut-out device, *Figure 28–61,* consisting of two pieces of metal separated by an insulator designed to puncture at a predetermined voltage. As long as the lamp is in operation, the voltage drop across the cut-out device is not sufficient to cause the film to puncture. If the lamp should burn out, producing an open circuit, the entire circuit voltage will appear across the cut-out device and cause it to short circuit.

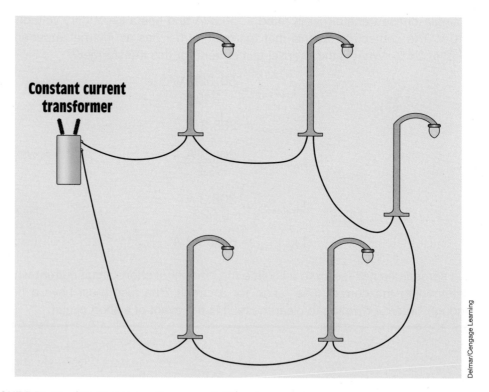

Constant current transformer

Delmar/Cengage Learning

FIGURE 28–59   Street lamps are often connected in series.

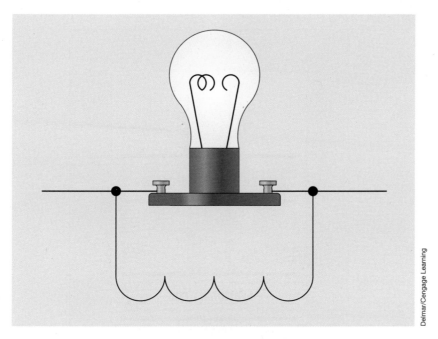

FIGURE 28–60 An inductor maintains the circuit if the lamp should fail.

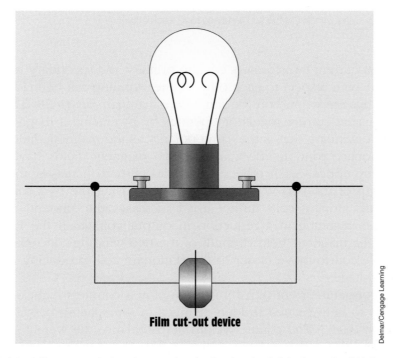

Film cut-out device

FIGURE 28–61 A film cut-out device shorts and maintains the circuit if the lamp should fail.

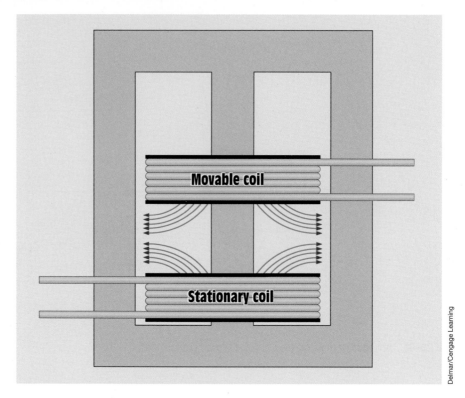

Delmar/Cengage Learning

FIGURE 28–62 Magnetic flux of the two windings repel each other.

Constant-current transformers contain primary and secondary winding that are movable with respect to each other. Either winding can be made movable. Both windings are wound on the same core material, *Figure 28–62*.

The constant-current regulator operates by producing a magnetic field in the movable winding that is the same polarity as the magnetic field produced in the stationary winding. Because the two magnetic fields have the same polarity, they oppose each other. If the load current increases, the magnetic field strength of the two windings increases, causing the two coils to move further apart. Moving coils farther apart increases the amount of magnetic flux leakage, resulting in a reduction in output voltage. If the load current decreases, the magnetic field strength of the two windings decreases, causing the movable coil to move closer to the stationary coil, producing an increase in output voltage.

Many constant-current transformers employ a counterweight and dashpot mechanism to help reduce sudden changes in the spacing between the two coils, *Figure 28–63*. The counterweight helps balance the weight of the movable coil, and the dashpot device helps reduce the "hunting" action between the two coils.

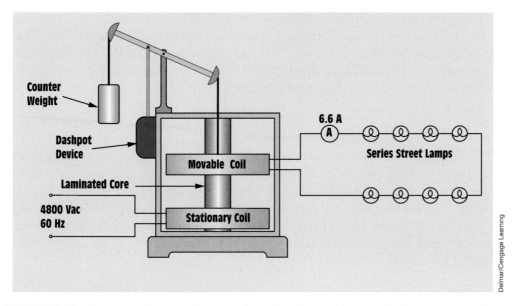

FIGURE 28–63 A counterweight and dashpot device help reduce sudden changes in the output current.

Series Connection of Transformer Secondaries

As a general rule, connecting the secondary windings of transformers in series does not present a problem. Because the current is the same in a series circuit, the problem of high circulating current does not exist in a series connection. The secondary windings can be connected in series to produce a higher output voltage. Assume that two transformers have a secondary voltage of 120 volts, *Figure 28–64*.

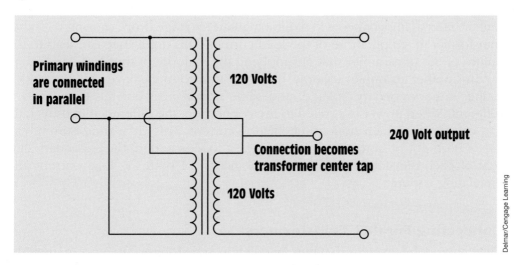

FIGURE 28–64 Transformer secondary windings connected in series.

The primary windings can be connected in parallel without a problem. When making this connection, the polarity of the two secondary windings must be connected additive of boost. The series connection of the two secondary windings will produce an output of 240 volts center tapped. If the polarity is not correct, the output voltage will be zero (0).

Parallel Transformer Connections

It is sometimes necessary to connect the secondary windings of transformers in parallel to increase the current capacity, but generally it is not done unless there is no other alternative, *Figure 28–65*. Connecting transformer primary windings in parallel is not a problem, but connecting the secondary windings in parallel can cause high circulating currents or extremely unbalance currents that can lead to transformer failure. Transformers that are to be connected in parallel must have

- the same kVA rating.
- the same turns ratio,
- the same impedance, and
- the same secondary voltage;
- and the secondary windings must have the same polarity.

If any of these factors are different, it can cause failure of one or both transformers. Assume for example, that two transformers have the same kVA rating, same secondary voltage, and same turns ratio, but the impedance is not the same. When load is added to the parallel connection, the transformer with the higher internal impedance will exhibit a greater voltage drop, causing the other transformer to supply more of the load current. This unbalance can lead to the failure of the transformer that is supplying the majority of the load current.

In another example, assume that two transformers have the same kVA rating, same secondary voltage, and same impedance, but the turns ratio is different. When power is applied to the connection, the difference in turns ratio can cause a very high no-load circulating current. This circulating current can cause burn-out of both transformers. When load is added, the secondary current of each transformer will be a combination of both the load current and circulating current.

Connecting Parallel Transformers

Care should be exercised when paralleling transformers to ensure that the polarity is correct. If the polarity is incorrect, it produces a short circuit. In this

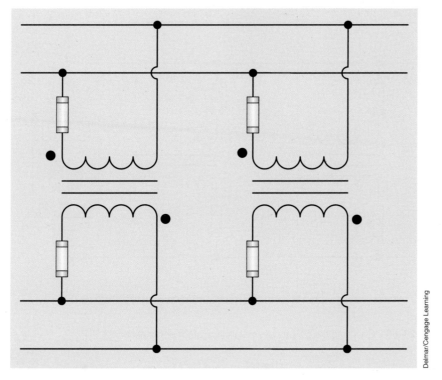

FIGURE 28–65   Transformers are sometimes connected in parallel to increase the current capacity of the circuit.

example, assume that one transformer is already connected to the line. Also assumed that the primary voltage is 4160 volts and the secondary voltage is 120 volts. Connect one of the secondary leads to the line and then energize the primary winding, *Figure 28–66*.

Connect a voltmeter between the secondary lead that has not been connected and the line of the intended connection. If the polarity of the two transformers is correct, the voltmeter should indicate zero (0) volts. If the polarity is not correct, the voltmeter will indicate double the amount of secondary voltage. In this example, the voltmeter would indicate 240 volts if the polarity were not correct.

Precautions When Servicing Parallel Transformers

If it should become necessary to remove one of the transformers for service, the secondary should be disconnected first. This can generally be accomplished by removing the secondary fuse, *Figure 28–67*.

In this example, the primary is connected to 4160 volts. If the secondary winding is not disconnected first, the transformer becomes a step-up

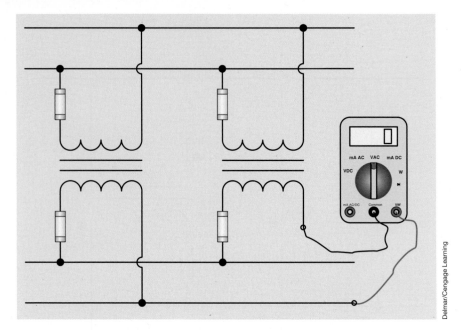

FIGURE 28–66   Checking the polarity of a parallel transformer connection.

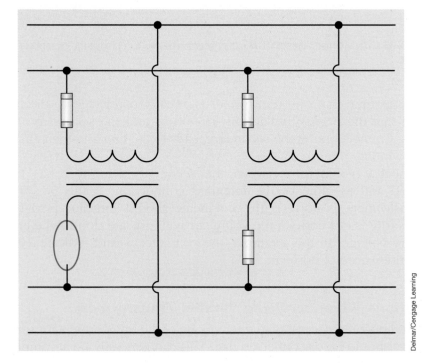

FIGURE 28–67   The secondary winding should be disconnected before disconnecting the primary winding.

transformer. The primary winding will still have a voltage of 4160 even if it is disconnected from the power line. Anytime parallel transformers are employed, a sign reading **CAUTION FEEDBACK VOLTAGE** should be located at each primary fuse.

Summary

- All values of voltage, current, and impedance in a transformer are proportional to the turns ratio.

- Transformers can change values of voltage, current, and impedance but cannot change the frequency.

- The primary winding of a transformer is connected to the powerline.

- The secondary winding is connected to the load.

- A transformer that has a lower secondary voltage than primary voltage is a step-down transformer.

- A transformer that has a higher secondary voltage than primary voltage is a step-up transformer.

- An isolation transformer has its primary and secondary windings electrically and mechanically separated from each other.

- When a coil induces a voltage into itself, it is known as self-induction.

- When a coil induces a voltage into another coil, it is known as mutual induction.

- Transformers can have very high inrush current when first connected to the powerline because of the magnetic domains in the core material.

- Inductors provide an air gap in their core material that causes the magnetic domains to reset to a neutral position.

- Autotransformers have only one winding, which is used as both the primary and secondary.

- Autotransformers have a disadvantage in that they have no line isolation between the primary and secondary winding.

- Isolation transformers help filter voltage and current spikes between the primary and secondary side.

- Polarity dots are often added to schematic diagrams to indicate transformer polarity.

- Transformers can be connected as additive or subtractive polarity.

- Constant-current transformers are also known as current regulators.

- Constant-current transformers are generally used to provide power to series-connected loads.

- As a general rule, transformer secondary windings should not be connected in parallel.

Review Questions

1. What is a transformer?

2. What are common efficiencies for transformers?

3. What is an isolation transformer?

4. All values of a transformer are proportional to its

   _____ _____.

5. What is an autotransformer?

6. What is a disadvantage of an autotransformer?

7. Explain the difference between a step-up and a step-down transformer.

8. A transformer has a primary voltage of 240 V and a secondary voltage of 48 V. What is the turns ratio of this transformer?

9. A transformer has an output of 750 VA. The primary voltage is 120 V. What is the primary current?

10. A transformer has a turns ratio of 1:6. The primary current is 18 A. What is the secondary current?

11. What do the dots shown beside the terminal leads of a transformer represent on a schematic?

12. A transformer has a primary voltage rating of 240 V and a secondary voltage rating of 80 V. If the windings were connected subtractive, what voltage would appear across the entire connection?

13. If the windings of the transformer in Question 12 were to be connected additive, what voltage would appear across the entire winding?

14. The primary leads of a transformer are labeled 1 and 2. The secondary leads are labeled 3 and 4. If polarity dots are placed beside leads 1 and 4, which secondary lead would be connected to terminal 2 to make the connection additive?

Practical Applications

You are working in an industrial plant. You must install a single-phase transformer. The transformer has the following information on the nameplate:

Primary voltage—13,800 V

Secondary voltage—240 V

Impedance—5%

150 kVA

The secondary fuse has a blow rating of 800 A and an interrupt rating of 10,000 A. Is this interrupt rating sufficient for this installation? ■

Practical Applications

You have been given a transformer to install on a 277-V line. The transformer nameplate is shown in *Figure 28–68*. The transformer must supply a 120-V, 20-A circuit. The transformer capacity should be not less than 115% of the rated load. Does the transformer you have been given have enough kVA capacity to supply the load? To which transformer terminals would you connect the incoming power? To which transformer terminals would you connect the load? ■

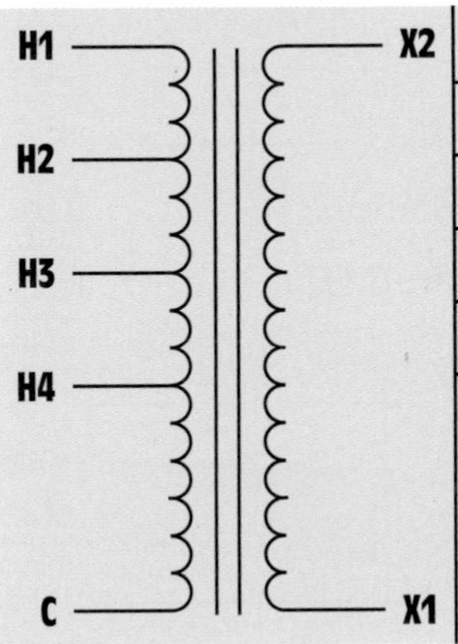

480	H1	Hz 60
277	H2	3 kVA
240	H3	40 °C
208	H4	Model #XXXXX
Com	C	Transformers Inc.
120	X1–X2	Dry type

Delmar/Cengage Learning

FIGURE 28–68 Transformer nameplate.

Practice Problems

Refer to *Figure 28–69* to answer the following questions. Find all the missing values.

1.

E_P 120 V	E_S 24 V
I_P ——————	I_S ——————
N_P 300 turns	N_S ——————
Ratio ——————	$Z = 3\ \Omega$

2.

E_P 240 V	E_S 320 V
I_P ——————	I_S ——————
N_P ——————	N_S 280
Ratio ——————	$Z = 500\ \Omega$

3.

E_P ——————	E_S 160 V
I_P ——————	I_S ——————
N_P ——————	N_S 80 turns
Ratio 1:2.5	$Z = 12\ \Omega$

4.

E_P 48	E_S 240 V
I_P ——————	I_S ——————
N_P 220 turns	N_S ——————
Ratio ——————	$Z = 360\ \Omega$

5.

E_P ——————	E_S ——————
I_P 16.5 A	I_S 3.25 A
N_P ——————	N_S 450 turns
Ratio ——————	$Z = 56\ \Omega$

6.

E_P 480 V	E_S ——————
I_P ——————	I_S ——————
N_P 275 turns	N_S 525 turns
Ratio ——————	$Z = 1.2\ k\Omega$

Refer to *Figure 28–70* to answer the following questions. Find all the missing values.

7.

E_P 208 V	E_{S1} 320 V	E_{S2} 120 V	E_{S3} 24 V
I_P ——————	I_{S1} ——————	I_{S2} ——————	I_{S3} ——————
NP 800 turns	N_{S1} ——————	N_{S2} ——————	N_{S3} ——————
	Ratio 1:	Ratio 2:	Ratio 3:
	R_1 12 kΩ	R_2 6 Ω	R_3 8 Ω

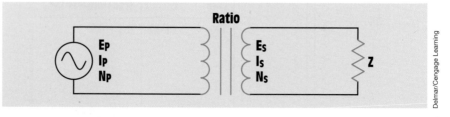

FIGURE 28–69 Isolation transformer practice problems.

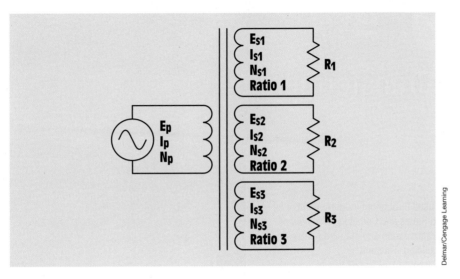

Delmar/Cengage Learning

FIGURE 28–70   Single-phase transformer with multiple secondaries.

8.

E_P 277 V	E_{S1} 480 V	E_{S2} 208 V	E_{S3} 120 V
I_P ———————	I_{S1} ———————	I_{S2} ———————	I_{S3} ———————
N_P 350 turns	N_{S1} ———————	N_{S2} ———————	N_{S3} ———————
	Ratio 1:	Ratio 2:	Ratio 3:
	R_1 200 Ω	R_2 60 Ω	R_3 24 Ω

Unit 29

Three-Phase Transformers

Why You Need to Know

*T*hree-phase transformers are used throughout industry. Almost all power generated in the United States and Canada is three phase. Transformers step up voltage for transmission and step it down again for use inside a plant or commercial building. This unit

- presents the difference between a true three-phase transformer and a three-phase transformer bank.
- determines different voltage and current values in a three-phase transformer.
- defines phase values in calculating the values of a transformer.
- explains how harmonics are identified and overcome.
- describes different three-phase transformer connections such as delta–wye, wye–delta, open-delta, T connected, and Scott connected.
- discusses installation and testing.

KEY TERMS

Closing a delta	Orange wire
Delta–wye	Single-phase loads
Dielectric oil	Tagging
High leg	Three-phase bank
One-line diagram	Wye–delta
Open-delta	

Objectives

After studying this unit, you should be able to

- discuss the operation of three-phase transformers.

- connect three single-phase transformers to form a three-phase bank.

- calculate voltage and current values for a three-phase transformer connection.

- connect two single-phase transformers to form a three-phase open-delta connection.

- discuss the characteristics of an open-delta connection.

- discuss different types of three-phase transformer connections and how they are used to supply single-phase loads.

- calculate values of voltage and current for a three-phase transformer used to supply both three-phase and single-phase loads.

- describe what a harmonic is.

- discuss the problems concerning harmonics.

- identify the characteristics of different harmonics.

- perform a test to determine if harmonic problems exist.

- discuss methods of dealing with harmonic problems.

Preview

Three-phase transformers are used throughout industry to change values of three-phase voltage and current. Because three-phase power is the most common way in which power is produced, transmitted, and used, an understanding of how three-phase transformer connections are made is essential. This unit discusses different types of three-phase transformer connections and presents examples of how values of voltage and current for these connections are calculated. ■

29–1   Three-Phase Transformers

A three-phase transformer is constructed by winding three single-phase transformers on a single core *(Figure 29–1)*. A photograph of a three-phase transformer is shown in *Figure 29–2*. The transformer is shown before it is mounted in an enclosure, which will be filled with a **dielectric oil.** The dielectric oil performs several functions. Because it is a dielectric, it provides electric insulation between the windings and the case. It is also used to help provide cooling and to prevent the formation of moisture, which can deteriorate the winding insulation.

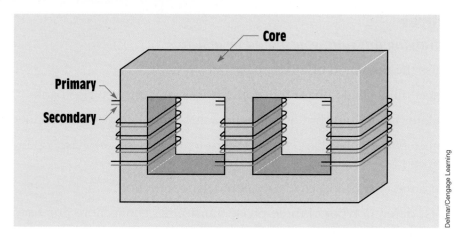

Delmar/Cengage Learning

FIGURE 29–1 Basic construction of a three-phase transformer.

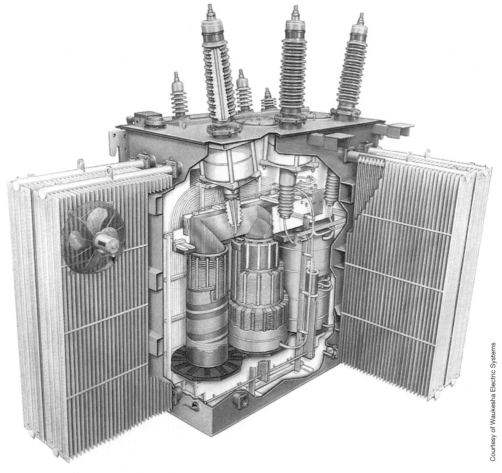

Courtesy of Waukesha Electric Systems

FIGURE 29–2 Three-phase transformer.

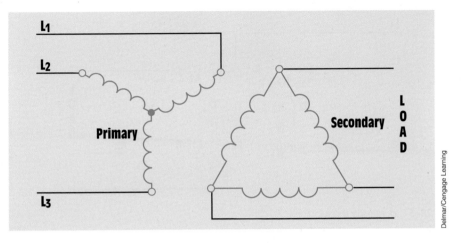

FIGURE 29–3 Wye–delta connected three-phase transformer.

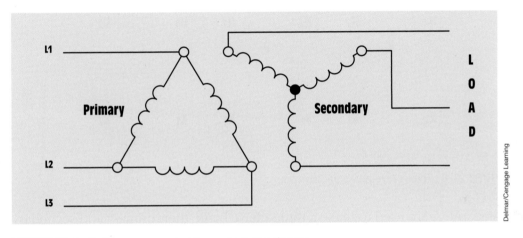

FIGURE 29–4 Delta–wye connected three-phase transformer.

Three-Phase Transformer Connections

Three-phase transformers are connected in delta or wye configurations. A **wye–delta** transformer, for example, has its primary winding connected in a wye and its secondary winding connected in a delta *(Figure 29–3)*. A **delta–wye** transformer would have its primary winding connected in delta and its secondary connected in wye *(Figure 29–4)*.

Connecting Single-Phase Transformers into a Three-Phase Bank

If three-phase transformation is needed and a three-phase transformer of the proper size and turns ratio is not available, three single-phase transformers can be connected to form a **three-phase bank.** When three single-phase

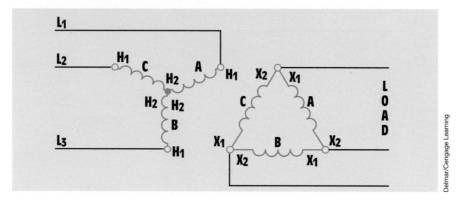

FIGURE 29–5 Identifying the windings.

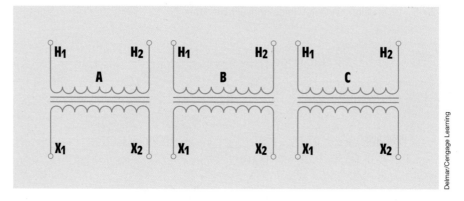

FIGURE 29–6 Three single-phase transformers.

transformers are used to make a three-phase bank, their primary and secondary windings are connected in a wye or delta connection. The three transformer windings in *Figure 29–5* are labeled A, B, and C. One end of each primary lead is labeled H_1, and the other end is labeled H_2. One end of each secondary lead is labeled X_1, and the other end is labeled X_2.

Figure 29–6 shows three single-phase transformers labeled A, B, and C. The primary leads of each transformer are labeled H_1 and H_2, and the secondary leads are labeled X_1 and X_2. The schematic diagram of *Figure 29–5* is used to connect the three single-phase transformers into a three-phase wye–delta connection, as shown in *Figure 29–7*.

The primary winding is first tied into a wye connection. The schematic in *Figure 29–5* shows that the H_2 leads of all three primary windings are connected together and the H_1 lead of each winding is open for connection to the incoming powerline. Notice in *Figure 29–7* that the H_2 leads of the primary windings are connected together and the H_1 lead of each winding has been connected to the incoming powerline.

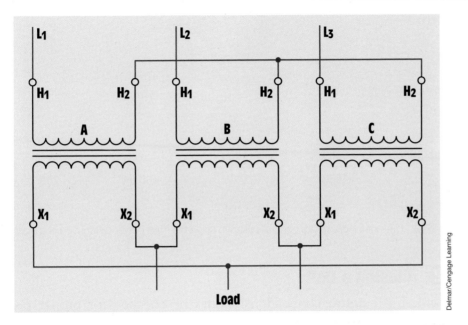

FIGURE 29–7 Connecting three single-phase transformers to form a wye–delta three-phase bank.

Figure 29–5 shows that the X_1 lead of Transformer A is connected to the X_2 lead of Transformer C. Notice that this same connection has been made in *Figure 29–7*. The X_1 lead of Transformer B is connected to the X_2 lead of Transformer A, and the X_1 lead of Transformer C is connected to the X_2 lead of Transformer B. The load is connected to the points of the delta connection.

Although *Figure 29–5* illustrates the proper schematic symbology for a three-phase transformer connection, some electrical schematics and wiring diagrams do not illustrate three-phase transformer connections in this manner. One type of diagram, called the **one-line diagram,** would illustrate a delta–wye connection, as shown in *Figure 29–8*. These diagrams are generally used to show the main power distribution system of a large industrial plant. The one-line diagram in *Figure 29–9* shows the main power to the plant and the transformation of voltages to different subfeeders. Notice that each transformer shows whether the primary and secondary are connected as a wye or delta and the secondary voltage of the subfeeder.

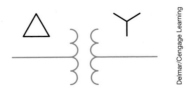

FIGURE 29–8 One-line diagram symbol used to represent a delta–wye three-phase transformer connection.

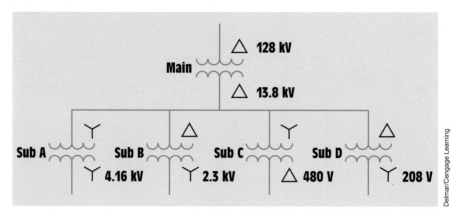

FIGURE 29–9   One-line diagrams are generally used to show the main power distribution of a plant.

29–2   Closing a Delta

When **closing a delta,** connections should be checked for proper polarity before making the final connection and applying power. If the phase winding of one transformer is reversed, an extremely high current will flow when power is applied. Proper phasing can be checked with a voltmeter, as shown in *Figure 29–10.* If power is applied to the transformer bank before the delta connection is closed, the voltmeter should indicate 0 volt. If one phase winding has been reversed, however, the voltmeter will indicate double the amount of voltage. For example, assume that the output voltage of a delta secondary is 240 volts. If the voltage is checked before the delta is closed, the voltmeter should indicate a voltage of 0 V if all windings have been phased properly. If one winding has been reversed, however, the voltmeter will indicate a voltage of 480 volts

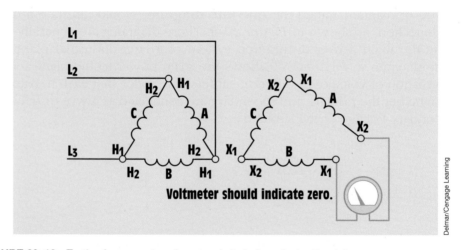

FIGURE 29–10   Testing for proper transformer polarity before closing the delta.

(240 V + 240 V). This test will confirm whether a phase winding has been reversed, but it will not indicate whether the reversed winding is located in the primary or secondary. If either primary or secondary windings have been reversed, the voltmeter will indicate double the output voltage.

Note, however, that a voltmeter is a high-impedance device. It is not unusual for a voltmeter to indicate some amount of voltage before the delta is closed, especially if the primary has been connected as a wye and the secondary as a delta. When this is the case, however, the voltmeter will generally indicate close to the normal output voltage if the connection is correct and double the output voltage if the connection is incorrect.

29–3 Three-Phase Transformer Calculations

To calculate the values of voltage and current for three-phase transformers, the formulas used for making transformer calculations and three-phase calculations must be followed. Another very important rule is that ***only phase values of voltage and current can be used when calculating transformer values.*** Refer to Transformer A in *Figure 29–6*. All transformation of voltage and current takes place between the primary and secondary windings. Because these windings form the phase values of the three-phase connection, only phase and not line values can be used when calculating transformed voltages and currents.

■ **EXAMPLE 29-1**

A three-phase transformer connection is shown in *Figure 29–11*. Three single-phase transformers have been connected to form a wye–delta bank. The primary is connected to a three-phase line of 13,800 V, and the secondary voltage is 480 V. A three-phase resistive load with an impedance of 2.77 Ω per phase is connected to the secondary of the transformer. Calculate the following values for this circuit:

$E_{P(PRIMARY)}$—phase voltage of the primary

$E_{P(SECONDARY)}$—phase voltage of the secondary

Ratio—turns ratio of the transformer

$E_{P(LOAD)}$—phase voltage of the load bank

$I_{P(LOAD)}$—phase current of the load bank

$I_{L(SECONDARY)}$—secondary line current

$I_{P(SECONDARY)}$—phase current of the secondary

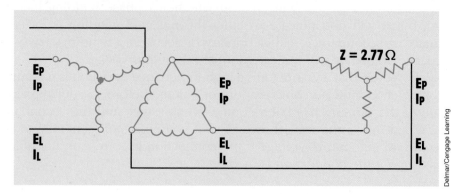

FIGURE 29–11   Example Circuit 1 three-phase transformer calculation.

$I_{P(PRIMARY)}$—phase current of the primary

$I_{L(PRIMARY)}$—line current of the primary

Solution

The primary windings of the three single-phase transformers have been connected to form a wye connection. In a wye connection, the phase voltage is less than the line voltage by a factor of 1.732 (the square root of 3). Therefore, the phase value of the primary voltage can be calculated using the formula

$$E_{P(PRIMARY)} = \frac{E_L}{1.732}$$

$$E_{P(PRIMARY)} = \frac{13,800 \text{ V}}{1.732}$$

$$E_{P(PRIMARY)} = 7967.667 \text{ V}$$

The secondary windings are connected as a delta. In a delta connection, the phase voltage and line voltage are the same:

$$E_{P(SECONDARY)} = E_{L(SECONDARY)}$$
$$E_{P(SECONDARY)} = 480 \text{ V}$$

The turns ratio can be calculated by comparing the phase voltage of the primary with the phase voltage of the secondary:

$$\text{Ratio} = \frac{\text{primary voltage}}{\text{secondary voltage}}$$

$$\text{Ratio} = \frac{7967.667 \text{ V}}{480 \text{ V}}$$

$$\text{Ratio} = 16.6:1$$

The load bank is connected in a wye connection. The voltage across the phase of the load bank will be less than the line voltage by a factor of 1.732:

$$E_{P(LOAD)} = \frac{E_{L\,(LOAD)}}{1.732}$$

$$E_{P(LOAD)} = \frac{480\ V}{1.732}$$

$$E_{P(LOAD)} = 277.136\ V$$

Now that the voltage across each of the load resistors is known, the current flow through the phase of the load can be calculated using Ohm's law:

$$I_{P(LOAD)} = \frac{E}{R}$$

$$I_{P(LOAD)} = \frac{277.136\ V}{2.77}$$

$$I_{P(LOAD)} = 100.049\ A$$

Because the load is connected as a wye connection, the line current is the same as the phase current:

$$I_{L(SECONDARY)} = 100.049\ A$$

The secondary of the transformer bank is connected as a delta. The phase current of the delta is less than the line current by a factor of 1.732:

$$I_{P(SECONDARY)} = \frac{I_L}{1.732}$$

$$I_{P(SECONDARY)} = \frac{100.049\ A}{1.732}$$

$$I_{P(SECONDARY)} = 57.765\ A$$

The amount of current flow through the primary can be calculated using the turns ratio. Because the primary has a higher voltage than the secondary, it will have a lower current. (Volts times amperes input must equal volts times amperes output.)

$$\text{Primary current} = \frac{\text{secondary current}}{\text{turns ratio}}$$

$$I_{P(PRIMARY)} = \frac{57.765\ A}{16.6}$$

$$I_{P(PRIMARY)} = 3.48\ A$$

Because all transformed values of voltage and current take place across the phases, the primary has a phase current of 3.48 A. In a wye connection, the phase current is the same as the line current:

$$I_{L(PRIMARY)} = 3.48 \text{ A}$$

The transformer connection with all calculated values is shown in *Figure 29–12*.

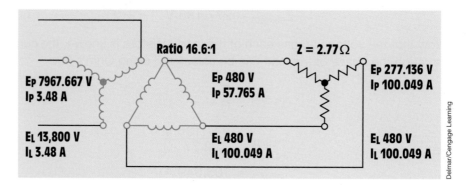

FIGURE 29–12 Example Circuit 1 with all missing values.

■ EXAMPLE 29-2

A three-phase transformer is connected in a delta–delta configuration *(Figure 29–13)*. The load is connected as a wye, and each phase has an impedance of 7 Ω. The primary is connected to a line voltage of 4160 V, and the secondary line voltage is 440 V. Find the following values:

$E_{P(PRIMARY)}$—phase voltage of the primary

$E_{P(SECONDARY)}$—phase voltage of the secondary

Ratio—turns ratio of the transformer

$E_{L(LOAD)}$—line voltage of the load

$E_{P(LOAD)}$—phase voltage of the load bank

$I_{P(LOAD)}$—phase current of the load bank

$I_{L(LOAD)}$—line current of the load

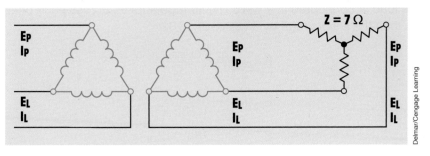

FIGURE 29–13   Example Circuit 2 three-phase transformer calculation.

$I_{L(SECONDARY)}$ — secondary line current

$I_{P(SECONDARY)}$ — phase current of the secondary

$I_{P(PRIMARY)}$ — phase current of the primary

$I_{L(PRIMARY)}$ — line current of the primary

Solution

The primary is connected as a delta. The phase voltage will be the same as the applied line voltage:

$$E_{P(PRIMARY)} = E_{L(PRIMARY)}$$
$$E_{P(PRIMARY)} = 4160 \text{ V}$$

The secondary of the transformer is connected as a delta also. Therefore, the phase voltage of the secondary will be the same as the line voltage of the secondary:

$$E_{P(SECONDARY)} = 440 \text{ V}$$

All transformer values must be calculated using phase values of voltage and current. The turns ratio can be found by dividing the phase voltage of the primary by the phase voltage of the secondary:

$$\text{Ratio} = \frac{E_{P(PRIMARY)}}{E_{P(SECONDARY)}}$$

$$\text{Ratio} = \frac{4160 \text{ V}}{440 \text{ V}}$$

$$\text{Ratio} = 9.45:1$$

The load is connected directly to the output of the secondary. The line voltage applied to the load must therefore be the same as the line voltage of the secondary:

$$E_{L(LOAD)} = 440 \text{ V}$$

The load is connected in a wye. The voltage applied across each phase will be less than the line voltage by a factor of 1.732:

$$E_{P(LOAD)} = \frac{E_{L(LOAD)}}{1.732}$$

$$E_{P(LOAD)} = \frac{440 \text{ V}}{1.732 \text{ V}}$$

$$E_{P(LOAD)} = 254.042 \text{ V}$$

The phase current of the load can be calculated using Ohm's law:

$$I_{P(LOAD)} = \frac{E_{P(LOAD)}}{Z}$$

$$I_{P(LOAD)} = \frac{254.042 \text{ V}}{7 \text{ }\Omega}$$

$$I_{P(LOAD)} = 36.292 \text{ A}$$

The amount of line current supplying a wye-connected load will be the same as the phase current of the load:

$$I_{L(LOAD)} = 36.292 \text{ A}$$

Because the secondary of the transformer is supplying current to only one load, the line current of the secondary will be the same as the line current of the load:

$$I_{L(SECONDARY)} = 36.292 \text{ A}$$

The phase current in a delta connection is less than the line current by a factor of 1.732:

$$I_{P(SECONDARY)} = \frac{I_{L(SECONDARY)}}{1.732}$$

$$I_{P(SECONDARY)} = \frac{36.292 \text{ A}}{1.732}$$

$$I_{P(SECONDARY)} = 20.954 \text{ A}$$

The phase current of the transformer primary can now be calculated using the phase current of the secondary and the turns ratio:

$$I_{P(PRIMARY)} = \frac{I_{P(SECONDARY)}}{\text{turns ratio}}$$

$$I_{P(PRIMARY)} = \frac{20.954 \text{ A}}{9.45}$$

$$I_{P(PRIMARY)} = 2.217 \text{ A}$$

In this example, the primary of the transformer is connected as a delta. The line current supplying the transformer will be higher than the phase current by a factor of 1.732:

$$I_{L(PRIMARY)} = I_{P(PRIMARY)} \times 1.732$$

$$I_{L(PRIMARY)} = 2.217 \text{ A} \times 1.732$$

$$I_{L(PRIMARY)} = 3.84 \text{ A}$$

The circuit with all calculated values is shown in *Figure 29–14*.

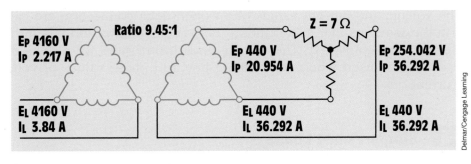

FIGURE 29–14   Example Circuit 2 with all missing values.

29–4   Open-Delta Connection

The **open-delta** transformer connection can be made with only two transformers instead of three *(Figure 29–15)*. This connection is often used when the amount of three-phase power needed is not excessive, such as in a small business. It should be noted that the output power of an open-delta connection is only 86.6% of the rated power of the two transformers. For example, assume two transformers, each having a capacity of 25 kilovolt-amperes, are connected

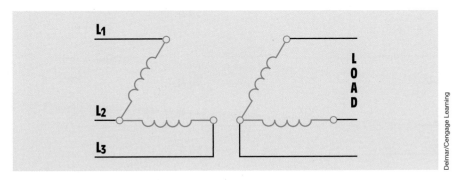

FIGURE 29–15   Open-delta connection.

in an open-delta connection. The total output power of this connection is 43.5 kilovolt-amperes (50 kVA × 0.866 = 43.3 kVA).

Another figure given for this calculation is 57.7%. This percentage assumes a closed-delta bank containing three transformers. If three 25-kilovolt-amperes transformers were connected to form a closed-delta connection, the total output power would be 75 kilovolt-amperes (3 × 25 kVA = 75 kVA). If one of these transformers were removed and the transformer bank operated as an open-delta connection, the output power would be reduced to 57.7 of its original capacity of 75 kilovolt-amperes. The output capacity of the open-delta bank is 43.3 kilovolt-amperes (75 kVA × 0.577 = 43.3 kVA).

The voltage and current values of an open-delta connection are calculated in the same manner as a standard delta–delta connection when three transformers are employed. The voltage and current rules for a delta connection must be used when determining line and phase values of voltage and current.

29–5   Single-Phase Loads

When true three-phase loads are connected to a three-phase transformer bank, there are no problems in balancing the currents and voltages of the individual phases. *Figure 29–16* illustrates this condition. In this circuit, a delta–wye three-phase transformer bank is supplying power to a wye-connected three-phase load in which the impedances of the three phases are the same. Notice that the amount of current flow in the phases is the same. This is the ideal condition and is certainly desired for all three-phase transformer loads. Although this is the ideal situation, it is not always possible to obtain a balanced load. Three-phase

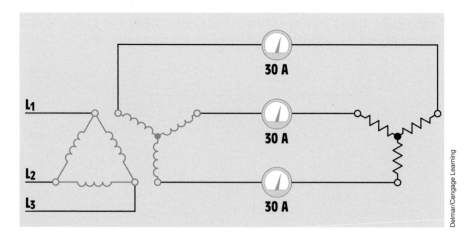

FIGURE 29–16   Three-phase transformer connected to a balanced three-phase load.

transformer connections are often used to supply **single-phase loads,** which tends to unbalance the system.

Open-Delta Connection Supplying a Single-Phase Load

The type of three-phase transformer connection used is generally determined by the amount of power needed. When a transformer bank must supply both three-phase and single-phase loads, the utility company often provides an open-delta connection with one transformer center-tapped as shown in *Figure 29–17*. In this connection, it is assumed that the amount of three-phase power needed is 20 kilovolt-amperes and the amount of single-phase power needed is 30 kilovolt-amperes. Notice that the transformer that has been center-tapped must supply power to both the three-phase and single-phase loads. Because this is an open-delta connection, the transformer bank can be loaded to only 86.6% of its full capacity when supplying a three-phase load. The rating of the three-phase transformer bank must therefore be 23 kilovolt-amperes (20 kVA/0.866 = 23 kVA). Because the rating of the two transformers can be added to obtain a total output power rating, one transformer is rated at only half the total amount of power needed, or 12 kilovolt-amperes (23 kVA/2 = 11.5 kVA). The transformer that is used to supply power to the three-phase load is only rated at 12 kilovolt-amperes. The transformer that has been center-tapped must supply power to both the single-phase and three-phase loads. Its capacity is therefore 42 kilovolt-amperes (12 kVA + 30 kVA). A 45-kilovolt-amperes transformer is used.

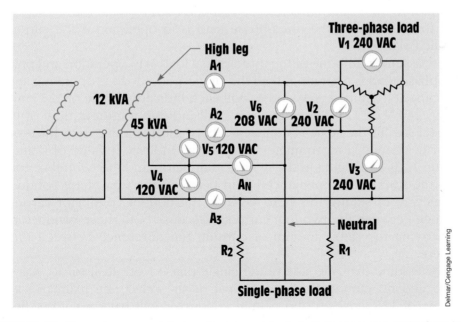

FIGURE 29–17   Three-phase open-delta transformer supplying both three-phase and single-phase loads.

Voltage Values

The connection shown in *Figure 29–17* has a line-to-line voltage of 240 volts. The three voltmeters V_1, V_2, and V_3 have all been connected across the three-phase lines and should indicate 240 volts each. Voltmeters V_4 and V_5 have been connected between the two lines of the larger transformer and its center tap. These two voltmeters will indicate a voltage of 120 volts each. Notice that it is these two lines and the center tap that are used to supply the single-phase power needed. The center tap of the larger transformer is used as a neutral conductor for the single-phase loads. Voltmeter V_6 has been connected between the center tap of the larger transformer and the line of the smaller transformer. This line is known as a **high leg** because the voltage between this line and the neutral conductor will be higher than the voltage between the neutral and either of the other two conductors. The high-leg voltage can be calculated by increasing the single-phase center-tapped voltage value by 1.732. In this case, the high-leg voltage will be 207.84 volts (120 V × 1.732 = 207.84 V). When this type of connection is employed, the *NEC* requires that the high leg be identified by connecting it to an **orange wire** or by **tagging** it at any point where a connection is made if the neutral conductor is also present.

Load Conditions

In the first load condition, it is assumed that only the three-phase load is in operation and none of the single-phase load is operating. If the three-phase load is operating at maximum capacity, Ammeters A_1, A_2, and A_3 will indicate a current flow of 48.114 amperes each [20 kVA/(240 V × 1.732) = 48.114 A]. Notice that only when the three-phase load is in operation is the current on each line balanced.

Now assume that none of the three-phase load is in operation and only the single-phase load is operating. If all the single-phase load is operating at maximum capacity, Ammeters A_2 and A_3 will each indicate a value of 125 amperes (30 kVA/240 V = 125 A). Ammeter A_1 will indicate a current flow of 0 ampere because all the load is connected between the other two lines of the transformer connection. Ammeter A_N will also indicate a value of 0 ampere. Ammeter A_N is connected in the neutral conductor, and the neutral conductor carries the sum of the unbalanced load between the two phase conductors. Another way of stating this is to say that the neutral conductor carries the difference between the two line currents. Because both these conductors are now carrying the same amount of current, the difference between them is 0 ampere.

Now assume that one side of the single-phase load, Resistor R_2, has been opened and no current flows through it. If the other line maintains a current flow of 125 amperes, the neutral conductor will have a current flow of 125 amperes also (125 A − 0 A = 125 A).

Now assume that Resistor R_2 has a value that will permit a current flow of 50 amperes on that phase. The neutral current will now be 75 amperes (125 A − 50 A = 75 A). Because the neutral conductor carries the sum of the unbalanced load, the neutral conductor never needs to be larger than the largest line conductor.

Now assume that both three-phase and single-phase loads are operating at the same time. If the three-phase load is operating at maximum capacity and the single-phase load is operating in such a manner that 125 amperes flow through Resistor R_1 and 50 amperes flow through Resistor R_2, the ammeters will indicate the following values:

$$A_1 = 48.1 \text{ A}$$
$$A_2 = 173.1 \text{ A } (48.1 \text{ A} + 125 \text{ A} = 173.1 \text{ A})$$
$$A_3 = 98.1 \text{ A } (48.1 \text{ A} + 50 \text{ A} = 98.1 \text{ A})$$
$$A_N = 75 \text{ A } (125 \text{ A} − 50 \text{ A} = 75 \text{ A})$$

Notice that the smaller of the two transformers is supplying current to only the three-phase load, but the larger transformer must supply current for both the single-phase and three-phase loads.

Although the circuit shown in *Figure 29–17* is the most common method of connecting both three-phase and single-phase loads to an open-delta transformer bank, it is possible to use the high leg to supply power to a single-phase load also. The circuit shown in *Figure 29–18* is a circuit of this type. Resistors R_1 and R_2 are connected to the lines of the transformer that has been center-tapped, and Resistor R_3 is connected to the line of the other transformer. If the line-to-line voltage is 240 volts, voltmeters V_1 and V_2 will each indicate a value of 120 volts across Resistors R_1 and R_2. Voltmeter V_3, however, will indicate that a voltage of 208 volts is applied across Resistor R_3.

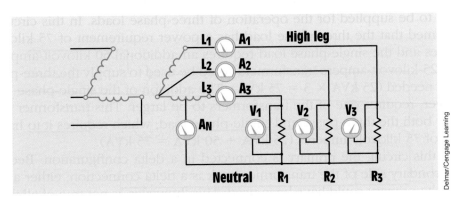

FIGURE 29–18 High leg supplies a single-phase load.

Harmonics with a positive sequence generally cause overheating of conductors, transformers, and circuit breakers. Negative-sequence harmonics can cause the same heating problems as positive harmonics plus additional problems with motors. Because the phasor rotation of a negative harmonic is opposite that of the fundamental frequency, it will tend to weaken the rotating magnetic field of an induction motor causing it to produce less torque. The reduction of torque causes the motor to operate below normal speed. The reduction in speed results in excessive motor current and overheating.

Although triplens do not have a phasor rotation, they can cause a great deal of trouble in a three-phase four-wire system, such as a 208/120-volt or 480/277-volt system. In a common 208/120-volt wye-connected system, the primary is generally connected in delta and the secondary is connected in wye *(Figure 29–32)*.

Single-phase loads that operate on 120 volts are connected between any phase conductor and the neutral conductor. The neutral current is the vector sum of the phase currents. In a balanced three-phase circuit (all phases having equal current), the neutral current is zero. Although single-phase loads tend to cause an unbalanced condition, the vector sum of the currents generally causes the neutral conductor to carry less current than any of the phase conductors. This is true for loads that are linear and draw a continuous sine wave current. When pulsating (nonlinear) currents are connected to a three-phase four-wire system, triplen harmonic frequencies disrupt the normal phasor relationship of the phase currents and can cause the phase currents to add in the neutral conductor instead of cancel. Because the neutral conductor is not protected by a fuse or circuit breaker, there is real danger of excessive heating in the neutral conductor.

Harmonic currents are also reflected in the delta primary winding where they circulate and cause overheating. Other heating problems are caused by

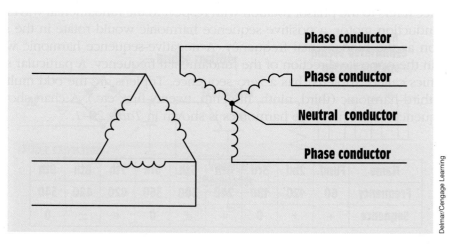

FIGURE 29–32 In a three-phase four-wire wye-connected system, the center of the wye-connected secondary is tapped to form a neutral conductor.

eddy current and hysteresis losses. Transformers are typically designed for 60-hertz operation. Higher harmonic frequencies produce greater core losses than the transformer is designed to handle. Transformers that are connected to circuits that produce harmonics must sometimes be derated or replaced with transformers that are specially designed to operate with harmonic frequencies.

Transformers are not the only electric component to be affected by harmonic currents. Emergency and standby generators can also be affected in the same way as transformers. This is especially true for standby generators used to power data-processing equipment in the event of a power failure. Some harmonic frequencies can even distort the zero crossing of the waveform produced by the generator.

Thermal-magnetic circuit breakers use a bimetallic trip mechanism that is sensitive to the heat produced by the circuit current. These circuit breakers are designed to respond to the heating effect of the true-RMS current value. If the current becomes too great, the bimetallic mechanism trips the breaker open. Harmonic currents cause a distortion of the RMS value, which can cause the breaker to trip when it should not or not to trip when it should. Thermal-magnetic circuit breakers, however, are generally better protection against harmonic currents than electronic circuit breakers. Electronic breakers sense the peak value of current. The peaks of harmonic currents are generally higher than the fundamental sine wave *(Figure 29–33)*. Although the peaks of harmonic currents are generally higher than the fundamental frequency, they can be lower. In some cases, electronic breakers may trip at low currents, and, in other cases, they may not trip at all.

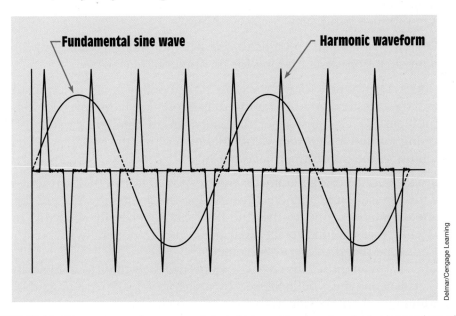

FIGURE 29–33   Harmonic waveforms generally have higher peak values than the fundamental waveform.

Triplen harmonic currents can also cause problems with neutral buss ducts and connecting lugs. A neutral buss is sized to carry the rated phase current. Because triplen harmonics can cause the neutral current to be higher than the phase current, it is possible for the neutral buss to become overloaded.

Electric panels and buss ducts are designed to carry currents that operate at 60 hertz. Harmonic currents produce magnetic fields that operate at higher frequencies. If these fields should become mechanically resonant with the panel or buss duct enclosures, the panels and buss ducts can vibrate and produce buzzing sounds at the harmonic frequency.

Telecommunications equipment is often affected by harmonic currents. Telecommunication cable is often run close to powerlines. To minimize interference, communication cables are run as far from phase conductors as possible and as close to the neutral conductor as possible. Harmonic currents in the neutral conductor induce high-frequency currents into the communication cable. These high-frequency currents can be heard as a high-pitch buzzing sound on telephone lines.

Determining Harmonic Problems on Single-Phase Systems

There are several steps to follow to determine if there is a problem with harmonics. One step is to do a survey of the equipment. This is especially important in determining if there is a problem with harmonics in a single-phase system.

1. Make an equipment check. Equipment such as personal computers, printers, and fluorescent lights with electronic ballasts are known to produce harmonics. Any piece of equipment that draws current in pulses can produce harmonics.

2. Review maintenance records to see whether there have been problems with circuit breakers tripping for no apparent reason.

3. Check transformers for overheating. If the cooling vents are unobstructed and the transformer is operating excessively hot, harmonics could be the problem. Check transformer currents with an ammeter capable of indicating a true-RMS current value. Make sure that the voltage and current ratings of the transformer have not been exceeded.

It is necessary to use an ammeter that responds to true RMS current when making this check. Some ammeters respond to the average value, not the RMS value. Meters that respond to the true-RMS value generally state this on the meter. Meters that respond to the average value are generally less expensive and do not state that they are RMS meters.

Meters that respond to the average value use a rectifier to convert the AC into DC. This value must be increased by a factor of 1.111 to change the average reading into the RMS value for a sine wave current. True-RMS responding meters calculate the heating effect of the current. The chart in *Figure 29–34*

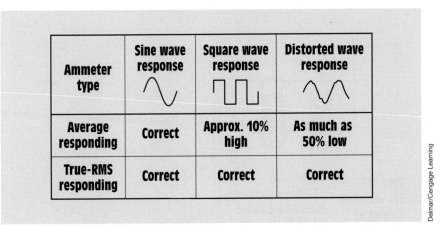

FIGURE 29–34 Comparison of average responding and true-RMS responding ammeters.

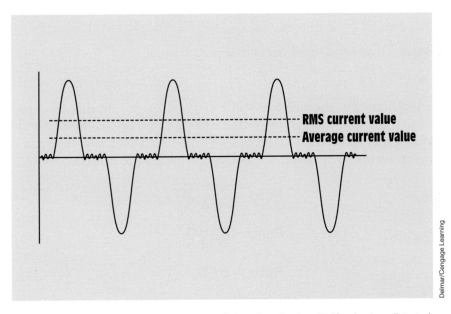

FIGURE 29–35 Average current values are generally less than the true-RMS value in a distorted waveform.

shows some of the differences between average-indicating meters and true-RMS meters. In a distorted waveform, the true-RMS value of current will no longer be *Average* × 1.111 *(Figure 29–35)*. The distorted waveform generally causes the average value to be as much as 50% less than the RMS value.

Another method of determining whether a harmonic problem exists in a single-phase system is to make two separate current checks. One check is made using an ammeter that indicates the true-RMS value and the other is made using

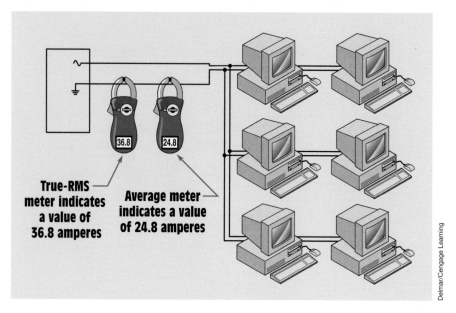

True-RMS meter indicates a value of 36.8 amperes

Average meter indicates a value of 24.8 amperes

Delmar/Cengage Learning

FIGURE 29–36 Determining harmonic problems using two ammeters.

a meter that indicates the average value *(Figure 29–36)*. In this example, it is assumed that the true-RMS ammeter indicates a value of 36.8 amperes and the average ammeter indicates a value of 24.8 amperes. Determine the ratio of the two measurements by dividing the average value by the true-RMS value:

$$\text{Ratio} = \frac{\text{Average}}{\text{RMS}}$$

$$\text{Ratio} = \frac{24.8 \text{ A}}{36.8 \text{ A}}$$

$$\text{Ratio} = 0.674$$

A ratio of 1 would indicate no harmonic distortion. A ratio of 0.5 would indicate extreme harmonic distortion. This method does not reveal the name or sequence of the harmonic distortion, but it does give an indication that there is a problem with harmonics. To determine the name, sequence, and amount of harmonic distortion present, a harmonic analyzer should be employed.

Determining Harmonic Problems on Three-Phase Systems

Determining whether a problem with harmonics exists in a three-phase system is similar to determining the problem in a single-phase system. Because harmonic problems in a three-phase system generally occur in a wye-connected four-wire system, this example asumes a delta-connected primary and wye-connected

Conductor	True-RMS responding ammeter	Average-responding ammeter
Phase 1	365 A	292 A
Phase 2	396 A	308 A
Phase 3	387 A	316 A
Neutral	488 A	478 A

TABLE 29-2   Three-Phase Four-Wire Wye-Connected System

secondary with a center-tapped neutral, as shown in *Figure 29–32*. To test for harmonic distortion in a three-phase four-wire system, measure all phase currents and the neutral current with both a true-RMS indicating ammeter and an average-indicating ammeter. It is assumed that the three-phase system being tested is supplied by a 200-kilovolt-ampere transformer, and the current values shown in *Table 29–2* were recorded. The current values indicate that a problem with harmonics does exist in the system. Note the higher current measurements made with the true-RMS indicating ammeter and also the fact that the neutral current is higher than any phase current.

Dealing with Harmonic Problems

After it has been determined that harmonic problems exist, something must be done to deal with the problem. It is generally not practical to remove the equipment causing the harmonic distortion, so other methods must be employed. It is a good idea to consult a power quality expert to determine the exact nature and amount of harmonic distortion present. Some general procedures for dealing with harmonics follow.

1. In a three-phase four-wire system, the 60-hertz part of the neutral current can be reduced by balancing the current on the phase conductors. If all phases have equal current flow, the neutral current would be zero.

2. If triplen harmonics are present on the neutral conductor, harmonic filters can be added at the load. These filters can help reduce the amount of harmonics on the line.

3. Pull extra neutral conductors. The ideal situation would be to use a separate neutral for each phase, instead of using a shared neutral.

4. Install a larger neutral conductor. If it is impractical to supply a separate neutral conductor for each phase, increase the size of the common neutral.

5. Derate or reduce the amount of load on the transformer. Harmonic problems generally involve overheating of the transformer. In many instances, it is necessary to derate the transformer to a point that it can handle the extra current caused by the harmonic distortion. When this is done, it is generally necessary to add a second transformer and divide the load between the two.

Determining Transformer Harmonic Derating Factor

Probably the most practical and straightforward method for determining the derating factor for a transformer is that recommended by the Computer & Business Equipment Manufacturers Association. To use this method, two ampere measurements must be made. One is the true-RMS current of the phases, and the second is the instantaneous peak phase current. The instantaneous peak current can be determined with an oscilloscope connected to a current probe or with an ammeter capable of indicating the peak value of current. Many of the digital clamp-on ammeters have the ability to indicate average, true-RMS, and peak values of current. For this example, it is assumed that peak current values are measured for the 200-kilovolt-ampere transformer discussed previously. These values are added to the previous data obtained with the true-RMS and average-indicating ammeters *(Table 29–3)*.

The formula for determining the transformer harmonic derating factor is

$$\text{THDF} = \frac{(1.414)(\text{RMS phase current})}{\text{instantaneous peak current}}$$

This formula produces a derating factor somewhere between 0 and 1.0. Because instantaneous peak value of current is equal to the RMS value × 1.414, if the current waveforms are sinusoidal (no harmonic distortion), the formula produces a derating factor of 1.0. Once the derating factor is determined, multiply the derating factor by the kilovolt-ampere capacity of the transformer. The product is the maximum load that should be placed on the transformer.

Conductor	True-RMS responding ammeter	Average-responding ammeter	Instantaneous peak current
Phase 1	365 A	292 A	716 A
Phase 2	396 A	308 A	794 A
Phase 3	387 A	316 A	737 A
Neutral	488 A	478 A	957 A

TABLE 29–3   Peak Currents Are Added to the Chart

If the phase currents are unequal, find an average value by adding the currents together and dividing by three:

$$\text{Phase (RMS)} = \frac{365\ A + 396\ A + 387\ A}{3}$$

$$\text{Phase (RMS)} = 382.7\ A$$

$$\text{Phase (Peak)} = \frac{716\ A + 794\ A + 737\ A}{3}$$

$$\text{Phase (Peak)} = 749$$

$$\text{THDF} = \frac{(1.414)(382.7\ A)}{749\ A}$$

$$\text{THDF} = 0.722$$

The 200-kilovolt-ampere transformer in this example should be derated to 144.4 kilovolt-amperes (200 kVA × 0.722).

Summary

- Three-phase transformers are constructed by winding three separate transformers on the same core material.
- Single-phase transformers can be used as a three-phase transformer bank by connecting their primary and secondary windings as either wyes or deltas.
- When calculating three-phase transformer values, the rules for three-phase circuits must be followed as well as the rules for transformers.
- Phase values of voltage and current must be used when calculating the values associated with the transformer.
- The total power output of a three-phase transformer bank is the sum of the rating of the three transformers.
- An open-delta connection can be made with the use of only two transformers.
- When an open-delta connection is used, the total output power is 86.6% of the sum of the power rating of the two transformers.
- It is common practice to center-tap one of the transformers in a delta connection to provide power for single-phase loads. When this is done, the remaining phase connection becomes a high leg.
- The *NEC* requires that a high leg be identified by an orange wire or by tagging.
- The center connection of a wye is often tapped to provide a neutral conductor for three-phase loads. This produces a three-phase four-wire system. Common voltages produced by this type of connection are 208/120 and 480/277.

- Transformers should not be connected as a wye–wye unless the incoming powerline contains a neutral conductor.

- T-connected transformers provide a better phase balance than open-delta connections.

- The T connection can be used to provide a three-phase four-wire connection with only two transformers.

- The Scott connection is used to change three-phase power into two-phase power.

- The zig-zag connection is primarily used for grounding purposes.

- Harmonics are generally caused by loads that pulse the powerline.

- Harmonic distortion on single-phase lines is often caused by computer power supplies, copy machines, fax machines, and light dimmers.

- Harmonic distortion on three-phase powerlines is generally caused by variable-frequency drives and electronic DC drives.

- Harmonics can have a positive rotation, negative rotation, or no rotation.

- Positive-rotating harmonics rotate in the same direction as the fundamental frequency.

- Negative-rotating harmonics rotate in the opposite direction of the fundamental frequency.

- Triplen harmonics are the odd multiples of the third harmonic.

- Harmonic problems can generally be determined by using a true-RMS ammeter and an average-indicating ammeter, or by using a true-RMS ammeter and an ammeter that indicates the peak value.

- Triplen harmonics generally cause overheating of the neutral conductor on three-phase four-wire systems.

Review Questions

1. How many transformers are needed to make an open-delta connection?

2. Two transformers rated at 100 kVA each are connected in an open-delta connection. What is the total output power that can be supplied by this bank?

3. How does the *NEC* specify that the high leg of a four-wire delta connection be marked?

4. An open-delta three-phase transformer system has one transformer center-tapped to provide a neutral for single-phase voltages. If the voltage from line to center tap is 277 V, what is the high-leg voltage?

5. If a single-phase load is connected across the two line conductors and neutral of the transformer in Question 4 and one line has a current of 80 A and the other line has a current of 68 A, how much current is flowing in the neutral conductor?

6. A three-phase transformer connection has a delta-connected secondary, and one of the transformers has been center-tapped to form a neutral conductor. The phase-to-neutral value of the center-tapped secondary winding is 120 V. If the high leg is connected to a single-phase load, how much voltage will be applied to that load?

7. A three-phase transformer connection has a delta-connected primary and a wye-connected secondary. The center tap of the wye is used as a neutral conductor. If the line-to-line voltage is 480 V, what is the voltage between any one phase conductor and the neutral conductor?

8. A three-phase transformer bank has the secondary connected in a wye configuration. The center tap is used as a neutral conductor. If the voltage across any phase conductor and neutral is 120 V, how much voltage would be applied to a three-phase load connected to the secondary of this transformer bank?

9. A three-phase transformer bank has the primary and secondary windings connected in a wye configuration. The secondary center tap is being used as a neutral to supply single-phase loads. Will connecting the center-tap connection of the secondary to the center-tap connection of the primary permit the secondary voltage to stay in balance when a single-phase load is added to the secondary?

10. Referring to the transformer connection in Question 9, if the center tap of the primary is connected to a neutral conductor on the incoming power, will it permit the secondary voltages to be balanced when single-phase loads are added?

11. What is the frequency of the second harmonic?

12. Which of the following are considered triplen harmonics: 3rd, 6th, 9th, 12th, 15th, and 18th?

13. Would a positive-rotating harmonic or a negative-rotating harmonic be more harmful to an induction motor? Explain your answer.

14. What instrument should be used to determine what harmonics are present in a power system?

15. A 22.5-kVA single-phase transformer is tested with a true-RMS ammeter and an ammeter that indicates the peak value. The true-RMS reading is 94 A. The peak reading is 204 A. Should this transformer be derated? If so, by how much?

Practical Applications

*Y*ou are working in an industrial plant. A three-phase transformer bank is connected wye–delta. The primary voltage is 12,470 V, and the secondary voltage is 480 V. The total capacity of the transformer bank is 450 kVA. One of the three transformers that form the three-phase bank develops a shorted primary winding and becomes unusable. A suggestion is made to reconnect the bank for operation as an open-delta. Can the two remaining transformers be connected open-delta? Explain your answer as to why they can or why they cannot be connected as an open-delta. If they can be reconnected open-delta, what would be the output capacity of the two remaining transformers? ■

Practical Applications

*Y*ou are a journeyman electrician working in an industrial plant. You are to install transformers that are to be connected in open-delta. Transformer A must supply its share of the three-phase load. Transformer B is to be center-tapped so it can provide power to single-phase loads as well as its share of the three-phase load. The total connected three-phase load is to be 40 kVA, and the total connected single-phase load is to be 60 kVA. The transformers are to have a capacity 115% greater than the rated load. What is the minimum kVA rating of each transformer? ■

Practice Problems

Refer to the transformer shown in *Figure 29–11* and find all the missing values.

1.

Primary	Secondary	Load
E_P _____	E_P _____	E_P _____
I_P _____	I_P _____	I_P _____
E_L 4160 V	E_L 440 V	E_L _____
I_L _____	I_L _____	I_L _____
Ratio	$Z = 3.5\ \Omega$	

2.

Primary	Secondary	Load
E_P _____	E_P _____	E_P _____
I_P _____	I_P _____	I_P _____
E_L 7200 V	E_L 240 V	E_L _____
I_L _____	I_L _____	I_L _____
Ratio	$Z = 4\ \Omega$	

Refer to the transformer connection shown in *Figure 29–37* and fill in the missing values.

3.

Primary	Secondary	Load
E_P _____	E_P _____	E_P _____
I_P _____	I_P _____	I_P _____
E_L 13,800 V	E_L 480 V	E_L _____
I_L _____	I_L _____	I_L _____
Ratio	$Z = 2.5\ \Omega$	

4.

Primary	Secondary	Load
E_P _____	E_P _____	E_P _____
I_P _____	I_P _____	I_P _____
E_L 23,000 V	E_L 208 V	E_L _____
I_L _____	I_L _____	I_L _____
Ratio	$Z = 3\ \Omega$	

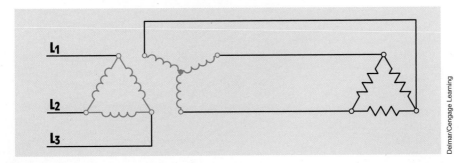

FIGURE 29–37   Practice problems circuit.

XI DC Machines

Unit 30
DC Generators

KEY TERMS

Armature
Armature reaction
Brushes
Commutator
Compound generators
Compounding
Countertorque
Cumulative compound
Differential-compound
Field excitation current
Frogleg-wound armatures
Generator
Lap-wound armatures
Pole pieces
Series field diverter
Series field windings
Series generator
Voltage regulator
Wave-wound armatures

Why You Need to Know

*D*C machines are still employed in many applications because of their variable speed and torque characteristics. This unit is devoted to the study of DC generators. Although many industries no longer use DC machinery, the study of DC generators is very important because it is the foundation for the study of AC generators or alternators. This unit

- describes the basic methods for generating electricity and how mechanical energy is converted to electric energy.
- explains many of the principles of induced voltage.
- describes how AC is produced in the armature of all rotating machines and how it is converted to DC before leaving the generator.
- discusses the various components of a DC generator and the principles of magnetic induction.
- explains the differences between series and shunt field windings and the characteristics of each.

Objectives

After studying this unit, you should be able to

- discuss the theory of operation of DC generators.

- list the factors that determine the amount of output voltage produced by a generator.

- list the three major types of DC generators.

- list different types of armature windings.

- describe the differences between series and shunt field windings.

- discuss the operating differences between different types of generators.

- draw schematic diagrams for different types of DC generators.

- set the brushes to the neutral plane position on the commutator of a DC machine.

Preview

Although most of the electric power generated throughout the world is AC, DC is used for some applications. Many industrial plants use DC generators to produce the power needed to operate large DC motors. DC motors have characteristics that make them superior to AC motors for certain applications. DC generators and motors are also used in diesel locomotives. The diesel engine in most locomotives is used to operate a large DC generator. The generator is used to provide power to DC motors connected to the wheels. ∎

30–1 What Is a Generator?

A **generator** is a device that converts mechanical energy into electric energy. DC generators operate on the principle of magnetic induction. In Unit 14 it was shown that a voltage is induced in a conductor when it cuts magnetic lines of flux *(Figure 30–1)*. In this example, the ends of the wire loop have been connected to two sliprings mounted on the shaft. Brushes are used to carry the current from the loop to the outside circuit.

In *Figure 30–2,* an end view of the shaft and wire loop is shown. At this particular instant, the loop of wire is parallel to the magnetic lines of flux and no cutting action is taking place. Because the lines of flux are not being cut by the loop, no voltage is induced in the loop.

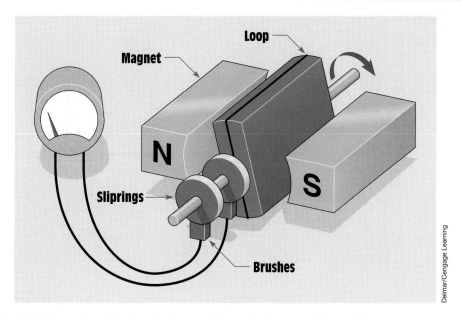

FIGURE 30–1   A voltage is induced in the conductor as it cuts magnetic lines of flux.

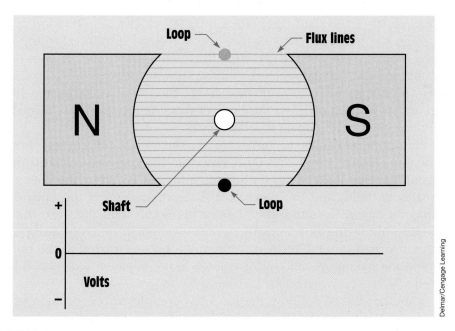

FIGURE 30–2   The loop is parallel to the lines of flux, and no cutting action is taking place.

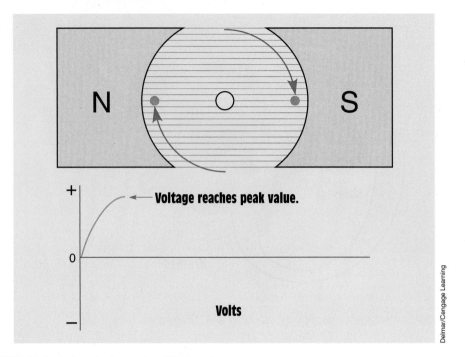

Delmar/Cengage Learning

FIGURE 30–3 Induced voltage after 90° of rotation.

In *Figure 30–3,* the shaft has been turned 90° clockwise. The loop of wire cuts through the magnetic lines of flux and a voltage is induced in the loop. When the loop is rotated 90°, it is cutting the maximum number of lines of flux per second and the voltage reaches its maximum, or peak, value.

After another 90° of rotation *(Figure 30–4),* the loop has completed 180° of rotation and is again parallel to the lines of flux. As the loop was turned, the voltage decreased until it again reached zero.

As the loop continues to turn, the conductors again cut the lines of magnetic flux *(Figure 30–5).* This time, however, the conductor that previously cut through the flux lines of the south magnetic field is cutting the lines of the north magnetic field, and the conductor that previously cut the lines of the north magnetic field is cutting the lines of the south field. Because the conductors are cutting the flux lines of opposite magnetic polarity, the polarity of voltage reverses. After 270° of rotation, the loop has rotated to the position shown and the maximum amount of voltage in the negative direction is being produced.

After another 90° of rotation, the loop has completed one rotation of 360° and returned to its starting position *(Figure 30–6).* The voltage decreased from its negative peak back to zero. Notice that the voltage produced in the **armature** (the rotating member of the machine) alternates polarity. ***The voltage produced in all rotating armatures is alternating voltage.***

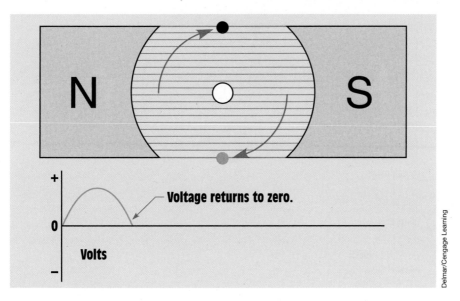

FIGURE 30–4 Induced voltage after 180° of rotation.

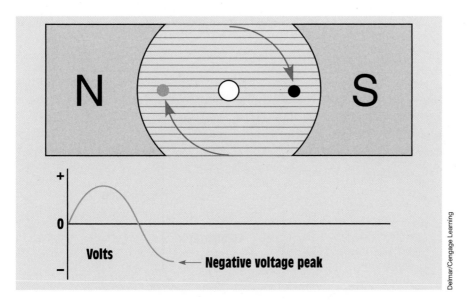

FIGURE 30–5 The negative voltage peak is reached after 270° of rotation.

Because DC generators must produce DC current instead of AC current, some device must be used to change the alternating voltage produced in the armature windings into direct voltage before it leaves the generator. This job is performed by the **commutator.** The commutator is constructed from a copper ring split into segments, with insulating material between the segments *(Figure 30–7)*. Brushes riding against the commutator segments carry the power to the outside circuit.

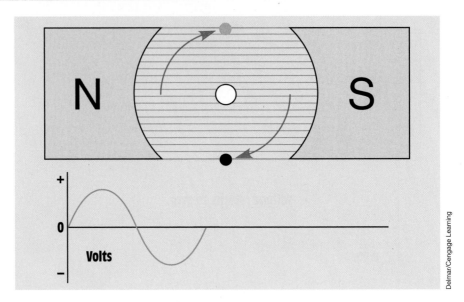

FIGURE 30–6 Voltage produced after 360° of rotation.

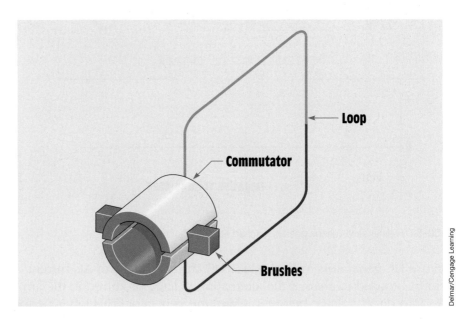

FIGURE 30–7 The commutator is used to convert the AC voltage produced in the armature into DC voltage.

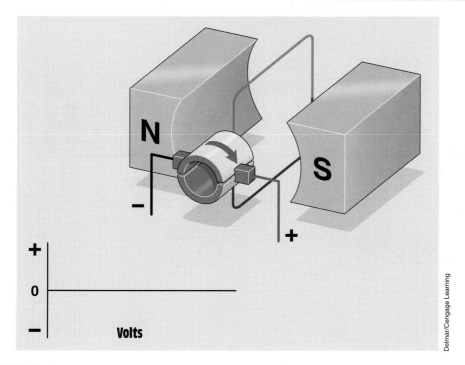

FIGURE 30–8   The loop is parallel to the lines of flux.

In *Figure 30–8,* the loop has been placed between the poles of two magnets. At this point in time, the loop is parallel to the magnetic lines of flux and no voltage is induced in the loop. Note that the brushes make contact with both of the commutator segments at this time. The position at which the windings are parallel to the lines of flux and there is no induced voltage is called the *neutral plane.* The brushes should be set to make contact between commutator segments when the armature windings are in the neutral plane position.

As the loop rotates, the conductors begin to cut through the magnetic lines of flux. The conductor cutting through the south magnetic field is connected to the positive brush, and the conductor cutting through the north magnetic field is connected to the negative brush *(Figure 30–9).* Because the loop is cutting lines of flux, a voltage is induced into the loop. After 90° of rotation, the voltage reaches its most positive point.

As the loop continues to rotate, the voltage decreases to zero. After 180° of rotation, the conductors are again parallel to the lines of flux and no voltage is induced in the loop. Note that the brushes again make contact with both segments of the commutator at the time when there is no induced voltage in the conductors *(Figure 30–10).*

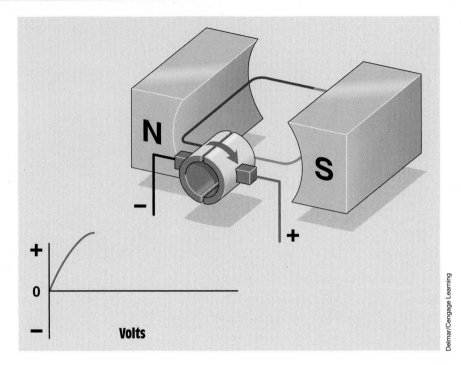

FIGURE 30–9 The loop has rotated 90°.

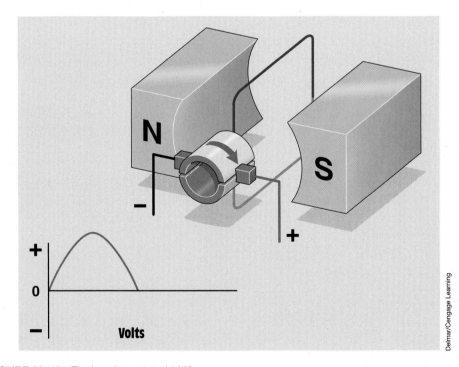

FIGURE 30–10 The loop has rotated 180°.

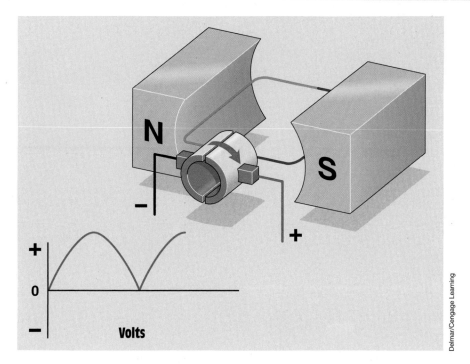

Delmar/Cengage Learning

FIGURE 30–11 The commutator maintains the proper polarity.

During the next 90° of rotation, the conductors again cut through the magnetic lines of flux. This time, however, the conductor that previously cut through the south magnetic field is now cutting the flux lines of the north magnetic field, and the conductor that previously cut the lines of flux of the north magnetic field is cutting the lines of flux of the south magnetic field *(Figure 30–11)*. Because these conductors are cutting the lines of flux of opposite magnetic polarities, the polarity of induced voltage is different for each of the conductors. The commutator, however, maintains the correct polarity to each brush. The conductor cutting through the north magnetic field will always be connected to the negative brush, and the conductor cutting through the south field will always be connected to the positive brush. Because the polarity at the brushes has remained constant, the voltage will increase to its peak value in the same direction.

As the loop continues to rotate *(Figure 30–12),* the induced voltage again decreases to zero when the conductors become parallel to the magnetic lines of flux. Notice that during this 360° rotation of the loop the polarity of voltage remained the same for both halves of the waveform. This is called *rectified* DC voltage. The voltage is pulsating or fluctuating. It does turn on and off, but it never reverses polarity. Because the polarity for each brush remains constant, the output voltage is DC.

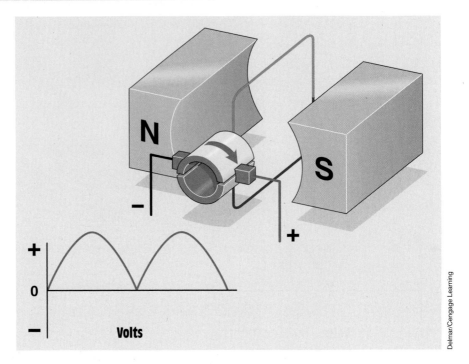

FIGURE 30–12   The loop completes one complete rotation.

To increase the amount of output voltage, it is common practice to increase the number of turns of wire for each loop *(Figure 30–13)*. If a loop contains 20 turns of wire, the induced voltage is 20 times greater than that for a single-loop conductor. The reason for this is that each loop is connected in series with the other loops. Because the loops form a series path, the voltage induced in the loops add. In this example, if each loop has an induced voltage of 2 volts, the total voltage for this winding would be 40 volts (2 V × 20 loops = 40 V).

It is also common practice to use more than one loop of wire *(Figure 30–14)*. When more than one loop is used, the average output voltage is higher and there is less pulsation of the rectified voltage. This pulsation is called *ripple*.

The loops are generally placed in slots of an iron core *(Figure 30–15)*. The iron acts as a magnetic conductor by providing a low-reluctance path for magnetic lines of flux to increase the inductance of the loops and provide a higher induced voltage. The commutator is connected to the slotted iron core. The entire assembly of iron core, commutator, and windings is called the armature *(Figure 30–16A* and *Figure 30–16B)*. The windings of armatures are connected in different ways depending on the requirements of the machine. The three basic types of armature windings are the *lap, wave,* and *frogleg*.

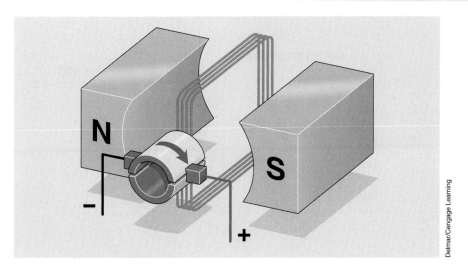

FIGURE 30–13 Increasing the number of turns increases the output voltage.

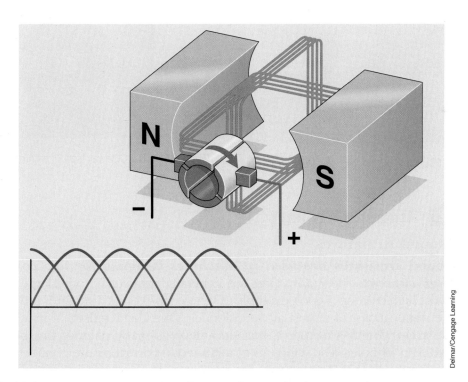

FIGURE 30–14 Increasing the number of loops produces a smoother output voltage.

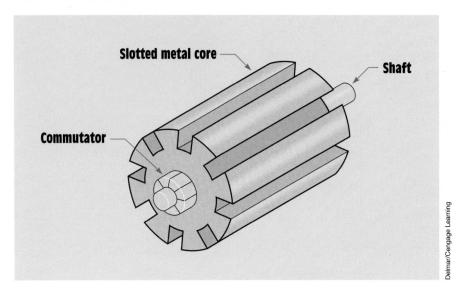

FIGURE 30–15   The loops of wire are wound around slots in a metal core.

FIGURE 30–16A   DC machine armature.

FIGURE 30–16B   Cutaway view of an armature.

30–2 Armature Windings

Lap-Wound Armatures

Lap-wound armatures *are used in machines designed for low voltage and high current.* These armatures are generally constructed with large wire because of high current. A good example of where lap-wound armatures are used is in the starter motor of almost all automobiles. One characteristic of machines that use a lap-wound armature is that they have as many pairs of brushes as there are pairs of poles. The windings of a lap-wound armature are connected in parallel *(Figure 30–17).* This permits the current capacity of each winding to be added and provides a higher operating current. Lap-wound armatures have as many parallel paths through the armature as there are pole pieces.

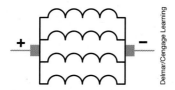

FIGURE 30–17   Lap-wound armatures have their windings connected in parallel. They are used in machines intended for high-current and low-voltage operation.

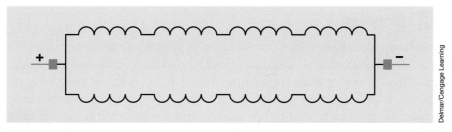

FIGURE 30–18   Wave-wound armatures have their windings connected in series. Wave windings are used in machines intended for high-voltage, low-current operation.

Wave-Wound Armatures

Wave-wound armatures *are used in machines designed for high voltage and low current.* These armatures have their windings connected in series as shown in *Figure 30–18.* When the windings are connected in series, the voltage of each winding adds, but the current capacity remains the same. A good example of where wave-wound armatures are used is in the small generator in hand-cranked megohmmeters. Wave-wound armatures never contain more than two parallel paths for current flow regardless of the number of pole pieces, and they never contain more than one set of brushes (a set being one brush or group of brushes for positive and one brush or group of brushes for negative).

Frogleg-Wound Armatures

Frogleg-wound armatures *are probably the most used. These armatures are used in machines designed for use with moderate current and moderate voltage.* The windings of a frogleg-wound armature are connected in series-parallel as shown in *Figure 30–19.* Most large DC machines use frogleg-wound armatures.

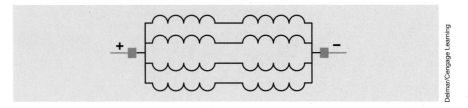

FIGURE 30–19 Frogleg-wound armatures are connected in series-parallel. These windings are generally used in machines intended for medium voltage and current operation.

30–3 Brushes

The **brushes** ride against the commutator segments and are used to connect the armature to the external circuit of the DC machine. Brushes are made from a material that is softer than the copper bars of the commutator. This permits the brushes, which are easy to replace, to wear instead of the commutator. The brush leads are generally marked A_1 and A_2 and are referred to as the *armature leads*.

30–4 Pole Pieces

The **pole pieces** are located inside the housing of the DC machine *(Figure 30–20)*. The pole pieces provide the magnetic field necessary for the operation of the machine. They are constructed of some type of good magnetic conductive material such as soft iron or silicon steel. Some DC generators use permanent magnets to provide the magnetic field instead of electromagnets. These machines are generally small and rated about one horsepower or less. A DC generator that uses permanent magnets as its field is referred to as a *magneto*.

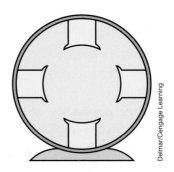

FIGURE 30–20 Pole pieces are constructed of soft iron and placed on the inside of the housing.

30–5   Field Windings

Most DC machines use wound electromagnets to provide the magnetic field. Two types of field windings are used. One is the series field and the other is the shunt field. **Series field windings** are made with relatively few turns of very large wire and have a very low resistance. They are so named because they are connected in series with the armature. The terminal leads of the series field are labeled S_1 and S_2. It is not uncommon to find the series field of large-horsepower machines wound with square or rectangular wire *(Figure 30–21)*. The use of square wire permits the windings to be laid closer together, which increases the number of turns that can be wound in a particular space. Additionally, smaller square and rectangular wire can be used to yield the same surface area of larger round wire *(Figure 30–22)*.

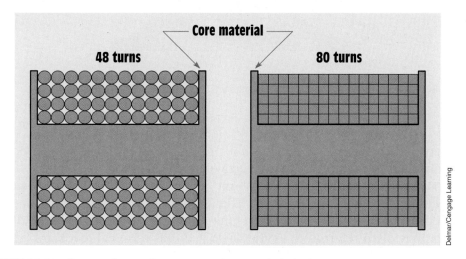

FIGURE 30–21   Square wire permits more turns than round wire in the same area.

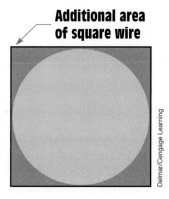

FIGURE 30–22   A square wire of equal size contains more surface area than round wire.

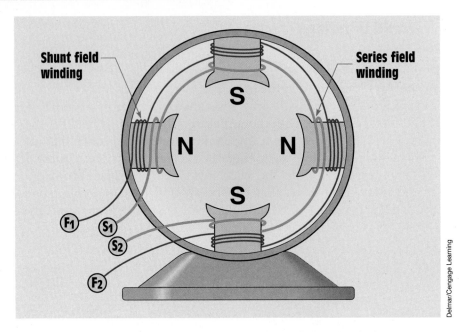

Shunt field winding

Series field winding

S

N   N

S

F₁ S₁ S₂ F₂

Delmar/Cengage Learning

FIGURE 30–23   Both series and shunt field windings are contained on each pole piece.

Shunt field windings are made with many turns of small wire. Because the shunt field is constructed with relatively small wire, it has a much higher resistance than the series field. The shunt field is intended to be connected in parallel with, or shunt, the armature. The resistance of the shunt field must be high because its resistance is used to limit current flow through the field. The shunt field is often referred to as the "field," and its terminal leads are labeled F_1 and F_2.

When a DC machine uses both series and shunt fields, each pole piece contains both windings *(Figure 30–23)*. The windings are wound on the pole pieces in such a manner that when current flows through the winding it produces alternate magnetic polarities. In *Figure 30–23,* two pole pieces form north magnetic polarities and two form south magnetic polarities. A DC machine with two field poles and one *interpole* is shown in *Figure 30–24.* Interpoles are discussed later in this unit.

30–6   Series Generators

There are three basic types of DC generators: the series, the shunt, and the compound. The type is determined by the arrangement and connection of field coils. The **series generator** contains only a series field connected in series with the armature *(Figure 30–25).* A schematic diagram used to represent a

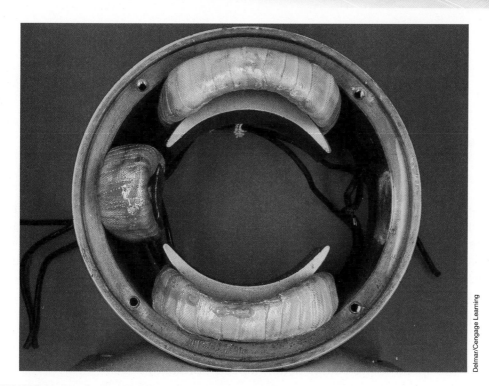

FIGURE 30–24   A two-pole DC machine with one interpole.

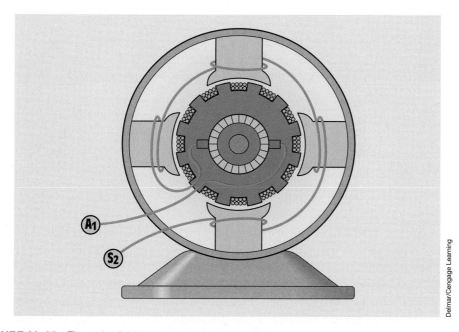

FIGURE 30–25   The series field is connected in series with the armature.

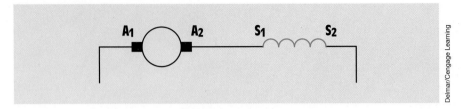

FIGURE 30–26   Schematic drawing of a series generator.

series-connected DC machine is shown in *Figure 30–26*. The series generator must be *self-excited,* which means that the pole pieces contain some amount of residual magnetism. This residual magnetism produces an initial output voltage that permits current to flow through the field if a load is connected to the generator. The amount of output voltage produced by the generator is proportional to three factors:

1. the number of turns of wire in the armature

2. the strength of the magnetic field of the pole pieces

3. the speed of the cutting action (speed of rotation)

To understand why these three factors determine the output voltage of a generator, recall from Unit 14 that 1 volt is induced in a conductor when it cuts magnetic lines of flux at a rate of 1 weber per second (1 weber = 100,000,000 lines of flux). When conductors are wound into a loop, each turn acts like a separate conductor. Because the turns are connected in series, the voltage induced into each conductor adds. If one conductor has an induced voltage of 0.5 volt and there are 20 turns, the total induced voltage would be 10 volts.

The second factor is the strength of the magnetic field. Flux density is a measure of the strength of a magnetic field. If the number of turns of wire in the armature remains constant and the speed remains constant, the output voltage can be controlled by the number of flux lines produced by the field poles. Increasing the lines of flux increases the number of flux lines cut per second and therefore the output voltage. The magnetic field strength can be increased until the iron of the pole pieces reaches saturation.

Induced voltage is proportional to the number of flux lines cut per second. If the strength of the magnetic field remains constant and the number of turns of wire in the armature remains constant, the output voltage is determined by the speed at which the conductors cut the flux lines. Increasing the speed of the armature increases the speed of the cutting action, which increases the output voltage. Likewise, decreasing the speed of the armature decreases the output voltage.

Connecting Load to the Series Generator

When a load is connected to the output of a series generator, the initial voltage produced by the residual magnetism of the pole pieces produces a current flow through the load *(Figure 30–27)*. Because the series field is connected in series with the armature, the current flowing through the armature and load must also flow through the series field. This causes the magnetism of the pole pieces to become stronger and produce more magnetic lines of flux. When the strength of the magnetic pole pieces increases, the output voltage increases also.

If another load is added *(Figure 30–28)*, more current flows and the pole pieces produce more magnetic lines of flux, which again increases the output

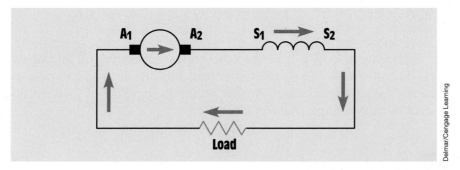

FIGURE 30–27 Residual magnetism produces an initial voltage used to provide a current flow through the load.

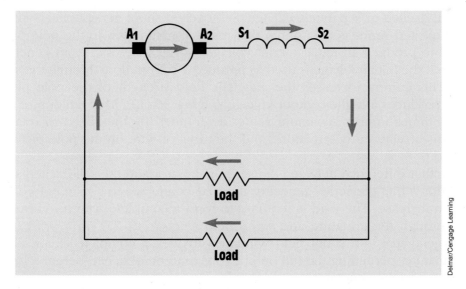

FIGURE 30–28 If more load is added, current flow and output voltage increase.

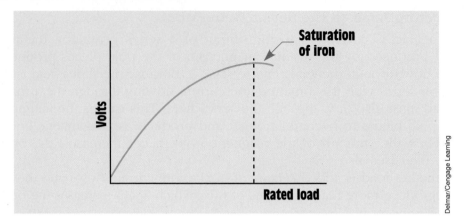

FIGURE 30–29   Characteristic curve of a series generator.

voltage. Each time a load is added to the series generator, its output voltage increases. This increase of voltage continues until the iron in the pole pieces and armature becomes saturated. At that point, an increase of load results in a decrease of output voltage *(Figure 30–29)*.

30–7   Shunt Generators

Shunt generators contain only a shunt field winding connected in parallel with the armature *(Figure 30–30)*. A schematic diagram used to represent a shunt-connected DC machine is shown in *Figure 30–31*. Shunt generators can be either self-excited or separately excited. Self-excited shunt generators are similar to self-excited series generators in that residual magnetism in the pole pieces is used to produce an initial output voltage. In the case of a shunt generator, however, the initial voltage is used to produce a current flow through the shunt field. This current increases the magnetic field strength of the pole pieces, which produces a higher output voltage *(Figure 30–32)*. This buildup of voltage continues until a maximum value, determined by the speed of rotation, the turns of wire in the armature, and the turns of wire on the pole pieces, is reached.

Another difference between the series generator and the self-excited shunt generator is that the series generator must be connected to a load before voltage can increase. The load is required to form a complete path for current to flow through the armature and series field *(Figure 30–27)*. In a self-excited shunt generator, the shunt field winding provides a complete circuit across the armature, permitting the full output voltage to be obtained before a load is connected to the generator.

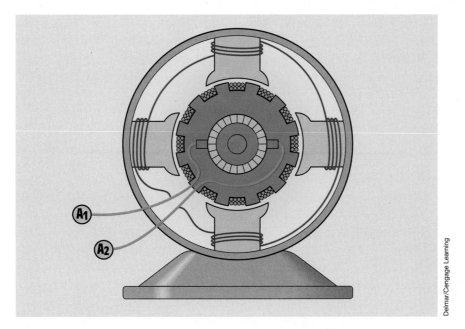

FIGURE 30–30 Shunt field windings are connected in parallel with the armature.

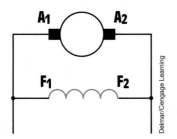

FIGURE 30–31 Schematic drawing of a shunt generator.

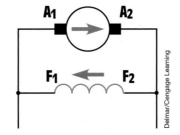

FIGURE 30–32 Residual magnetism in the pole pieces produces an initial voltage, which causes current to flow through the shunt field and the field flux to increase.

Separately excited generators have their fields connected to an external source of DC *(Figure 30–33)*. The advantages of the separately excited machine are that it gives better control of the output voltage and that its voltage drop is less when load is added. The characteristic curves of both self-excited and separately excited shunt generators are shown in *Figure 30–34*.

The self-excited generator exhibits a greater drop in voltage when load is added because the armature voltage is used to produce the current flow in the shunt field. Each time the voltage decreases, the current flow through the field decreases, causing a decrease in the amount of magnetic flux lines in

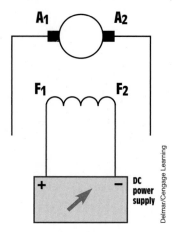

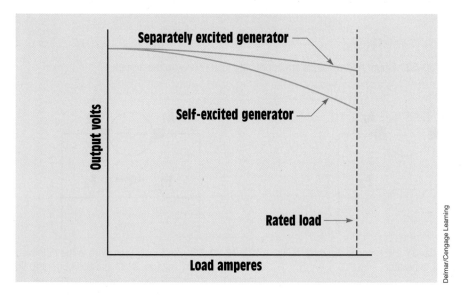

FIGURE 30–33   Separately excited shunt generators must have an external power source to provide excitation current for the shunt field.

FIGURE 30–34   Characteristic curves of self- and separately excited shunt generators.

the pole pieces. This decrease of flux in the pole pieces causes a further decrease of output voltage. The separately excited machine does not have this problem because the field flux is held constant by the external power source.

Field Excitation Current

Regardless of which type of shunt generator is used, the amount of output voltage is generally controlled by the amount of **field excitation current.** Field excitation current is the DC that flows through the shunt field winding.

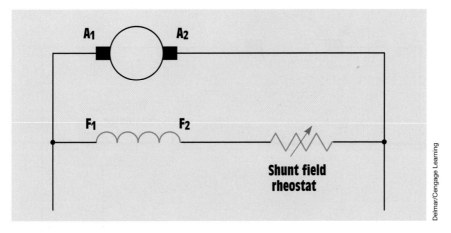

FIGURE 30–35 The shunt field rheostat is used to control the output voltage.

This current is used to turn the iron pole pieces into electromagnets. Because one of the factors that determines the output voltage of a DC generator is the strength of the magnetic field, the output voltage can be controlled by the amount of current flow through the field coils. A simple method of controlling the output voltage is by the use of a shunt field rheostat. The shunt field rheostat is connected in series with the shunt field winding *(Figure 30–35)*. By adding or removing resistance connected in series with the shunt field winding, the amount of current flow through the field can be controlled. This in turn controls the strength of the magnetic field of the pole pieces.

When it is important that the output voltage remain constant regardless of load, an electronic **voltage regulator** can be used to adjust the shunt field current *(Figure 30–36)*. The voltage regulator connects in series with the shunt field in a similar manner as the shunt field rheostat. The regulator, however, senses the amount of voltage across the load. If the output voltage drops, the regulator permits more current to flow through the shunt field. If the output voltage becomes too high, the regulator decreases the current flow through the shunt field.

Generator Losses

When load is added to the shunt generator, the output voltage drops. This voltage drop is due to losses that are inherent to the generator. The largest of these losses is generally caused by the resistance of the armature. In *Figure 30–37*, it is assumed that the armature has a wire resistance of 10 ohms. When a load is connected to the output of the generator, current flows from the armature, through the load, and back to the armature. As current flows through the armature, the resistance of the wire causes a voltage

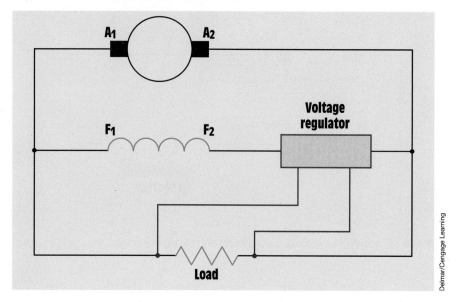

FIGURE 30–36 The voltage regulator controls the amount of shunt field current.

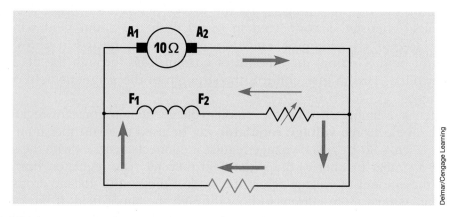

FIGURE 30–37 Armature resistance causes a drop in output voltage.

drop. Assume that the armature has a current flow of 2 amperes. If the resistance of the armature is 10 ohms, it will require 20 volts to push the current through the resistance of the armature.

Now assume that the armature has a resistance of 2 ohms. The same 2 amperes of current flow require only 4 volts to push the current through the armature resistance. A low-resistance armature is generally a very desirable characteristic for DC machines. ***In the case of a generator, the voltage***

regulation is determined by the resistance of the armature. Voltage regulation is measured by the amount that output voltage drops as load is added. A generator with good voltage regulation has a small amount of voltage drop as load is added.

Some other losses are I²R losses, eddy current losses, and hysteresis losses. Recall that I²R is one of the formulas for finding power, or watts. In the case of a DC machine, I²R describes the power loss associated with heat due to the resistance of the wire in both the armature and field windings.

Eddy currents are currents that are induced into the metal core material by the changing magnetic field as the armature spins through the flux lines of the pole pieces. Eddy currents are so named because they circulate around inside the metal in a manner similar to the swirling eddies in a river *(Figure 30–38)*. These swirling currents produce heat that heats the surrounding metal and causes a power loss. Many machines are constructed with laminated pole pieces and armature cores to help reduce eddy currents. The surface of each lamination forms a layer of iron oxide, which acts as an insulator to help prevent the formation of eddy currents.

Hysteresis losses are losses due to molecular friction. As discussed previously, AC is produced inside the armature. This reversal of the direction of current flow causes the molecules of iron in the core to realign themselves each time the current changes direction. The molecules of iron are continually rubbing against each other as they realign magnetically. The friction of the molecules rubbing together causes heat, which is a power loss. Hysteresis loss is proportional to the speed of rotation of the armature. The faster the armature rotates, the more current reversals there are per second and the more heat is produced because of friction.

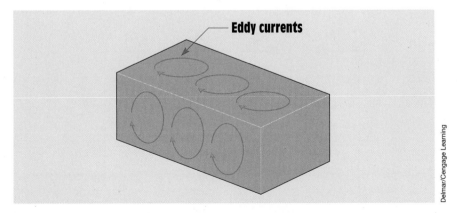

Delmar/Cengage Learning

FIGURE 30–38 Eddy currents heat the metal and cause power loss.

30–8 Compound Generators

Compound generators contain both series and shunt fields. Most large DC machines are compound wound. The series and shunt fields can be connected in two ways. One connection is called *long shunt (Figure 30–39)*. The long shunt connection has the shunt field connected in parallel with both the armature and series field. This is the most used of the two connections.

The second connection is called *short shunt (Figure 30–40)*. The short shunt connection has the shunt field connected in parallel with the armature. The series field is connected in series with the armature. This is a very common connection for DC generators that must be operated in parallel with each other.

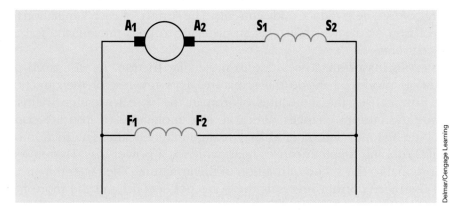

FIGURE 30–39 Schematic drawing of a long shunt compound generator.

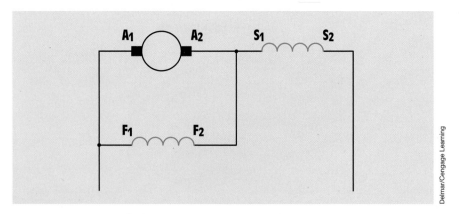

FIGURE 30–40 Schematic drawing of a short shunt compound generator.

30–9 Compounding

The relationship of the strengths of the two fields in a generator determines the amount of **compounding** for the machine. A machine is *overcompounded* when the series field has too much control and the output voltage increases each time a load is added to the generator. Basically, the generator begins to take on the characteristics of a series generator. Overcompounding is characterized by the fact that the output voltage at full load is greater than the output voltage at no load *(Figure 30–41)*.

When the generator is *flat compounded,* the output voltage is the same at full load as it is at no load. Flat compounding is accomplished by permitting the series field to increase the output voltage by an amount that is equal to the losses of the generator.

If the series field is too weak, however, the generator becomes *undercompounded*. This condition is characterized by the fact that the output voltage is less at full load than it is at no load. When a generator is undercompounded, it has characteristics similar to those of a shunt generator.

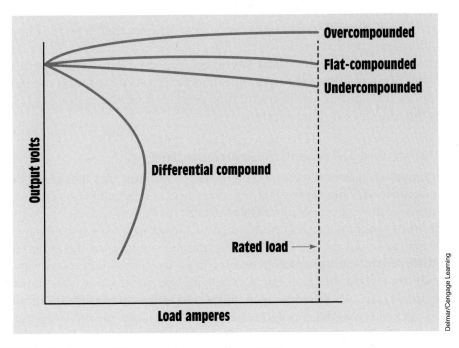

FIGURE 30–41   Characteristic curves of compound generators.

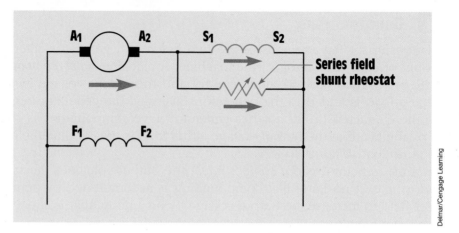

FIGURE 30–42 The series field shunt rheostat controls the amount of compounding.

Controlling Compounding

Most DC machines are constructed in such a manner that they are overcompounded if no control is used. This permits the series field strength to be weakened and thereby control the amount of compounding. The amount of compounding is controlled by connecting a low-value variable resistor in parallel with the series field *(Figure 30–42)*. This resistor is known as the series field shunt rheostat, or the **series field diverter.** The rheostat permits part of the current that normally flows through the series field to flow through the resistor. This reduces the amount of magnetic flux produced by the series field, which reduces the amount of compounding.

Cumulative and Differential Compounding

DC generators are generally connected in such a manner that they are a **cumulative compound.** This means that the shunt and series fields are connected in such a manner that when current flows through them they aid each other in the production of magnetism *(Figure 30–43)*. In the example shown, each of the field windings would produce the same magnetic polarity for the pole piece.

A **differential-compound** generator has its fields connected in such a manner that they oppose each other in the production of magnetism *(Figure 30–44)*. In this example, the shunt and series fields are attempting to produce opposite magnetic polarities for the same pole piece. This results in the magnetic field becoming weaker as current flow through the series field increases. Although there are some applications for a differential-compound machine, they are very limited.

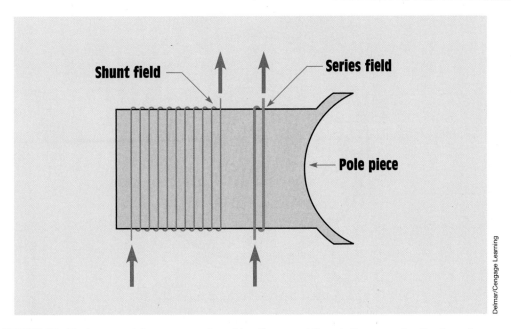

FIGURE 30–43   In a cumulative-compound machine, the current flows in the same direction through both the series and the shunt field.

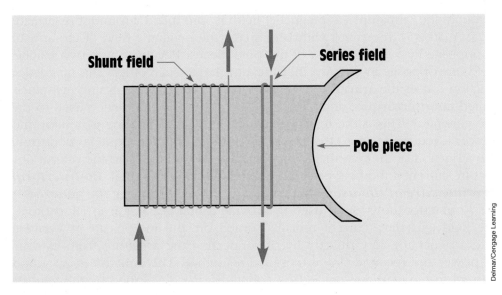

FIGURE 30–44   In a differential-compound machine, the current flows through the shunt field in a direction opposite that of the current flow through the series field.

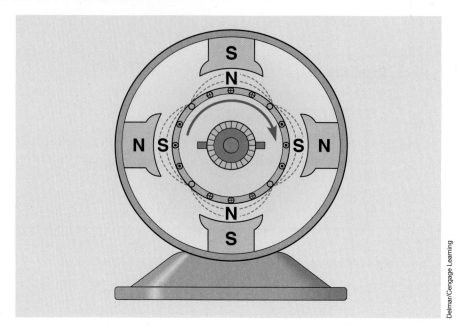

Delmar/Cengage Learning

FIGURE 30–45   A magnetic field is produced around the armature.

30–10   Countertorque

When a load is connected to the output of a generator, current flows from the armature, through the load, and back to the armature. As current flows through the armature, a magnetic field is produced around the armature *(Figure 30–45)*. In accord with Lenz's law, the magnetic field of the armature is opposite in polarity to that of the pole pieces. Because these two magnetic fields are opposite in polarity, they are attracted to each other. This magnetic attraction causes the armature to become hard to turn. This turning resistance is called **countertorque,** and it must be overcome by the device used to drive the generator. This is the reason that, as load is added to the generator, more power is required to turn the armature. Because countertorque is produced by the attraction of the two magnetic fields, it is proportional to the output or armature current if the field excitation current remains constant. ***Countertorque is a measure of the useful electric energy produced by the generator.***

Countertorque is often used to provide a braking action in DC motors. If the field excitation current remains turned on, the motor can be converted into a generator very quickly by disconnecting the armature from its source of power and reconnecting it to a load resistance. The armature now supplies current to the load resistance. The countertorque developed by the generator action causes the armature to decrease in speed. When this type of braking action is used, it is referred to as *dynamic braking* or *regenerative braking*.

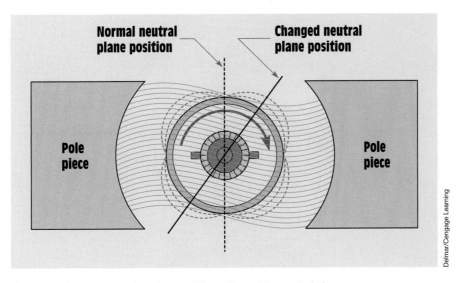

Normal neutral
plane position

Changed neutral
plane position

Pole
piece

Pole
piece

Delmar/Cengage Learning

FIGURE 30–46   Armature reaction changes the position of the neutral plane.

30–11   Armature Reaction

Armature reaction is the twisting or bending of the magnetic lines of flux of the pole pieces. It is caused by the magnetic field produced around the armature as it supplies current to the load *(Figure 30–46)*. This distortion of the main magnetic field causes the position of the neutral plane to change position. When the neutral plane changes, the brushes no longer make contact between commutator segments at a time when no voltage is induced in the armature. This results in power loss and arcing and sparking at the brushes, which can cause overheating and damage to both the commutator and brushes. The amount of armature reaction is proportional to armature current.

Correcting Armature Reaction

Armature reaction can be corrected in several ways. One method is to rotate the brushes an equal amount to the shift of the neutral plane *(Figure 30–47)*. This method is only satisfactory, however, if the generator delivers a constant current. Because the distortion of the main magnetic field is proportional to armature current, the brushes have to be adjusted each time the load current changes. In the case of a generator, the brushes are rotated in the direction of rotation of the armature. In the case of a motor, the brushes are rotated in a direction opposite that of armature rotation.

Another method that is used often is to insert small pole pieces, called interpoles or commutating poles, between the main field poles *(Figure 30–24)*. The interpoles are sometimes referred to as the *commutating winding* because they are wound with a few turns of large wire similar to the series field

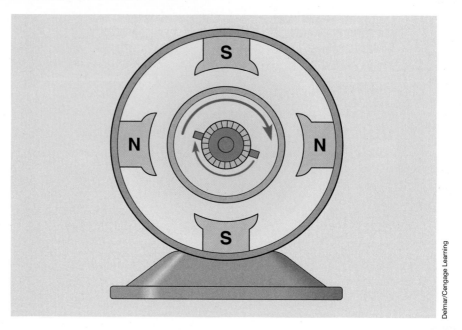

Delmar/Cengage Learning

FIGURE 30–47 In a generator, the brushes are rotated in the direction of armature rotation to correct armature reaction.

winding. The interpoles are connected in series with the armature, which permits their strength to increase with an increase of armature current *(Figure 30–48)*. Interpole connections are often made inside the housing of the machine. When the interpole connection is made internally, the A_1 lead is actually connected to one end of the interpole winding. When the interpole windings are brought out of the machine separately, they are generally labeled C_1 and C_2, which stands for commutating field. It is not unusual, however, to find them labeled S_3 and S_4.

In a generator, the magnetic field of the armature tends to bend the main magnetic field upward *(Figure 30–46)*. In a motor, the armature field bends the main field downward. The function of the interpoles is to restore the field

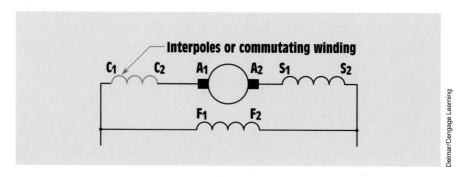

Delmar/Cengage Learning

FIGURE 30–48 Interpoles are connected in series with the armature.

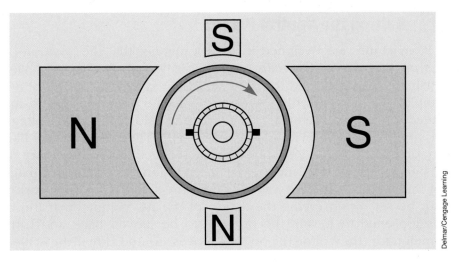

FIGURE 30–49   Interpoles must have the same polarity as the pole piece directly ahead of them.

to its normal condition. When a DC machine is used as a generator, the interpoles have the same polarity as the main field pole directly ahead of them (ahead in the sense of the direction of rotation of the armature) *(Figure 30–49)*. When a DC machine is used as a motor, the interpoles have the same polarity as the pole piece behind them in the sense of direction of rotation of the armature.

Interpoles do have one disadvantage. They restore the field only in their immediate area and are not able to overcome all the field distortion. Large DC generators use another set of windings called *compensating windings* to help restore the main magnetic field. Compensating windings are made by placing a few large wires in the face of the pole piece parallel to the armature windings *(Figure 30–50)*. The compensating winding is connected in series with the armature so that its strength increases with an increase of output current.

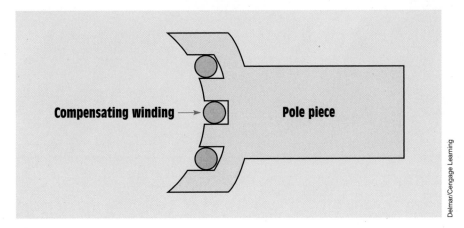

FIGURE 30–50   Compensating winding helps correct armature reaction.

30–12 Setting the Neutral Plane

Most DC machines are designed in such a manner that the position of the brushes on the commutator can be set or adjusted. An exposed view of the brushes and brush yoke of a DC machine is shown in *Figure 30–51*. The simplest method of setting the brushes to the neutral plane position is to connect an AC voltmeter across the shunt field leads. Low-voltage AC is then applied to the armature *(Figure 30–52)*. The armature acts like the primary of a transformer, and the shunt field acts like the secondary. If the brushes are not set at the neutral plane position, the changing magnetic field of the armature induces a voltage into the shunt field. The brush position can be set by observing the action of the AC voltmeter. If the brush yoke is loosened to permit the brushes to be moved back and forth on the commutator, the voltmeter pointer moves up and down the scale. The brushes are set to the neutral plane position when the voltmeter is at its lowest possible reading.

FIGURE 30–51 Exposed view of a DC machine.

Courtesy of GE Industrial Systems

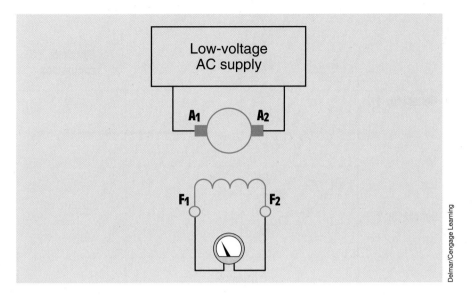

Low-voltage
AC supply

A₁   A₂

F₁   F₂

Delmar/Cengage Learning

FIGURE 30–52   Setting the brushes at the neutral plane.

30–13   Paralleling Generators

There may be occasions when one DC generator cannot supply enough current to operate the connected load. In such a case, another generator is connected in parallel with the first. DC generators should never be connected in parallel without an equalizer connection *(Figure 30–53)*. The equalizer connection is used to connect the series fields of the two machines in parallel with each other. This arrangement prevents one machine from taking the other over as a motor.

Assume that two generators are to be connected in parallel and the equalizing connection has not been made. Unless both machines are operating with identical field excitation when they are connected in parallel, the machine with the greatest excitation takes the entire load and begins operating the other machine as a motor. The series field of the machine that accepts the load is strengthened, and the series field of the machine that gives up the load is weakened. The machine with the stronger series field takes even greater load, and the machine with the weaker series field reduces load even further.

The generator that begins motoring has the current flow through its series field reversed, which causes it to operate as a differential-compound motor *(Figure 30–54)*. If the motoring generator is not removed from the line, the magnetic field strength of the series field will become greater than the field strength of the shunt field. This causes the polarity of the residual magnetism in the pole pieces to reverse. This is often referred to as *flashing the field*. Flashing

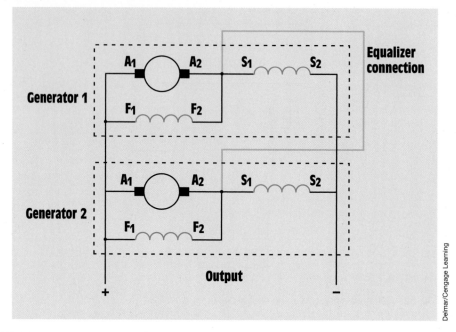

FIGURE 30–53   The equalizer connection is used to connect the series fields in parallel with each other.

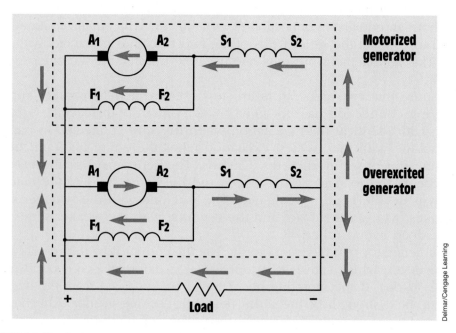

FIGURE 30–54   One generator takes all the load, and the other becomes a motor.

the field results in the polarity of the output voltage being reversed when the machine is restarted as a generator. The equalizer connection prevents field reversal even if the generator becomes a motor.

The resistance of the equalizer cable should not exceed 20% of the resistance of the series field winding of the smallest paralleled generator. This ensures that the current flow provided to the series fields will divide in the approximate inverse ratio of the respective series field winding.

Summary

- A generator is a machine that converts mechanical energy into electric energy.

- Generators operate on the principle of magnetic induction.

- AC is produced in all rotating armatures.

- The commutator changes the AC produced in the armature into DC.

- The brushes are used to make contact with the commutator and to carry the output current to the outside circuit.

- The position at which there is no induced voltage in the armature is called the neutral plane.

- Lap-wound armatures are used in machines designed for low-voltage and high-current operation.

- Wave-wound armatures are used in machines designed for high-voltage and low-current operation.

- Frogleg-wound armatures are the most used and are intended for machines designed for moderate voltages and current.

- The loops of wire, iron core, and commutator are made as one unit and are referred to as the armature.

- The armature connection is marked A_1 and A_2.

- Series field windings are made with a few turns of large wire and have a very low resistance.

- Series field windings are connected in series with the armature.

- Series field windings are marked S_1 and S_2.

- Shunt field windings are made with many turns of small wire and have a high resistance.

- Shunt field windings are connected in parallel with the armature.

- The shunt field windings are marked F_1 and F_2.
- Three factors that determine the voltage produced by a generator are

  a. the number of turns of wire in the armature.

  b. the strength of the magnetic field of the pole pieces.

  c. the speed of the armature.
- Series generators increase their output voltage as load is added.
- Shunt generators decrease their output voltage as load is added.
- The voltage regulation of a DC generator is proportional to the resistance of the armature.
- Compound generators contain both series and shunt field windings.
- A long shunt compound generator has the shunt field connected in parallel with both the armature and series field.
- A short shunt compound generator has the shunt field connected in parallel with the armature but in series with the series field.
- When a generator is overcompounded, the output voltage is higher at full load than it is at no load.
- When a generator is flat-compounded, the output voltage is the same at full load and no load.
- When a generator is undercompounded, the output voltage is less at full load than it is at no load.
- Cumulative-compound generators have their series and shunt fields connected in such a manner that they aid each other in the production of magnetism.
- Differential-compound generators have their series and shunt field winding connected in such a manner that they oppose each other in the production of magnetism.
- Armature reaction is the twisting or bending of the main magnetic field.
- Armature reaction is caused by the interaction of the magnetic field produced in the armature.
- Armature reaction is proportional to armature current.
- Interpoles are small pole pieces connected between the main field poles used to help correct armature reaction.
- Interpoles are connected in series with the armature.

- Interpoles used in a generator must have the same polarity as the main field pole directly ahead of them in the sense of rotation of the armature.

- Interpoles used in a motor must have the same polarity as the main field pole directly behind them in the sense of rotation of the armature.

- Interpole leads are sometimes marked C_1 and C_2 or S_3 and S_4.

- Interpole leads are not always brought out of the machine.

- The neutral plane can be set by connecting an AC voltmeter to the shunt field and a source of low-voltage AC to the armature. The brushes are then adjusted until the voltmeter indicates the lowest possible voltage.

- When a generator supplies current to a load, countertorque is produced, which makes the armature harder to turn.

- Countertorque is proportional to the armature current if the field excitation current remains constant.

- Countertorque is a measure of the useful electric energy produced by the generator.

Review Questions

1. What is a generator?

2. What type of voltage is produced in all rotating armatures?

3. What are the three types of armature winding?

4. What type of armature winding would be used for a machine intended for high-voltage, low-current operation?

5. What are interpoles, and what is their purpose?

6. How are interpoles connected in relation to the armature?

7. What type of field winding is made with many turns of small wire?

8. How is the series field connected in relation to the armature?

9. How is the shunt field connected in relation to the armature?

10. What is armature reaction?

11. What is armature reaction proportional to?

12. What are eddy currents?

13. What condition characterizes overcompounding?

14. What is the function of the shunt field rheostat?

15. What is used to control the amount of compounding for a generator?

16. Explain the difference between cumulative- and differential-compounded connections.

17. What three factors determine the amount of output voltage for a DC generator?

18. What is the voltage regulation of a DC generator proportional to?

19. What is countertorque proportional to?

20. What is countertorque a measure of?

Practical Applications

You are working as an electrician in a large steel manufacturing plant, and you are in the process of doing preventive maintenance on a large DC generator. You have megged both the series and shunt field windings and found that each has over 10 MΩ to ground. Your ohmmeter, however, indicates a resistance of 1.5 Ω across terminals S_1 and S_2. The ohmmeter indicates a resistance of 225 Ω between terminals F_1 and F_2. Are these readings normal for this type machine, or is there a likely problem? Explain your answer. ■

Unit 31
DC Motors

KEY TERMS

Brushless DC
motors
Compound motor
Constant-speed
motors
Counter-
electromotive
force (CEMF)
(back-EMF)
Cumulative-
compound
motors
Differential-
compound
motors

Field-loss relay
(FLR)
Motor
Permanent magnet
(PM) motors
Printed-circuit motor
Pulse-width
modulation
Series motor
ServoDisc motor
Servomotors
Shunt motor
Speed regulation
Torque

Why You Need to Know

*D*C motors are used in many applications that require variable speed. They are also known for their ability to produce a large amount of torque at low speed. DC motors are commonly employed in the automotive industry to start vehicles and operate such devices as electric seats, electric windows, and blowers. The air-conditioning industry employs brushless DC motors in many air handlers for central heating and cooling systems. This unit

- introduces many different types of DC motors. Some of these types include series, shunt, compound, brushless, permanent magnet, and disc servo.
- explains how torque is produced and how it is used in HVAC air handlers and explains when variable speeds are required.
- discusses the different methods of connecting DC motors based on required rotation and speed control requirements.

Objectives

After studying this unit, you should be able to

- discuss the principle of operation of DC motors.

- discuss different types of DC motors.

- draw schematic diagrams of different types of DC motors.

- be able to connect a DC motor for a particular direction of rotation.

- discuss counter-electromotive force (CEMF).

- describe methods for controlling the speed of DC motors.

Preview

DC motors are used throughout industry in applications where variable speed is desirable. The speed-torque characteristic of DC motors makes them desirable for many uses. A good example is the San Francisco cable car system. It is powered by four 550 horsepower direct current motors. The automotive industry uses DC motors to start internal combustion engines and to operate blower fans, power seats, and other devices where a small motor is needed. ■

31–1   DC Motor Principles

DC motors operate on the principle of repulsion and attraction of magnetism. Motors use the same types of armature windings and field windings as the generators discussed in Unit 30. A **motor,** however, performs the opposite function of a generator. A *motor* is a device used to convert electric energy into mechanical energy. DC motors were, in fact, the first electric motors to be invented. For many years, it was not believed possible to make a motor that could operate using AC.

DC motors are the same basic machines as DC generators. In the case of a generator, some device is used to turn the shaft of the armature and the power produced by the turning armature is supplied to a load. In the case of a motor, power connected to the armature causes it to turn. To understand why the armature turns when current is applied to it, refer to the simple one-loop armature in *Figure 31–1*. Electrons enter the loop through the negative brush, flow around the loop, and exit through the positive brush. As current flows through the loop, a magnetic field is created around the loop. An end view, illustrating the pole pieces and the two conductors of a single loop, is shown in *Figure 31–2*. The X indicates electrons moving away from the observer like the back of an arrow moving away. The dot represents electrons moving toward the observer like the point of an approaching arrow. The left-hand rule for magnetism can be used to check the direction of the magnetic field around the conductors.

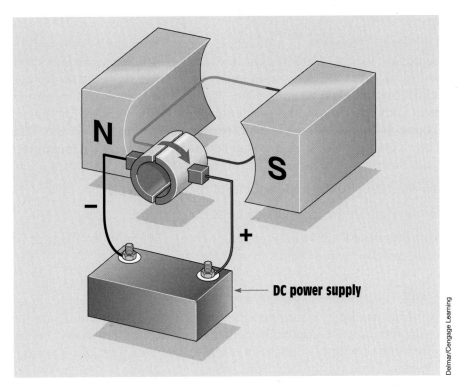

FIGURE 31–1 DC is supplied to the loop.

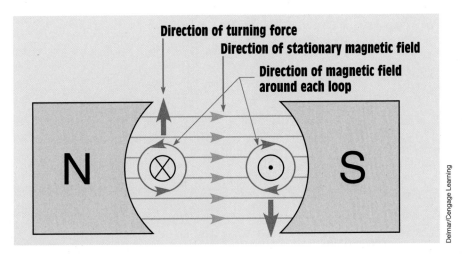

FIGURE 31–2 Flux lines in the same direction repel each other, and flux lines in the opposite direction attract each other.

Torque

Magnetic lines of force flow in a direction of north to south between the poles of the stationary magnet (left to right in *Figure 31–2*). When magnetic lines of flux flow in the same direction, they repel each other. When they flow in opposite directions, they attract each other. The magnetic lines of flux around the conductors cause the loop to be pushed in the direction shown by the arrows. This pushing or turning force is called **torque** and is created by the magnetic field of the pole pieces and the magnetic field of the loop or armature. Two factors that determine the amount of torque produced by a DC motor are

1. the strength of the magnetic field of the pole pieces.

2. the strength of the magnetic field of the armature.

Notice that there is no mention of speed or cutting action. One characteristic of a DC motor is that it can develop maximum torque at 0 rpm.

Increasing the Number of Loops

In the previous example, a single-loop armature was used to illustrate the operating principle of a DC motor. In actual practice, armatures are constructed with many turns of wire per loop and many loops. This provides a strong continuous turning force for the armature *(Figure 31–3)*.

The Commutator

When a DC machine is used as a generator, the commutator performs the function of a mechanical rectifier to change the AC produced in the armature

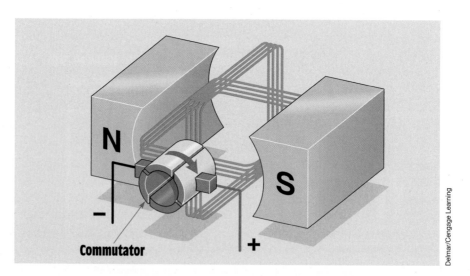

Commutator

FIGURE 31–3   Increasing the number of loops and turns increases the torque.

Delmar/Cengage Learning

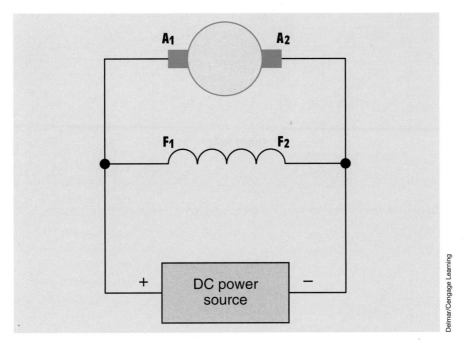

FIGURE 31–4 Brushes provide power connection from an outside source to the armature.

into DC before it exits the machine through the brushes. When a DC machine is used as a motor, the commutator performs the function of a rotary switch and maintains the correct direction of current flow through the armature windings. In order for the motor to develop a turning force, the magnetic field polarity of the armature must remain constant in relation to the polarity of the pole pieces. The commutator forces the direction of current flow to remain constant through certain sections of the armature as it rotates. The brushes are used to provide power to the armature from an external power source *(Figure 31–4)*.

31–2 Shunt Motors

The three basic types of DC motors are the shunt, the series, and the compound. The DC **shunt motor** has the shunt field connected in parallel with the armature *(Figure 31–5)*. This permits an external power source to supply current to the shunt field and maintain a constant magnetic field. The shunt motor has very good speed characteristics. The full-load speed generally remains within 10% of the no-load speed. Shunt motors are often referred to as **constant-speed motors.** Characteristic curves for shunt, series, and compound motors are shown in *Figure 31–6*. Note that the shunt

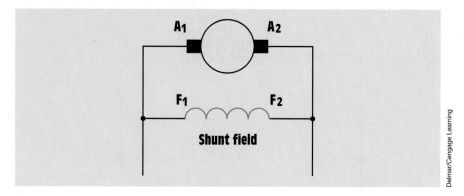

FIGURE 31–5 The shunt motor has the shunt field connected in parallel with the armature.

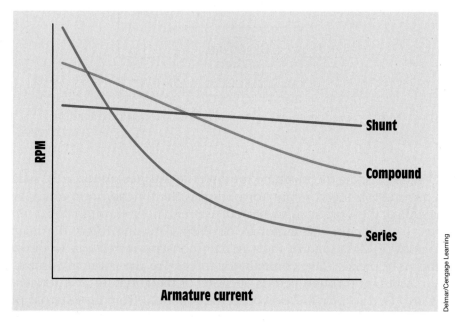

FIGURE 31–6 Characteristic speed curves for different DC motors.

motor maintains the most constant speed as load is added and armature current increases.

Counter-Electromotive Force

When the windings of the armature spin through the magnetic field produced by the pole pieces, a voltage is induced into the armature. This induced voltage is opposite in polarity to the applied voltage and is known as **counter-electromotive force (CEMF)** or **back-EMF** *(Figure 31–7)*. It is the CEMF that

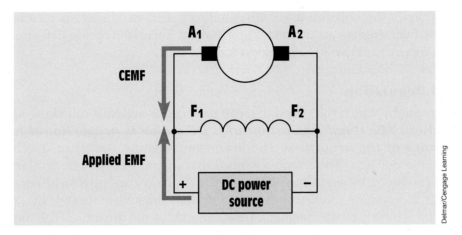

FIGURE 31–7   CEMF limits the flow of current through the armature.

limits the flow of current through the armature when the motor is in operation. The amount of CEMF produced in the armature is proportional to three factors:

1. the number of turns of wire in the armature

2. the strength of the magnetic field of the pole pieces

3. the speed of the armature

Speed-Torque Characteristics

When a DC motor is first started, the inrush of current can be high because no CEMF is being produced by the armature. The only current-limiting factor is the amount of wire resistance in the windings of the armature. When current flows through the armature, a magnetic field is produced and the armature begins to turn. As the armature windings cut through the magnetic field of the pole pieces, CEMF is induced in the armature. The CEMF opposes the applied voltage, causing current flow to decrease. If the motor is not connected to a load, the armature continues to increase in speed until the CEMF is almost the same value as the applied voltage. At this point, the motor produces enough torque to overcome its own losses. Some of these losses are

1. I^2R loss in the armature windings.

2. windage loss.

3. bearing friction.

4. brush friction.

When a load is added to the motor, the torque is not sufficient to support the load at the speed at which the armature is turning. The armature therefore slows down. When the armature slows down, CEMF is reduced and more current flows through the armature windings. This produces an increase in magnetic field strength and an increase in torque. This is the reason that armature current increases when load is added to the motor.

Speed Regulation

The amount by which the speed decreases as load is added is called the **speed regulation.** *The speed regulation of a DC motor is proportional to the resistance of the armature.* The lower the armature resistance, the better the speed regulation. The reason for this is that armature current determines the torque produced by the motor if the field excitation current is held constant. In order to produce more torque, more current must flow though the armature, which increases the magnetic field strength of the armature. The amount of current flowing through the armature is determined by the CEMF and the armature resistance. If the field excitation current is constant, the amount of CEMF is proportional to the speed of the armature. The faster the armature turns, the higher the CEMF. When the speed of the armature decreases, the CEMF decreases also.

Assume an armature has a resistance of 6 ohms. Now assume that when load is added to the motor, an additional 3 amperes of armature current are required to produce the torque necessary to overcome the added load. In this example, a voltage of 18 volts is required to increase the armature current by 3 amperes (3 A × 6 Ω = 18 V). This means that the speed of the armature must drop enough so that the CEMF is 18 volts less than it was before. The reduction in CEMF permits the applied voltage to push more current through the resistance of the armature.

Now assume that the armature has a resistance of 1 ohm. If a load is added that requires an additional 3 amperes of armature current, the speed of the armature must drop enough to permit a 3-volt reduction in CEMF (3 A × 1 Ω = 3 V). The armature does not have to reduce speed as much to cause a 3-volt reduction in CEMF as it does for a reduction of 18 volts.

31–3   Series Motors

The operating characteristics of the DC **series motor** are very different from those of the shunt motor. The reason is that the series motor has only a series field connected in series with the armature *(Figure 31–8)*. The armature current, therefore, flows through the series field. The speed of the series motor is controlled by the amount of load connected to the motor. When load is increased, the speed of the motor decreases. This causes a reduction in the amount of CEMF produced in the armature and an increase in armature and series field current. Because the current increases in both the armature and series field, the torque increases by the square of the current. In other words,

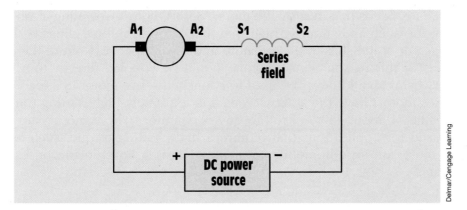

FIGURE 31–8   Series motor connection.

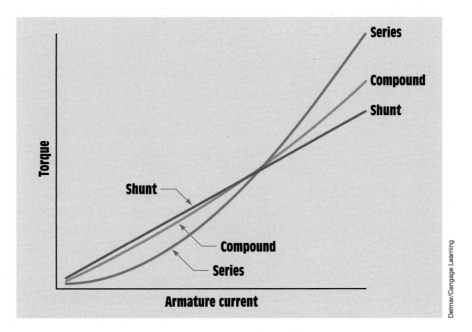

FIGURE 31–9   Torque curves of DC motors.

if the current doubles, the torque increases four times. Characteristic curves showing the relationship of torque and armature current for the three main types of DC motors are shown in *Figure 31–9*. Notice that the series motor produces the most torque of the three motors.

Series Motor Speed Characteristics

Series motors have no natural speed limit and should therefore never be operated in a no-load condition. Large series motors that suddenly lose their load race to speeds that destroy the motor. Series motors operating at no load can develop such an extremely high rpm that centrifugal force slings both the windings out of the slots in the armature and the copper bars out of the commutator. For this reason, series motors should be coupled directly to a load. Belts or chains should never be used to connect a series motor to a load.

Series motors have the ability to develop extremely high starting torques. An average of about 450% of full torque is common. These motors are generally used for applications that require a high starting torque, such as the starter motor on an automobile, cranes and hoists, and electric buses and streetcars.

31–4   Compound Motors

The **compound motor** uses both a series field and a shunt field. This motor is used to combine the operating characteristics of both the series and the shunt motor. The series field of the compound motor permits the motor to develop high torque, and the shunt field permits speed control and regulation. The compound motor is used more than any other type of DC motor in industry. The compound motor does not develop as much torque as the series motor, but it does develop more than the shunt motor. The speed regulation of a compound motor is not as good as a shunt motor, but it is much better than a series motor *(Figure 31–6)*.

Compound motors can be connected as short shunt or long shunt just as compound generators can *(Figure 31–10)*. The long shunt connection is more common because it has superior speed regulation. Compound motors can

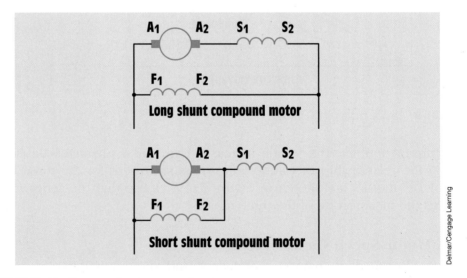

Long shunt compound motor

Short shunt compound motor

FIGURE 31–10   Compound motor connections.

also be **cumulative-compound motors** or **differential-compound motors.** Although there are some applications for differential-compound motors, it is generally a connection to be avoided. If a generator is inadvertently connected as a differential-compound machine, the greatest consequence is that the output voltage drops rapidly as load is added. This is not the case with a motor, however. When a motor is connected differential compound, the shunt field determines the direction of rotation of the motor at no load or light load. When load is added to the motor, the series field becomes stronger. If enough load is added, the magnetic field of the pole pieces reverses polarity and the motor suddenly stops, reverses direction, and begins operating like a series motor. This can damage the motor and the equipment to which the motor is connected.

If a compound motor is to be connected, the following steps can be followed to prevent the motor from being accidentally connected differential compound:

1. Disconnect the motor from the load.

2. Connect the series field and armature windings together to form a series motor connection, leaving the shunt field disconnected. Connect the motor to the power source *(Figure 31–11)*.

3. Turn the power on momentarily to determine the direction of rotation. This application of power must be of very short duration because the motor is now being operated as a series motor with no load. The idea is to "bump" the motor just to check for direction of rotation. If the motor turns in the opposite direction than is desired, reverse the connection of the armature leads. This reverses the direction of rotation of the motor.

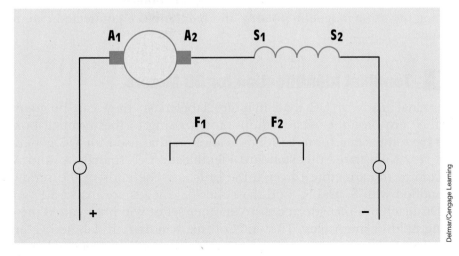

FIGURE 31–11   The motor is first connected as a series motor.

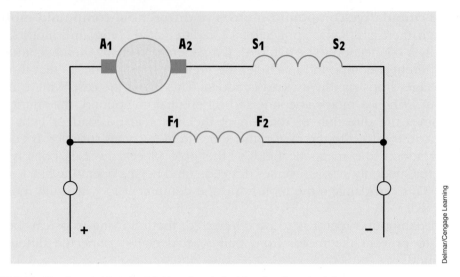

Delmar/Cengage Learning

FIGURE 31–12   Connect the shunt field and again test for direction of rotation.

4. Connect the shunt field leads to the incoming power *(Figure 31–12)*. Again turn on the power and check the direction of rotation. If the motor operates in the desired direction, it is connected cumulative compound. If the motor turns in the opposite direction, it is connected differential compound and the shunt field leads should be reversed.

This test can be used to check for a differential- or cumulative-compound connection because the shunt field controls the direction of rotation at no load. If the motor operates in the same direction as both a series and compound motor, the magnetic polarity of the pole pieces must be the same for both connections. This indicates that both the series and shunt field windings must be producing the same magnetic polarity and are therefore connected cumulative compound.

31–5   Terminal Identification for DC Motors

The terminal leads of DC machines are labeled so they can be identified when they are brought outside the motor housing to the terminal box. DC motors have the same terminal identification as that used for DC generators. *Figure 31–13* illustrates this standard identification. Terminals A_1 and A_2 are connected to the armature through the brushes. The ends of the series field are identified with S_1 and S_2, and the shunt field leads are marked F_1 and F_2. Some DC machines provide access to another set of windings called the commutating field or interpoles. The ends of this winding are labeled C_1 and C_2 or S_3 and S_4. It is common practice to provide access to the interpole winding on machines designed to be used as motors or generators.

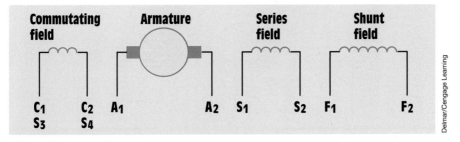

FIGURE 31–13   Terminal identification for DC machines.

31–6   Determining the Direction of Rotation of a DC Motor

The direction of rotation of a DC motor is determined by facing the commutator end of the motor. This is generally the back or rear of the motor. If the windings have been labeled in a standard manner, it is possible to determine the direction of rotation when the motor is connected. *Figure 31–14* illustrates the standard connections for a series motor. The standard connections for a shunt motor are illustrated in *Figure 31–15*, and the standard connections for a compound motor are shown in *Figure 31–16*.

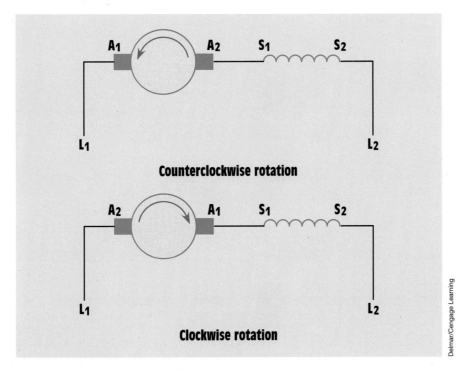

FIGURE 31–14   Series motors.

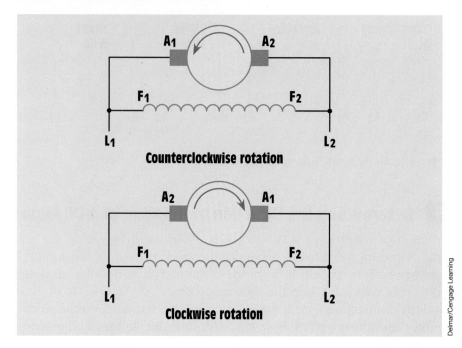

FIGURE 31–15 Shunt motors.

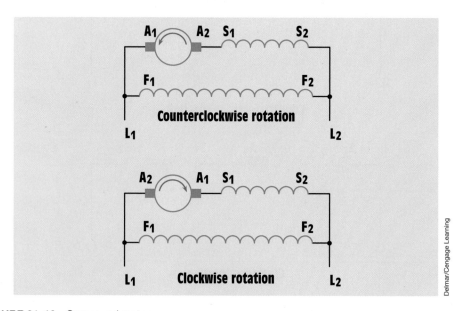

FIGURE 31–16 Compound motors.

The direction of rotation of a DC motor can be reversed by changing the connections of the armature leads or the field leads. It is common practice to change the connection of the armature leads. This is done to prevent changing a cumulative-compound motor into a differential-compound motor.

Although it is standard practice to change the connection of the armature leads to reverse the direction of rotation, it is not uncommon to reverse the rotation of small shunt motors by changing the connection of the field leads. If a motor contains only a shunt field, there is no danger of changing the motor from a cumulative- to a differential-compound motor. The shunt field leads are often changed on small motors because the amount of current flow through the field is much less than the current flow through the armature. This permits a small double-pole double-throw (DPDT) switch to be used as a control for reversing the direction of rotation *(Figure 31–17)*.

Large compound motors often use a control circuit similar to the one shown in *Figure 31–18* for reversing the direction of rotation. This control circuit uses magnetic contactors to reverse the flow of current through the armature. If the circuit is traced, it will be seen that when the forward or reverse direction is chosen, only the current through the armature changes direction. The current flow through the shunt and series fields remains the same.

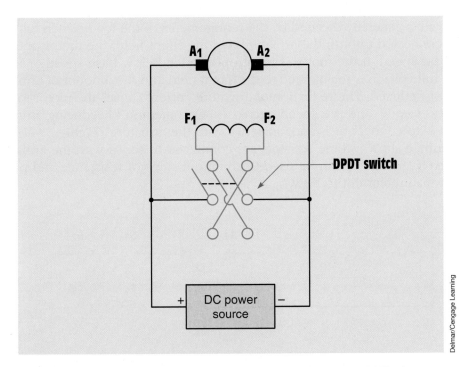

FIGURE 31–17   Reversing the rotation of a shunt motor by changing the shunt field leads.

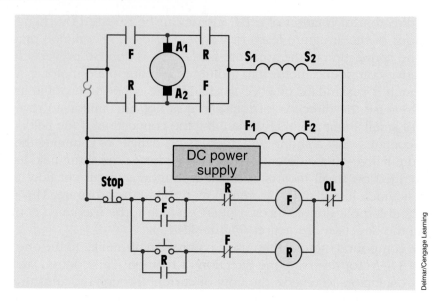

FIGURE 31–18   Forward-reverse control for a compound motor.

31–7   Speed Control

There are several ways of controlling the speed of DC motors. The method employed is generally dictated by the requirements of the load. When full voltage is connected to both the armature and the shunt field, the motor operates at its base speed. If the motor is to be operated at below base speed or under speed, full voltage is maintained to the shunt field and the amount of armature current is reduced. The reduction of armature current causes the motor to produce less torque, and the speed decreases. One method of reducing armature current is to connect resistance in series with the armature *(Figure 31–19)*. In the example shown, three resistors are connected in series with the armature. Contacts S_1, S_2, and S_3 can be used to shunt out steps of resistance and permit more armature current to flow.

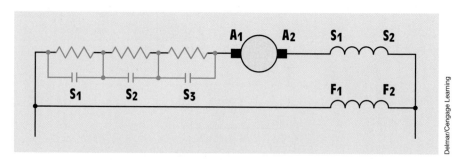

FIGURE 31–19   Resistors limit armature current.

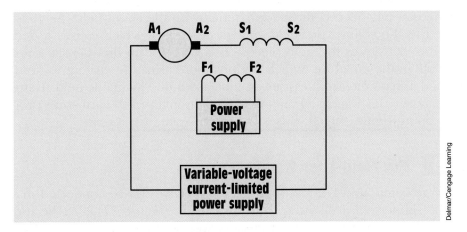

FIGURE 31–20   Armature current is controlled by a variable-voltage power supply.

Although adding resistance in the armature circuit does permit the motor to be underspeeded, it has several disadvantages. As current flows through the resistors, they waste power in the form of heat. Also, the speed of the motor can only be controlled in steps. There is no smooth increase or decrease in speed. Most large DC motors use an electronic controller to supply variable voltage to the armature circuit separately from the field *(Figure 31–20)*. This permits continuous adjustment of the speed from zero to full RPM. Electronic power supplies can also sense the speed of the motor and maintain a constant speed as load is changed. Most of these power supplies are current-limited. The maximum current output can be set to a value that does not permit the motor to be harmed if it stalls or if the load becomes too great.

A DC motor can be overspeeded by connecting full voltage to the armature and reducing the current flow through the shunt field. In *Figure 31–21,* a shunt

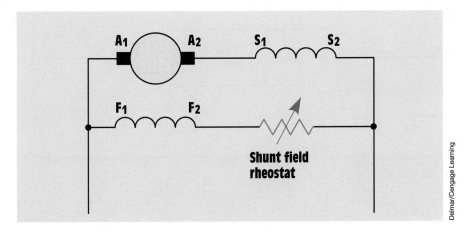

FIGURE 31–21   The shunt field rheostat can be used to overspeed the motor.

field rheostat has been connected in series with the shunt field. As resistance is added to the shunt field, field current decreases, which causes a decrease in the flux density of the pole pieces. This decrease in flux density produces less CEMF in the armature, which permits more armature current to flow. The increased armature current causes an increase in the magnetic field strength of the armature. This increased magnetic field strength of the armature produces a net gain in torque, which causes the motor speed to increase.

31-8 The Field-Loss Relay

Most large compound DC motors have a protective device connected in series with the shunt field, called the **field-loss relay** *(Figure 31–22)*. The function of the field-loss relay is to disconnect power to the armature if current flow through the shunt field decreases below a certain level. If the shunt field current stops completely, the compound motor becomes a series motor and increases rapidly in speed. This can cause damage to both the motor and the load.

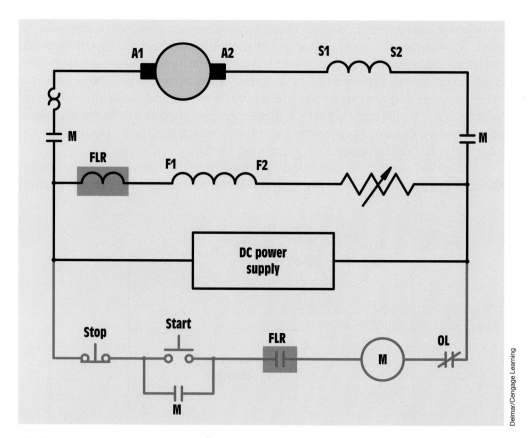

FIGURE 31–22 The shunt field relay disconnects power to the armature if shunt field current stops.

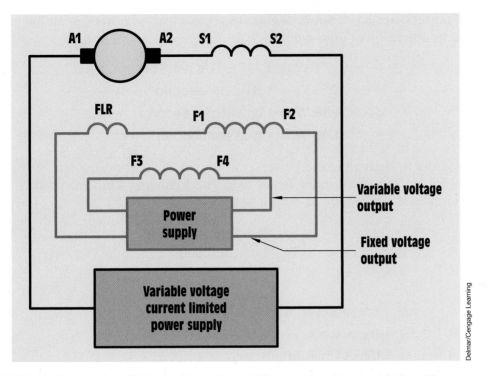

FIGURE 31–23 One shunt field is used to provide a stable speed; the other shunt field provides overspeed control.

Many large DC compound motors intended to operate in an overspeed condition actually contain two separate shunt fields *(Figure 31–23)*. One shunt field is connected to a fixed voltage and maintains a constant field to provide an upper limit to motor speed. This shunt field is connected to the field-loss relay. The second shunt field is connected to a source of variable voltage. This shunt field is used to increase speed above the base speed. For this type of motor, base speed is achieved by applying full voltage to the armature and both shunt fields. Note that most large DC motors have voltage applied to the shunt field at all times, even when the motor is not in operation. The resistance of the winding produces heat, which is used to prevent any formation of moisture inside the motor.

31–9 Horsepower

When James Watt first began marketing steam engines, he found that he needed a way to compare them to the horses they were to replace. After conducting experiments, Watt found that the average horse could do work at a rate of 550 foot-pounds per second. This became the basic horsepower measurement. Horsepower can also be expressed in the basic electric unit for power, which is the watt:

$$1 \text{ hp} = 746 \text{ W}$$

Once horsepower has been converted to a basic unit of power, it can be converted to other power units such as

$$1 \text{ W} = 3.412 \text{ BTUs per hour}$$
$$1055 \text{ W} = 1 \text{ BTU per second}$$
$$4.18 \text{ W} = 1 \text{ calorie per second}$$
$$1.36 \text{ W} = 1 \text{ ft-lb per second}$$

In order to determine the horsepower output of a motor, the rate at which the motor is doing work must be known. The following formula can be used to determine the horsepower output of a motor:

$$hp = \frac{(1.59)(torque)(rpm)}{100,000}$$

where

$$hp = \text{horsepower}$$
$$1.59 = \text{a constant}$$
$$torque = \text{torque in lb-in.}$$
$$rpm = \text{speed (angular velocity in revolutions per minute)}$$
$$100,000 = \text{a constant}$$

■ EXAMPLE 31–1

How much horsepower is being produced by a motor turning a load of 350 lb-in. at a speed of 1375 rpm?

Solution

$$hp = \frac{1.59 \times 350 \text{ lb-in.} \times 1375 \text{ rpm}}{100,000}$$

$$hp = 7.652$$

Once the output horsepower is known, it is possible to determine the efficiency of the motor by using the formula

$$Eff. = \frac{\text{power out}}{\text{power in}} \times 100$$

■ EXAMPLE 31-2

A DC motor is connected to a 120-VDC line and has a current draw of 1.3 A. The motor is operating a load that requires 8 lb-in. of torque and is turning at a speed of 1250 rpm. What is the efficiency of the motor?

Solution

The first step is to determine the horsepower output of the motor:

$$hp = \frac{1.59 \times 8 \text{ lb-in.} \times 1250 \text{ rpm}}{100,000}$$

$$hp = 0.159$$

Now that the output horsepower is known, horsepower can be changed into W using the formula

$$1 \text{ hp} = 746 \text{ W}$$

$$0.159 \times 746 = 118.614 \text{ W}$$

The amount of input power can be found by using the formula

$$W = V \times A$$

$$W = 120 \text{ V} \times 1.3 \text{ A}$$

$$W = 156$$

The efficiency of the motor can be found by using the formula

$$\text{Eff.} = \frac{\text{power out}}{\text{power in}} \times 100$$

The answer is multiplied by 100 to change it to a percent:

$$\text{Eff.} = \frac{118.614 \text{ W}}{156 \text{ W}} \times 100$$

$$\text{Eff.} = 76\%$$

   Torque is often measured in lb-ft instead of lb-in. Another formula often used to determine horsepower when the torque is measured in lb-ft is

$$hp = \frac{(2\pi)(\text{torque})(\text{rpm})}{33,000}$$

The speed of the rotating magnetic field for the motor just described is

$$S = \frac{120 \times 60}{64}$$

$$S = 112.5 \text{ rpm}$$

The use of this many poles to provide low speed and high torque is often referred to as *magnetic gearing* because it eliminates the need for mechanical gears and other speed-reducing equipment. This in turn eliminates friction and backlash associated with mechanical gears. Motors that use a high pole count for low-speed high-torque applications are often called *ring motors* or *ring torquers*.

Inside-Out Motors

Another type of brushless DC motor uses a rotor shaped like a hollow cylinder or cup. The stator windings are wound inside the rotor *(Figure 31–27)*. The permanent magnet rotor is mounted on the outside of the stator windings. The size and shape of the rotor cause it to act like a flywheel, which gives these motors a large amount of inertia. ***Motors with a large amount of inertia exhibit superior speed-regulation characteristics.*** Inside-out motors are often used to drive the hard disks in computers, to operate tape

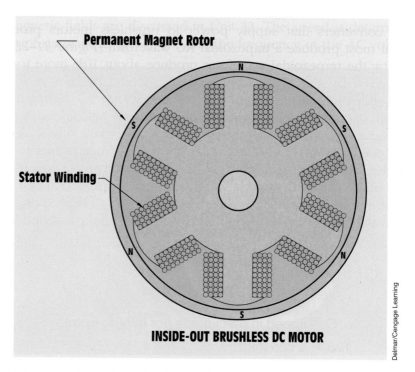

FIGURE 31–27   Inside-out brushless DC motor.

cartridge drives, to provide power for robots, and to operate fans and blowers in high-speed air-conditioning systems.

Differences between Brush-Type and Brushless Motors

Because brushless motors do not have a commutator or brushes, they are generally smaller and cost less than brush-type motors. The added expense of the converter, however, makes the cost about the same as a brush-type motor. Brushless motors dissipate heat more quickly because the stator windings can dissipate heat faster than a wound armature. They are more efficient than brush-type motors, require less maintenance, and as a general rule have less downtime.

31–11   Converters

As stated previously, converters are used to change the DC into multiphase AC to produce a rotating magnetic field in the stationary armature or stator of the brushless DC motor. An example circuit for a DC-to-three-phase converter is shown in *Figure 31–28*. In this example, Transistor Q_1 through Transistor Q_6 are used as switches. By turning the transistors on or off in the proper sequence, current can be routed through the stator windings in such a manner

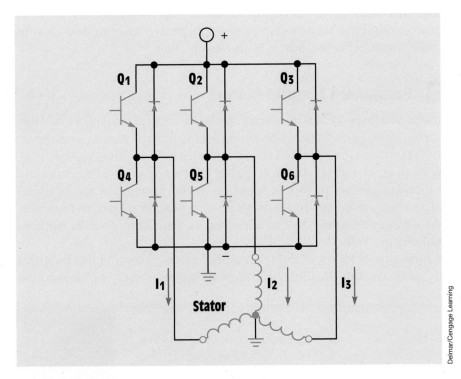

FIGURE 31–28   Typical circuit to change DC voltage into three-phase AC voltage.

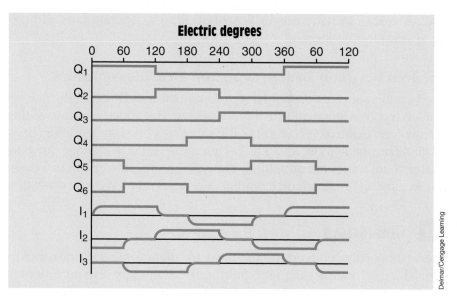

FIGURE 31–29 *Converter commutation sequence.*

that it produces three ACs 120° out of phase with each other. The commutation sequence provides an action similar to that of the commutator in a brush-type motor. A chart illustrating the firing order of the transistors to produce this commutation sequence is shown in *Figure 31–29*. The current flow through the three stator windings is illustrated in the same chart.

31–12 Permanent Magnet Motors

Permanent magnet (PM) motors contain a wound armature and brushes like a conventional DC motor. The pole pieces, however, are permanent magnets. This eliminates the need for shunt or series field windings *(Figure 31–30)*. Permanent magnet motors have a higher efficiency than conventional field-wound motors because power is supplied to the armature circuit only. These motors have been popular for many years in applications where batteries must be used to supply power to the motor, such as trolling motors on fishing boats and small electric vehicles.

The horsepower rating of PM (permanent magnet) motors has increased significantly since the introduction of rare-earth magnets such as Samarium-cobalt

GREEN TIPS: Direct current motors that contain permanent magnet field poles are more efficient because no power is required to produce the magnetic fields of the pole pieces. ■

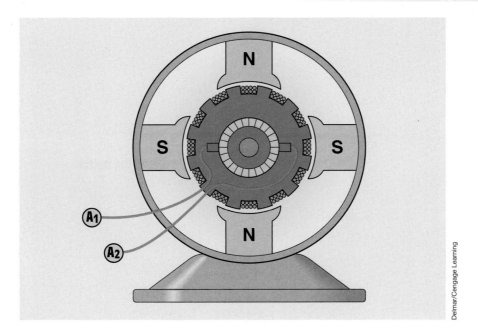

FIGURE 31–30 Permanent magnet motor.

Delmar/Cengage Learning

and Neodymium in the mid-1970s. These materials have replaced Alnico and ferrite magnets in most PM motors. The increased strength of rare-earth magnets permits motors to produce torques that range from 7.0 ounces per inch to 4500 foot-pounds. Permanent magnet motors with horsepower ratings of over 15 horsepower are now available. The torque-to-weight ratios of PM motors equipped with rare-earth magnets can exceed those of conventional field-wound motors by 40% to 90%. The power-to-weight ratios can exceed conventional motors by 50% to 200%. Permanent magnet motors of comparable horsepower ratings are smaller and lighter in weight than conventional field-wound motors.

Operating Characteristics

Because the fields are permanent magnets, the field flux of PM motors remains constant at all times. This gives the motor operating characteristics very similar to those of conventional separately excited shunt motors. The speed can be controlled by the amount of voltage applied to the armature, and the direction of rotation is reversed by reversing the polarity of voltage applied to the armature leads *(Figure 31–31)*.

DC Servomotors

Small PM motors are used as **servomotors.** These motors have small, lightweight armatures that contain very little inertia. This permits servomotors to be operated at high speed and then stopped or reversed very quickly. Servomotors generally contain from two to six poles. They are used to operate tape

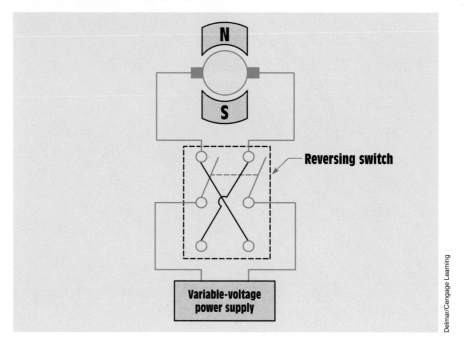

FIGURE 31–31   PM motors can be reversed by reversing the polarity to the armature, and the speed can be controlled by variable voltage.

drives on computers and to power the spindles on numerically controlled (NC) machines such as milling machines and lathes.

DC ServoDisc Motors

Another type of DC servomotor that is totally different in design is the **ServoDisc motor.** This motor uses permanent magnets to provide a constant magnetic field like conventional servomotors, but the design of the armature is completely different. In a conventional servomotor, the armature is constructed in the same way as other DC motors by cutting slots in an iron core and winding wire through the slots. A commutator is then connected to one end to provide commutation, which switches the current path as the armature turns. This maintains a constant magnetic polarity for different sections of the armature. In the conventional servomotor, the permanent magnets are mounted on the motor housing in such a manner that they create a radial magnetic field that is perpendicular to the windings of the armature *(Figure 31–32)*. The armature core is made of iron because it must conduct the magnetic lines of flux between the two pole pieces.

The armature of the ServoDisc motor does not contain any iron. It is made of two to four layers of copper conductors formed into a thin disc. The conductors are "printed" on a fiberglass material in much the same way as a printed

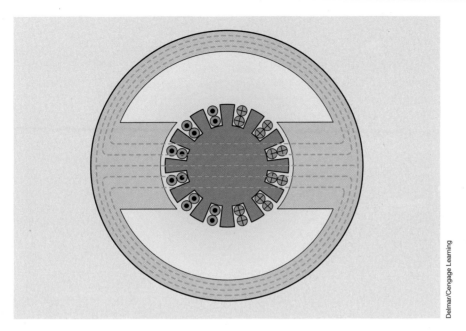

Delmar/Cengage Learning

FIGURE 31–32   Magnetic lines of flux must travel a great distance between the poles of the permanent magnets.

circuit. For this reason, the ServoDisc motor is often called a **printed-circuit motor.** Because the disc armature is very thin, it permits the permanent magnets to be mounted on either side of the disc and parallel to the shaft of the motor *(Figure 31–33)*. Because the disc too is very thin, the air gap between the two magnets is small.

Torque is produced when current flowing though the copper conductors of the disc produces a magnetic field that reacts with the magnetic field of the permanent magnets. The permanent magnet pairs are arranged around the circumference of the motor housing in such a manner that they provide alternate magnetic fields *(Figure 31–34)*. The conductors on each side of the armature, upper and lower, are arranged in such a manner that when current flows through them, a force tangent to the pole piece is produced. This tangential force is the vector sum of the forces produced by the upper and lower conductors *(Figure 31–35)*. The ServoDisc motor produces a relatively strong torque for its size and weight.

The conductors of the armature are so arranged on the fiberglass disc that they form a commutator on one side of the armature. Brushes riding against this commutator supply DC to the armature conductors. Because the armature contains no iron, it has almost no inductance. This greatly reduces any arcing at the brushes, which results in extremely long brush life. A typical ServoDisc motor is shown in *Figure 31–36.*

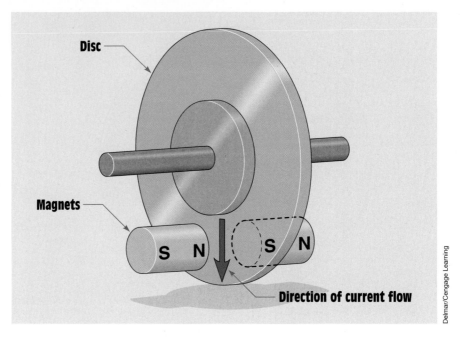

FIGURE 31–33 Basic construction of a ServoDisc motor.

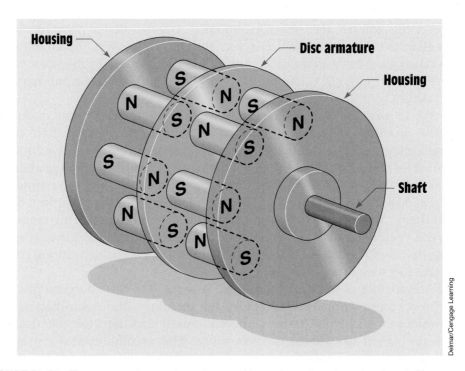

FIGURE 31–34 The permanent magnets are arranged to produce alternate magnetic polarities.

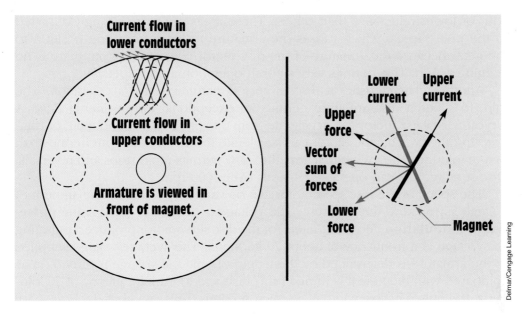

Delmar/Cengage Learning

FIGURE 31–35 A force tangent to the magnet is produced.

Courtesy of Danaher Motion Technologies

FIGURE 31–36 DC ServoDisc motors.

Characteristics of ServoDisc Motors

The unique construction of the disc-type servomotor gives it some operating characteristics that are different from those of other types of permanent magnet DC motors. Permanent magnet DC motors with iron armatures have a problem with cogging at low speeds. Cogging is caused by the reaction of the iron armature with the field of the permanent magnets. If the armature is turned by hand, it jumps, or "cogs," from one position to another. As the armature is turned, the magnetic flux lines of the pole pieces pass through the core of the armature *(Figure 31–32)*. Because the armature is slotted, certain areas offer a path of lower reluctance than others. These areas are more strongly attracted to the pole pieces, which causes the armature to jump from one position to the next. Because the armature of the disc motor contains no iron, there is no cogging action. This permits very smooth operation at low speeds.

Another characteristic of disc motors is extremely fast acceleration. The thin, low-inertia disc armature permits an exceptional torque–inertia ratio. A typical ServoDisc motor can accelerate from 0 to 3000 rpm in about 60° of rotation. In other words, the motor can accelerate from 0 to 3000 rpm in one-sixth of a revolution. The low-inertia armature also permits rapid stops and reversals. Disc servomotors can operate at speeds over 4000 rpm.

The speed of the disc servomotor can be varied by changing the amount of voltage supplied to the armature. The voltage is generally varied using **pulse-width modulation.** Most amplifiers for the disc servomotor produce a pulsating DC voltage at a frequency of about 20 kilohertz. The average voltage supplied to the armature is determined by the length of time the voltage is turned on as compared with the time it is turned off (pulse width). At a frequency of 20 kilohertz, the pulses have a width of 50 microseconds (pulse width = 1/frequency). Assume that the pulses have a peak voltage of 24 volts and are turned on for 40 microseconds and off for 10 microseconds during each pulse *(Figure 31–37)*. In this example, 24 volts are supplied to the motor for 80% of the time, producing an average voltage of 19.2 volts (24 V × 0.80 = 19.2 V). Now assume that the pulse width is changed so that the on time is 10 microseconds and the off time is 40 microseconds *(Figure 31–38)*. A voltage of 24 volts is now supplied to the motor 20% of the time, resulting in an average value of 4.8 volts (24 V × 0.20 = 4.8 V).

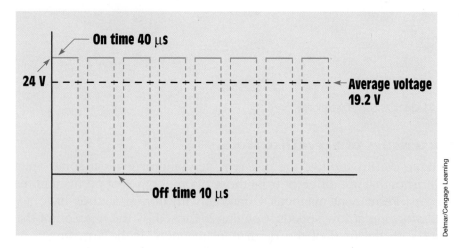

FIGURE 31–37 The amount of average voltage is determined by the peak voltage and the amount of time it is turned on or off.

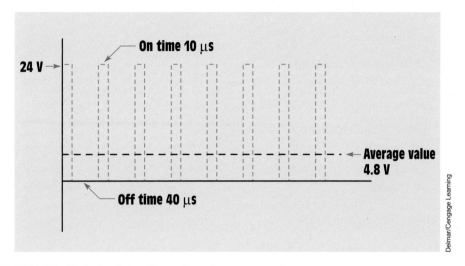

24 V →

On time 10 μs

Average value 4.8 V

Off time 40 μs

Delmar/Cengage Learning

FIGURE 31–38 Reducing the on time reduces the average voltage.

Courtesy of Danaher Motion Technologies

FIGURE 31–39 The DC ServoDisc motor can be constructed with small, thin cases.

Because of the basic design of the motor, the ServoDisc motor can be constructed in a very small, thin case *(Figure 31–39)*. This makes it very useful in some applications where other types of servomotors could not be used.

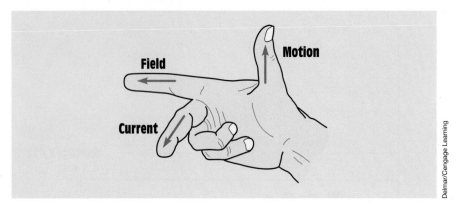

FIGURE 31–40 Right-hand motor rule.

31–13 The Right-Hand Motor Rule

In Unit 14, it was shown that it is possible to determine the direction of current flow through the armature of a generator using the fingers of the left hand when the polarity of the field poles and the direction of rotation of the armature are known. Similarly, the fingers of the right hand can be used to determine the direction of rotation of the armature when the magnetic field polarity of the pole pieces and the direction of current flow through the armature are known *(Figure 31–40)*. The thumb indicates the direction of thrust or movement of the armature. The forefinger indicates the direction of the field flux assuming that flux lines are in a direction of north to south, and the center finger indicates the direction of current flow through the armature. A simple method of remembering which finger represents which quantity is

THumb = THrust (direction of armature rotation)

Forefinger = Field (direction of magnetic field)

Center finger = Current (direction of armature current)

Summary

- A motor is a machine that converts electric energy into mechanical energy.
- DC motors operate on the principle of attraction and repulsion of magnetism.
- Two factors that determine the torque produced by a motor are
  a. the strength of the magnetic field of the pole pieces.
  b. the strength of the magnetic field of the armature.

- Torque is proportional to armature current if field excitation current remains constant.

- Current flow through the armature is limited by CEMF and armature resistance.

- Three factors that determine the amount of CEMF produced by a motor are

    a. the strength of the magnetic field of the pole pieces.

    b. the number of turns of wire in the armature.

    c. the speed of the armature.

- Three basic types of DC motors are the series, the shunt, and the compound.

- Shunt motors are sometimes known as constant-speed motors.

- Series motors must never be operated under a no-load condition.

- Series motors can develop extremely high starting torque.

- Compound motors contain both series and shunt field windings.

- When full voltage is applied to both the armature and shunt field, the motor operates at base speed.

- When full voltage is applied to the field and reduced voltage is applied to the armature, the motor operates below base speed.

- When full voltage is applied to the armature and reduced voltage is applied to the shunt field, the motor operates above base speed.

- The direction of rotation of a DC motor can be changed by reversing the connection of either the armature or the field leads.

- It is common practice to reverse the connection of the armature leads to prevent changing a compound motor from a cumulative- to a differential-compound connection.

- The shunt field relay is used to disconnect power to the armature if shunt field current drops below a certain level.

- Brushless DC motors do not contain a wound armature, commutator, or brushes.

- The stator windings for brushless motors are generally three phase but can be four phase or two phase.

- Brushless DC motors require a converter to change the DC into multi-phase AC.

- Brushless motors generally require less maintenance and have less down-time than conventional DC motors.

- PM motors use permanent magnets as the pole pieces and do not require series or shunt field windings.

- PM motors are more efficient than conventional DC motors.

- The operating characteristics of a PM motor are similar to those of a shunt motor with external excitation.

- Servomotors have lightweight armatures with low inertia.

- Disc servomotors contain an armature that is made of several layers of copper conductors.

- Because there is no iron in the armature of a disc servomotor, it does not have a problem with cogging.

- Disc servomotors can be accelerated very rapidly.

Review Questions

1. What is the principle of operation of a DC motor?

2. What is a motor?

3. What is the function of the commutator in a DC motor?

4. What two factors determine the amount of torque developed by a DC motor?

5. What type of motor is known as a constant-speed motor?

6. What is CEMF?

7. What three factors determine the amount of CEMF produced in the armature?

8. What limits the amount of current flow through the armature when power is first applied to the motor?

9. What factor determines the speed regulation of a DC motor?

10. What type of motor should never be operated at no load?

11. In general, what type of compound-motor connection should be avoided?

12. What is the most common way of changing the direction of rotation of a compound motor?

13. How can a DC motor be made to operate at its base speed?

14. How can a DC motor be made to operate above its base speed?

15. What device is used to disconnect power to the armature if the shunt field current drops below a certain level?

16. Why do many industries leave power connected to the shunt field at all times even when the motor is not operating?

17. Who was the first person to establish a measurement for hp?

18. One horsepower is equal to how many W?

19. A motor is operating a load that requires a torque of 750 lb-in. and is turning at a speed of 1575 rpm. How much hp is the motor producing?

20. The motor in Question 19 is connected to a 250-VDC line and has a current draw of 80 A. What is the efficiency of this motor?

Practical Applications

You are working as a plant electrician. You are told that a large DC motor was returned from the rewind shop and was installed on the previous shift. The motor seemed to operate properly until load was added. The motor then stopped, reversed direction, and began to run at high speed. What is the probable cause of trouble, and what would you do to correct the problem? Explain your answer. ∎

XII AC Machines

Unit 32

Three-Phase Alternators

KEY TERMS

Alternators

Brushless exciter

Field-discharge
    resistor

Hydrogen

Parallel alternators

Phase rotation

Revolving-armature-
    type alternator

Revolving-field-type
    alternator

Rotor

Sliprings

Stator

Synchroscope

Why You Need to Know

Alternators produce most of the electric power in the world. Some applications for small single-phase alternators are used as portable generators for home emergency or to provide the power for portable power tools on a work site, but most alternators are three phase. This unit

- explains the principles of operation for almost all alternators regardless of size.
- discusses what determines output frequency and how output voltage is controlled.
- explains how alternators are connected in parallel to provide more power when needed.
- discusses the different types of alternators and the operation of each.
- explains how to interpret the *NEC* when determining how to connect and determine protective devices.

Objectives

After studying this unit, you should be able to

- discuss the operation of a three-phase alternator.

- explain the effect of speed of rotation on frequency.

- explain the effect of field excitation on output voltage.

- connect a three-phase alternator and make measurements using test instruments.

Preview

Most of the electric power in the world today is produced by AC generators or alternators. Electric power companies use alternators rated in gigawatts (1 gigawatt = 1,000,000,000 W) to produce the power used throughout the United States and Canada. The entire North American continent is powered by AC generators connected together in parallel. These alternators are powered by steam turbines. The turbines, called prime movers, are powered by oil, coal, natural gas, or nuclear energy. ■

32–1   Three-Phase Alternators

Alternators operate on the same principle of electromagnetic induction as DC generators, but they have no commutator to change the AC produced in the armature into DC. There are two basic types of alternators: the revolving-armature type and the revolving-field type. Although there are some single-phase alternators that are used as portable power units for emergency home use or to operate power tools in a remote location, most alternators are three phase.

Revolving-Armature-Type Alternators

The **revolving-armature-type alternator** is the least used of the two basic types. This alternator uses an armature similar to that of a DC machine with the exception that the loops of wire are connected to **sliprings** instead of to a commutator *(Figure 32–1)*. Three separate windings are connected in either delta or wye. The armature windings are rotated inside a magnetic field *(Figure 32–2)*. Power is carried to the outside circuit via brushes riding against the sliprings. This alternator is the least used because it is very limited in the amount of output voltage and kilovolt-ampere (kVA) capacity it can develop.

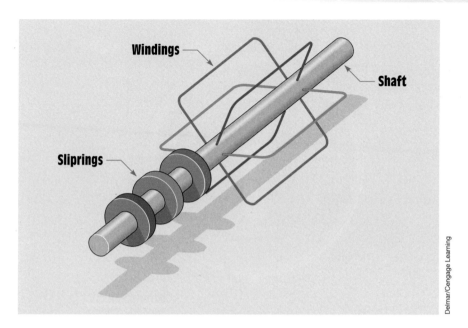

FIGURE 32-1   Basic design of a three-phase armature.

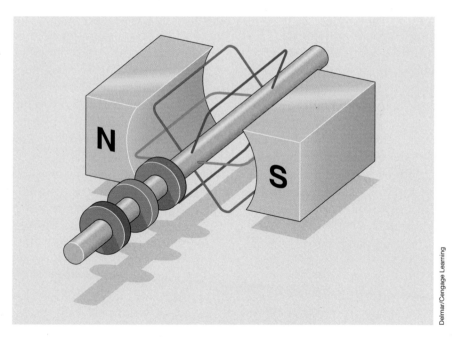

FIGURE 32-2   The armature conductors rotate inside a magnetic field.

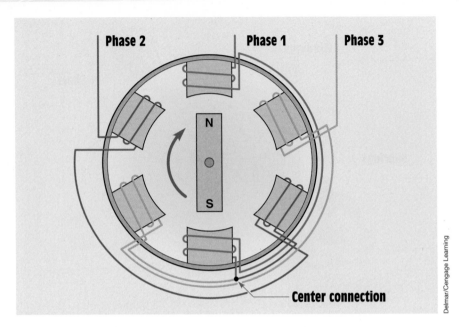

Delmar/Cengage Learning

FIGURE 32–3   Basic design of a three-phase alternator.

Revolving-Field-Type Alternators

The **revolving-field-type alternator** uses a stationary armature called the **stator** and a rotating magnetic field. This design permits higher voltage and kilovolt-ampere ratings because the outside circuit is connected directly to the stator and is not routed through sliprings and brushes. This type of alternator is constructed by placing three sets of windings 120° apart *(Figure 32–3)*. In *Figure 32–3,* the winding of Phase 1 winds around the top center pole piece. It then proceeds 180° around the stator and winds around the opposite pole piece in the opposite direction. The second phase winding winds around the top pole piece directly to the left of the top center pole piece. The second phase winding is wound in an opposite direction to the first. It then proceeds 180° around the stator housing and winds around the opposite pole piece in the opposite direction. The finish end of Phase 2 connects to the finish end of Phase 1. The start end of Phase 3 winds around the top pole piece to the right of the top center pole piece. This winding is wound in a direction opposite to Phase 1 also. The winding then proceeds 180° around the stator frame to its opposite pole piece and winds around it in an opposite direction. The finish end of Phase 3 is then connected to the finish ends of Phases 1 and 2. This forms a wye connection for the stator winding. When the magnet is rotated, voltage is induced in the three windings. Because these windings are spaced 120° apart, the induced voltages are 120° out of phase with each other *(Figure 32–4)*.

FIGURE 32–4 The alternator produces three sine wave voltages 120° out of phase with each other.

The stator shown in *Figure 32–3* is drawn in a manner to aid in understanding how the three phase windings are arranged and connected. In actual practice, the stator windings are placed in a smooth cylindrical core without projecting pole pieces *(Figure 32–5)*. This design provides a better path for magnetic lines of flux and increases the efficiency of the alternator.

FIGURE 32–5 Wound stator.

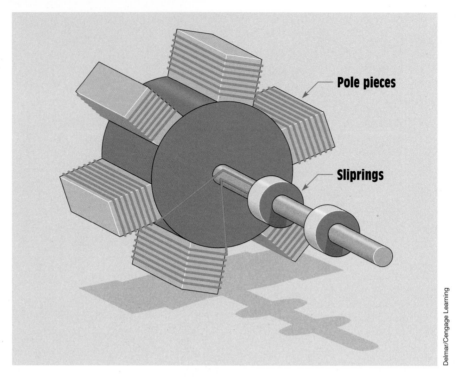

Pole pieces

Sliprings

Delmar/Cengage Learning

FIGURE 32–6   The rotor contains pole pieces that become electromagnets.

32–2   The Rotor

The **rotor** is the rotating member of the machine. It provides the magnetism needed to induce voltage into the stator windings. The magnets of the rotor are electromagnets and require some source of external DC to excite the alternator. This DC is known as excitation current. The alternator cannot produce an output voltage until the rotor has been excited. Some alternators use sliprings and brushes to provide the excitation current to the rotor *(Figure 32–6)*. A good example of this type of rotor can be found in the alternator of most automobiles. The DC excitation current can be varied in order to change the strength of the magnetic field. A rotor with salient (projecting) poles is shown in *Figure 32–7*.

32–3   The Brushless Exciter

Most large alternators use an exciter that contains no brushes. This is accomplished by adding a separate small alternator of the armature type on the same shaft of the rotor of the larger alternator. The armature rotates between wound

Courtesy of Cutler Hammer

FIGURE 32–7 Rotor of the salient pole type.

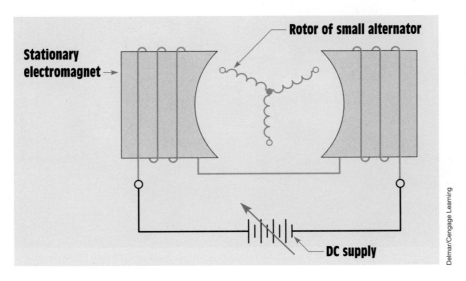

Delmar/Cengage Learning

FIGURE 32–8 The brushless exciter uses stationary electromagnets.

electromagnets. The DC excitation current is connected to the wound stationary magnets *(Figure 32–8)*. The amount of voltage induced in the rotor can be varied by changing the amount of excitation current supplied to the electromagnets. The output voltage of the armature is connected to a three-phase bridge rectifier mounted on the rotor shaft *(Figure 32–9)*. The bridge rectifier converts the three-phase AC voltage produced in the armature into DC voltage

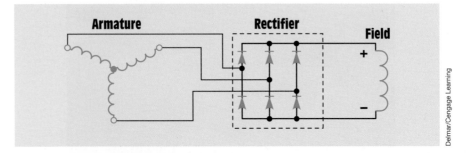

FIGURE 32–9   Basic brushless exciter circuit.

FIGURE 32–10   Brushless exciter assembly.

before it is applied to the main rotor windings. Because the armature, rectifier, and rotor winding are connected to the main rotor shaft, they all rotate together and no brushes or sliprings are needed to provide excitation current for the large alternator. A photograph of the **brushless exciter** assembly is shown in *Figure 32–10*. The field winding is placed in slots cut in the core material of the rotor *(Figure 32–11)*.

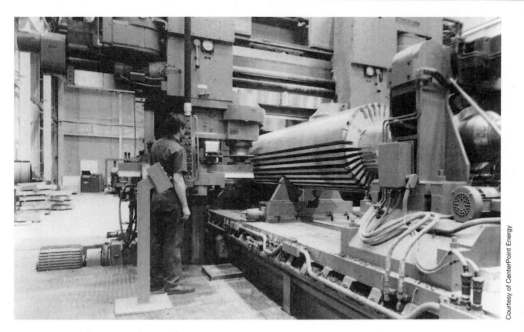

Courtesy of CenterPoint Energy

FIGURE 32–11   Two-pole rotor slotting.

32–4   Alternator Cooling

There are two main methods of cooling alternators. Alternators of small kilovolt-ampere rating are generally air-cooled. Open spaces are left in the stator windings, and slots are often provided in the core material for the passage of air. Air-cooled alternators have a fan attached to one end of the shaft that circulates air through the entire assembly.

Large-capacity alternators are often enclosed and operate in a **hydrogen** atmosphere. There are several advantages in using hydrogen. Hydrogen is less dense than air at the same pressure. The lower density reduces the windage loss of the spinning rotor. A second advantage in operating an alternator in a hydrogen atmosphere is that hydrogen has the ability to absorb and remove heat much faster than air. At a pressure of one atmosphere, hydrogen has a specific heat of approximately 3.42. The specific heat of air at a pressure of 1 atmosphere is approximately 0.238. This means that hydrogen has the ability to absorb approximately 14.37 times more heat than air. A cutaway drawing of an alternator intended to operate in a hydrogen atmosphere is shown in *Figure 32–12*.

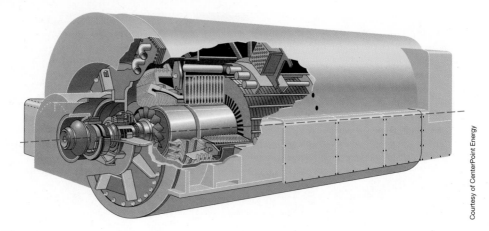

Courtesy of CenterPoint Energy

FIGURE 32–12   Two-pole, turbine-driven, hydrogen-cooled alternator.

32–5   Frequency

The output frequency of an alternator is determined by two factors:

1. the number of stator poles

2. the speed of rotation of the rotor

Because the number of stator poles is constant for a particular machine, the output frequency is controlled by adjusting the speed of the rotor. The following chart shows the speed of rotation needed to produce 60 hertz for alternators with different numbers of poles.

rpm	Stator Poles
3600	2
1800	4
1200	6
900	8

The following formula can also be used to determine the frequency when the poles and revolutions per minute (rpm) are known:

$$f = \frac{PS}{120}$$

where

$$f = \text{frequency in hertz}$$
$$P = \text{number of poles per phase}$$
$$S = \text{speed in rpm}$$
$$120 = \text{a constant}$$

■ EXAMPLE 32-1

What is the output frequency of an alternator that contains six poles per phase and is turning at a speed of 1000 rpm?

Solution

$$f = \frac{6 \times 1000 \text{ rpm}}{120}$$

$$f = 50 \text{ Hz}$$

32–6   Output Voltage

Three factors determine the amount of output voltage of an alternator:

1. the length of the armature or stator conductors (number of turns)

2. the strength of the magnetic field of the rotor

3. the speed of rotation of the rotor

The following formula can be used to calculate the amount of voltage induced in the stator winding:

$$E = \frac{BLV}{10^8}$$

where

$$10^8 = \text{flux lines equal to 1 weber}$$
$$E = \text{induced voltage (in volts)}$$
$$B = \text{flux density in gauss}$$
$$L = \text{length of the conductor (in cm)}$$
$$v = \text{velocity (in cm/s)}$$

One of the factors that determines the amount of induced voltage is the length of the conductor. This factor is often stated as number of turns of wire in the stator because the voltage induced in each turn adds. Increasing the number of turns of wire has the same effect as increasing the length of one conductor.

Controlling Output Voltage

The number of turns of wire in the stator cannot be changed in a particular machine without rewinding the stator, and the speed of rotation is generally maintained at a certain level to provide a constant output frequency. Therefore, the output voltage is controlled by increasing or decreasing the strength of the magnetic field of the rotor. The magnetic field strength can be controlled by controlling the DC excitation current to the rotor.

32–7   Paralleling Alternators

Because one alternator cannot produce all the power that is required, it often becomes necessary to use more than one machine. When more than one alternator is to be used, they are connected in parallel with each other. Several conditions must be met before **parallel alternators** can be used:

1. The phases must be connected in such a manner that the phase rotation of all the machines is the same.

2. Phases A, B, and C of one machine must be in sequence with Phases A, B, and C of the other machine. For example, Phase A of Alternator 1 must reach its positive peak value of voltage at the same time Phase A of Alternator 2 does *(Figure 32–13)*.

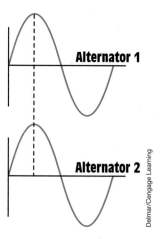

FIGURE 32–13   The voltages of both alternators must be in phase with each other.

3. The output voltage of the two alternators should be the same.

4. The frequency should be the same.

Determining Phase Rotation

The most common method of detecting when the **phase rotation** (the direction of magnetic field rotation) of one alternator is matched to the phase rotation of the other is with the use of three lights *(Figure 32–14)*. In *Figure 32–14,* the two alternators that are to be paralleled are connected together through a synchronizing switch. A set of lamps acts as a resistive load between the two machines when the switch contacts are in the open position. The voltage developed across the lamps is proportional to the difference in voltage between the two alternators. The lamps are used to indicate two conditions:

1. The lamps indicate when the phase rotation of one machine is matched to the phase rotation of the other. When both alternators are operating, both are producing a voltage. The lamps blink on and off when the phase rotation of one machine is not synchronized to the phase rotation of the other machine. If all three lamps blink on and off at the same time, or in unison, the phase rotation of Alternator 1 is correctly matched to the phase rotation of Alternator 2. If the lamps blink on and off but not in unison, the phase rotation between the two machines is not correctly matched, and two lines of Alternator 2 should be switched.

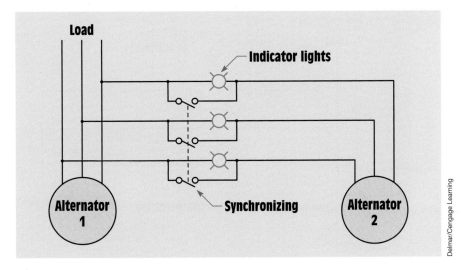

FIGURE 32–14   Determining phase rotation using indicator lights.

2. The lamps also indicate when the phase of one machine is synchronized with the phase of the other machine. If the positive peak of Alternator 1 does not occur at the same time as the positive peak of Alternator 2, there is a potential between the two machines. This permits the lamps to glow. The brightness of the lamps indicates how far out of synchronism the two machines are. When the peak voltages of the two alternators occur at the same time, there is no potential difference between them. The lamps should be off at this time. The synchronizing switch should never be closed when the lamps are glowing.

The Synchroscope

Another instrument often used for paralleling two alternators is the **synchroscope** *(Figure 32–15)*. The synchroscope measures the difference in voltage and frequency of the two alternators. The pointer of the synchroscope is free to rotate in a 360° arc. The alternator already connected to the load is considered to be the base machine. The synchroscope indicates whether the frequency of the alternator to be parallel to the base machine is fast or slow. When the voltages of the two alternators are in phase, the pointer covers the shaded area on the face of the meter. When the two alternators are synchronized, the synchronizing switch is closed.

If a synchroscope is not available, the two alternators can be paralleled using three lamps, as described earlier. If the three-lamp method is used, an AC voltmeter connected across the same phase of each machine indicates when the potential difference between the two machines is zero *(Figure 32–16)*. That is the point at which the synchronizing switch should be closed.

FIGURE 32–15 Synchroscope.

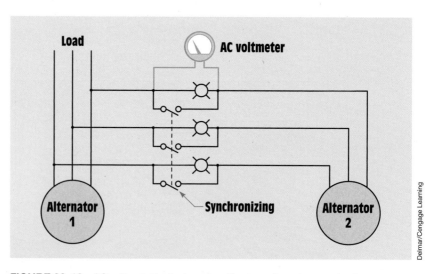

FIGURE 32–16 AC voltmeter indicates when the two alternators are in phase.

32–8 Sharing the Load

After the alternators have been paralleled, the power input to Alternator 2 must be increased to permit it to share part of the load. For example, if the alternator is being driven by a steam turbine, the power of the turbine would have to be increased. When this is done, the power to the load remains constant. The power output of the base alternator decreases, and the power output of the second alternator increases.

32–9 Field-Discharge Protection

When the DC excitation current is disconnected, the collapsing magnetic field can induce a high voltage in the rotor winding. This voltage can be high enough to arc contacts and damage the rotor winding or other circuit components. One method of preventing the induced voltage from becoming excessive is with the use of a **field-discharge resistor.** A special double-pole single-throw switch with a separate blade is used to connect the resistor to the field before the switch contacts open. When the switch is closed and DC is connected to the field, the circuit connecting the resistor to the field is open *(Figure 32–17)*. When the switch is opened, the special blade connects the resistor to the field before the main contacts open *(Figure 32–18)*.

Another method of preventing the high voltage discharge is to connect a diode in parallel with the field *(Figure 32–19)*. The diode is connected in such a manner that, when excitation current is flowing, the diode is reverse-biased and no current flows through the diode.

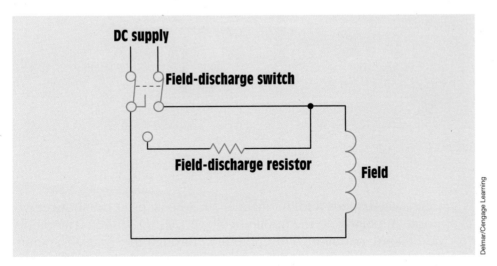

FIGURE 32–17 Switch in closed position.

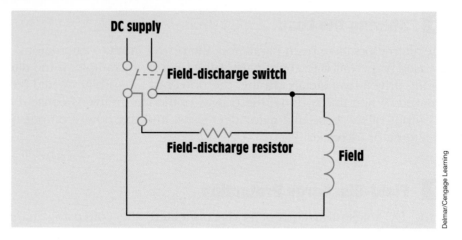

Delmar/Cengage Learning

FIGURE 32–18   Switch in open position.

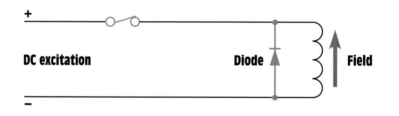

Delmar/Cengage Learning

FIGURE 32–19   Normal current flow.

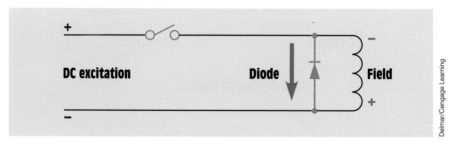

Delmar/Cengage Learning

FIGURE 32–20   Induced current flow.

When the switch opens and the magnetic field collapses, the induced volt-age is opposite in polarity to the applied voltage *(Figure 32–20)*. The diode is now forward-biased, permitting current to flow through the diode. The energy contained in the magnetic field is dissipated in the form of heat by the diode and field winding.

Summary

- There are two basic types of three-phase alternators: the revolving-armature type and the revolving-field type.

- The rotating-armature type is the least used because of its limited voltage and power rating.

- The rotor of the revolving-field-type alternator contains electromagnets.

- DC must be supplied to the field before the alternator can produce an output voltage.

- The DC supplied to the field is called excitation current.

- The output frequency of an alternator is determined by the number of stator poles and the speed of rotation.

- Three factors that determine the output voltage of an alternator are

  a. the length of the conductor of the armature or stator winding.

  b. the strength of the magnetic field of the rotor.

  c. the speed of the rotor.

- The output voltage is controlled by the amount of DC excitation current.

- Before two alternators can be connected in parallel, the output voltage of the two machines should be the same, the phase rotation of the machines must be the same, and the output voltages of the two machines must be in phase.

- Three lamps connected between the two alternators can be used to test for phase rotation.

- A synchroscope can be used to determine phase rotation and difference of frequency between two alternators.

- Two devices used to prevent a high voltage being induced in the rotor when the DC excitation current is stopped are a field-discharge resistor and a diode.

- Many large alternators use a brushless exciter to supply DC to the rotor winding.

Review Questions

1. What conditions must be met before two alternators can be paralleled together?

2. How can the phase rotation of one alternator be changed in relationship to the other alternator?

3. What is the function of the synchronizing lamps?

4. What is a synchroscope?

5. Assume that Alternator A is supplying power to a load and that Alternator B is to be paralleled to A. After the paralleling has been completed, what must be done to permit Alternator B to share the load with Alternator A?

6. What two factors determine the output frequency of an alternator?

7. At what speed must a six-pole alternator turn to produce 60 Hz?

8. What three factors determine the output voltage of an alternator?

9. What are sliprings used for on a revolving-field-type alternator?

10. Is the rotor excitation current AC or DC?

11. When a brushless exciter is used, what converts the AC produced in the armature winding into DC before it is supplied to the field winding?

12. What two devices are used to eliminate the induced voltage produced in the rotor when the field excitation current is stopped?

Unit 33
Three-Phase Motors

KEY TERMS

Amortisseur winding
Code letter
Differential selsyn
Direction of rotation
Dual-voltage motors
Percent slip
Phase rotation meter
Rotating magnetic field
Rotor frequency
Selsyn motors
Single-phasing
Squirrel-cage rotor
Synchronous condenser
Synchronous speed
Wound-rotor motor

Why You Need to Know

*T*hree-phase motors are the backbone of industry. They range in size from fractional horsepower to several thousand horsepower. It is imperative that anyone working in the electrical field have a thorough understanding of the different types of three-phase motors and their operating characteristics. This unit

- describes the three basic types of three-phase motors. Some are designed to operate on a single voltage, and others can be made to operate on two different voltages. Some motors have their stator windings connected in wye, and others are connected in delta.
- explains how stator windings are numbered and how to connect them to operate on the voltage supplied to the motor.
- describes how to calculate protective devices, different wiring methods, and operational principles.

Objectives

After studying this unit, you should be able to

- discuss the basic operating principles of three-phase motors.

- list factors that produce a rotating magnetic field.

- list different types of three-phase motors.

- discuss the operating principles of squirrel-cage motors.

- connect dual-voltage motors for proper operation on the desired voltage.

- discuss the operation of consequent-pole motors.

- discuss the operation of wound-rotor motors.

- discuss the operation of synchronous motors.

- determine the direction of rotation of a three-phase motor using a phase rotation meter.

Preview

Three-phase motors are used throughout the United States and Canada as the prime mover for industry. These motors convert the three-phase AC into mechanical energy to operate all types of machinery. Three-phase motors are smaller and lighter and have higher efficiencies per horsepower than single-phase motors. They are extremely rugged and require very little maintenance. Many of these motors are operated 24 hours a day, 7 days a week for many years without problem. ∎

33–1   Three-Phase Motors

The three basic types of three-phase motors are

1. the squirrel-cage induction motor.

2. the wound-rotor induction motor.

3. the synchronous motor.

All three motors operate on the same principle, and they all use the same basic design for the stator windings. The difference among them is the type of rotor used. Two of the three motors are induction motors and operate on the principle of electromagnetic induction in a manner similar to that of transformers. In fact, AC induction motors were patented as rotating transformers by Nikola Tesla. The stator winding of a motor is often referred to as the *motor primary,* and the rotor is referred to as the *motor secondary.*

33–2 The Rotating Magnetic Field

The operating principle for all three-phase motors is the **rotating magnetic field.** There are three factors that cause the magnetic field to rotate. These are

1. the fact that the voltages in a three-phase system are 120° out of phase with each other.

2. the fact that the three voltages change polarity at regular intervals.

3. the arrangement of the stator windings around the inside of the motor.

Figure 33–1 shows three AC sine waves 120° out of phase with each other, and the stator winding of a three-phase motor. The stator illustrates a two-pole three-phase motor. Two pole means that there are two poles per phase. AC motors do not generally have actual pole pieces as shown in *Figure 33–1,*

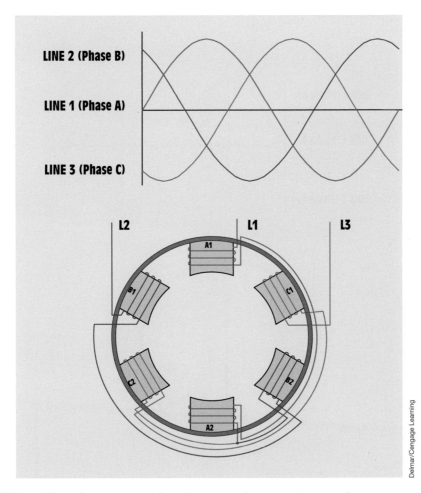

FIGURE 33–1 Three-phase stator and three sine wave voltages.

but they will be used here to aid in understanding how the rotating magnetic field is created in a three-phase motor. Notice that pole pieces A1 and A2 are located opposite each other. The same is true for poles B1 and B2 and C1 and C2. Pole pieces A1 and A2 are wound in such a manner that when current flows through the winding, they will develop opposite magnetic polarities. This is also true for poles B1 and B2 and C1 and C2. The windings of poles B1 and C1 are wound in the same direction in relation to each other, but in an opposite direction from the winding of pole A1. The start end of the winding for poles A1 and A2 is connected to Line 1, the start end of the winding for poles B1 and B2 is connected to Line 2, and the start end of the winding for poles C1 and C2 is connected to Line 3. The finish ends of all three windings are joined to form a wye connection for the stator.

To understand how the magnetic field rotates around the inside of the stator, refer to *Figure 33–2*. A dashed line labeled A has been drawn through the three sine waves of the three-phase system. This line is used to illustrate the

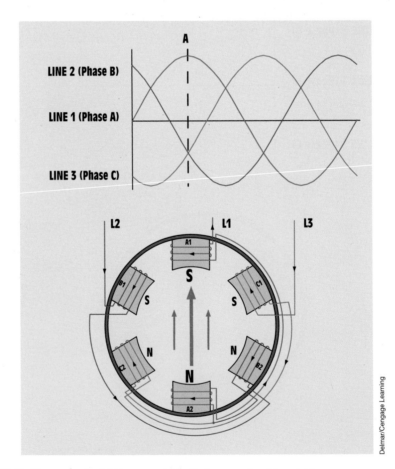

FIGURE 33–2 The magnetic field is concentrated between poles A1 and A2.

Delmar/Cengage Learning

condition of the three voltages at this point in time. The arrows drawn inside the motor indicate the greatest concentration of magnetic lines of flux; the arrows are pointing in the direction that indicates magnetic lines of flux from north to south. Line 1 has reached its maximum peak voltage in the positive direction, and Lines 2 and 3 are less than maximum and in the negative direction. The magnetic field is concentrated between poles A1 and A2. Weaker lines of magnetic flux also exist between poles B1 and B2 and C1 and C2. Also note that poles A1, B1, and C1 all form a south magnetic polarity. Poles A2, B2, and C2 form a north magnetic polarity.

In *Figure 33–3*, line B is drawn at a point in time when the voltage of Line 3 is zero and the voltages of Lines 1 and 2 are less than maximum but opposite in polarity. The magnetic field is now concentrated between the pole pieces of phases A and B. Phase C has no current flow at this time and therefore no magnetic field.

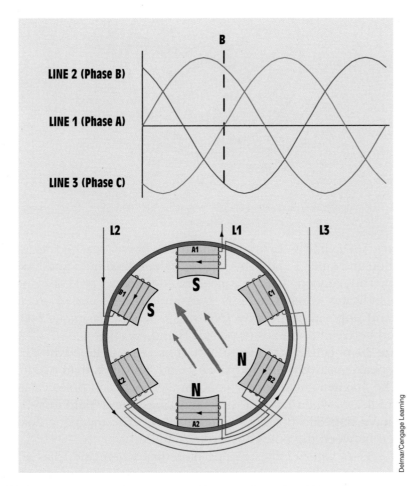

FIGURE 33–3   The magnetic field is concentrated between phases A and B.

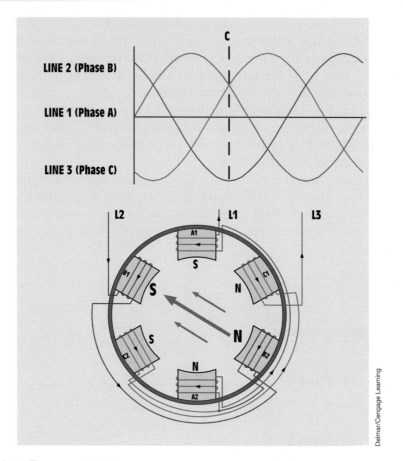

FIGURE 33–4   The magnetic field is concentrated between poles B1 and B2.

In *Figure 33–4,* line C is drawn when Line 2 has reached its maximum negative value and Lines 1 and 3 are both less than maximum and have a positive polarity. The magnetic field is concentrated between poles B1 and B2.

Line D indicates when Line 1 is zero and Lines 2 and 3 are less than maximum and opposite in polarity *(Figure 33–5).* The magnetic field is now concentrated between the poles of phases B and C.

In *Figure 33–6,* Line E is drawn to indicate a point in time when Line 3 has reached its peak positive point and Lines 1 and 2 are less than maximum and negative. The magnetic field is now concentrated between poles C1 and C2.

Line F indicates when Line 2 is zero and Lines 1 and 3 are less than maximum and have opposite polarities *(Figure 33–7).* The magnetic field is now concentrated between the poles of phases A and C.

In *Figure 33–8,* Line G indicates a point in time when Line 1 has reached its maximum negative value and Lines 2 and 3 are less than maximum and have a positive polarity. The magnetic field is again concentrated between poles A1 and A2. However, pole A1 has a north magnetic polarity instead of pole A2.

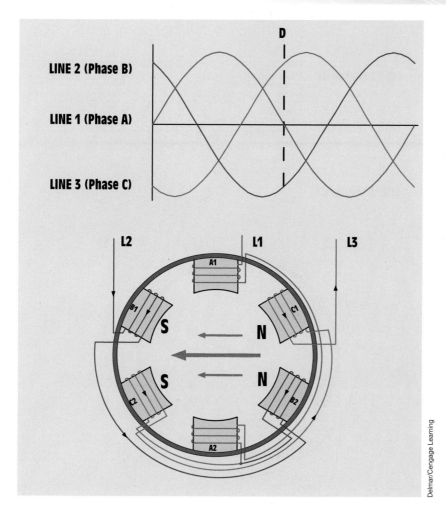

FIGURE 33–5 The magnetic field is concentrated between phases B and C.

Figure 33–9 shows the position of the magnetic at position H. The field has now rotated 270°. At the end of one complete cycle, the magnetic field completes a full 360° of rotation *(Figure 33–10)*. The speed of the rotating magnetic field is 3600 rpm in a two-pole motor connected to a 60-Hz line.

Synchronous Speed

The speed at which the magnetic field rotates is called the **synchronous speed.** Two factors that determine the synchronous speed of the rotating magnetic field are

1. the number of stator poles (per phase).

2. the frequency of the applied voltage.

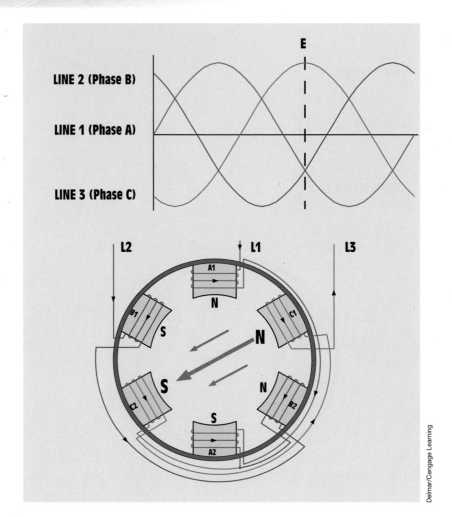

FIGURE 33–6   The magnetic field is concentrated between poles C1 and C2.

The following chart shows the synchronous speed at 60 hertz for different numbers of stator poles:

rpm	Stator poles
3600	2
1800	4
1200	6
900	8

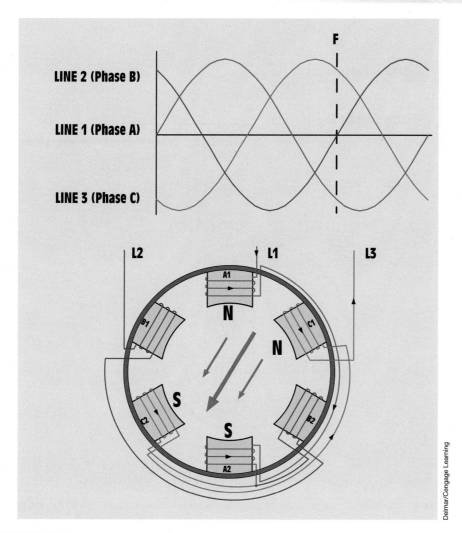

FIGURE 33–7 The magnetic field is concentrated between phases A and C.

The stator winding of a three-phase motor is shown in *Figure 33–11*. The synchronous speed can be calculated using the formula

$$S = \frac{120\,f}{P}$$

where

S = speed in rpm

f = frequency in Hz

P = number of stator poles (per phase)

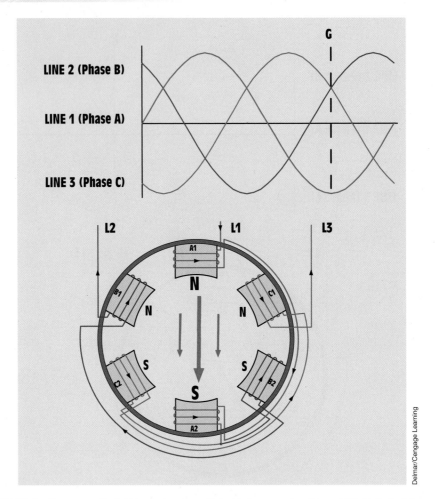

LINE 2 (Phase B)

LINE 1 (Phase A)

LINE 3 (Phase C)

G

L2   L1   L3

A1
N
N   N
B1   C1
S   S
C2   B2
S
A2

Delmar/Cengage Learning

FIGURE 33–8   The magnetic field is again concentrated between poles A1 and A2. Note that the polarity of the magnetic fields has reversed. The magnetic field has rotated 180° during one half-cycle.

■ **EXAMPLE 33-1**

What is the synchronous speed of a four-pole motor connected to 50 hertz?

Solution

$$S = \frac{120 \times 50 \text{ Hz}}{4}$$

$$S = 1500 \text{ rpm}$$

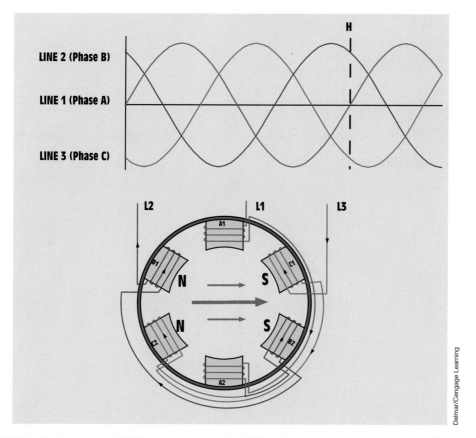

LINE 2 (Phase B)

LINE 1 (Phase A)

LINE 3 (Phase C)

Delmar/Cengage Learning

FIGURE 33–9   The magnetic field has rotated a total of 270° and is concentrated between phases B and C.

■ EXAMPLE 33-2

What frequency should be applied to a six-pole motor to produce a synchronous speed of 400 rpm?

Solution

First change the base formula to find frequency. Once that is done, known values can be substituted in the formula

$$f = \frac{PS}{120}$$

$$f = \frac{6 \times 400 \text{ Hz}}{120}$$

$$f = 20 \text{ Hz}$$

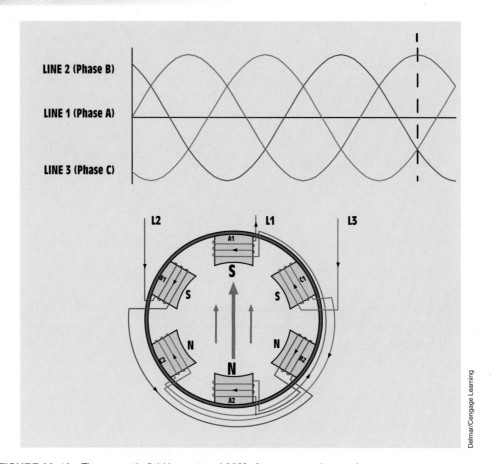

FIGURE 33–10 The magnetic field has rotated 360° after one complete cycle.

FIGURE 33–11 Stator of a three-phase motor.

Determining the Direction of Rotation for Three-Phase Motors

On many types of machinery, the direction of rotation of the motor is critical. ***The direction of rotation of any three-phase motor can be changed by reversing two of its stator leads.*** This causes the direction of the rotating magnetic field to reverse. When a motor is connected to a machine that will not be damaged when its direction of rotation is reversed, power can be momentarily applied to the motor to observe its direction of rotation. If the rotation is incorrect, any two line leads can be interchanged to reverse the motor's rotation.

When a motor is to be connected to a machine that can be damaged by incorrect rotation, however, the direction of rotation must be determined before the motor is connected to its load. The **direction of rotation** can be determined in two basic ways. One way is to make an electric connection to the motor before it is mechanically connected to the load. The direction of rotation can then be tested by momentarily applying power to the motor before it is coupled to the load.

There may be occasions when it is not practical or is very inconvenient to apply power to the motor before it is connected to the load. In such a case, a **phase rotation meter** can be used *(Figure 33–12)*. The phase rotation meter compares the phase rotation of two different three-phase connections. The meter contains six terminal leads. Three of the leads are connected to one side of the meter and are labeled MOTOR. These three motor leads are labeled A, B, or C. The LINE leads are located on the other side of the meter and are labeled A, B, or C.

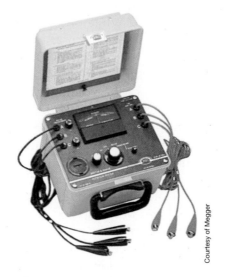

Courtesy of Megger

FIGURE 33–12   Phase rotation meter.

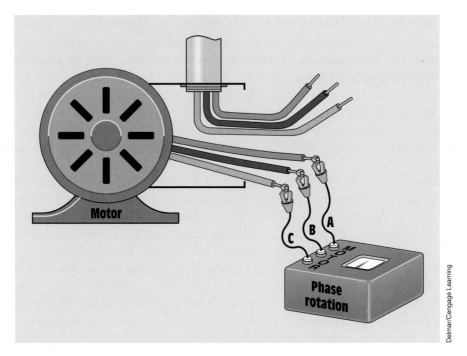

Delmar/Cengage Learning

FIGURE 33–13   Connecting the phase rotation meter to the motor.

To determine the direction of rotation of the motor, first zero the meter by following the instructions provided by the manufacturer. Then set the meter selector switch to MOTOR, and connect the three MOTOR leads of the meter to the "T" leads of the motor *(Figure 33–13)*. The phase rotation meter contains a zero-center voltmeter. One side of the voltmeter is labeled INCORRECT, and the other side is labeled CORRECT. While observing the zero-center voltmeter, manually turn the motor shaft in the direction of desired rotation. The zero-center voltmeter will immediately swing in the CORRECT or INCORRECT direction. When the motor shaft stops turning, the needle may swing in the opposite direction. It is the *first* indication of the voltmeter that is to be used.

If the voltmeter needle indicates CORRECT, label the motor T leads A, B, or C to correspond with the MOTOR leads from the phase rotation meter. If the voltmeter needle indicates INCORRECT, change any two of the MOTOR leads from the phase rotation meter and again turn the motor shaft. The voltmeter needle should now indicate CORRECT. The motor T leads can now be labeled to correspond with the MOTOR leads from the phase rotation meter.

After the motor T leads have been labeled A, B, or C to correspond with the leads of the phase rotation meter, the rotation of the line supplying power to the motor must be determined. Set the selector switch on the phase rotation meter to the LINE position. After making certain the power has been turned off, connect the three LINE leads of the phase rotation meter to the incoming

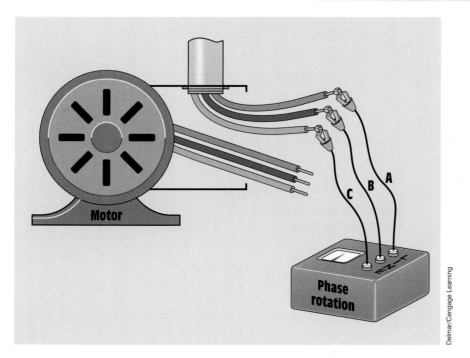

Delmar/Cengage Learning

FIGURE 33–14 Connecting the phase rotation meter to the line.

powerline *(Figure 33–14)*. Turn on the power and observe the zero-center voltmeter. If the meter is pointing in the CORRECT direction, turn off the power and label the line leads A, B, or C to correspond with the LINE leads of the phase rotation meter.

If the voltmeter is pointing in the INCORRECT direction, turn off the power and change any two of the leads from the phase rotation meter. When the power is turned on, the voltmeter should point in the CORRECT direction. Turn off the power and label the line leads A, B, or C to correspond with the leads from the phase rotation meter.

Now that the motor T leads and the incoming power leads have been labeled, connect the line lead labeled A to the T lead labeled A, the line lead labeled B to the T lead labeled B, and the line lead labeled C to the T lead labeled C. When power is connected to the motor, it will operate in the proper direction.

33–3 Connecting Dual-Voltage Three-Phase Motors

Many of the three-phase motors used in industry are designed to be operated on two voltages, such as 240 volts and 480 volts. Motors of this type, called **dual-voltage motors,** contain two sets of windings per phase. Most dual-voltage motors bring out nine T leads at the terminal box. A standard method

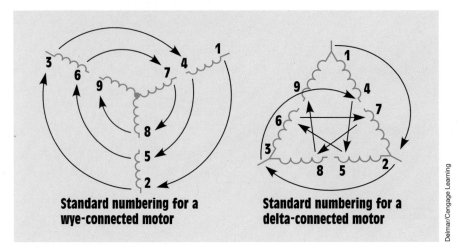

Standard numbering for a
wye-connected motor

Standard numbering for a
delta-connected motor

Delmar/Cengage Learning

FIGURE 33–15   Standard numbering for three-phase motors.

used to number these leads is shown in *Figure 33–15*. Starting with Terminal 1, the leads are numbered in a decreasing spiral as shown. Another method of determining the proper lead numbers is to add three to each terminal. For example, starting with Lead 1, add three to one. Three plus one equals four. The phase winding that begins with 1 ends with 4. Now add three to four. Three plus four equals seven. The beginning of the second winding for phase one is seven. This method will work for the windings of all phases. If in doubt, draw a diagram of the phase windings and number them in a spiral.

High-Voltage Connections

Three-phase motors can be constructed to operate in either wye or delta. If a motor is to be connected to high voltage, the phase windings are connected in series. In *Figure 33–16,* a schematic diagram and terminal connection chart for high voltage are shown for a wye-connected motor. In *Figure 33–17,* a schematic diagram and terminal connection chart for high voltage are shown for a delta-connected motor. Notice that in both cases the windings are connected in series.

Low-Voltage Connections

When a motor is to be connected for low-voltage operation, the phase windings must be connected in parallel. *Figure 33–18* shows the basic schematic diagram for a wye-connected motor with parallel phase windings. In actual practice, however, it is not possible to make this exact connection with a nine-lead motor. The schematic shows that Terminal 4 connects to the other end of the phase winding that starts with Terminal 7. Terminal 5 connects to the other

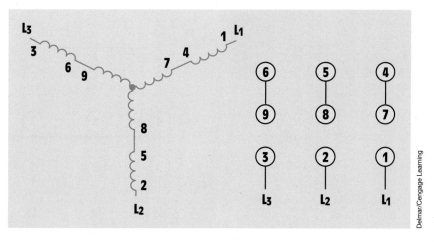

FIGURE 33–16 High-voltage wye connection.

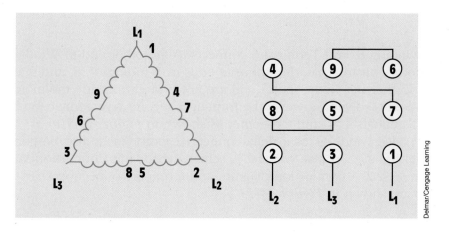

FIGURE 33–17 High-voltage delta connection.

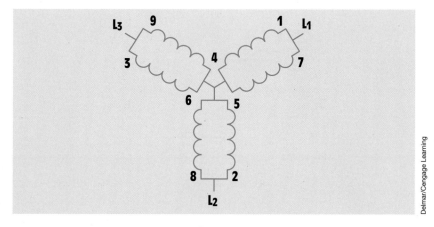

FIGURE 33–18 Stator windings connected in parallel.

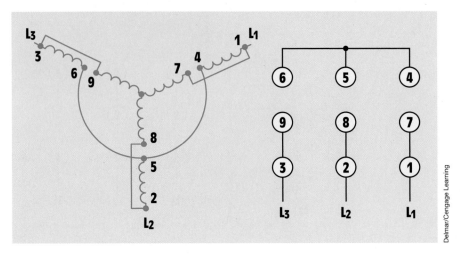

FIGURE 33–19 Low-voltage wye connection.

end of Winding 8, and Terminal 6 connects to the other end of Winding 9. In actual motor construction, the opposite ends of Windings 7, 8, and 9 are connected together inside the motor and are not brought outside the motor case. The problem is solved, however, by forming a second wye connection by connecting Terminals 4, 5, and 6 together as shown in *Figure 33–19*.

The phase windings of a delta-connected motor must also be connected in parallel for use on low voltage. A schematic for this connection is shown in *Figure 33–20*. A connection diagram and terminal connection chart for this hook-up are shown in *Figure 33–21*.

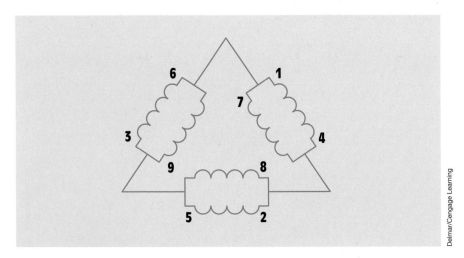

FIGURE 33–20 Parallel delta connection.

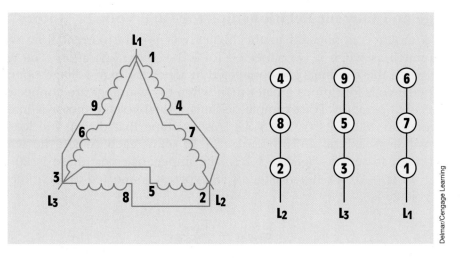

FIGURE 33–21   Low-voltage delta connection.

Some dual-voltage motors contain 12 T leads instead of 9. In this instance, the opposite ends of Terminals 7, 8, and 9 are brought out for connection. *Figure 33–22* shows the standard numbering for both delta- and wye-connected motors. Twelve leads are brought out if the motor is intended to be used for wye-delta starting. When this is the case, the motor must be designed for normal operation with its windings connected in delta. If the windings are connected in wye during starting, the starting current of the motor is reduced to one third of what it is if the motor is started as a delta.

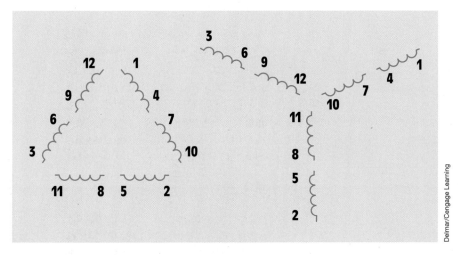

FIGURE 33–22   A 12-lead motor.

Voltage and Current Relationships for Dual-Voltage Motors

When a motor is connected to the higher voltage, the current flow will be half as much as when it is connected for low-voltage operation. The reason is that, when the windings are connected in series for high-voltage operation, the impedance is four times greater than when the windings are connected for low-voltage operation. For example, assume a dual-voltage motor is intended to operate on 480 volts or 240 volts. Also assume that during full load, the motor windings exhibit an impedance of 10 ohms each. When the winding is connected in series *(Figure 33–23),* the impedance per phase is 20 ohms (10 Ω + 10 Ω = 20 Ω). If a voltage of 480 volts is connected to the motor, the phase voltage is

$$E_{PHASE} = \frac{E_{LINE}}{1.732}$$

$$E_{PHASE} = \frac{480}{1.732}$$

$$E_{PHASE} = 277 \text{ V}$$

The amount of current flow through the phase can be calculated using Ohm's law:

$$I = \frac{E}{Z}$$

$$I = \frac{277 \text{ V}}{20 \ \Omega}$$

$$I = 13.85 \text{ A}$$

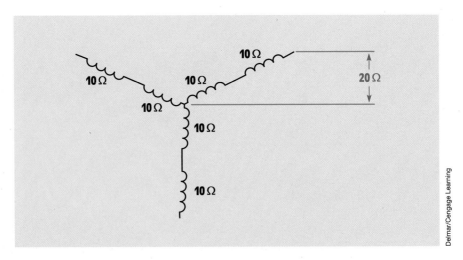

FIGURE 33–23   Impedance adds in series.

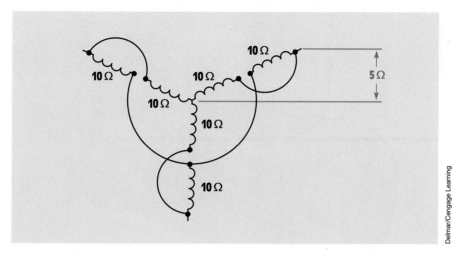

FIGURE 33–24 Impedance is less in parallel.

If the stator windings are connected in parallel, the total impedance is found by adding the reciprocals of the impedances of the windings *(Figure 33–24)*:

$$Z_T = \frac{1}{\frac{1}{Z_1} + \frac{1}{Z_2}}$$

$$Z_T = 5 \ \Omega$$

If a voltage of 240 volts is connected to the motor, the voltage applied across each phase is 138.6 volts (240 V/1.732 = 138.6 V). The amount of phase current can now be calculated using Ohm's law:

$$I = \frac{E}{Z}$$

$$I = \frac{138.6 \ V}{5 \ \Omega}$$

$$I = 27.7 \ A$$

33–4 Squirrel-Cage Induction Motors

The **squirrel-cage rotor** induction motor receives its name from the type of rotor used in the motor. A squirrel-cage rotor is made by connecting bars to two end rings. If the metal laminations were removed from the rotor, the result would look very similar to a squirrel cage *(Figure 33–25)*. A squirrel cage is a cylindrical device constructed of heavy wire. A shaft placed through the center of the cage

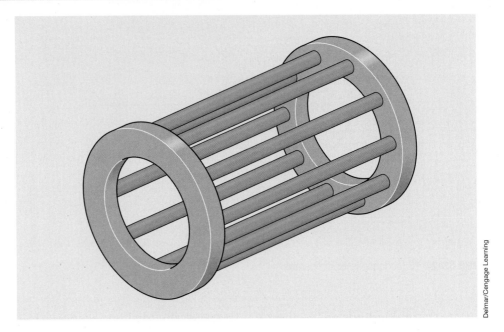

Delmar/Cengage Learning

FIGURE 33–25 Basic squirrel-cage rotor without laminations.

Courtesy of Electric Machinery Corp.

FIGURE 33–26 Squirrel-cage rotor.

permits the cage to spin around the shaft. A squirrel cage is placed inside the cage of small pets such as squirrels and hamsters to permit them to exercise by running inside of the squirrel cage. A squirrel-cage rotor is shown in *Figure 33–26*.

Principle of Operation

The squirrel-cage motor is an induction motor. That means that the current flow in the rotor is produced by induced voltage from the rotating magnetic field of the stator. In *Figure 33–27*, a squirrel-cage rotor is shown inside the stator of a three-phase motor. It will be assumed that the motor shown in *Figure 33–27* contains four poles per phase, which produces a rotating magnetic field with a synchronous speed of 1800 rpm when the stator is connected to a 60-hertz

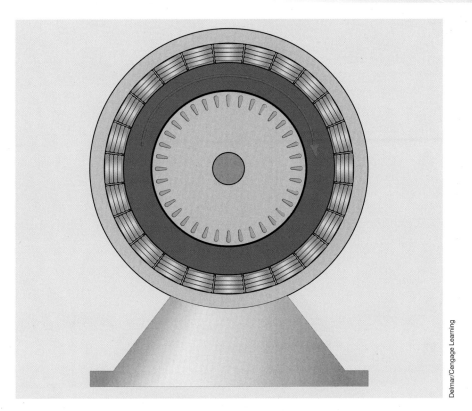

FIGURE 33–27   Voltage is induced into the rotor by the rotating magnetic field.

line. When power is first connected to the stator, the rotor is not turning. The magnetic field of the stator cuts the rotor bars at a rate of 1800 rpm. This cutting action induces a voltage into the rotor bars. This induced voltage will be the same frequency as the voltage applied to the stator. The amount of induced voltage is determined by three factors:

1. The strength of the magnetic field of the stator

2. The number of turns of wire cut by the magnetic field (in the case of a squirrel-cage rotor, this will be the number of bars in the rotor)

3. The speed of the cutting action

Because the rotor is stationary at this time, maximum voltage is induced into the rotor. The induced voltage causes current to flow through the rotor bars. As current flows through the rotor, a magnetic field is produced around each bar *(Figure 33–28)*.

The magnetic field of the rotor is attracted to the magnetic field of the stator, and the rotor begins to turn in the same direction as the rotating magnetic field.

As the speed of the rotor increases, the rotating magnetic field cuts the rotor bars at a slower rate. For example, assume the rotor has accelerated to

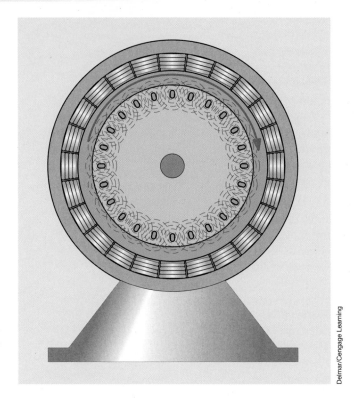

Delmar/Cengage Learning

FIGURE 33–28 A magnetic field is produced around each rotor bar.

a speed of 600 rpm. The synchronous speed of the rotating magnetic field is 1800 rpm. Therefore, the rotor bars are being cut at a rate of 1200 rpm (1800 rpm − 600 rpm = 1200 rpm). Because the rotor bars are being cut at a slower rate, less voltage is induced in the rotor, reducing rotor current. When the rotor current decreases, the stator current decreases also.

As the rotor continues to accelerate, the rotating magnetic field cuts the rotor bars at a decreasing rate. This reduces the amount of induced voltage and therefore the amount of rotor current. If the motor is operating without a load, the rotor continues to accelerate until it reaches a speed close to that of the rotating magnetic field.

Torque

The amount of torque produced by an AC induction motor is determined by three factors:

1. The strength of the magnetic field of the stator

2. The strength of the magnetic field of the rotor

3. The phase angle difference between rotor and stator fields

$$T = K_T \times \varphi_S \times I_R \times \cos \theta_R$$

where

$$T = \text{torque in lb-ft}$$
$$K_T = \text{torque constant}$$
$$\varphi_S = \text{stator flux (constant at all speeds)}$$
$$I_R = \text{rotor current}$$
$$\cos \theta_R = \text{rotor power factor}$$

Notice that one of the factors that determines the amount of torque produced by an induction motor is the strength of the magnetic field of the rotor. ***An induction motor can never reach synchronous speed.*** If the rotor were to turn at the same speed as the rotating magnetic field, there would be no induced voltage in the rotor and consequently no rotor current. Without rotor current, there could be no magnetic field developed by the rotor and therefore no torque or turning force. A motor operating at no load will accelerate until the torque developed is proportional to the windage and bearing friction losses.

If a load is connected to the motor, it must furnish more torque to operate the load. This causes the motor to slow down. When the motor speed decreases, the rotating magnetic field cuts the rotor bars at a faster rate. This causes more voltage to be induced in the rotor and therefore more current. The increased current flow produces a stronger magnetic field in the rotor, which causes more torque to be produced. The increased current flow in the rotor causes increased current flow in the stator. This is why motor current increases as load is added.

Another factor that determines the amount of torque developed by an induction motor is the phase angle difference between stator and rotor field flux. ***Maximum torque is developed when the stator and rotor flux are in phase with each other.*** Note in the preceding formula that one of the factors that determines the torque developed by an induction motor is the cosine of the rotor power factor. The cosine function reaches its maximum value of 1 when the phase angle is 0 ($\cos 0° = 1$).

Starting Characteristics

When a squirrel-cage motor is first started, it has a current draw several times greater than its normal running current. The actual amount of starting current is determined by the type of rotor bars, the horsepower rating of the motor, and the applied voltage. The type of rotor bars is indicated by the code letter found on the nameplate of a squirrel-cage motor. *Table 430.7(B)* of the *NEC* can be used to calculate the locked rotor current (starting current) of a squirrel-cage motor when the applied voltage, horsepower, and code letter are known.

■ EXAMPLE 33-3

An 800-hp, three-phase squirrel-cage motor is connected to 2300 V. The motor has a code letter of J. What is the starting current of this motor?

Solution

Table 430.7(B) of the *NEC* gives a value of 7.1 to 7.99 kVAs per hp as the locked-rotor current of a motor with a code letter J *(Figure 33–29)*. An average value of 7.5 is used for this calculation. The apparent power can be calculated by multiplying the 7.5 times the hp rating of the motor:

$$kVA = 7.5 \text{ kVA/hp} \times 800 \text{ hp}$$
$$kVA = 6000$$

The line current supplying the motor can now be calculated using the formula

$$I_{(LINE)} = \frac{VA}{E_{(LINE)} \times 1.732}$$

$$I_{(LINE)} = \frac{6,000,000 \text{ VA}}{2300 \text{ V} \times 1.732}$$

$$I_{(LINE)} = 1506.175 \text{ A}$$

This large starting current is caused by the fact that the rotor is not turning when power is first applied to the stator. Because the rotor is not turning, the squirrel-cage bars are cut by the rotating magnetic field at a fast rate. Remember that one of the factors that determines the amount of induced voltage is speed of the cutting action. This high induced voltage causes a large amount of current to flow in the rotor. The large current flow in the rotor causes a large amount of current flow in the stator. Because a large amount of current flows in both the stator and rotor, a strong magnetic field is established in both.

It would first appear that the starting torque of a squirrel-cage motor is high because the magnetic fields of both the stator and rotor are strong at this point. Recall that the third factor for determining the torque developed by an induction motor is the difference in phase angle between stator flux and rotor flux. Because the rotor is being cut at a high rate of speed by the rotating stator field, the bars in the squirrel-cage rotor appear to be very inductive at this point because of the high frequency of the induced voltage. This causes the phase angle difference between the induced voltage in the rotor and rotor current to

Table 430.7(B) Locked-Rotor Indicating Code Letters

Code Letter	Kilovolt-Amperes per Horsepower with Locked Rotor
A	0 – 3.14
B	3.15 – 3.54
C	3.55 – 3.99
D	4.0 – 4.49
E	4.5 – 4.99
F	5.0 – 5.59
G	5.6 – 6.29
H	6.3 – 7.09
J	7.1 – 7.99
K	8.0 – 8.99
L	9.0 – 9.99
M	10.0 – 11.19
N	11.2 – 12.49
P	12.5 – 13.99
R	14.0 – 15.99
S	16.0 – 17.99
T	18.0 – 19.99
U	20.0 – 22.39
V	22.4 and up

FIGURE 33–29 *NEC Table 430.7(B). (Reprinted with permission from NFPA 70–2011)*

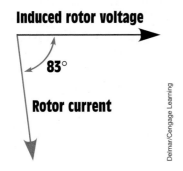

FIGURE 33–30 Rotor current is almost 90° out of phase with the induced voltage at the moment of starting.

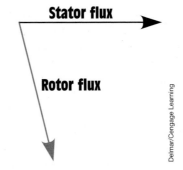

FIGURE 33–31 Rotor flux lags the stator flux by a large amount during starting.

be almost 90° out of phase with each other, producing a lagging power factor for the rotor *(Figure 33–30)*. This causes the rotor flux to lag the stator flux by a large amount; consequently, a relatively weak starting torque, per A of starting current when compared with other types of three-phase motors, is developed *(Figure 33–31)*.

Percent Slip

The speed performance of an induction motor is measured in **percent slip.** The percent slip can be determined by subtracting the synchronous speed from the speed of the rotor. For example, assume an induction motor has a synchronous speed of 1800 rpm and at full load the rotor turns at a speed of 1725 rpm. The difference between the two speeds is 75 rpm (1800 rpm − 1725 rpm = 75 rpm). The percent slip can be determined using the formula

$$\text{Percent slip} = \frac{\text{synchronous speed} - \text{rotor speed}}{\text{synchronous speed}} \times 100$$

$$\text{Percent slip} = \frac{75 \text{ rpm}}{1800 \text{ rpm}} \times 100$$

$$\text{Percent slip} = 4.16\%$$

A rotor slip of 2% to 5% is common for most squirrel-cage induction motors. The amount of slip for a particular motor is greatly affected by the type of rotor bars used in the construction of the rotor. Squirrel-cage motors are considered to be constant-speed motors because there is a small difference between noload speed and full-load speed.

Rotor Frequency

In the previous example, the rotor slips behind the rotating magnetic field by 75 rpm. This means that at full load, the bars of the rotor are being cut by magnetic lines of flux at a rate of 75 rpm. Therefore, the voltage being induced in the rotor at this point in time is at a much lower frequency than when the motor was started. The **rotor frequency** can be determined using the formula

$$f = \frac{P \times S_R}{120}$$

where

$$f = \text{frequency in Hz}$$
$$P = \text{number of stator poles}$$
$$S_R = \text{rotor slip in rpm}$$
$$f = \frac{4 \times 75 \text{ rpm}}{120}$$
$$f = 2.5 \text{ Hz}$$

Because the frequency of the current in the rotor decreases as the rotor approaches synchronous speed, the rotor bars become less inductive. The current flow through the rotor becomes limited more by the resistance of the bars and less by inductive reactance. The current flow in the rotor becomes more

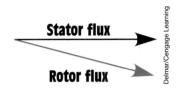

FIGURE 33–32 Rotor and stator flux become more in phase with each other as motor speed increases.

in phase with the induced voltage, which causes less phase angle shift between stator and rotor flux *(Figure 33–32)*. This is the reason that squirrel-cage motors generally have a relatively poor starting torque per ampere of starting current when compared with other types of three-phase motors but a good running torque. Although the starting torque per ampere of starting current is lower than other types of three-phase motors, the starting torque can be high because of the large amount of inrush current.

Reduced Voltage Starting

Because many squirrel-cage motors require a large amount of starting current, it is sometimes necessary to reduce the voltage during the starting period. When the voltage is reduced, the starting torque is reduced also. If the applied voltage is reduced to 50% of its normal value, the magnetic fields of both the stator and rotor are reduced to 50% of normal. The 50% reduction of the magnetic fields causes the starting torque to be reduced to 25% of normal. A chart showing a typical torque curve for a squirrel-cage motor is shown in *Figure 33–33*.

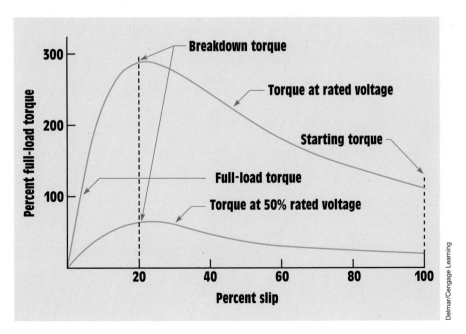

FIGURE 33–33 Typical torque curves for a squirrel-cage motor.

The torque formula given earlier can be used to show why this large reduction of torque occurs. Both the stator flux, φ_S, and the rotor current, I_R, are reduced to half their normal value. The product of these two values, torque, is reduced to one fourth. The torque varies as the square of the applied voltage for any given value of slip.

Code Letters

Squirrel-cage rotors are not all the same. Rotors are made with different types of bars. The type of rotor bars used in the construction of the rotor determines the operating characteristics of the motor. AC squirrel-cage motors are given a **code letter** on their nameplate. The code letter indicates the type of bars used in the rotor. *Figure 33–34* shows a rotor with type A bars. A type A rotor has the highest resistance of any squirrel-cage rotor. This means that the starting torque per ampere of starting current will be high because the rotor current is closer to being in phase with the induced voltage than on any other type of rotor. Also, the high resistance of the rotor bars limits the amount of current flow in the rotor when starting. This produces a low starting current for the motor. A rotor with type A bars has very poor running characteristics, however. Because the bars are resistive, a large amount of voltage will have to be induced into the rotor to produce an increase in rotor current and therefore an increase in the rotor magnetic field. This means that when load is added to the motor, the rotor must slow down a great amount to produce enough current in the rotor to increase the torque. Motors with type A rotors have the highest percent slip of any squirrel-cage motor. Motors with type A rotors are generally used in applications where starting is a problem, such as a motor that must accelerate a large flywheel from 0 rpm to its full speed. Flywheels can have a

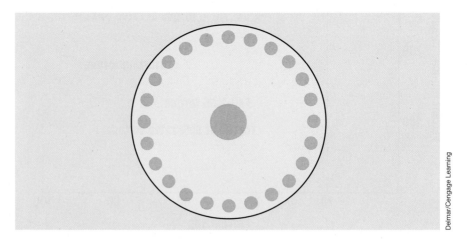

FIGURE 33–34   Type A rotor.

Delmar/Cengage Learning

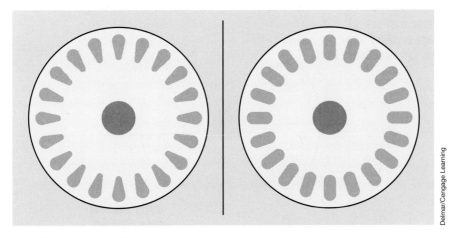

FIGURE 33–35   Type B–E rotor.   **FIGURE 33–36**   Type F–V rotor.

very large amount of inertia, which may require several minutes to accelerate them to their running speed when they are started.

Figure 33–35 shows a rotor with bars similar to those found in rotors with code letters B through E. These rotor bars have lower resistance than the type A rotor. Rotors of this type have fair starting torque, low starting current, and fair speed regulation.

Figure 33–36 shows a rotor with bars similar to those found in rotors with code letters F through V. This rotor has low starting torque per ampere of starting current, high starting current, and good running torque. Motors containing rotors of this type generally have very good speed regulation and low percent slip.

The Double-Squirrel-Cage Rotor

Some motors use a rotor that contains two sets of squirrel-cage windings *(Figure 33–37)*. The outer winding consists of bars with a relatively high resistance located close to the top of the iron core. Because these bars are located close to the surface, they have a relatively low reactance. The inner winding consists of bars with a large cross-sectional area, which gives them a low resistance. The inner winding is placed deeper in the core material, which causes it to have a much higher reactance.

When the double-squirrel-cage motor is started, the rotor frequency is high. Because the inner winding is inductive, its impedance will be high compared with the resistance of the outer winding. During this period of time, most of the rotor current flows through the outer winding. The resistance of the outer winding limits the current flow through the rotor, which limits the starting current to a relatively low value. Because the current is close to being in phase

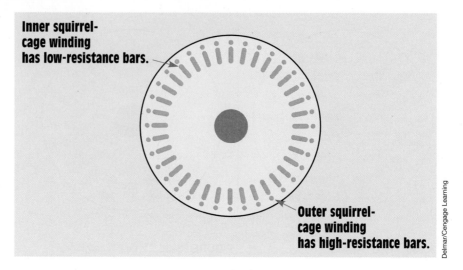

Inner squirrel-cage winding has low-resistance bars.

Outer squirrel-cage winding has high-resistance bars.

Delmar/Cengage Learning

FIGURE 33–37 Double-squirrel-cage rotor.

with the induced voltage, the rotor flux and stator flux are close to being in phase with each other and a strong starting torque is developed. The starting torque of a double-squirrel-cage motor can be as high as 250% of rated full-load torque.

When the rotor reaches its full-load speed, rotor frequency decreases to 2 or 3 hertz. The inductive reactance of the inner winding has now decreased to a low value. Most of the rotor current now flows through the low-resistance inner winding. This type of motor has good running torque and excellent speed regulation.

Power Factor of a Squirrel-Cage Induction Motor

At no load, most of the current is used to magnetize the stator and rotor. Because most of the current is magnetizing current, it is inductive and lags the applied voltage by close to 90°. A very small resistive component is present, caused mostly by the resistance of the wire in the stator and the power needed to overcome bearing friction and windage loss. At no load, the motor appears to be a resistive-inductive series circuit with a large inductive component as compared with resistance *(Figure 33–38)*. A power factor of about 10% is common for a squirrel-cage motor at no load.

As load is added, electric energy is converted into mechanical energy and the in-phase component of current increases. The circuit now appears to contain more resistance than inductance *(Figure 33–39)*. This causes the phase angle between applied voltage and motor current to decrease, causing the power factor to increase. In practice, the power factor of an induction motor at full load is from about 85% to 90% lagging.

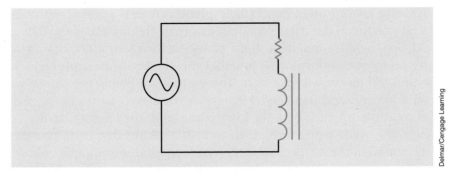

FIGURE 33–38   At no load, the motor appears to have a large amount of inductance and a very small resistance.

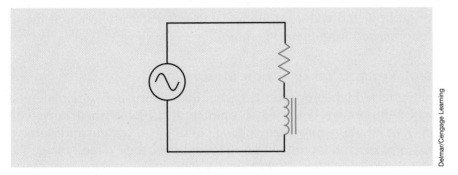

FIGURE 33–39   At full load, the resistive component of the circuit appears to be greater than the inductive component.

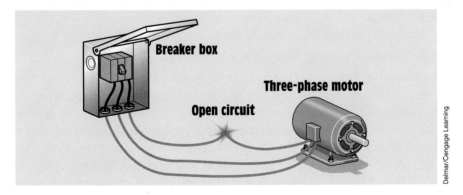

FIGURE 33–40   Single-phasing occurs when one line of a three-phase system is open.

Single-Phasing

Three lines supply power to a three-phase motor. If one of these lines should open, the motor will be connected to single-phase power *(Figure 33–40)*. This condition is known as **single-phasing.**

If the motor is not running and single-phase power is applied to the motor, the induced voltage in the rotor sets up a magnetic field in the rotor. This magnetic field opposes the magnetic field of the stator (Lenz's law). As a result, practically no torque is developed in either the clockwise or counterclockwise direction and the motor will not start. The current supplying the motor will be excessive, however, and damage to the stator windings can occur.

If the motor is operating under load at the time the single-phasing condition occurs, the rotor will continue to turn at a reduced speed. The moving bars of the rotor cut the stator field flux, which continues to induce voltage and current in the bars. Due to reduced speed, the rotor has high-reactive and low-resistive components, causing the rotor current to lag the induced voltage by almost 90°. This lagging current creates rotor fields midway between the stator poles, resulting in greatly reduced torque. The reduction in rotor speed causes high current flow and will most likely damage the stator winding if the motor is not disconnected from the powerline.

Effects of Voltage Variation on Motors

Motors are affected when operated at other than their rated nameplate voltage. NEMA rated motors are designed to operate at ±10% of their rated voltage. *Table 33–1* shows the approximate change in full-load current and starting current for typical electric motors when operated over their rated voltage (110%) and under their rated voltage (90%).

Motors are intended to operate on systems with balanced voltage (the voltage is the same between all phases). Unbalanced voltage is one of the leading causes of motor failure. Unbalanced voltage is generally caused when single-phase loads are supplied by three-phase systems.

Determining the Amount of Voltage Unbalance

The values listed in *Table 33–1* assume that the voltages across the phase conductors as measured between phases AB, BC, and AC are balanced. In other words, the table indicates the effect on motor current when voltage is greater or less than the motor nameplate rating in a balanced system. Greater harm is caused when the voltages are unbalanced. The greatest example of voltage unbalance occurs when one phase of a three-phase system is lost and the motor begins single-phasing. This causes a 173% increase of current in two of the motor windings.

Voltage Variation	Full-Load Current	Starting Current
110%	7% increase	10–12% increase
90%	11% increase	10–12% decrease

TABLE 33–1   Effects of voltage variation

If the normal full-load current of a motor is 20 amperes, the two windings still connected to power will have a current of 34.6 amperes and one winding that has lost power will have a current of 0 ampere. NEMA recommends that the unbalanced voltage not exceed ±1%. The following steps illustrate how to determine the percent of voltage unbalance in a three-phase system:

1. Take voltage measurements between all phases. In this example assume the voltage between AB = 496 volts, BC = 460 volts, and AC = 472 volts.

2. Find the average voltage.

$$
\begin{array}{r}
496 \\
460 \\
\underline{472} \\
1428
\end{array}
\quad 1428 / 3 = 476 \text{ volts}
$$

3. Subtract the average voltage from the voltage reading that results in the greatest difference:

$$496 - 476 = 20 \text{ volt}$$

4. Determine the percent difference:

$$\frac{100 \times \text{Greatest voltage difference}}{\text{Average voltage}}$$

$$\frac{100 \times 20}{476} = 4.2\% \text{ voltage unbalance}$$

Heat Rise

The percent of heat rise in the motor caused by the voltage unbalance is equal to twice the percent squared [2 × (percent voltage unbalance)^2]:

$$2 \times 4.2 \times 4.2 = 35.28\% \text{ temperature increase in the winding with the highest current.}$$

The Nameplate

Electric motors have nameplates that give a great deal of information about the motor. *Figure 33–41* illustrates the nameplate of a three-phase squirrel-cage induction motor. The nameplate shows that the motor is 10 horsepower, is a three-phase motor, and operates on 240 or 480 volts. The full-load running current of the motor is 28 amperes when operated on 240 volts and 14 amperes when operated on 480 volts. The motor is designed to be operated on a 60-hertz AC voltage and has a full-load speed of 1745 rpm. This speed indicates that the motor has four poles per phase. Because the full-load speed is 1745 rpm, the

Manufacturer	
HP 10	**Phase** 3
Volts 240/480	**Amps** 28/14
Hz 60	**FL Speed** 1745 RPM
Code J	**SF** 1.25
Temp 40 °C	**NEMA Code** B
FRAME XXXX	**MODEL NO.** XXXX

FIGURE 33–41 Motor nameplate.

synchronous speed would be 1800 rpm. The motor contains a type J squirrel-cage rotor and has a service factor of 1.25. The code letter indicating the type of rotor bars should not be confused with the NEMA code letter. In this example, the code letter used to determine locked-rotor current is J. The NEMA code letter is B. The *NEC* requires the NEMA code to be placed on the nameplate of squirrel-cage motors. It is used to determine fuse size when installing the motor. The service factor is used to determine the amperage rating of the overload protection for the motor. Some motors indicate a marked temperature rise in Celsius degrees instead of a service factor. The frame number indicates the type of mounting the motor has. *Figure 33–42* shows the schematic symbol used to represent a three-phase squirrel-cage motor.

Consequent-Pole Squirrel-Cage Motors

Consequent-pole squirrel-cage motors permit the synchronous speed to be changed by changing the number of stator poles. If the number of poles is

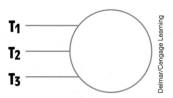

FIGURE 33–42 Schematic symbol of a three-phase squirrel-cage induction motor.

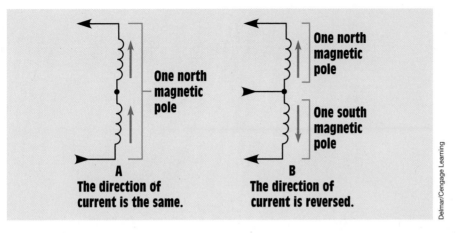

One north
magnetic
pole

One north
magnetic
pole

One south
magnetic
pole

A
The direction of
current is the same.

B
The direction of
current is reversed.

Delmar/Cengage Learning

FIGURE 33–43 The number of poles can be changed by reversing the current flow through alternate poles.

doubled, the synchronous speed is reduced by one-half. A two-pole motor has a synchronous speed of 3600 rpm when operated at 60 hertz. If the number of poles is doubled to four, the synchronous speed becomes 1800 rpm. The number of stator poles can be changed by changing the direction of current flow through alternate pairs of poles.

Figure 33–43 illustrates this concept. In *Figure 33–43A,* two coils are connected in such a manner that current flows through them in the same direction. Both poles produce the same magnetic polarity and are essentially one pole. In *Figure 33–43B,* the coils have been reconnected in such a manner that current flows through them in opposite directions. The coils now produce the opposite magnetic polarities and are essentially two different poles.

Consequent-pole motors with one stator winding bring out six leads labeled T_1 through T_6. Depending on the application, the windings will be connected as a series delta or a parallel wye. If it is intended that the motor maintain the same horsepower rating for both high and low speed, the high-speed connection will be a series delta *(Figure 33–44).* The low-speed connection will be a parallel wye *(Figure 33–45).*

If it is intended that the motor maintain constant torque for both low and high speeds, the series-delta connection will provide low speed, and the parallel wye will provide high speed.

Because the speed range of a consequent-pole motor is limited to a 1:2 ratio, motors intended to operate at more than two speeds contain more than one stator winding. A consequent-pole motor with three speeds, for example, has one stator winding for one speed only and a second winding with taps. The tapped winding may provide synchronous speeds of 1800 and 900 rpm, and the separate winding may provide a speed of 1200 rpm. Consequent-pole motors with four speeds contain two separate stator windings with taps. If the

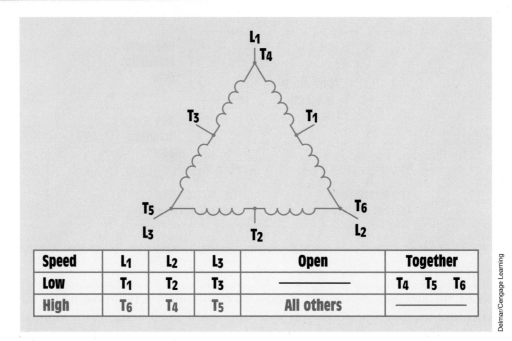

Speed	L₁	L₂	L₃	Open	Together
Low	T₁	T₂	T₃	—————	T₄ T₅ T₆
High	T₆	T₄	T₅	All others	—————

FIGURE 33–44 High-speed series-delta connection.

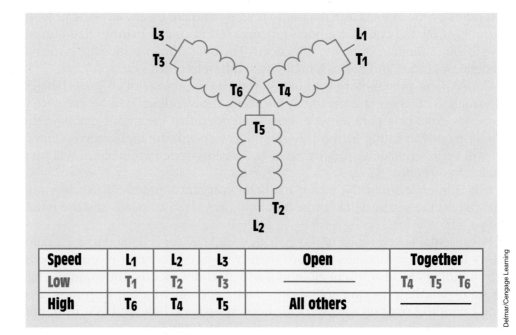

Speed	L₁	L₂	L₃	Open	Together
Low	T₁	T₂	T₃	—————	T₄ T₅ T₆
High	T₆	T₄	T₅	All others	—————

FIGURE 33–45 Low-speed parallel-wye connection.

second stator winding of the motor in this example were to be tapped, the motor would provide synchronous speeds of 1800, 1200, 900, and 600 rpm.

Motor Calculations

In the following example, output horsepower and motor efficiency are calculated. It is assumed that a 1/2-horsepower squirrel-cage motor is connected to a load. A wattmeter is connected to the motor, and the load torque measurement is calibrated in pound-inches. The motor is operating at a speed of 1725 rpm and producing a torque of 16 lb-in. The wattmeter is indicating an input power of 500 watts.

The actual amount of horsepower being produced by the motor can be calculated by using the formula

$$hp = \frac{6.28 \times rpm \times L \times P}{33,000}$$

where

$$hp = horsepower$$
$$6.28 = a\ constant$$
$$rpm = speed\ in\ revolutions\ per\ minute$$
$$L = distance\ in\ feet$$
$$P = pounds$$
$$33,000 = a\ constant$$

Because the formula uses feet for the distance and the torque of the motor is rated in pound-inches, L will be changed to 1/12 of a foot, or 1 inch. To simplify the calculation, the fraction 1/12 will be changed into its decimal equivalent (0.08333). To calculate the output horsepower, substitute the known values in the formula

$$hp = \frac{6.28 \times 1725\ rpm \times 0.08333\ ft/in \times 16\ lb\text{-}in.}{33,000}$$

$$hp = \frac{14,444\ ft\text{-}lb/min}{33,000}$$

$$hp = 0.438$$

One horsepower is equal to 746 watts. The output power of the motor can be calculated by multiplying the output horsepower by 746:

$$Power\ out = 746\ W/hp \times 0.438\ hp$$
$$Power\ out = 326.5\ W$$

The efficiency of the motor can be calculated by using the formula:

$$Eff. = \frac{power\ out}{power\ in} \times 100$$

$$Eff. = \frac{326.5\ W}{500\ W} \times 100$$

$$Eff. = 0.653 \times 100$$

$$Eff. = 65.3\%$$

33–5   Wound-Rotor Induction Motors

The **wound-rotor motor** induction motor is very popular in industry because of its high starting torque and low starting current. The stator winding of the wound-rotor motor is the same as the squirrel-cage motor. The difference between the two motors lies in the construction of the rotor. Recall that the squirrel-cage rotor is constructed of bars connected together at each end by a shorting ring as shown in *Figure 33–25*.

The rotor of a wound-rotor motor is constructed by winding three separate coils on the rotor 120° apart. The rotor will contain as many poles per phase as the stator winding. These coils are then connected to three sliprings located on the rotor shaft *(Figure 33–46)*. Brushes, connected to the sliprings, provide external connection to the rotor. This permits the rotor circuit to be connected to a set of resistors *(Figure 33–47)*.

The stator terminal connections are generally labeled T_1, T_2, and T_3. The rotor connections are commonly labeled M_1, M_2, and M_3. The M_2 lead is generally connected to the middle slip ring, and the M_3 lead is connected close to the rotor windings. The direction of rotation for the wound-rotor motor is reversed

Delmar/Cengage Learning

FIGURE 33–46   Rotor of a wound-rotor induction motor.

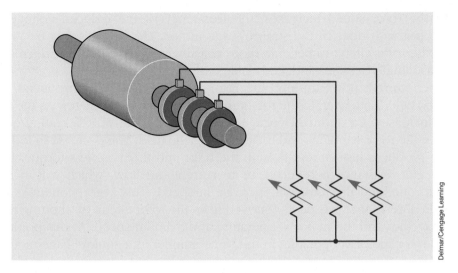

FIGURE 33–47 The rotor of a wound-rotor motor is connected to external resistors.

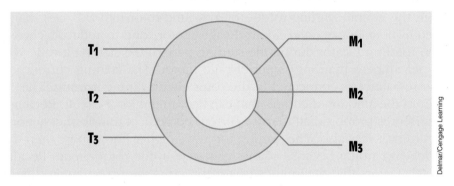

FIGURE 33–48 Schematic symbol for a wound-rotor induction motor.

by changing any two stator leads. Changing the M leads will have no effect on the direction of rotation. The schematic symbol for a wound-rotor motor is shown in *Figure 33–48*.

Principles of Operation

When power is applied to the stator winding, a rotating magnetic field is created in the motor. This magnetic field cuts through the windings of the rotor and induces a voltage into them. The amount of current flow in the rotor is determined by the amount of induced voltage and the total impedance of the rotor circuit ($I = E/Z$). The rotor impedance is a combination of inductive reactance created in the rotor windings and the external resistance. The impedance could be calculated using the formula for resistance and inductive reactance connected in series:

$$Z = \sqrt{R^2 + X_L^2}$$

As the rotor speed increases, the frequency of the induced voltage decreases just as it does in the squirrel-cage motor. The reduction in frequency causes the rotor circuit to become more resistive and less inductive, decreasing the phase angle between induced voltage and rotor current.

When current flows through the rotor, a magnetic field is produced. This magnetic field is attracted to the rotating magnetic field of the stator. As the rotor speed increases, the induced voltage decreases because of less cutting action between the rotor windings and rotating magnetic field. The decrease in induced voltage produces less current flow in the rotor and therefore less torque. If the rotor circuit resistance is reduced, more current can flow, which will increase motor torque, and the rotor will increase in speed. This action continues until all external resistance has been removed from the rotor circuit by shorting the M leads together and the motor is operating at maximum speed. At this point, the wound-rotor motor is operating in the same manner as a squirrel-cage motor.

Starting Characteristics of a Wound-Rotor Motor

Although the overall starting torque of a wound-rotor motor is less than that of an equivalent horsepower squirrel-cage motor, due to reduced current in both the rotor and stator during the starting period, the starting torque will be higher per ampere than in a squirrel-cage motor. The starting current is less because resistance is connected in the rotor circuit during starting. This resistance limits the amount of current that can flow in the rotor circuit. Because the stator current is proportional to rotor current because of transformer action, the stator current is less also. The starting torque is higher per ampere than that of a squirrel-cage motor because of the resistance in the rotor circuit. Recall that one of the factors that determines motor torque is the phase angle difference between stator flux and rotor flux. Because resistance is connected in the rotor circuit, stator and rotor flux are close to being in phase with each other producing a high starting torque for the wound-rotor induction motor. The wound-rotor motor generally exhibits a higher starting torque than an equivalent size squirrel-cage motor connected to a reduced-voltage starter.

If an attempt is made to start the motor with no circuit connected to the rotor, the motor cannot start. If no resistance is connected to the rotor circuit, there can be no current flow and consequently no magnetic field can be developed in the rotor.

Speed Control

The speed of a wound-rotor motor can be controlled by permitting resistance to remain in the rotor circuit during operation. When this is done, the rotor and stator current is limited, which reduces the strength of both magnetic fields. The reduced magnetic field strength permits the rotor to slip behind the rotating magnetic field of the stator. The resistors of speed controllers must have higher power ratings than the resistors of starters because they operate for extended periods of time.

The operating characteristics of a wound-rotor motor with the sliprings shorted are almost identical to those of a squirrel-cage motor. The percent slip, power factor, and efficiency are very similar for motors of equal horsepower rating.

33–6 Synchronous Motors

The three-phase synchronous motor has several characteristics that separate it from the other types of three-phase motors. Some of these characteristics follow.

1. The synchronous motor is not an induction motor. It does not depend on induced current in the rotor to produce a torque.

2. It will operate at a constant speed from full load to no load.

3. The synchronous motor must have DC excitation to operate.

4. It will operate at the speed of the rotating magnetic field (synchronous speed).

5. It has the ability to correct its own power factor and the power factor of other devices connected to the same line.

Rotor Construction

The synchronous motor has the same type of stator windings as the other two three-phase motors. The rotor of a synchronous motor has windings similar to the rotor of an alternator *(see Figure 32–6)*. Wound pole pieces become electromagnets when DC is applied to them. The excitation current can be applied to the rotor through two sliprings located on the rotor shaft or by a brushless exciter. The brushless exciter for a synchronous motor is the same as that used for the alternator discussed in Unit 32.

Starting a Synchronous Motor

The rotor of a synchronous motor also contains a set of squirrel-cage bars similar to those found in a type A rotor. This set of squirrel-cage bars is used to start the motor and is known as the **amortisseur winding** *(Figure 33–49)*. When power is first connected to the stator, the rotating magnetic field cuts through the squirrel-cage bars. The cutting action of the field induces a current into the squirrel-cage winding. The current flow through the amortisseur winding produces a rotor magnetic field that is attracted to the rotating magnetic field of the stator. This causes the rotor to begin turning in the direction of rotation of the stator field. When the rotor has accelerated to a speed that is close to the synchronous speed of the field, DC is connected to the rotor through the sliprings on the rotor shaft or by a brushless exciter *(Figure 33–50)*.

When DC is applied to the rotor, the windings of the rotor become electromagnets. The electromagnetic field of the rotor locks in step with the rotating magnetic field of the stator. The rotor will now turn at the same speed as the

Courtesy of Electric Machinery Corp.

FIGURE 33–49 Synchronous-motor rotor with amortisseur winding.

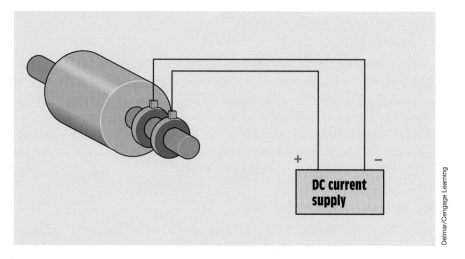

Delmar/Cengage Learning

FIGURE 33–50 DC excitation current supplied through sliprings.

rotating magnetic field. When the rotor turns at the synchronous speed of the field, there is no more cutting action between the stator field and the amortisseur winding. This causes the current flow in the amortisseur winding to cease.

Notice that the synchronous motor starts as a squirrel-cage induction motor. Because the rotor uses bars that are similar to those used in a type A rotor, they have a relatively high resistance, which gives the motor good starting torque and low starting current. ***A synchronous motor must never be started with DC connected to the rotor.*** If DC is applied to the rotor, the field poles of the rotor become electromagnets. When the stator is energized, the rotating magnetic field begins turning at synchronous speed. The electromagnets are alternately attracted and repelled by the stator field. As a result, the rotor does not turn. The rotor and power supply can be damaged by high induced voltages, however.

The Field-Discharge Resistor

When the stator winding is first energized, the rotating magnetic field cuts through the rotor winding at a fast rate of speed. This causes a large amount of voltage to be induced into the winding of the rotor. To prevent this from becoming excessive, a resistor is connected across the winding. This resistor is known as the *field-discharge resistor (Figure 33–51)*. It also helps to reduce the voltage induced into the rotor by the collapsing magnetic field when the DC is disconnected from the rotor. The field-discharge resistor is connected in parallel with the rotor winding during starting. If the motor is manually started, a field-discharge switch is used to connect the excitation current to the rotor. If the motor is automatically started, a special type of relay is used to connect excitation current to the rotor and disconnect the field-discharge resistor.

Constant-Speed Operation

Although the synchronous motor starts as an induction motor, it does not operate as one. After the amortisseur winding has been used to accelerate the rotor to about 95% of the speed of the rotating magnetic field, DC is connected to the rotor and the electromagnets lock in step with the rotating field. Notice that the synchronous motor does not depend on induced voltage from the stator field to produce a magnetic field in the rotor. The magnetic field of the rotor is produced by external DC applied to the rotor. This is the reason that the synchronous motor has the ability to operate at the speed of the rotating magnetic field.

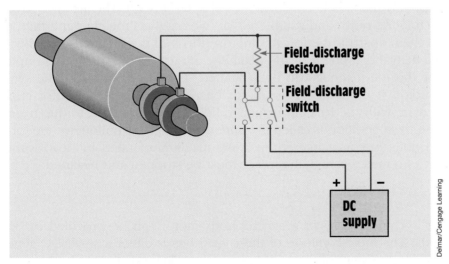

FIGURE 33–51   The field-discharge resistor is connected in parallel with the rotor winding during starting.

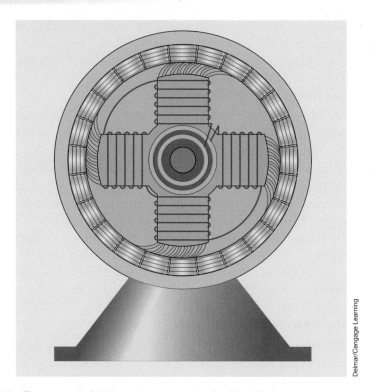

Delmar/Cengage Learning

FIGURE 33–52   The magnetic field becomes stressed as load is added.

As load is added to the motor, the magnetic field of the rotor remains locked with the rotating magnetic field of the stator and the rotor continues to turn at the same speed. The added load, however, causes the magnetic fields of the rotor and stator to become stressed *(Figure 33–52)*. The action is similar to connecting the north and south ends of two magnets together and then trying to pull them apart. If the force being used to pull the magnets apart becomes greater than the strength of the magnetic attraction, the magnetic coupling is broken and the magnets can be separated. The same is true for the synchronous motor. If the load on the motor becomes too great, the rotor is pulled out of sync with the rotating magnetic field. The amount of torque necessary to cause this condition is called the *pullout torque*. The pullout torque for most synchronous motors ranges from 150% to 200% of rated full-load torque. If pullout torque is reached, the motor must be stopped and restarted.

The Power Supply

The DC power supply of a synchronous motor can be provided by several methods. The most common of these methods is either a small DC generator mounted to the shaft of the motor or an electronic power supply that converts the AC line voltage into DC voltage.

Power Factor Correction

The synchronous motor has the ability to correct its own power factor and the power factor of other devices connected to the same line. The amount of power factor correction is controlled by the amount of excitation current in the rotor. If the rotor of a synchronous motor is underexcited, the motor has a lagging power factor like a common induction motor. As rotor excitation current is increased, the synchronous motor appears to be more capacitive. When the excitation current reaches a point that the power factor of the motor is at unity or 100%, it is at the normal excitation level. At this point, the current supplying the motor drops to its lowest value.

If the excitation current is increased above the normal level, the motor has a leading power factor and appears as a capacitive load. When the rotor is overexcited, the current supplying the motor increases due to the change in power factor. The power factor at this point, however, is leading and not lagging. Because capacitance has now been added to the line, it corrects the lagging power factor of other inductive devices connected to the same line. Changes in the amount of excitation current do not affect the speed of the motor.

Interaction of the DC and AC Fields

Figure 33–53 illustrates how the magnetic flux of the AC field aids or opposes the DC field. In this example, it is assumed that the DC field is held stationary and the rotating armature is connected to the AC source. Although most synchronous motors have a stationary AC field and a rotating DC field, the principle of operation is the same. When the excitation DC is less than the amount required for normal excitation, the AC must supply some portion of the magnetizing current to aid the weak DC *(Figure 33–53A)*. This portion of magnetizing current lags the applied voltage by 90°. The current waveform shown in *Figure 33–53A* depicts only the portion of magnetizing current that is out of phase with the voltage. The remaining part of the AC is used to produce the torque necessary to operate the load. The synchronous motor has a lagging power factor at this time.

In *Figure 33–53B,* the excitation DC has been increased to the normal excitation value. All the AC is now used to produce the torque necessary to operate the load. Because the AC no longer supplies any of the magnetizing current, it is in phase with the voltage and the motor power factor is at unity or 100%. The amount of AC supplied to the motor is at its lowest value during this period.

In *Figure 33–53C,* the excitation DC is greater than that needed for normal excitation. The AC now supplies a demagnetizing component of current. The portion of AC used to demagnetize the overexcited DC field will lead the applied voltage by 90°. The current waveform shown in *Figure 33–53C* illustrates only the portion of AC used to demagnetize the DC field and does not take into account the amount of AC used to produce torque for the load. The synchronous motor now has a leading power factor.

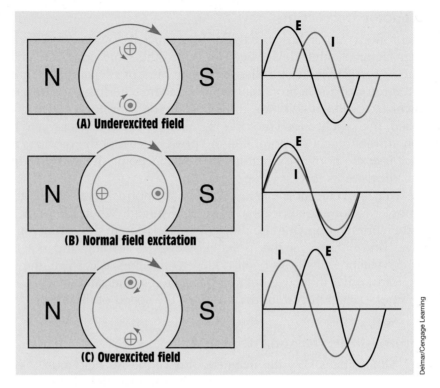

FIGURE 33–53 Field excitation in a synchronous motor.

Synchronous Motor Applications

Synchronous motors are very popular in industry, especially in the large horse-power ratings (motors up to 5000 hp are not uncommon). They have a low starting current per horsepower and a high starting torque. They operate at a constant speed from no load to full load and maintain maximum efficiency. Synchronous motors are used to operate DC generators, fans, blowers, pumps, and centrifuges. They correct their own power factor and can correct the power factor of other inductive loads connected to the same feeder *(Figure 33–54)*. Synchronous motors are sometimes operated at no load and are used for power factor correction only. When this is done, the motor is referred to as a **synchronous condenser.**

Advantages of the Synchronous Condenser

The advantage of using a synchronous condenser over a bank of capacitors for power factor correction is that the amount of correction is easily controlled. When a bank of capacitors is used for correcting power factor, capacitors must be added to or removed from the bank if a change in the amount of correction is needed. When a synchronous condenser is used, only the excitation current must be changed to cause an alteration of power factor. The schematic symbol for a synchronous motor is shown in *Figure 33–55.*

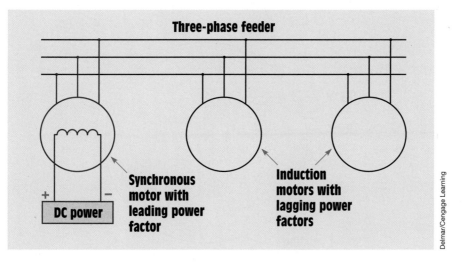

FIGURE 33–54   Synchronous motor used to correct the power factor of other motors.

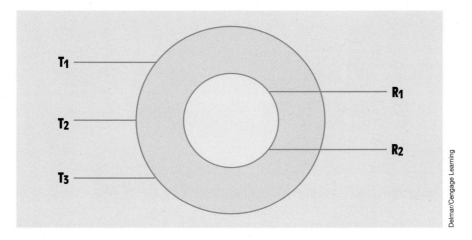

FIGURE 33–55   Schematic symbol for a synchronous motor.

33-7   Selsyn Motors

The word *selsyn* is a contraction derived from *self-synchronous*. **Selsyn motors** are used to provide position control and angular feedback information in industrial applications. Although selsyn motors are actually operated on single-phase AC, they do contain three-phase windings *(Figure 33–56)*. The schematic symbol for a selsyn motor is shown in *Figure 33–57*. This symbol is very similar to the symbol used to represent a three-phase synchronous motor. The stator windings are labeled S_1, S_2, and S_3. The rotor leads are labeled R_1 and R_2. The rotor leads are connected to the rotor winding by means of sliprings and brushes.

When selsyn motors are employed, at least two are used together. One motor is referred to as the *transmitter* and the other is called the *receiver*. It

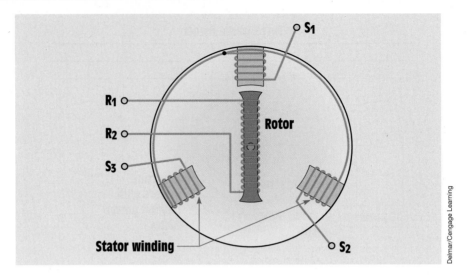

FIGURE 33–56 Selsyn motor.

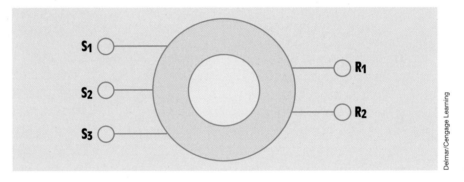

FIGURE 33–57 Schematic symbol for a selsyn motor.

makes no difference which motor acts as the transmitter and which acts as the receiver. Connection is made by connecting S_1 of the transmitter to S_1 of the receiver, S_2 of the transmitter to S_2 of the receiver, and S_3 of the transmitter to S_3 of the receiver. The rotor leads of each motor are connected to a source of single-phase AC *(Figure 33–58)*. If the stator-winding leads of the two selsyn motors are connected improperly, the receiver rotates in a direction opposite that of the transmitter. If the rotor leads are connected improperly, the rotor of the transmitter and the rotor of the receiver have an angle difference of 180°.

Selsyn Motor Operation

Selsyn motors actually operate as transformers. The rotor winding is the primary, and the stator winding is the secondary. In *Figure 33–58*, the rotor of the transmitter is in line with stator winding S_1. Because the rotor is connected to a source of AC, an alternating magnetic field exists in the rotor. This alternating

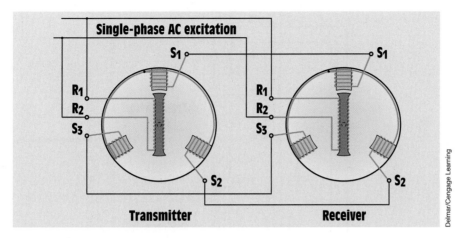

FIGURE 33–58 Connection of two selsyn motors.

magnetic field induces a voltage into the windings of the stator. Because the rotors of both motors are connected to the same source of AC, magnetic fields of identical strength and polarity exist in both motors.

Because the rotor of the transmitter is in line with stator winding S_1, maximum voltage and current are being induced in stator S_1 and less than maximum voltage and current are being induced in stator windings S_2 and S_3. Because the stator windings of the receiver are connected to the stator windings of the transmitter, the same current will flow through the receiver, producing a magnetic field in the receiver. This magnetic field attracts or repels the magnetic field of the rotor depending on the relative polarity of the two fields. When the rotor of the receiver is in the same position as the rotor of the transmitter, an equal amount of voltage is induced in the stator windings of the receiver, causing stator winding current to become zero.

If the rotor of the transmitter is turned to a different position, the magnetic field of the stator changes, resulting in a change of the magnetic field in the stator of the receiver. This causes the rotor of the receiver to rotate to a new position, where the two stator magnetic fields again cancel each other. Each time the rotor position of the transmitter is changed, the rotor of the receiver changes the same amount.

The Differential Selsyn

The **differential selsyn** is used to produce the algebraic sum of the rotation of two other selsyn units. Differential selsyns are constructed in a manner different from other selsyn motors. The differential selsyn contains three rotor windings connected in wye as well as three stator windings connected in wye. The rotor windings are brought out through three sliprings and brushes in a manner very similar to a wound-rotor induction motor. The differential selsyn

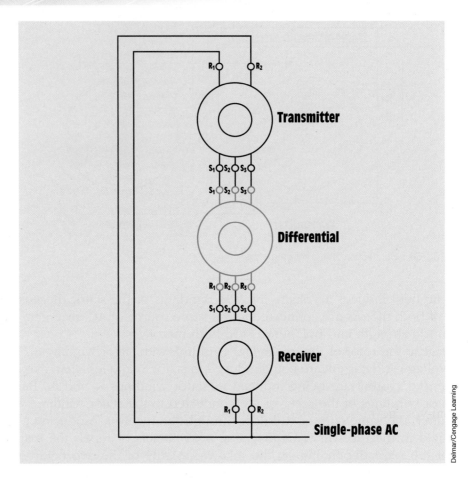

FIGURE 33–59   Differential selsyn connection.

is not connected to a source of power. Power must be provided by one of the other selsyn motors connected to it *(Figure 33–59)*.

If any one of the selsyn units is held in place and a second unit is turned, the third turns by the same amount. If any two of the selsyn units are turned at the same time, the third turns an amount equal to the sum of the angle of rotation of the other two.

Summary

- Three basic types of three-phase motors are

    a. squirrel-cage induction motor.

    b. wound-rotor induction motor.

    c. synchronous motor.

- All three-phase motors operate on the principle of a rotating magnetic field.

- Three factors that cause a magnetic field to rotate are

  a. the fact that the voltages of a three-phase system are 120° out of phase with each other.

  b. the fact that voltages change polarity at regular intervals.

  c. the arrangement of the stator windings.

- The speed of the rotating magnetic field is called the synchronous speed.

- Two factors that determine the synchronous speed are

  a. number of stator poles per phase.

  b. frequency of the applied voltage.

- The direction of rotation of any three-phase motor can be changed by reversing the connection of any two stator leads.

- The direction of rotation of a three-phase motor can be determined with a phase rotation meter before power is applied to the motor.

- Dual-voltage motors have 9 or 12 leads brought out at the terminal connection box.

- Dual-voltage motors intended for high-voltage connection have their phase windings connected in series.

- Dual-voltage motors intended for low-voltage connection have their phase windings connected in parallel.

- Motors that bring out 12 leads are generally intended for wye–delta starting.

- Three factors that determine the torque produced by an induction motor are

  a. the strength of the magnetic field of the stator.

  b. the strength of the magnetic field of the rotor.

  c. the phase angle difference between rotor and stator flux.

- Maximum torque is developed when stator and rotor flux are in phase with each other.

- The code letter on the nameplate of a squirrel-cage motor indicates the type of rotor bars used in the construction of the rotor.

- The type A rotor has the lowest starting current, highest starting torque, and poorest speed regulation of any type squirrel-cage rotor.

- The double-squirrel-cage rotor contains two sets of squirrel-cage windings in the same rotor.

- Consequent-pole squirrel-cage motors change speed by changing the number of stator poles.

- Wound-rotor induction motors have wound rotors that contain three-phase windings.

- Wound-rotor motors have three sliprings on the rotor shaft to provide external connection to the rotor.

- Wound-rotor motors have higher starting torque and lower starting current than squirrel-cage motors of the same horsepower.

- The speed of a wound-rotor motor can be controlled by permitting resistance to remain in the rotor circuit during operation.

- Synchronous motors operate at synchronous speed.

- Synchronous motors operate at a constant speed from no load to full load.

- When load is connected to a synchronous motor, stress develops between the magnetic fields of the rotor and stator.

- Synchronous motors must have DC excitation from an external source.

- DC excitation is provided to some synchronous motors through two sliprings located on the rotor shaft, and other motors use a brushless exciter.

- Synchronous motors have the ability to produce a leading power factor by overexcitation of the DC supplied to the rotor.

- Synchronous motors have a set of type A squirrel-cage bars used for starting. This squirrel-cage winding is called the amortisseur winding.

- A field-discharge resistor is connected across the rotor winding during starting to prevent high voltage in the rotor due to induction.

- Changing the DC excitation does not affect the speed of the motor.

- Selsyn motors are used to provide position control and angular feedback information.

- Although selsyn motors contain three-phase windings, they operate on single-phase AC.

- A differential selsyn unit can be used to determine the algebraic sum of the rotation of two other selsyn units.

Review Questions

1. What are the three basic types of three-phase motors?

2. What is the principle of operation of all three-phase motors?

3. What is synchronous speed?

4. What two factors determine synchronous speed?

5. Name three factors that cause the magnetic field to rotate.

6. Name three factors that determine the torque produced by an induction motor.

7. Is the synchronous motor an induction motor?

8. What is the amortisseur winding?

9. Why must a synchronous motor never be started when DC excitation is applied to the rotor?

10. Name three characteristics that make the synchronous motor different from an induction motor.

11. What is the function of the field-discharge resistor?

12. Why can an induction motor never operate at synchronous speed?

13. A squirrel-cage induction motor is operating at 1175 rpm and producing a torque of 22 lb-ft. What is the horsepower output of the motor?

14. A wattmeter measures the input power of the motor in Question 13 to be 5650 W. What is the efficiency of the motor?

15. What is the difference between a squirrel-cage motor and a wound-rotor motor?

16. What is the advantage of the wound-rotor motor over the squirrel-cage motor?

17. Name three factors that determine the amount of voltage induced in the rotor of a wound-rotor motor.

18. Why will the rotor of a wound-rotor motor not turn if the rotor circuit is left open with no resistance connected to it?

19. Why is the starting torque per A of starting current of a wound-rotor motor higher than that of a squirrel-cage motor although the starting current is less?

20. When is a synchronous motor a synchronous condenser?

21. What determines when a synchronous motor is at normal excitation?

22. How can a synchronous motor be made to have a leading power factor?

23. Is the excitation current of a synchronous motor AC or DC?

24. How is the speed of a consequent-pole squirrel-cage motor changed?

25. A three-phase squirrel-cage motor is connected to a 60-Hz line. The full-load speed is 870 rpm. How many poles per phase does the stator have?

Practical Applications

*Y*ou are working as a plant electrician. It is your job to install a 300-hp three-phase squirrel-cage induction motor. The supply voltage is 480 V. The power company has determined that the maximum amount of starting current that can be permitted by any motor in the plant is 3000 A. The motor nameplate is as follows:

Phase: 3	FLA: 352
Volts: 480	RPM: 1755
Frame: XXX	Code: L

Will it be possible to start this motor across the line, or will it be necessary to use a reduced-voltage starter to reduce starting current? ■

Practical Applications

*Y*ou have been given the task of connecting a nine-lead three-phase dual-voltage motor to a 240-V line. You discover that the nameplate has been painted and you cannot see the connection diagram. To make the proper connections, you must know if the motor stator winding is connected wye or delta. How could you determine this using an ohmmeter? Explain your answer. ■

Practical Applications

*Y*ou are an electrician working in an industrial plant. A 30-hp three-phase squirrel-cage motor keeps tripping out on overload. The motor is connected to a 240-volt line. Voltage measurements indicate the following voltages between the different phases: A–B = 276 volts, B–C = 221 volts, and A–C = 267 volts. Determine the percentage of heat rise in the winding with the highest current. ■

KEY TERMS

Centrifugal switch
Compensating winding
Conductive compensation
Consequent-pole motor
Holtz motor
Inductive compensation
Multispeed motors
Neutral plane
Repulsion motor
Run winding
Shaded-pole induction motor
Shading coil
Split-phase motors
Start winding
Stepping motors
Synchronous motors
Two-phase
Universal motor
Warren motor

Unit 34
Single-Phase Motors

Why You Need to Know

*S*ingle-phase motors are used almost exclusively in residential applications and to operate loads that require fractional horsepower motors in industrial and commercial locations. Many of these motors you will recognize from everyday life and may have wondered how they work. Unlike three-phase motors, there are many different types of single-phase motors and they do not all operate on the same principle. There are some that operate on the principle of a rotating magnetic field, but others do not. Some single-phase motors are designed to operate at more than one speed. This unit

- presents several different types of single-phase motors and explains how they operate.
- explains how to determine the appropriate motor to be used under a given situation by evaluating the operating principles of each.

Objectives

After studying this unit, you should be able to

- list the different types of split-phase motors.

- discuss the operation of split-phase motors.

- reverse the direction of rotation of a split-phase motor.

- discuss the operation of multispeed split-phase motors.

- discuss the operation of shaded-pole-type motors.

- discuss the operation of repulsion-type motors.

- discuss the operation of stepping motors.

- discuss the operation of universal motors.

Preview

Although most of the large motors used in industry are three phase, at times single-phase motors must be used. Single-phase motors are used almost exclusively to operate home appliances such as air conditioners, refrigerators, well pumps, and fans. They are generally designed to operate on 120 volts or 240 volts. They range in size from fractional horsepower to several horsepower, depending on the application. ■

34–1 Single-Phase Motors

In Unit 33, it was stated that there are three basic types of three-phase motors and that all operate on the principle of a rotating magnetic field. Although that is true for three-phase motors, it is not true for single-phase motors. There are not only many different types of single-phase motors, but they also have different operating principles.

34–2 Split-Phase Motors

Split-phase motors fall into three general classifications:

1. The resistance-start induction-run motor

2. The capacitor-start induction-run motor

3. The capacitor-start capacitor-run motor

Although all these motors have different operating characteristics, they are similar in construction and use the same operating principle. Split-phase motors

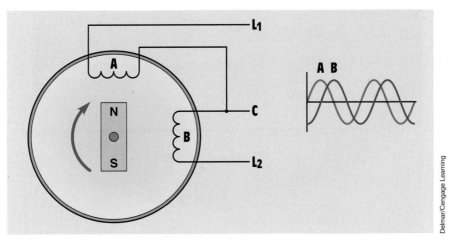

FIGURE 34–1 A two-phase alternator produces voltages that are 90° out of phase with each other.

receive their name from the manner in which they operate. Like three-phase motors, split-phase motors operate on the principle of a rotating magnetic field. A rotating magnetic field, however, cannot be produced with only one phase. Split-phase motors therefore split the current flow through two separate windings to simulate a two-phase power system. A rotating magnetic field can be produced with a two-phase system.

The Two-Phase System

In some parts of the world, **two-phase** power is produced. A two-phase system is produced by having an alternator with two sets of coils wound 90° apart *(Figure 34–1)*. The voltages of a two-phase system are therefore 90° out of phase with each other. These two out-of-phase voltages can be used to produce a rotating magnetic field in a manner similar to that of producing a rotating magnetic field with the voltages of a three-phase system. Because there have to be two voltages or currents out of phase with each other to produce a rotating magnetic field, split-phase motors use two separate windings to create a phase difference between the currents in each of these windings. These motors literally split one phase and produce a second phase, hence the name split-phase motor.

Stator Windings

The stator of a split-phase motor contains two separate windings, the **start winding** and the **run winding.** The start winding is made of small wire and is placed near the top of the stator core. The run winding is made of relatively large wire and is placed in the bottom of the stator core. *Figures 34–2A and B*

FIGURE 34–2A   Stator winding of a resistance-start induction-run motor. The start winding contains much smaller wire than the run winding.

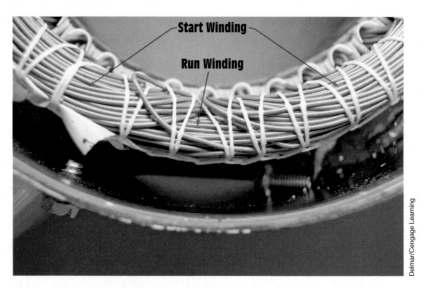

FIGURE 34–2B   Stator winding of a capacitor-start capacitor-run motor. The wire size is the same for both start and run windings.

are photographs of two split-phase stators. The stator in *Figure 34–2A* is used for a resistance-start induction-run motor or a capacitor-start induction-run motor. The stator in *Figure 34–2B* is used for a capacitor-start capacitor-run motor. Both stators contain four poles, and the start winding is placed at a 90° angle from the run winding.

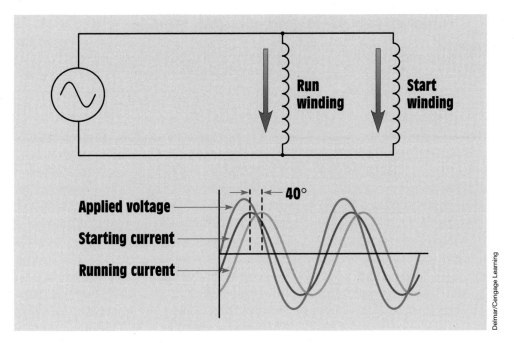

FIGURE 34–3 The start and run windings are connected in parallel with each other.

Notice the difference in size and position of the two windings of the stator shown in *Figure 34–2A*. The start winding is made from small wire and placed near the top of the stator core. This causes it to have a higher resistance than the run winding. The start winding is located between the poles of the run winding. The run winding is made with larger wire and placed near the bottom of the core. This gives it higher inductive reactance and less resistance than the start winding. These two windings are connected in parallel with each other *(Figure 34–3)*.

When power is applied to the stator, current flows through both windings. Because the start winding is more resistive, the current flow through it is more in phase with the applied voltage than the current flow through the run winding. The current flow through the run winding lags the applied voltage due to inductive reactance. These two out-of-phase currents are used to create a rotating magnetic field in the stator. The speed of this rotating magnetic field is called synchronous speed and is determined by the same two factors that determined the synchronous speed for a three-phase motor:

1. Number of stator poles per phase

2. Frequency of the applied voltage

34–3   Resistance-Start Induction-Run Motors

The resistance-start induction-run motor receives its name from the fact that the out-of-phase condition between start and run winding current is caused by the start winding being more resistive than the run winding. The amount of starting torque produced by a split-phase motor is determined by three factors:

1. The strength of the magnetic field of the stator

2. The strength of the magnetic field of the rotor

3. The phase angle difference between current in the start winding and current in the run winding (Maximum torque is produced when these two currents are 90° out of phase with each other.)

Although these two currents are out of phase with each other, they are not 90° out of phase. The run winding is more inductive than the start winding, but it does have some resistance, which prevents the current from being 90° out of phase with the voltage. The start winding is more resistive than the run winding, but it does have some inductive reactance, preventing the current from being in phase with the applied voltage. Therefore, a phase angle difference of 35° to 40° is produced between these two currents, resulting in a rather poor starting torque *(Figure 34–4)*.

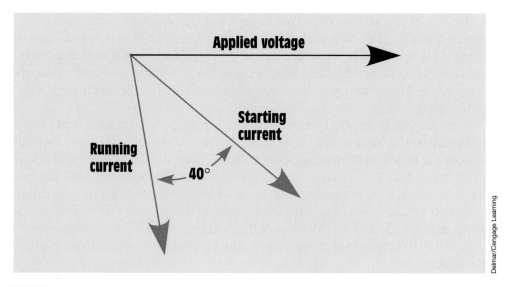

FIGURE 34–4   Running current and starting current are 35° to 40° out of phase with each other.

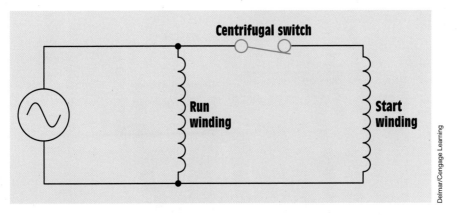

FIGURE 34–5   A centrifugal switch is used to disconnect the start winding from the circuit.

Disconnecting the Start Winding

A stator rotating magnetic field is necessary only to start the rotor turning. Once the rotor has accelerated to approximately 75% of rated speed, the start winding can be disconnected from the circuit and the motor will continue to operate with only the run winding energized. Motors that are not hermetically sealed (most refrigeration and air-conditioning compressors are hermetically sealed) use a **centrifugal switch** to disconnect the start windings from the circuit. The contacts of the centrifugal switch are connected in series with the start winding *(Figure 34–5)*. The centrifugal switch contains a set of spring-loaded weights. When the shaft is not turning, the springs hold a fiber washer in contact with the movable contact of the switch *(Figure 34–6)*. The fiber washer causes the movable contact to complete a circuit with a stationary contact.

When the rotor accelerates to about 75% of rated speed, centrifugal force causes the weights to overcome the force of the springs. The fiber washer retracts and permits the contacts to open and disconnect the start winding from the circuit *(Figure 34–7)*. The start winding of this type motor is intended to be energized only during the period of time that the motor is actually starting. If the start winding is not disconnected, it will be damaged by excessive current flow.

Starting Relays

Resistance-start induction-run and capacitor-start induction-run motors are sometimes hermetically sealed, such as with air-conditioning and refrigeration compressors. When these motors are hermetically sealed, a centrifugal switch cannot be used to disconnect the start winding. Some device that can be mounted externally must be used to disconnect the start windings from

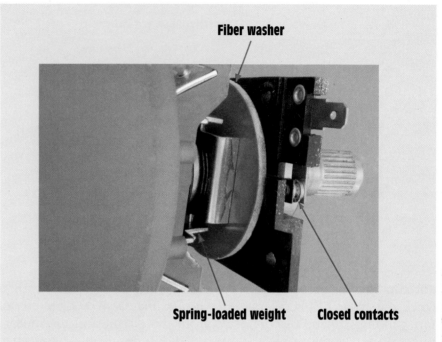

Delmar/Cengage Learning

FIGURE 34–6 The centrifugal switch is closed when the rotor is not turning.

Delmar/Cengage Learning

FIGURE 34–7 The contact opens when the rotor reaches about 75% of rated speed.

the circuit. Starting relays are used to perform this function. There are three basic types of starting relays used with the resistance-start and capacitor-start motors:

1. Hot-wire relay

2. Current relay

3. Solid-state starting relay

The *hot-wire relay* functions as both a starting relay and an overload relay. In the circuit shown in *Figure 34–8*, it is assumed that a thermostat controls the operation of the motor. When the thermostat closes, current flows through a resistive wire and two normally closed contacts connected to the start and run windings of the motor. The high starting current of the motor rapidly heats the resistive wire, causing it to expand. The expansion of the wire causes the spring-loaded start winding contact to open and disconnect the start winding from the circuit, reducing motor current. If the motor is not overloaded, the resistive wire never becomes hot enough to cause the overload contact to open and the motor continues to run. If the motor should become overloaded, however, the resistive wire expands enough to open the overload contact and

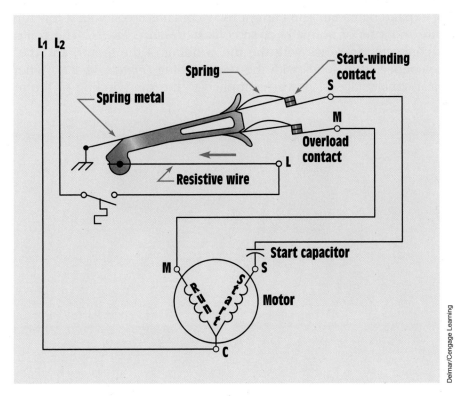

Delmar/Cengage Learning

FIGURE 34–8   Hot-wire relay connection.

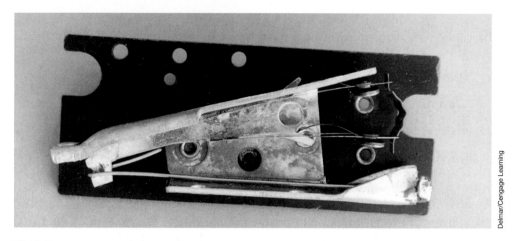

FIGURE 34–9   Hot-wire type of starting relay.

disconnect the motor from the line. A photograph of a hot-wire starting relay is shown in *Figure 34–9*.

The *current relay* also operates by sensing the amount of current flow in the circuit. This type of relay operates on the principle of a magnetic field instead of expanding metal. The current relay contains a coil with a few turns of large wire and a set of normally open contacts *(Figure 34–10)*. The coil of the relay is connected in series with the run winding of the motor, and the contacts are connected in series with the start winding *(Figure 34–11)*. When the thermostat contact closes, power is applied to the run winding of the motor.

FIGURE 34–10   Current type of starting relay.

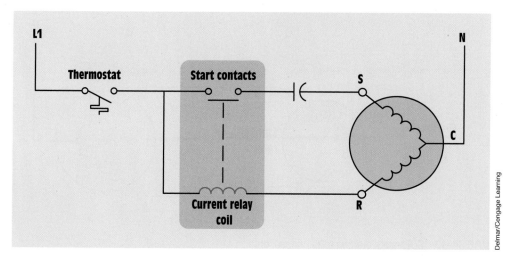

FIGURE 34–11 Current relay connection.

Because the start winding is open, the motor cannot start, causing a high current to flow in the run winding circuit. This high current flow produces a strong magnetic field in the coil of the relay, causing the normally open contacts to close and connect the start winding to the circuit. When the motor starts, the run-winding current is greatly reduced, permitting the start contacts to reopen and disconnect the start winding from the circuit.

The *solid-state starting relay (Figure 34–12)* performs the same basic function as the current relay and in many cases is replacing both the current relay and the centrifugal switch. The solid-state starting relay is generally more reliable and less expensive than the current relay or the centrifugal switch. The solid-state

FIGURE 34–12 Solid-state starting relay.

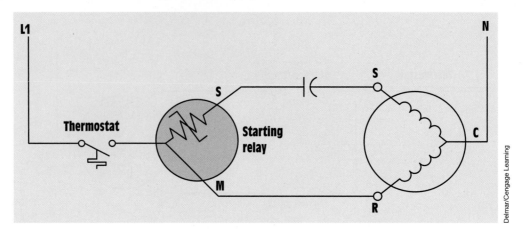

FIGURE 34–13   Solid-state starting relay connection.

starting relay is actually an electronic component known as a *thermistor*. A thermistor is a device that exhibits a change of resistance with a change of temperature. This particular thermistor has a positive coefficient of temperature, which means that when its temperature increases, its resistance increases also. The schematic diagram in *Figure 34–13* illustrates the connection of the solid-state starting relay. The thermistor is connected in series with the start winding of the motor. When the motor is not in operation, the thermistor is at a low temperature and its resistance is low, typically 3 or 4 ohms. When the thermostat contact closes, current flows to both the run and start windings of the motor. The current flowing through the thermistor causes an increase in temperature. This increased temperature causes the resistance of the thermistor to suddenly change to a high value of several thousand ohms. The change of temperature is so sudden that it has the effect of opening a set of contacts. Although the start winding is never completely disconnected from the powerline, the amount of current flow though it is very small, typically 0.03 to 0.05 amperes, and does not affect the operation of the motor. This small amount of *leakage current* maintains the temperature of the thermistor and prevents it from returning to a low value of resistance. After the motor is disconnected from the powerline, a cooldown time of two to three minutes should be allowed to permit the thermistor to return to a low resistance before the motor is restarted.

Relationship of Stator and Rotor Fields

The split-phase motor contains a squirrel-cage rotor very similar to those used with three-phase squirrel-cage motors *(Figure 34–14)*. When power is connected to the stator windings, the rotating magnetic field induces a voltage into the bars of the squirrel-cage rotor. The induced voltage causes current to flow in the rotor, and a magnetic field is produced around the rotor bars. The magnetic field of the rotor is attracted to the stator field, and the rotor begins to

FIGURE 34–14   Squirrel-cage rotor used in a split-phase motor.

Courtesy of Bodine Electric Co.

turn in the direction of the rotating magnetic field. After the centrifugal switch opens, only the run winding induces voltage into the rotor. This induced voltage is in phase with the stator current. The inductive reactance of the rotor is high, causing the rotor current to be almost 90° out of phase with the induced voltage. This causes the pulsating magnetic field of the rotor to lag the pulsating magnetic field of the stator by 90°. Magnetic poles, located midway between the stator poles, are created in the rotor *(Figure 34–15)*. These two pulsating magnetic fields produce a rotating magnetic field of their own, and the rotor continues to rotate.

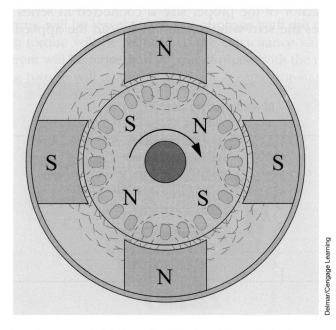

FIGURE 34–15   A rotating magnetic field is produced by the stator and rotor flux.

Delmar/Cengage Learning

34–5   Dual-Voltage Split-Phase Motors

Many split-phase motors are designed for operation on 120 or 240 volts. *Figure 34–19* shows the schematic diagram of a split-phase motor designed for dual-voltage operation. This particular motor contains two run windings and two start windings. The lead numbers for single-phase motors are numbered in a standard manner. One of the run windings has lead numbers of T_1 and T_2. The other run winding has its leads numbered T_3 and T_4. This particular motor uses two different sets of start-winding leads. One set is labeled T_5 and T_6, and the other set is labeled T_7 and T_8.

If the motor is to be connected for high-voltage operation, the run windings and start windings are connected in series, as shown in *Figure 34–20*. The start windings are then connected in parallel with the run windings. If the opposite direction of rotation is desired, T_5 and T_8 are changed.

For low-voltage operation, the windings must be connected in parallel, as shown in *Figure 34–21*. This connection is made by first connecting the run windings in parallel by hooking T_1 and T_3 together and T_2 and T_4 together. The start windings are paralleled by connecting T_5 and T_7 together and T_6 and T_8 together. The start windings are then connected in parallel with the run windings. If the opposite direction of rotation is desired, T_5 and T_6 should be reversed along with T_7 and T_8.

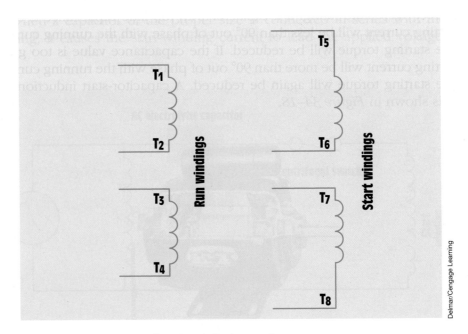

FIGURE 34–19   Dual-voltage windings for a split-phase motor.

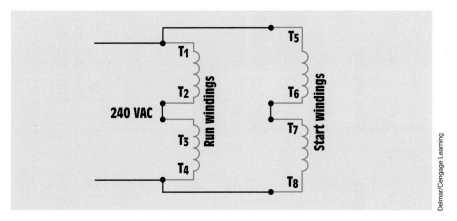

FIGURE 34–20 High-voltage connection for a split-phase motor with two run and two start windings.

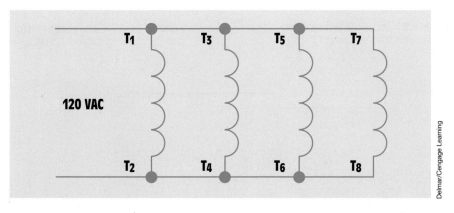

FIGURE 34–21 Low-voltage connection for a split-phase motor with two run and two start windings.

Not all dual-voltage single-phase motors contain two sets of start windings. *Figure 34–22* shows the schematic diagram of a motor that contains two sets of run windings and only one start winding. In this illustration, the start winding is labeled T_5 and T_6. Some motors, however, identify the start winding by labeling it T_5 and T_8, as shown in *Figure 34–23*.

Regardless of which method is used to label the terminal leads of the start winding, the connection is the same. If the motor is to be connected for high-voltage operation, the run windings are connected in series and the start winding is connected in parallel with one of the run windings (*Figure 34–24*). In this type of motor, each winding is rated at 120 volts. If the run windings are connected in series across 240 volts, each winding has a voltage drop of 120 volts. By connecting the start winding in parallel across only one run winding, it

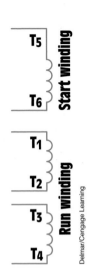

FIGURE 34–22   Dual-voltage motor with one start winding labeled T_5 and T_6.

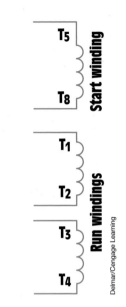

FIGURE 34–23   Dual-voltage motor with one start winding labeled T_5 and T_8.

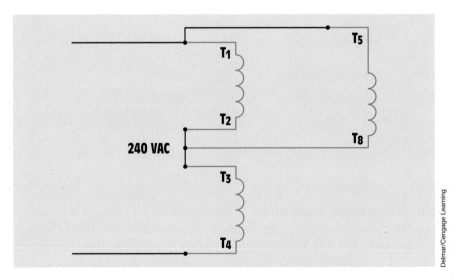

FIGURE 34–24   High-voltage connection with one start winding.

receives only 120 volts when power is applied to the motor. If the opposite direction of rotation is desired, T_5 and T_8 should be changed.

If the motor is to be operated on low voltage, the windings are connected in parallel as shown in *(Figure 34–25)*. Because all windings are connected in parallel, each receives 120 volts when power is applied to the motor.

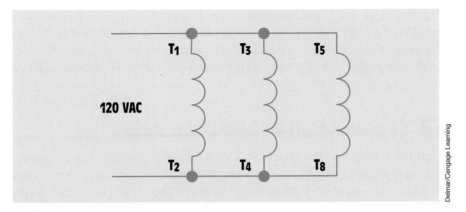

FIGURE 34–25   Low-voltage connection for a split-phase motor with one start winding.

34–6   Determining the Direction of Rotation for Split-Phase Motors

The direction of rotation of a single-phase motor can generally be determined when the motor is connected. The direction of rotation is determined by facing the back or rear of the motor. *Figure 34–26* shows a connection diagram for rotation. If clockwise rotation is desired, T_5 should be connected to T_1. If counterclockwise rotation is desired, T_8 (or T_6) should be connected to T_1. This connection diagram assumes that the motor contains two sets of run and two sets of start windings. The type of motor used determines the actual connection. For example, *Figure 34–24* shows the connection of a motor with two

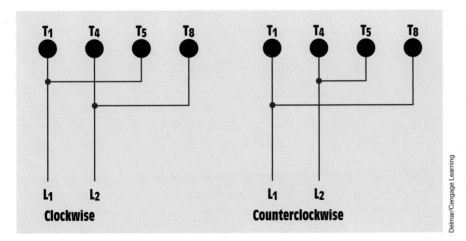

FIGURE 34–26   Determining direction of rotation for a split-phase motor.

run windings and only one start winding. If this motor were to be connected for clockwise rotation, terminal T_5 would have to be connected to T_1, and terminal T_8 would have to be connected to T_2 and T_3. If counterclockwise rotation is desired, terminal T_8 would have to be connected to T_1, and terminal T_5 would have to be connected to T_2 and T_3.

34–7   Capacitor-Start Capacitor-Run Motors

Although the capacitor-start capacitor-run motor is a split-phase motor, it operates on a different principle than the resistance-start induction-run motor or the capacitor-start induction-run motor. The capacitor-start capacitor-run motor is designed in such a manner that its start winding remains energized at all times. A capacitor is connected in series with the winding to provide a continuous leading current in the start winding *(Figure 34–27)*. Because the start winding remains energized at all times, no centrifugal switch is needed to disconnect the start winding as the motor approaches full speed. The capacitor used in this type of motor is generally of the oil-filled type because it is intended for continuous use. An exception to this general rule is small fractional-horsepower motors used in reversible ceiling fans. These fans have a low current draw and use an AC electrolytic capacitor to help save space.

The capacitor-start capacitor-run motor actually operates on the principle of a rotating magnetic field in the stator. Because both run and start windings remain energized at all times, the stator magnetic field continues to rotate and the motor operates as a two-phase motor. This motor has excellent starting and running torque. It is quiet in operation and has a high efficiency. Because the capacitor remains connected in the circuit at all times, the motor power factor is close to unity.

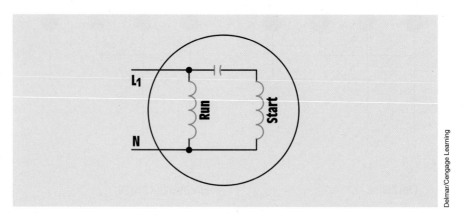

FIGURE 34–27   A capacitor-start capacitor-run motor.

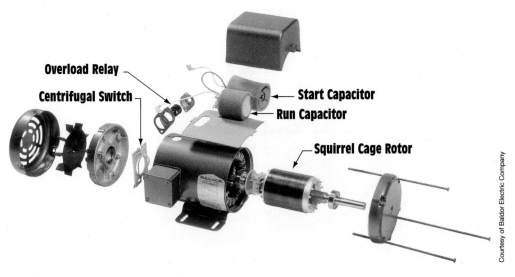

Overload Relay

Centrifugal Switch

Start Capacitor

Run Capacitor

Squirrel Cage Rotor

Courtesy of Baldor Electric Company

FIGURE 34–28 Capacitor-start capacitor-run motor with additional starting capacitor.

Although the capacitor-start capacitor-run motor does not require a centrifugal switch to disconnect the capacitor from the start winding, some motors use a second capacitor during the starting period to help improve starting torque *(Figure 34–28)*. A good example of this can be found on the compressor of a central air-conditioning unit designed for operation on single-phase power. If the motor is not hermetically sealed, a centrifugal switch will be used to disconnect the start capacitor from the circuit when the motor reaches approximately 75% of rated speed. Hermetically sealed motors, however, must use some type of external switch to disconnect the start capacitor from the circuit.

The capacitor-start capacitor-run motor, or permanent split-capacitor motor as it is generally referred to in the air-conditioning and refrigeration industry, generally employs a potential starting relay to disconnect the starting capacitor when a centrifugal switch cannot be used. The potential starting relay *(Figure 34–29A and B)* operates by sensing an increase in the voltage developed in the start winding when the motor is operating. A schematic diagram of a potential starting relay circuit is shown in *Figure 34–30*. In this circuit, the potential relay is used to disconnect the starting capacitor from the circuit when the motor reaches about 75% of its full speed. The starting-relay coil, SR, is connected in parallel with the start winding of the motor. A normally closed SR contact is connected in series with the starting capacitor. When the thermostat contact closes, power is applied to both the run and start windings. At this point in time, both the start and run capacitors are connected in the circuit.

As the rotor begins to turn, its magnetic field induces a voltage into the start winding, producing a higher voltage across the start winding than the applied voltage. When the motor has accelerated to about 75% of its full

A

Delmar/Cengage Learning

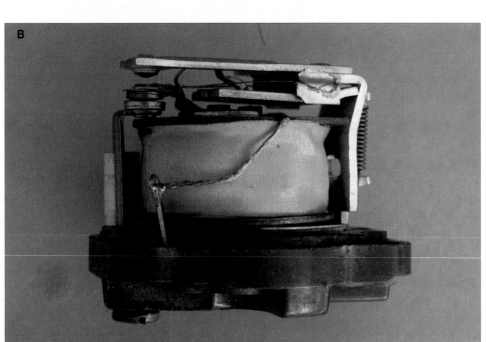

B

Delmar/Cengage Learning

FIGURE 34–29A AND B Potential starting relays.

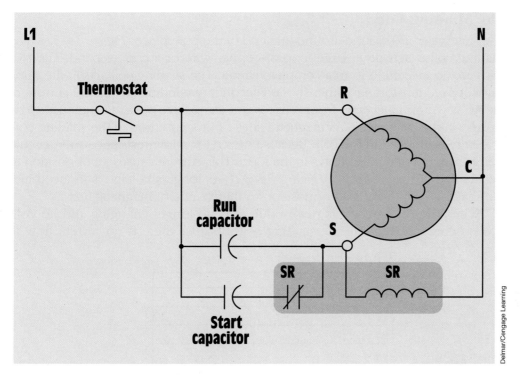

FIGURE 34–30 Potential relay connection.

speed, the voltage across the start winding is high enough to energize the coil of the potential relay. This causes the normally closed SR contact to open and disconnect the start capacitor from the circuit. Because the start winding of this motor is never disconnected from the powerline, the coil of the potential starting relay remains energized as long as the motor is in operation.

34–8 Shaded-Pole Induction Motors

The **shaded-pole induction motor** is popular because of its simplicity and long life. This motor contains no start windings or centrifugal switch. It contains a squirrel-cage rotor and operates on the principle of a rotating magnetic field. The rotating magnetic field is created by a **shading coil** wound on one side of each pole piece. Shaded-pole motors are generally fractional-horsepower motors and are used for low-torque applications such as operating fans and blowers.

The Shading Coil

The shading coil is wound around one end of the pole piece *(Figure 34–31)*. The shading coil is actually a large loop of copper wire or a copper band. The two ends are connected to form a complete circuit. The shading coil acts in the same manner as a transformer with a shorted secondary winding. When the current of the AC waveform increases from zero toward its positive peak, a magnetic field is created in the pole piece. As magnetic lines of flux cut through the shading coil, a voltage is induced in the coil. Because the coil is a low-resistance short circuit, a large amount of current flows in the loop. This current causes an opposition to the change of magnetic flux *(Figure 34–32)*. As long as voltage is induced into the shading coil, there is an opposition to the change of magnetic flux.

When the AC reaches its peak value, it is no longer changing and no voltage is being induced into the shading coil. Because there is no current flow in

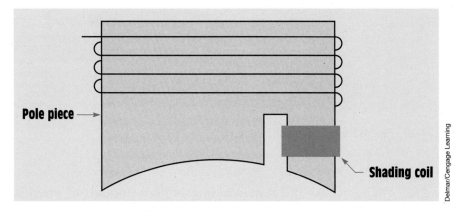

FIGURE 34–31   A shaded pole.

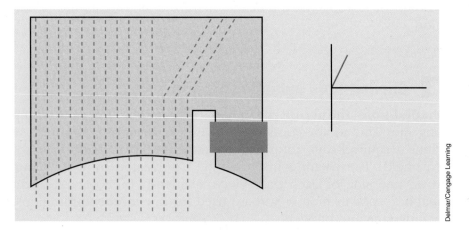

FIGURE 34–32   The shading coil opposes a change of flux as current increases.

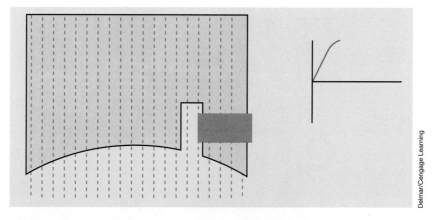

FIGURE 34–33 There is opposition to magnetic flux when the current is not changing.

the shading coil, there is no opposition to the magnetic flux. The magnetic flux of the pole piece is now uniform across the pole face *(Figure 34–33)*.

When the AC begins to decrease from its peak value back toward zero, the magnetic field of the pole piece begins to collapse. A voltage is again induced into the shading coil. This induced voltage creates a current that opposes the change of magnetic flux *(Figure 34–34)*. This causes the magnetic flux to be concentrated in the shaded section of the pole piece.

When the AC passes through zero and begins to increase in the negative direction, the same set of events happens except that the polarity of the magnetic field is reversed. If these events were to be viewed in rapid order, the magnetic field would be seen to rotate across the face of the pole piece. A pole piece with a shading coil is shown in *Figure 34–35*.

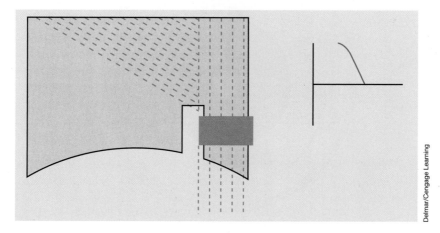

FIGURE 34–34 The shading coil opposes a change of flux when the current decreases.

FIGURE 34–35 The shading coil is a large copper conductor wound around one side of the pole piece.

Speed

The speed of the shaded-pole induction motor is determined by the same factors that determine the synchronous speed of other induction motors: frequency and number of stator poles. Shaded-pole motors are commonly wound as four- or six-pole motors. *Figure 34–36* shows a drawing of a four-pole shaded-pole induction motor.

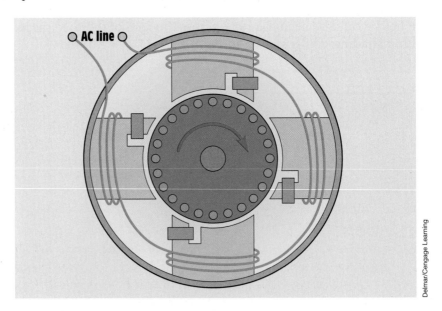

FIGURE 34–36 Four-pole shaded-pole induction motor.

FIGURE 34–37 Stator winding and rotor of a shaded-pole induction motor.

General Operating Characteristics

The shaded-pole motor contains a standard squirrel-cage rotor. The amount of torque produced is determined by the strength of the magnetic field of the stator, the strength of the magnetic field of the rotor, and the phase angle difference between rotor and stator flux. The shaded-pole induction motor has low starting and running torque.

The direction of rotation is determined by the direction in which the rotating magnetic field moves across the pole face. The rotor turns in the direction shown by the arrow in *Figure 34–36*. The direction can be changed by removing the stator winding and turning it around. This is not a common practice, however. As a general rule, the shaded-pole induction motor is considered to be nonreversible. *Figure 34–37* shows a photograph of the stator winding and rotor of a shaded-pole induction motor.

34–9 Multispeed Motors

There are two basic types of single-phase **multispeed motors.** One is the consequent-pole type and the other is a specially wound *capacitor-start* capacitor-run motor or shaded-pole induction motor. The single-phase **consequent-pole motor** operates in the same basic way as the three-phase consequent pole discussed in Unit 33. The speed is changed by reversing the current flow through alternate poles and increasing or decreasing the total number of stator poles. The consequent-pole motor is used where high running torque must be maintained at different speeds. A good example of where this type of motor is used is in two-speed compressors for central air-conditioning units.

Multispeed Fan Motors

Multispeed fan motors have been used for many years. These motors are generally wound for two to five steps of speed and operate fans and squirrel-cage blowers. A schematic drawing of a three-speed motor is shown in *Figure 34–38*. Notice that the run winding has been tapped to produce low, medium, and high speed. The start winding is connected in parallel with the run-winding section. The other end of the start-winding lead is connected to an external oil-filled capacitor. This motor obtains a change of speed by inserting inductance in series with the run winding. The actual run winding for this motor is between the terminals marked *high* and *common*. The winding shown between *high* and *medium* is connected in series with the main run winding. When the rotary switch is connected to the medium speed position, the inductive reactance of this coil limits the amount of current flow through the run winding. When the current of the run winding is reduced, the strength of the magnetic field of the run winding is reduced and the motor produces less torque. This causes a greater amount of slip and the motor speed to decrease.

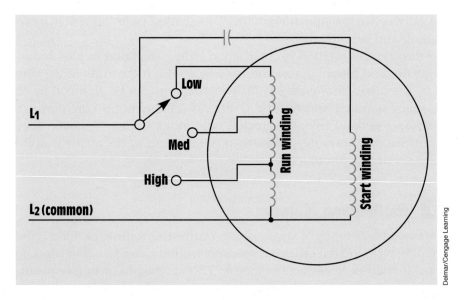

FIGURE 34–38   A three-speed motor.

If the rotary switch is changed to the *low* position, more inductance is inserted in series with the run winding. This causes less current to flow through the run winding and another reduction in torque. When the torque is reduced, the motor speed decreases again.

Common speeds for a four-pole motor of this type are 1625, 1500, and 1350 rpm. Notice that this motor does not have wide ranges between speeds as would be the case with a consequent-pole motor. Most induction motors would overheat and damage the motor winding if the speed were reduced to this extent. This type of motor, however, has much higher impedance windings than most other motors. The run windings of most split-phase motors have a wire resistance of 1 to 4 ohms. This motor generally has a resistance of 10 to 15 ohms in its run winding. It is the high impedance of the windings that permits the motor to be operated in this manner without damage.

Because this motor is designed to slow down when load is added, it is not used to operate high-torque loads. This type of motor is generally used to operate only low-torque loads such as fans and blowers.

34–10   Repulsion-Type Motors

There are three basic repulsion-type motors:

1. The repulsion motor

2. The repulsion-start induction-run motor

3. The repulsion-induction motor

Each of these three types has different operating characteristics.

34–11   Construction of Repulsion Motors

A **repulsion motor** operates on the principle that like magnetic poles repel each other, not on the principle of a rotating magnetic field. The stator of a repulsion motor contains only a run winding very similar to that used in the split-phase motor. Start windings are not necessary. The rotor is actually called an armature because it contains a slotted metal core with windings placed in the slots. The windings are connected to a commutator. A set of brushes makes contact with the surface of the commutator bars. The entire assembly looks very much like a DC armature and brush assembly. One difference, however, is that the brushes of the repulsion motor are shorted together. Their function is to provide a current path through certain parts of the armature, not to provide power to the armature from an external source.

Operation

Although the repulsion motor does not operate on the principle of a rotating magnetic field, it is an induction motor. When AC power is connected to the stator winding, a magnetic field with alternating polarities is produced in the poles. This alternating field induces a voltage into the windings of the armature. When the brushes are placed in the proper position, current flows through the armature windings, producing a magnetic field of the same polarity in the armature. The armature magnetic field is repelled by the stator magnetic field, causing the armature to rotate. Repulsion motors contain the same number of brushes as there are stator poles. Repulsion motors are commonly wound for four, six, or eight poles.

Brush Position

The position of the brushes is very important. Maximum torque is developed when the brushes are placed 15° on either side of the pole pieces. *Figure 34–39* shows the effect of having the brushes placed at a 90° angle to the pole pieces. When the brushes are in this position, a circuit is completed between the coils located at a right angle to the poles. In this position, there is no induced voltage in the armature windings and no torque is produced by the motor.

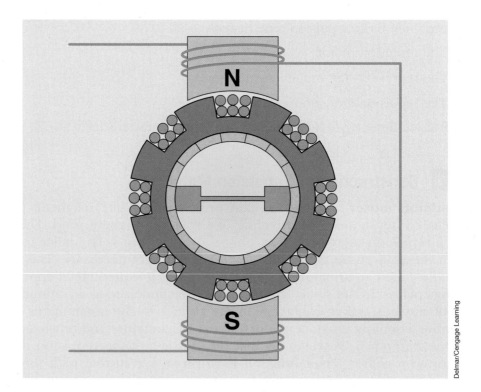

FIGURE 34–39   Brushes are placed at a 90° angle to the poles.

Delmar/Cengage Learning

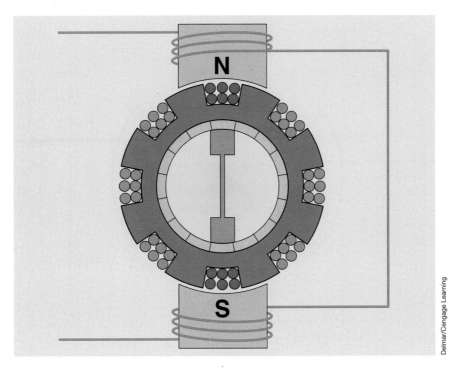

FIGURE 34–40   The brushes are set at a 0° angle to the pole pieces.

In *Figure 34–40*, the brushes have been moved to a position so that they are in line with the pole pieces. In this position, a large amount of current flows through the coils directly under the pole pieces. This current produces a magnetic field of the same polarity as the pole piece. Because the magnetic field produced in the armature is at a 0° angle to the magnetic field of the pole piece, no twisting or turning force is developed and the armature does not turn.

In *Figure 34–41*, the brushes have been shifted in a clockwise direction so that they are located 15° from the pole piece. The induced voltage in the armature winding produces a magnetic field of the same polarity as the pole piece. The magnetic field of the armature is repelled by the magnetic field of the pole piece, and the armature turns in the clockwise direction.

In *Figure 34–42*, the brushes have been shifted counterclockwise to a position 15° from the center of the pole piece. The magnetic field developed in the armature again repels the magnetic field of the pole piece, and the armature turns in the counterclockwise direction.

The direction of armature rotation is determined by the setting of the brushes. The direction of rotation for any type of repulsion motor is changed by setting the brushes 15° on either side of the pole pieces. Repulsion-type motors have the highest starting torque of any single-phase motor. The speed of a repulsion motor, not to be confused with the repulsion-start induction-run

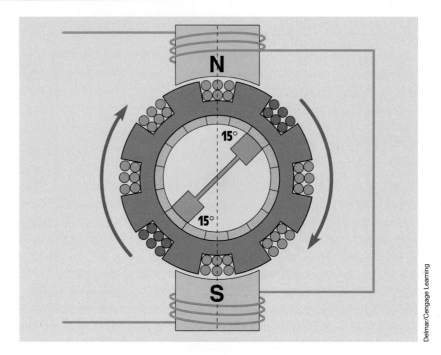

FIGURE 34–41   The brushes have been shifted clockwise 15°.

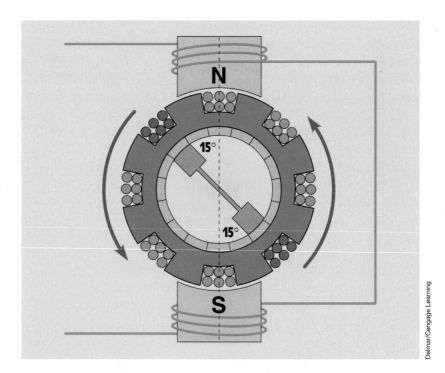

FIGURE 34–42   The brushes have been shifted counterclockwise 15°.

motor or the repulsion-induction motor, can be varied by changing the AC voltage supplying power for the motor. The repulsion motor has excellent starting and running torque but can exhibit unstable speed characteristics. The repulsion motor can race to very high speed if operated with no mechanical load connected to the shaft.

34–12   Repulsion-Start Induction-Run Motors

The repulsion-start induction-run motor starts as a repulsion motor but runs like a squirrel-cage motor. There are two types of repulsion-start induction-run motors:

1. The brush-riding type

2. The brush-lifting type

The brush-riding type uses an axial commutator *(Figure 34–43).* The brushes ride against the commutator segments at all times when the motor is in operation. After the motor has accelerated to approximately 75% of its full-load speed, centrifugal force causes copper segments of a short-circuiting ring to overcome the force of a spring *(Figure 34–44).* The segments sling out and make contact with the segments of the commutator. This effectively short-circuits all the commutator segments together, and the motor operates in the same manner as a squirrel-cage motor.

   The brush-lifting-type motor uses a radial commutator *(Figure 34–45).* Weights are mounted at the front of the armature. When the motor reaches about 75% of full speed, these weights swing outward due to centrifugal force and cause two push rods to act against a spring barrel and short-circuiting necklace. The weights overcome the force of the spring and cause the entire spring barrel and brush holder assembly to move toward the back of the motor

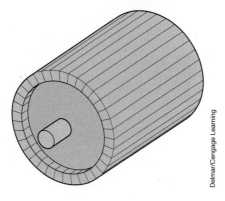

FIGURE 34–43   Axial commutator.

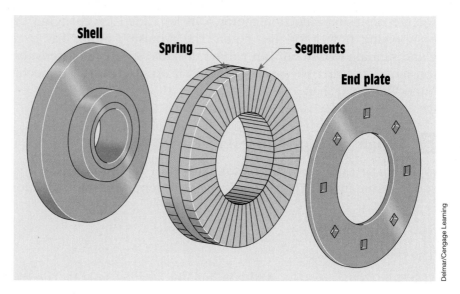

FIGURE 34–44 Short-circuiting ring for brush-riding-type repulsion-start induction-run motor.

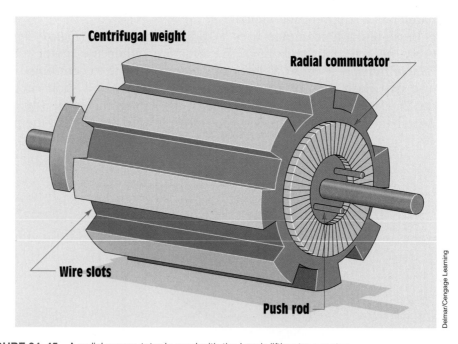

FIGURE 34–45 A radial commutator is used with the brush-lifting-type motor.

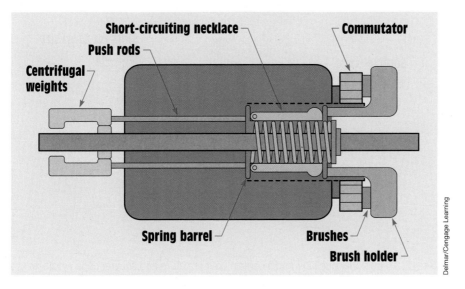

FIGURE 34–46   Brush-lifting-type repulsion-start induction-run motor.

(Figure 34–46). The motor is so designed that the short-circuiting necklace will short-circuit the commutator bars before the brushes lift off the surface of the radial commutator. The motor will now operate as a squirrel-cage induction motor. The brush-lifting motor has several advantages over the brush-riding motor. Because the brushes lift away from the commutator surface during operation, wear on both the commutator and brushes is greatly reduced. Also, the motor does not have to overcome the friction of the brushes riding against the commutator surface during operation. As a result, the brush-lifting motor is quieter in operation.

34–13   Repulsion-Induction Motors

The repulsion-induction motor is basically the same as the repulsion motor except that a set of squirrel-cage windings are added to the armature *(Figure 34–47)*. This type of motor contains no centrifugal mechanism or short-circuiting device. The brushes ride against the commutator at all times. The repulsion-induction motor has very high starting torque because it starts as a repulsion motor. The squirrel-cage winding, however, gives it much better speed characteristics than a standard repulsion motor. This motor has very good speed regulation between no load and full load. Its running characteristics are similar to a DC compound motor. The schematic symbol for a repulsion motor is shown in *Figure 34–48*.

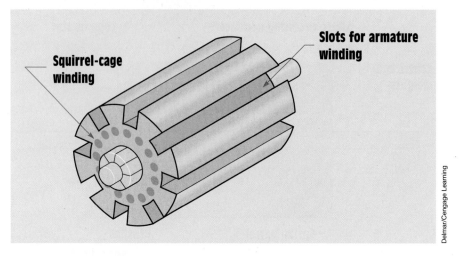

FIGURE 34–47 Repulsion-induction motors contain both armature and squirrel-cage windings.

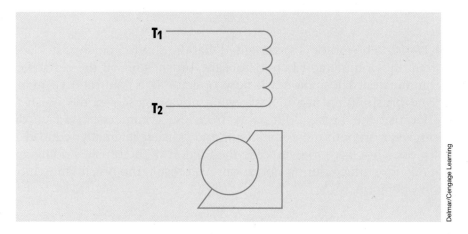

FIGURE 34–48 Schematic symbol for a repulsion motor.

34–14 Single-Phase Synchronous Motors

Single-phase **synchronous motors** are small and develop only fractional horsepower. They operate on the principle of a rotating magnetic field developed by a shaded-pole stator. Although they will operate at synchronous speed, they do not require DC excitation. They are used in applications where constant speed is required such as clock motors, timers, and recording instruments. They also are used as the driving force for small fans because they are small and inexpensive to manufacture. There are two basic types of synchronous motor: the Warren, or General Electric motor, and the Holtz motor. These motors are also referred to as hysteresis motors.

Warren Motors

The **Warren motor** is constructed with a laminated stator core and a single coil. The coil is generally wound for 120-VAC operation. The core contains two poles, which are divided into two sections each. One half of each pole piece contains a shading coil to produce a rotating magnetic field *(Figure 34–49)*. Because the stator is divided into two poles, the synchronous field speed is 3600 rpm when connected to 60 hertz.

The difference between the Warren and Holtz motor is the type of rotor used. The rotor of the Warren motor is constructed by stacking hardened steel laminations onto the rotor shaft. These disks have high hysteresis loss. The laminations form two crossbars for the rotor. When power is connected to the motor, the rotating magnetic field induces a voltage into the rotor and a strong starting torque is developed causing the rotor to accelerate to near-synchronous

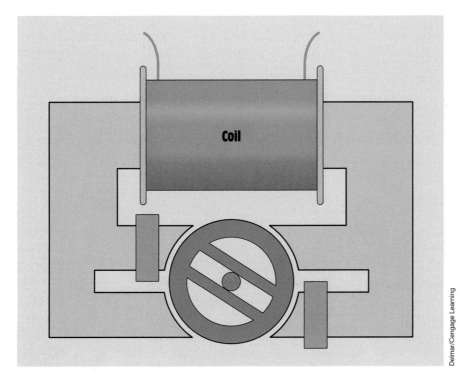

FIGURE 34–49   A Warren motor.

speed. Once the motor has accelerated to near-synchronous speed, the flux of the rotating magnetic field follows the path of minimum reluctance (magnetic resistance) through the two crossbars. This causes the rotor to lock in step with the rotating magnetic field, and the motor operates at 3600 rpm. These motors are often used with small geartrains to reduce the speed to the desired level.

Holtz Motors

The **Holtz motor** uses a different type of rotor *(Figure 34–50)*. This rotor is cut in such a manner that six slots are formed. These slots form six salient (projecting or jutting) poles for the rotor. A squirrel-cage winding is constructed by inserting a metal bar at the bottom of each slot. When power is connected to the motor, the squirrel-cage winding provides the torque necessary to start the rotor turning. When the rotor approaches synchronous speed, the salient poles lock in step with the field poles each half cycle. This produces a rotor speed of 1200 rpm (one-third of synchronous speed) for the motor.

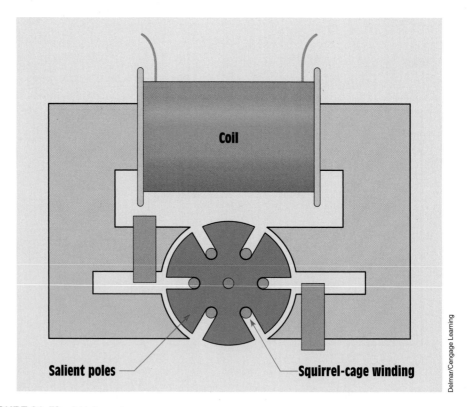

FIGURE 34–50   A Holtz motor.

34-15 Stepping Motors

Stepping motors are devices that convert electric impulses into mechanical movement. Stepping motors differ from other types of DC or AC motors in that their output shaft moves through a specific angular rotation each time the motor receives a pulse. The stepping motor allows a load to be controlled as to speed, distance, or position. These motors are very accurate in their control performance. There is generally less than 5% error per angle of rotation, and this error is not cumulative regardless of the number of rotations. Stepping motors are operated on DC power but can be used as a two-phase synchronous motor when connected to AC power.

Theory of Operation

Stepping motors operate on the theory that like magnetic poles repel and unlike magnetic poles attract. Consider the circuit shown in *Figure 34–51*. In this illustration, the rotor is a permanent magnet and the stator windings consist of two electromagnets. If current flows through the winding of stator pole A in such a direction that it creates a north magnetic pole and through B in such a direction that it creates a south magnetic pole, it is impossible to determine the direction of rotation. In this condition, the rotor could turn in either direction.

Now consider the circuit shown in *Figure 34–52*. In this circuit, the motor contains four stator poles instead of two. The direction of current flow through stator pole A is still in such a direction as to produce a north magnetic field,

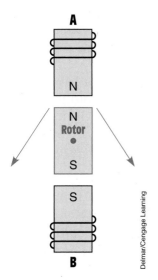

FIGURE 34–51 The rotor could turn in either direction.

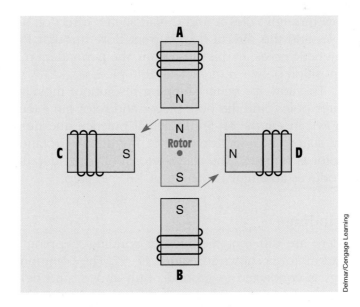

FIGURE 34–52 The direction of rotation is known.

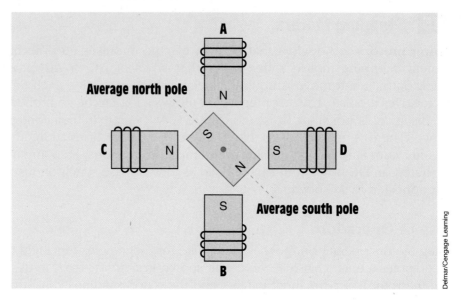

FIGURE 34–53 The magnet aligns with the average magnetic pole.

and the current flow through pole B produces a south magnetic field. The current flow through stator pole C, however, produces a south magnetic field and the current flow through pole D produces a north magnetic field. In this illustration, there is no doubt as to the direction or angle of rotation. In this example, the rotor shaft turns 90° in a counterclockwise direction.

Figure 34–53 shows yet another condition. In this example, the current flow through Poles A and C is in such a direction as to form a north magnetic pole, and the direction of current flow through Poles B and D forms south magnetic poles. In this illustration, the permanent magnetic rotor has rotated to a position between the actual pole pieces.

To allow for better stepping resolution, most stepping motors have eight stator poles, and the pole pieces and rotor have teeth machined into them, as shown in *Figure 34–54*. In actual practice, the number of teeth machined in the stator and rotor determines the angular rotation achieved each time the motor is stepped. The stator-rotor tooth configuration shown in *Figure 34–54* produces an angular rotation of 1.8° per step.

Windings

There are different methods of winding stepping motors. A standard three-lead motor is shown in *Figure 34–55*. The common terminal of the two windings is connected to ground of an above- and below-ground power supply. Terminal 1 is connected to the common of a single-pole double-throw switch (Switch 1), and Terminal 3 is connected to the common of another single-pole

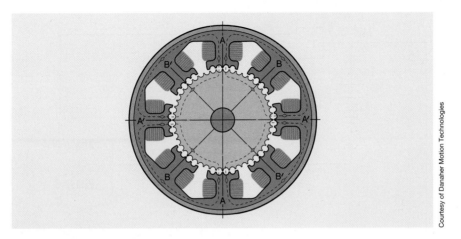

FIGURE 34–54   Construction of a stepping motor.

Courtesy of Danaher Motion Technologies

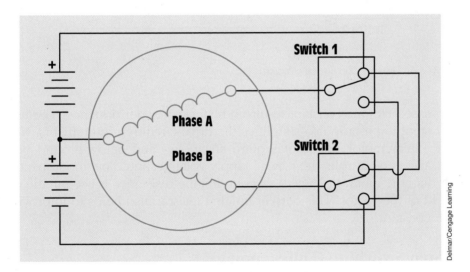

FIGURE 34–55   A standard three-lead motor.

Delmar/Cengage Learning

double-throw switch (Switch 2). One of the stationary contacts of each switch is connected to the positive, or above-ground, voltage, and the other stationary contact is connected to the negative, or below-ground, voltage. The polarity of each winding is determined by the position setting of its control switch.

Stepping motors can also be wound bifilar, as shown in *Figure 34–56*. The term *bifilar* means that two windings are wound together. This is similar to a transformer winding with a center-tap lead. Bifilar stepping motors have twice as many windings as the three-lead type, which makes it necessary to use smaller wire in the windings. This results in higher wire resistance in the winding, producing a better inductive-resistive (LR) time constant for the bifilar-wound motor.

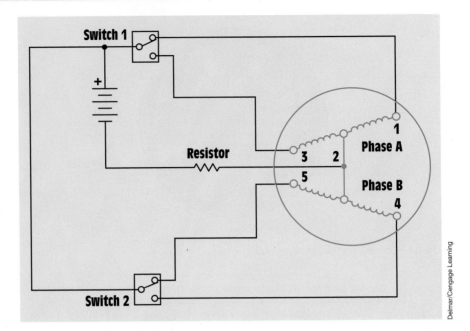

FIGURE 34–56 Bifilar-wound stepping motor.

The increased LR time constant results in better motor performance. The use of a bifilar stepping motor also simplifies the drive circuitry requirements. Notice that the bifilar motor does not require an above- and below-ground power supply. As a general rule, the power supply voltage should be about five times greater than the motor voltage. A current-limiting resistance is used in the common lead of the motor. This current-limiting resistor also helps to improve the LR time constant.

Four-Step Switching (Full-Stepping)

The switching arrangement shown in *Figure 34–56* can be used for a four-step switching sequence (full-stepping). Each time one of the switches changes position, the rotor advances one-fourth of a tooth. After four steps, the rotor has turned the angular rotation of one "full" tooth. If the rotor and stator have 50 teeth, it will require 200 steps for the motor to rotate one full revolution. This corresponds to an angular rotation of 1.8° per step (360°/200 steps = 1.8° per step). The chart shown in *Table 34–1* illustrates the switch positions for each step.

Eight-Step Switching (Half-Stepping)

Figure 34–57 illustrates the connections for an eight-step switching sequence (half-stepping). In this arrangement, the center-tap leads for Phases A and B

Step	Switch 1	Switch 2
1	1	5
2	1	4
3	3	4
4	3	5
1	1	5

TABLE 34–1 Four-Step Switching Sequence

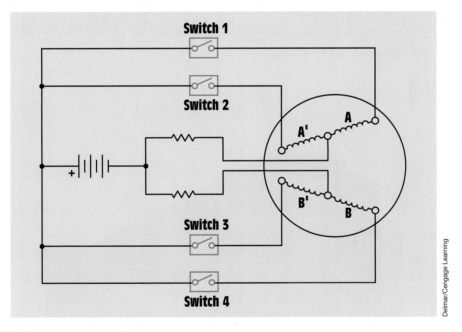

FIGURE 34–57 Eight-step switching.

Delmar/Cengage Learning

are connected through their own separate current-limiting resistors back to the negative of the power supply. This circuit contains four separate single-pole switches instead of two switches. The advantage of this arrangement is that each step causes the motor to rotate one-eighth of a tooth instead of one-fourth of a tooth. The motor now requires 400 steps to produce one revolution, which produces an angular rotation of 0.9° per step. This results in better stepping resolution and greater speed capability. The chart in *Table 34–2* illustrates the switch position for each step. A stepping motor is shown in *Figure 34–58*.

Step	Switch 1	Switch 2	Switch 3	Switch 4
1	On	Off	On	Off
2	On	Off	Off	Off
3	On	Off	Off	On
4	Off	Off	Off	On
5	Off	On	Off	On
6	Off	On	Off	Off
7	Off	On	On	Off
8	Off	Off	On	Off
1	On	Off	On	Off

TABLE 34–2 Eight-Step Switching Sequence

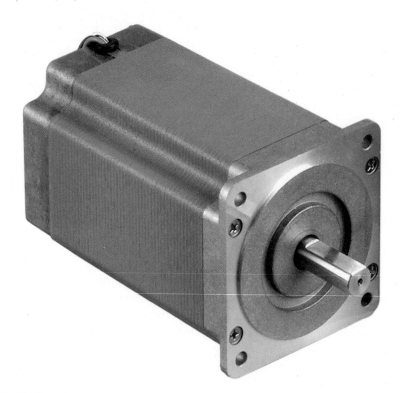

FIGURE 34–58 Stepping motor.

Courtesy of Danaher Motion Technologies

AC Operation

Stepping motors can be operated on AC voltage. In this mode of operation, they become two-phase, AC, synchronous, constant-speed motors and are classified as *permanent magnet induction motors*. Refer to the cutaway of a stepping motor shown in *Figure 34–59*. Notice that this motor has no brushes, sliprings, commutator, gears, or belts. Bearings maintain a constant air gap between the permanent magnet rotor and the stator windings. A typical eight-stator-pole stepping motor will have a synchronous speed of 72 rpm when connected to a 60-hertz, two-phase AC powerline.

A resistive-capacitive network can be used to provide the 90° phase shift needed to change single-phase AC into two-phase AC. A simple forward-off-reverse switch can be added to provide directional control. A sample circuit

Courtesy of Danaher Motion Technologies

FIGURE 34–59 Cutaway of a stepping motor.

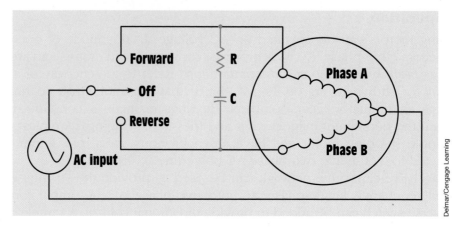

FIGURE 34–60 Phase-shift circuit converts single-phase into two-phase.

of this type is shown in *Figure 34–60*. The correct values of resistance and capacitance are necessary for proper operation. Incorrect values can result in random direction of rotation when the motor is started, change of direction when the load is varied, erratic and unstable operation, and failure to start. The correct values of resistance and capacitance will be different with different stepping motors. The manufacturer's recommendations should be followed for the particular type of stepping motor used.

Stepping Motor Characteristics

When stepping motors are used as two-phase synchronous motors, they can start, stop, or reverse direction of rotation virtually instantly. The motor will start within about 1-1/2 cycles of the applied voltage and will stop within 5 to 25 milliseconds. The motor can maintain a stalled condition without harm to the motor. Because the rotor is a permanent magnet, there is no induced current in the rotor. There is no high inrush of current when the motor is started. The starting and running currents are the same. This simplifies the power requirements of the circuit used to supply the motor. Due to the permanent magnetic structure of the rotor, the motor does provide holding torque when turned off. If more holding torque is needed, DC voltage can be applied to one or both windings when the motor is turned off. An example circuit of this type is shown in *Figure 34–61*. If DC is applied to one winding, the holding torque will be approximately 20% greater than the rated torque of the motor. If DC is applied to both windings, the holding torque will be about 1 to 1/2 times greater than the rated torque.

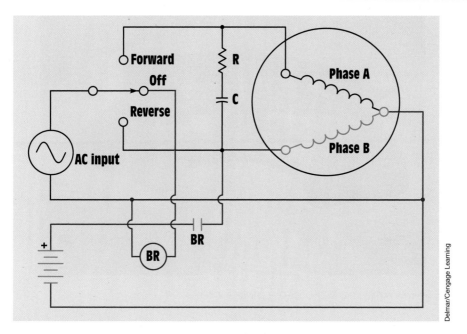

FIGURE 34–61 Applying DC voltage to increase holding torque.

34–16 Universal Motors

The **universal motor** is often referred to as an AC series motor. This motor is very similar to a DC series motor in its construction in that it contains a wound armature and brushes *(Figure 34–62)*. The universal motor, however, has the addition of a **compensating winding.** If a DC series motor is connected to AC, the motor operates poorly for several reasons. The armature windings have a large amount of inductive reactance when connected to AC. Another reason for poor operation is that the field poles of most DC machines contain solid metal pole pieces. If the field is connected to AC, a large amount of power is lost to eddy current induction in the pole pieces. Universal motors contain a laminated core to help prevent this problem. The compensating winding is wound around the stator and functions to counteract the inductive reactance in the armature winding.

The universal motor is so named because it can be operated on AC or DC voltage. When the motor is operated on DC, the compensating winding is connected in series with the series field winding *(Figure 34–63)*.

Connecting the Compensating Winding for AC

When the universal motor is operated with AC power, the compensating winding can be connected in two ways. If it is connected in series

Delmar/Cengage Learning

FIGURE 34–62   Armature and brushes of a universal motor.

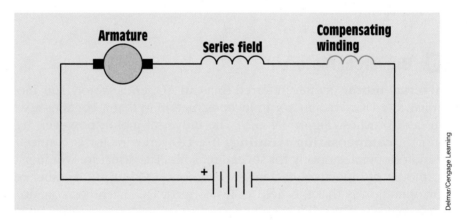

Delmar/Cengage Learning

FIGURE 34–63   The compensating winding is connected in series with the series field winding.

with the armature, as shown in *Figure 34–64*, it is known as **conductive compensation.**

The compensating winding can also be connected by shorting its leads together, as shown in *Figure 34–65*. When connected in this manner, the winding acts like a shorted secondary winding of a transformer. Induced current permits the winding to operate when connected in this manner. This connection is known as **inductive compensation.** Inductive compensation cannot be used when the motor is connected to DC.

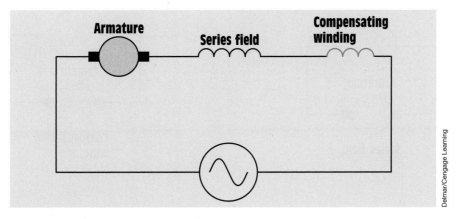

FIGURE 34–64 Conductive compensation.

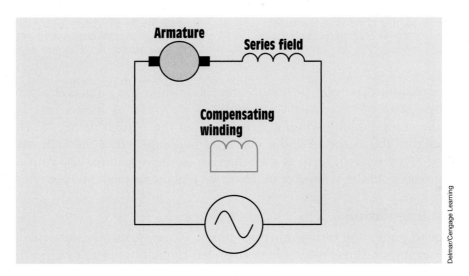

FIGURE 34–65 Inductive compensation.

The Neutral Plane

Because the universal motor contains a wound armature, commutator, and brushes, the brushes should be set at the **neutral plane** position. This can be done in the universal motor in a manner similar to that of setting the neutral plane of a DC machine. When setting the brushes to the neutral plane position in a universal motor, either the series or compensating winding can be used. To set the brushes to the neutral plane position using the series winding *(Figure 34–66)*, AC is connected to the armature leads. A voltmeter is connected to the series winding. Voltage is then applied to the armature. The brush position is then moved until the voltmeter connected to the series field reaches

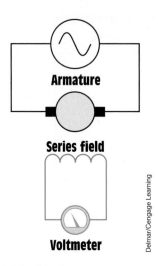

FIGURE 34–66   Using the series field to set the brushes at the neutral plane position.

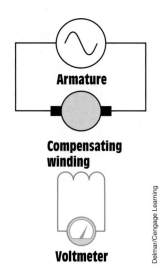

FIGURE 34–67   Using the compensating winding to set the brushes to the neutral plane position.

a null position. (The null position is reached when the voltmeter reaches its lowest point.)

If the compensating winding is used to set the neutral plane, AC is again connected to the armature and a voltmeter is connected to the compensating winding *(Figure 34–67)*. AC is then applied to the armature. The brushes are then moved until the voltmeter indicates its highest or peak voltage.

Speed Regulation

The speed regulation of the universal motor is very poor. Because this motor is a series motor, it has the same poor speed regulation as a DC series motor. If the universal motor is connected to a light load or no load, its speed is almost unlimited. It is not unusual for this motor to be operated at several thousand revolutions per minute. Universal motors are used in a number of portable appliances where high horsepower and light weight are needed, such as drill motors, skill saws, and vacuum cleaners. The universal motor is able to produce a high horsepower for its size and weight because of its high operating speed.

Changing the Direction of Rotation

The direction of rotation of the universal motor can be changed in the same manner as changing the direction of rotation of a DC series motor. To change the direction of rotation, change the armature leads with respect to the field leads.

Summary

- Not all single-phase motors operate on the principle of a rotating magnetic field.

- Split-phase motors start as two-phase motors by producing an out-of-phase condition for the current in the run winding and the current in the start winding.

- The resistance of the wire in the start winding of a resistance-start induction-run motor is used to produce a phase angle difference between the current in the start winding and the current in the run winding.

- The capacitor-start induction-run motor uses an AC electrolytic capacitor to increase the phase angle difference between starting and running current. This causes an increase in starting torque.

- Maximum starting torque for a split-phase motor is developed when the start-winding current and run-winding current are 90° out of phase with each other.

- Most resistance-start induction-run motors and capacitor-start induction-run motors use a centrifugal switch to disconnect the start windings when the motor reaches approximately 75% of full-load speed.

- The capacitor-start capacitor-run motor operates like a two-phase motor because both the start and run windings remain energized during motor operation.

- Most capacitor-start capacitor-run motors use an AC oil-filled capacitor connected in series with the start winding.

- The capacitor of the capacitor-start capacitor-run motor does help to correct the power factor.

- Shaded-pole induction motors operate on the principle of a rotating magnetic field.

- The rotating magnetic field of a shaded-pole induction motor is produced by placing shading loops or coils on one side of the pole piece.

- The synchronous-field speed of a single-phase motor is determined by the number of stator poles and the frequency of the applied voltage.

- Consequent-pole motors are used when a change of motor speed is desired and high torque must be maintained.

- Multispeed fan motors are constructed by connecting windings in series with the main run winding.

- Multispeed fan motors have high-impedance stator windings to prevent them from overheating when their speed is reduced.

- There are three basic repulsion-type motors: the repulsion motor, the repulsion-start induction-run motor, and the repulsion-induction motor.

- Repulsion motors have the highest starting torque of any single-phase motor.

- The direction of rotation of repulsion motors is changed by setting the brushes 15° on either side of the pole pieces.

- The direction of rotation for split-phase motors is changed by reversing the start winding in relation to the run winding.

- Shaded-pole motors are generally considered to be nonreversible.

- There are two types of repulsion-start induction-run motors: the brush-riding type and the brush-lifting type.

- The brush-riding type of motor uses an axial commutator and a short-circuiting device, which short-circuits the commutator segments when the motor reaches approximately 75% of full-load speed.

- The brush-lifting type of repulsion-start induction-run motor uses a radial commutator. A centrifugal device causes the brushes to move away from the commutator and a short-circuiting necklace to short-circuit the commutator when the motor reaches about 75% of full-load speed.

- The repulsion-induction motor contains both a wound armature and squirrel-cage windings.

- There are two types of single-phase synchronous motor: the Warren and the Holtz.

- Single-phase synchronous motors are sometimes called hysteresis motors.

- The Warren motor operates at a speed of 3600 rpm.

- The Holtz motor operates at a speed of 1200 rpm.

- Stepping motors generally operate on DC and are used to produce angular movements in steps.

- Stepping motors are generally used for position control.

- Stepping motors can be used as synchronous motors when connected to two-phase AC.

- Stepping motors operate at a speed of 72 rpm when connected to 60-hertz power.

- Stepping motors can produce a holding torque when DC is connected to their windings.

- Universal motors operate on DC or AC.

- Universal motors contain a wound armature and brushes.

- Universal motors are also called AC series motors.

- Universal motors have a compensating winding that helps overcome inductive reactance.

- The direction of rotation for a universal motor can be changed by reversing the armature leads with respect to the field leads.

Review Questions

1. What are the three basic types of split-phase motors?

2. The voltages of a two-phase system are how many degrees out of phase with each other?

3. How are the start and run windings of a split-phase motor connected in relation to each other?

4. In order to produce maximum starting torque in a split-phase motor, how many degrees out of phase should the start- and run-winding currents be with each other?

5. What is the advantage of the capacitor-start induction-run motor over the resistance-start induction-run motor?

6. On the average, how many degrees out of phase with each other are the start- and run-winding currents in a resistance-start induction-run motor?

7. What device is used to disconnect the start windings for the circuit in most nonhermetically sealed capacitor-start induction-run motors?

8. Why does a split-phase motor continue to operate after the start windings have been disconnected from the circuit?

9. How can the direction of rotation of a split-phase motor be reversed?

10. If a dual-voltage split-phase motor is to be operated on high voltage, how are the run windings connected in relation to each other?

11. When determining the direction of rotation for a split-phase motor, should you face the motor from the front or from the rear?

12. What type of split-phase motor does not generally contain a centrifugal switch?

13. What type of single-phase motor develops the highest starting torque?

14. What is the principle of operation of a repulsion motor?

15. What type of commutator is used with a brush-lifting-type repulsion-start induction-run motor?

16. When a repulsion-start induction-run motor reaches about 75% of rated full-load speed, it stops operating as a repulsion motor and starts operating as a squirrel-cage motor. What must be done to cause the motor to begin operating as a squirrel-cage motor?

17. What is the principle of operation of a capacitor-start capacitor-run motor?

18. What causes the magnetic field to rotate in a shaded-pole induction motor?

19. How can the direction of rotation of a shaded-pole induction motor be changed?

20. How is the speed of a consequent-pole motor changed?

21. Why can a multispeed fan motor be operated at lower speed than most induction motors without harm to the motor windings?

22. What is the speed of operation of the Warren motor?

23. What is the speed of operation of the Holtz motor?

24. Explain the difference in operation between a stepping motor and a common DC motor.

25. What is the principle of operation of a stepping motor?

26. What does the term *bifilar* mean?

27. Why do stepping motors have teeth machined in the stator poles and rotor?

28. When a stepping motor is connected to AC power, how many phases must be applied to the motor?

29. What is the synchronous speed of an eight-pole stepping motor when connected to a two-phase, 60-Hz AC line?

30. How can the holding torque of a stepping motor be increased?

31. Why is the AC series motor often referred to as a universal motor?

32. What is the function of the compensating winding?

33. How is the direction of rotation of the universal motor reversed?

34. When the motor is connected to DC voltage, how must the compensating winding be connected?

35. Explain how to set the neutral plane position of the brushes using the series field.

36. Explain how to set the neutral plane position using the compensating winding.

Practical Applications

You are an electrical contractor, and you have been called to a home to install a well pump. The homeowner has purchased the pump but does not know how to connect it. You open the connection terminal cover and discover that the motor contains eight terminal leads marked T_1 through T_8. The motor is to be connected to 240 V. At present, the T leads are connected as follows: T_1, T_3, T_5, and T_7 are connected together; and T_2, T_4, T_6, and T_8 are connected together. L_1 is connected to the group of terminals with T_1, and L_2 is connected to the group of terminals with T_2. Is it necessary to change the leads for operation on 240 V? If so, how should they be connected? ∎

Identifying the Leads of a Three-Phase, Wye-Connected, Dual-Voltage Motor

The terminal markings of a three-phase motor are standardized and used to connect the motor for operation on 240 or 480 volts. *Figure 1* shows these terminal markings and their relationship to the other motor windings. If the motor is connected to a 240-volt line, the motor windings are connected parallel to each other, as shown in *Figure 2*. If the motor is to be operated on a 480-volt line, the motor windings are connected in series, as shown in *Figure 3*.

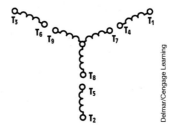

FIGURE 1 Standard terminal markings for a 3-phase motor.

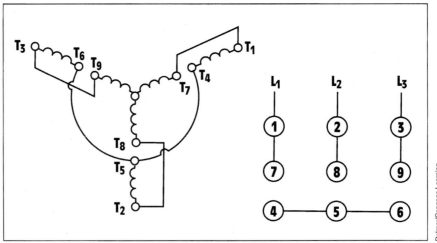

FIGURE 2 Low-voltage connection.

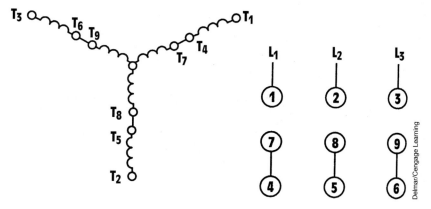

FIGURE 3   High-voltage connection.

As long as these motor windings remain marked with proper numbers, connecting the motor for operation on a 240- or 480-volt powerline is relatively simple. If these numbers are removed or damaged, however, the lead must be reidentified before the motor can be connected. The following procedure can be used to identify the proper relationship of the motor windings:

1. Using an ohmmeter, divide the motor windings into four separate circuits. One circuit will have continuity to three leads, and the other three circuits will have continuity between only two leads (see *Figure 1*).

    Caution: The circuits that exhibit continuity between two leads must be identified as pairs, but do not let the ends of the leads touch anything.

2. Mark the three leads that have continuity with each other as T_7, T_8, and T_9. Connect these three leads to a 240-volt, three-phase power source *(Figure 4)*. (Note: Because these windings are rated at 240 volts each, the

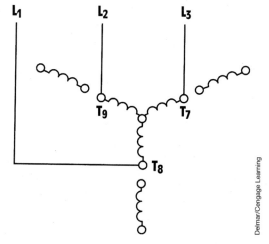

FIGURE 4 T_7, T_8, and T_9 connected to a 3-phase, 240-volt line.

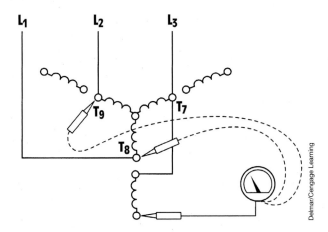

FIGURE 5 Measure voltage from unconnected paired leads to T_8 and T_9.

motor can be safely operated on one set of windings as long as it is not connected to a load.)

3. With the power turned off, connect one end of one of the paired leads to the paired leads to the terminal marked T_7. Turn the power on, and using an AC voltmeter set for a range not less than 480 volts, measure the voltage from the unconnected end of the paired lead to Terminals T_8 and T_9 *(Figure 5)*. If the measured voltages are unequal, the wrong paired lead is connected to Terminal T_7. Turn the power off, and connect another paired lead to T_7. When the correct set of paired leads is connected to T_7, the voltage readings to T_8 and T_9 are equal.

4. After finding the correct pair of leads, a decision must be made as to which lead should be labeled T_4 and which should be labeled T_1. Because an induction motor is basically a transformer, the phase windings act very similar to a multiwinding autotransformer. If Terminal T_1 is connected to Terminal T_7, it will operate similarly to a transformer with its windings connected to form subtractive polarity. If an AC voltmeter is connected to T_4, a voltage of about 140 volts should be seen between T_4 and T_8, or T_4 and T_9 *(Figure 6)*.

   If Terminal T_4 is connected to T_7, the winding will operate similarly to a transformer with its windings connected for additive polarity. If an AC voltmeter is connected to T_1, a voltage of about 360 volts will be indicated when the other lead of the voltmeter is connected to T_8 or T_9 *(Figure 7)*.

   Label leads T_1 and T_4, using the preceding procedure to determine which lead is correct. Then disconnect and separate T_1 and T_4.

5. To identify the other leads, follow the same basic procedure. Connect one end of the remaining pairs to T_8. Measure the voltage between the unconnected lead and T_7 and T_9 to determine whether it is the correct

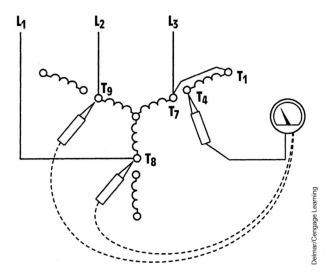

FIGURE 6 T_1 connected to T_7.

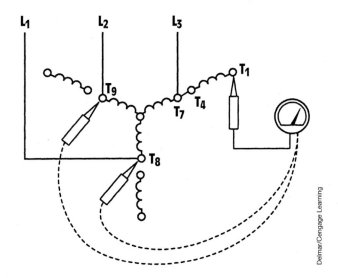

FIGURE 7 T_4 connected to T_7.

lead pair for Terminal T_8. When the correct lead pair is connected to T_8, the voltage between the unconnected terminal and T_7 or T_9 will be equal. Then determine which is T_5 or T_2 by measuring for a high or low voltage. When T_5 is connected to T_8, about 360 volts can be measured between T_2 and T_7, or T_2 and T_9.

6. The remaining pair can be identified as T_3 or T_6. When T_6 is connected to T_9, voltage of about 360 volts can be measured between T_3 and T_7, or T_3 and T_8.

AC Formulas

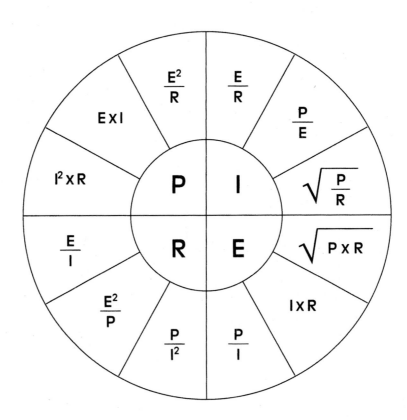

Instantaneous and Maximum Values

The instantaneous value of voltage and current for a sine wave is equal to the peak, or maximum, value of the waveform times the sine of the angle. For example, a waveform has a peak value of 300 volts. What is the voltage at an angle of 22°?

$$E_{(INST)} = E_{(MAX)} \times \sin \angle$$
$$E_{(INST)} = 300 \times 0.3746 \ (\sin \text{ of } 22°)$$
$$E_{(INST)} = 112.328 \text{ V}$$

$$E_{MAX} = \frac{E_{INST}}{\sin \angle}$$

$$\sin \angle = \frac{E_{INST}}{E_{MAX}}$$

Changing Peak, RMS, and Average Values

To change	To	Multiply by
Peak	RMS	0.707
Peak	Average	0.637
Peak	Peak-to-Peak	2
RMS	Peak	1.414
Average	Peak	1.567
RMS	Average	0.9
Average	RMS	1.111

Pure Resistive Circuit

$E = I \times R$ $I = \dfrac{E}{R}$ $R = \dfrac{E}{I}$ $P = E \times I$

$E = \dfrac{P}{I}$ $I = \dfrac{P}{R}$ $R = \dfrac{E^2}{P}$ $P = I^2 \times R$

$E = \sqrt{P \times R}$ $I = \sqrt{\dfrac{P}{R}}$ $R = \dfrac{P}{I^2}$ $P = \dfrac{E^2}{R}$

Series Resistive Circuits

$R_T = R_1 + R_2 + R_3$
$I_T = I_1 = I_2 = I_3$
$E_T = E_1 + E_2 + E_3$
$P_T = P_1 + P_2 + P_3$

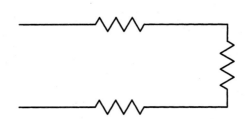

Parallel Resistive Circuits

$R_T = \dfrac{1}{\dfrac{1}{R_1} + \dfrac{1}{R_2} + \dfrac{1}{R_3}}$

$R_T = \dfrac{R_1 \times R_2}{R_1 + R_2}$

$I_T = I_1 + I_2 + I_3$

$E_T = E_1 = E_2 = E_3$

$P_T = P_1 = P_2 = P_3$

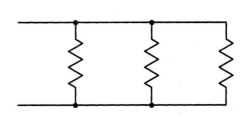

Pure Inductive Circuits

In a pure inductive circuit, the current lags the voltage by 90°. Therefore, there is no true power, or watts, and the power factor is 0. VARs is the inductive equivalent of watts.

$$L = \frac{X_L}{2\pi f}$$

$$X_L = 2\pi fL$$

$$E_L = I_L \times X_L$$

$$I_L = \frac{E_L}{X_L}$$

$$X_L = \frac{E_L}{I_L}$$

$$VARs_L = E_L \times I_L$$

$$E_L = \sqrt{VARs_L \times X_L}$$

$$I_L = \frac{VARs_L}{E_L}$$

$$X_L = \frac{E_L^2}{VARs_L}$$

$$VARs_L = I_L^2 \times X_L$$

$$E_L = \frac{VARs_L}{I_L}$$

$$I_L = \sqrt{\frac{VARs_L}{X_L}}$$

$$X_L = \frac{VARs_L}{I_L^2}$$

$$VARs_L = \frac{E_L^2}{X_L}$$

Series Inductive Circuits

$$E_{LT} = E_{L1} + E_{L2} + E_{L3}$$

$$X_{LT} = X_{L1} + X_{L2} + X_{L3}$$

$$I_{LT} = I_{L1} = I_{L2} = I_{L3}$$

$$VARs_{LT} = VARs_{L1} + VARs_{L2} + VARs_{L3}$$

$$L_T = L_1 + L_2 + L_3$$

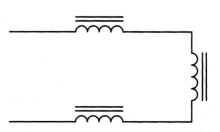

Parallel Inductive Circuits

$$E_{LT} = E_{L1} = E_{L2} = E_{L3}$$

$$I_{LT} = I_{L1} + I_{L2} + I_{L3}$$

$$X_{LT} = \frac{X_{L1} \times X_{L2}}{X_{L1} + X_{L2}}$$

$$X_{LT} = \frac{1}{\dfrac{1}{X_{L1}} + \dfrac{1}{X_{L2}} + \dfrac{1}{X_{L3}}}$$

$$L_T = \frac{1}{\dfrac{1}{L_1} + \dfrac{1}{L_2} + \dfrac{1}{L_3}}$$

$$L_T = \frac{L_1 \times L_2}{L_1 + L_2}$$

$$VARs_{LT} = VARs_{L1} + VARs_{L2} + VARs_{L3}$$

Pure Capacitive Circuits

In a pure capacitive circuit, the current leads the voltage by 90°. For this reason there is no true power, or watts, and no power factor. VARs is the equivalent of watts in a pure capacitive circuit.

The value of C in the formula for finding capacitance is in farads and must be changed into the capacitive units being used.

$$C = \frac{1}{2\pi f X_C} \qquad\qquad X_C = \frac{1}{2\pi f C}$$

$$E_C = I_C \times X_C \qquad I_C = \frac{E_C}{X_C} \qquad X_C = \frac{E_C}{I_C} \qquad VARs_C = E_C \times I_C$$

$$E_C = \sqrt{VARs_C \times X_C} \qquad I_C = \frac{VARs_C}{E_C} \qquad X_C = \frac{E_C^2}{VARs_C} \qquad VARs_C = I_C^2 \times X_C$$

$$E_C = \frac{VARs_C}{I_C} \qquad I_C = \sqrt{\frac{VARs_C}{X_C}} \qquad X_C = \frac{VARs_C}{I_C^2} \qquad VARs_C = \frac{W_C^2}{X_C}$$

Series Capacitive Circuits

$$E_{CT} = E_{C1} + E_{C2} + E_{C3}$$

$$I_{CT} = I_{C1} = I_{C2} = I_{C3}$$

$$X_{CT} = X_{C1} + X_{C2} + X_{C3}$$

$$VARs_{CT} = VARs_{C1} + VARs_{C2} + VARs_{C3}$$

$$C_T = \frac{1}{\dfrac{1}{C_1} + \dfrac{1}{C_2} + \dfrac{1}{C_3}} \qquad C_T = \frac{C_1 \times C_2}{C_1 + C_2}$$

Parallel Capacitive Circuits

$$E_{CT} = E_{C1} = E_{C2} = E_{C3} \qquad\qquad I_{CT} = I_{C1} + I_{C2} + I_{C3}$$

$$X_{CT} = \frac{1}{\dfrac{1}{X_{C1}} + \dfrac{1}{X_{C2}} + \dfrac{1}{X_{C3}}} \qquad X_{CT} = \frac{X_{C1} \times X_{C2}}{X_{C1} + X_{C2}}$$

$$C_T = C_1 + C_2 + C_3 \qquad\qquad VARs_{CT} = VARs_{C1} + VARs_{C2} + VARs_{C3}$$

Resistive-Inductive Series Circuits

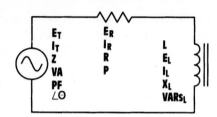

To find values for the resistor, use the formulas in the Pure Resistive section.

To find values for the inductor, use the formulas in the Pure Inductive section.

$$E_T = \sqrt{E_R^2 + E_L^2}$$

$$E_T = I_T \times Z$$

$$E_T = \frac{VA}{I_T}$$

$$E_T = \frac{E_R}{PF}$$

$$E_T = \sqrt{VA \times Z}$$

$$PF = \frac{R}{Z}$$

$$PF = \frac{P}{VA}$$

$$PF = \frac{E_R}{E_T}$$

$$PF = COS \angle\theta$$

$$R = \sqrt{Z^2 - X_L^2}$$

$$R = \frac{E_R}{I_R}$$

$$R = \frac{E_R^2}{P}$$

$$R = \frac{P}{I_R^2}$$

$$R = Z \times PF$$

$$VARs_L = \sqrt{VA^2 - P^2}$$

$$Z = \sqrt{R^2 + X_L^2}$$

$$Z = \frac{E_T}{I_T}$$

$$Z = \frac{VA}{I_T^2}$$

$$Z = \frac{R}{PF}$$

$$Z = \frac{E_T^2}{VA}$$

$$P = E_R \times I_R$$

$$P = \sqrt{VA^2 - VARs_L^2}$$

$$P = \frac{E_R^2}{R}$$

$$P = I_R^2 \times R$$

$$P = VA \times PF$$

$$E_L = I_L \times X_L$$

$$E_L = \sqrt{E_T^2 - E_R^2}$$

$$E_L = \sqrt{VARs_L \times X_L}$$

$$E_L = \frac{VARs_L}{I_L}$$

$$VARs_L = E_L \times I_L$$

$$VA = E_T \times I_T$$

$$VA = I_T^2 \times Z$$

$$VA = \frac{E_T^2}{Z}$$

$$VA = \sqrt{P^2 + VARs_L^2}$$

$$VA = \frac{P}{PF}$$

$$E_R = I_R \times R$$

$$E_R = \sqrt{P \times R}$$

$$E_R = \frac{P}{I_R}$$

$$E_R = \sqrt{E_T^2 - E_L^2}$$

$$E_R = E_T \times PF$$

$$I_L = I_R = I_T$$

$$I_L = \frac{E_L}{X_L}$$

$$I_L = \frac{VARs_L}{E_L}$$

$$I_L = \sqrt{\frac{VARs_L}{X_L}}$$

$$VARs_L = \frac{E_L^2}{X_L}$$

$$L = \frac{X_L}{2\pi f}$$

$$I_T = I_R = I_L$$

$$I_T = \frac{E_T}{Z}$$

$$I_T = \frac{VA}{E_T}$$

$$I_T = \sqrt{\frac{VA}{Z}}$$

$$I_R = I_T = I_L$$

$$I_R = \frac{E_R}{R}$$

$$I_R = \frac{P}{E_R}$$

$$I_R = \sqrt{\frac{P}{R}}$$

$$X_L = \sqrt{Z^2 - R^2}$$

$$X_L = \frac{E_L}{I_L}$$

$$X_L = \frac{E_L^2}{VARs_L}$$

$$X_L = \frac{VARs_L}{I_L^2}$$

$$X_L = 2\pi f L$$

$$VARs_L = I_L^2 \times X_L$$

Resistive-Inductive Parallel Circuits

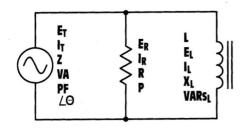

$$Z = \frac{1}{\sqrt{\left(\frac{1}{R}\right)^2 + \left(\frac{1}{X_L}\right)^2}}$$

$$Z = \frac{E_T}{I_T} \quad Z = \frac{E_T^2}{VA}$$

$$VA = E_T \times I_T$$

$$VA = I_T^2 \times Z$$

$$VA = \frac{E_T^2}{Z}$$

$$VA = \sqrt{P^2 + VARs_L^2}$$

$$VA = \frac{P}{PF}$$

$$VARs_L = I_L^2 \times X_L$$

$$X_L = \frac{E_L}{I_L}$$

$$X_L = \frac{E_L^2}{VARs_L}$$

$$X_L = \frac{VARs_L}{I_L^2}$$

$$Z = \frac{VA}{I_T^2}$$

$$Z = R \times PF$$

$$PF = \frac{Z}{R}$$

$$PF = \frac{P}{VA}$$

$$PF = \frac{I_R}{I_T}$$

$$PF = COS \angle\theta$$

$$VARs_L = \sqrt{VA^2 - P^2}$$

$$X_L = \frac{1}{\sqrt{\left(\frac{1}{Z}\right)^2 - \left(\frac{1}{R}\right)^2}}$$

$$X_L = 2\pi fL$$

$$R = \frac{E_R}{I_R}$$

$$E_T = E_R = E_L$$

$$E_T = I_T \times Z$$

$$L = \frac{X_L}{2\pi f}$$

$$I_T = \sqrt{I_R^2 + I_L^2}$$

$$I_T = \frac{E_T}{Z}$$

$$E_L = I_L \times X_L$$

$$E_L = E_T = E_R$$

$$E_L = \sqrt{VARs_L \times X_L}$$

$$E_L = \frac{VARs_L}{I_L}$$

$$VARs_L = E_L \times I_L$$

$$E_L = I_R \times R$$

$$E_R = \sqrt{P \times R}$$

$$E_R = \frac{P}{I_R}$$

$$E_R = E_T = E_L$$

$$R = \frac{P}{I_R^2}$$

$$E_T = \frac{VA}{I_T}$$

$$E_T = \sqrt{VA \times Z}$$

$$I_T = \frac{VA}{E_T} \quad I_T = \sqrt{\frac{VA}{Z}}$$

$$I_T = \frac{I_R}{PF}$$

$$I_L = \sqrt{I_T^2 - I_R^2}$$

$$I_L = \frac{E_L}{X_L}$$

$$I_L = \frac{VARs_L}{E_L}$$

$$I_L = \sqrt{\frac{VARs_L}{X_L}}$$

$$VARs_L = \frac{E_L^2}{X_L}$$

$$I_R = \sqrt{I_T^2 - I_L^2}$$

$$I_R = \frac{E_R}{R}$$

$$I_R = \frac{P}{E_R}$$

$$I_R = \sqrt{\frac{P}{R}}$$

$$I_R = I_T \times PF$$

$$R = \frac{1}{\sqrt{\left(\frac{1}{Z}\right)^2 - \left(\frac{1}{X_L}\right)^2}}$$

$$P = E_R \times I_R$$

$$R = \frac{E_R^2}{P}$$

$$P = \frac{E_R^2}{R}$$

$$R = \frac{Z}{PF}$$

$$P = I_R^2 \times R$$

$$P = \sqrt{VA^2 - VARs_L^2}$$

$$P = VA \times PF$$

Resistive-Capacitive Series Circuits

$$P = I_R^2 \times R$$

$$P = VA \times PF$$

$$E_C = I_C \times X_C$$

$$E_C = \sqrt{E_T^2 - E_R^2}$$

$$I_T = I_R = I_C$$

$$I_T = \frac{E_T}{Z}$$

$$I_T = \frac{VA}{E_T}$$

$$C = \frac{1}{2\pi f X_C}$$

$$E_R = I_R \times R$$

$$E_R = \sqrt{P \times R}$$

$$E_R = \frac{P}{I_R}$$

$$E_R = \sqrt{E_T^2 - E_C^2}$$

$$E_R = E_T \times PF$$

$$I_C = I_R = I_T$$

$$I_C = \frac{E_C}{X_C}$$

$$I_R = \frac{P}{E_R}$$

$$I_R = \sqrt{\frac{P}{R}}$$

$$X_C = \sqrt{Z^2 - R^2}$$

$$X_C = \frac{E_C}{I_C}$$

$$VARs_C = \sqrt{VA^2 - P^2}$$

$$R = \frac{P}{I_R^2}$$

$$R = Z \times PF$$

$$X_C = \frac{E_C^2}{VARs_C}$$

$$X_C = \frac{VARs_C}{I_C^2}$$

$$E_C = \sqrt{VARs_C \times X_C}$$

$$E_C = \frac{VARs_C}{I_C}$$

$$X_C = \frac{1}{2\pi fC}$$

$$VARs_C = E_C \times I_C$$

$$I_C = \frac{VARs_C}{E_C}$$

$$I_C = \sqrt{\frac{VARs_C}{X_C}}$$

$$VARs_C = I_C^2 \times X_C$$

$$VARs_C = \frac{E_C^2}{X_C}$$

$$Z = \frac{E_T^2}{VA}$$

$$Z = \sqrt{R^2 + X_C^2}$$

$$Z = \frac{E_T}{I_T}$$

$$VA = E_T \times I_T$$

$$VA = I_T^2 \times Z$$

$$VA = \frac{E_T^2}{Z}$$

$$VA = \sqrt{P^2 + VARs_C^2}$$

$$VA = \frac{P}{PF}$$

$$I_R = I_T = I_C$$

$$I_R = \frac{E_R}{R}$$

$$Z = \frac{VA}{I_T^2}$$

$$Z = \frac{R}{PF}$$

$$PF = \frac{R}{Z}$$

$$PF = \frac{P}{VA}$$

$$PF = \frac{E_R}{E_T}$$

$$PF = \cos \angle\theta$$

$$R = \sqrt{Z^2 - X_C^2}$$

$$R = \frac{E_R}{I_R}$$

$$R = \frac{E_R^2}{P}$$

$$E_T = \sqrt{E_R^2 + E_C^2}$$

$$E_T = I_T \times Z$$

$$E_T = \frac{VA}{I_T}$$

$$E_T = \frac{E_R}{PF}$$

$$P = E_R \times I_R$$

$$P = \sqrt{VA^2 - VARs_C^2}$$

$$P = \frac{E_R^2}{R}$$

Resistive-Capacitive Parallel Circuits

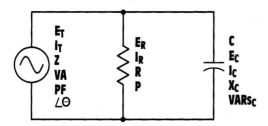

$$Z = \frac{1}{\sqrt{\left(\frac{1}{R}\right)^2 + \left(\frac{1}{X_C}\right)^2}}$$

$$I_T = \sqrt{I_R^2 + I_C^2}$$

$$I_T = \frac{E_T}{Z}$$

$$I_T = \frac{VA}{E_T}$$

$$I_T = \frac{I_R}{PF}$$

$$I_T = \sqrt{\frac{VA}{Z}}$$

$$Z = \frac{VA}{I_T^2}$$

$$VA = E_T \times I_T$$

$$VA = I_T^2 \times Z$$

$$VA = \frac{E_T^2}{Z}$$

$$VA = \sqrt{P^2 + VARs_C^2}$$

$$VA = \frac{P}{PF}$$

$$E_T = E_R = E_C$$

$$E_T = I_T \times Z$$

$$E_T = \frac{VA}{I_T}$$

$$C = \frac{1}{2\pi f X_C}$$

$$Z = \frac{E_T}{I_T} \quad Z = \frac{E_T^2}{VA}$$

$$PF = \frac{Z}{R}$$

$$PF = \frac{P}{VA}$$

$$PF = \frac{I_R}{I_T}$$

$$PF = \cos \angle\theta$$

$$E_T = \sqrt{VA \times Z}$$

$$Z = R \times PF$$

$$P = E_R \times I_R$$

$$P = \sqrt{VA^2 - VARs_C^2}$$

$$P = \frac{E_R^2}{R}$$

$$P = I_R^2 \times R$$

$$P = VA \times PF$$

$$E_R = I_R \times R$$

$$E_R = \sqrt{P \times R}$$

$$E_R = \frac{P}{I_R}$$

$$E_R = E_T = E_C$$

$$I_C = \sqrt{I_T^2 - I_R^2}$$

$$I_C = \frac{E_C}{X_C}$$

$$I_C = \frac{VARs_C}{E_C}$$

$$I_C = \sqrt{\frac{VARs_C}{X_C}}$$

$$I_R = \sqrt{I_T^2 - I_C^2}$$

$$I_R = \frac{E_R}{R}$$

$$I_R = \frac{P}{E_R}$$

$$I_R = \sqrt{\frac{P}{R}}$$

$$X_C = \frac{1}{\sqrt{\left(\frac{1}{Z}\right)^2 - \left(\frac{1}{R}\right)^2}}$$

$$X_C = \frac{E_C}{I_C}$$

$$R = \frac{1}{\sqrt{\left(\frac{1}{Z}\right)^2 - \left(\frac{1}{X_C}\right)^2}}$$

$$R = \frac{E_R}{I_R}$$

$$R = \frac{E_R^2}{P}$$

$$R = \frac{P}{I_R^2}$$

$$R = \frac{Z}{PF}$$

$$X_C = \frac{E_C^2}{VARs_C}$$

$$X_C = \frac{VARs_C}{I_C^2}$$

$$X_C = \frac{1}{2\pi f C}$$

$$E_C = I_C \times X_C$$

$$E_C = E_T = E_R$$

$$E_C = \sqrt{VARs_C \times X_C}$$

$$E_C = \frac{VARs_C}{I_C}$$

$$VARs_C = I_C^2 \times X_C$$

$$VARs_C = \frac{E_C^2}{X_C}$$

$$VARs_C = E_C \times I_C$$

$$VARs_C = \sqrt{VA^2 - P^2}$$

Resistive-Inductive-Capacitive Series Circuits

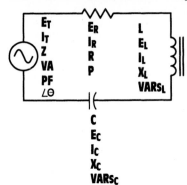

$$R = \sqrt{Z^2 - (X_L - X_C)^2}$$

$$R = \frac{E_R}{I_R}$$

$$R = \frac{E_R^2}{P}$$

$$X_C = \frac{1}{2\pi fC}$$

$$X_C = \frac{E_C}{VARs_C}$$

$$X_C = \frac{E_C}{I_C}$$

$$VARs_C = I_C^2 \times X_C$$

$$VARs_C = \frac{E_C^2}{X_C}$$

$$VARs_C = E_C \times I_C$$

$$I_T = I_R = I_L = I_C$$

$$I_T = \frac{E_T}{Z}$$

$$VA = E_T \times I_T$$

$$VA = I_T^2 \times Z$$

$$VA = \sqrt{P^2 + (VARs_L - VARs_C)^2}$$

$$R = \frac{P}{I_R^2}$$

$$X_C = \frac{VARs_C}{I_C^2}$$

$$L = \frac{X_L}{2\pi f}$$

$$X_L = \frac{E_L}{I_L}$$

$$X_L = \frac{E_L^2}{VARs_L}$$

$$I_T = \frac{VA}{E_T}$$

$$I_T = \sqrt{\frac{VA}{Z}}$$

$$VA = \frac{P}{PF}$$

$$VA = \frac{E_T^2}{Z}$$

$$E_R = I_R \times R$$

$$E_R = \frac{P}{I_R}$$

$$E_R = \sqrt{E_T^2 - (E_L - E_C)^2}$$

$$E_T = \sqrt{E_R^2 (E_L - E_C)^2}$$

$$E_T = I_T \times Z$$

$$E_T = \frac{VA}{I_T}$$

$$E_T = \frac{E_R}{PF}$$

$$Z = \sqrt{R^2 + (X_L - X_C)^2}$$

$$Z = \frac{E_T}{I_T}$$

$$Z = \frac{VA}{I_T^2}$$

$$Z = \frac{R}{PF}$$

$$PF = \frac{R}{Z}$$

$$PF = \frac{P}{VA}$$

$$PF = \frac{E_R}{E_T}$$

$$PF = \cos \angle\theta$$

$$I_R = I_T = I_C = I_L$$

$$I_R = \frac{E_R}{R}$$

$$I_R = \frac{P}{E_R}$$

$$I_R = \sqrt{\frac{P}{R}}$$

$$I_C = I_R = I_T = I_L$$

$$I_C = \frac{E_C}{X_C}$$

$$I_C = \frac{VARs_C}{E_C}$$

$$I_C = \sqrt{\frac{VARs_C}{X_C}}$$

$$P = E_R \times I_R$$

$$P = \sqrt{VA^2 - (VARs_L - VARs_C)^2}$$

$$P = \frac{E_R^2}{R}$$

$$P = I_R^2 \times R$$

$$P = VA \times PF$$

$$E_R = E_T \times PF$$

$$E_C = I_C \times X_C$$

$$E_C = \sqrt{VARs_C \times X_C}$$

$$E_R = \sqrt{P \times R}$$

$$R = Z \times PF$$

Resistive-Inductive-Capacitive Series Circuits (continued)

$$E_C = \frac{VARs_C}{I_C}$$

$$C = \frac{1}{2\pi f X_C}$$

$$X_L = 2\pi f L$$

$$X_L = \frac{VARs_L}{I_L^2}$$

$$E_L = I_L \times X_L$$

$$E_L = \sqrt{VARs_L \times X_L}$$

$$E_L = \frac{VARs_L}{I_L}$$

$$I_L = I_R = I_T = I_C$$

$$I_L = \frac{E_L}{X_L}$$

$$I_L = \frac{VARs_L}{E_L}$$

$$I_L = \sqrt{\frac{VARs_L}{X_L}}$$

$$VARs_L = E_L \times I_L$$

$$VARs_L = \frac{E_L^2}{X_L}$$

$$VARs_L = I_L^2 \times X_L$$

Resistive-Inductive-Capacitive Parallel Circuits

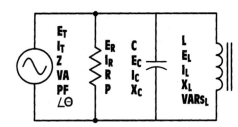

$$Z = \frac{1}{\sqrt{\left(\frac{1}{R}\right)^2 + \left(\frac{1}{X_L} - \frac{1}{X_C}\right)^2}}$$

$$Z = \frac{E_T}{I_T} \qquad Z = \frac{E_T^2}{VA} \qquad I_L = \frac{E_L}{X_L}$$

$$VA = E_T \times I_T \qquad\qquad Z = \frac{VA}{I_T^2}$$

$$VA = I_T^2 \times Z \qquad\qquad Z = R \times PF$$

$$VA = \frac{E_T^2}{Z}$$

$$VA = \sqrt{P^2 + (VARs_L - VARs_C)^2}$$

$$VA = \frac{P}{PF} \qquad\qquad PF = \frac{Z}{R}$$

$$PF = \frac{P}{VA}$$

$$PF = \frac{I_R}{I_T}$$

$$PF = \cos \angle \theta$$

$$X_L = \frac{E_L}{I_L}$$

$$I_L = \frac{VARs_L}{E_L}$$

$$I_L = \sqrt{\frac{VARs_L}{X_L}}$$

$$E_T = E_R = E_L = E_C$$

$$E_T = I_T \times Z$$

$$E_T = \frac{VA}{I_T}$$

$$E_T = \sqrt{VA \times Z}$$

$$I_T = \sqrt{I_R^2 + (I_L - I_C)^2}$$

$$I_T = \frac{E_T}{Z}$$

$$E_L = I_L \times X_L$$

$$E_L = \frac{VARs_L}{I_L}$$

$$X_L = 2\pi f L$$

$$X_L = \frac{E_L^2}{VARs_L}$$

$$X_L = \frac{VARs_L}{I_L^2}$$

$$I_T = \frac{VA}{E_T} \qquad I_T = \sqrt{\frac{VA}{Z}}$$

$$I_T = \frac{I_R}{PF}$$

$$E_L = E_T = E_R = E_C$$

$$E_L = \sqrt{VARs_L \times X_L}$$

$$L = \frac{X_L}{2\pi f}$$

Resistive-Inductive-Capacitive Parallel Circuits (continued)

$$VARS_L = \frac{E_L^2}{X_L}$$

$$I_R = \sqrt{I_T^2 - (I_L - I_C)^2}$$

$$I_R = I_T \times PF$$

$$R = \frac{1}{\sqrt{\left(\frac{1}{Z}\right)^2 - \left(\frac{1}{X_L} - \frac{1}{X_C}\right)^2}}$$

$$VARS_L = I_L^2 \times X_L$$

$$I_R = \frac{E_R}{R}$$

$$R = \frac{E_R}{I_R}$$

$$VARS_L = E_L \times I_L$$

$$R = \frac{E_R^2}{P}$$

$$I_R = \frac{P}{E_R}$$

$$R = \frac{P}{I_R^2} \quad R = \frac{Z}{PF}$$

$$E_R = I_R \times R$$

$$P = I_R^2 \times R$$

$$E_R = \sqrt{P \times R}$$

$$I_R = \sqrt{\frac{P}{R}}$$

$$P = E_R \times I_R$$

$$E_R = \frac{P}{I_R}$$

$$X_C = \frac{1}{2\pi fC}$$

$$P = VA \times PF$$

$$P = \frac{E_R^2}{R}$$

$$E_R = E_T = E_L = E_C$$

$$VARS_C = E_C \times I_C$$

$$X_C = \frac{E_C}{I_C}$$

$$P = \sqrt{VA^2 - (VARS_L - VARS_C)^2}$$

$$VARS_C = I_C^2 \times X_C$$

$$E_C = \frac{VARS_C}{I_C}$$

$$I_C = \frac{E_C}{X_C}$$

$$X_C = \frac{E_C^2}{VARS_C}$$

$$VARS_C = \frac{E_C^2}{X_C}$$

$$E_C = I_C \times X_C$$

$$I_C = \frac{VARS_C}{E_C}$$

$$X_C = \frac{VARS_C}{I_C^2}$$

$$E_C = E_T = E_R = E_L$$

$$E_C = \sqrt{VARS_C \times X_C}$$

$$I_C = \sqrt{\frac{VARS_C}{X_C}}$$

Transformers

$$\frac{E_P}{E_S} = \frac{N_P}{N_S} \qquad \frac{E_P}{E_S} = \frac{I_S}{I_P} \qquad \frac{N_P}{N_S} = \frac{I_S}{I_P} \qquad \frac{Z_P}{Z_S} = \left(\frac{N_P}{N_S}\right)^2 \qquad Z_P = Z_S\left(\frac{N_P}{N_S}\right)^2 \qquad Z_S = Z_P\left(\frac{N_S}{N_P}\right)^2$$

E_P—Voltage of the primary

E_S—Voltage of the secondary

I_P—Current of the primary

I_S—Current of the secondary

N_P—Number of turns of the primary

N_S—Number of turns of the secondary

Z_P—Impedance of the primary

Z_S—Impedance of the secondary

Three-Phase Connections

Wye Connection

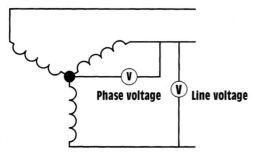

Phase voltage   Line voltage

Delta Connection

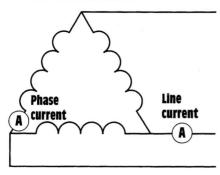

Phase current   Line current

Open Delta Connection

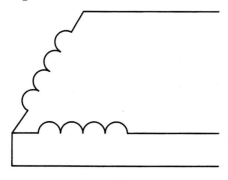

Apparent and True Power

$$VA = \sqrt{3} \times E_{LINE} \times I_{LINE}$$

$$VA = 3 \times E_{PHASE} \times I_{PHASE}$$

In a **wye** connection, the line current and phase current are the same:

$$I_{LINE} = I_{PHASE}$$

In a **wye** connection, the line voltage is higher than the phase voltage by a factor of the square root of 3:

$$E_{LINE} = E_{PHASE} \times \sqrt{3}$$

$$E_{PHASE} = \frac{E_{LINE}}{\sqrt{3}}$$

In a **delta**-connected system, the line voltage and phase voltage are the same:

$$E_{LINE} = E_{PHASE}$$

In a **delta**-connected system, the line current is higher than the phase current by a factor of the square root of 3:

$$I_{LINE} = I_{PHASE} \times \sqrt{3}$$

$$I_{PHASE} = \frac{I_{LINE}}{\sqrt{3}}$$

Open-delta connections provide 86.6% of the sum of the power rating of the two transformers. Example: The two transformers shown are rated at 60 kilovolt-amperes each. The total power rating for this connection would be

60 + 60 = 120 kVA

120 kVA × 0.866 = 103.92 kVA

All values of voltage and current are computed in the same manner as for a closed-delta connection.

$$P = \sqrt{3} \times E_{LINE} \times I_{LINE} \times PF$$

$$P = 3 \times E_{PHASE} \times I_{PHASE} \times PF$$

Greek Alphabet

Name of letter	Uppercase	Lowercase	Designates
Alpha	A	α	Angles
Beta	B	β	Angles; flux density
Gamma	Γ	γ	Conductivity; photon
Delta	Δ	δ	Change of a quantity
Epsilon	E	ε	Permittivity (dielectric constant)
Zeta	Z	ζ	Impedance; coefficients, coordinates
Eta	H	η	Hysteresis coefficient; efficiency
Theta	Θ	θ	Phase angle
Iota	I	ι	
Kappa	K	κ	Dielectric constant; coefficient of coupling; susceptibility
Lambda	Λ	λ	Wavelength
Mu	M	μ	Permeability; the prefix micro-, amplification factor
Nu	N	ν	Change of quantity
Xi	Ξ	ξ	
Omicron	O	o	
Pi	Π	π	3.1416
Rho	P	ρ	Resistivity
Sigma	Σ	σ	Conductivity
Tau	T	τ	Time constant; time phase displacement
Upsilon	Υ	υ	Reluctivity
Phi	Φ	φ	Angles; magnetic flow
Chi	X	χ	Susceptibility
Psi	Ψ	ψ	Dielectric flux; phase difference
Omega	Ω	ω	Ohm, unit of resistance angular velocity

Metals

Metal	Sym.	Spec. grav.	Melt Point °C	Melt Point °F	Elec. Cond. % copper	Lb/in.³
Aluminum	Al	2.710	660	1220	64.9	0.0978
Antimony	Sb	6.620	631	1167	4.42	0.2390
Arsenic	As	5.730			4.90	0.2070
Beryllium	Be	1.830	1287	2336	9.32	0.0660
Bismuth	Bi	9.800	271	512	1.50	0.3540
Brass (70–30)		8.510	900	1652	28.0	0.3070
Bronze (5% Sn)		8.870	1000	1382	18.0	0.3200
Cadmium	Cd	8.650	321	610	22.7	0.3120
Calcium	Ca	1.550	842	1562	50.1	0.0560
Cobalt	Co	8.900	1495	2723	17.8	0.3210
Copper	Cu	8.900	1085	1981	100.0	0.3210
Gold	Au	19.30	1064	1945	71.2	0.6970
Graphite		2.250	3500	6332	0.001	0.0812
Indium	In	7.300	156	311	20.6	0.2640
Iridium	Ir	22.40	2446	4442	32.5	0.8090
Iron	Fe	7.200	1400 to 1500	2552 to 2732	17.6	0.2600
Malleable		7.200	1600 to 1500	2912 to 2732	10.0	0.2600
Wrought		7.700	1600	2912	10.0	0.2780
Lead	Pb	11.40	327	621	8.35	0.4120
Magnesium	Mg	1.740	650	1204	38.7	0.0628
Manganese	Mn	7.200	1246	2273	0.90	0.2600
Mercury	Hg	13.65	−38.8	−37.7	1.80	0.4930
Molybdenum	Mo	10.20	2623	4748	36.1	0.3680

Metal	Sym.	Spec. grav.	Melt Point °C	Melt Point °F	Elec. Cond. % copper	Lb/in.³
Monel (63–37)		8.870	1300	2372	3.00	0.3200
Nickel	Ni	8.90	1455	2646	25.0	0.3210
Phosphorus	P	1.82	44.2	111.6	10^{-17}	0.0657
Platinum	Pt	21.46	1768	3214	17.5	0.0657
Potassium	K	0.860	63.0	146.1	28.0	0.0310
Selenium	Se	4.81	221	430	14.4	0.1740
Silicon	Si	2.40	1414	2577	10^{-5}	0.0866
Silver	Ag	10.50	962	1764	106	0.3790
Steel (carbon)		7.84	1330 to 1380	2436 to 2516	10.0	0.2830
Stainless						
(18–8)		7.92	1500	2732	2.50	0.2860
(13–Cr)		7.78	1520	2768	3.50	0.2810
(18–Cr)		7.73	1500	2732	3.00	0.2790
Tantalum	Ta	16.6	3017	5462	13.9	0.5990
Tellurium	Te	6.20	450	846	10^{-5}	0.2240
Thorium	Th	11.7	1750	3182	9.10	0.4220
Tin	Sn	7.30	232	449	15.0	0.2640
Titanium	Ti	4.50	1668	3034	2.10	0.1620
Tungsten	W	19.3	3422	6181	31.5	0.6970
Uranium	U	18.7	1135	2075	2.80	0.6750
Vanadium	V	5.96	1910	3470	6.63	0.2150
Zinc	Zn	7.14	420	788	29.1	0.2580
Zirconium	Zr	6.40	1855	3371	4.20	0.2310

Scientific Notation

In scientific notation, the coefficient of the power of 10 is always expressed with one digit to the left of the decimal point and the required power of 10.

Example:

$$350,000 = 3.5^{\times 10^5}$$
$$875 = 8.75^{\times 10^2}$$
$$0.002 = 2^{\times 10^{-2}}$$
$$0.000,045 = 4.5^{\times 10^{-5}}$$

Notice that in each example there is one digit to the left of the decimal place. $9500 = 95^{\times 10^{-2}}$ and $0.0056 = 0.56^{\times 10^{-2}}$ are not examples of scientific notation.

Scientific notation is used in almost all scientific calculations. It was first used for making calculations with a slide rule. A slide rule is a tool that can perform mathematical operations such as multiplication and division; find square roots; find logs of numbers; and find sines, cosines, and tangents of angles. When finding a number on a slide rule, only the actual digits are used, not decimal points or zeros (unless they come between two other digits such as 102). To a slide rule, the numbers 0.000012, 0.0012, 0.12, 1.2, 12, 120, 12000, or 12000000 are all the same number: 12. Since the slide rule recognizes only the basic digits of any number, imagine the problem of determining where to place a decimal point in an answer. As long as only simple calculations are done, there is no problem in determining where the decimal point should be placed.

■ EXAMPLE 1

Multiply 12 × 20. It is obvious where the decimal point should be placed in this problem:

$$240.00$$

■ EXAMPLE 2

Now assume that the following numbers are to be multiplied together:

$$0.000041 \times 380{,}000 \times 0.19 \times 720 \times 0.0032$$

In this problem, it is not obvious where the decimal point should be placed in the answer. Scientific notation can be used to simplify the numbers so that an estimated answer can be obtained. Scientific notation is used to change any number to a simple whole number by dividing or multiplying the number by a power of 10. Any number can be multiplied by 10 by moving the decimal point one place to the right. Any number can be divided by 10 by moving the decimal point one place to the left. For example, the number 0.000041 can be changed to a whole number of 4.1 by multiplying it by 10 five times. Therefore, if the number 4.1 is divided by 10 five times it will be the same as the original number, 0.000041. In this problem, the number 0.000041 will be changed to 4.1 by multiplying it by 10 five times. The new number is 4.1 times 10 to the negative fifth:

$$4.1 \times 10^{-5}$$

Because super- and subscripts are often hard to produce in printed material, it is a common practice to use the letter E to indicate the exponent in scientific notation. The preceding notation can also be written

$$4.1E - 05$$

The number 380,000 can be reduced to a simple whole number of 3.8 by dividing it by 10 five times. The number 3.8 must, therefore, be notated to indicate that the original number is actually 3.8 multiplied by 10 five times:

$$3.8 \times 10^{+5}$$

or

$$3.8E05$$

The other numbers in the problem can also be changed to simple whole numbers using scientific notation:

0.19 becomes 1.9E − 01

720 becomes 7.2E02

0.0032 becomes 3.2E − 03

Now that the numbers have been simplified using scientific notation, an estimate can be obtained by multiplying these simple numbers together and adding the exponents: 4.1 is about 4; 3.8 is about 4; 1.9 is about 2; 7.2 is about 7; and 3.2 is about 3. Therefore:

$$4 \times 4 \times 2 \times 7 \times 3 = 672$$

When the exponents are added:

$$(E - 05) + (E05) + (E - 01) + (E02) + (E - 03) = E - 02$$

The estimated answer becomes 672E − 02. When the calculation is completed, the actual answer becomes:

$$682.03E - 02$$

or

$$6.8203$$

Using Scientific Notation with Calculators

In the early 1970s, scientific calculators, often referred to as *slide-rule calculators,* became common place. Most of these calculators have the ability to display from 8 to 10 digits, depending on the manufacturer. Scientific calculations, however, often involve numbers that contain more than eight or ten digits. To overcome the limitation of an 8- or 10- digit display, slide-rule calculators depend on scientific notation. When a number becomes too large for the calculator to display, scientific notation is used automatically. Imagine, for example, that it became necessary to display the distance (in kilometers) that light travels in one year (approx. 9,460,800,000,000 km). This number contains 13 digits. The calculator would display this number as shown:

$$\boxed{9.4608 \ 12}$$

The number 12 shown to the right of 9.4608 is the scientific notation exponent. This number could be written 9.4608E12, indicating that the decimal point should be moved to the right 12 places.

　If a minus sign should appear ahead of the scientific notation exponent, it indicates that the decimal point should be moved to the left. The number on the following display contains a negative scientific notation exponent:

$$\boxed{7.5698 - 06}$$

This number could be rewritten as:

$$0.0000075698$$

Entering Numbers in Scientific Notation

Slide-rule calculators also have the ability to enter numbers in scientific notation. To do this, the exponent key must be used. There are several ways in which manufacturers mark the exponent key. Some are marked **EXP** and others are marked **EE.** Some manufacturers use **E** and **EXX.** Regardless of how the manufacturer chooses to indicate the functions, they work the same.

Assume that the number 549E08 is to be entered. The following key strokes are used:

$$\boxed{5} \quad \boxed{4} \quad \boxed{9} \quad \boxed{EE} \quad \boxed{8}$$

There are also different ways in which scientific notation may appear on the display. Several are shown:

$$\boxed{5.49 \ 10} \qquad \boxed{5.49 \ _{\times 10^{10}}} \qquad \boxed{5.49 \ E10}$$

Notice that although the number is entered as 549 EE 8, the calculator automatically changes the number so there is only one digit to the left of the decimal place.

If a number with a negative exponent is to be entered, the change sign (+/−) key should be used. Assume that the number 1.276E − 4 is to be entered. The following key strokes would be used:

$$\boxed{1} \quad \boxed{\bullet} \quad \boxed{2} \quad \boxed{7} \quad \boxed{6} \quad \boxed{EXP} \quad \boxed{+/-} \quad \boxed{4}$$

The display would show the following:

$$\boxed{1.276 - 04}$$

Setting the Display

Some calculators permit the answer to be displayed in any of three different ways. One of these ways is with *floating decimals (FD)*. When the calculator is set for this mode of operation, the answers are displayed with the decimal point appearing in the normal position. The only exception to this is if the number to be displayed is too large. In this case, the calculator automatically displays the number in scientific notation. In the *scientific mode (Sci),* the calculator displays all entries and answers in scientific notation.

When set in the *engineering mode (Eng),* the calculator displays all entries and answers in *engineering notation*. Engineering notation is very similar to scientific notation except that all answers are displayed in steps of thousand instead of ten. This simply means that the scientific notation units are in steps of three instead of one. When in engineering notation, the calculator does not display exponents of 1, 2, 4, 5, 7, 8, and so on. When displayed in steps of three, the notation corresponds to standard engineering notation units such as kilo, mega, giga, milli, micro, and so on. For example, assume a calculator is set in the scientific mode and displays the number shown:

$$\boxed{5.69836 \ 05}$$

Now assume that the calculator is reset to the engineering mode. The number would now be displayed as shown:

$$\boxed{569.836 \ 03}$$

The number could now be read as "569.836 kilo," because kilo means 1000, or 1E03.

Answers to Practice Problems

Unit 2 Electrical Quantities and Ohm's Law

Volts (E)	Amperes (I)	Ohms (R)	Watt (P)
153	0.056	2732.143 Ω	8.568
305.5	0.65	470 Ω	198.575
24	5.167	4.645 Ω	124
3.59	0.00975	368.179 Ω	0.035
76.472	0.0112	6.8 kΩ	0.86
460	6.389	72 Ω	2938.889
48	1.2	40 Ω	57.6
123.2	154	0.8 Ω	18,972.8
277	2.744	100.959 Ω	760
14.535	0.0043	3380.206 Ω	0.0625
54.083	0.000416	130 kΩ	0.0225
96	0.0436	2.2 kΩ	4.189

Unit 5 Resistors

1st Band	2nd Band	3rd Band	4th Band	Value	%Tol
Red	Yellow	Brown	Silver	240 Ω	10
Blue	Gray	Red	Gold	6800	5
Orange	Orange	Orange	Gold	33 kΩ	5
Brown	Red	Black	Red	12 Ω	2
Brown	Green	Silver	Silver	0.15 Ω	10
Brown	Gray	Green	Silver	1.8 MΩ	10
Brown	Black	Yellow	None	100 kΩ	20
Brown	Black	Orange	Gold	10 kΩ	5
Violet	Green	Black	Red	75 Ω	2
Yellow	Violet	Red	None	4.7 kΩ	20
Gray	Red	Green	Red	8.2 MΩ	2
Green	Blue	Gold	Red	5.6 Ω	2

Unit 6 Series Circuits

1.

E_T120 V	E_1 16.34 V	E_2 13.68 V	E_3 28.5 V	E_4 38 V	E_5 23.56 V
I_T 0.038 A	I_1 0.038 A	I_2 0.038 A	I_3 0.038 A	I_4 0.038 A	I_5 0.038 A
R_T 3160 Ω	R_1 430 Ω	R_2 360 Ω	R_3 750 Ω	R_4 1000 Ω	R_5 620 Ω
P_T 4.56 W	P_1 0.621 W	P_2 0.52 W	P_3 1.083 W	P_4 1.444 W	P_5 0.894 W

E_T 50 V	E_1 6 V	E_2 16.5 V	E_3 11 V	E_4 9 V	E_5 7.5 V
I_T 0.005 A	I_1 0.005 A	I_2 0.005 A	I_3 0.005A	I_4 0.005 A	I_5 0.005 A
R_T 10 kΩ	R_1 1.2 kΩ	R_2 3.3 kΩ	R_3 2.2 kΩ	R_4 1.8 kΩ	R_5 1.5 kΩ
P_T 0.25 W	P_1 0.03 W	P_2 0.0825 W	P_3 0.055 W	P_4 0.045 W	P_5 0.0375 W

E_T 340 V	E_1 44 V	E_2 94 V	E_3 60 V	E_4 40 V	E_5 102 V
I_T 0.002 A	I_1 0.002 A	I_2 0.002 A	I_3 0.002 A	I_4 0.002 A	I_5 0.002 A
R_T 170 kΩ	R_1 22 kΩ	R_2 47 kΩ	R_3 30 kΩ	R_4 20 kΩ	R_5 51 kΩ
P_T 0.68 W	P_1 0.088 W	P_2 0.188 W	P_3 0.12 W	P_4 0.08 W	P_5 0.204 W

2.

$E_1 = 18.46$ V $E_2 = 40.62$ V $E_3 = 33.23$ V $E_4 = 27.69$ V

Unit 7 Parallel Circuits

1.

E_T 120.03 V	E_1 120.03 V	E_2 120.03 V	E_3 120.03 V	E_4 120.03 V
I_T 0.942 A	I_1 0.177 A	I_2 0.146 A	I_3 0.255 A	I_4 0.364 A
R_T 127.42 Ω	R_1 680 Ω	R_2 820 Ω	R_3 470 Ω	R_4 330 Ω
P_T 113.068 W	P_1 21.425 W	P_2 17.542 W	P_3 27.007 W	P_4 43.619 W

2.

E_T 277.056 V	E_1 277.056 V	E_2 277.056 V	E_3 277.056 V	E_4 277.056 V
I_T 0.00639 A	I_1 0.00231 A	I_2 0.00139 A	I_3 0.00154 A	I_4 0.00115 A
R_T 43,357.746 Ω	R_1 119,937.662 Ω	R_2 199,320.863 Ω	R_3 179,906.494 Ω	R_4 240,918.261 Ω
P_T 1.771 W	P_1 0.640 W	P_2 0.384 W	P_3 0.427 W	P_4 0.319 W

3.

E_T 47.994 V	E_1 47.994 V	E_2 47.994 V	E_3 47.994 V	E_4 47.994 V
I_T 13.399 A	I_1 3 A	I_2 4.799 A	I_3 3.2 A	I_4 2.4 A
R_T 3.582 Ω	R_1 16 Ω	R_2 10 Ω	R_3 15 Ω	R_4 20 Ω
P_T 643.072 W	P_1 143.982 W	P_2 230.323 W	P_3 153.581 W	P_4 115.186 W

4.

E_T 240.78 V	E_1 240.78 V	E_2 240.78 V	E_3 240.78 V	E_4 240.78 V
I_T 0.0143 A	I_1 0.00293 A	I_2 0.0032 A	I_3 0.00429 A	I_4 0.00387 A
R_T 16,802.591 Ω	R_1 82 Ω	R_2 75 kΩ	R_3 56 Ω	R_4 62 Ω
P_T 3.436 W	P_1 0.704 W	P_2 0.769 W	P_3 1.031 W	P_4 0.932 W

5.

$R_T = 83.735\ \Omega$ $I_1 = 0.0116\ A$ $I_2 = 0.00891\ A$
$I_3 = 0.0140\ A$ $I_4 = 0.0155\ A$

6.

$R_T = 53{,}171.513\ \Omega$ $I_1 = 1.18\ mA$ $I_2 = 0.886\ mA$
$I_3 = 0.742\ mA$ $I_4 = 3.19\ mA$

Unit 8 Combination Circuits

1.

| I_T 0.0241 A | E_1 36.15 V | E_2 12.103 V | E_3 26.6 V | E_4 38.782 V |
| R_T 3109.216 Ω | I_1 0.0241 A | I_2 0.0133 A | I_3 0.0133 A | I_4 0.0107 A |

2.

| I_T 0.00946 A | E_1 208.12 V | E_2 85.14 V | E_3 56.76 V | E_4 141.9 V |
| R_T 37 kΩ | I_1 0.00946 A | I_2 0.00473 A | I_3 0.00473 A | I_4 0.00473 A |

3.

| I_T 0.0616 A | E_1 5.051 V | E_2 5.456 V | E_3 7.502 V | E_4 12.943 V |
| R_T 292.118 | I_1 0.0616 | I_2 0.0341 | I_3 0.0341 | I_4 0.0275 |

4.

| E_T 712.95 V | E_1 712.95 V | E_2 295.478 V | E_3 417.472 V | E_4 417.472 V |
| R_T 891.187 Ω | I_1 0.475 A | I_2 0.325 A | I_3 0.209 A | I_4 0.116 A |

5.

| E_T 8033.068 V | E_1 8033.068 V | E_2 4275 V | E_3 3762 V | E_4 3762 V |
| R_T 12,358.566 Ω | I_1 0.365 A | I_2 0.285 A | I_3 0.171 A | I_4 0.144 A |

6.

| E_T 58.55 V | E_1 58.55 V | E_2 19.693 V | E_3 38.813 V | E_4 38.813 V |
| R_T 48.792 Ω | I_1 0.781 A | I_2 0.419 A | I_3 0.176 A | I_4 0.243 A |

7.

E_T 250 V	E_1 55 V	E_2 50 V	E_3 18.8 V	E_4 11.2 V
I_T 0.25 A	I_1 0.25 A	I_2 0.1 A	I_3 0.04 A	I_4 0.04 A
R_T 1000 Ω	R_1 220 Ω	R_2 500 Ω	R_3 470 Ω	R_4 280 Ω
P_T 62.5 W	P_1 13.75 W	P_2 5 W	P_3 0.752 W	P_4 0.448 W

E_5 40 V	E_6 30 V	E_7 52.5 V	E_8 67.5 V	E_9 75 V
I_5 0.1 A	I_6 0.06 A	I_7 0.15 A	I_8 0.15 A	I_9 0.25 A
R_5 400 Ω	R_6 500 Ω	R_7 350 Ω	R_8 450 Ω	R_9 300 Ω
P_5 4 W	P_6 1.8 W	P_7 7.875 W	P_8 10.125 W	P_9 18.75 W

8.

E_T 24 V	E_1 3.358 V	E_2 4.797	E_3 0.478 V	E_4 1.248 V
I_T 0.024 A	I_1 0.024 A	I_2 0.00961 A	I_3 0.00385 A	I_4 0.00385 A
R_T 1000 Ω	R_1 139.917 Ω	R_2 499.167 Ω	R_3 124.156 Ω	R_4 324.156 Ω
P_T 0.576 W	P_1 0.0806 W	P_2 0.0461 W	P_3 0.00184 W	P_4 0.00481 W
E_5 2.112 V	E_6 1.726 V	E_7 3.597 V	E_8 5.042 V	E_9 12 V
I_5 0.0096 A	I_6 0.00576 A	I_7 0.0144 A	I_8 0.0144 A	I_9 0.024 A
R_5 219.771 Ω	R_6 299.653 Ω	R_7 249.792 Ω	R_8 350.139 Ω	R_9 500 Ω
P_5 0.0203 W	P_6 0.00995 W	P_7 0.0518 W	P_8 0.0726 W	P_9 0.288 W

Unit 9 Kirchhoff's Laws, Thevenin's, Norton's, and Superposition Theorems

To solve the following Kirchhoff's law problems, refer to the circuit shown in *Figure 9–35*.

1.

$E_{S1} = 12$ V	$E_1 = 0.898$ V	$E_2 = 20.9$ V	$E_3 = 11.1$ V
$E_{S2} = 32$ V	$I_1 = 0.00132$ A	$I_2 = 0.0209$ A	$I_3 = 0.0222$ A
$R_1 = 680$ Ω	$R_2 = 1000$ Ω	$R_3 = 500$ Ω	

2.

$E_{S1} = 3$ V	$E_1 = 1.822$ V	$E_2 = 0.322$ V	$E_3 = 1.18$ V
$E_{S2} = 1.5$ V	$I_1 = 0.00911$ A	$I_2 = 0.00268$ A	$I_3 = 0.0118$ A
$R_1 = 200$ Ω	$R_2 = 120$ Ω	$R_3 = 100$ Ω	

3.

$E_{S1} = 6$ V	$E_1 = 22.64$ V	$E_2 = 31.333$ V	$E_3 = 28.8$ V
$E_{S2} = 60$ V	$I_1 = -0.0141$ A	$I_2 = 0.0261$ A	$I_3 = 0.012$ A
$R_1 = 1.6$ KΩ	$R_2 = 1.2$ KΩ	$R_3 = 2.4$ KΩ	

To answer the following questions, refer to the circuit shown in *Figure 9–36*.

4. Find the Thevenin equivalent voltage and resistance across terminals A and B.
   $E_S = 32$ V $R_1 = 4$ Ω $R_2 = 6$ Ω $E_{TH} = 19.2$ V $R_{TH} = 2.4$ Ω

5.
   $E_S = 18$ V $R_1 = 2.5$ Ω $R_2 = 12$ Ω $E_{TH} = 14.896$ V $R_{TH} = 2.07$ Ω

6. Find the Norton equivalent current and resistance across terminals A and B.
   $E_S = 10$ V $R_1 = 3$ Ω $R_2 = 7$ Ω $I_N = 3.33$ A $R_N = 2.1$ Ω

7.
   $E_S = 48$ V $R_1 = 12$ Ω $R_2 = 64$ Ω $I_N = 4$ A $R_N = 10.1$ Ω

Unit 10 Measuring Instruments

1. First find the voltage necessary to deflect the meter movement full scale.

$$E = 0.000100 \text{ A} \times 5000 \ \Omega$$

$$E = 0.5 \text{ V}$$

The series resistor must have a full-scale voltage value of 9.5 V with a current flow of 100 μA.

$$R = \frac{9.5}{0.000100}$$
$$R = 95 \text{ k}\Omega$$

2.

$$R_1 = \frac{14.5 \text{ V}}{0.000100 \text{ A}} = 145 \text{ k}\Omega$$

$$R_2 = \frac{45 \text{ V}}{0.000100 \text{ A}} = 450 \text{ k}\Omega$$

$$R_3 = \frac{90 \text{ V}}{0.000100 \text{ A}} = 900 \text{ k}\Omega$$

$$R_4 = \frac{150 \text{ V}}{0.000100 \text{ A}} = 1.5 \text{ k}\Omega$$

3. First find the amount of current that must flow through the shunt.

$$R_S = 2 \text{ A } 0.000500 \text{ A}$$

$$R_S = 1.9995 \text{ A}$$

Next find the amount of resistance necessary to produce a voltage drop of 50 mV when that amount of current flows through it.

$$R_S = \frac{0.050 \text{ V}}{1.9995 \text{ A}}$$
$$R_S = 0.025 \ \Omega$$

4. The first step is to find the resistance of the meter movement.

$$R_M = \frac{0.050 \text{ V}}{0.000500 \text{ A}}$$
$$R_M = 100 \ \Omega$$

The resistance of the total shunt is found using the formula

$$R_S = \frac{I_M \times R_M}{I_T}$$

Note: When the ammeter is set for a value of 0.5 A full scale, the total shunt is connected across the meter movement.

$$R_S = \frac{0.000500 \text{ A} \times 100 \text{ } \Omega}{0.5 \text{ A}}$$

$$R_S = 0.1 \text{ } \Omega$$

The value of Resistor R_1 will be computed next. Resistor R_1 is used to produce the 5-A scale. To find this value, use the formula

$$R_1 = \frac{I_M \times +R_{SUM}}{I_T}$$

$$R_1 = \frac{0.000500 \text{ A} \times 100.1 \text{ } \Omega}{5 \text{ A}}$$

$$R_1 = 0.01 \text{ } \Omega$$

When the meter is connected to the 1-A range, the Resistors R_1 and R_2 are connected in series with each other. The value of these two resistors can be found using the preceding formula:

$$R_1 + R_2 = \frac{0.000500 \text{ A} \times 100.1 \text{ } \Omega}{1 \text{ A}}$$

$$R_1 + R_2 = 0.05 \text{ } \Omega$$

Because the value of Resistor R_1 is known, the value of R_2 can be found by subtracting the total of $R_1 + R_2$ from the value of R_1:

$$R_2 = 0.05 \text{ } \Omega - 0.01 \text{ } \Omega = 0.04 \text{ } \Omega$$

The value of Resistor R_3 can be found by subtracting the value of Resistors R_1 and R_2 from the total value of the shunt:

$$R_3 = 0.1 \text{ } \Omega - 0.04 \text{ } \Omega - 0.01 \text{ } \Omega = 0.05 \text{ } \Omega$$

5.

$$R_S = \frac{2.5 \text{ V}}{0.000010 \text{ A}} = 250 \text{ k}\Omega$$

Unit 11 Using Wire Tables and Determining Conductor Sizes

1. $10.4 \text{ } \Omega \text{ cm/ft} \times 450 \text{ ft} = 4680 \text{ } \Omega \text{ cm}$

$$\frac{4680 \text{ } \Omega \text{ cm}}{66.370 \text{ cm}} = .0705 \text{ } \Omega$$

2.
$$1.817 \ \Omega \times 16{,}510 \ \text{cm} = 29{,}998.67 \ \Omega \ \text{cm}$$

$$\frac{29{,}998.67 \ \Omega \ \text{cm}}{500 \ \text{ft}} = 60 \ \frac{\Omega \ \text{cm}}{\text{ft}} \ (\text{Iron})$$

3. **430 A** (from *310.15(B)(16)*) × **0.71** (correction factor [*Table 310.15(B)(2)(a)*]) = 305.3 A

4. **30 A** × 0.70 = 21 A [*Table 310.15(B)(3)(a)*]

5. Max voltage drop: 480 V × 0.05 = 24 V
   Length of conductor: 300 ft × 2 × 0.866 = 519.6 ft

   Max resistance of conductors: $\dfrac{24 \ \text{V}}{522 \ \text{A}} = 0.046 \ \Omega$

   $$\text{CM} = 10.4 \times \frac{519.6 \ \text{ft}}{0.046 \ \Omega} \qquad \text{CM} = 117{,}474.78$$

   Wire size = 2/0 AWG

6. **4/0 AWG** (746 × 50 = 37,300 watts) (37,300/250 = 149.2 amps) (149.2 × 125% = 186.5 amps) (Correct for ambient temperature: 186.5/0.82 = 227.4 amps.) (Conductor is chosen from the 75°C column of *Table 310.15(B)(16)*.)

7. **2/0 AWG** (25,000/480 = 52.1 amperes) (52.1 × 125% = 65.1 amperes) (Because the initial current demand is 65.1 amperes, the conductors must be chosen from the 60°C column of *Table 310.15(B)(16)*.) (Adjust for ambient temperature: 65.1/0.58 = 112.2 amps.) (Adjust for number of conductors in the raceway: 112.2/0.80 = 140.2 amps.)

8. **#6 AWG** (15 × 746 = 11,190 watts) (11,190/240 = 46.6 amps) (46.6 × 125% = 58.25 amps) (Because the motor contains a NEMA code rating, the conductors are chosen from the 75°C column.) (No adjustment is necessary for ambient temperature or number of conductors in the raceway.)

9. **R = 1.27 Ω**

   $$R = R_{\text{ref}}[1 + \alpha(T - T_{\text{ref}})]$$

   $$R = 10.4[1 + 0.0039(87 - 20)]$$

   $$R = 13.12$$

   $$R = \frac{K \times L}{\text{CM}}$$

   $$R = \frac{13.12 \times 250}{2583}$$

   $$R = 1.27\Omega$$

10. $R = 18.36\ \Omega$

$$R = R_{ref}[1 + \alpha(T - T_{ref})]$$
$$R = 17[1 + 0.004(40 - 20)]$$
$$R = 18.36\ \Omega$$

Use the *NEC* to determine the ampacity of the following conductors:

Size	Material	Insulation	Amb. Temp.	Conductors in Raceway	Amps
#10 AWG	Copper	RHW	44°C	3	28.7 amps
350 kcmil	Copper	XHH	128°F	6	212.8 amps
#2 AWG	Aluminum	TW	86°F	2	75 amps
3/0 AWG	Aluminum	XHHW-2	38°C	9	111.48 amps
500 kcmil	Copper	THWN	48°C	6	228 amps
#6 AWG	Copper	THW-2	150°F	3	24.75 amps
2/0 AWG	Aluminum	UF	86°F	12	57.5 amps
750 kcmil	Aluminum	RHW-2	34°C	6	334.08 amps

Unit 15 Basic Trigonometry and Vectors

∠X	∠Y	Side A	Side B	Hyp.
40°	50°	5.142	6.128	8
57°	33°	72	46.759	85.85
37.1°	52.9°	38	50.249	63
52°	38°	17.919	14	22.739
48°	42°	173.255	156	233.138

Unit 16 Alternating Current

Sine Wave Values

Peak volts	Inst. volts	Degrees	Peak volts	Inst. volts	Degrees
347	208	36.829	87.2	23.7	15.771
780	536.64	43.5	156.9	155.488	82.3
80.464	24.3	17.6	110.387	62.7	34.6
224	5.65	1.327	1256	400	18.571
48.7	43.976	64.6	15,720	3269.76	12
339.463	240	45	126.795	72.4	34.8

Peak, RMS, and Average Values

Peak volts	RMS	Average	Peak volts	RMS	Average
12.7	8.979	8.09	339.7	240.168	216.389
76.073	53.8	48.42	17.816	12.6	11.34
257.794	182.426	164.2	14.13	9.999	9
1235	873.145	786.695	123.7	87.456	78.797
339.36	240	216	105.767	74.8	67.32
26.06	18.443	16.6	169.56	119.988	108

Unit 17 Inductance in AC Circuits

Inductive Circuits

1.

Inductance (H)	Frequency (Hz)	Induct. Rct. (Ω)	Inductance (H)	Frequency (Hz)	Induct. Rct. (Ω)
1.2	60	452.4	0.5	60	188.5
0.085	400.052	213.628	0.85	1199.934	6408.849
0.75	1000	4712.389	1.6	20	201.062
0.65	600	2450.448	0.45	400	1130.976
3.6	29.999	678.584	4.8	80.001	2412.743
2.619	25	411.459	0.0065	1000	40.841

2. 480 Hz $\left(X_L = \dfrac{99.526}{2 \times 3.1416 \times 0.033} \right)$

3. 0.0729 henrys $(Z = 120/4)$ $(Z = 30\ \Omega)$ $(X_L = \sqrt{30^2 - 12^2})$

4. 296.722 Ω $(X_L = 2 \times 3.1416 \times 60)$ $(X_L = 282.744\ \Omega)$ $(Z = \sqrt{90^2 + 282.722^2})$

5. $Q = 5.239$ $(Z = 240/3)$ $(Z = 80\ \Omega)$ $(X_L = \sqrt{80^2 - 15^2})$ $(X_L = 78.581\ \Omega)$ $(Q = 78.581/15)$

Unit 18 Resistive-Inductive Series Circuits

1.

E_T 480 V	E_R 247.128 V	E_L 411.489 V
I_T 20.594 A	I_R 20.594 A	I_L 20.594 A
Z 23.308 Ω	R 12 Ω	X_L 19.981 Ω
VA 9885.12	P 5089.354 W	$VARS_L$ 8474.204
PF 51.485%	$\angle\theta$ 59.003°	L 0.053 H

2.

E_T 130 V	E_R 78 V	E_L 104 V
I_T 6.5 A	I_R 6.5 A	I_L 6.5 A
Z 100 Ω	R 12 Ω	X_L 16 Ω
VA 845	P 507 W	$VARs_L$ 676
PF 60%	∠θ 53.13°	L 0.0424 H

3.

E_T 120 V	E_R 96 V	E_L 72 V
I_T 1.2 A	I_R 1.2 A	I_L 1.2 A
Z 100 Ω	R 80 Ω	X_L 60 Ω
VA 144	P 115.2 W	$VARs_L$ 86.4
PF 80%	∠θ 36.87°	L 0.15915

4.

E_T 239.806 V	E_R 187.049 V	E_L 150.059 V
I_T 1.563 A	I_R 1.563 A	I_L 1.563 A
Z 153.427 Ω	R 119.673 Ω	X_L 96.007 Ω
VA 374.817	P 292.357 W	$VARs_L$ 234.553
PF 78%	∠θ 38.739°	L 0.0382

5. 187.4 volts

6. 84.8 PF

7. 195.86 VARs

8. 439.38 VARs

9. 10.03 Ω

10. 81.98 Ω

11. 390.3 watts

12. 37.98 VARs

13. 807.77 Ω

14. 77.9% PF

15. 90.05 volts

16. 154.89 volts

Unit 19 Resistive-Inductive Parallel Circuits

1.

E_T 240.005 V	E_R 240.005 V	E_L 240.005 V
I_T 34.553 A	I_R 17.143 A	I_L 30 A
Z 6.946 Ω	R 14 Ω	X_L 8 Ω
VA 8292.893	P 4114.406 W	$VARs_L$ 7200.15
PF 49.61%	∠θ 60.25°	L 0.02122 H

2.

E_T 277 V	E_R 277 V	E_L 277 V
I_T 39 A	I_R 15 A	I_L 36 A
Z 7.103 Ω	R 18.467 Ω	X_L 7.694 Ω
VA 10,803	P 4155 W	$VARs_L$ 9972
PF 38.46%	∠θ 67.36°	L 0.0204

3.

E_T 72 V	E_R 72 V	E_L 72 V
I_T 2 A	I_R 1.6 A	I_L 1.2 A
Z 36 Ω	R 45 Ω	X_L 60 Ω
VA 144	P 115.2 W	$VARs_L$ 86.4
PF 80%	∠θ 36.87°	L 0.15915

4.

E_T 150.062 V	E_R 150.062 V	E_L 150.062 V
I_T 2.498 A	I_R 1.948 A	I_L 1.563 A
Z 60.073 Ω	R 77.034 Ω	X_L 96.007 Ω
VA 374.817	P 292.357 W	$VARs_L$ 234.553
PF 78%	∠θ 38.74°	L 0.0382

5. 199.69 Ω
6. 1600 volts
7. 0.8 amps
8. 240 Ω
9. 174.5 VARs
10. 202.6 VA
11. 0.144 amps
12. 83.2% PF
13. 0.8 amps
14. 7.09 VA
15. 0.127 henrys
16. 1.632 amps

Unit 20 Capacitors

RC Time Constants

1.

Resistance	Capacitance	Time constant	Total time
150 kΩ	100 μF	15 s	75 s
350 kΩ	20 μF	7 s	35 s
142,857,142.9	350 pF	0.05 s	0.25 s
40 MΩ	0.05 μF	2 s	10 s
1.2 MΩ	0.47 μF	0.564 s	2.82 s
4166.67 Ω	12 μF	0.05 s	0.25 s
86 kΩ	3.488 μF	0.3 s	1.5 s
120 kΩ	470 pF	56.4 μs	282 μs
80 kΩ	250 nF	20 ms	100 ms
3.75 Ω	8 μF	30 μs	150 μs
100 kΩ	1.5 μF	150 ms	750 ms
33 kΩ	4 μF	132 ms	660 ms

2. 34.28 μF   3. 200 μF   4. 2.48047 μF   5. 84 seconds

Unit 21 Capacitance in AC Circuits

Capacitive Circuits

1.

Capacitance	X_C	Frequency
38 μF	69.803 Ω	60 Hz
5.049 μF	78.8 Ω	400 Hz
250 pF	4.5 kΩ	141.471 kHz
234 μF	0.068 Ω	10 kHz
13.263 μF	240 Ω	50 Hz
10 μF	36.8 Ω	432.486 Hz
560 nF	0.142 Ω	2 MHz
176.835 nF	15 kΩ	60 Hz
75 μF	560 Ω	3.789 Hz
470 pF	1.693 kΩ	200 kHz
58.513 nF	6.8 kΩ	400 Hz
34 μF	450 Ω	10.402 Hz

2. 564.37 Ω   3. 3.57 μF   4. 282.19 Hz

5. 3.127 amps (C = 20 + 40 + 60) (C = 110 μF) (X_C = 24.114 Ω) (E = 24.114 × 8.6) (E = 207.38 volts) (X_C for 40 μF capacitor = 66.313 Ω) (I_C = 207.38/66.313)

6. 180 Ω (Find C at 60 Hz.) (Use the value of C to find X_C at 100 Hz.)

7. The existing capacitor has a value of 132.6 μF. The needed capacitance is 262.2 μF. (480/24 = 20 Ω) (Use the value of X_C to determine existing capacitor value.) (Determine the value of X_C needed to limit the current to 16 amperes.) (480/16 = 30 Ω) (Determine the total capacitance needed to produce 30 Ω of X_C.) (C = 88.4 μF) (Convert the formula for total capacitance of series-connected capacitors to determine the value of the second capacitor:

$$\left(C_{Needed} = \frac{1}{\dfrac{1}{88.4} - \dfrac{1}{132.6}} \right) (C_{Needed} = 265.2 \ \mu F)$$

Unit 22 Resistive-Capacitive Series Circuits

1.

E_T 480 V	E_R 288 V	E_C 384 V
I_T 24 A	I_R 24 A	I_C 24 A
Z 20 Ω	R 12 Ω	X_C 16 Ω
Va 11,520	P 6912 W	$VARs_C$ 9216
PF 60%	$\angle\theta$ 53.13°	C 165.782 μF

2.

E_T 130 V	E_R 78 V	E_C 104 V
I_T 6.5 A	I_R 6.5 A	I_C 6.5 A
Z 20 Ω	R 12 Ω	X_C 16 Ω
VA 845	P 507 W	$VARs_C$ 676
PF 60%	$\angle\theta$ 53.13°	C 165.782 μF

3.

E_T 346.709 V	E_R 277.368 V	E_C 207.975 V
I_T 1.246 A	I_R 1.246 A	I_C 1.246 A
Z 278.258 Ω	R 222.607 Ω	X_C 166.914 Ω
VA 432	P 345.6 W	$VARs_C$ 259.2
PF 80%	$\angle\theta$ 36.87°	C 15.8915 μF

4.

E_T 186.335 V	E_R 126.708 V	E_C 135.936 V
I_T 1.61 A	I_R 1.61 A	I_C 1.61 A
Z 115.736 Ω	R 78.701 Ω	X_C 84.432 Ω
VA 300	P 204 W	$VARs_C$ 219.964
PF 68%	$\angle\theta$ 47.156°	C 4.7125 μF

5. 196.07 volts   7. 51.9% PF   9. 9.2 μF

6. 47.3 VARs   8. 330.9 volts   10. 3,680.6 watts

Unit 23 Resistive-Capacitive Parallel Circuits

1.

E_T 122.689 V	E_R 122.689 V	E_C 122.689 V
I_T 10.463 A	I_R 8.764 A	I_C 5.718 A
Z 11.725 Ω	R 14 Ω	X_C 21.456 Ω
VA 1283.695	P 1075.246 W	$VARs_C$ 701.536
PF 83.76%	$\angle\theta$ 33.07°	C 132.626 μF

2.

E_T 54 V	E_R 54 V	E_C 54 V
I_T 2.5 A	I_R 1.5 A	I_C 2 A
Z 21.6 Ω	R 36 Ω	X_C 27 Ω
VA 135	P 81 W	$VARs_C$ 108
PF 60%	∠θ 53.13°	C 14.736 μF

3.

E_T 307.202 V	E_R 307.202 V	E_C 307.202 V
I_T 140.991 A	I_R 65.6 A	I_C 124.8 A
Z 2.17888 Ω	R 4.683 Ω	X_C 2.462 Ω
VA 43,312.717 VA	P 20,152.451	$VARs_C$ 38,338.81 VARs
PF 46.53%	∠θ 62.29°	C 107.653 μF

4.

E_T 69.549 V	E_R 69.549 V	E_C 69.549 V
I_T 7.5 A	I_R 6.999 A	I_C 2.696 A
Z 9.273 Ω	R 9.937 Ω	X_C 25.797 Ω
VA 521.618	P 486.75 W	$VARs_C$ 187.5
PF 93.31%	∠θ 21.09°	C 6.169 μF

5. 2,857.96 Ω   7. 110 Ω   9. 81.2% PF

6. 0.436 amps   8. 2,883.4 VARs   10. 24.99 μF

Unit 24 Resistive-Inductive-Capacitive Series

1.

E_T 120 V	E_R 72 V	E_L 200 V	E_C 104 V
I_T 2 A	I_R 2 A	I_L 2 A	I_C 2 A
Z 60 Ω	R 36 Ω	X_L 100 Ω	X_C 52 Ω
VA 240	P 144 W	$VARs_L$ 400	$VARs_C$ 208
PF 60%	∠θ 53.13°	L 0.265	C 51.01 μF

2.

E_T 35.678 V	E_R 24 V	E_L 21.6 V	E_C 48 V
I_T 0.6 A	I_R 0.6 A	I_L 0.6 A	I_C 0.6 A
Z 59.464 Ω	R 40 Ω	X_L 36 Ω	X_C 80 Ω
VA 21.407	P 14.4 W	$VARs_L$ 12.96	$VARs_C$ 28.8
PF 67.27%	∠θ 47.7°	L 0.0143	C 4.974 μF

3.

E_T 23.988 V	E_R 14.993 V	E_L 75 V	E_C 56.25 V
I_T 1.25 A	I_R 1.25 A	I_L 1.25 A	I_C 1.25 A
Z 19.19	R 11.994 Ω	X_L 60 Ω	X_C 45 Ω
VA 29.985	P 18.741 W	$VARs_L$ 93.75	$VARs_C$ 70.313
PF 62.5%	∠θ 51.32°	L 0.159 H	C 58.945 μF

4.

E_T 349.057 V	E_R 185 V	E_L 740 V	E_C 444 V
I_T 0.148 A	I_R 0.148 A	I_L 0.148 A	I_C 0.148 A
Z 2358.493 Ω	R 1250 Ω	X_L 5000 Ω	X_C 3000 Ω
VA 51.8	P 27.38 W	VARs$_L$ 109.52	VARs$_C$ 65.712
PF 52.86%	∠θ 58.06°	L 0.796 H	C 0.053 μF

5. 101.24 volts

6. 87.4 % PF

7. The power factor is leading because there are more capacitive VARs than inductive VARs.

8. 35.6 volts   9. 45.07 Ω   10. 283.15 Ω

Unit 25 Resistive-Inductive-Capacitive Parallel Circuits

1.

E_T 120 V	E_R 120 V	E_L 120 V	E_C 120 V
I_T 3.387 A	I_R 3.333 A	I_L 3 A	I_C 2.4 A
Z 35.431 Ω	R 36 Ω	X_L 40 Ω	X_C 50 Ω
VA 406.44	P 400 W	VARs$_L$ 360	VARs$_C$ 288
PF 98.41%	∠θ 10.2°	L 0.106 H	C 53.05 μF

2.

E_T 243.885 V	E_R 243.885 V	E_L 243.885 V	E_C 243.885 V
I_T 22.267 A	I_R 15.745 A	I_L 7.873 A	I_C 23.618 A
Z 10.953 Ω	R 15.49 Ω	X_L 30.977 Ω	X_C 10.326 Ω
VA 5430.58	P 3840 W	VARs$_L$ 1920	VARs$_C$ 5760
PF 70.71%	∠θ 45.01°	L 0.0123 H	C 38.532 μF

3.

E_T 23.999 V	E_R 23.999 V	E_L 23.999 V	E_C 23.999 V
I_T 2.004 A	I_R 2 A	I_L 0.4 A	I_C 0.533 A
Z 11.973 Ω	R 12 Ω	X_L 60 Ω	X_C 45 Ω
VA 48.106	P 48 W	VARs$_L$ 9.6	VARs$_C$ 12.791
PF 99.78%	∠θ 3.62°	L 0.159	C 58.94 μF

4.

E_T 480 V	E_R 480 V	E_L 480 V	E_C 480 V
I_T 100 A	I_R 60 A	I_L 150 A	I_C 70 A
Z 4.8 Ω	R 8 Ω	X_L 3.2 Ω	X_C 6.857 Ω
VA 48 kVA	P 28.8 kW	VARs$_L$ 72 kVARs	VARs$_C$ 33.6 VARs
PF 60%	∠θ 53.13°	L 0.509 mH	C 23.211 μF

5. 16,571.4 Ω   7. ∠θ = 56.98°   9. 480 volts

6. 92.3% PF   8. 312.2 watts   10. 3.33 amperes

Unit 27 Three-Phase Circuits

1.

$E_{P(A)}$ 138.568 V	$E_{P(L)}$ 240 V
$I_{P(A)}$ 34.64 A	$I_{P(L)}$ 20 A
$E_{L(A)}$ 240 V	$E_{L(L)}$ 240 V
$I_{L(A)}$ 34.64 A	$I_{L(L)}$ 34.64 A
P 14,399.155 W	Z_{PHASE} 12 Ω

2.

$E_{P(A)}$ 4160 V	$E_{P(L)}$ 2401.849 V
$E_{P(a)}$ 23.112 A	$E_{P(L)}$ 40.031 A
$E_{L(A)}$ 4160 V	$E_{L(L)}$ 4160 V
$I_{L(A)}$ 40.031 A	$I_{L(L)}$ 40.031 A
P 288,428.159 W	Z_{PHASE} 60 Ω

3.

$E_{P(A)}$ 323.326 V	$E_{P(L1)}$ 323.326 V	$E_{P(L2)}$ 560 V
$I_{P(A)}$ 185.91 A	$I_{P(L1)}$ 64.665 A	$I_{P(L2)}$ 70 A
$E_{L(A)}$ 560 V	$E_{L(L1)}$ 560 V	$E_{L(L2)}$ 560 V
$I_{L(A)}$ 185.905 A	$I_{L(L1)}$ 64.665 A	$I_{L(L2)}$ 121.24 A
P 180,317.83 W	Z_{PHASE} 5 Ω	Z_{PHASE} 8 Ω

4.

$E_{P(A)}$ 277.136 V	$E_{P(L1)}$ 277.136 V	$E_{P(L2)}$ 480 V	$E_{P(L3)}$ 277.136 V
$I_{P(A)}$ 34.485 A	$I_{P(L1)}$ 23.095 A	$I_{P(L2)}$ 30 A	$I_{P(L3)}$ 27.714 A
$E_{L(A)}$ 480 V	$E_{L(L1)}$ 480 V	$E_{L(L2)}$ 480 V	$E_{L(L3)}$ 480 V
$I_{L(A)}$ 34.049 A	$I_{L(L1)}$ 23.095 A	$I_{L(L2)}$ 51.96 A	$I_{L(L3)}$ 27.714 A
VA 27,838.09	Z_{PHASE} 12 Ω	$Z_{L(PHASE)}$ 16 Ω	$X_{C(PHASE)}$ 10 Ω
	P 19,200.259 W	$VARs_L$ 43,197.466	$VARs_C$ 23,040.311

Unit 28 Single-Phase Transformers

1.

E_P 120 V	E_S 24 V
I_P 1.6 A	I_S 8 A
N_P 300 turns	N_S 60 turns
Ratio 5:1	Z = 3 Ω

2.

E_P 240 V E_S 320 V
I_P 0.853 A I_S 0.643 A
N_P 210 turns N_S 280 turns
Ratio 1:1.333 Z = 500 Ω

3.

E_P 64 V E_S 160 V
I_P 33.333 A I_S 13.333 A
N_P 32 turns N_S 80 turns
Ratio 1:2.5 Z = 12 Ω

4.

E_P 48 V E_S 240 V
I_P 3.333 A I_S 0.667 A
N_P 220 turns N_S 1100 turns
Ratio 1:5 Z = 360 Ω

5.

E_P 35.848 V E_S 182 V
I_P 16.5 A I_S 3.25 A
N_P 87 turns N_S 450 turns
Ratio 1:5.077 Z = 56 Ω

6.

E_P 480 V E_S 916.346 V
I_P 1.458 A I_S 0.764 A
N_P 275 turns N_S 525 turns
Ratio 1:1.909 Z = 1.2 kΩ

7.

E_P 208 V E_S 1 320 V E_{S2} 120 V E_{S3} 24 V
I_P 11.93 A I_{S1} 0.0267 A I_{S2} 20 A I_{S3} 3 A
N_P 800 turns N_{S1} 1231 turns N_{S2} 462 turns N_{S3} 92 turns
                   Ratio 1 1:1.54   Ratio 2 1.73:1   Ratio 3 1:8.67
                   R_1 12 kΩ R_2 6 Ω R_3 8 Ω

8.

E_P 277 V E_{S1} 480 V E_{S2} 208 V E_{S3} 120 V
I_P 8.93 A I_{S1} 2.4 A I_{S2} 3.47 A I_{S3} 5 A
N_P 350 turns N_{S1} 606 turns N_{S2} 263 turns N_{S3} 152 turns
                   Ratio 1 1:1.73   Ratio 2 1.33:1   Ratio 3 2.31:1
                   R_1 200 Ω R_2 60 Ω R_3 24 Ω

Unit 29 Three-Phase Transformers

1.

E_P 2401.8 V	E_P 440 V	E_P 254.04 V
I_P 7.67 A	I_P 41.9 A	I_P 72.58 A
E_L 4160 V	E_L 440 V	E_L 440 V
I_L 7.67 A	I_L 72.58 A	I_L 72.58 A
Ratio 5.46:1	Z 3.5 Ω	

2.

E_P 4157.04 V	E_P 240 V	E_P 138.57 V
I_P 1.15 A	I_P 20 A	I_P 34.64 A
E_L 7200 V	E_L 240 V	E_L 240 V
I_L 1.15 A	I_L 34.64 A	I_L 34.64 A
Ratio 17.32:1	Z 4 Ω	

3.

E_P 13,800 V	E_P 277 V	E_P 480 V
I_P 6.68 A	I_P 332.54 A	I_P 192 A
E_L 13,800 V	E_L 480 V	E_L 480 V
I_L 11.57 A	I_L 332.54 A	I_L 332.54 A
Ratio 49.76:1	Z 2.5 Ω	

4.

E_P 23,000 V	E_P 120 V	E_P 208 V
I_P 0.626 A	I_P 120.08 A	I_P 69.33 A
E_L 23,000 V	E_L 208 V	E_L 208 V
I_L 1.08 A	I_L 120.08 A	I_L 120.08 A
Ratio 191.66:1	Z 3 Ω	

GLOSSARY

A

Acids any of a large group of substances having a pH less than 7.

Across-the-line a method of motor starting that connects the motor directly to the supply line on starting or running. Also known as full-voltage starting.

Adjacent next to or beside.

Air gap the space between two magnetically related components.

Alkalies substances having a pH greater than 7.

Alternating current (AC) current that reverses its direction of flow periodically. Reversals generally occur at regular intervals.

Alternator a machine used to generate AC by rotating conductors through a magnetic field.

Ambient air temperature the temperature surrounding a device.

American Wire Gauge (AWG) a measurement of the diameter of a wire. The gauge scale was formerly known as the Brown and Sharp scale. The scale has a fixed constant of 1.123 between gauge sizes.

Ammeter an instrument used to measure the flow of current.

Ammeter shunt a device used to change the full scale value of an ammeter by providing another path for current low. Most ammeter shunts produce a voltage drop of 50 millivolts when rated current flows through them.

Amortisseur winding a squirrel-cage winding on the rotor of a synchronous motor used for starting purposes only.

Ampacity the maximum current-carrying capacity of a wire or device.

Ampere (A) a unit of measure for the rate of current flow. One ampere equals 1 coulomb per second.

Ampere-hour (A-hr) a unit of measure for describing the current capacity of a battery or a cell.

Ampere-turns a basic unit for measurement of magnetism. The product of number of turns of wire times current flow.

Amplifier a device used to increase a signal.

Amplitude the highest value reached by a signal, voltage, or current.

Analog meters meters that rely on a mechanical pointer and employ a scale to measure indicated quantities.

Analog voltmeter a voltmeter that uses a meter movement to indicate the voltage value. Analog meters use a pointer and scale.

Angle theta ($\angle\theta$) generally used to express the phase angle difference between voltage and current in a reactive circuit.

Anode the positive terminal of an electric device.

Apparent power (VA) the value found by multiplying the applied voltage by the total current of an AC circuit. Apparent power is measured in volt-amperes (VA) and should not be confused with true power, measured in watts.

Appears to flow current only appears to flow through a capacitor when it is connected to AC current. Current cannot actually flow through it because of an insulator separating the plates.

Applied voltage the amount of voltage connected to a circuit or device.

Arc a luminous discharge of electricity through gas.

Armature the rotating member of a motor or generator. The armature generally contains windings and a commutator.

Armature reaction the twisting or bending of the main magnetic field of a motor or generator. Armature reaction is proportional to armature current.

Artificial respiration providing a person with an external method of breathing. Generally employing the mouth-to-mouth method.

ASA American Standards Association.

Atom the smallest part of an element that contains all the properties of that element.

Atomic number the atomic number is equal to the number of protons (positively charged particles) in the nucleus of an atom.

Attenuator a device that decreases the amount of signal voltage or current.

Attraction the force that causes one object to pull another object to itself.

Automatic self-acting, operation by its own mechanical or electric mechanism.

Autotransformers transformers that use only one winding for both primary and secondary.

Average the value of DC voltage when it is rectified from AC voltage. The ratio is dependent on the type of AC waveform that has been rectified.

Ayrton shunt an ammeter shunt that permits multiple range values.

B

Back-EMF see counter-EMF

Back-voltage the induced voltage in the coil of an inductor or generator that opposes the applied voltage.

Bandpass filters pass a range of frequencies.

Band-rejection filters reject a range of frequencies.

Bandwidth a range of frequencies.

Base the semiconductor region between the collector and emitter of a transistor. The base controls the current flow through the collector-emitter circuit.

Battery a device used to convert chemical energy into electric energy. A group of voltaic cells connected together in a series or parallel connection.

Bias a DC applied to the base of a transistor to protect its operating point.

Bidirectional an object that moves in two directions.

Bimetallic strip a strip made by bonding two unlike metals together. The metals expand at different temperatures when heated, causing a bending or warping action.

Branch circuit that portion of a wiring system that extends beyond the circuit protective device, such as a fuse or circuit breaker.

Breakdown torque the maximum amount of torque that can be developed by a motor at rated voltage and frequency before an abrupt change in speed occurs.

Bridge circuit a circuit that consists of four sections connected in series to form a closed loop.

Bridge rectifier a device constructed with four diodes that converts both positive and negative cycles of AC voltage into DC voltage. The bridge rectifier is one type of full-wave rectifier.

British thermal unit (Btu) the amount of heat necessary to raise or lower 1 pound of water 1° Fahrenheit.

Broadband circuits that can tune over a wide range of frequencies.

Brushes sliding contacts, generally made of carbon, used to provide connection to rotating parts of machines.

Brushless DC motors DC motors that do not depend on a commutator and brushes to operate.

Brushless exciter a device used to excite the rotor of an alternator or synchronous motor that does not employ the used of brushes or slip rings.

Busway an enclosed system used for power transmission that is voltage- and current-rated.

C

Capacitance (C) the electrical size of a capacitor.

Capacitive reactance (X_C) the current-limiting property of a capacitor in an AC circuit.

Capacitor a device made with two conductive plates separated by an insulator or dielectric.

Capacitor-start induction-run motor a single-phase induction motor that uses a capacitor connected in series with the start winding to increase starting torque.

Carbon film resistor a resistor that is produced by coating a nonconductive material with a film of carbon.

Cardiopulmonary respiration (CPR) a method of alternating closed heart massage and artificial respiration for a person who is not breathing and has no pulse.

Cathode the negative terminal of an electric device.

Cell a single unit that converts chemical energy into electrical energy.

Center-tapped transformer a transformer that has a wire connected to the electric midpoint of its winding. Generally the secondary winding is tapped.

Centrifugal switch a switch sensitive to speed of rotation, generally used to disconnect the start windings of a split-phase motor.

Charging current the current flowing from an electric source to a capacitor.

Chassis ground ground connection to the case of a piece of equipment.

Choke an inductor designed to present an impedance to AC or to be used as the current filter of a DC power supply.

Circuit an electrical path between two points.

Circuit branch a point in an electric circuit where current can flow in more than one direction.

Circuit breakers devices designed to open under an abnormal amount of current flow. These devices are not damaged and may be used repeatedly. Rated by voltage, current, and horsepower.

Circuit impedance the total current limiting property of an AC circuit.

Circular mil an area measure of wire determined by squaring the diameter of the wire in mils (1 mil = 0.001 inch). Example: A wire with a diameter of 0.020 inch would have a circular mil area of 400 cmil (20 × 20 = 400).

Clamp-on ammeter an ammeter that has a movable jaw that can be clamped around a conductor.

Clock timer a time-delay device that uses an electric clock to measure the delay time.

Closing a delta making the final connection to complete a delta connection.

Code letter a letter found on the nameplate of a motor that describes the types of bars used in the squirrel-cage rotor.

Coercive force the magnetizing force necessary to reduce the flux density in a magnetic material to zero.

Collapse of a magnetic field occurs when a magnetic field suddenly changes from its maximum value to a zero value.

Collector a semiconductor region of a transistor that must be connected to the same polarity as the base.

Color code a system where colors are used to represent numbers.

Combination circuit a circuit that contains both series and parallel paths.

Commutating field a field used in DC machines to help overcome the problems of armature reaction. The commutating field connects in series with the armature and is also known as the interpole winding.

Commutator strips or bars of metal insulated from each other and arranged around one end of an armature. They provide connection between the armature windings and the brushes. The commutator is used to ensure proper direction of current flow through the armature windings.

Comparator a device or circuit that compares two like quantities such as voltage levels.

Compensating winding the winding embedded in the main field poles of a DC machine. The compensating winding is used to help overcome armature reaction.

Complete path the path that exists in a circuit when it can be traced from one terminal of a power source back to the other terminal without a break.

Composition carbon resistors resistors that are made by mixing carbon with other compounds to produce a substance that has a known resistance.

Compound DC machine a generator or motor that uses both series and shunt fields windings. DC machines may be connected long-shunt compound, short-shunt compound, cumulative compound, or differential compound.

Compound generators generators that contain both series and shunt field windings.

Compound motor a motor that contains both series and shunt field windings.

Compounding generally indicates the voltage increase caused by the series field as compared to the voltage drop of the generator.

Conduction level the point at which an amount of voltage or current causes a device to conduct.

Conductive compensation a method of connecting the compensating winding in a universal motor.

Conductor a device or material that permits current to flow through it easily.

Confined spaces spaces that have very limited entrance and exit.

Consequent-pole motor a type of motor that changes speeds by changing the number of stator poles.

Constant-current transformer a transformer that supplies a constant amount of current to a varying load.

Constant-speed motors motors that run at the same approximate speed from no load to full load.

Contact a conducting part of a relay that acts as a switch to connect or disconnect a circuit or component.

Continuity a complete path for current flow.

Control transformer a common type of isolation transformer generally used to change line voltage of 240 or 480 volts into 120 volts to operate motor- control equipment.

Conventional current flow theory a theory that considers current to flow from the most positive source to the most negative source.

Copper losses power loss due to current flowing through wire. Copper loss is proportional to the resistance of the wire and the square of the current.

Copper sulfate a chemical compound formed by combining copper, sulfur, and oxygen ($CuSO_4$)

Core magnetic material used to form the center of a coil or transformer. The core may be made of a nonmagnetic conductor (air core), iron, or some other magnetic material.

Core losses the power loss in the core material caused by eddy current induction and hysteresis loss.

Correction factor as pertaining to conductors, the de-rating of the current capacity of conductors due to ambient temperature.

Cosine in trigonometry, it is the ratio of the adjacent side of the angle and the hypotenuse.

Coulomb a quantity of electrons equal to 6.25×10^{18}.

Counter-electromotive force (CEMF) the voltage induced in the armature of a DC motor that opposes the applied voltage and limits armature current; also called back-EMF.

Countertorque the magnetic force developed in the armature of a generator that makes the shaft hard to turn. Countertorque is proportional to armature current and is a measure of the electric energy produced by the generator.

Cumulative compound when the series and shunt fields are connected in such a manner as to aid each other in the production of magnetism.

Cumulative-compound motors motors that are so connected that their series and shunt fields aid each other in the production of magnetism.

Cuprous cyanide a chemical compound used to electroplate copper to iron.

Current the rate of flow of electrons.

Current capacity the amount of current a power source is capable of supplying.

Current dividers a circuit that is constructed to produce a specific amount of current flow through a branch.

Current flow through the capacitor the amount of current flowing through a capacitor.

Current flow through the inductor (I_L) the amount of current flowing through an inductor.

Current flow through the resistor (I_R) amount of current flowing through a resistor.

Current lags voltage the condition that occurs in an AC circuit containing inductance.

Current rating the amount of current flow a device is designed to withstand.

Current relay a relay that is operated by a predetermined amount of current flow. Current relays are often used as one type of starting relay for air-conditioning and refrigeration equipment.

Current transformer a transformer used for metering alternating current. Current transformers generally have a standard output current value of 5 amps at rated primary current.

Cycle one complete AC waveform.

D

D'Arsonval meter a meter movement using a permanent magnet and a coil of wire. The basic meter movement used in many analog type voltmeters, ammeters, and ohmmeters.

Damp locations locations that are protected from the weather and are not subject to saturation with water or other liquids but are subject to moderate degrees of moisture.

De-energized circuit a circuit that has no power applied to it.

Delta connection a circuit formed by connecting three electric devices in series to form a closed loop. It is used most often in three-phase connections.

Delta-wye a three-phase transformer connection with a delta-connected primary and a wye-connected secondary.

Demagnetize to remove magnetism from an object.

Diac a bidirectional diode.

Diamagnetic a material that will not conduct magnetic lines of flux. Diamagnetic materials have a permeability rating less than that of air (1).

Dielectric an electric insulator.

Dielectric breakdown the point at which the insulating material separating two electric charges permits current to flow between the two charges. Dielectric breakdown is often caused by excessive voltage, excessive heat, or both.

Dielectric constant a method of determining the insulation properties of a dielectric using air as a reference.

Dielectric oil a type of insulating oil often used to cool transformers.

Dielectric stress the molecular stress placed on a dielectric material proportional to the level of applied voltage.

Differential compound when the series and shunt field are connected in such a manner as to oppose each other in the production of magnetism.

Differential compound motors motors that are so connected that their series and shunt fields oppose each other in the production of magnetism.

Differential selsyn a special type of selsyn motor.

Digital device a device that has only two states of operation, on or off.

Digital logic circuit elements connected in such a manner as to solve problems using components that have only two states of operation.

Digital voltmeter a voltmeter that uses direct-reading numerical display as opposed to a meter movement.

Diode a two-element device that permits current to flow through it in only one direction.

Direct current (DC) current that does not reverse its direction of flow.

Direction of rotation the direction, clockwise or counterclockwise, that the rotor of a motor turns.

Distribution transformer a common transformer used by utility companies to change their line voltage into the 240/120 volts necessary to operate homes and small businesses.

Disconnect to physically remove an object from a circuit or a switch that is employed to remove power for a circuit.

Disconnecting means (disconnect) a device or group of devices used to disconnect a circuit or device from its source of supply.

Domain a group of atoms aligning themselves north and south to create a magnetic material.

Dot notation dots placed beside transformer windings on a schematic to indicate relative polarity between different windings.

Dry locations locations not normally subject to dampness or wetness.

Dual-voltage motors motors that can be connected to operate on either of two voltages.

DVM abbreviation for digital voltmeter.

Dynamic braking (1) Using a DC motor as a generator to produce countertorque and thereby produce a braking action. (2) Applying DC to the stator winding of an AC induction motor or cause a magnetic braking action.

E

Earth ground a ground established by driving a metal rod into the earth.

Eddy current circular induced current contrary to the main currents. Eddy currents are a source of heat and power loss in magnetically operated devices.

Effective often referred to as RMS voltage or current. The amount of AC voltage or current that will produce the same heating effects as a like amount of DC voltage or current.

Electric controller a device or group of devices used to govern in some predetermined manner the operation of a circuit or piece of electric apparatus.

Electrical interlock when the contacts of one device or circuit prevent the operation of the some other device or circuit.

Electrodynamometer a machine used to measure the torque developed by a motor or engine for the purpose of determining output horsepower.

Electrolysis the decomposition of a chemical compound or metals caused by an electric current.

Electrolyte a chemical compound capable of conducting electric current by being broken down into ions. Electrolytes can be acids or alkalis.

Electrolytic a type of capacitor that is polarity sensitive.

Electromagnetic induction voltage produced by lines of magnetic flux cutting a conductor.

Electromagnets magnets that relay on electricity to become magnetized.

Electromotive force (EMF) electrical pressure, the force that pushes electrons through a wire; the force that causes electrons to flow through a circuit. The most common definition of voltage.

Electromotive series of metals a list of metals that can be employed for constructing voltaic cells.

Electron one of the three major parts of an atom. The electron carries a negative charge.

Electron flow theory the most popular theory of electric current flow that states electricity flows from the more negative part of the power source to the more positive.

Electron impact the collision that occurs when an electron strikes another electron or object.

Electron orbit the shell in which an electron moves around the nucleus of an atom.

Electron spin patterns the pattern or direction in which electrons spin as they orbit the nucleus of an atom.

Electronic control a control circuit that uses solid-state devices as control components.

Electroplating the act of depositing the atoms of one type metal onto another.

Electroscope a device used to determine whether an object contains a positive or negative charge.

Electrostatic charge the static charge that is developed by adding or removing electrons from an object.

Electrostatic charges voltage developed by an excess or lack of electrons on an object.

Electrostatic field the field of force that surrounds a charged object. The term is often used to describe the force of a charged capacitor.

Element (1) one of the basic building blocks of nature. An atom is the smallest part of an element; (2) one part of a group of devices.

Emitter the semiconductor region of a transistor that must be connected to a polarity different from that of the base.

Enclosure mechanical, electric, or environmental protection for components used in a system.

Energized circuit a circuit that has power applied to it.

Eutectic alloy a metal with a low and sharp melting point used in thermal overload relay.

Excitation current (1) the DC used to produce electromagnetism in the fields of a DC motor or generator, or in the rotor of an alternator or synchronous motor; (2) the AC that flows in the primary of a transformer with no load connected.

Exponential the rate at which current rises and falls in inductor or the rate at which capacitors charge and discharge.

Exponential curve a natural curve that is the rate at which current can rise or fall in an inductor or the rate of charge and discharge for a capacitor.

F

Farad (F) the basic unit of capacitance.

Feeder the circuit conductor between either the service equipment or the generator switchboard of an isolated plant and the branch circuit overcurrent protective device.

Femto- a metric prefix corresponding to 10^{-15}.

Ferromagnetic a metal that will conduct magnetic lines of force easily, such as iron (ferrum). Ferromagnetic materials have a permeability much greater than that of air (1).

Fibrillation the state in which the heart quivers and does not pump blood through the body.

Field-discharge resistor a resistor used to reduce the voltage spike caused by a collapsing magnetic field when the DC power supplied to the rotor of a synchronous motor or alternator is disconnected.

Field excitation current the current used to produce a magnetic field in a pole piece.

Field-loss relay (FLR) a current relay connected in series with the shunt field of a DC motor. The relay causes power to be disconnected from the armature in the event that field current should drop below a certain level.

Filter (1) a device used to remove the ripple produced by a rectifier; (2) a frequency selective device.

Filter circuits circuits designed to block or pass particular frequencies.

Fire-retardant clothing clothing that is treated with chemicals that reduce its ability to burn.

Fixed resistors resistors that have a value that cannot be changed.

Flashing the field method used to produce residual magnetism in the pole pieces of a DC machine. It is done by applying full voltage to the field winding for a period of not less than 30 s.

Flat-compounding setting the strength of the series field in a DC generator so that the output voltage will be the same at full load as it is at no load.

Flux magnetic lines of force.

Flux density the number of magnetic lines contained in a certain area. The area measurement depends on the system of measurement.

Flux leakage the number of magnetic flux lines that radiate into space and do not cut through the conductors of an inductor or secondary winding of a transformer.

Frequency the number of complete cycles of AC voltage that occur in 1 s.

Frog leg-wound armatures armatures that have their windings connected in such a way as to form a series-parallel connection.

Full-load torque the amount of torque necessary to produce the full horsepower of a motor at rated speed.

Fuses devices used to protect a circuit or electric device from excessive current. Fuses operate by melting a metal link when current becomes excessive.

G

Gain the increase in signal power produced by an amplifier.

Galvanometer a meter movement requiring microamperes to cause a full-scale deflection. Many galvanometers have a zero center, which permits them to measure both positive and negative values.

Gate (1) a device that has multiple inputs and a single output. There are five basic types of gates: *and, or, nand, nor,* and *inverter;* (2) one terminal of some electronic devices such as SCRs, Triacs, and field-effect transistors (FETs).

Gauss a unit of measure in the CGS system. One gauss equals 1 maxwell per square centimeter.

Generator a device used to convert mechanical energy into electric energy.

Giga- a metric prefix meaning one billion (10^9).

Gilbert a basic unit of magnetomotive force in the CGS system.

Grounding conductor the conductor that is used to force a piece of equipment to exist at ground potential. It is generally green in color or bare.

Ground point the point at which a grounding conductor is attached.

H

Heat sink a metallic device designed to increase the surface area of an electronic component for the purpose of removing heat at a faster rate.

Henry (H) the basic unit of inductance.

Hermetic completely enclosed. Airtight.

Hertz (Hz) the international unit of frequency.

High leg a certain type of three-phase power connection that exhibits a higher voltage on one leg when measured to ground than on the other legs.

High-pass filters pass high frequencies and reject low frequencies.

HIPOT a high voltage tester often used to test the dielectric of a capacitor.

Holding contacts contacts used for the purpose of maintaining current flow to the coil of a relay.

Holding current the amount of current needed to keep an SCR or Triac turned on.

Holtz motor a type of single-phase synchronous motor.

Horseplay the act of playing jokes on others. Generally through some physical act.

Horsepower a measure of power for electric and mechanical devices.

Hydrogen a gas.

Hydrometer a device used to measure the specific gravity of a fluid, such as the electrolyte used in a battery.

Hypotenuse the longest side of a right triangle.

Hysteresis loop a graphic curve that shows the value of magnetizing force for a particular type of material.

Hysteresis loss power loss in a conductive material caused by molecular friction. Hysteresis loss is proportional to frequency.

I

Idiot proofing designing a piece of equipment so that it cannot be misused or connected in an incorrect manner.

Impedance (Z) the total opposition to current flow in an electric circuit.

Incandescence the ability to produce light as a result of heating.

Induced current current produced in a conductor by the cutting action of a magnetic field.

Induced voltage voltage that is produced in a conductor by magnetic lines of flux cutting the conductor.

Inductance (L) property of a circuit that tends to oppose any change of current.

Inductive compensation a method of connecting the compensating windings in a universal motor.

Inductive reactance (X_L) the current-limiting property of an inductor in an AC circuit.

Inductor a coil.

In phase a condition that occurs when voltage and current reach all peak values at the same time.

Input voltage the amount of voltage connected to a device or circuit.

Inrush current the amount of current that is drawn when an inductive device is first connected to the power line.

Insulation a resistive coating that prevents the flow of electrons.

Insulator a material used to electrically isolate two conductive surfaces.

Interlock a device used to prevent some action from taking place in a piece of equipment or circuit until some other action has occurred.

Internal resistance the resistance inherent to a power source.

Interpole small pole piece placed between the main field poles of a DC machine to reduce armature reaction.

Ion a charged atom.

Ionization potential the amount of voltage or potential that must be applied to cause ionization of a gas.

Isolation transformers transformers in which the secondary winding is electrically isolated from the primary winding.

J

JAN (Joint Army-Navy) standard Joint Army Navy standard.

Joule (J) a basic unit of electric energy. A joule is the amount of worked done when 1 A flows through 1 Ω for 1 s. A joule is equal to 1 W-s.

Jumper a short piece of conductor used to make connection between components or a break in a circuit.

Junction diode a diode that is made by joining together two pieces of semiconductor material.

K

Kick-back diode a diode used to eliminate the voltage spike induced in a coil by the collapse of a magnetic field.

Kilo- a metric prefix meaning 1000 (10^3).

Kinetic energy the energy of a moving object, such as the energy of a flywheel in motion.

Kirchhoff's current law the sum of the current entering and leaving a point must equal zero.

Kirchhoff's laws a set of electrical laws dealing with voltage and current.

Kirchhoff's voltage law the sum of the voltages around a closed loop must equal zero.

L

Lagging power factor power factor caused by current lagging the voltage in an inductive circuit.

Laminated something made of thin sheets stacked together.

Lamination one thickness of the sheet material used to construct the core material for transformers, inductors, and AC motors.

Lap-wound armatures armatures that have their windings connected in such a way as to form a parallel connection.

Leading power factor power factor caused by current leading the voltage in a capacitive circuit.

Leakage current the amount of current that flows through the dielectric separating the plates of a capacitor.

Left-hand generator rule a rule for determining the direction of current flow through an armature when the direction of motion and polarity of the magnetic field are known.

Left-hand-rule a rule used for determining the polarity of an electromagnet when the direction of current flow is known.

Lenz's law a law that explains the relationships of movement, voltage polarity, and magnetic field polarity that pertain to magnetic induction.

Leyden jar a glass jar used to store electric charges in the very early days of electric experimentation. The Leyden jar was constructed by lining the inside and outside of the jar with metal foil. The Leyden jar was a basic capacitor.

light-emitting diode (LED) a diode that will produce light when current flows through it.

Lightning a natural static charge that can produce extremely high voltage arcs.

Lightning arrestor a device designed to help prevent damage to equipment by transferring the energy of a lightning strike safely to ground.

Lightning bolts high voltage arcs caused by lightning discharge. Lighting bolts can flow from cloud to cloud, cloud to ground, or ground to cloud.

Lightning rods metal poles that are generally pointed on one end. They are mounted above the structure that is to be protected from lightning.

Limit switch a mechanically operated switch that detects the position or movement of an object.

Linear when used in comparing electric devices or quantities, signifies that one unit is equal to another.

Linear wave a waveform that is produced when voltage or current rises or falls in direct proportion to time.

Line current the current flow in the line supplying a three-phase load.

Lines of flux lines of magnetic force.

Line voltage the voltage as measured between the lines of a three-phase system.

Load the power consumed by a piece of equipment or a circuit while performing its function.

Load center generally the service entrance. A point from which branch circuits originate.

Load test a test for determining the condition of charge for a battery.

Locked-rotor current the amount of current produced when voltage is applied to a motor and the rotor is not turning.

Locked-rotor torque the amount of torque produced by a motor at the time of starting.

Lockout a mechanical device used to prevent the operation of some other component.

Lockout and tagout the act of placing a lock and warning tag on the disconnect of a piece of equipment that is not to be energized.

Lodestones natural magnets.

Long-shunt compound the connection of field windings in a DC machine where the shunt field is connected in parallel with both the armature and series field.

Low-pass filters pass low frequencies and reject high frequencies.

Low-voltage protection a magnetic relay circuit so connected that a drop in voltage causes the motor starter to disconnect the motor from the line.

M

Magnetic contactor a contactor operated electromechanically.

Magnetic domains a section within a magnetic material where atoms are aligned to produce a north or south polarity.

Magnetic field the space in which a magnetic force exists.

Magnetic molecules see magnetic domains.

Magnetomotive force (mmf) the magnetic force produced by current flowing through a conductor or coil.

Maintaining contacts used to maintain the coil circuit in a relay control circuit. The contact is connected in parallel with the start push button. Also known as holding or sealing contacts.

Manual controller a controller operated by hand at the location of the controller.

Material safety data sheets (MSDS) written material supplied by various manufacturers that lists safety information about a particular product.

Matter anything that occupies space.

Maximum operating temperature the highest operating temperature of a conductor. Generally determined by the type of insulation around the conductor.

Maxwell a measure of magnetic flux in the CGS system.

Mega- a metric prefix meaning 1,000,000 (10^6).

Megger a registered name by Biddle Industries. A high-voltage tester used for testing conductor insulation.

Metal film resistors resistors that are produced by coating a nonconductive material with a film of metal.

Metal glaze resistors resistors that are produced by applying a mixture of metal and glass to a nonconductive material.

Metallic salt a chemical compound that is generally in a crystal state and exhibits the properties of a metal.

Metal oxide varistor (MOV) a semiconductor device that changes its resistance in accord with the voltage applied across it.

Meter a device for measuring some quantity such as voltage, current, resistance, etc.

Mica a mineral used as an electric insulator.

Micro- a metric prefix meaning $1/1,000,000$ (10^{-6}).

Microprocessor a small computer. The central processing unit is generally made from a single integrated circuit.

Mil a unit for measuring the diameter of a wire equal to $1/1000$ of an inch.

Mil-foot a standard for measuring the resistivity of wire. A mil-foot is the resistance of a piece of wire 1 mil in diameter and 1 foot in length.

Milli- a metric prefix for $1/1000$ (10^{-3}).

Milliampere $1/1000$ of an ampere (0.001 ampere).

Mode a state or condition.

Molecules the smallest part of a compound that is made of individual atoms.

Motor a device used to convert electric energy into mechanical energy.

Motor controller a device used to control the operation of a motor.

Moving-coil meter a meter that is constructed by connecting a point to a coil that is permitted to move when current flows through it.

Moving iron meter a meter that operates by connecting a pointer to a piece of iron that is attracted or repelled in proportion to the strength of a magnetic field.

Multirange ammeters ammeters that can be set for more than one full range value.

Multirange voltmeters voltmeters that can be set for more than one full range value.

Multispeed motor a motor that can be operated at more than one speed.

Multiturn variable resistors variable resistors that require multiple turns to change their resistance value. They are generally employed where precise control of the resistance value is critical.

N

Nano- metric prefix meaning one-billionth (10^{-9}).

Narrow band passes or rejects a very select range of frequencies.

National Electrical Code® (NEC®) a national code that gives specifications for the installation of electrical equipment.

Negative one polarity of a voltage, current, or charge.

NEMA National Electrical Manufacturers Association.

NEMA ratings electric control device ratings of voltage, current, horsepower, and interrupting capability given by NEMA.

Neutral conductor the neutral conductor is most often part of a 120-volt AC power source. It is connected to ground potential and is considered a current-carrying conductor. The neutral is generally white or gray in color.

Neutral conductor the grounded conductor of a power system.

Neutral plane the point at which there is no voltage induced in the armature of a motor or generator.

Neutron one of the principal parts of an atom. The neutron has no charge and is part of the nucleus.

Nickel-cadmium (nicad) cell a voltaic cell using nickel and cadmium as cell plates with an electrolyte of potassium hydroxide producing a voltage of approximately 1.2 volts.

Node a joining point where electric connections are made.

Noninductive load an electric load that does not have induced voltages caused by a coil. Noninductive loads are generally considered to be resistive, but they can be capacitive.

Nonpolarized capacitors capacitors that can be connected to either polarity of voltage. Generally used in alternating current circuits.

Nonreversing a device that can be operated in only one direction.

Normally closed the contact of a relay that is closed when the coil is deenergized.

Normally open the contact of a relay that is open when the coil is de-energized.

Norton's theorem is used to reduce a circuit network into a simple current source and single parallel resistor.

Notch filters filters designed to reject a particular frequency with sharp cutoff at each end of the frequency.

Nucleus the center or middle.

Nuisance static charges static electric charges that develop on objects when it is not desirable for the object to become charged.

Occupational Safety and Health Administration (OSHA) a government organization responsible for maintaining a safe work environment.

Off-delay timer a timer that delays changing its contacts back to their normal position when the coil is de-energized.

Ohm (Ω) the unit of measure for electrical resistance.

Ohmmeter a device used to measure resistance.

Ohm's law a set of electrical laws developed by Georg S. Ohm that permits electrical quantities to be calculated mathematically.

On-delay timer a timer that delays changing the position of its contacts when the coil is energized.

Unity 100%, or 1.

Unity power factor a power factor of 1 (100%). Unity power factor is accomplished when the applied voltage and circuit current are in phase with each other.

Universal motor a type of motor that can be operated on AC or DC power.

Useful static charges static electric charges that are intentionally placed on objects.

V

Vacuum-tube voltmeter (VTVM) a voltmeter that uses the grid of a vacuum tube to produce a very high input impedance. VTVMs are used in electronic circuits to prevent loading the circuit that is being tested. Modern VTVMs replace the vacuum tubes with field-effect transistors (FETs).

Valence electrons electrons located in the outer orbit of an atom.

Variable capacitors capacitors whose value can be changed.

Variable resistor a resistor whose resistance value can be varied between its minimum and maximum values.

Varistor a resistor that changes its resistance value with a change of voltage.

Vector addition a method of adding angles.

Vector a line having a specific length and direction.

Volt (V) a measure of electromotive force. The potential necessary to cause 1 coulomb to produce 1 joule of work.

Voltage an electrical measurement of potential difference, electrical pressure, or electromotive force (EMF).

Voltage divider a circuit that generally is constructed by connecting resistors in series to produce desired voltage drops across the resistors.

Voltage drop the amount of voltage required to cause an amount of current to flow through a certain resistance.

Voltage drop across the capacitor the amount of voltage measured across a capacitor.

Voltage drop across the resistor the amount of voltage measured across a resistor.

Voltage polarity the polarity of the voltage at a particular point in a circuit. Generally of concern in direct current circuits.

Voltage rating a rating that indicates the amount of voltage that can be safely connected to a device.

Voltage regulator a device or circuit that maintains a constant value of voltage.

Voltage spike a momentary increase of voltage.

Voltaic cell a device that converts chemical energy into electric energy.

Voltaic pile a group of voltaic cells connected together to form a battery.

Voltmeter an instrument used to measure a level of voltage.

Volt-ohm-milliammeter (VOM) a test instrument designed to measure voltage, resistance, or milliamperes.

Volts-per-turn ratio determined by dividing the voltage applied to the primary of a transformer by the number of turns of wire in the primary.

VTVM abbreviation for vacuum-tube voltmeter.

W

Warren motor a type of single-phase synchronous motor.

Watt (W) a measure of true power.

Watt-hour (W-hr) a unit of measure for describing the current capacity of a battery or a cell.

Wattless power see reactive power.

Wattmeter a meter used to measure true power or watts.

Waveform the shape of a wave as obtained by plotting a graph with respect to voltage and time.

Wave-wound armatures armatures that have their windings connected in such a way as to form a series connection.

Weber (Wb) a measure of magnetic lines of flux in the MKS system. 1 Wb = 100 million lines of flux.

Wet locations areas that are subject to saturation of water or other liquids.

Wheatstone bridge a circuit used to accurately measure resistance.

Windage loss the losses encountered by the armature or rotor of a rotating machine caused by the friction of the surrounding air.

Wire-wound resistors resistors that are constructed by winding resistive wire around a nonconductive material.

Wiring diagram an electrical diagram used to show components in their approximate physical location with connecting wires.

Wound rotor a type of three-phase rotor that contains phase winding and slip rings.

Wound-rotor motor a three-phase motor containing a rotor with windings and sliprings. This rotor permits control of rotor current by connecting external resistance in series with the rotor winding.

Wye connection a connection of three components made in such a manner that one end of each component is connected. This connection is generally used to connect devices to a three-phase power system.

Wye-delta a three-phase transformer connection that has its primary connected in wye and a secondary connected in delta.

X

X-rays beams of high-energy protons.

Z

Zener diode a diode that has a constant voltage drop when operated in the reverse direction. Zener diodes are commonly used as voltage regulators in electronic circuits.

Zener region the region of a semiconductor device that is reached when current flows through it in the reverse direction.

Zenith a culmination or high point.

Zone an area, region, or division distinguished from adjacent parts or objects by some feature, characteristic, or barrier.

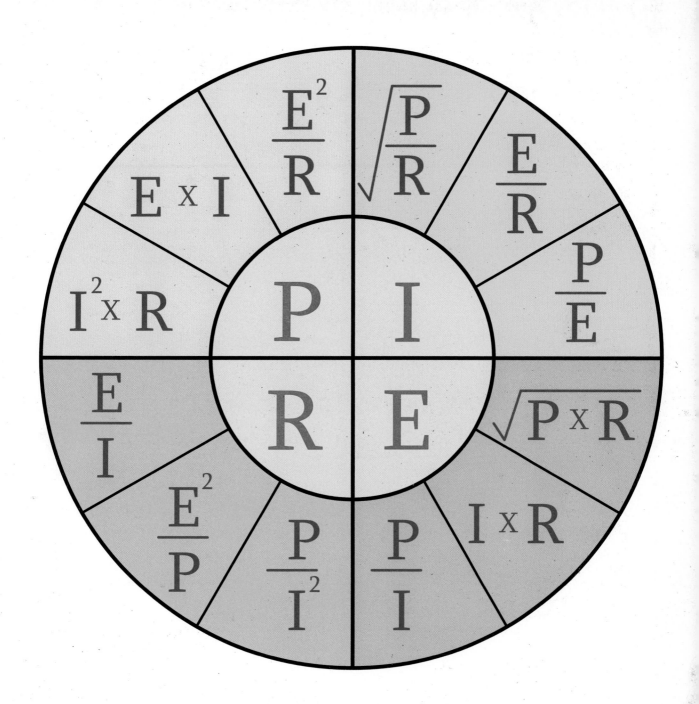